The Microbiota in Gastrointestinal Pathophysiology

The Microbiota in Gastrointestinal Pathophysiology

Implications for Human Health, Prebiotics, Probiotics, and Dysbiosis

Edited by

Martin H. Floch, MD, MACG, FACP, AGAF
Clinical Professor of Medicine
Department of Internal Medicine
Section of Digestive Diseases
Yale University School of Medicine
New Haven, CT
United States

Yehuda Ringel, MD, AGAF, FACG
Department of Gastroenterology
Rabin Medical Center
Petach Tikva, Israel;
Adjunct Professor
Department of Medicine
University of North Carolina
School of Medicine at Chapel Hill
Chapel Hill, NC, United States

W. Allan Walker, MD, AGAF, FACG
Conrad Taff Professor of Nutrition and
Professor of Pediatrics
Harvard Medical School
Mucosal Immunology Laboratory
Pediatric Gastroenterology and Nutrition
Massachusetts General Hospital for Children
Boston, MA, United States

AMSTERDAM • BOSTON • HEIDELBERG • LONDON • NEW YORK • OXFORD • PARIS
SAN DIEGO • SAN FRANCISCO • SINGAPORE • SYDNEY • TOKYO

Academic Press is an imprint of Elsevier

Academic Press is an imprint of Elsevier
125 London Wall, London EC2Y 5AS, United Kingdom
525 B Street, Suite 1800, San Diego, CA 92101-4495, United States
50 Hampshire Street, 5th Floor, Cambridge, MA 02139, United States
The Boulevard, Langford Lane, Kidlington, Oxford OX5 1GB, United Kingdom

Library of Congress Cataloging-in-Publication Data
A catalog record for this book is available from the Library of Congress

British Library Cataloguing-in-Publication Data
A catalogue record for this book is available from the British Library

ISBN: 978-0-12-804024-9

For information on all Academic Press publications
visit our website at https://www.elsevier.com/

Working together
to grow libraries in
developing countries

www.elsevier.com • www.bookaid.org

Publisher: Mica Haley
Acquisition Editor: Stacy Masucci
Editorial Project Manager: Sam Young
Production Project Manager: Julia Haynes
Designer: Greg Harris

Typeset by Thomson Digital

We would like to dedicate this book to our wives,
Gladys Floch, Tammy Ringel-Kulka, and Ann Sattler-Walker.
We are forever grateful for their guidance, patience, and understanding
for the time we spent working on this project.

Contents

23. Prebiotics: Inulin and Other Oligosaccharides

S. Mitmesser and M. Combs

24. The Benefits of Yogurt, Cultures, and Fermentation

M. Freitas

Part D
Basic Physiologic Effects of Microbiota

25. Dysbiosis

W.A. Walker

26. Immunologic Response in the Host

K. Madsen and H. Park

39. Celiac Disease, the Microbiome, and Probiotics

S. Krishnareddy and P.H.R. Green

40. Probiotics for the Treatment of Liver Disease

C. Punzalan and A. Qamar

41. The Prevention and Treatment of Radiation and Chemotherapy-Induced Intestinal Mucositis

M.A. Ciorba

42. The Role of the Brain–Gut–Microbiome in Mental Health and Mental Disorders

G.J. Treisman

43. Management of Disease and Disorders by Prebiotics and Probiotic Therapy: Probiotics in Bacterial Vaginosis

B. Vitali, A. Abruzzo and P. Mastromarino

Contributors

A. Abruzzo, PhD Department of Pharmacy and Biotechnology, University of Bologna, Bologna, Italy

A. Alahmadi, MBBS Department of Medicine, Division of Gastroenterology, University of Alberta, Edmonton, AB, Canada

J.C. Arthur, PhD Department of Microbiology and Immunology, Center for Gastrointestinal Biology and Disease, University of North Carolina at Chapel Hill, Chapel Hill, NC, United States

G.J. Bakker, MD Department of Vascular Medicine, Academic Medical Center, Amsterdam, The Netherlands

P. Bercik, MD Department of Medicine, Farncombe Family Digestive Health Research Institute, McMaster University, Hamilton, ON, Canada

R.J. Bertelsen, MSc, PhD Department of Clinical Science, University of Bergen; Department of Occupational Medicine, Haukeland University Hospital, Bergen, Norway

R.A. Britton, PhD Department of Molecular Virology and Microbiology, Center for Metagenomics and Microbiome Research, Baylor College of Medicine, Houston, TX, United States

J.P. Burton, PhD Division of Urology, Department of Surgery/Department of Microbiology & Immunology, Western University; Lawson Health Research Institute and Canadian Centre for Human Microbiome and Probiotics, London, ON, Canada

J.S. Carr, MD Department of Surgery, University of North Carolina at Chapel Hill, Chapel Hill, NC, United States

I.M. Carroll, PhD Department of Medicine, University of North Carolina at Chapel Hill, Chapel Hill, NC, United States

R.M. Chanyi, PhD Department of Microbiology & Immunology, Western University; Lawson Health Research Institute, Canadian Centre for Human Microbiome and Probiotics Research, London, ON, Canada

R. Chibbar, MD Department of Medicine, Division of Gastroenterology, University of Alberta, Edmonton, AB, Canada

M.G. Cifone, PhD Department of Life, Health & Environmental Sciences, University of L'Aquila, L'Aquila, Italy

B. Cinque, PhD Department of Life, Health & Environmental Sciences, University of L'Aquila, L'Aquila, Italy

M.A. Ciorba, MD Division of Gastroenterology, Washington University in Saint Louis, Saint Louis, MO, United States

M. Combs, MS Department of Nutrition & Scientific Affairs, NBTY, Inc., Ronkonkoma, NY, United States

E.M. Comelli, PhD Department of Nutritional Sciences and Center for Child Nutrition and Health, University of Toronto, Toronto, ON, Canada

T.M. Darby, PhD Department of Pediatrics, Emory University School of Medicine, Atlanta, GA, United States

L.A. Dieleman, MD, PhD Department of Medicine, Division of Gastroenterology, University of Alberta, Edmonton, AB, Canada

N.S. Ding, MBBS (Hons.), FRACP Inflammatory Bowel Disease Unit, St Mark's Hospital, Imperial College London, London, United Kingdom

G. Donelli, PhD Microbial Biofilm Laboratory, IRCCS Fondazione Santa Lucia, Rome, Italy

S. Doron, MD Department of Medicine, Tufts University School of Medicine; Division of Geographic Medicine and Infectious Diseases, Tufts Medical Center, Boston, MA, United States

M. Ellermann, PhD Department of Microbiology and Immunology, University of North Carolina at Chapel Hill, Chapel Hill, NC; Department of Microbiology, University of Texas Southwestern Medical Center, Dallas, TX, United States

M. Eloit, DVM, PhD Institut Pasteur, Biology of Infection Unit, Inserm U1117, Laboratory of Pathogen Discovery, Paris; Ecole Nationale Vétérinaire d'Alfort, Maisons-Alfort, France

S. Fine, MD, MS Department of Gastroenterology, Warren Alpert School of Medicine, Brown University, Providence, RI, United States

M.H. Floch, MD, MACG, FACP, AGAF Department of Internal Medicine, Section of Digestive Diseases, Yale University School of Medicine, New Haven, CT, United States

N. Floch, MD, FACS Department of Surgery, Norwalk Hospital, University of Vermont School of Medicine, and Quinnipiac School of Medicine Fairfield County Bariatrics and Surgical Specialists, P.C, Norwalk, CT, United States

A.A. Fodor, PhD Department of Bioinformatics and Genomics, University of North Carolina at Chapel Hill, Chapel Hill, NC, United States

M. Freitas, PhD Health Affairs, The Dannon Company Inc., White Plains, NY, United States

S. Gorbach, MD Department of Public Health & Medicine, Tufts University School of Medicine; Division of Infectious Diseases, Tufts Medical Center, Boston, MA, United States

P.H.R. Green, MD Celiac Disease Center, Columbia University, NY, United States

S. Guandalini, MD Section of Gastroenterology, Hepatology and Nutrition, University of Chicago Medicine—Comer Children's Hospital, Chicago, IL, United States

M. Guslandi, MD Clinical Hepato-Gastroenterology Unit, Division of Gastroenterology & Digestive Endoscopy, S. Raffaele University Hospital, Milan, Italy

A. Hart, PhD Inflammatory Bowel Disease Unit, St Mark's Hospital, Imperial College London, London, United Kingdom

R.A. Hickman, MD Department of Pathology, New York University School of Medicine, New York, NY, United States

K. Hod, RD, MSc Department of Epidemiology and Preventive Medicine, School of Public Health, Sackler Faculty of Medicine, Tel Aviv University; Research Division, Epidemiological Service, Assuta Medical Center, Tel Aviv, Israel

E. Holmes, PhD Department of Surgery and Cancer, Division of Computational Systems Medicine, Imperial College London, London, United Kingdom

D.J.A. Jenkins, MD, PhD Clinical Nutrition & Risk Factor Modification Center, Li Ka Shing Knowledge Institute and Division of Endocrinology and Metabolism, St. Michael's Hospital; Departments of Nutritional Sciences and Medicine, Faculty of Medicine, University of Toronto, Toronto, ON, Canada

E.T. Jensen, MPH, PhD Department of Epidemiology and Prevention, Division of Public Health Sciences, Wake Forest School of Medicine, Winston-Salem; Department of Medicine, Division of Gastroenterology and Hepatology, University of North Carolina at Chapel Hill, Chapel Hill, NC, United States

R.M. Jones, PhD Department of Pediatrics, Emory University School of Medicine, Atlanta, GA, United States

C.R. Kelly, MD Warren Alpert Medical School of Brown University, Miriam Hospital, and Lifespan Hospital System, Providence, RI, United States

C.W.C. Kendall, PhD Clinical Nutrition & Risk Factor Modification Center, St. Michael's Hospital; Department of Nutritional Sciences, University of Toronto, Toronto, ON; College of Pharmacy and Nutrition, University of Saskatchewan, Saskatoon, SK, Canada

A. Khan, MD Department of Medicine, New York University School of Medicine, New York, NY, United States

S. Krishnareddy, MD Celiac Disease Center, Columbia University, NY, United States

K.I. Kroeker, MD Department of Medicine, Division of Gastroenterology, University of Alberta, Edmonton, AB, Canada

C. La Torre, PhD Department of Life, Health & Environmental Sciences, University of L'Aquila, L'Aquila, Italy

M. Lecuit, MD, PhD Institut Pasteur, Biology of Infection Unit, Inserm U1117, Laboratory of Pathogen Discovery; Paris Descartes University, Sorbonne Paris Cité; Necker-Enfants Malades University Hospital, Division of Infectious Diseases and Tropical Medicine, Paris, France

F. Lombardi, PhD Department of Life, Health & Environmental Sciences, University of L'Aquila, L'Aquila, Italy

C. Lu, MD Department of Medicine, Division of Gastroenterology, University of Alberta, Edmonton, AB, Canada

K. Madsen, PhD Division of Gastroenterology, The Center of Excellence for Gastrointestinal Inflammation and Immunity Research, University of Alberta, Edmonton, AB, Canada

F. Magro, MD Department of Medicine, Tufts University School of Medicine; Division of Geographic Medicine and Infectious Diseases, Tufts Medical Center, Boston, MA, United States

P. Mastromarino, PhD Department of Public Health and Infectious Diseases, Sapienza University, Rome, Italy

L.V. McFarland, PhD Department of Medicinal Chemistry, School of Pharmacy, University of Washington, Seattle, WA, United States

S. Mitmesser, PhD Department of Nutrition & Scientific Affairs, NBTY, Inc., Ronkonkoma, NY, United States

A. S. Neish, MD Department of Pathology, Emory University School of Medicine, Atlanta, GA, United States

A.G. Neto, MD Department of Pathology, New York University School of Medicine, New York, NY, United States

M. Nieuwdorp, MD, PhD Department of Vascular Medicine, Academic Medical Center; Department of Internal Medicine, VUmc Diabetes Center, Free University Medical Center, Amsterdam, The Netherlands; Wallenberg Laboratory, University of Gothenberg, Gothenberg, Sweden

J.C. Noronha, HBSc Department of Nutritional Sciences, University of Toronto; Clinical Nutrition & Risk Factor Modification Center, St. Michael's Hospital, Toronto, ON, Canada

C. Nossa, MD Research & Development, Gene By Gene, Ltd., Houston, TX, United States

A. Nowak-Wegrzyn, MD Pediatrics, Allergy and Immunology, Icahn School of Medicine at Mount Sinai, Jaffe Food Allergy Institute, New York, NY, United States

P. Palumbo, PhD Department of Life, Health & Environmental Sciences, University of L'Aquila, L'Aquila, Italy

H. Park, PhD Faculty of Medicine and Dentistry, University of Alberta, Edmonton, AB, Canada

Z. Pei, MD Department of Pathology; Department of Medicine; Department of Veterans Affairs New York Harbor Healthcare System, New York University School of Medicine, New York, NY, United States

V. Philip, PhD Candidate Department of Medicine, Farncombe Family Digestive Health Research Institute, McMaster University, Hamilton, ON, Canada

C.S. Pitchumoni, MD Saint Peter's University Hospital, New Brunswick, NJ, United States

C. Punzalan, MD Lahey Hospital and Medical Center, Burlington, MA, United States

A. Qamar, MD Lahey Hospital and Medical Center, Burlington; Tufts University School of Medicine, Boston, MA, United States

E.M.M. Quigley, MD, FRCP, FACP, MACG, FRCPI Lynda K and David M Underwood Center for Digestive Disorders, Houston Methodist Hospital and Weill Cornell Medical College, Houston, TX, United States

Y. Ringel, MD, AGAF, FACG Department of Gastroenterology, Rabin Medical Center, Petach Tikva, Israel; Department of Medicine, University of North Carolina School of Medicine at Chapel Hill, Chapel Hill, NC, United States

T. Ringel-Kulka, MD, MPH Department of Maternal and Child Health, UNC Gillings School of Global Public Health, University of North Carolina at Chapel Hill, Chapel Hill, NC, United States

N. Sansotta, MD Section of Gastroenterology, Hepatology and Nutrition, University of Chicago Medicine—Comer Children's Hospital, Chicago, IL, United States

M.H. Sarafian, MD Department of Surgery and Cancer, Division of Computational Systems Medicine, Imperial College London, London, United Kingdom

A. Sarkar, MD Rutgers, Robert Wood Johnson Medical School, New Brunswick, NJ, United States

M. Schultz, PhD Department of Medicine, Dunedin School of Medicine, University of Otago, Dunedin, New Zealand

J.L. Sievenpiper, MD, PhD Clinical Nutrition & Risk Factor Modification Center, Li Ka Shing Knowledge Institute and Division of Endocrinology and Metabolism, St. Michael's Hospital; Department of Nutritional Sciences, University of Toronto, Toronto, ON, Canada

H. Szajewska, MD Department of Paediatrics, The Medical University of Warsaw, Warsaw, Poland

G.J. Treisman, MD, PhD Departments of Psychiatry and Behavioral Sciences and Internal Medicine, Johns Hopkins University School of Medicine, Baltimore, MD, United States

M.E. van den Rest, PhD Biovisible, Groningen, The Netherlands

B. Vitali, PhD Department of Pharmacy and Biotechnology, University of Bologna, Bologna, Italy

C. Vuotto, PhD Microbial Biofilm Laboratory, IRCCS Fondazione Santa Lucia, Rome, Italy

W.A. Walker, MD, AGAF, FACG Harvard Medical School; Mucosal Immunology and Biology Research Center, Pediatric Gastroenterology and Nutrition, Massachusetts General Hospital for Children, Boston, MA, United States

J.M.W. Wong, PhD, RD Clinical Nutrition & Risk Factor Modification Center and Li Ka Shing Knowledge Institute, St. Michael's Hospital, Toronto, ON, Canada; New Balance Obesity Foundation Obesity Prevention Center, Boston Children's Hospital, Boston, MA, United States

About the Editors

M.H. Floch, MD, MACG, FACP, AGAF

Martin Floch is a graduate of New York University, has a Master's Degree from the University of New Hampshire in Durham and received his MD from New York Medical College. He completed his internal medicine residency at Beth Israel Hospital in New York and his gastroenterology training at the former Seton Hall College of Medicine in South Orange, New Jersey.

He is a Master of the American College of Gastroenterology and holds their Samuel Weiss Award for service to the college. He was Associate Editor for the American Journal of Nutrition before he became Editor for the American College of Gastroenterology. Following that, he became Editor of The Gastroenterologist and is now completing his 16th year as Editor of The Journal of Clinical Gastroenterology.

Dr. Floch was Chairman of the Department of Medicine at Norwalk Hospital, a major affiliate of Yale University, from 1970 to 1994 and also had a consultation practice. He served as Director of the Yale GI clinics for a term of 5 years. While at Yale, Dr. Howard Spiro was his mentor and colleague. Dr. Floch has had several NIH Grants and has done research on Irritable Bowel Syndrome, Diverticulitis, and Probiotics. During the past decade, he has held four international Yale CME Workshop Programs on the Recommendations for Clinical Use of Probiotics, the results of which were published.

W.A. Walker, MD

Allan Walker is the Conrad Taff Professor of Nutrition and Professor of Pediatrics at Harvard Medical School. After completing medical school at Washington University of St. Louis, he trained in Pediatrics and Immunology with Dr. Robert Good at the University of Minnesota before coming to Massachusetts General Hospital (Harvard Medical School) to train in Gastroenterology and Nutrition with Dr. Kurt Isselbacher. Dr. Walker joined the faculty of Pediatrics in 1971 and became Professor in 1982. He established the first Pediatric Gastroenterology Unit at Massachusetts General Hospital in 1973 and became Director of a Combined Program in Pediatric Gastroenterology at Boston Children's Hospital and Massachusetts General Hospital in 1982. He is currently Director of Nutrition at Harvard Medical School and an Investigator in the Mucosal Immunology and Biology Research Center at Massachusetts General Hospital for Children. His research interests include the development of mucosal immune function and the role of breastfeeding, protective nutrients, and initial colonization in the ontogeny of intestinal host defense. He has received numerous honors in recognition of his research including the Shwachman Award from NASPGHAN and the Hugh Butt Award from the AGA for mucosal immunologic research and a R-37 (MERIT) Award from the National Institutes of Health for outstanding investigation in breast milk immunology. Recently, Dr. Walker's laboratory has defined the cellular mechanism of probiotic protection in necrotizing enterocolitis.

Y. Ringel, MD, AGAF, FACG

Yehuda Ringel is an Adjunct Professor of Medicine from the University of North Carolina (UNC) School of Medicine at Chapel Hill. Dr. Ringel earned his medical degree *cum laude* at The Technion—The Israel Institute of Technology and completed his medical residency and fellowship in Internal Medicine and Gastroenterology at Tel Aviv Medical Center in Israel. Additional training includes a Master in Internal Medicine at Tel Aviv University and a postdoctoral fellowship in functional gastrointestinal (GI) and motility disorders with Drs Douglas Drossman and William Whitehead at UNC—Chapel Hill.

Dr. Ringel has been the Director of the UNC Clinic for Functional GI Disorders and the Associate Director of the UNC Center for Functional GI and Motility Disorders and Chief of GI Division at Rabin Medical Center in Israel. Dr. Ringel has been involved in clinical and translational research for over 20 years. His research focuses on intestinal physiology and functional GI disorders and in the past few years, on the role of the intestinal microbiome and pre- and probiotics in these conditions. He was awarded grants from the *National Institute of Health* (NIDDK) to examine the role of the intestinal microbiome and immune function in the pathogenesis of GI symptoms. Dr. Ringel's team has been involved with some of the largest studies characterizing the intestinal microbiota and investigating the effects of interventions targeting the microbiota in patients with functional GI disorders. He is a recipient of several prestigious awards including from the *American College of Gastroenterology* (ACG), the *American Gastroenterology Association* (AGA), and the *Functional Brain–Gut Research Group* (FBG).

Introduction

M.H. Floch*, Y. Ringel**,† and W.A. Walker‡,§

*Department of Internal Medicine, Section of Digestive Diseases, Yale University School of Medicine, New Haven, CT, United States;
**Department of Gastroenterology, Rabin Medical Center, Petach Tikva, Israel; †Department of Medicine, University of North Carolina School of
Medicine at Chapel Hill, Chapel Hill, NC, United States; ‡Harvard Medical School, Boston, MA, United States; §Mucosal Immunology Laboratory,
Pediatric Gastroenterology and Nutrition, Massachusetts General Hospital for Children, Boston, MA, United States

The modern history of the microbiota of the intestinal/gut microbiome is fascinating. The scientific world did not intensively develop its interest in this world of organisms until the national committee stimulated by Dr. Jeffrey Gordon published the paper Turnbaugh et al. that clearly showed the microbiota of the gastrointestinal tract were numerous at approximately 10^{14} and, in effect, were more in number than cells in the human body [1]. With this information, huge interest began to develop and it also stimulated the production and use of probiotics and prebiotics. The supplement industry is approximately \$6 billion annually and it continues to grow worldwide throughout North and South America, Europe, and Asia. However, detailed and useful publications on the scientific basis and the clinical use of prebiotics and probiotics are limited. Although many of the prebiotic and probiotic industries market their products, there are few that make recommendations for use in humans. We began the Yale University Workshops which brought together noted scientists who were studying the use of prebiotics and probiotics in human conditions [2–4].

One of us, Floch, was invited by the US Army early in the 1970s to study the anaerobic flora of the gastrointestinal tract [5]. The government was concerned about what would happen, in space, to the organisms. We had laboratories financed at both Norwalk Hospital and Yale University and studied the anaerobes in various physiologic states and diseases [6].

During this period of time, several microbiologists were working in the field. Drs Renee DuBois and Russell Shadler of the Rockefeller were leaders. Drs Moore and Holdeman in Blacksburg, Virginia, developed anaerobic techniques that were used by many [7]. Initial leaders in the field were Sydney Finegold [8], Donald Luckey [9], and many others. Many publications came forth, but early in the 1980s interest decreased. It wasn't until the 1990s and 2000s that a huge interest developed in using live organisms as probiotics. The probiotic industry became huge and is now an important industry in the world.

This book attempts to bring forward the latest information on the microbiota in humans. It is divided into four parts: (1) the general disposition of the organisms throughout the gut, (2) a detailed description of the most common probiotics organisms used today as described by Andrew Neish, Michael Schultz, Eamonn Quigley, and Sherwood Gorbach, (3) the careful functions of the microbiota and their effect on the immunopathology of the gut and the metabolism of foods by the microbiota and its use of the various nutrients on the host; and finally, (4) is devoted to the recommendations made for use of supplements in patients with possible dysbiosis.

As we noted previously, Martin Floch has worked intensively in identification of the flora. The second Editor, Yehuda Ringel, is a gastrointestinal physiologist who has intensively investigated the interaction between the gut microbiota and the neurobiology and function of the gastrointestinal tract. He also has interest and published extensively on the use of prebiotics and probiotics in patients with functional GI symptoms.

Allan Walker is renowned for his interest and knowledge in the immunologic response of the gut to the microbiota and the effect of all of its nutrients on the host responses. He brings his interest in this field to the editing of the book. He feels that early appropriate colonization when the immune system is developing is important in the expression of chronic disease, later in life.

We are extremely appreciative to the authors who have helped put these chapters together. Most of their work is unique and it adds to science of understanding the microbiome.

We are deeply grateful to the producers of the book, Elsevier, including Samuel Young and Stacy Masucci. We are also very appreciative to the persons who have helped in the administration of the book, particularly Candace Peabody who acted as the coordinator. We hope this book will be helpful as you continue to gain knowledge of the microbiome and its importance.

REFERENCES

[1] Turnbaugh PJ, Ley RE, Hamady M, Fraser-Liggett CM, Knight R, Gordon JI. The human microbiome project. Nature 2007;449:804–10.

[2] Floch MH, Walker WA, Guandalini S, et al. Recommendations for probiotic use—2008. J Clin Gastroenterol 2008;42:S104–8.

[3] Floch MH, Walker WA, Madsen K, et al. Recommendations for probiotic use—2011 update. J Clin Gastroenterol 2011;45:S168–71.

[4] Floch MH, Walker WA, Sanders ME, et al. Recommendations for probiotic use—2015 update: proceedings and consensus opinion. J Clin Gastroenterol 2015;49:S69–73.

[5] Floch MH, Gorbach SL, Luckey TD. Intestinal microflora. Am J Clin Nutr 1970;23:1425–6.

[6] Floch MH, Wolfman M, Doyle R. Fiber and gastrointestinal microecology. J Clin Gastroenterol 1980;2:175–84.

[7] Moore WEC, Holdeman LV. Human fecal flora: the normal flora of 20 Japanese-Hawaiians. Appl Microbiol 1974;27:961–79.

[8] Finegold SM, Sutter VL, Sugihara PT, Elder HA, Lehmann SM, Phillips RL. Fecal microbial flora in Seventh Day Adventist population and control subjects. Am J Clin Nutr 1977;30:1781–92.

[9] Luckey TD. Introduction to the ecology of the intestinal flora. Am J Clin Nutr 1970;23:1430–2.

Part A

The Microbiota of the Gastrointestinal Tract

Chapter 1

The Upper Gastrointestinal Tract—Esophagus and Stomach

A.G. Neto*, R.A. Hickman*, A. Khan**, C. Nossa‡ and Z. Pei*,**,†

*Department of Pathology, New York University School of Medicine, New York, NY, United States; **Department of Medicine, New York University School of Medicine, New York, NY, United States; †Department of Veterans Affairs New York Harbor Healthcare System, New York University School of Medicine, New York, NY, United States; ‡Research & Development, Gene By Gene, Ltd., Houston, TX, United States

INTRODUCTION

The human body is host to an abundant and diverse population of microbes. Every external and internal surface of the human body that is exposed to the environment is inhabited by its own unique population of microbes. Some parts of the body that are not in direct contact with the environment are also populated with microbes. The role of the human microbiome in both health and disease ranges from negligible to crucial, and much remains undiscovered.

Some microbes inhabiting the human body are true commensals and have no noticeable effect on host physiology. Other members of the human microbiota are beneficial, providing the host with nutrients or defenses against other microbes. For example, beneficial microbes serve as a *de facto* metabolic organ within the host colon. Conversely, other microbes can be harmful, either by acting as acute pathogens with a range of virulence or by promoting chronic disease. While obligate and opportunistic pathogenic microbes have long been studied for their specific influence on disease, commensal microbes as causative agents of chronic disease have only recently been recognized as an important field of study.

THE HUMAN MICROBIOME PROJECTS

The microbiome had not been a main focus of study until the launch of the human microbiome projects in the last decade, specifically the National Institutes of Health (NIH) Human Microbiome Project and the International Human Microbiome Consortium [1,2]. The primary goals of these initiatives are: to determine a core human microbiome, common to all healthy individuals; to assess how this core microbiome changes during disease; and to establish whether a change in the microbiome, termed dysbiosis, is associated with disease. It was anticipated that the knowledge gained from human microbiome research would offer insight into the etiology and pathogenesis of idiopathic chronic diseases that have a serious impact on human health. Further associated with the overall aim is the characterization of all of the microbial genes within the human body, not only the species constituting the microbiome. Characterization of the human microbial metagenome has the potential to uncover the metabolic processes and pathways of microbes, decipher how they contribute to or perturb normal human health, and ultimately lead to novel treatments and techniques.

After establishing a normal core microbiota, the secondary goal of human microbiome research is to discover correlations between diseases and microbial dysbiosis. Dysbiosis is any perturbation of the normal microbiome content that could disrupt the symbiotic relationship between the host and associated microbes, a disruption that can result in diseases, such as inflammatory bowel disease and other gastrointestinal (GI) disorders, including gastritis, peptic ulcer disease, irritable bowel syndrome, and even gastric and colon cancer [3–6]. The purpose of this research is first to confirm that the correlation between dysbiosis and disease is meaningful in regard to the etiology of disease, and, if so, to then determine the molecular mechanisms of disease causation in order to tailor therapies that target the microbial dysbiosis.

METHODS FOR MICROBIAL ANALYSIS AND ADVANCES IN SEQUENCING TECHNOLOGY

Cultivation-based methods have been the gold standard for bacterial detection, bacterial identification, and disease diagnosis, but these methods are strongly biased toward microbes that can easily be cultured in laboratory settings. The vast majority of microbes are fastidious and therefore invisible to culture-dependent analysis, because they can exist in vivo in a viable but nonculturable state, or as dormant vegetative cells with low metabolic activity [7,8]. While these techniques have advanced tremendously, cultivation-based methods are still inadequate to fully define microbial populations [9]. Cultivation-independent techniques have been developed to attempt to identify both cultivable and fastidious microorganisms by direct sequencing of the microbial DNA and then searching databases of known microbes for genetically similar organisms.

The microbial populations so far defined by cultivation-independent techniques are far more diverse than previously suspected. Thousands of bacterial species have been revealed in specific environments of microbiomes previously assumed to be dominated by a few cultivable species. Currently, most of the research into the gut microbiome focuses on members of the kingdom Eubacteria. This is due to the initial use of the 16S rRNA gene marker using primers that are specifically designed with conserved Eubacteria sequences for phylogenetic analysis. The 16S rRNA subunit sequence was chosen because it not only provides regions that serve as markers all the way down to the species level, but also provides highly conserved regions that serve as primer target regions [10–13]. While 5S and 23S rRNA subunit sequences have been evaluated as possible species-level markers, 16S has proven the most popular bacterial marker [14–16].

Along with the impetus provided by the NIH Human Microbiome Project, the introduction and commercial availability of next-generation sequencing (NGS), namely Roche 454 and Illumina pyrosequencing, contributed to the phenomenal increase in microbiome research. The sequencing capacity and reduced cost of NGS allowed researchers to increase the size and scope of relevant research projects. Previous researchers, constrained by cost, manpower, and logistics, had to limit sample sizes to at most tens of samples; NGS's massive parallel sequencing allows for the analysis of hundreds to thousands of samples in both less time and at a lower cost. In the past decade, since the inception of the NIH Human Microbiome Project, the capacity of NGS has continued to increase, allowing for in-depth research focusing on microbial identities as well as whole microbial genomic contents. Eventually, the price of sequencing was reduced and capacity increased to the point that specific markers were no longer necessary. It is now cost-effective to sequence the entire metagenome of microbial samples, not only for microbial composition but also for gene pathways and metabolic networks. This is creating a trove of information, allowing for more insight into viral, fungal, and protozoan contributions to the microbiome.

ADVANCES IN MICROBIOME RESEARCH IN THE PROXIMAL VERSUS DISTAL GUT

Among the sites within the human microbiome, research focused on the GI tract has been the most successful in uncovering microbiome-associated disease. This is especially true in the distal gut, most likely due to the relatively high biomass present as well as the easy access for sampling. This microbiome has been shown to contribute not only to potential disease, but also to normal human physiologic functioning.

Mouth to colon analyses of the gut microbiome have been published and have established the continuity and progression of the gut microbiota through the GI tract. As the microbiome was characterized from mouth to anus, it was deciphered that there are certain phyla of bacteria common throughout the GI tract, but not necessarily in constant ratios. Species diversity also varies greatly along the gut; from the proximal to the distal gut, there is a marked difference in number and diversity of bacteria [17].

Study of the proximal gut microbiome has been greatly overshadowed by the investigations of the distal gut microbiome, with the exception being the documented consequences of *Helicobacter pylori* colonization in the stomach. This is due partly to past findings and assumptions regarding the sterility and/or lack of diversity within the proximal gut, as well as the laborious and invasive nature of collecting samples in the esophagus and stomach, which involves endoscopic probing in a specific manner to avoid cross-site contamination. On embryological grounds, the proximal gut comprises the mouth through the esophagus, stomach and extends to the major papillae of the duodenum. The accessory organs that are regarded as being associated with the proximal gut include the liver, gallbladder, and pancreas.

Characterization of the proximal gut microbiome has improved dramatically as the use of culture-independent microbiome analysis has been facilitated by NGS technologies. Although there was already significant research on the oral microbiome, studies involving the microbiome of the esophagus and stomach were sparse; most involved few samples with very shallow characterizations. Recent publications have contributed much more in-depth characterization of the proximal gut microbiome as interest has grown about the nature of the entire human gut microbiome, providing information that is not simply extrapolated from distal gut analysis.

The esophageal and gastric microbiomes share characteristics with the well-described distal gut microbiome and with the gut microbiome as a whole. They also each have their own distinct properties. Currently, associations exist between both the esophageal and the gastric microbiomes and several chronic diseases, some of which are systemic or not localized to the upper GI tract itself. In most cases, it has not yet been determined whether these associations are causative or correlative, as research remains in the discovery phase.

THE ESOPHAGEAL MICROBIOME

Early studies of the normal esophageal microflora were undertaken to explore the role of bacteria in postoperative infection [18–20]. Cultivation yielded only a few bacterial species and, in many specimens, researchers were unable to detect any bacteria at all. The few microbes isolated were suspected of having been transiently deposited from either the mouth or stomach rather than being intrinsic to the esophagus [21].

This ambiguity regarding the esophageal microbiome was not sufficiently assessed until two important advances were made in early 2000s: studies of the microbiome by cultivation-independent technology, and visualization of bacteria on the esophageal mucosa using in situ staining. The first broad-range 16S rRNA gene survey revealed the consistent presence of a complex microbiota in the esophagus represented by six major phyla (Firmicutes, Bacteroidetes, Actinobacteria, Proteobacteria, Fusobacteria, and *TM7*) and dominated by the genus *Streptococcus* [22]. Another study successfully detected bacteria in all esophageal specimens obtained from 24 subjects [23]. Furthermore, two studies visually demonstrated the presence of bacteria in the mucosal surface using tissue Gram staining and fluorescence in situ hybridization [22,24]. These findings provided a firm foundation for the existence of a complex bacterial community in the esophagus as well as a logical basis for further exploration of its role in disease.

The recent interest in the esophageal microbiome was largely inspired by the rise in the incidence of esophageal adenocarcinoma (EAC). In the United States and Europe, squamous cell carcinoma of the esophagus comprised the vast majority of esophageal cancers until the 1970s, when the incidence of adenocarcinoma of the esophagus began to gradually increase and ultimately surpass squamous cell carcinoma as the most common esophageal cancer [25,26]. EAC is the end point of a sequence of diseases in the distal esophagus, including reflux esophagitis (RE) and Barrett's esophagus (BE).

The incidence of EAC has increased sixfold in the United States since the 1970s, parallel to a significant increase in the prevalence of RE. Although specific host factors might predispose to disease risk, the startling increase in EAC incidence suggests a predominantly environmental cause. Our understanding of EAC is mainly derived from studies of epidemiological risk factors [27,28]. Cigarette smoking, obesity, gastroesophageal reflux disorder (GERD), and insufficient consumption of fruits and vegetables account for 39.7, 41.1, 29.7, and 15.3% of EAC, respectively, with a combined population attributable risk of 78.7% [29]. However, these factors cannot consistently explain the recent increase in EAC incidence. For instance, tobacco smoking is unlikely to have a role in more recent increases of EAC in view of the decreasing prevalence of smoking (reduced from 37.1% in 1974 to 20.1% in 2004) [30].

THE MICROBIOME IN ESOPHAGEAL DISEASES

Given the lack of a known common causative risk factor for GERDs, it was hypothesized that the esophageal microbiome may be an overlooked contributor to the rise of EAC as well as RE and BE [31]. The esophageal microbiome can be classified into two types (Table 1.1). The type I microbiome, mainly associated with the normal esophagus, is a bacterial community dominated by Gram-positive bacteria from the Firmicutes phylum or *Streptococcus* at the genus level. The type II microbiome is found more often in patients with RE or BE. In the type II microbiome, there is a shift from Gram-positive species to Gram-negative species, specifically due to the diminishment of *Streptococcus* species. The decrease in *Streptococcus* organisms correlates with an increase in the relative abundance of 24 other genera. The most prominent increases involve *Veillonella, Prevotella, Haemophilus, Neisseria, Rothia, Granulicatella, Campylobacter, Porphyromonas, Fusobacterium*, and *Actinomyces*, many of which are Gram-negative anaerobes or microaerophiles, and are generally considered to be pathogens for periodontal disease as well. Overall, Gram-negative bacteria constitute 53.4% of the type II microbiome but only 14.9% of the type I microbiome. These observations

TABLE 1.1 Main Difference Between Type I and Type II Esophageal Microbiomes at the Genus Level by Relative Abundance[a]

Type I (%)	Genus	Type II (%)
78.75	*Streptococcus*	29.96
4.30	*Prevotella*	12.93
3.08	*Veillonella*	8.61
1.93	*Haemophilus*	5.75
1.03	*Rothia*	3.86
0.90	*Neisseria*	5.39
0.48	*Fusobacterium*	1.50
0.45	*Actinomyces*	1.75
0.43	*Porphyromonas*	1.43
0.35	*Granulicatella*	2.39
0.20	*Campylobacter*	1.14

[a]*Genera with relative abundance >1% in either type I or type II microbiome are listed [31].*

support the hypothesis that the esophageal microbiome is altered in reflux disorders [31,32].

In one study, bacterial quantification was performed by staining specimens from esophageal disease biopsies. It was subsequently found that bacteria were detected more often in BE than in non-BE specimens. It was also discovered that metaplasia and dysplasia were both associated with increasing bacterial stain scores [31,33,34]. Remarkably, Macfarlane et al. found that the esophagus in the majority of patients with BE was colonized with *Campylobacter* species, specifically *Campylobacter concisus* and *Campylobacter rectus*, whereas *Campylobacter* was not identified in any of the control subjects in their study. Furthermore, this result was confirmed by fluorescence in situ hybridization using specific 16S rRNA oligonucleotide probes, which demonstrated that BE mucosae are colonized by *Campylobacter* species [24,31]. It has also been documented that *C. concisus* can cause potentially DNA-damaging nitrosative and oxidative stress, and when coupled with the induction of interleukin-8 (IL-8, which induces natural killer cell activity and apoptosis), this species' colonization can increase in reflux-related esophageal disease [35]. These results further validated the previous findings by Yang et al. that the esophageal microbiome is altered in reflux disorders, specifically by an increase in Gram-negative bacteria [31]

The type II microbiome, with its larger content of Gram-negative bacteria, may also engage innate immune functions of the epithelial cells differently than the type I microbiome. A relevant explanation is that the type II microbiome produces larger amounts of Gram-negative microbe components, such as lipopolysaccharides and endotoxin [36,37]. The emptying of the stomach after a meal reduces gastric pressure, thus decreasing the opportunity for reflux. In mice treated with lipopolysaccharides, it was found that nearly 80% of food intake remains in the stomach 4 h after a meal; gastric emptying is significantly longer than in untreated mice (25% longer) [38]. The delay in gastric emptying induced by endotoxin can be blocked by pretreatment with either the nonselective cyclooxygenase (COX)-1 and -2 inhibitor indomethacin or selective COX-2 inhibition. Thus, the Gram-negative enriched type II microbiome might contribute to the development of gastroesophageal reflux by an increase in intragastric pressure caused by delayed gastric emptying.

Another factor that determines the development of pathologic reflux is the muscle tone of the lower esophageal sphincter, which serves as a gatekeeper against reflux. In a mouse model, lipopolysaccharides caused a dose-dependent fall in the basal tone. This effect can be prevented by pretreatment with the inducible nitric oxide synthase (iNOS) inhibitor L-canavanine [39]. Thus, it is speculated that the

type II microbiome causes an abnormal relaxation of the lower esophageal sphincter and contributes to the etiology of GERD and EAC. The pathway involved in the induction of iNOS and COX-2 expression by Gram-negative bacteria might start with triggering Toll-like receptors (TLRs), which are expressed in human esophageal epithelial cells. TLRs are a class of membrane-spanning proteins that recognize structurally conserved molecules derived from microbes. Activation of TLRs induces nuclear translocation of nuclear factor kappa light-chain-enhancer of activated B cells (NF-κB). This, in turn, leads to the upregulation of expression of its downstream target genes involved in inflammation, including genes for iNOS and COX-2, IL-1b and IL-8, innate immune responses, adaptive immune responses, apoptosis blocking, cell proliferation, and cell differentiation [40].

Helicobacter pylori INFECTION AND ITS EFFECT ON THE ESOPHAGUS

Before discussing the effects of *H. pylori* in gastric conditions, it is imperative to understand its influence on the esophageal mucosa, as microbes from the stomach and esophagus have been implicated in esophageal disease. While bacterial involvement has a potentially important role in inflammation and dysplasia in the esophagus as discussed previously (i.e., *Campylobacter* involvement), with exposure to bacterial products eliciting inflammation through TLR signaling, gastric acid and bile acids in the reflux contents are considered the primary factors contributing to progressive esophageal inflammation [41].

An inverse association between *H. pylori* colonization and the risk of EAC and BE has been observed in several studies [34,41]. This has been interpreted as *H. pylori* having a protective function, with the assumption that the eradication of *H. pylori* may predispose to EAC [42]. However, an alternative explanation is that the increased use of proton pump inhibitors (PPIs), which are also part of anti-*Helicobacter* treatment regimens, can modify gastric and/or esophageal microbiota toward a more carcinogenic bacterial community, and the presence or absence of *H. pylori* is not, in itself, directly implicated in these changes. Ironically, acid-suppressive therapy with PPIs has been effective in healing gastritis and esophagitis, concurrently with the increase in EAC incidence. Gastric acid is a relevant component in the consideration of chemoprevention of EAC because (1) in gastroesophageal reflux, it causes esophageal tissue damage, which is repaired by intestinal metaplasia/dysplasia; (2) acid and bile exposure have also been shown to upregulate COX-2 expression in BE, which has been implicated in esophageal carcinogenesis; and (3) the hostile gastric juice might convert the type I microbiome into the abnormal type II microbiome by killing sensitive bacteria in the esophagus [43–46].

Conversely, complete acid suppression over a period of 6 months, as measured by 24-h pH monitoring, has been shown to decrease markers of epithelial proliferation and increase cell differentiation markers in patients with BE [46]. As well as their acid secretion–suppressive effects in patients with esophagitis and GERD, which may result in gastric bacterial overgrowth and enhanced deconjugation of bile acids, PPIs also have antiinflammatory properties [41,47]. These properties, which are independent of their acid-suppressive effects, may also contribute to their chemopreventive effect against EAC [43–46]. In patients with BE, the formation of esophageal squamous islands may be induced by long-term aggressive PPI therapy and may reduce the length of the BE segment [43–46]. However, clinical studies have been inconclusive as to whether PPIs cause regression of established dysplasia in patients with BE. These studies are limited due to a short duration of PPI therapy (<6 months), small numbers of patients with dysplasia, and short-term follow-up for disease progression [43–46].

In addition to reflux disorders, the microbiome has been examined in other diseases involving the esophagus. In eosinophilic esophagitis, *Neisseria* and *Haemophilus* in the phylum Proteobacteria, as well as *Corynebacterium* in the Actinobacteria phylum are enriched, while Firmicutes organisms are depleted [48,49]. Consumption of highly allergenic foods enriches *Granulicatella* and *Campylobacter* in the esophagus, suggesting that microbes may contribute to inflammation in the esophagus [48]. In HIV infection, Firmicutes species are depleted, while Proteobacteria species are enriched. Low CD4 T-cell counts are associated with invasion of the proximal gut, in particular in the stomach and duodenum, by nitrogen-fixing bacteria *Burkholderia fungorum* and *Bradyrhizobium pachyrhizi* [50]. Since these bacteria are not detected in normal controls or in HIV-infected patients with normal CD4 counts, it was suggested that the presence of these environmental bacteria reflects a loss of colonization resistance in HIV infection and that compromised immunity could be responsible for the observed invasion by the exogenous microbes.

Although microbiomes may be altered in different esophageal conditions, esophageal pathology across a variety of disease types appears to be associated with common changes in the microbiome: depletion of the dominant phylum Firmicutes and enrichment of the proinflammatory phylum Proteobacteria. To elucidate the etiological impact of these stereotypic changes on disease progression, a potential therapy could be designed to normalize the microbiome in diseased patients by increasing the relative abundance of *Streptococcus* [34]. Therapeutic manipulation of the altered microbiome could be achieved via the use of TLR inhibitors, probiotics, prebiotics, or antibiotics.

THE GASTRIC MICROBIOME

Much like the esophageal microbiome, the gastric microbiome was long considered to be barren. The basic explanation was that a vibrant microbiome would struggle to exist in the hostile, acidic nature of the gastric lumen [51]. Additional factors that were perceived to be adverse to bacterial growth included the retrograde duodenal reflux of bile as well as toxic metabolites that are produced by oral bacteria [52–57]. For these reasons, which were further compounded by practical difficulties in sample collection and a lack of reliable diagnostic tests, this historical dogma persisted for many years.

With the discovery of *H. pylori* (a spiral shaped, ureolytic Gram-negative bacterium) colonization, it was shown that certain specialized microbes could inhabit the gastric niche. All known gastric *Helicobacter* species are urease positive and have flagella which provide motility. *H. pylori* escapes gastric elimination by two principal methods. First, *H. pylori* produces urease enzyme, which converts urea to ammonia, thereby neutralizing gastric acid and affording an alkaline micromilieu that can allow bacterial survival. Second, the bacterium penetrates the alkaline viscous mucous layer through its urea- and bicarbonate-mediated chemotactic motility. Through these processes, *H. pylori* can colonize and persist on the surface of the gastric epithelium, inducing a chronic inflammatory response within the gastric mucosa. The presence of *H. pylori,* which is assumed to be the major constituent of the gastric microbiota in more than half of the human population, has been shown to be important in several disease states, such as peptic ulcer disease, gastric cancer, and esophageal disease, as mentioned previously. With a primarily acidic environment, it was previously assumed that gastric species diversity would be low. However, as research advanced, it became increasingly apparent that although the overall biomass may be relatively small, the gastric microbiome is surprisingly diverse and includes numerous non-*Helicobacter* organisms. However, the exception is in subjects with active *H. pylori* colonization, wherein this scenario, most species are outcompeted by the predominant *H. pylori* bacteria.

Interestingly, non-*Helicobacter* species were discovered in the gastric lumen prior to the landmark finding of *H. pylori* by Warren and coworkers [58]. In particular, several acid-resistant strains of bacteria, notably *Streptococcus, Neisseria,* and *Lactobacillus* species are now considered to have originated from the oral cavity [59,60]. Currently, it is realized that at least two thirds of phylotypes found within the gastric lumen are also present within the oral cavity. Furthermore, changes in the microbiota of the oral cavity can also result in transient changes in the composition of the gastric microbiome, although the influences of these fleeting changes are generally not perceived to be significant to the host due to insufficient time to colonization within the gastric mucosa, which minimizes interaction with the host [59,61,62].

IMPACT OF *Helicobacter pylori* INFECTION ON THE COMPOSITION OF GASTRIC MICROBIOTA

The first comprehensive gastric microbiome analysis by cloning and Sanger sequencing of bacterial 16S rRNA genes showed that most of the subjects had *H. pylori* present, even though not all had clinically tested positive for *H. pylori* [60]. In subjects who had clinically tested positive for *H. pylori*, it was the predominant species (72%), while in those testing negative for *H. pylori* its relative abundance was only 11%. The core phyla of the gastric microbiome identified in this study were Proteobacteria (representing *H. pylori*; 50%), Firmicutes (30%), Bacteroidetes (10%), Actinobacteria (9%), and Fusobacteria (5%). Later studies showed similar results among the core phyla [51,60]. *H. pylori* was found to be the most abundant phylotype (42% of all sequences obtained in the first analysis), with the next most dominant genera *Prevotella* (8%) and *Streptococcus* (6%), which are also abundant in upstream oral and esophageal microbiomes. In contrast, the healthy human stomach is dominated by *Streptococcus, Prevotella, Veillonella, and Rothia* followed by *Fusobacterium, Haemophilus, Neisseria, Porphyromonas, Pasturellaceae, Propionobacterium,* and *Lactobacillus* [60,63–65]. In another study, *H. pylori* gastritis was found to be associated with higher abundance of families Bradyrhizobiaceae, Caulobacteraceae, Lactobacillaceae, and Burkholderiaceae [66]. In patients with atrophic gastritis, there is an increased abundance of *Streptococcus* species and fewer *Prevotella* species [64]. This implies that *H. pylori* is the determining factor, and in the absence of it, the core stomach microbiome is more representative of those upstream in the mouth and esophagus [51].

Bacterial diversity within the stomach determined by 16S rRNA gene survey by cloning and Sanger sequencing identified more than 100 bacterial species that reside within the normal stomach [60]. A more complete survey using NGS found 276 phylotypes representing 13 phyla. This included Chlamydiae and Cyanobacteria phyla, in addition to the core phyla stated previously [51,67]. Most of the phyla within the stomach are represented by only one or two genera, which may explain the decrease of species diversity down the GI tract from the mouth. This may constitute a selection of more specialized genera for the more demanding niches found within the GI tract. As bacteria-bearing materials (i.e., food, saliva) pass through the alimentary canal, those species that cannot survive the relatively harsh environments of the esophagus and stomach may perish, while those that have evolved specializations to survive may

remain and colonize, outcompeting the less well-equipped commensals [51].

COFACTORS IN THE DEVELOPMENT OF *Helicobacter pylori*-ASSOCIATED GASTRIC ADENOCARCINOMA

Gastric cancer is a leading cause of infection-associated cancer [68]. Although its mortality has been steadily declining since 1930, gastric cancer continues to be common in Asian countries where nearly 60% of new cases occur [69]. In the United States, its incidence rates are nearly twice in nonwhites, including African Americans, Hispanics, and Asians [70]. Among whites, rates of noncardia gastric cancer are declining in older adults but unexpectedly increasing in persons born since 1952 [71]. Colonization of *H. pylori* is a major risk factor of intestinal type noncardia gastric adenocarcinoma [72–74]. While nearly all patients diagnosed with gastric adenocarcinoma are infected by *H. pylori* some time in their life, only less than 1% of *H. pylori*-infected subjects eventually develop cancer for reasons unclear [75–77]. Potential cofactors include male gender, cigarette smoking, high intake of preserved food, meats, and salty food [69,78–83].

Several factors that mediate carcinogenesis in certain strains of *H. pylori* infection include the secretion of VacA, type IV cag secretion, and upregulation of beta-catenin within the gastric epithelium [84,85]. Host genetics also play a role. Notably, in regions of the world where *H. pylori* prevalence rates are high, such as Japan, the incidence of gastric cancer in patients with familial adenomatous polyposis (FAP, an autosomal dominant disorder due to germline *APC* mutation) ranges from 39% to 50%. In this setting, coexisting *H. pylori* infection increases the risk of gastric adenomas further [85–87].

H. pylori-associated gastric cancer is characterized by a continuum of progression from chronic superficial gastritis to chronic atrophic gastritis, intestinal metaplasia, dysplasia, and adenocarcinoma [88,89]. This sequence of events may span several decades. Although present in gastric tissue in 80–100% of patients with active acute gastritis, *H. pylori* is absent in gastric tissues in the large majority of patients diagnosed with advanced atrophic body gastritis, intestinal metaplasia or gastric cancer even when serology is positive, suggesting the disappearance of active *H. pylori* infection during later continuum of gastric cancer development [90–94]. The loss of *H. pylori* and impairment of acid secretion in these lesions may facilitate the colonization of other bacteria into the stomach, including those with nitrogen-reducing ability that are able to produce carcinogenic *N*-nitroso compounds through conversion of nitrates or nitrites [95,96]. In a study that investigated the constituents of the microbiome in chronic gastritis, intestinal metaplasia and gastric cancer, it was found that the abundance of

Helicobacteraceae was significantly reduced in the cancer group compared to the chronic gastritis and intestinal metaplasia groups, whereas Streptococcaceae were significantly increased [51,67]. Mechanistic studies in a mouse model indicate that *H. pylori* can act synergistically with even a limited gastric microbiota to promote gastric neoplasia and that a community of bacteria contribute to cancer risk [97]. These data have culminated in an alternative model for *H. pylori* promoted carcinogenesis, that is, long-term *H. pylori* infection serves as an initiator to deteriorate the gastric compartment that results in disappearing *H. pylori* and progression to cancer is determined by dysbiosis with the expansion of cancer promoting microorganisms in the absence of *H. pylori*.

THERAPEUTIC INTERVENTIONS AND THEIR EFFECT ON THE GASTRIC PHYSIOLOGY AND MICROBIOME

It has been suggested that the use of PPIs increases the rate of development of atrophic gastritis within the gastric corpus, which may inadvertently increase the risk of gastric cancer [98,99]. However, several recent studies have found no evidence of acceleration of corpus atrophy in *H. pylori*-infected subjects receiving PPI therapy [99–103]. Certainly, PPI therapy does appear to alter the distribution of *H. pylori*-induced gastritis, prompting it to move from the antrum into the more proximal corpus mucosa of the stomach. In this way, it induces what is referred to as a corpus- or body-predominant gastritis. This may have significance with respect to the subsequent risk of *H. pylori*-associated gastric cancer [99]. Whether or not PPI therapy accelerates atrophy, it is concerning that it induces the distribution of gastritis most associated with an increased risk of gastric cancer: corpus-predominant gastritis. However, the association of these two factors does not prove a cause and effect relationship. It has not been determined whether corpus gastritis by itself increases the risk of cancer or whether it is just an epiphenomenon resulting from an underlying factor that represents the link with cancer [99]. Eradication of *H. pylori* can also remodel and affect the equilibrium of the GI ecosystem. In addition, long-term PPI therapy can block microbiota-modulating agents [3,104]. It should be noted that the therapeutic eradication of *H. pylori* by no means guarantees the prevention of gastric cancer in patients with premalignant lesions, including atrophy and intestinal metaplasia [66,105].

Alternative therapies have been developed, such as probiotic agents, which may reduce the side effects of *H. pylori* eradication treatment and increase tolerability. In addition, probiotic therapy may increase the overall efficacy of the eradication. Yet, the results may vary for a number of reasons: variability in treatment protocol, different types of probiotics, and the fact that different geographical areas are likely to have discrete *H. pylori* strains, as well as differing host susceptibility. Probiotics include several microorganisms, mostly within the *Lactobacillus* or *Bifidobacterium* genus, which can be grouped under the current definition of living microorganisms and which, after being ingested in sufficient numbers, may confer health benefits beyond inherent basic nutrition [106,107]. The beneficial effects of probiotics on GI diseases, including antibiotic-associated diarrhea, have been widely described [106,108]. Numerous in vitro studies demonstrating bacterial destruction or inhibition were reproduced by preclinical and clinical studies [106]. These studies indicated that probiotics, administered alone, were only partly efficacious against *H. pylori*, but an increase of efficacy and/or reduction of side effects was evident when they were administered together with the eradication drugs [109,110]. Antibiotics are known to cause diarrhea by eliminating a significant percentage of intestinal microflora, leading to a proliferation of resistant bacterial strains as well as impairing the fermentation processes carried out by intestinal microorganisms [111]. The beneficial effects of probiotics may be explained by their ability to stimulate mucosal immune mechanisms (e.g., activation of local macrophages to increase antigen presentation and modulation of cytokine profiles). For instance, administration of probiotics containing yogurt to *H. pylori*-infected children was shown to restore the normal *Bifidobacterium* spp./*Escherichia coli* ratio, increase serum immunoglobulin A, and reduce serum IL-6 [106]. Probiotics may also produce antioxidants and antimicrobial substances, alter local pH, stimulate mucin production, strengthen the barrier function of the intestines, neutralize pathogen-derived toxins, and affect colonization by competing with pathogens for nutrients and for the binding to the host cell surface [108,112]. All of these general actions of probiotics have been proposed as operations that contribute to their efficacy in increasing *H. pylori* eradication and decreasing side effects when used together with eradication therapy.

There has been no evidence that probiotics pose a significant danger to humans as generally administered, and several strains are being used in specific disorders [108]. In one study by Pan et al., the probiotic candidate, *Lactobacillus plantarum* ZDY 2013, was investigated for this species' potential use as a protective agent against the gastric mucosal inflammation and alteration of gastric microbiota induced by *H. pylori* infection in a mouse model [113]. By using quantitative real-time PCR and high-throughput 16S rRNA gene amplicon sequencing, the authors showed that *L. plantarum* ZDY 2013 pretreatment prevented an increase in inflammatory cytokines (e.g., IL-1β and gamma interferon) and inflammatory cell infiltration in gastric lamina propria induced by *H. pylori* infection. *L. plantarum* ZDY 2013 pretreatment also prevented the alteration in gastric microbiota following *H. pylori* infection. Twenty-two bacterial taxa (e.g., Pasteurellaceae, Erysipelotrichaceae, Halomonadaceae, Helicobacteraceae, and Spirochaetaceae) overgrew in the gastric microbiota of *H. pylori*-infected mice. Most of these belonged to the Proteobacteria phylum. *L. plantarum* ZDY 2013 pretreatment prevented this alteration; only six taxa (e.g., *Lachnospiraceae,* Ruminococcaceae, and Clostridiaceae), mainly from the taxa of Firmicutes and Bacteroidetes, were dominant in the gastric microbiota of the mice pretreated with *L. plantarum* ZDY 2013. Administration of *L. plantarum* ZDY 2013 for 3 weeks led to an increase in several bacterial taxa (e.g., *Rikenella, Staphylococcus, Bifidobacterium*), although a nonsignificant alteration was found in the gastric microbiota. The authors concluded that *L. plantarum* ZDY 2013 pretreatment had an important role in preventing gastric mucosal inflammation and gastric microbiota alteration induced by *H. pylori* infection [113].

CONCLUSIONS AND PERSPECTIVES

Traditionally, study of the proximal gut microbiome has been overshadowed by the research devoted to the microbiome of the distal gut, with the exception of the consequences of *H. pylori* colonization in the stomach. However, in the past decade the knowledge regarding the relevancy of both the esophageal and gastric microbiome on human health and disease has been significantly advancing. Specific changes have now been described in the esophageal microbiome, along with the related and potentially important impact on GERD and its downstream complications of BE and EAC. *H. pylori* remains the main species of interest when discussing the gastric microbiome, due to its widespread colonization and infection rate, even in asymptomatic individuals, and its known association with peptic ulcer disease and gastric malignancy. *H. pylori* has a role in cancer initiation but colonizes gastric precancerous lesions and cancer poorly. The loss of *H. pylori* and impairment of acid secretion in these lesions might facilitate colonization of the stomach by other bacteria that promote gastric carcinogenesis. Studies investigating the influence of other gastric bacteria on the progression from intestinal metaplasia to gastric cancer are needed.

Currently, with the exception of *H. pylori*, true causative studies linking the upper GI microbiome to human disease are lacking. There are several steps needed to prove that changes in the proximal gut can alter the natural course of diseases, such as GERD, BE, and EAC. Technological advances beyond specific markers, such as 16S must continue, as metagenomics, transcriptomics, proteomics, metabolomics, and immunological profiling can help characterize the functional potential of the microbiome as well as cross talk between the microbiome and its host. Overall, the composition of the Archaeome, the Mycobiome, the Protistome and the Virome needs further attention as well, and may eventually uncover contributions from these overlooked microbes to the core microbiome or certain relevant pathological states. However, major pathophysiology in the proximal gut, such as disease progression from GERD to EAC, must be studied mechanistically, as the key mutations in carcinogenesis and specific relevance of the microbiome are still unclear. To accomplish this, animal models are needed, despite the inherent disadvantage of known animal models being unable to naturally progress from GERD to EAC. With surgical manipulation and careful selection of the appropriate animal model, findings from hypothesis generating studies can be evaluated to potentially show causality between changes in the proximal gut microbiome and associated esophageal diseases [114]. Then, novel therapies targeting the microbiome by the use of prebiotics, probiotics, and antibiotics may be developed to prevent and treat major esophageal and gastric disease.

ACKNOWLEDGMENTS

This chapter was supported in part by the Department of Pathology, New York University Langone Medical Center, the National Cancer Institute, the National Institute of Allergy and Infectious Diseases, and the National Institute of Dental and Craniofacial Research of the National Institutes of Health under award numbers UH3CA140233, U01CA182370, R01CA204113, R01CA159036, R01AI110372, and R21DE025352. ZP is a Staff Physician at the Department of Veterans Affairs New York Harbor Healthcare System. The content is solely the responsibility of the authors and does not necessarily represent the official views of the National Institutes of Health, the US Department of Veterans Affairs, or the United States government.

REFERENCES

[1] Peterson J, Garges S, Giovanni M, et al. The NIH human microbiome project. Genome Res 2009;19(12):2317–23.

[2] The International Human Microbiome Consortium, Available from: http://www.human-microbiome.org; 2008.

[3] Lopetuso LR, Scaldaferri F, Franceschi F, Gasbarrini A. The gastrointestinal microbiome–functional interference between stomach and intestine. Best Prac Res Clin gastroenterol 2014;28(6): 995–1002.

[4] Frank DN, St Amand AL, Feldman RA, Boedeker EC, Harpaz N, Pace NR. Molecular-phylogenetic characterization of microbial community imbalances in human inflammatory bowel diseases. Proc Natl Acad Sci USA 2007;104(34):13780–5.

[5] Parashar UD, Gibson CJ, Bresee JS, Glass RI. Rotavirus and severe childhood diarrhea. Emerg Infect Diseases 2006;12(2):304–6.

[6] Sartor RB. Microbial influences in inflammatory bowel diseases. Gastroenterology 2008;134(2):577–94.

[7] Li L, Mendis N, Trigui H, Oliver JD, Faucher SP. The importance of the viable but non-culturable state in human bacterial pathogens. Frontiers Microbiol 2014;5:258.

[8] Costerton JW, Stewart PS, Greenberg EP. Bacterial biofilms: a common cause of persistent infections. Science 1999;284(5418): 1318–22.

[9] Carlos N, Tang YW, Pei Z. Pearls and pitfalls of genomics-based microbiome analysis. Emerg Microbes Infect 2012;1(12):e45.

[10] Pei AY, Oberdorf WE, Nossa CW, et al. Diversity of 16S rRNA genes within individual prokaryotic genomes. Appl Environ Microbiol 2010;76(12):3886–97.

[11] Doolittle WF. Phylogenetic classification and the universal tree. Science 1999;284(5423):2124–9.

[12] Nossa CW, Oberdorf WE, Yang L, et al. Design of 16S rRNA gene primers for 454 pyrosequencing of the human foregut microbiome. World J Gastroenterol 2010;16(33):4135–44.

[13] Woese CR. Bacterial evolution. Microbiol Rev 1987;51(2):221–71.

[14] Pei A, Nossa CW, Chokshi P, et al. Diversity of 23S rRNA genes within individual prokaryotic genomes. PLoS one 2009;4(5):e5437.

[15] Hunt DE, Klepac-Ceraj V, Acinas SG, Gautier C, Bertilsson S, Polz MF. Evaluation of 23S rRNA PCR primers for use in phylogenetic studies of bacterial diversity. Appl Environ Microbiol 2006;72(3):2221–5.

[16] Pei A, Li H, Oberdorf WE, et al. Diversity of 5S rRNA genes within individual prokaryotic genomes. FEMS Microbiol Lett 2012;335(1):11–8.

[17] O'Hara AM, Shanahan F. The gut flora as a forgotten organ. EMBO Rep 2006;7(7):688–93.

[18] Lau WF, Wong J, Lam KH, Ong GB. Oesophageal microbial flora in carcinoma of the oesophagus. Aust N Z J Surg 1981;51(1):52–5.

[19] Finlay IG, Wright PA, Menzies T, McArdle CS. Microbial flora in carcinoma of oesophagus. Thorax 1982;37(3):181–4.

[20] Mannell A, Plant M, Frolich J. The microflora of the oesophagus. Ann R Coll Surg Engl 1983;65(3):152–4.

[21] Gagliardi D, Makihara S, Corsi PR, et al. Microbial flora of the normal esophagus. Dis Esophagus 1998;11(4):248–50.

[22] Pei Z, Bini EJ, Yang L, Zhou M, Francois F, Blaser MJ. Bacterial biota in the human distal esophagus. Proc Natl Acad Sci USA 2004;101(12):4250–5.

[23] Pei Z, Yang L, Peek RM Jr, Levine SM, Pride DT, Blaser MJ. Bacterial biota in reflux esophagitis and Barrett's esophagus. World J Gastroenterol 2005;11(46):7277–83.

[24] Macfarlane S, Furrie E, Macfarlane GT, Dillon JF. Microbial colonization of the upper gastrointestinal tract in patients with Barrett's esophagus. Clin Infect Dis 2007;45(1):29–38.

[25] El-Serag HB. Time trends of gastroesophageal reflux disease: a systematic review. Clin Gastroenterol Hepatol 2007;5(1):17–26.

[26] Pohl H, Welch HG. The role of overdiagnosis and reclassification in the marked increase of esophageal adenocarcinoma incidence. J Natl Cancer Inst 2005;97(2):142–6.

[27] Ong CA, Lao-Sirieix P, Fitzgerald RC. Biomarkers in Barrett's esophagus and esophageal adenocarcinoma: predictors of progression and prognosis. World J Gastroenterol 2010;16(45):5669–81.

[28] Reid BJ, Li X, Galipeau PC, Vaughan TL. Barrett's oesophagus and oesophageal adenocarcinoma: time for a new synthesis. Nat Rev Cancer 2010;10(2):87–101.

[29] Engel LS, Chow WH, Vaughan TL, et al. Population attributable risks of esophageal and gastric cancers. J Natl Cancer Inst 2003;95(18):1404–13.

[30] Levy DT, Mumford EA, Gerlowski DA. Examining trends in quantity smoked. Nicotine Tob Res 2007;9(12):1287–96.

[31] Yang L, Lu X, Nossa CW, Francois F, Peek RM, Pei Z. Inflammation and intestinal metaplasia of the distal esophagus are associated with alterations in the microbiome. Gastroenterology 2009;137(2):588–97.

[32] Shaheen N, Ransohoff DF. Gastroesophageal reflux, barrett esophagus, and esophageal cancer: scientific review. Jama 2002;287(15):1972–81.

[33] Osias GL, Bromer MQ, Thomas RM, et al. Esophageal bacteria and Barrett's esophagus: a preliminary report. Dig Dis Sci 2004;49(2):228–36.

[34] Neto AG, Whitaker A, Pei Z, editors. Microbiome and Potential Targets for Chemoprevention of Esophageal Adenocarcinoma. Seminars in Oncology. New York: Elsevier; 2015.

[35] Blackett KL, Siddhi SS, Cleary S, et al. Oesophageal bacterial biofilm changes in gastro-oesophageal reflux disease, Barrett's and oesophageal carcinoma: association or causality? Aliment Pharmacol Ther 2013;37(11):1084–92.

[36] Yang L, Chaudhary N, Baghdadi J, Pei Z. Microbiome in reflux disorders and esophageal adenocarcinoma. Cancer J 2014;20(3):207–10.

[37] Yang L, Francois F, Pei Z. Molecular pathways: pathogenesis and clinical implications of microbiome alteration in esophagitis and Barrett esophagus. Clin Cancer Res 2012;18(8):2138–44.

[38] Calatayud S, Garcia-Zaragoza E, Hernandez C, et al. Downregulation of nNOS and synthesis of PGs associated with endotoxin-induced delay in gastric emptying. Am J Physiol Gastrointest Liver Physiol 2002;283(6):G1360–7.

[39] Fan YP, Chakder S, Gao F, Rattan S. Inducible and neuronal nitric oxide synthase involvement in lipopolysaccharide-induced sphincteric dysfunction. Am J Physiol Gastrointest Liver Physiol 2001;280(1):G32–42.

[40] Rogers CJ, Prabhu KS, Vijay-Kumar M. The microbiome and obesity-an established risk for certain types of cancer. Cancer J 2014;20(3):176–80.

[41] Amir I, Konikoff FM, Oppenheim M, Gophna U, Half EE. Gastric microbiota is altered in oesophagitis and Barrett's oesophagus and further modified by proton pump inhibitors. Environ Microbiol 2014;16(9):2905–14.

[42] Blaser MJ. Disappearing microbiota: *Helicobacter pylori* protection against esophageal adenocarcinoma. Cancer Prev Res (Phila) 2008;1(5):308–11.

[43] Subramanian CR, Triadafilopoulos G. Endoscopic treatments for dysplastic Barrett's esophagus: resection, ablation, what else? World J Surg 2015;39(3):597–605.

[44] Hillman LC, Chiragakis L, Shadbolt B, Kaye GL, Clarke AC. Effect of proton pump inhibitors on markers of risk for high-grade dysplasia and oesophageal cancer in Barrett's oesophagus. Aliment Pharmacol Ther 2008;27(4):321–6.

[45] Cooper BT, Chapman W, Neumann CS, Gearty JC. Continuous treatment of Barrett's oesophagus patients with proton pump inhibitors up to 13 years: observations on regression and cancer incidence. Aliment Pharmacol Ther 2006;23(6):727–33.

[46] Singh S, Garg SK, Singh PP, Iyer PG, El-Serag HB. Acid-suppressive medications and risk of oesophageal adenocarcinoma in patients with Barrett's oesophagus: a systematic review and meta-analysis. Gut 2014;63(8):1229–37.

[47] Theisen J, Nehra D, Citron D, et al. Suppression of gastric acid secretion in patients with gastroesophageal reflux disease results in gastric bacterial overgrowth and deconjugation of bile acids. J Gastrointest Surg 2000;4(1):50–4.

[48] Benitez AJ, Hoffmann C, Muir AB, et al. Inflammation-associated microbiota in pediatric eosinophilic esophagitis. Microbiome 2015;3:23.

[49] Harris JK, Fang R, Wagner BD, et al. Esophageal microbiome in eosinophilic esophagitis. PLoS one 2015;10(5):e0128346.

[50] Yang L, Poles MA, Fisch GS, et al. HIV-induced immunosuppression is associated with colonization of the proximal gut by environmental bacteria. AIDS 2016;30(1):19–29.

[51] Nossa CW, Yang L, Pei Z. Microbiome, foregut. Encyclopedia of metagenomics: environmental metagenomics. NY, USA: Springer; 2015. p. 403–11.

[52] Duncan C, Dougall H, Johnston P, et al. Chemical generation of nitric oxide in the mouth from the enterosalivary circulation of dietary nitrate. Nat Med 1995;1(6):546–51.

[53] McKnight GM, Smith LM, Drummond RS, Duncan CW, Golden M, Benjamin N. Chemical synthesis of nitric oxide in the stomach from dietary nitrate in humans. Gut 1997;40(2):211–4.

[54] Xu J, Xu X, Verstraete W. The bactericidal effect and chemical reactions of acidified nitrite under conditions simulating the stomach. J Appl Microbiol 2001;90(4):523–9.

[55] Yang I, Nell S, Suerbaum S. Survival in hostile territory: the microbiota of the stomach. FEMS Microbiol Rev 2013;37(5):736–61.

[56] Wu WM, Yang YS, Peng LH. Microbiota in the stomach: new insights. J Dig Dis 2014;15(2):54–61.

[57] Giannella RA, Broitman SA, Zamcheck N. Gastric acid barrier to ingested microorganisms in man: studies in vivo and in vitro. Gut 1972;13(4):251–6.

[58] Bacteria in the stomach. Lancet 1981;2(8252):906–7.

[59] Kazor CE, Mitchell PM, Lee AM, et al. Diversity of bacterial populations on the tongue dorsa of patients with halitosis and healthy patients. J Clin Microbiol 2003;41(2):558–63.

[60] Bik EM, Eckburg PB, Gill SR, et al. Molecular analysis of the bacterial microbiota in the human stomach. Proc Natl Acad Sci USA 2006;103(3):732–7.

[61] Nardone G, Compare D. The human gastric microbiota: is it time to rethink the pathogenesis of stomach diseases? United European Gastroenterol J 2015;3(3):255–60.

[62] Zilberstein B, Quintanilha AG, Santos MA, et al. Digestive tract microbiota in healthy volunteers. Clinics (Sao Paulo) 2007;62(1):47–54.

[63] Li XX, Wong GL, To KF, et al. Bacterial microbiota profiling in gastritis without Helicobacter pylori infection or non-steroidal antiinflammatory drug use. PLoS one 2009;4(11):e7985.

[64] Engstrand L, Lindberg M. Helicobacter pylori and the gastric microbiota. Best practice & research Clinical gastroenterology 2013;27(1):39–45.

[65] Delgado S, Cabrera-Rubio R, Mira A, Suarez A, Mayo B. Microbiological survey of the human gastric ecosystem using culturing and pyrosequencing methods. Microb Ecol 2013;65(3):763–72.

[66] Eun CS, Kim BK, Han DS, et al. Differences in gastric mucosal microbiota profiling in patients with chronic gastritis, intestinal metaplasia, and gastric cancer using pyrosequencing methods. Helicobacter 2014;19(6):407–16.

[67] Andersson AF, Lindberg M, Jakobsson H, Backhed F, Nyren P, Engstrand L. Comparative analysis of human gut microbiota by barcoded pyrosequencing. PLoS one 2008;3(7):e2836.

[68] de Martel C, Ferlay J, Franceschi S, et al. Global burden of cancers attributable to infections in 2008: a review and synthetic analysis. Lancet Oncol 2012;13(6):607–15.

[69] Crew KD, Neugut AI. Epidemiology of gastric cancer. World journal of gastroenterology 2006;12(3):354–62.

[70] Watabe K, Nishi M, Miyake H, Hirata K. Lifestyle and gastric cancer: a case-control study. Oncol Rep 1998;5(5):1191–4.

[71] Anderson WF, Camargo MC, Fraumeni JF Jr, Correa P, Rosenberg PS, Rabkin CS. Age-specific trends in incidence of noncardia gastric cancer in US adults. Jama 2010;303(17):1723–8.

[72] Nomura A, Stemmermann GN, Chyou PH, Kato I, Perez-Perez GI, Blaser MJ. Helicobacter pylori infection and gastric carcinoma among Japanese Americans in Hawaii. N Engl J Med 1991;325(16):1132–6.

[73] Parsonnet J, Friedman GD, Vandersteen DP, et al. Helicobacter pylori infection and the risk of gastric carcinoma. N Engl J Med 1991;325(16):1127–31.

[74] Forman D, Newell DG, Fullerton F, et al. Association between infection with Helicobacter pylori and risk of gastric cancer: evidence from a prospective investigation. BMJ 1991;302(6788):1302–5.

[75] Chen Y, Segers S, Blaser MJ. Association between Helicobacter pylori and mortality in the NHANES III study. Gut 2013;62(9):1262–9.

[76] Ernst PB, Peura DA, Crowe SE. The translation of Helicobacter pylori basic research to patient care. Gastroenterology 2006;130(1):188–206. [quiz 12–3].

[77] Kusters JG, van Vliet AH, Kuipers EJ. Pathogenesis of Helicobacter pylori infection. Clin Microbiol Rev 2006;19(3):449–90.

[78] Terry MB, Gaudet MM, Gammon MD. The epidemiology of gastric cancer. Semin Radiat Oncol 2002;12(2):111–27.

[79] Correa P, Piazuelo MB, Camargo MC. The future of gastric cancer prevention. Gastric Cancer 2004;7(1):9–16.

[80] Gonzalez CA, Jakszyn P, Pera G, et al. Meat intake and risk of stomach and esophageal adenocarcinoma within the European Prospective Investigation Into Cancer and Nutrition (EPIC). J Natl Cancer Inst 2006;98(5):345–54.

[81] Ji BT, Chow WH, Yang G, et al. Dietary habits and stomach cancer in Shanghai, China. Int J Cancer 1998;76(5):659–64.

[82] Ward MH, Lopez-Carrillo L. Dietary factors and the risk of gastric cancer in Mexico City. Am J Epidemiol 1999;149(10):925–32.

[83] Lee JK, Park BJ, Yoo KY, Ahn YO. Dietary factors and stomach cancer: a case-control study in Korea. Int J Epidemiol 1995;24(1):33–41.

[84] Cao L, Yu J. Effect of Helicobacter pylori infection on the composition of gastric microbiota in the development of gastric cancer. Gastrointest Tumors 2015;2(1):14–25.

[85] Polk DB, Peek RM Jr. Helicobacter pylori: gastric cancer and beyond. Nat Rev Cancer 2010;10(6):403–14.

[86] Iida M, Yao T, Itoh H, et al. Natural history of gastric adenomas in patients with familial adenomatosis coli/Gardner's syndrome. Cancer 1988;61(3):605–11.

[87] Nakamura S, Matsumoto T, Kobori Y, Iida M. Impact of Helicobacter pylori infection and mucosal atrophy on gastric lesions in patients with familial adenomatous polyposis. Gut 2002;51(4):485–9.

[88] Correa PA. human model of gastric carcinogenesis. Cancer Res 1988;48(13):3554–60.

[89] Correa P, Haenszel W, Cuello C, Tannenbaum S, Archer M. A model for gastric cancer epidemiology. Lancet 1975;2(7924):58–60.

[90] Hirschl A, Potzi R, Stanek G, et al. Occurrence of campylobacter pyloridis in patients from Vienna with gastritis and peptic ulcers. Infection 1986;14(6):275–8.

[91] Marshall BJ, Warren JR. Unidentified curved bacilli in the stomach of patients with gastritis and peptic ulceration. Lancet 1984;1(8390):1311–5.

[92] Kwak HW, Choi IJ, Cho SJ, et al. Characteristics of gastric cancer according to Helicobacter pylori infection status. J Gastroenterol Hepatol 2014;29(9):1671–7.

[93] Karnes WE Jr, Samloff IM, Siurala M, et al. Positive serum antibody and negative tissue staining for *Helicobacter pylori* in subjects with atrophic body gastritis. Gastroenterology 1991;101(1):167–74.

[94] Galiatsatos P, Wyse J, Szilagyi A. Accuracy of biopsies for *Helicobacter pylori* in the presence of intestinal metaplasia of the stomach. Turk J Gastroenterol 2014;25(1):19–23.

[95] Ziebarth D, Spiegelhalder B, Bartsch H. *N*-nitrosation of medicinal drugs catalysed by bacteria from human saliva and gastrointestinal tract, including *Helicobacter pylori*. Carcinogenesis 1997;18(2):383–9.

[96] Dicksved J, Lindberg M, Rosenquist M, Enroth H, Jansson JK, Engstrand L. Molecular characterization of the stomach microbiota in patients with gastric cancer and in controls. J Med Microbiol 2009;58(Pt 4):509–16.

[97] Lertpiriyapong K, Whary MT, Muthupalani S, et al. Gastric colonisation with a restricted commensal microbiota replicates the promotion of neoplastic lesions by diverse intestinal microbiota in the *Helicobacter pylori* INS-GAS mouse model of gastric carcinogenesis. Gut 2014;63(1):54–63.

[98] Kuipers EJ. *Helicobacter pylori* and the risk and management of associated diseases: gastritis, ulcer disease, atrophic gastritis and gastric cancer. Aliment Pharmacol Ther 1997;11(Suppl. 1):71–88.

[99] McColl KE. *Helicobacter pylori* infection and long term proton pump inhibitor therapy. Gut 2004;53(1):5–7.

[100] Uemura N, Okamoto S, Yamamoto S, et al. Changes in *Helicobacter pylori*-induced gastritis in the antrum and corpus during long-term acid-suppressive treatment in Japan. Aliment Pharmacol Ther 2000;14(10):1345–52.

[101] Geboes K, Dekker W, Mulder CJ, Nusteling K, Dutch Study G. Long-term lansoprazole treatment for gastro-oesophageal reflux disease: clinical efficacy and influence on gastric mucosa. Aliment Pharmacol Ther 2001;15(11):1819–26.

[102] Stolte M, Meining A, Schmitz JM, Alexandridis T, Seifert E. Changes in *Helicobacter pylori*-induced gastritis in the antrum and corpus during 12 months of treatment with omeprazole and lansoprazole in patients with gastro-oesophageal reflux disease. Aliment Pharmacol Ther 1998;12(3):247–53.

[103] Singh P, Indaram A, Greenberg R, Visvalingam V, Bank S. Long term omeprazole therapy for reflux esophagitis: follow-up in serum gastrin levels, EC cell hyperplasia and neoplasia. World J Gastroenterol 2000;6(6):789–92.

[104] Merli M, Lucidi C, Di Gregorio V, et al. The chronic use of beta-blockers and proton pump inhibitors may affect the rate of bacterial infections in cirrhosis. Liver Int 2015;35(2):362–9.

[105] Wong BC, Lam SK, Wong WM, et al. *Helicobacter pylori* eradication to prevent gastric cancer in a high-risk region of China: a randomized controlled trial. Jama 2004;291(2):187–94.

[106] Ruggiero P. Use of probiotics in the fight against *Helicobacter pylori*. World J Gastrointest Pathophysiol 2014;5(4):384–91.

[107] Guarner F, Schaafsma GJ. Probiotics. International journal of food microbiology 1998;39(3):237–8.

[108] Guarner F, Khan AG, Garisch J, et al. World Gastroenterology Organisation Global Guidelines: probiotics and prebiotics October 2011. J Clin Gastroenterol 2012;46(6):468–81.

[109] Wilhelm SM, Johnson JL, Kale-Pradhan PB. Treating bugs with bugs: the role of probiotics as adjunctive therapy for *Helicobacter pylori*. Ann Pharmacother 2011;45(7–8):960–6.

[110] Patel A, Shah N, Prajapati JB. Clinical application of probiotics in the treatment of *Helicobacter pylori* infection—a brief review. J Microbiol Immunol Infect 2014;47(5):429–37.

[111] Sarowska J, Choroszy-Krol I, Regulska-Ilow B, Frej-Madrzak M, Jama-Kmiecik A. The therapeutic effect of probiotic bacteria on gastrointestinal diseases. Adv Clin Exp Med 2013;(5):759–66.

[112] Ljungh A, Wadstrom T. Lactic acid bacteria as probiotics. Curr Issues Intest Microbiol 2006;7(2):73–89.

[113] Pan M, Wan C, Xie Q, et al. Changes in gastric microbiota induced by *Helicobacter pylori* infection and preventive effects of *Lactobacillus plantarum* ZDY 2013 against such infection. J Dairy Sci 2016;99(2):970–81.

[114] Kapoor H, Lohani KR, Lee TH, Agrawal DK, Mittal SK. Animal models of Barrett's esophagus and esophageal adenocarcinoma-past, present, and future. Clin Transl Sci 2015;8(6):841–7.

Chapter 2

Characterizing and Functionally Defining the Gut Microbiota: Methodology and Implications

M. Ellermann*,¶, J.S. Carr**, A.A. Fodor†, J.C. Arthur‡ and I.M. Carroll§

*Department of Microbiology and Immunology, University of North Carolina at Chapel Hill, Chapel Hill, NC, United States; **Department of Surgery, University of North Carolina at Chapel Hill, Chapel Hill, NC, United States; †Department of Bioinformatics and Genomics, University of North Carolina at Chapel Hill, Chapel Hill, NC, United States; ‡Department of Microbiology and Immunology, Center for Gastrointestinal Biology and Disease, University of North Carolina at Chapel Hill, Chapel Hill, NC, United States; §Department of Medicine, University of North Carolina at Chapel Hill, Chapel Hill, NC, United States; ¶Department of Microbiology, University of Texas Southwestern Medical Center, Dallas, TX, United States

INTRODUCTION

The intestinal microbiota is a complex microbial community consisting of eukaryota, prokaryota, and viruses [1]. This community colonizes the lumen and mucosa of the entire alimentary canal from mouth to anus, with differences in diversity and abundance between locations [2]. Traditional culturing techniques have been used to understand which microbes survive and persist in the intestine [3]; however, these techniques are limited because a significant portion of this complex microbial community cannot be cultured [4]. Therefore, molecular techniques have been routinely applied to characterize the intestinal microbiota (Table 2.1) in healthy individuals, patients at specific disease states, and mouse models of diseases.

The depth and cost of characterizing enteric microbial communities have rapidly improved over recent years. This change is largely due to advances in sequencing technologies that allow more efficient, rapid data collection with a lower cost of materials. Specifically, clone libraries (a collection of DNA fragments stored within microbial plasmids) coupled with Sanger sequencing (sequencing based on incorporating chain-terminating dideoxynucleotides during in vitro DNA replication) would yield anywhere up to 300 16S rRNA gene reads per sample (depending on the number of clones sequenced) [5], while the new Illumina platforms (HiSeq 2500 or MiSeq) can yield up to one million 16S rRNA gene reads per sample, depending on the number of multiplexed samples included in a flow-cell. Additionally, novel creative techniques have emerged that do not simply describe the composition of the microbial community, but instead allow us to move toward understanding the functional impact of the microbiota on specific phenotypes. For example, colonizing germ free mice (described later) with human enteric microbial communities has demonstrated the functional impact of the intestinal microbiota from obese individuals on the accumulation of fat [6]. This chapter will thus review and discuss both the methods used to characterize microbial communities in the gastrointestinal tract, as well as the methods used to investigate the functional impact of microbial communities on the host.

Employing novel approaches to elucidate the functional and mechanistic impact of complex microbial communities on host physiology and disease have the potential to identify safe and effective therapies for enteric microbe-associated diseases. This approach has led to breakthroughs in utilizing fecal microbial transplants (FMT) to treat *Clostridium difficile* infections [7], metabolic syndrome [8], and ulcerative colitis (UC) [9]. As the intestinal microbiota encompasses anywhere from 500 to 1000 bacterial species [10], the next task will be to identify the microbes, or combination of microbes, within this complex community that are responsible for the beneficial outcomes of FMT. Since these microbes are effective in treating specific diseases (e.g. *C. difficile*), they have the potential to be effective probiotic organisms in prophylaxis or treatment of a number of other disorders. Thus, this chapter will also review and discuss the clinical relevance of intestinal niches where microbes reside, the known impact of the microbiota on health and disease, and the potential of microbes to act as effective probiotic therapies.

TABLE 2.1 Advantages and Disadvantages of Intestinal Microbiota Characterization Techniques

Techniques	Advantages	Disadvantages
Microbial culture	Investigate live microbial cells Characterize morphological differences between microbial colonies	A significant amount of species in the intestinal microbiota are nonculturable Subject to contamination
16S rRNA gene sequencing	Determine the abundance of taxa and diversity of microbes within a sample Multiplexing of samples makes this method cost effect	Limited taxonomic resolution Analysis is based on one gene Cannot differentiate between live and dead microbes
Metagenomic sequencing	Capable of determining the functional capacity of microbial communities	More sequences are needed per sample, therefore it is not as cost effective as 16S rRNA gene sequencing. Gene presence does not mean the gene is active
Metatranscriptomics	Capable of determining which genes are active/turned on	As some proteins are regulated in a posttranslational manner, RNA abundance can sometimes be uninformative
Proteomics/ metabolomics	Potentially identify microbial molecules responsible for host physiological status	Cannot readily differentiate between host and microbial proteins/metabolites
Gnotobiotics	Determine the functional impact of one or a community of microbes on specific host phenotypes	Expensive. Germ free mice have physiological differences compared to mice with a normal gut microbiota, which may influence outcomes

MICROBIAL NICHES OF THE INTESTINES

There are two major niches where complex microbial communities reside within the gut, namely the lumen and mucosa [11–13]. In human studies, fecal samples are routinely evaluated as a proxy for the luminal microbiota, and intestinal biopsies are evaluated as a proxy for the mucosal-associated microbiota. Since they are easily collected, microbiota studies often use fecal samples. Conversely, there are a limited number of studies characterizing the human mucosal-associated microbiota due to the difficulty and expense associated with obtaining this type of sample [14–19]. Unfortunately, the fecal microbiota often does not fully reflect the composition and metabolic function of the mucosal-associated microbiota or longitudinal distribution along the length of the intestines.

Given the differences in microbial densities and environmental factors encountered from the mouth to anus, it is not surprising that studies comparing human fecal and mucosal microbial ecosystems have reported diversity and compositional differences between the fecal and mucosal niches [17,19–23]. For example, a multicenter study identified common compositional changes in the intestinal microbiota within the mucosa of treatment naïve pediatric Crohn's disease (CD) patients that were absent in noninflamed controls [17]. Interestingly, changes to the fecal microbial community are only weakly associated with CD, highlighting the importance of niche-specificity in correlating compositional changes to the microbiota with specific diseases states. Another study investigated healthy individuals and irritable bowel syndrome (IBS) patients who were carefully screened in order to exclude GI diseases, chronic

symptoms, and recent use of interventions (e.g., antibiotics, probiotics, antidiarrheal, promotility and antiinflammatory agents) that can alter the intestinal microbiota. Analysis of the microbiota in healthy individuals using high throughput sequencing of the 16S rRNA revealed distinct differences in composition and diversity between the fecal and mucosal microbiota, with lower microbial richness and a higher abundance of Proteobacteria associated with the mucosal environment. This suggests different roles for these two distinct microbial populations within the intestinal microbiota ecosystem [23].

As the mucosal-associated microbiota is in closer proximity to host cells (e.g. epithelial and immune cells), it can be argued that examining this niche is more relevant than fecal characterization in human studies. Indeed, a study describing the intestinal microbiotas from drug-naïve newly diagnosed CD patients demonstrated that the disease associations were greater with the mucosal-associated microbiota than the fecal microbiota [17]. Of note, this study collected intestinal mucosal biopsies from patients following a bowel preparation. One concern in human microbiota studies is whether bowel preparation impacts the luminal or mucosal microbiota. Many studies investigating the mucosal microbiota collect biopsies during a screening colonoscopy where patients have undergone a bowel preparation [14–19]. Bowel preparation appears to affect both the luminal and mucosal microbiota, but these differences are far outweighed by interindividual variations [24].

In addition to the differences demonstrated in microbiotas from mucosal and fecal niches, the complexity of microbial populations is complicated by the fact that the mucus layer that lines the intestinal wall is not uniform throughout the alimentary canal, and is instead separated into two layers [25]. In the healthy colon, the inner layer is relatively sterile and the outer layer contains microbes [26]. Additionally, the mucus layer in the small intestine is irregular, but is continuous in the large bowel and stomach. Moreover, it has been reported that the interactions between enteric microbes and their hosts are likely to vary not only by the composition of the microbial community, but also with the location within the intestinal tract [27].

Understanding the differences in microbial richness, composition, and function within different intestinal niches in humans is a fundamental consideration for designing probiotics and other microbial therapies. Specifically, as the composition of microbial communities changes along the alimentary canal, the functional impact of a microbial therapy on the host may differ depending on which part of the intestines is targeted. Ultimately, with this knowledge, it may be possible to design probiotics that will have a beneficial interaction within a targeted region of the gut.

THE 16S rRNA GENE

The prokaryotic 16S ribosomal rRNA gene, which encodes rRNA found within ribosomes, is the basis for many molecular microbiota techniques. In addition to its presence within all bacterial genomes, the prokaryotic 16S rRNA gene is distinct from the eukaryotic 18S rRNA gene, and therefore serves as an ideal molecular marker for distinguishing bacteria within a given sample. The 16S rRNA gene contains 10 highly conserved regions and 9 regions that are variable between bacterial taxa as a result of evolutionary forces. This unique structure of the 16S rRNA gene is well suited for classifying bacteria using molecular methods. For example, the conserved regions enable the development of universal primers that can amplify the 16S rRNA gene from bacterial taxa using polymerase chain reaction (PCR). The unique nucleotide sequence within the variable region, which spans the two universal primer annealing sites, can then be sequenced to define bacterial identity using reference databases. Primers complementary to the variable regions can also be designed to identify and quantify specific bacteria taxa of interest using more targeted approaches, such as quantitative PCR. The 16S rRNA gene has become the basis for both targeted and high throughput sequencing techniques that have enabled the comprehensive characterization of endogenous human microbial communities and has been instrumental in highlighting their importance to human health and disease.

High Throughput Sequencing of the 16S rRNA Gene

Sequencing the 16S rRNA gene in a high throughput manner provides a broad and unbiased view of microbial community composition. Universal primers that are utilized for high throughput sequencing contain unique barcodes for multiplex community analysis, enabling simultaneous sequencing and analysis of numerous samples. The amplified 16S rRNA gene can then be sequenced on one of numerous platforms including 454-pyrosequencing [28], Illumina Hiseq, and Miseq [29], and Ion Torrent [30]. Computational platforms are subsequently applied to the 16S rRNA sequence data in order to identify bacterial taxa within each sample, to quantify the relative abundances of these taxa, and to calculate microbial ecological measures, such as the diversity or microbial richness of a community.

THE HUMAN MICROBIOTA IN HEALTH AND DISEASE

High throughput sequencing enables the analysis of numerous communities at once and the relative relationship between these communities. One analysis method evaluates the extent of similarities and differences in microbial community structure between groups, such as, disease cases versus controls. This method is called beta diversity. Another method quantifies the number of unique bacterial taxa and relative distribution within one community. This method is called alpha diversity, which can describe the diversity metrics of richness and evenness. Such analyses have been instrumental in demonstrating the potential role of resident bacteria in a wide range of pathologies and have correlated compositional changes with specific diseases—a state known as dysbiosis.

Although numerous clinical and animal studies have identified common microbial signatures that correlate with disease, one fundamental challenge has been identifying the microbial characteristics that comprise a "healthy" microbiome—the collection of bacteria and their genes within a given community. The Human Microbiome Project (HMP), funded by the National Institutes of Health, was established in 2007 to identify core microbiomes across numerous body sites and determine how microbiome changes correlate with health and disease [31]. Through the HMP, up to 18 body sites (including the nasal cavity, oral cavity, skin, stool and urogenital tract), were sampled in 242 healthy adults and underwent 16S rRNA gene sequencing [31]. The data obtained from this and other studies demonstrated that interindividual gut microbiota variability was greater than intraindividual variability over time [32,33]. This suggests that composition of the intestinal microbiota is relatively distinct between individuals, but maintains long-term compositional stability within each healthy individual. In addition, it appears possible to distinguish individuals by their gut microbiota.

The HMP revealed that defining a "core" microbiota based upon community composition will be difficult. Correlations observed between microbial composition and host genetic variation in both healthy individuals and in disease states has indicated that host genetics influence composition of the microbiota [34–36]. A further complication is the impact of diet in mediating transient changes to the intestinal microbiota. Indeed, even short-term dietary interventions appear to result in reversible compositional changes. For example, switching healthy human adults to a high animal protein diet resulted in rapid, but transient, changes that were abrogated upon termination of the dietary intervention [37]. Together, these findings highlight the difficulty in defining the compositional characteristics of a healthy microbiota within a population.

In addition to the limitations of 16S rRNA high throughput sequencing for defining a "core" microbiota, the modality lacks resolution in distinguishing compositional differences at lower taxonomies, including the species, strain and substrain levels. High functional heterogeneity can exist between different bacterial species and strains within a particular genus. For example, the *Escherichia* genus encompasses pathogenic, commensal and probiotic members that exhibit marked genetic diversity, including the presence or absence of toxins, pathogenicity islands, and other virulence factors, all of which impact their functional potential [38,39]. Relying on 16S rRNA sequencing alone ignores the functional and genetic diversity within a given bacterial family or genus, which can have important implications for microbial community structure and interactions with the host.

METAGENOMICS OF THE HUMAN MICROBIOME

Although defining a core human microbiota has thus far been elusive, the HMP and other studies have begun to characterize a core set of microbial genes and metabolic pathways present within the healthy human microbiota [31,40]. Indeed, data from the HMP demonstrated that despite high interindividual variability with microbiota composition, metagenomic analysis revealed more consistency in the bacterial genes and pathways present between individuals, across different body sites [31]. Metagenomic characterization of a microbial community utilizes shotgun high throughput sequencing to read all bacterial DNA sequences within a given sample. This contrasts with 16S rRNA sequencing, which targets only the 16S gene. Instead of defining only which microbial groups are present, metagenomic sequencing can ascertain the abundance of various bacterial genes and functional pathways within a community, and thus offers insight into the functional capacity of the microbiota. For example, metagenomics was instrumental in identifying gut bacterial enzymes required for digesting seaweed, which are present in *Bacteroides* species of the Japanese but not the North American intestinal microbiota [41]. Seaweeds are ubiquitous in the Japanese diet, therefore suggesting that coevolutionary forces and selection

pressure from a seaweed-containing diet favored the acquisition and retention of this function by the Japanese microbiota [41].

Metagenomic studies have also associated disease states with specific bacterial pathways in cases versus healthy controls. For example, both compositional shifts and bacterial metagenomes of the oral microbiota differed between periodontitis patients and healthy controls [42]. This included the enrichment of genes involved in parasitic metabolic relationships with the host, as well as resistance against xenobiotic and metal stress [42]. Similarly, another study demonstrated more consistent changes in the metagenome of CD patients rather than the abundance of particular bacterial genera. This included an increased presence of pathways involved in nutrient import, resistance to oxidative stress and virulence, the latter of which included gene clusters that enable enhanced association with host cells [43]. Therefore, when investigating the intestinal microbiota with respect to disease, the microbial taxa present may not be as important as the collective genes this community possesses.

Metagenomics Versus 16S rRNA Gene Sequencing

Metagenomics has the primary advantage of functionally—rather than just taxonomically—characterizing the capabilities of a given community. This is because the entire collection of genomes is sequenced in metagenomics, revealing the entire gene repertoire of the community. In contrast, 16S sequencing assigns taxonomy based upon a region of one conserved gene.

Another advantage of metagenomics is that it can provide higher taxonomical resolution than 16S rRNA sequencing, reaching the species and strain levels. For example, data obtained from metagenomic sequencing has identified pathogens and opportunistic pathogens in samples that would otherwise be indistinguishable using 16S rRNA sequencing methods alone [31]. Nonetheless, 16S rRNA sequencing is more efficient and cost-effective for general community characterization. Furthermore, 16S rRNA sequencing can easily be applied to microbial communities intimately associated with host tissues—such as the microbiota of intestinal tissue biopsies—because PCR amplification of the 16S rRNA gene eliminates mammalian DNA contamination. Sample preparation for metagenomic sequencing does not remove mammalian DNA, and thus greater read depth is required to generate statistically sound comparisons between microbial metagenomes.

Despite the advantages of metagenomic approaches over high throughput 16S rRNA sequencing, advancements in utilizing 16S rRNA sequencing to predict the functional capacity of a bacterial community have been achieved via design of complex computer algorithms. One such program, the Phylogenetic Investigation of Communities by Reconstruction of Unobserved States (PICRUSt.), has been applied to the 16S rRNA sequencing data from the HMP, and recapitulated the metagenomic sequence data with approximately 80% accuracy [44]. PICRUSt. has also been applied to other datasets generated through 16S rRNA sequencing to infer functional changes to the intestinal microbiota during disease states [17,45]. For example, the application of PICRUSt. on 16S rRNA sequencing data obtained from intestinal mucosal samples distinguished patients with CD versus UC, as well as inflamed versus noninflamed mucosal samples obtained from individual UC patients [45]. Significant changes to the abundance of predicted bacterial pathways present in the mucosa of CD patients relative to healthy controls identified putative mucosally-associated bacterial functions associated with CD [17].

Although PICRUSt. is simply a predictive computational tool that utilizes 16S rRNA sequencing data, this and similar approaches have some advantages over metagenomic sequencing. For example, PICRUSt. may be more feasible for sample types with high levels of contaminating host DNA (such as mucosal biopsies) that can interfere with metagenomic sequencing and analysis. PICRUSt. can also be applied to past studies that have solely utilized 16S rRNA sequencing to gain insight into functional changes that occur within the microbiota before investing in metagenomic sequencing.

TRANSCRIPTOMICS AND PROTEOMICS OF THE HUMAN MICROBIOME

While data obtained from metagenomic studies can link particular microbial community functions to disease or healthy states, the mere presence of particular genes or pathways does not necessarily signify they are actively transcribed, or are even functional. Point mutations within a gene or within distant genetic loci can render complete pathways inactive. For example, most *Escherichia coli* strains harbor a gene cluster that enables biosynthesis of curli fimbriae, an extracellular component of biofilms [46]. However, some point mutations have inactivated this gene cluster in select strains, while other mutations have restricted curli production to ambient temperatures, indicating that curli expression is likely inactive in these strains when within the host environment [46–48]. Such information is not possible to glean from metagenomic studies. Thus, although metagenomic sequencing does provide insight into the functional capacity of a particular bacterial community, other approaches, including metatranscriptomics, metaproteomics, and metabolomics, are then required to inform us how bacteria in the community are functioning in real-time.

Metatranscriptomics, or community-wide bacterial RNA-seq, uses high throughput RNA sequencing to identify and quantify active transcription of genes present within a bacterial community. Nonetheless, since many microbial functions are regulated at the posttranscriptional level, comparing bacterial proteins (metaproteomics) and metabolites (metabolomics) can provide even further insight into the functional activities of different communities. Metaproteomics incorporates chromatography and shotgun mass spectroscopy to quantify proteins present within a sample. When this approach was applied to human fecal samples, for example, bacterial and human proteins could be distinguished from one another, thus producing a nonbiased view of host–microbial interactions on a community-wide scale [49]. This approach also detected plant proteins derived from the human diet, which comprised approximately 2–9% of the proteins within the fecal sample, whereas 30–35% of proteins were of host (human) origin [49].

To compliment metaproteomics, analytical chemistry tools like mass spectrometry and nuclear magnetic resonance (NMR) spectrometry can be utilized in an approach collectively known as metabolomics [50,51]. Metabolomics compares the relative abundances of various small molecules and metabolites, such as fatty acids, amino acids, and their intermediates. However, unlike proteomics, results obtained through metabolomics currently cannot distinguish whether metabolites are of human or microbial origin. All three of these approaches have been used to correlate alterations in microbial function with specific diseases states [52–55].

16S rRNA GENE SEQUENCING DATA ANALYSIS

Data analysis of metagenomic sequences can be challenging. We therefore simplify this topic and focus on the analysis of 16S rRNA sequences for this particular section. While there are a number of popular software packages that can be used to analyze 16S rRNA sequences (including the popular programs Qiime [56] and Mothur [57]), nearly every step of the data analysis pipeline has been the subject of some controversy in the literature. Typically, the initial processing of sequence data removes low quality sequences, and then classifies the remaining sequences by comparison to databases of 16S rRNA sequences such as Ribosomal Database Project (RDP) [58] or Greengenes [59]. This has been criticized because different taxonomies can be inconsistent, especially at more refined levels of the phylogenetic tree.

Another consistent source of controversy in the literature focuses on the compositional nature of sequence data. All count-based sequence data is compositional, which means that different samples have different numbers of sequences. Before statistical analysis can commence, therefore some normalization must be applied to correct for variable sequencing depths. The correct normalization scheme to apply has been particularly contentious. The Qiime pipeline often utilizes rarefaction, in which sequences for downstream analysis are chosen at random from each sample until each sample in the analysis has the same number of sequences. It has been argued that rarefaction reduces the accuracy and power of downstream analyses, as information is removed when many of the sequences in a sequencing run are ignored [60]. This assertion is controversial, however, and no clear consensus has been reached in the literature regarding the best normalization scheme. After normalization has been achieved, the next requirement of most microbiome studies is the application of a statistical model to identify taxa that are different between biological conditions, such as healthy and disease states. As with other areas of data analysis, there are many different competing statistical models and there is little consensus as to which group of models yields the best results. While the consequences of using these different methods will undoubtedly continue to be the focus of research, given the contentious nature of existing literature, it seems unlikely that we can expect the emergence of a consensus pipeline in the near future.

As a result, it may be useful to run datasets through multiple pipelines and then examine the results for consistency. If associations of the microbiome with health and disease phenotypes are robust, then it should be true that significant results can be observed across multiple pipelines. Indeed, an early result of the Microbiome Quality Control Project is that while steps taken during "wet-lab" sample processing (especially specific extraction procedures) can have an enormous impact on biasing the final results, different procedures performed during informatics processing often have much more modest impacts on the outcome. It may be the case that despite attracting a good deal of controversy, the choices we make in data analysis end up mattering less than choices made in the design and execution of our experiments, and that the bias associated with different informatics pipelines is often small when compared to the biological noise that all of these pipelines must evaluate.

DEFINING MICROBIOTA FUNCTION THROUGH GNOTOBIOTICS

As discussed previously, typical approaches to identify specific members of the intestinal microbiota within a disease-state have included high throughput sequencing methods to correlate the abundance of a microbe, or composition of a microbial community with and within a particular host phenotype. The success of these approaches is rare, as the causal relationship between a microbe and a specific host phenotype may be dependent on the activity of the microbe rather than the abundance of microbes or the expression of genes, gene transcription, and proteins. Thus, alternate methods must be used to determine the functional impact of a particular microbe, or community of microbes, on a host's phenotype.

Gnotobiotics is a powerful tool enabling the study of host/microbe interactions, and offers an investigator a platform to determine the functional impact of a complex microbial community on host physiology. Interestingly, gnotobiotic is derived from the Greek words "gnotos" meaning known and "bios" meaning life. Gnotobiotic animals live in the absence or in association with known viable microorganisms. This is achieved by rearing animals in sterile Trexler isolators where they can be germ free (GF), mono-associated with a pure culture of a microbial strain, dual-associated with two pure cultures of microbial strains, or associated with a consortium of known or unknown microbes. The animals are then observed to determine the impact of each microbial state on a particular health or disease outcome.

The concept of gnotobiotics stems from Pasteur's work in 1885 [61] where he speculated that life would become impossible with the elimination of microbial associates (i.e. a GF state). In contrast, Elle Metchnikoff a few years later declared microbes to be antagonists for the well-being of the host [62]. It was not until the 1940's when rats and chickens raised in a GF state emerged as definitive proof that life could exist in the absence of microorganisms [63,64]. Subsequent work with gnotobiotic animals quickly revealed several characteristics associated with the GF state: (1) underdevelopment of the immune system, (2) increased resistance to radiation, (3) and an enlarged cecum [65].

Given that life is possible in a GF state, and that life without microbes endows abnormal physiological characteristics on the host, Pasteur and Metchnikoff's initial hypotheses were not entirely incorrect. It could be concluded that intestinal microbiotas or the products they produce found on and in an animal host are needed for normal physiological development. Nonetheless, the ability to control the microbial state of an animal has placed gnotobiotics as an effective tool for establishing whether structural or functional configurations of enteric microbial communities are causally related to a given physiologic disease or phenotype.

Gnotobiotics, Enteric Microbes and Host Adiposity

Consistent evidence implicates the intestinal microbiota in the extraction and storage of fat (adiposity). However, specific microbial taxa and the degree to which they contribute to extraction remains controversial. Initial studies reported an association between murine obesity and compositional alterations in the intestinal microbiota. Specifically, the enteric microbiota in genetically obese mice exhibited a 50% reduction in the abundance of the Bacteroidetes phylum and a similar increase in the Firmicutes phylum [66]. A subsequent study revealed that, in a human population, fewer Bacteroidetes and more Firmicutes were present in the intestinal microbiota of obese individuals when compared to their lean controls [5]. Additionally, a 1-year low-calorie diet (either low-fat or low-carbohydrate) was associated with a rise in the relative level of Bacteroidetes and a decrease in the relative level of Firmicutes in the obese group. Subsequent research both confirmed [67–69] and refuted [70–73] the association between obesity and these trends in Firmicutes and Bacteroidetes phyla in human populations. Irrespective of the specific taxa that may be associated with obesity, these studies were fundamentally descriptive and could not establish causality.

To establish causality between the intestinal microbiota and obesity, investigators turned to gnotobiotic experiments, where GF mice received transplants of intact uncultured fecal microbiotas from mouse and human sources. It was found that the intestinal microbiota in genetically obese mice was more effective at extracting calories from food than that of lean mice, and that this phenotype could be passed to GF mice via fecal transplant, resulting in increased adiposity [74]. This study established that an obesity-like phenotype was transmissible between mice via the intestinal microbiota.

To establish causality between obesity and enteric microbial communities from a human source, investigators used a similar gnotobiotic study design with fecal microbiotas from twins discordant for obesity. This population and study design offered an opportunity to examine interrelations between obesity, diet, and the intestinal microbiota. Twins discordant for obesity provide an attractive model for studying host-gut microbiota interactions, as the healthy twin is a valuable reference control to contrast with the cotwin's obese-associated enteric microbial community. Transferring fecal samples from obese adult females into GF mice resulted in increased body fat, fat mass, and obesity-associated metabolic phenotypes. Significantly, cohousing wild type mice harboring an obese twin's microbiota with wild type mice containing the lean cotwin's microbiota prevented the development of increased body mass and obesity-associated metabolic phenotypes in the mice colonized with the obese-associated gut microbiota [6]. This suggests that there are enteric microbes within a lean individual that can prevent obesity-related traits. Therefore, identifying and culturing these microbes in a pure form could lead to novel probiotics that treat obesity in a human population. Indeed, the authors of this study identify specific *Bacteroides* species that are associated with the prevention of increased adiposity.

Gnotobiotics can Demonstrate the Functional Impact of Cultured Enteric Microbes

The intestinal microbiotas within a population is estimated to encompass up to 1000 different bacterial species, whereas an average individual's intestine harbors approximately 100–200 unique bacterial species [75,76]. This high level of diversity makes it extremely difficult to identify the microbe, or combination of microbes, that functionally impact specific aspects of the host (e.g. intestinal physiology, absorption, or immune function). In order to isolate and determine the impact of enteric microbes within a diverse microbial community, the microbes that influence the host phenotype must be identified and cultured. Using a single rich medium and anaerobic growth conditions, it is possible to generate culture collections of bacterial strains that originated from an individual's gut. In fact, at least 50% of the taxonomic species identified by sequencing methods can be cultured from frozen fecal samples [4,6,76].

Determining the effect of a cultured microbe (or a combination of microbes) on a host, even with 50% of the members of the intestinal microbiota, would require testing an inordinate number of combinations. Therefore, one group developed a creative method to identify enteric microbial effector strains that influenced immune responses [77]. Based on the concept that microbe-driven immune phenotypes are likely to be additive and involve two or more specific microbes eliciting a response, it was postulated that the complexity of determining host-microbe interactions could be dramatically reduced. The essential task was to determine which culturable components of the intestinal microbiota could transmit the same host response seen with fecal colonization in GF mice. A system for generating arrayed collections of cultured anaerobic bacterial members from a human intestinal microbiota sample in a multiwell system was established, in which each well contained a cultured strain from an intestinal sample [4]. Combining culturable strains into subsets with overlapping strains in a range of sizes (e.g. 3, 5, or 10 members), and then transferring the subsets into GF mice, allowed the number of strains necessary to influence the host's phenotype of interest to be determined. Bacterial strains that are present in a tested subset that best explain an observed phenotypic influence on the host were then identified and validated alone or in combination with other microbes in a GF recipient. This approach led to the identification of specific enteric microbial strains that influence colonic regulatory T cell accumulation, adiposity, and cecal metabolites [78]. While no single approach is devoid of limitations, this method outlines a manner in which microbes that impact specific components of the host can be cultured and further characterized.

Gnotobiotics, Enteric Microbes and Host Behavior

A link between the enteric microbiota, the intestine, and the brain has been reported [79,80]. This collective body of evidence suggests a bidirectional communication between the intestinal microbiota and brain function both in times of homeostasis and disease [81]. Communication appears to occur via complex interactions with the hypothalamic–pituitary–adrenal (HPA) axis and structures in the central nervous system that can affect cognition, mood, and emotion. For example, investigators found that inducing stress (also implicated in the pathogenesis of depression and composition changes in the gut microbiota) in GF mice caused an exaggerated response in the HPA axis that was reversed by *Bifidobacterium infantis*, a normal resident of the gut microbiota [82]. Furthermore, GF BALB/c mice colonized with an NIH Swiss intestinal microbiota showed reduced anxiety-like behavior compared to GF NIH Swiss mice colonized with a BALB/c microbiota, indicating that gut microbe-associated changes in anxiety are transmissible and are affected by the composition of the intestinal microbiota [83].

To address whether enteric microbes can beneficially impact mood and behavior, a research group showed that a probiotic originally isolated from the intestinal microbiota of a healthy individual (*Lactobacillus rhamnosus* JB-1) was able to reduce anxiety- and stress-related behavior in mice with a normal gut microbiota via modulation of the expression of gamma-aminobutyric acid (GABA) in the brain [84]. These biological and behavioral effects were not seen in vagotomized mice, illustrating the critical role of microbe–gut–brain communication [84]. This study suggests that "psychobiotics" (probiotics that effect mood) could potentially benefit behavior in humans. Indeed, a placebo controlled study was designed where healthy female volunteers consumed a fermented milk product (FMP) containing a probiotic cocktail, a non-FMP, or received no intervention. After 4 weeks, their intrinsic connectivity and responses to emotional attention tasks were assessed [85]. This revealed that ingestion of a FMP with probiotics was associated with changes in midbrain connectivity. Although further studies are needed to elucidate the complex interactions between the intestinal microbiota and mood/behavior, this field is particularly rich for exploration and has the potential to treat mood disorders and cognition.

INTEGRATING METAOMIC APPROACHES TO ASSESS THE EFFICACY OF PREBIOTIC AND PROBIOTIC INTERVENTIONS

The enrichment of particular bacterial taxa, genes or pathways, as well as overall changes to community structure or function, have been linked to a wide variety of pathological states. This provides rationale to mine the human microbiome for reliable diagnostic and prognostic markers to monitor susceptibility, development, progression, and resolution of these diseases. Prebiotic and probiotic interventions can potentially be utilized to manage or treat diseases associated with dysbiosis, perhaps by preventing further detrimental changes to the community or reverting the community back to a more stable and symbiotic state. High throughput sequencing and analytical chemistry techniques can provide the means to determine the impact of prebiotic and

probiotic interventions on the human microbiome in a non-biased manner. Indeed, several clinical studies have investigated the impact of prebiotics and probiotics on the human microbiome. Two studies reported minimal effects on the composition of the salivary and intestinal microbiotas with the consumption of probiotics by healthy adults [86]. Similarly, consumption of a FMP containing 5 probiotic strains, or administration of *L. rhamnosus* GG, did not significantly alter the intestinal microbiota or metagenome in healthy adults [87,88]. Instead, significant changes were observed within the fecal metatranscriptome during the intervention period, suggesting that the functional capabilities of the microbiota had changed [87,88]. These functional changes to the fecal microbiome are likely transient changes that require regular probiotic consumption. In contrast to healthy adults, in IBS patients, FMP, and probiotic strains impacted the intestinal metagenome [89]. FMP increased endogenous butyrate producers and decreased known pathobionts such as *Bilophila*, which corresponded with an improvement of IBS-related symptoms [89]. Subsequent studies using a colonic fermenter inoculated with human feces demonstrated that exposure to FMP stimulated butyrate production [89], although metabolomic studies are needed to demonstrate the same effect in vivo. Taken together, the impact of prebiotic and probiotic interventions on the microbiome depends on numerous factors including the health status of the individual, the body region sampled and the type of readouts analyzed. Thus, future animal and clinical studies that integrate several metaomic approaches are needed to characterize the impact and mechanism of action of resident microbes as well as probiotic treatments on the health status of the host.

CONCLUSIONS

The mutualistic relationship between the human host and the microbial communities that reside in the gut has been well documented. Investigation of this relationship has been made possible via the continuing advancement of high throughput sequencing technologies and the creative methods employed to characterize and interpret the resulting data. Investigations regarding the role the intestinal microbiota plays in disease is moving from cross-sectional observational studies to more mechanistic approaches where the functional impact of a microbial community on the host can be accessed via gnotobiotics. Specifically, gnotobiotic experimental approaches have the potential to determine whether a complex microbial community, or a member of this community, from an individual with an active disease state (e.g. inflammatory bowel diseases, obesity, or IBS) can initiate the disease, or exacerbate disease symptoms. These investigations have led to the identification of effector strains that in the future may serve as novel next-generation probiotics, or targets for probiotics, that are better suited for therapy as their effect on a

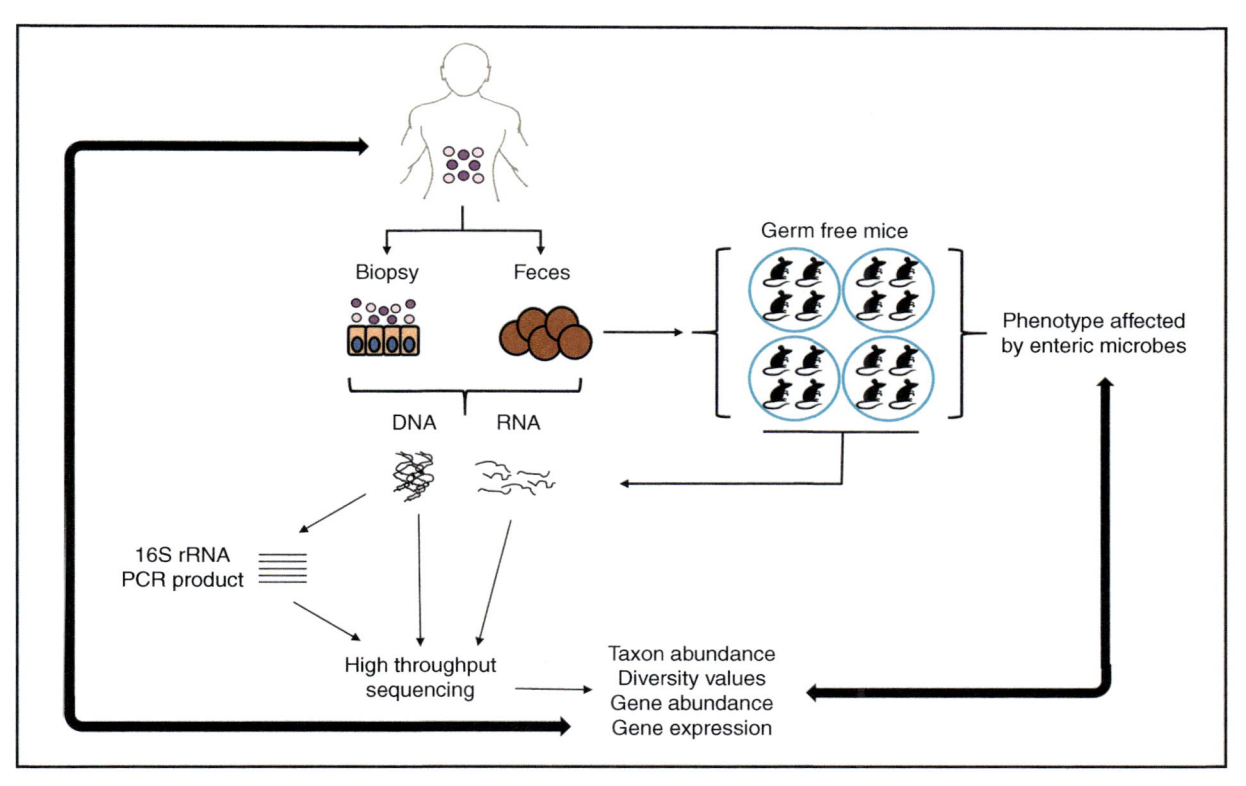

FIGURE 2.1 **Schematic diagram of methods used to characterize and determine the functional impact of enteric microbial communities on host phenotypes.**

host phenotype or cell type has been characterized. However, the exact mechanism(s) in which specific effector microbes initiate, perturb, or maintain a disease state remains to be elucidated. Additionally, given the enormous surface area of the intestine, the niche where these microbes exert their effect remains to be determined.

As modulation of the enteric microbiota through FMT has become an effective therapy for *C. difficile* infection, the next step will be to identify the microbes, or the products that they produce, that are associated with the beneficial effects of this treatment. This has the potential to lead to more targeted approaches for intestinal microbiota-based therapies and next generation probiotics. We have rapidly gained a considerable amount of knowledge regarding the relationship between our enteric microbes and health. However, there is still much we do not know about these microbial communities, and future research has the potential to generate novel microbe-based therapies for dysbiotic-associated diseases (Fig. 2.1).

REFERENCES

[1] Backhed F, Ley RE, Sonnenburg JL, et al. Host-bacterial mutualism in the human intestine. Science 2005;307(5717):1915–20.

[2] Zhang Z, Geng J, Tang X, et al. Spatial heterogeneity and co-occurrence patterns of human mucosal-associated intestinal microbiota. Isme J 2014;8(4):881–93.

[3] McBain AJ, Macfarlane GT. Investigations of bifidobacterial ecology and oligosaccharide metabolism in a three-stage compound continuous culture system. Scand J Gastroenterol Suppl 1997;222: 32–40.

[4] Goodman AL, Kallstrom G, Faith JJ, et al. Extensive personal human gut microbiota culture collections characterized and manipulated in gnotobiotic mice. Proc Natl Acad Sci USA 2011;108(15):6252–7.

[5] Ley RE, Turnbaugh PJ, Klein S, et al. Microbial ecology: human gut microbes associated with obesity. Nature 2006;444(7122):1022–3.

[6] Ridaura VK, Faith JJ, Rey FE, et al. Gut microbiota from twins discordant for obesity modulate metabolism in mice. Science 2013;341(6150):1241214.

[7] van Nood E, Vrieze A, Nieuwdorp M, et al. Duodenal infusion of donor feces for recurrent *Clostridium* difficile. N Engl J Med 2013;368(5):407–15.

[8] Vrieze A, Van Nood E, Holleman F, et al. Transfer of intestinal microbiota from lean donors increases insulin sensitivity in individuals with metabolic syndrome. Gastroenterology 2012;143(4): 913–6. e7.

[9] Moayyedi P, Surette MG, Kim PT, et al. Fecal microbiota transplantation induces remission in patients with active ulcerative colitis in a randomized controlled trial. Gastroenterology 2015;149(1):102–9. e6.

[10] Ley RE, Hamady M, Lozupone C, et al. Evolution of mammals and their gut microbes. Science 2008;320(5883):1647–51.

[11] Eckburg PB, Bik EM, Bernstein CN, et al. Diversity of the human intestinal microbial flora. Science 2005;308(5728):1635–8.

[12] Van den Abbeele P, Van de Wiele T, Verstraete W, et al. The host selects mucosal and luminal associations of coevolved gut microorganisms: a novel concept. FEMS Microbiol Rev 2011;35(4):681–704.

[13] Walker AW, Sanderson JD, Churcher C, et al. High-throughput clone library analysis of the mucosa-associated microbiota reveals dysbiosis and differences between inflamed and non-inflamed regions of the intestine in inflammatory bowel disease. BMC Microbiol 2011; 11:7.

[14] Sokol H, Lepage P, Seksik P, et al. Molecular comparison of dominant microbiota associated with injured versus healthy mucosa in ulcerative colitis. Gut 2007;56(1):152–4.

[15] Hong PY, Croix JA, Greenberg E, et al. Pyrosequencing-based analysis of the mucosal microbiota in healthy individuals reveals ubiquitous bacterial groups and micro-heterogeneity. PLoS One 2011;6(9):e25042.

[16] Parkes GC, Rayment NB, Hudspith BN, et al. Distinct microbial populations exist in the mucosa-associated microbiota of subgroups of irritable bowel syndrome. Neurogastroenterol Motil 2012;24(1):31–9.

[17] Gevers D, Kugathasan S, Denson LA, et al. The treatment-naive microbiome in new-onset Crohn's disease. Cell Host Microbe 2014;15(3):382–92.

[18] Frank DN, St Amand AL, Feldman RA, et al. Molecular-phylogenetic characterization of microbial community imbalances in human inflammatory bowel diseases. Proc Natl Acad Sci USA 2007;104(34):13780–5.

[19] Lyra A, Forssten S, Rolny P, et al. Comparison of bacterial quantities in left and right colon biopsies and faeces. World J Gastroenterol 2012;18(32):4404–11.

[20] Carroll IM, Chang YH, Park J, et al. Luminal and mucosal-associated intestinal microbiota in patients with diarrhea-predominant irritable bowel syndrome. Gut Pathog 2010;2(1):19.

[21] Carroll IM, Ringel-Kulka T, Keku TO, et al. Molecular analysis of the luminal- and mucosal-associated intestinal microbiota in diarrhea-predominant irritable bowel syndrome. Am J Physiol Gastrointest Liver Physiol 2011;301(5):G799–807.

[22] Araujo-Perez F, McCoy AN, Okechukwu C, et al. Differences in microbial signatures between rectal mucosal biopsies and rectal swabs. Gut Microbes 2012;3(6):530–5.

[23] Ringel Y, Maharshak N, Ringel-Kulka T, et al. High throughput sequencing reveals distinct microbial populations within the mucosal and luminal niches in healthy individuals. Gut Microbes 2015;6(3):173–81.

[24] Shobar RM, Velineni S, Keshavarzian A, et al. The effects of bowel preparation on microbiota-related metrics differ in health and in inflammatory bowel disease and for the mucosal and luminal microbiota compartments. Clin Transl Gastroenterol 2016;7:e143.

[25] Atuma C, Strugala V, Allen A, et al. The adherent gastrointestinal mucus gel layer: thickness and physical state in vivo. Am J Physiol Gastrointest Liver Physiol 2001;280(5):G922–9.

[26] Johansson ME, Phillipson M, Petersson J, et al. The inner of the two Muc2 mucin-dependent mucus layers in colon is devoid of bacteria. Proc Natl Acad Sci USA 2008;105(39):15064–9.

[27] Li H, Limenitakis JP, Fuhrer T, et al. The outer mucus layer hosts a distinct intestinal microbial niche. Nat Commun 2015;6:8292.

[28] Hamady M, Walker JJ, Harris JK, et al. Error-correcting barcoded primers for pyrosequencing hundreds of samples in multiplex. Nat Methods 2008;5(3):235–7.

[29] Bartram AK, Lynch MD, Stearns JC, et al. Generation of multimillion-sequence 16S rRNA gene libraries from complex microbial communities by assembling paired-end illumina reads. Appl Environ Microbiol 2011;77(11):3846–52.

[30] Whiteley AS, Jenkins S, Waite I, et al. Microbial 16S rRNA Ion Tag and community metagenome sequencing using the Ion Torrent (PGM) Platform. J Microbiol Methods 2012;91(1):80–8.

[31] Human Microbiome Project Consortium. Structure, function and diversity of the healthy human microbiome. Nature 2012;486(7402): 207–14.

[32] Flores GE, Caporaso JG, Henley JB, et al. Temporal variability is a personalized feature of the human microbiome. Genome Biol 2014;15(12):531.

[33] Caporaso JG, Lauber CL, Costello EK, et al. Moving pictures of the human microbiome. Genome Biol 2011;12(5):R50.

[34] Blekhman R, Goodrich JK, Huang K, et al. Host genetic variation impacts microbiome composition across human body sites. Genome Biol 2015;16:191.

[35] Frank DN, Robertson CE, Hamm CM, et al. Disease phenotype and genotype are associated with shifts in intestinal-associated microbiota in inflammatory bowel diseases. Inflamm Bowel Dis 2011;17(1):179–84.

[36] Knights D, Silverberg MS, Weersma RK, et al. Complex host genetics influence the microbiome in inflammatory bowel disease. Genome Med 2014;6(12):107.

[37] David LA, Maurice CF, Carmody RN, et al. Diet rapidly and reproducibly alters the human gut microbiome. Nature 2014;505(7484): 559–63.

[38] Johnson TJ, Nolan LK. Pathogenomics of the virulence plasmids of *Escherichia coli*. Microbiol Mol Biol Rev 2009;73(4):750–74.

[39] Lukjancenko O, Wassenaar TM, Ussery DW. Comparison of 61 sequenced *Escherichia coli* genomes. Microb Ecol 2010;60(4):708–20.

[40] Turnbaugh PJ, Quince C, Faith JJ, et al. Organismal, genetic, and transcriptional variation in the deeply sequenced gut microbiomes of identical twins. Proc Natl Acad Sci USA 2010;107(16):7503–8.

[41] Hehemann JH, Correc G, Barbeyron T, et al. Transfer of carbohydrate-active enzymes from marine bacteria to Japanese gut microbiota. Nature 2010;464(7290):908–12.

[42] Liu Y, Zhang Y, Wang L, et al. Prevalence of Porphyromonas gingivalis four rag locus genotypes in patients of orthodontic gingivitis and periodontitis. PLoS One 2013;8(4):e61028.

[43] Morgan XC, Tickle TL, Sokol H, et al. Dysfunction of the intestinal microbiome in inflammatory bowel disease and treatment. Genome Biol 2012;13(9):R79.

[44] Langille MG, Zaneveld J, Caporaso JG, et al. Predictive functional profiling of microbial communities using 16S rRNA marker gene sequences. Nat Biotechnol 2013;31(9):814–21.

[45] Davenport M, Poles J, Leung JM, et al. Metabolic alterations to the mucosal microbiota in inflammatory bowel disease. Inflamm Bowel Dis 2014;20(4):723–31.

[46] Romling U, Sierralta WD, Eriksson K, et al. Multicellular and aggregative behaviour of *Salmonella* typhimurium strains is controlled by mutations in the agfD promoter. Mol Microbiol 1998;28(2): 249–64.

[47] Viazis N, Rekoumis G, Vlachogiannakos J, et al. Effect of octreotide and corticosteroids on human sphincter of oddi motility. J Gastroenterol Hepatol 2004;19(1):116–7.

[48] Uhlich GA, Keen JE, Elder RO. Mutations in the csgD promoter associated with variations in curli expression in certain strains of *Escherichia coli* O157:H7. Appl Environ Microbiol 2001;67(5):2367–70.

[49] Verberkmoes NC, Russell AL, Shah M, et al. Shotgun metaproteomics of the human distal gut microbiota. Isme J 2009;3(2): 179–89.

[50] Kolmeder CA, de Been M, Nikkila J, et al. Comparative metaproteomics and diversity analysis of human intestinal microbiota testifies for its temporal stability and expression of core functions. PLoS One 2012;7(1):e29913.

[51] Wu GD, Compher C, Chen EZ, et al. Comparative metabolomics in vegans and omnivores reveal constraints on diet-dependent gut microbiota metabolite production. Gut 2016;65(1):63–72.

[52] Kang D, Shi B, Erfe MC, et al. Vitamin B_{12} modulates the transcriptome of the skin microbiota in acne pathogenesis. Sci Transl Med 2015;7(293):293ra103.

[53] Ponnusamy K, Choi JN, Kim J, et al. Microbial community and metabolomic comparison of irritable bowel syndrome faeces. J Med Microbiol 2011;60(Pt 6):817–27.

[54] Haberman Y, Tickle TL, Dexheimer PJ, et al. Pediatric Crohn disease patients exhibit specific ileal transcriptome and microbiome signature. J Clin Invest 2014;124(8):3617–33.

[55] Le Gall G, Noor SO, Ridgway K, et al. Metabolomics of fecal extracts detects altered metabolic activity of gut microbiota in ulcerative colitis and irritable bowel syndrome. J Proteome Res 2011;10(9): 4208–18.

[56] Caporaso JG, Kuczynski J, Stombaugh J, et al. QIIME allows analysis of high-throughput community sequencing data. Nat Methods 2010;7(5):335–6.

[57] Schloss PD, Westcott SL, Ryabin T, et al. Introducing mothur: open-source, platform-independent, community-supported software for describing and comparing microbial communities. Appl Environ Microbiol 2009;75(23):7537–41.

[58] Cole JR, Chai B, Farris RJ, et al. The ribosomal database project (RDP-II): introducing *myRDP* space and quality controlled public data. Nucleic Acids Res 2007;35(Database issue):D169–72.

[59] McDonald D, Price MN, Goodrich J, et al. An improved Greengenes taxonomy with explicit ranks for ecological and evolutionary analyses of bacteria and archaea. ISME J 2012;6(3):610–8.

[60] McMurdie PJ, Holmes S. Waste not, want not: why rarefying microbiome data is inadmissible. PLoS Comput Biol 2014;10(4):e1003531.

[61] Pasteur L. Observations relatives a la note precedente de M. Declaux. Acad Sci Paris 1885;200:68.

[62] Metchnikoff E. Sur la flore du corps humain. Manchester Lit Philos Soc 1901;45:1–38.

[63] Reyniers JA, Trexler PC, Ervin RF. Rearing germ-free albino rats. Lobund Reports 1946;(1):1–84.

[64] Reyniers JA, Trexler PC, et al. A complete life-cycle in the germ-free bantam chicken. Nature 1948;162(4132):67.

[65] Gordon HA, Pesti L. The gnotobiotic animal as a tool in the study of host microbial relationships. Bacteriol Rev 1971;35(4):390–429.

[66] Ley RE, Backhed F, Turnbaugh P, et al. Obesity alters gut microbial ecology. Proc Natl Acad Sci USA 2005;102(31):11070–5.

[67] Armougom F, Henry M, Vialettes B, et al. Monitoring bacterial community of human gut microbiota reveals an increase in *Lactobacillus* in obese patients and Methanogens in anorexic patients. PLoS One 2009;4(9):e7125.

[68] Turnbaugh PJ, Hamady M, Yatsunenko T, et al. A core gut microbiome in obese and lean twins. Nature 2009;457(7228):480–4.

[69] Zuo HJ, Xie ZM, Zhang WW, et al. Gut bacteria alteration in obese people and its relationship with gene polymorphism. World J Gastroenterol 2011;17(8):1076–81.

[70] Collado MC, Isolauri E, Laitinen K, et al. Distinct composition of gut microbiota during pregnancy in overweight and normal-weight women. Am J Clin Nutr 2008;88(4):894–9.

[71] Fernandes J, Su W, Rahat-Rozenbloom S, et al. Adiposity, gut microbiota and faecal short chain fatty acids are linked in adult humans. Nutr Diabetes 2014;4:e121.

[72] Mai V, McCrary QM, Sinha R, et al. Associations between dietary habits and body mass index with gut microbiota composition and fecal water genotoxicity: an observational study in African American and Caucasian American volunteers. Nutr J 2009;8:49.

[73] Schwiertz A, Taras D, Schafer K, et al. Microbiota and SCFA in lean and overweight healthy subjects. Obesity (Silver Spring) 2010;18(1):190–5.

[74] Turnbaugh PJ, Ley RE, Mahowald MA, et al. An obesity-associated gut microbiome with increased capacity for energy harvest. Nature 2006;444(7122):1027–31.

[75] Qin J, Li R, Raes J, et al. A human gut microbial gene catalogue established by metagenomic sequencing. Nature 2010;464(7285):59–65.

[76] Faith JJ, Guruge JL, Charbonneau M, et al. The long-term stability of the human gut microbiota. Science 2013;341(6141):1237439.

[77] Ahern PP, Faith JJ, Gordon JI. Mining the human gut microbiota for effector strains that shape the immune system. Immunity 2014;40(6):815–23.

[78] Faith JJ, Ahern PP, Ridaura VK, et al. Identifying gut microbe-host phenotype relationships using combinatorial communities in gnotobiotic mice. Sci Transl Med 2014;6(220):220ra11.

[79] Forsythe P, Bienenstock J, Kunze WA. Vagal pathways for microbiome-brain-gut axis communication. Adv Exp Med Biol 2014;817:115–33.

[80] Foster JA, McVey Neufeld KA. Gut-brain axis: how the microbiome influences anxiety and depression. Trends Neurosci 2013;36(5):305–12.

[81] Mayer EA. Gut feelings: the emerging biology of gut-brain communication. Nat Rev Neurosci 2011;12(8):453–66.

[82] Sudo N, Chida Y, Aiba Y, et al. Postnatal microbial colonization programs the hypothalamic-pituitary-adrenal system for stress response in mice. J Physiol 2004;558(Pt 1):263–75.

[83] Bercik P, Denou E, Collins J, et al. The intestinal microbiota affect central levels of brain-derived neurotropic factor and behavior in mice. Gastroenterology 2011;141(2):599–609. 609.e1–609.e3.

[84] Bravo JA, Forsythe P, Chew MV, et al. Ingestion of *Lactobacillus* strain regulates emotional behavior and central GABA receptor expression in a mouse via the vagus nerve. Proc Natl Acad Sci USA 2011;108(38):16050–5.

[85] Tillisch K, Labus J, Kilpatrick L, et al. Consumption of fermented milk product with probiotic modulates brain activity. Gastroenterology 2013;144(7):1394–401. 1401.e1–1401.e4.

[86] Kim SW, Suda W, Kim S, et al. Robustness of gut microbiota of healthy adults in response to probiotic intervention revealed by high-throughput pyrosequencing. DNA Res 2013;20(3):241–53.

[87] McNulty NP, Yatsunenko T, Hsiao A, et al. The impact of a consortium of fermented milk strains on the gut microbiome of gnotobiotic mice and monozygotic twins. Sci Transl Med 2011;3(106):106ra06.

[88] Eloe-Fadrosh EA, Brady A, Crabtree J, et al. Functional dynamics of the gut microbiome in elderly people during probiotic consumption. MBio 2015;6(2.).

[89] Veiga P, Pons N, Agrawal A, et al. Changes of the human gut microbiome induced by a fermented milk product. Sci Rep 2014;4:6328.

Chapter 3

Microbiota of the Gastrointestinal Tract in Infancy

E.T. Jensen[*,**], R.J. Bertelsen[†,‡] and T. Ringel-Kulka[§]

*Department of Epidemiology and Prevention, Division of Public Health Sciences, Wake Forest School of Medicine, Winston-Salem, NC, United States;

**Department of Medicine, Division of Gastroenterology and Hepatology, University of North Carolina at Chapel Hill, Chapel Hill, NC, United States;

†Department of Clinical Science, University of Bergen, Bergen, Norway; ‡Department of Occupational Medicine, Haukeland University Hospital,

Bergen, Norway; §Department of Maternal and Child Health, UNC Gillings School of Global Public Health, University of North Carolina at Chapel Hill,

Chapel Hill, NC, United States

HUMAN MICROBIOTA

Humans have more than 10^{14} symbiotic organisms in the distal small intestine and colon, and 10–100 times more bacterial cells than body cells. While the human gastrointestinal (GI) tract consists primarily of a bacterial-dominant ecosystem, it also contains eukaryotic viruses, bacterial viruses (bacteriophages), fungi, and archaea [1]. At the individual level, a unique composition of bacterial species exists, but across the various body habitats, for example, saliva, skin, and gut, specific bacteria will be found among all individuals, although with varying abundance [2].

Previously, our capacity to characterize the bacterial species residing in the human gut and elsewhere on the body was limited by the technologies available for identifying bacteria. Specifically, we were limited to identifying bacteria species that could be cultured in the laboratory. More recently, high-throughput sequencing techniques have provided the capacity to identify species previously unrecognized, thus growing our appreciation of the vast abundance of bacterial species inhabiting the gut. Most phylotypes residing in the gut belong to two divisions (superkingdoms) of bacteria—the Bacteroidetes and Firmicutes. Other phylotypes are distributed among the Proteobacteria, Verrucomicrobia, Fusobacteria, Cyanobacteria, Spirochaetes, and VandinBE97 divisions [3,4].

The microbiota shares a symbiotic relationship with the human host, including supporting metabolic function, trophic activity, such as stimulation of angiogenesis, maintaining intestinal motility function through support of motor and sensory functions, contributing to the development of immune tolerance, and preserving mucosal barrier function [5]. The human gut microbiome is considered to be critically important in maintaining health.

Epidemiologic studies have identified associations between changes to the diversity, structure, or function of the gut microbiota (dysbiosis) in early life and development of disease later in life (described in subsequent sections). These observations, together with the established relationship between host–microbiome interactions and immune development, suggest that the period of early colonization is a period of developmental susceptibility (Fig. 3.1). Understanding when and how the microbiome is established offers not only the opportunity to identify mechanisms for pathology, but also opportunities for prevention through modification of microbiota during these developmental periods.

Fetal Microbiota

Until recently, it was believed that colonization of the human gut microbiome was initiated at birth, with the fetus maintaining a sterile gut environment until exposure to maternal vaginal and fecal bacteria at birth (for vaginal deliveries) or to maternal skin microbiome (for cesarean deliveries). More recent studies, reviewed by Koleva et al. [6], have identified the presence of bacteria in the meconium of both term and preterm newborn infants, indicating that colonization initiates in utero. Studies evaluating differences in microbiota in term and preterm infants have identified reduced microbial diversity in the meconium of infants born preterm [7,8]. The source of this first inoculation of bacteria to the fetus has not been elucidated, although fetal membranes, placental tissue, newborn umbilical cord, and

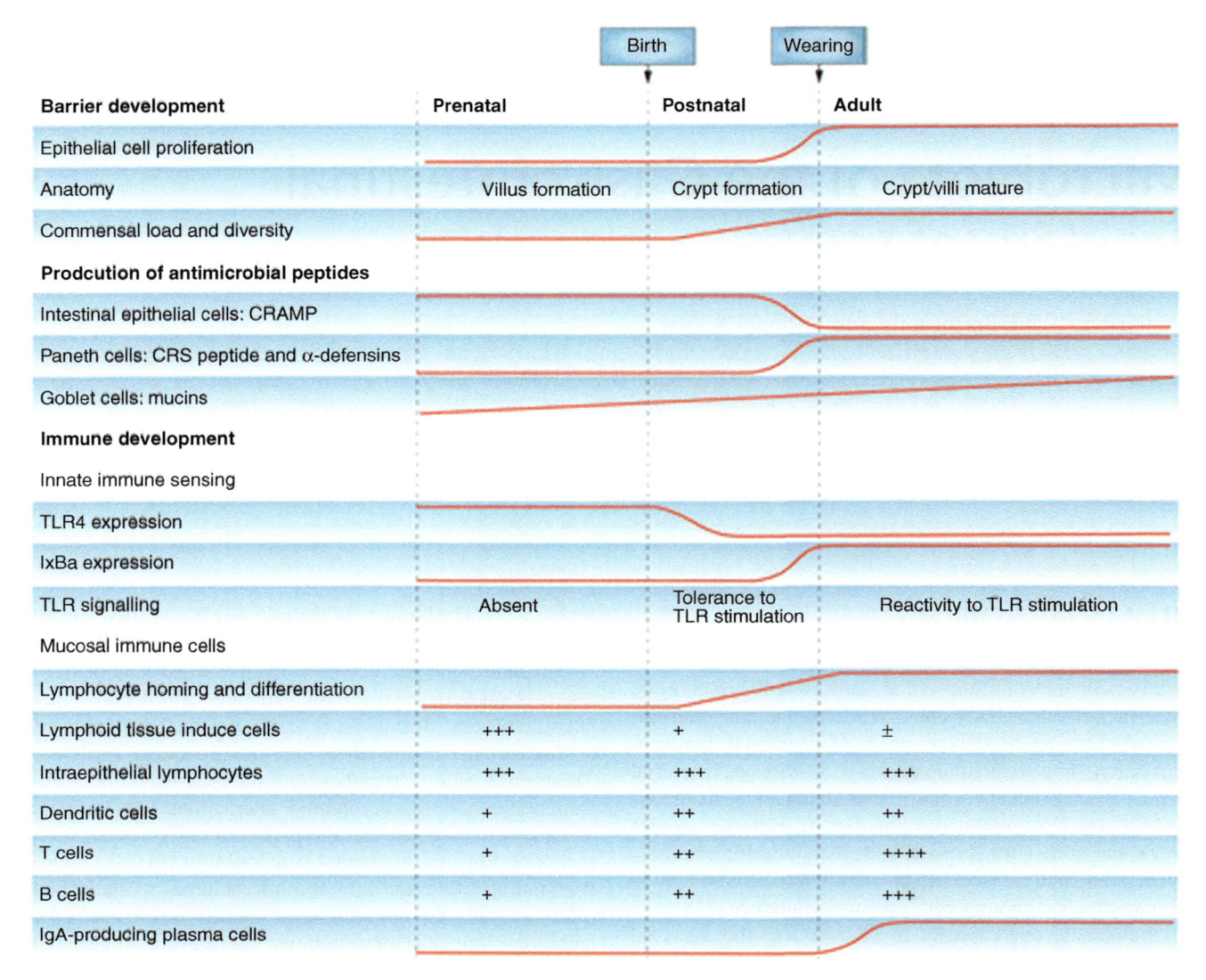

FIGURE 3.1 **Development and maturation of the intestinal mucosal barrier and mucosal immune system.** *Jain N, Walker WA. Diet and host–microbial crosstalk in postnatal intestinal immune homeostasis. Nat Rev Gastroenterol Hepatol 2015;12(1):14–25.*

amniotic fluid have all been identified to contain bacteria and may contribute. Studies examining the microbiota in amniotic fluid and meconium have identified bacteria common to both sites. In midpregnancy the fetus is swallowing amniotic fluid and excreting fluid to maintain amniotic fluid balance and some amniotic fluid reaches the fetal lumen [9,10] and thus this is believed to be a possible source of early exposure for the fetus.

Development and Colonization of the Microbiota

Previously, it was believed that establishment of the human gut microbiome was complete by the end of the first year of life. However, more recent studies employing new methods

for characterizing the microbiota have identified differences between microbiota of young children and adults [11]. At least one study has observed differences in the microbiota in preadolescence and adulthood [12]. The process of colonization is best characterized as a sequence of successive exposures that add to the complexity in the microbiota as illustrated by measures of diversity and relative abundance of species. This progression of increasing richness in the microbiota in early life is well described. It is unclear as to what constitutes the ideal microbiota profile, although some studies have suggested that the ratio of Enterobacteriaceae to Bacteroidaceae is relevant to disease associations [13]. Other studies suggest that reduced microbial diversity may be implicated in disease [14,15]. Still, other studies have identified associations between the abundance of specific bacteria and disease. For example, while bifidobacteria and lactobacilli are considered beneficial to the human host, staphylococci and clostridia may be pathogenic [2,16,17]. Studies characterizing the longitudinal progression of microbiota abundance and diversity have found significant intersubject variation and there is significant variation between studies as well. While some of the differences between studies may be attributable to the methods used for identifying bacteria, with newer methods offering improved capacity to identify and characterize bacteria at the subspecies level, some differences between and within studies may be attributable to the differences in the exposures of the children studied. The composition of the microbiota in early life has been demonstrated to be influenced by a host of varying factors. The association between early life perturbations that may alter the diversity, structure, or function of the gut microbiota (dysbiosis) and later development of disease is the subject of many research studies.

Factors Influencing the Microbiota

Numerous factors have demonstrated to alter the establishment of the human microbiome in early life. Prenatal, antenatal, and postnatal factors have been identified and indicate that these periods are developmentally sensitive periods where perturbations leading to certain microbiota profiles

could lead to dysregulation of the immune response with future implications on health. These developmentally sensitive periods may also offer opportunities for modification of the microbiome to support healthier colonization patterns, ostensibly to prevent future disease (Fig. 3.2). Although less well-studied, given the differences in microbiota composition observed in older children [18], there may also be the possibility of altering microbiota establishment through exposures experienced in childhood.

Genetic Factors

While most factors suggested to influence the establishment of the infant microbiome reflect exogenous factors, studies have examined the potential that an individual's genotype may contribute to his or her microbiome. Many of these studies have been conducted in mouse models and have identified interaction between host genotype and gut microbiota [19–21]. Fewer studies have been conducted in humans, but these are suggestive that the gut microbiome reflects a complex interplay between host genotype and environment. In a study of monozygotic and dizygotic twins ($n = 416$ twin pairs), researchers identified stronger concordance in relative abundance of gut bacteria in monozygotic twins as compared to dizygotic twins. In this study, host genetics were observed to influence colonization of microbiota that have influence in host metabolism [22]. In

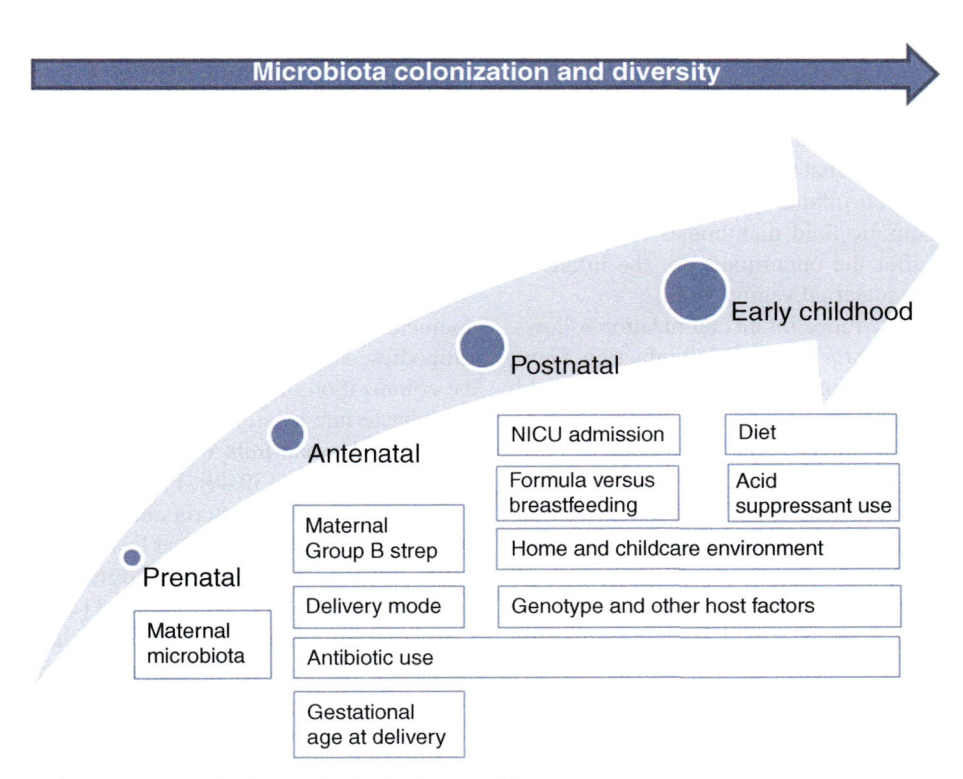

FIGURE 3.2 **Influential factors on gut microbiota colonization in early life.**

another study, researchers estimated the degree to which genetic variation explained the proportion of variance in the microbiota in a Hutterite population ($n = 127$) where, presumably, environmental factors would be less variable. While this study was underpowered for hypothesis testing, their results were suggestive that a portion ($\sim$10–15%) of the bacterial taxa interrogated had at least some evidence of heritability [23].

Pre- and Antenatal Factors
Cesarean Delivery

Many studies have examined and described differences in microbiota colonization in infants born via cesarean delivery versus vaginally [24,25]. In general, infants born via cesarean are more likely to be colonized with bacteria identified from maternal skin. Infants born vaginally are more likely to be colonized with bacteria found in the maternal vaginal and GI tracts. Bifidobacteria, believed to be one of the most important and beneficial bacteria to colonize the infant gut, has been isolated in stool from infants born vaginally, with corresponding identification of the same strains of bifidobacterium in their mothers. In contrast, infants born by cesarean did not have the same monophyletic strains of bifidobacteria as identified in their mothers [26]. In a study of microbiota composition in infant stool from 102 term infants at 6 weeks of age in the United States and controlling for infant diet, vaginally delivered infants had an increased abundance of Bacteroides and Pretobacterium and a decreased abundance of *Staphylococcus*, *Rothia*, and *Propionibacterium* [25]. A limitation to the body of research exploring associations between mode of delivery and infant microbiota has been the fact that many of these studies have not distinguished between infants born via cesarean after the rupture of the amniotic fluid membranes. The rupture of membranes may offer the opportunity for the infant to experience exposure to maternal vaginal flora.

A few studies have examined infant colonization following maternal Group B *Streptococcus* (GBS) infection. Generally, these studies have focused on differences that could be attributable to prophylactic antibiotic use in the prevention of GBS. At least one study focused on whether GBS is associated with differences in microbiota colonization independent of antibiotic use. In a study of 298 infants in the United States, maternal GBS status was not associated with composition differences at 1 month of age, however, differences were observed, in adjusted models, at 6 months of age. No differences were observed in alpha diversity metrics at 1 or 6 months of age [27]. Differences in colonization patterns in the infant have been observed between infants exposed and unexposed to intrapartum prophylactic use of antibiotics during delivery to prevent GBS in the infant, a standard treatment practice in the United States [28,29]. In well-controlled study of healthy, term infants

($n = 84$) in Italy born to women who were GBS positive at 34–37 weeks gestation, 35 infants were exposed to intrapartum antibiotics (intravenous ampicillin administered every 4 h until delivery, as per the Centers for Disease Control Guidelines) [30] and 49 were unexposed. At 7 days postpartum, the infants exposed to the antibiotics had a significantly reduced count of *Bifidobacterium* spp. as compared to the unexposed infants. However, no differences were observed at 30 days postpartum [31]. Longer term differences (up to 12 months of age in infants that were not breastfed) in bacterial abundance were observed in another Canadian study examining microbiota alterations following intrapartum antibiotic prophylaxis. In this study, intrapartum antibiotic use, either for GBS or for maternal cesarean delivery, was associated with reduction in microbiota richness and depletion of Bacteroides and increased relative abundance of Firmicutes. The influence of antibiotics on microbiota was modified by breastfeeding [32].

Postnatal Factors
Prematurity

Premature infants show a delayed colonization with a limited number of bacterial species, as compared to term infants. These differences may be attributable to differences in delivery practices [33,34], the aseptic neonatal intensive care environment, diet, immaturity in the preterm infants gut, and intrapartum or postnatal medical treatments [35,36]. While most healthy full-term neonates are colonized with a heterogeneous bacterial flora dominated by bifidobacteria in breastfed infants, the gut of preterm neonates are dominated by a progression of bacteria colonization from the classes Bacilli, Gammaproteobacteria, and Clostridia [37].

Infant Nutrition

Breastfeeding is a critically important factor in the colonization of the infant gut microbiome. Breastmilk contains nonmetabolized oligosaccharides, lysozomes, lactoferrin, antibodies, and cytokines and these components promote the colonization of the gut with beneficial bacteria, such as bifidobacterium by stimulating certain gut bacteria to proliferate [38]. Human milk oligosaccharides assist the growth of bifidobacterium [39,40]. Furthermore, human breastmilk also contains viable bacteria and a large range of bacterial DNA signatures, which during lactation is transported from gut-associated lymphoid tissue to the mammary glands via the lymphatics and peripheral blood [41–43]. This suggests a bacterial imprinting of the neonatal immune system, and highlights the importance of breastfeeding infants born by vagina and caesarean delivery and who has not been exposed to maternal bacterial gut flora. Whether or not the infant is exclusively fed breastmilk may also be important. Infants fed both breastmilk and infant formula (mixed feeding) have been demonstrated to have a microbiota composition similar to that of exclusively formula-fed infants at 6 weeks of age [25]. Exclusively breastfed infants have been demonstrated to have reduced relative abundance of *Lactococcus* as compared to exclusively formula-fed infants [25].

Antibiotics

Antibiotics administered in infancy and early childhood is designed to eradicate specific strains of bacteria. However, the eradication of bacteria is generally not specific to only the bacteria eliciting harm. Numerous studies have examined the effect of antibiotics on gut microflora and whether antibiotics confer long term alterations. While many of these studies suggest that the changes to gut microbiota do not persist overtime, some suggest that these temporary changes, during developmental susceptible periods of time in immune development, may result in immune dysregulatory effects that persist over time [44]. Numerous observational studies have identified associations between antibiotic use in infancy and later development of immune-mediated diseases including asthma, inflammatory bowel disease, and eosinophilic esophagitis, with some studies noting differences in associations by class and frequency of dosing that the antibiotics were administered [45–49].

Acid Suppressants

The use of acid suppressants (i.e., proton pump inhibitors and H2-receptor antagonists) in infants to treat reflux or colics has increased significantly [50,51] and use of these agents has been associated with increased risk for respiratory infections, gastroenteritis, community-acquired pneumonia, pharyngitis, and *Clostridium difficile* colitis [52–54]. Use of acid suppressants has been demonstrated to alter gastric microflora, contributing to bacterial overgrowth of *Staphylococcus* [55,56] and *Streptococcus* [55].

Geographic Factors

Significant differences have been observed between studies carried out in different countries and it has become clear that country of residence is associated with microbiota abundance and diversity. In a study of 606 infants from five European centers (United Kingdom, Italy, Sweden, Germany, and Spain), fecal samples at 6 weeks, and questionnaire data on prenatal and postnatal factors were analyzed. While differences were observed between infants who were breastfed versus bottlefed, vaginally versus cesarean delivered, and exposed to antibiotics versus not, the greatest differences were based on geography. This study suggested that there was a north-south gradient in the distribution of bifidobacteria and Bacteroides, with the northern areas characterized by greater abundance of bifidobacteria and the southern regions characterized by greater abundance of Bacteroides. Southern centers also demonstrated greater diversity in microbiota. Principal Components Analysis was used to explore these differences and account for regional variation in breastfeeding behaviors [57]. Similarly, in a study of 90 infants in Finland, Sweden, Germany, and the United States (three sites—Colorado, Washington State, and Florida/Georgia), there was significant variability between sites in colonization and diversity of gut microbiomes [58].

Evidence for Association Between the Microbiota and Disease Conditions in Later Life

Immune Dysregulatory Effects

There is mounting evidence that the microbiota colonization observed in early life is associated with disease conditions in infancy and later in life. Most microbiota species are commensal bacteria contributing to host–bacterial interactions necessary for immune development and homeostasis, maintenance of the intestinal barrier function, promotion of tolerance, and protection against pathogens. Dysbiosis, or the disruption of this commensal relationship, may contribute to the pathogenesis of many disease states [59].

The microbiota has also been demonstrated to influence immune response to infectious conditions [60]. For the preterm newborn, an overgrowth of *Proteobacteria* (e.g., *Escherichia* sp.) at the expense of Firmicutes has been associated with necrotizing enterocolitis (NEC) [61,62]. Microbiota colonization and/or diversity has also been associated with changes to immune function and to immune-mediated diseases that may be chronic, including asthma and allergic disease [13,14,63,64], inflammatory bowel disease [65], and type 1 diabetes [66]. In a study of 30 children with fecal samples obtained longitudinally through the first 2 months of life and mononuclear cells obtained from venous blood at the age of 2, it was identified that colonization with *Lactobacillus casei*, *Lactobacillus paracasei*, and *Lactobacillus rhamnosus* at 2 weeks of age trended toward an inverse association with IL-4-, IL-10-, and IFN-gamma-producing cells at the age of 2 years. Colonization with *Bifidobacterium adolescentis*, *Bifidobacterium bifidum*, and *Bifidobacterium breve* was not associated with cytokine-producing cell counts. *Staphylococcus aureus* colonization at 2 weeks of age was significantly associated with increased counts of IL-4-, IL-10-producing cells at the age of 2. The relationship between *S. aureus* and counts of IL-4-, IL-10-producing cells at the age of 2 was modified by the presence of *Lactobacillus* spp. [63]. *S. aureus* colonization has been associated with development of allergic disease [67].

Metabolic Dysfunction

Studies have identified associations between dysbiosis and obesity, hypothesizing that dysbiosis contributes to differences in metabolic function. For instance, exposure to broad-spectrum antibiotics before 2 years of age have been associated with childhood obesity [68], implying a potential lasting effect of antibiotics on the gut microbiota and its metabolic modulation. Gut colonizing bacteria influence gut permeability and metabolic regulation and could contribute to not only obesity, but also other pathogenic metabolic conditions, such as type 2 diabetes [69,70]. In infants, higher bifidobacterial abundance in fecal samples during infancy was identified in children who remained normal weight than in children who became overweight at the age of 7 [71]. In a recent review on microbiome and overweight, the authors concluded that lower Bacteroides colonization of the gut microbiota within 3 months of birth predicts risk for infant and child overweight [72]. In addition, pregnancy overweight may influence the compositional structure of gut microbiota in infants through vertical transfer of microbiota and/or their metabolites during pregnancy, delivery, and breastfeeding [73].

Neurodevelopmental Dysfunction

Microbiota colonization has also been associated with neurological conditions in children, including autism spectrum disorders [74,75] and attention-deficit/hyperactivity disorders [76]. For example, children with autism spectrum disorders have been demonstrated to have higher relative abundance of Proteobacteria and Bacteroides and reduced abundance of Firmicutes and bifidobacteria as compared to healthy controls [77]. With dysbiosis and proliferation of pathogenic bacteria, the intestinal barrier of the gut may be impaired and pathogenic bacterial production of neuro- and endotoxins may result in infiltration of the mucosa and submucosa with bacteria, triggering immune cell activation and a proinflammatory response. The proinflammatory response leads to further intestinal permeability and maintenance of the inflammatory state and subsequent activation of the hypothalamic–pituitary–adrenal axis stress response [77,78].

Manipulation of the Intestinal Microbiota in Early Life

Pre- and Probiotics

Given the observation that colonization of the gut in early life can be altered through various environmental factors and that alteration of the gut microbiome during the developmentally sensitive period of early life may confer increased risk of developing disease, there have been numerous studies examining the potential that the gut microbiota could be manipulated in early life to confer a more beneficial microbiota profile, thus ameliorating the effects of certain exposures (i.e., antibiotics, cesarean delivery, use of infant formula) or preventing disease in infants who may be genetically predisposed. One approach to manipulating the microbiota is through introduction of pre- and probiotics. They have been used to manipulate the microbiota of

infants, including administering them to women prenatally, to premature infants for prevention of NEC, and to infants through supplementation of infant formula [79,80].

The strongest evidence for use of prebiotics and probiotics is for treatment/prevention of NEC. NEC typically develops in the first few weeks after birth, with the age of onset inversely related to gestational age at birth. It is characterized by intestinal tract damage, specifically mucosal injury in the mildest of cases, up to full-thickness necrosis and perforation in the more severe cases. Use of probiotics in infants at risk for NEC has reduced or lessened the severity of NEC in multiple observational and clinical trials [81], however, not all studies have suggested an association [82], thus indicating that results may be strain specific [83].

Probiotics are sometimes recommended in children who receive antibiotics as a means of preventing antibiotic-associated diarrhea. Probiotics have been evaluated for use in prevention of other pediatric conditions, including allergic diseases, obesity, inflammatory bowel disease, colic, and infectious disease [84], with mixed evidence as to their effectiveness.

The use of pre- and probiotics in infant formula is growing, however, the studies in support of their use remain incomplete. The Committee on Nutrition of the European Society for Paediatric Gastroenterology, Hepatology, and Nutrition (ESPGHAN) conducted a systematic review on the safety and health effects of pre- and probiotic supplementation of infant formula and identified no safety concerns for use. Additionally, there was insufficient data to recommend use of infant formula supplemented with pre- and probiotics [85].

Fecal Microbial Transplant

Fecal microbial transplant (FMT) is an emerging, potentially therapeutic tool for repopulating the gut microbiome in diseases associated with dysbiosis. While considerable evidence exists in support of FMT for treatment of C. difficile infection [86,87], and the body of literature on use of FMT in other disease conditions is growing [88], little is known about use of FMT in the pediatric population. In a small study of just four adolescents, a single dose of FMT via nasogastric tube in pediatric ulcerative colitis was not associated with clinical or laboratory benefit [89].

CONCLUSIONS

Our capacity to characterize the human microbiome is rapidly evolving and our understanding of how the microbiome impacts human health and disease will likely change significantly in the years to come. The present state of knowledge indicates that colonization of the human microbiome begins very early in life, before birth, and that there are numerous factors that influence the microbiome. Some of these factors are intrinsic; others may be modifiable and offer opportunities for promoting health through prevention. Manipulation of the intestinal microbiota through pre- and probiotics, or through fecal microbiota transfer, may offer incredible therapeutic potential as we continue to improve our understanding of the complex relationship between the human host and their GI microbiome.

COI STATEMENT

No conflict of interest has been declared by the authors.

ACKNOWLEDGMENTS

This work was supported in part by the Research Council of Norway, Grant number 230827 (RJB—Lead PI, TRK—UNC PI).

REFERENCES

[1] Lim ES, Zhou Y, Zhao G, et al. Early life dynamics of the human gut virome and bacterial microbiome in infants. Nat Med 2015;21(10):1228–34.

[2] Walker A. Intestinal colonization and programming of the intestinal immune response. J Clin Gastroenterol 2014;48(Suppl. 1): S8–S11.

[3] Backhed F, Ley RE, Sonnenburg JL, et al. Host-bacterial mutualism in the human intestine. Science 2005;307(5717):1915–20.

[4] Eckburg PB, Bik EM, Bernstein CN, et al. Diversity of the human intestinal microbial flora. Science 2005;308(5728):1635–8.

[5] Round JL, Mazmanian SK. The gut microbiota shapes intestinal immune responses during health and disease. Nat Rev Immunol 2009;9(5):313–23.

[6] Koleva PT, Kim JS, Scott JA, et al. Microbial programming of health and disease starts during fetal life. Birth Defects Res C Embryo Today 2015;105(4):265–77.

[7] Mshvildadze M, Neu J. The infant intestinal microbiome: friend or foe? Early Hum Dev 2010;86(Suppl. 1):67–71.

[8] Ardissone AN, de la Cruz DM, Davis-Richardson AG, et al. Meconium microbiome analysis identifies bacteria correlated with premature birth. PLoS One 2014;9(3):e90784.

[9] Curran MA, Nijland MJ, Mann SE, et al. Human amniotic fluid mathematical model: determination and effect of intramembranous sodium flux. Am J Obstet Gynecol 1998;178(3):484–90.

[10] Brace RA. Physiology of amniotic fluid volume regulation. Clin Obstet Gynecol 1997;40(2):280–9.

[11] Ringel-Kulka T, Cheng J, Ringel Y, et al. Intestinal microbiota in healthy U.S. young children and adults—a high throughput microarray analysis. PLoS One 2013;8(5):e64315.

[12] Hollister EB, Riehle K, Luna RA, et al. Structure and function of the healthy pre-adolescent pediatric gut microbiome. Microbiome 2015;3:36.

[13] Azad MB, Konya T, Guttman DS, et al. Infant gut microbiota and food sensitization: associations in the first year of life. Clin Exp Allergy 2015;45(3):632–43.

[14] Abrahamsson TR, Jakobsson HE, Andersson AF, et al. Low gut microbiota diversity in early infancy precedes asthma at school age. Clin Exp Allergy 2014;44(6):842–50.

[15] Forsberg A, Abrahamsson TR, Bjorksten B, et al. Pre- and post-natal lactobacillus reuteri supplementation decreases allergen responsiveness in infancy. Clin Exp Allergy 2013;43(4):434–42.

[16] Rastall RA. Bacteria in the gut: friends and foes and how to alter the balance. J Nutr 2004;134(8 Suppl.):2022s–6s.

[17] Walker WA, Iyengar RS. Breast milk, microbiota, and intestinal immune homeostasis. Pediatr Res 2015;77(1–2):220–8.

[18] Cheng J, Ringel-Kulka T, Heikamp-de Jong I, et al. Discordant temporal development of bacterial phyla and the emergence of core in the fecal microbiota of young children. ISME J 2016;10: 1002–14.

[19] Zhang C, Zhang M, Wang S, et al. Interactions between gut microbiota, host genetics and diet relevant to development of metabolic syndromes in mice. ISME J 2010;4(2):232–41.

[20] Kovacs A, Ben-Jacob N, Tayem H, et al. Genotype is a stronger determinant than sex of the mouse gut microbiota. Microb Ecol 2011;61(2):423–8.

[21] Benson AK, Kelly SA, Legge R, et al. Individuality in gut microbiota composition is a complex polygenic trait shaped by multiple environmental and host genetic factors. Proc Natl Acad Sci USA 2010;107(44):18933–8.

[22] Goodrich Julia K, Waters Jillian L, Poole Angela C, et al. Human genetics shape the gut microbiome. Cell 2014;159(4):789–99.

[23] Davenport ER, Cusanovich DA, Michelini K, et al. Genome-wide association studies of the human gut microbiota. PLoS One 2015;10(11):e0140301.

[24] Azad MB, Konya T, Maughan H, et al. Gut microbiota of healthy Canadian infants: profiles by mode of delivery and infant diet at 4 months. CMAJ 2013;185(5):385–94.

[25] Madan JC, Hoen AG, Lundgren SN, et al. Association of cesarean delivery and formula supplementation with the intestinal microbiome of 6-week-old infants. JAMA Pediatr 2016;170:212–9.

[26] Makino H, Kushiro A, Ishikawa E, et al. Mother-to-infant transmission of intestinal bifidobacterial strains has an impact on the early development of vaginally delivered infant's microbiota. PLoS One 2013;8(11):e78331.

[27] Cassidy-Bushrow AE, Sitarik A, Levin AM, et al. Maternal group B streptococcus and the infant gut microbiota. J Dev Orig Health Dis 2016;7(1):45–53.

[28] Arboleya S, Sanchez B, Milani C, et al. Intestinal microbiota development in preterm neonates and effect of perinatal antibiotics. J Pediatr 2015;166(3):538–44.

[29] Keski-Nisula L, Kyynarainen HR, Karkkainen U, et al. Maternal intrapartum antibiotics and decreased vertical transmission of lactobacillus to neonates during birth. Acta Paediatr 2013;102(5):480–5.

[30] Verani JR, McGee L, Schrag SJ. Prevention of perinatal group B streptococcal disease—revised guidelines from CDC. MMWR Recomm Rep 2010;59(RR-10):1–36.

[31] Corvaglia L, Tonti G, Martini S, et al. Influence of intrapartum antibiotic prophylaxis for group B streptococcus on gut microbiota in the first month of life. J Pediatr Gastroenterol Nutr 2016;62(2): 304–8.

[32] Azad MB, Konya T, Persaud RR, et al. Impact of maternal intrapartum antibiotics, method of birth and breastfeeding on gut microbiota during the first year of life: a prospective cohort study. BJOG 2016;123:983–93.

[33] Gregory KE, LaPlante RD, Shan G, et al. Mode of birth influences preterm infant intestinal colonization with bacteroides over the early neonatal period. Adv Neonatal Care 2015;15(6):386–93.

[34] Gewolb IH, Schwalbe RS, Taciak VL, et al. Stool microflora in extremely low birthweight infants. Arch Dis Child Fetal Neonatal Ed 1999;80(3):F167–73.

[35] Westerbeek EA, van den Berg A, Lafeber HN, et al. The intestinal bacterial colonisation in preterm infants: a review of the literature. Clin Nutr 2006;25(3):361–8.

[36] Gomez M, Moles L, Melgar A, et al. Early gut colonization of preterm infants: effect of enteral feeding tubes. J Pediatr Gastroenterol Nutr 2016;62:893–900.

[37] La Rosa PS, Warner BB, Zhou Y, et al. Patterned progression of bacterial populations in the premature infant gut. Proc Natl Acad Sci USA 2014;111(34):12522–7.

[38] Ballard O, Morrow AL. Human milk composition: nutrients and bioactive factors. Pediatr Clin North Am 2013;60(1):49–74.

[39] Fanaro S, Marten B, Bagna R, et al. Galacto-oligosaccharides are bifidogenic and safe at weaning: a double-blind randomized multicenter study. J Pediatr Gastroenterol Nutr 2009;48(1):82–8.

[40] Newburg DS. Neonatal protection by an innate immune system of human milk consisting of oligosaccharides and glycans. J Anim Sci 2009;87(13 Suppl.):26–34.

[41] Donnet-Hughes A, Perez PF, Dore J, et al. Potential role of the intestinal microbiota of the mother in neonatal immune education. Proc Nutr Soc 2010;69(3):407–15.

[42] Perez PF, Dore J, Leclerc M, et al. Bacterial imprinting of the neonatal immune system: lessons from maternal cells? Pediatrics 2007;119(3):e724–32.

[43] Fernandez L, Langa S, Martin V, et al. The human milk microbiota: origin and potential roles in health and disease. Pharmacol Res 2013;69(1):1–10.

[44] Vangay P, Ward T, Gerber JS, et al. Antibiotics, pediatric dysbiosis, and disease. Cell Host Microbe 2015;17(5):553–64.

[45] Hviid A, Svanstrom H, Frisch M. Antibiotic use and inflammatory bowel diseases in childhood. Gut 2011;60(1):49–54.

[46] Jensen ET, Kappelman MD, Kim HP, et al. Early life exposures as risk factors for pediatric eosinophilic esophagitis. J Pediatr Gastroenterol Nutr 2013;57(1):67–71.

[47] Kozyrskyj AL, Ernst P, Becker AB. Increased risk of childhood asthma from antibiotic use in early life. Chest 2007;131(6):1753–9.

[48] Kronman MP, Zaoutis TE, Haynes K, et al. Antibiotic exposure and IBD development among children: a population-based cohort study. Pediatrics 2012;130(4):e794–803.

[49] Risnes KR, Belanger K, Murk W, et al. Antibiotic exposure by 6 months and asthma and allergy at 6 years: findings in a cohort of 1,401 US children. Am J Epidemiol 2011;173(3):310–8.

[50] Kothari S, Nelson SP, Wu EQ, et al. Healthcare costs of gerd and acid-related conditions in pediatric patients, with comparison between histamine-2 receptor antagonists and proton pump inhibitors. Curr Med Res Opin 2009;25(11):2703–9.

[51] Nelson SP, Kothari S, Wu EQ, et al. Pediatric gastroesophageal reflux disease and acid-related conditions: trends in incidence of diagnosis and acid suppression therapy. J Med Econ 2009;12(4):348–55.

[52] Canani RB, Cirillo P, Roggero P, et al. Therapy with gastric acidity inhibitors increases the risk of acute gastroenteritis and community-acquired pneumonia in children. Pediatrics 2006;117(5):e817–20.

[53] Turco R, Martinelli M, Miele E, et al. Proton pump inhibitors as a risk factor for paediatric *Clostridium difficile* infection. Aliment Pharmacol Ther 2010;31(7):754–9.

[54] Cohen S, Bueno de Mesquita M, Mimouni FB. Adverse effects reported in the use of gastroesophageal reflux disease treatments

[55] Rosen R, Amirault J, Liu H, et al. Changes in gastric and lung microflora with acid suppression: acid suppression and bacterial growth. JAMA Pediatr 2014;168(10):932–7.

[56] Rosen R, Hu L, Amirault J, et al. 16S community profiling identifies proton pump inhibitor related differences in gastric, lung, and oropharyngeal microflora. J Pediatr 2015;166(4):917–23.

[57] Fallani M, Young D, Scott J, et al. Intestinal microbiota of 6-week-old infants across europe: geographic influence beyond delivery mode, breast-feeding, and antibiotics. J Pediatr Gastroenterol Nutr 2010;51(1):77–84.

[58] Kemppainen KM, Ardissone AN, Davis-Richardson AG, et al. Early childhood gut microbiomes show strong geographic differences among subjects at high risk for type 1 diabetes. Diabetes Care 2015;38(2):329–32.

[59] Goulet O. Potential role of the intestinal microbiota in programming health and disease. Nutr Rev 2015;73(Suppl. 1):32–40.

[60] Ichinohe T, Pang IK, Kumamoto Y, et al. Microbiota regulates immune defense against respiratory tract influenza a virus infection. Proc Natl Acad Sci USA 2011;108(13):5354–9.

[61] Stewart CJ, Marrs EC, Nelson A, et al. Development of the preterm gut microbiome in twins at risk of necrotising enterocolitis and sepsis. PLoS One 2013;8(8):e73465.

[62] Claud EC, Keegan KP, Brulc JM, et al. Bacterial community structure and functional contributions to emergence of health or necrotizing enterocolitis in preterm infants. Microbiome 2013;1(1):20.

[63] Johansson MA, Sjogren YM, Persson JO, et al. Early colonization with a group of lactobacilli decreases the risk for allergy at five years of age despite allergic heredity. PLoS One 2011;6(8):e23031.

[64] Bisgaard H, Li N, Bonnelykke K, et al. Reduced diversity of the intestinal microbiota during infancy is associated with increased risk of allergic disease at school age. J Allergy Clin Immunol 2011;128(3):646–52. e1–e5.

[65] Oberc A, Coombes BK. Convergence of external Crohn's disease risk factors on intestinal bacteria. Front Immunol 2015;6:558.

[66] Kostic AD, Gevers D, Siljander H, et al. The dynamics of the human infant gut microbiome in development and in progression toward type 1 diabetes. Cell Host Microbe 2015;17(2):260–73.

[67] Pastacaldi C, Lewis P, Howarth P. Staphylococci and staphylococcal superantigens in asthma and rhinitis: a systematic review and meta-analysis. Allergy 2011;66(4):549–55.

[68] Bailey LC, Forrest CB, Zhang P, et al. Association of antibiotics in infancy with early childhood obesity. JAMA Pediatr 2014;168(11):1063–9.

[69] Everard A, Matamoros S, Geurts L, et al. *Saccharomyces boulardii* administration changes gut microbiota and reduces hepatic steatosis, low-grade inflammation, and fat mass in obese and type 2 diabetic db/db mice. MBio 2014;5(3):e01011–4.

[70] Geurts L, Neyrinck AM, Delzenne NM, et al. Gut microbiota controls adipose tissue expansion, gut barrier and glucose metabolism: novel insights into molecular targets and interventions using prebiotics. Benef Microbes 2014;5(1):3–17.

[71] Kalliomaki M, Collado MC, Salminen S, et al. Early differences in fecal microbiota composition in children may predict overweight. Am J Clin Nutr 2008;87(3):534–8.

[72] Kozyrskyj AL, Kalu R, Koleva PT, et al. Fetal programming of overweight through the microbiome: boys are disproportionately affected. J Dev Orig Health Dis 2016;7(1):25–34.

[73] Galley JD, Bailey M, Kamp Dush C, et al. Maternal obesity is associated with alterations in the gut microbiome in toddlers. PLoS One 2014;9(11):e113026.

[74] Finegold SM, Downes J, Summanen PH. Microbiology of regressive autism. Anaerobe 2012;18(2):260–2.

[75] Mezzelani A, Landini M, Facchiano F, et al. Environment, dysbiosis, immunity and sex-specific susceptibility: a translational hypothesis for regressive autism pathogenesis. Nutr Neurosci 2015;18(4):145–61.

[76] Partty A, Kalliomaki M, Wacklin P, et al. A possible link between early probiotic intervention and the risk of neuropsychiatric disorders later in childhood: a randomized trial. Pediatr Res 2015;77(6):823–8.

[77] Ghaisas S, Maher J, Kanthasamy A. Gut microbiome in health and disease: linking the microbiome-gut-brain axis and environmental factors in the pathogenesis of systemic and neurodegenerative diseases. Pharmacol Ther 2016;158:52–62.

[78] Petra AI, Panagiotidou S, Hatziagelaki E, et al. Gut-microbiota-brain axis and its effect on neuropsychiatric disorders with suspected immune dysregulation. Clin Ther 2015;37(5):984–95.

[79] Thum C, Cookson AL, Otter DE, et al. Can nutritional modulation of maternal intestinal microbiota influence the development of the infant gastrointestinal tract? J Nutr 2012;142(11):1921–8.

[80] Rutten NB, Gorissen DM, Eck A, et al. Long term development of gut microbiota composition in atopic children: impact of probiotics. PLoS One 2015;10(9):e0137681.

[81] Olsen R, Greisen G, Schroder M, et al. Prophylactic probiotics for preterm infants: a systematic review and meta-analysis of observational studies. Neonatology 2016;109(2):105–12.

[82] Costeloe K, Hardy P, Juszczak E, et al. *Bifidobacterium breve* BBG-001 in very preterm infants: a randomised controlled phase 3 trial. Lancet 2016;387(10019):649–60.

[83] Abrahamsson TR. Not all probiotic strains prevent necrotising enterocolitis in premature infants. Lancet 2016;387(10019):624–5.

[84] Ringel-Kulka T, Kotch JB, Jensen ET, et al. Randomized, double-blind, placebo-controlled study of synbiotic yogurt effect on the health of children. J Pediatr 2015;166(6):1475–81. e1–e3.

[85] Braegger C, Chmielewska A, Decsi T, et al. Supplementation of infant formula with probiotics and/or prebiotics: a systematic review and comment by the ESPGHAN committee on nutrition. J Pediatr Gastroenterol Nutr 2011;52(2):238–50.

[86] Brandt LJ. American Journal of Gastroenterology lecture: intestinal microbiota and the role of fecal microbiota transplant (FMT) in treatment of *C. difficile* infection. Am J Gastroenterol 2013;108(2):177–85.

[87] Aroniadis OC, Brandt LJ, Greenberg A, et al. Long-term follow-up study of fecal microbiota transplantation for severe and/or complicated *Clostridium difficile* infection: a multicenter experience. J Clin Gastroenterol 2016;50:398–402.

[88] Smits LP, Bouter KE, de Vos WM, et al. Therapeutic potential of fecal microbiota transplantation. Gastroenterology 2013;145(5):946–53.

[89] Suskind DL, Singh N, Nielson H, et al. Fecal microbial transplant via nasogastric tube for active pediatric ulcerative colitis. J Pediatr Gastroenterol Nutr 2015;60(1):27–9.

Chapter 4

Identification of the Microbiota in the Aging Process

A. Sarkar* and C.S. Pitchumoni**

*Rutgers, Robert Wood Johnson Medical School, New Brunswick, NJ, United States; **Saint Peter's University Hospital, New Brunswick, NJ, United States

INTRODUCTION

Most westernized countries have arbitrarily accepted the chronological age of 65 years as a definition of "elderly." The period of old age has been further divided into three subgroups. Using the classification of young–old (65–74), middle–old (75–84), and oldest–old (85+), the number of oldest–old is recognized now as the fastest growing segment of many populations. The US Census Bureau used the term "century club" to define the group of people aged over 100 years, numbering over 50,000 in 2004 [1]. Chronic noncommunicable diseases are now the major cause of death among older people in both developed and less developed countries. In 2006, almost 500 million people worldwide were 65 and older. By 2030, it is projected to increase to one billion, one in every eight of the world's population, and approximately two billion by 2050. While the growing number of older adult population reflects a crowning achievement of the scientific progress made in the last century, it does not come without a significant challenge.

Gerontologists are looking for what distinguishes normal aging from disease to answer why older adults are increasingly vulnerable to disease and disability. A scientific definition of aging is difficult but aging has been defined as the regression of physiological function accompanied by advancement of age [2] which is associated with physiological changes in the gastrointestinal tract (GIT), as well as changes in dietary patterns and immune function (Fig. 4.1) [3].

The association between healthy aging and microbes was first made by the Nobel prize winner Elie Metchnikoff from Kharkov/Ukraine along with Paul Ehrlich as early as 1908 in a book impressively titled *The Prolongation of Life*. An excellent review paper published in early 1990s on intestinal flora and ageing brought to light that many factors, in addition to ageing, altered the composition of the gut microbiota (GM) (i.e., physiological state, drugs, disease,

diet and stress) [4]. Although subsequent studies confirmed that the process of aging influences the GM and vice versa, research focusing on GM in the older adult and their possible consequences on the health status of the host is just evolving [5].

Metagenomic studies have highlighted variations of the GM as a function of age and diet. Physiological changes in the GIT influenced by environmental factors and the host immune system affect the composition of the GM. In this chapter, therefore, we briefly point out age-related physiological changes in the GIT, GM, and the clinical implications. The postulated roles of GM in a number of medical problems of the older adult are wide and include functional dyspepsia, altered immunity, metabolic syndrome, insulin resistance, obesity and related comorbidities, brain health, many autoimmune diseases, *Clostridium difficile* colitis, inflammatory bowel diseases, diverticulitis, irritable bowel syndrome, celiac disease, sclerosing cholangitis and primary biliary cirrhosis, colon cancer, pancreatitis, and the list is growing. Although many of the above are multifactorial, a central role may exist with GM. This chapter is an extension of many recent excellent articles on related topics [2,5–17].

GUT MICROBIOTA, THE HIDDEN AND FORGOTTEN ORGAN

With considerable metabolic activity GM is a virtual organ within the host with profound influence on normal structural and functional development of the GIT mucosal immune system with a powerful influence on health and disease. Since the bacterial flora has a combined metabolic activity equal to a well-formed organ within the host, but not so far well recognized, it is a "forgotten organ" within the human body [18]. The human GIT is a nutrient-rich environment packed with up to 100 trillion microbes, approximately

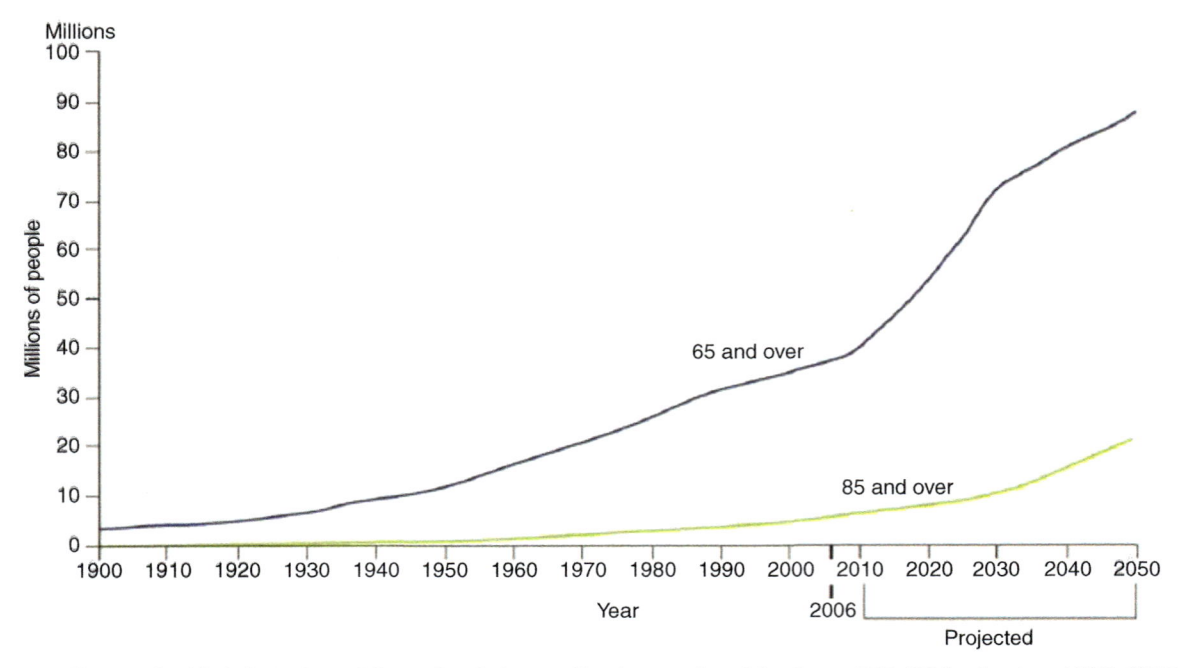

FIGURE 4.1 **Increase in elderly in the United States.** Population age 65 and over and age 85 and over, 1900–2006 and projected 2010–50 [3].

weighing 1.5 kg constituting the GM a complex ecosystem. The specialized microbes inhabiting the surfaces and GIT of a human adult outnumber the human cells by a factor of 10 [19,20]. The vast majority reside in our colon where densities approach 10^{11}–10^{12} cells/mL. Approximately 400–500 bacterial species make up the GM [5]. Human beings have been recently reviewed as "metaorganisms" as a result of a close symbiotic relationship with the GM [21]. The GM contributes to net metabolic transformations in the GIT, and to a diverse range of health-related activities not only in the elderly but in all age groups [22].

AGE-RELATED GASTROENTEROLOGICAL CHANGES

Aging related GIT changes are decremental and progressive. Overall the functional reserve decreases when a person enters the "middle old" age of 75–85 years. There is bidirectional relationship; the GM may contribute to the physiological changes and vice versa. Normal aging is associated with gradual and subtle changes in structure and function. The specific microbial diversity of the GM plays a vital role in maintaining immune homeostasis suggesting that microbiotal alterations are related to many GI and systemic diseases [23].

The GIT has diverse cell types [24] with absorptive, secretory, immunological, and endocrine functions. It contains the largest number of neurons next only to the nervous system. The gut mucosa represents the single largest immunological organ with much of the immunoglobulin producing cells [25]. The GIT is a lymphoid organ and the gut-associated lymphoid tissue (GALT) is an important immune organ in the body. The enterochromaffin cells constitute the largest endocrine cell population in the GIT secreting a large number of hormones: gastrin, somatostatin, ghrelin, motilin, secretin, cholecystokinin, peptide YY, and other not so well recognized hormones, influencing many physiological functions within the GIT and elsewhere. A brief presentation of the pathophysiological changes in the GIT will help in evaluating the role of GM in the pathogenesis of some of the age-related disorders. The factors include an interplay of comorbidity, diet, GI side effects of medications, antibiotics and acid reducing agents, chemo- and radiotherapy, impact of repeated and invasive diagnostic tests and hospitalizations. The age-related physiological changes in the GIT are tabulated (Table 4.1) [26].

As age advances, mechanisms regulating energy homeostasis change. Age-related weight gain with increase in BMI is attributed to hunger signals prevailing over satiety signals [27]. Anorexia of aging and protein energy malnutrition are related to conditions, such as chronic obstructive pulmonary disease (COPD), Parkinson's disease, and arthritis-associated inflammation. Peptic ulcer disease (often due to use of NSAIDs) and malabsorptive disorders contribute to decreased food intake and unintentional weight loss. Cognitive impairment can also lead to weight loss. A change in the secretion of satiety hormone CCK has been proposed as an explanation for anorexia of aging [28,29]. Studies looking at factors regulating appetite in the older adult are limited. Ghrelin, CCK, GLP1, and leptin are currently being

TABLE 4.1 Age-Related GI Tract Changes [26]

Location	Age-Related Changes
Oral cavity	• Soreness (i.e., ill-fitting dentures, vitamin deficiencies, reduced salivary secretion) • Oropharyngeal dysphagia
Esophagus	• Esophageal dysphagia (i.e., achalasia, diffuse esophageal spasm, tumors, strictures, Schatzki's ring, extrinsic compression) • Odynophagia (i.e. pill-induced esophagitis, *Candida*) • Gastroesophageal reflux disease • Barrett's esophagus • Esophagitis • Malignancy
Stomach	• Gastric atrophy • Decreased gastric acid secretion • Decreased pepsin • Decreased mucus production and mucosal prostaglandin levels • Reduced gastric emptying owing to slowing of transit time • Reduced blood supply possibly exacerbated by atherosclerotic disease • Peptic ulcer disease (i.e. NSAID use) • *H. pylori* infection
Small intestine	• Malabsorption (i.e., pancreatic insufficiency, cholestasis, parasitic infections, Whipple's disease (rare), Celiac disease, Crohn's disease, bacterial overgrowth) • Diarrhea (i.e. achlorhydria, exposure to enteropathogens, luminal stasis, decreased mucosal immunity, antibiotic use) • Mesenteric ischemia
Large intestine	• Constipation (i.e. decrease in transit time, inactivity, poor hydration, depression, medication) • Irritable bowel syndrome • Diverticular disease • Colorectal cancer

studied in relation to appetite and satiety regulation in all age groups.

Malnutrition is a classical feature of old age and is linked to physiological changes that impact food digestion and absorption. Some of the oral changes in the older adult contribute to alteration in food habits and GM. Problems with teeth, chewing and dry mouth (xerostomia) caused by side effects of most medications encourage individuals to eat less nutritious soft foods. Clinically significant changes in the motor function of the esophagus are rare although motility disorders, such as decrease in the amplitude of contractions and increase in diffuse contractions (diffuse esophageal spasm) occur. Dysphagia is common due to neurologically related oropharyngeal problems or esophageal disorders. Secondary peristalsis may be absent or decreased in amplitude. The myenteric interstitial cells of Cajal (ICC-MY)

serve as a pacemaker, which creates the bioelectrical slow wave potential that leads to contraction of the smooth muscle. ICC-MY decline with age, influencing motor functions. Decreased gastric emptying has been demonstrated along with impaired mucosal protective mechanism because of a decline in mucosal prostaglandins [30]. Small bowel and colonic motility in the older adult is affected more often by medications than by the process of aging.

Blind loop syndrome (small intestinal bacterial overgrowth, SIBO) can occur after gastric surgery or with jejunal diverticulosis. It is also important to note SIBO occurs in the elderly without anatomic blind loops due to decreased intestinal motility. Normally the proximal small intestine contains less than 10^4 bacteria per milliliter and in SIBO, the small bowel is colonized with increase in species diversity as a result of decrease in gastric acid secretion or prolonged use of proton pump inhibitors. Malabsorption of fat and bacterial utilization of B_{12} is clinical features. The serum folate and biotin levels are increased because of synthesis of the vitamins by enteric bacteria.

Large intestine may show delay in transit time. The prevalence of diverticular disease progressively increases with age, attributed to a dietary fiber deficiency. In many western of countries more than half the population may have the disorder, which is often silent and rarely presents with complications.

The increased incidence of *C. difficile*-associated (CDA) colitis poses an epidemiological risk to the institutionalized and free-living elderly and as discussed later in another chapter, may be due to age-related changes in fecal flora, immune senescence, or the presence of other underlying diseases, or medications [31,32]. The increased incidence is not solely due to greater exposure to antibiotics [33]. Inflammatory bowel disorders (IBD), Crohn's as well as ulcerative colitis (UC) have dual peaks of onset and the second peak is suspected to be related to GM. Ischemic bowel disorders, lymphocytic or collagenous colitis, drug-induced colitis and infectious colitis occur in greater frequency and are in the differential diagnosis. Benign and malignant diseases of the colon and liver pose serious threat to the elderly. There is an age-associated insulin resistance and glucose intolerance and increasing incidence of pancreatic cancer. An explosion of literature in the field of GM overall raises the possibility of changing GM in the GIT of elderly and its relationship to the special problems of the older adult and quality of living (Fig. 4.2).

GUT MICROBIOTA IN THE ELDERLY

For many reasons the older adults are often excluded from clinical research and are generally underrepresented in intervention studies [35]. A major source of data for the topic under discussion is from the ELDERMET group of scientists working at University College Cork, Cork University

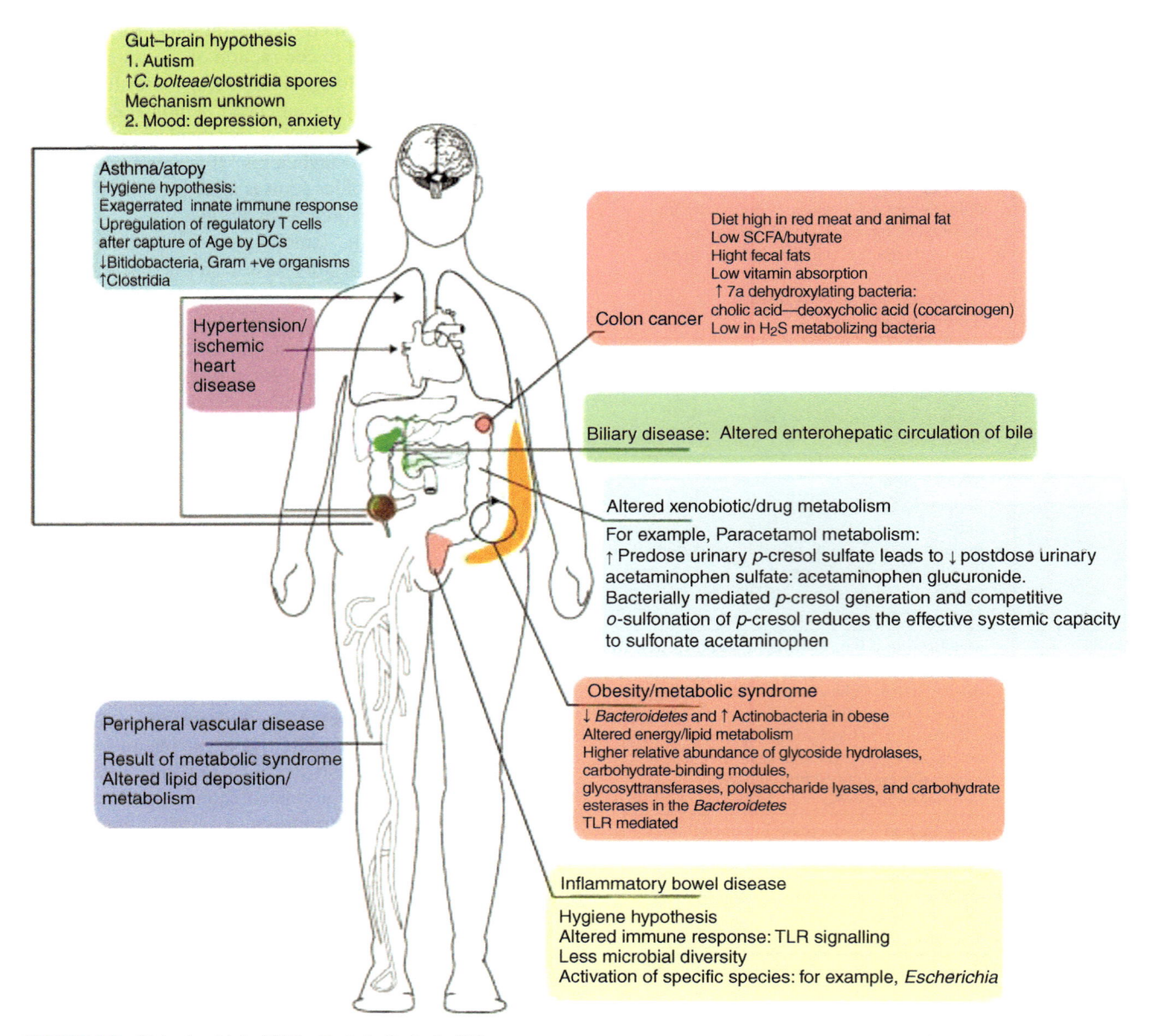

Gut–brain hypothesis
1. Autism
↑*C. bolteae*/clostridia spores
Mechanism unknown
2. Mood: depression, anxiety

Asthma/atopy
Hygiene hypothesis:
Exagerrated innate immune response
Upregulation of regulatory T cells
after capture of Age by DCs
↓Bitidobacteria, Gram +ve organisms
↑Clostridia

Hypertension/ ischemic heart disease

Diet high in red meat and animal fat
Low SCFA/butyrate
Hight fecal fats
Low vitamin absorption
↑ 7a dehydroxylating bacteria:
Colon cancer cholic acid—deoxycholic acid (cocarcinogen)
Low in H_2S metabolizing bacteria

Biliary disease: Altered enterohepatic circulation of bile

Altered xenobiotic/drug metabolism

For example, Paracetamol metabolism:
↑ Predose urinary *p*-cresol sulfate leads to ↓ postdose urinary
acetaminophen sulfate: acetaminophen glucuronide.
Bacterially mediated *p*-cresol generation and competitive
o-sulfonation of *p*-cresol reduces the effective systemic capacity
to sulfonate acetaminophen

Obesity/metabolic syndrome
↓ *Bacteroidetes* and ↑ Actinobacteria in obese
Altered energy/lipid metabolism
Higher relative abundance of glycoside hydrolases,
carbohydrate-binding modules,
glycosyttransferases, polysaccharide lyases, and carbohydrate
esterases in the *Bacteroidetes*
TLR mediated

Peripheral vascular disease
Result of metabolic syndrome
Altered lipid deposition/ metabolism

Inflammatory bowel disease
Hygiene hypothesis
Altered immune response: TLR signalling
Less microbial diversity
Activation of specific species: for example, *Escherichia*

FIGURE 4.2 **Gut microbiota (GM) affects in the body [34].**

Hospital, an impressive project funded by the Government of Ireland [36]. Although the topic of microbial evolution in the gut from infancy to adulthood has been discussed in another chapter in the book, briefly we have summarized the milestones in microbiotal changes up to the level of centenarians.

Establishment of the intestinal microbiota is a progressive process [37]. The major functions of GM begin to manifest at the end of the second year of life and comprise nutrient absorption and food fermentation [38], stimulation of the host immune system [39], and barrier effects against pathogens [40]. Once a stable composition is achieved near the end of adolescence, this ecosystem displays a high stability in healthy adults [41]. Although there are about 500 species resident at any one time in the GIT, 4 divisions of bacteria predominate: Firmicutes, Bacteroidetes, Proteobacteria and Actinobacteria. Firmicutes alone constitute ~64% of the microbiota, whereas Bacteroidetes account for ~23% of the normal microbiota, suggesting that these are especially important in the functioning of the host. Most bacteria belong to the genera *Bacteroides, Clostridium, Faecalibacterium, Eubacterium, Ruminococcus, Peptococcus, Peptostreptococcus,* and *Bifidobacterium* [42–44]. Other genera, such as *Escherichia* and *Lactobacillus,* are present to a lesser extent [42]. Although the intestinal microbiota is relatively stable throughout adult life, recent studies indicate

TABLE 4.2 Microbiota in Age Groups [49]

| | | TaqMan Detection | | | | | | SYBR Green Detection | |
| | | Firmicutes | | | | | | | Firmicutes |
	n	All Bacteria (a)	*Clostridium leptum* Group (b)	*Clostridium coccoides* Group (b)	*Bacteroides/ Prevotella* Group	*Bifidobacterium* Genus (b)	*Escherichia coli* (b)	*Lactobacillus /Leuconostoc /Pediococcus* Group (b)
Infant	21	10.7 ± 0.1 (A)	−3.2 ± 0.4 (A)	−3.2 ± 0.4 (A)	−1.5 ± 0.3 (A)	−0.6 ± 0.2 (A)	−1.5 ± 0.3 (A)	−3 ± 0.2 (A)
Adult	21	11.5 ± 0.1 (B)	−0.7 ± 0.1 (B)	−1.2 ± 0.1 (B)	−1.5 ± 0.1 (AB)	−2.3 ± 0.3 (B)	−3.8 ± 0.1 (B)	−3.9 ± 0.3 (AB)
Elder	20	11.4 ± 0.1 (B)	−1.1 ± 0.1 (C)	−1.8 ± 0.1 (A)	−1 ± 0.1 (A)	−2.3 ± 0.3 (B)	−2.4 ± 0.2 (C)	−4.2 ± 0.2 (B)

that modifications occur in the composition in elderly individuals. A reduction in the numbers of bifidobacteria and *Bacteroides* has been observed, accompanied by a decrease of Lactobacilli. A commensurate increase in the number of facultative anaerobes highlights the variation between adults and elderly individuals [2,45,46]. In elderly healthy adults, 80% of the identified fecal microbiota can be classified into three dominant phyla: Bacteroidetes, Firmicutes, and Actinobacteria [47]. In general terms the Firmicutes to Bacteroidetes ratio is regarded to be of significant relevance in human GM composition [48,49] (Table 4.2).

GM benefits the host by gleaning the energy from the fermentation of undigested carbohydrates and the subsequent absorption of short-chain fatty acids (SCFA). The most important of these are butyrates, metabolized by the colonic epithelium, propionates by the liver, and acetates by the muscle tissue. Intestinal bacteria also play a role in synthesizing vitamin B_{12} and vitamin K as well as metabolizing bile acids, sterols, and xenobiotics. Influenced by many factors, the human GM varies between individuals, population groups, and in different age groups. Also the composition of the microbiota varies within the GI tract depending on segments of intestine, and within the segments, for example, the mucus harbors different bacteria compared to the lumen [11]. The microbes in the GM of an individual may be continuously present (individual core) and microbes that can be detected in most people described as "collective core."

Another important role of gut flora is that they prevent species that would harm the host from colonizing the gut through competitive exclusion, an activity termed the "barrier effect." Harmful yeasts and bacterial species, such as *C. difficile* [the overgrowth of which can cause *C. difficile*-associated diseases (CDAD) like pseudomembranous colitis] are unable to grow excessively due to competition from helpful gut flora adhering to the mucosal lining of the intestine. The barrier effect protects from both invading species and species normally present in the gut in low numbers [42]. Symbiotic bacteria are more at home in this ecological niche and are thus more successful in the competition.

Indigenous gut flora also produce bacteriocins, which are proteinaceous toxins that inhibit growth of similar bacterial strains, substances that kill harmful microbes and the levels of which can be regulated by enzymes produced by the host.

Our ability to study the microbiota has substantially improved with the development of culture-independent methods for studying the microbes circumventing the poor laboratory culturability [50]. Microarray for the characterization of the human GM (human intestinal tract chip, HITChip) has been recently developed and validated [51]. Using the new tool, Biagi et al. addressed the age-related differences in both the GM and the inflammatory status among different stages of adult life, starting from young adults (∼30 years old), through elderly (∼70 years old), to the extreme limit of the human lifespan, represented by a group of centenarians [5]. The authors provided evidence that the aging process profoundly affects the structure of the human GM, as well as its homeostasis with the host's immune system [10,52,53]. The composition of the GM in older people (>65 years) is extremely variable between individuals, and differs from the core microbiota and diversity levels of younger adults [54,55]. The diversity is the result of coevolution between microbial communities and their hosts [56–59]. The characteristic change is an alteration of the relative proportions of the Firmicutes and the *Bacteroides*. The elderly has a higher number of *Bacteroides* while younger individuals a higher proportion of Firmicutes [14,49]. The Firmicutes/Bacteroidetes ratio undergoes an increase from birth to adulthood and is further altered with advanced age. Infants and elderly have a Firmicutes/Bacteroidetes ratio <1 and adults have a ratio >1 [48,49]. A reduction in the numbers and diversity of many protective commensal anaerobes occur in the elderly microbiome, such as *Bacteroides* and bifidobacteria [48,49] as well as an increase in the total number of facultative anaerobes and shifts in the dominant species within several bacterial groups.

There is a compromised microbiota in the centenarians associated with an increased inflammatory status, known as inflammaging (discussed later) [5]. After 100 years of symbiotic association with the human host, the microbiota was

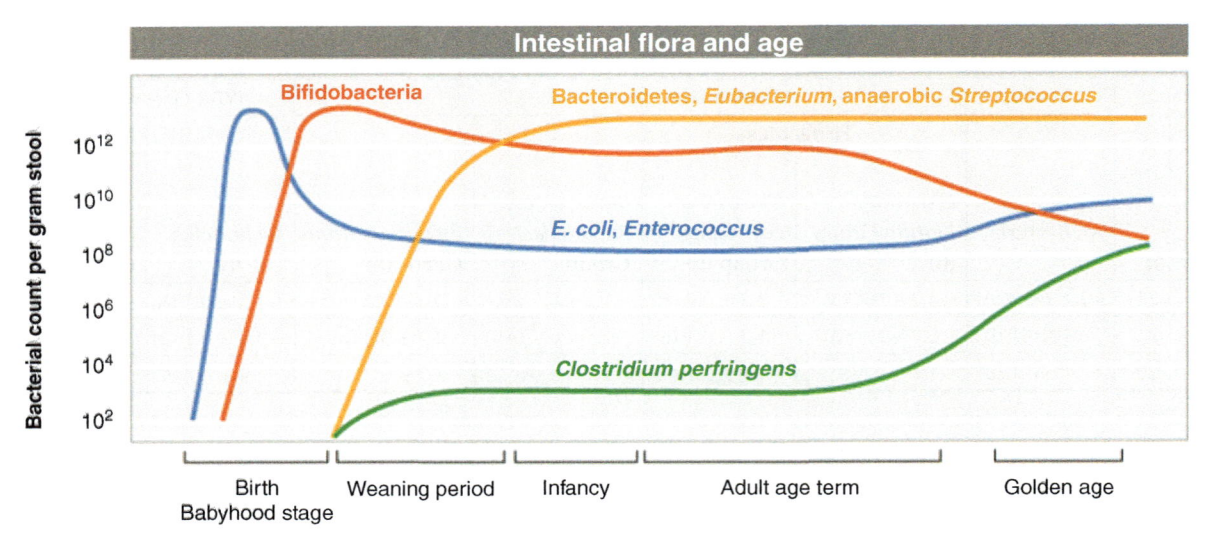

FIGURE 4.3 Age-related microbiota changes.

characterized by a rearrangement in the Firmicutes population and an enrichment in facultative anaerobes, notably pathobionts [60] (Fig. 4.3; Table 4.3).

DISCREPANCIES IN ELDERLY MICROBIOTA

The effect of age on the dominant components of the GM, Firmicutes and Bacteroidetes varies according to nationality and age [21]. Firmicutes, members of the *Clostridium* cluster XIVa (a dominant group in the intestinal microbiota, which includes among others the species *Eubacterium rectale*, *Eubacterium hallii*, *Eubacterium ventriosum*, *C. coccoides*, *Clostridium symbiosum*, *Ruminococcus gnavus*, *Ruminococcus obeum*, and the genera *Dorea*, *Roseburia*, *Lachnospira*, *Butyrivibrio*) were found to decrease in Japanese, Finnish, and Italian elderly and centenarians [5,57,61,62], whereas an inverse trend was found in German old adults [57]. The species *F. prausnitzii*, belonging to the *Clostridium* cluster IV (part of Firmicutes phylum), was markedly decreased in Italian elderly and centenarians [5,57], but not confirmed in other European populations [57,63]. However, it is well established that a decline in this antiinflammatory Firmicutes member of the GM is typical of frail, hospitalized, antibiotic- and antiinflammatory-treated elderly [52,64–66]. Conversely, an age related increase in Bacteroidetes was found in German, Austrian, Finnish and Irish elderly [57,61,63,64], but not noted in Italian elderly and centenarians [5,57]. In the case of Irish elderly, Bacteroidetes were found to be the dominant phylum instead of Firmicutes, which has always been regarded as the most abundant in healthy adults [63]. Health-promoting bacteria, such as bifidobacteria, were commonly regarded as decreasing along with ageing [10,57], but the most recent studies do not completely support this [5,51,67]. Much less controversial is the commonly reported age-related increase in facultative anaerobes, including streptococci, staphylococci, enterococci, and enterobacteria [5,49,51,57,61], often classified as pathobionts (bacteria present in the healthy GM in low concentration, which are able to thrive in inflamed conditions) [68].

TABLE 4.3 Age-Related Microbiome Changes [60]

Human Groups	Intestinal Microbiota Alterations Identified	Impact on Extraintestinal Locations
Healthy elderly	Country dependent changes in Firmicutes/Bacteroidetes ratio	Decline in the normal function of the immune system
	Elevated facultative anaerobes	Risk of autoimmune and chronic diseases
	Reduced level SCFA	Poor response to vaccination
	Lower diversity	Increased vulnerability to infection
Elderly under medical treatment	Lower levels of Bacteroidetes, *Bifidobacterium*, *Clostridium* cluster XIVa, *Faecalibacterium prausnitzii* members (cluster IV) Increased levels of *Lactobacillus* after antibiotic therapy Lower diversity	Increased susceptibility to infections by *C. difficile*

SCFA, Short-chain fatty acid.

TABLE 4.4 Impact of Age-Related GM Changes [70]

Age-Related Modifications in Gut Microbiota Composition	Health Outcome	Mechanism
Reduced biodiversity	Increased risk for CDAD	Reduction of colonization resistance to *C. difficile*
Proliferation of Enterobacteriaceae	Stimulation of inflammatory response	Excessive endotoxin production
	Increased probability to develop metastatic CRC	
Decrease of butyrate-producing bacterial groups	Weakening of the colonic epithelium	Reduction of the protective and trophic function of butyrate on the colonic epithelium
	Stimulation of inflammatory response	Reduction of the antiinflammatory effect of butyrate
	Increased risk for CRC	Reduction of the antineoplastic effect of butyrate
Colonization by toxin-producing *E. coli*, *H. pylori*, *B. fragilis*	Increased risk for CRC	Perturbation of enterocytes cell cycle regulation and growth control, and DNA damage by toxins

CDAD, *Clostridium difficile*-associated diarrhea; CRC, colorectal cancer.

INFLAMM-AGING

A marked impairment of the immune response in GALT, has been reported by several studies conducted mostly in animal models [3,69–71].

Microbiota components account for the production of SCFA and, in particular butyrate, acetate, and propionate. SCFAs are a subset of fatty acids that are produced by the GM during the fermentation of partially and nondigestible polysaccharides. SCFA are endowed with antiinflammatory (inhibition of NF-κB) and antineoplastic activities, also exerting a protective function in favor of intestinal epithelia [72]. In aged people, a reduction of butyrate levels depends on the decreased number of *F. prausnitzii*, *E. hallii*, and *E. rectal/Roseburia* group [5]. Therefore, SCFA decrease may lead to an impaired secretion of mucins by the intestinal epithelial cells promoting, easier entry of pathogens into the intestinal mucosa, especially Enterobacteriaceae. These Gram-negative bacteria are able to release lipopolysaccharides or endotoxins, which, in turn, aggravate the inflammatory condition [73].

In a healthy GI microenvironment the largest part of the resident GM is compartmentalized to the lumen, away from the epithelial surface [74]. In order to minimize the contact of the GM with the enterocytes, human beings have evolved an immune apparatus, where innate and adaptive immune systems cooperate to keep microbiota under control by a physiological low grade inflammatory status [75]. Capable to sense microorganisms, enterocytes tune the immune response of the GALT depending on the perceived degree of threat [76]. The impairment of the GALT to efficiently

synthesize strain-specific secretory IgA, together with the reduced efficiency of the innate immune defences, such as alpha-defensins, antimicrobial peptides and mucus secretion, results in the failure to control the resident microbiota, allowing an uncontrolled microbial growth on the enterocyte surface [70]. In this context enterocytes could engage the activation of inflammatory cytokines and chemokines, forcing dendritic cells in the underlying GALT to drive the differentiation of effector T_H1, T_H2, and T_H17 cells that induce a strong proinflammatory response [77]. Immunosenescence is also accompanied by a chronic, low grade overall inflammatory condition named "inflamm-aging" [71,78]. Inflammation aides growth of pathobionts, a minor component of the healthy intestinal microbiota that in an inflamed GI ecosystem can overtake mutualistic symbionts and support inflammation [70,79] (Table 4.4).

FACTORS INFLUENCING CHANGES IN GM IN THE ELDERLY

Diet

The diet of the elderly undergoes a series of changes influenced by age-related factors, such as changes in socioeconomic status, mobility, dentition, taste, smell, digestion, appetite, and conditions, such as depression and dementia [21,54,80–84]. The composition of the GM is modulated by dietary components including prebiotics [85–88].

The ELDERMET group of scientists observed the relationship between diet, gut bacteria, and health status in a large number of elderly (>65 years) Irish subjects. The knowledge is contributing to the NU-AGE project, a large multidisciplinary consortium with 30 partners, involving nutritionists, biogerontologists, immunologists, and molecular biologists in Europe. The NU-AGE project aims to explore how diet can help the elderly to live a healthier, longer life and to design new dietary strategies to address the specific needs of elderly.

Wu et al. [89] assessed the microbiota by pyrosequencing of 16S rDNA gene segments in 98 subject undergoing different diets. Whereas short-term diets had no influence on enterotypes, long-term diets were able to influence and affect the enterotype of individuals. Whereas diets enriched in protein and animal fat favored the *Bacteroides* enterotype, a carbohydrate enriched diet supported the *Prevotella* enterotype. Long-term change in diet is relevant in influencing the microbiota. The fecal microbiota of vegetarians was found to be different from omnivorous samples. A strict vegan or vegetarian diet results in a significant shift in the microbiota [90].

Claesson et al. in an elegant study from Ireland looked at the fecal microbiota composition from 178 elderly individuals formed four groups based on their diet: group 1 consuming low fat high fiber, group 2 moderate fat fiber, group 3 moderate fat and low fiber, and finally group 4 consuming high fat moderate fiber diet. They also correlated their findings based on community living or short or long term stay in a long term residential care facility [54]. The separation of microbiota composition significantly correlated with measures of frailty, comorbidity, nutritional status, markers

of inflammation and with metabolites in fecal water. The individual microbiota of people in long-stay care showed significant diversity of GM. Also the GM of community living elderly was more capable of producing SCFAs. As mentioned previously, SCFAs play an important role in the maintenance of health and the development of disease. The findings are in strong support of a relationship between diet, microbiota, and health status, a scientific basis not only for a prudent diet, but also for probiotic therapy. Also given the vast beneficial effects of SCFAs, and that their levels are regulated by diet, there is a new basis (i.e., reduced SCFA in the colon) to explain the increased prevalence of inflammatory disease in Westernized countries [2,18,55,91] (Table 4.5).

SPECIAL PROBLEMS OF THE OLDER ADULT PROBABLY RELATED TO GM

Clostridium difficile-Associated Diarrhea in the Elderly

Susceptibility to gastrointestinal infections in the older adult is correlated with these microbial changes in the gut. *C. difficile*-associated diarrhea one of the most common nosocomial infections in the elderly, has a profound effect on morbidity, mortality and health care costs. Risk factors CDAD are increasing age, severity of comorbid conditions, nonsurgical gastrointestinal procedures, tube feedings, alimentary surgery, PPI therapy, H2 antagonists and methotrexate, stay in ICU, duration of hospital stay, duration of antibiotic course, administration of multiple antibiotics, gastrointestinal problems, such as inflammatory bowel disease *C. difficile* is a Gram-positive, spore-forming anaerobic bacterium and is the agent that causes *C. difficile* infection (CDI).

Total enteral nutrition induces an imbalance between anaerobes and aerobes. Compared with controls, the number of anaerobic bacteria was significantly lower in patients on total enteral nutrition [45,92]. Since healthy GM is able to control the pathogen infection, restoration of the GM with diet, prebiotic and probiotic therapy appears to be a promising approach in the treatment of recurrent cases [45].

A recent spurt of publications on fecal microbiota transplantation (FMT) describes transferring feces from a healthy donor to a patient with recurrent *C. difficile* colitis. To correct the dysbiosis in the colon, which may be a predisposing factor, FMT is being currently used in patients who have had at least three recurrences of CDI and have failed all the conventional therapies, including a pulsed, tapered regimen of vancomycin [93,94]. FMT has been very successful, restoring the intestinal microbial diversity in up to 95% of patients [46,95–97].

Recent findings suggest that probiotics may have an immunoenhancing effect in the elderly. The beneficial role

TABLE 4.5 Key Functions of the GM [2,18,55]

Key Functions of the GM
Protective (production of antimicrobial factors, e.g. bacteriocins and lactic acids)
Structural (barrier fortification, induction of IgA, immune system development, synthesis of vitamins biotin and folate, fermentation of nondigestible dietary carbohydrates and endogenous epithelial-derived mucus, ion absorption, salvage of energy)
Metabolic functions
Intestinal motility
Synthesis of vitamins K and folic acid
Enterohepatic circulation of bile acids
Cholesterol metabolism
Role in gut homeostasis, adaptive immune system linked to immunosenescence
Disorders including obesity, cardio metabolic disorders, diabetes mellitus, inflammatory bowel disease, Parkinsonism and other neurological issues
Carcinogenesis

of lactobacilli and *Saccharomyces boulardii* in *C. difficile-*associated diarrhea has been well established in a recent metaanalysis [45]. Studies on probiotics to treat CDI have shown conflicting results [98].

GM and Cancer, Chemotherapy-Associated Clinical Problems

The incidence of chemotherapy-induced GI diseases may be as high as 50–80% of patients based on the agent and is associated with gastrointestinal mucositis and/or ulcers of the GIT clinically manifested as diarrhea, abdominal pain, bleeding, fatigue, malnutrition, dehydration, electrolyte imbalance, and infections. The prudent management often is to reduce the dose of chemotherapeutic agents, which may be suboptimal [99,100]. The GM plays a major role in the maintenance of intestinal homoeostasis and integrity. The beneficial role of prebiotic and probiotic therapies seen in a few clinical studies needs to be confirmed despite good experimental evidence. In rats treated with a single intraperitoneal dose of irinotecan, VSL#3 significantly prevented moderate or severe diarrhea, weight loss, prevented irinotecan-induced increase in goblet cells and apoptosis of intestinal crypt cells [101,102]. As compared with placebo, supplementation with Yakult (*Bifidobacterium breve* strain Yakult, 10/9 living bacteria given 3 times daily, started 2 weeks before the first day of chemotherapy, and continued for 6 weeks) resulted in a decreased incidence of fever and prevented some modifications in GM, such as an increase in Enterobacteriaceae. Caution must be exercised in using probiotics in immunosuppressed cancer patients receiving chemotherapy [103–105].

Obesity, Type II Diabetes, NAFLD, and Cardiovascular Diseases

Obesity is increasing in the elderly [106]. The link between the GM and the development of obesity, cardiovascular disease and metabolic syndromes, such as type 2 diabetes, is becoming clearer [107–112]. Some of the factors related to obesity and GM include synthesis of micronutrients, fermentation of indigestible food substances, assistance in the absorption of certain electrolytes and trace minerals [113]. A number of thought-provoking studies has shown that the GM flora of obese mice and humans include fewer Bacteroidetes and correspondingly more Firmicutes than that of their lean counterparts, suggesting that differences in caloric extraction of ingested food substances varies with the composition of the GM. Bacterial lipopolysaccharide derived from the GM may act as a triggering factor linking inflammation to high-fat diet-induced metabolic syndrome [114–116]. With the increase of Firmicutes in obesity there is also further aggravation of inflamm-ageing, as previously mentioned [48,69,112]. Analysis of the fecal microbiota of

morbidly obese subjects before and after Rou-en-Y gastric bypass surgery reported lower numbers of the *Bacteroides–Prevotella* group in obese subjects than in control subjects at baseline, while the numbers increased after surgery [117]. Another study, including normal weight, obese and anorexic subjects, also reported lower Bacteroidetes gene copy numbers in obese human subjects compared to control subjects [118]. What's more, a recent study found a decrease in specific species (*Bifidobacterium animalis* and *Methanobrevibacter smithii*) and an increase in others, such as *Staphylococcus aureus*, *E. coli*, and *Lactobacillus reuteri*, which have been associated with obesity [119]. The mechanisms rely either on the "energy harvest" hypothesis (an increased utilization of indigestible carbohydrates by microbial enzymes) or on the impact of microbial metabolites or cell-derived signals on the host pathways that regulate energy intake and/or fat deposition [115].

One out of four Americans who are 65 or older have type 2 diabetes. There is a relationship between the composition of the GM and diabetes [120]. Several studies on mice models and in humans have shown increase in body weight associated with a larger proportion of Firmicutes and relatively less Bacteroidetes [114,121,122].

In type 2 diabetes, there was a decrease in Firmicutes, while the proportion of Bacteroidetes and Proteobacteria was somewhat higher in diabetics persons compared to their nondiabetic counterparts [120].

Atherogenesis and Lipid Metabolism

Tang et al. [123] identified a biochemical pathway linking host and GM with atherogenesis [124]. Dietary choline metabolized by GM to trimethylamine (TMA), which on absorption into blood stream is converted to trimethylamine-*N*-oxide (TMAO), promotes atherogenesis.

INFLAMMATORY BOWEL DISEASE

Crohn's disease and UC preferentially occur in the colon and/or distal ileum, structures associated with the highest intestinal bacterial concentrations, suggesting a possible relationship between GM and disease [125]. Enteric bacteria may cause IBD in a genetically susceptible person [126]. Defects in the innate immune response to commensal intestinal bacteria, result in an exaggerated adaptive immune response to these organisms [127]. Rutgeerts et al. [128] are the first to provide evidence that GM play a role in the pathogenesis of IBD from clinical studies. Diversion of fecal stream from a segment of inflamed small bowel in patients with Crohn's disease showed histologic resolution. Studies demonstrated marked alterations in the GM, decreased microbial diversity, especially in numbers within the Firmicutes and *Bacteroides* phyla [129]. In general terms, patients with IBD exhibit an abnormal microbiota with instability of dominant species which is higher than in healthy controls. In particular, *F. prausnitzii* (butyrate-producing bacteria) is severely reduced in Crohn's disease and in UC with an increased prevalence of adherent-invasive *E. coli* strains [69]. Deviations from the healthy-like intestinal microbiota profile similar to that associated with the ageing process has been reported to accompany inflammatory disorders, such as inflammatory bowel diseases [70,130,131]. The increase in proinflammatory pathobionts and the decrease in immune modulatory species belonging to the *Clostridium* clusters IV and XIVa are hypothesized to be involved in the proinflammatory loop that promotes and sustains the inflammatory disorders [70]. In most patients with Crohn's disease there is a substantial antibody response to bacterial and fungal antigens [127].

Potential benefits of prebiotics, probiotics, and symbiotics are being extensively evaluated in many areas of elderly medicine [132]. Prebiotics as functional foods induce the growth of microorganisms in the GM providing benefits to the host. Providing a variety of food with plenty of dietary fiber is a good recommendation. Selected probiotic strains have been shown to beneficially influence the GM in the elderly by increasing the bifidobacteria or lactobacilli levels as evidenced by fecal analysis.

Modifying the bacterial flora with probiotics may attenuate the inflammatory process and prevent relapses in UC [133]. Three trials compared probiotics to mesalamine and one trial compared probiotics with placebo. The studies ranged in length from 3 to 12 months. The risk of bias was high in two studies due to incomplete outcome data and lack of blinding. There is insufficient evidence to make conclusions about the efficacy of probiotics for maintenance of remission in UC [133]. A metaanalysis evaluated 13 randomized controlled studies on remission induction and maintenance effect of probiotics in UC. Compared with the placebo group and nonprobiotics group, there was a higher the remission rate for UC patients who received probiotics and a lesser degree of recurrence. Probiotic treatment was more effective than placebo in maintaining remission in UC [134].

CANCERS

GM seems to affect intestinal and extraintestinal cancer incidences. Recent data implicate dysregulated host responses to enteric bacteria leading to cancers in extraintestinal sites.

The intestinal microbiota, which is composed of bacteria, viruses, and microeukaryotes, acts as an accessory organ system with distinct functions along the intestinal tract that are critical for health.

Colorectal Cancer

The incidence of colorectal cancer (CRC) increases in the elderly: about 50% of the western population develops colorectal polyps at the age of 70 and 5% of these polyps progress to cancer [135]. Aging produces a proinflammatory dysbiosis and together with the age associated decreased butyrate production in the intestine, this may lead to an increased risk of CRC [136,137]. There is increasing evidence that the colonic microbiota plays an important role in the cause of sporadic CRC [17,136]. Metabolic profiling studies of human fecal water extracts demonstrated a profound decrease of SCFA content in CRC. Colonic bacteria affect the neoplastic process by induction of mucosal inflammation in the GI tract [137]. In turn, chronic inflammation can support carcinogenesis by inducing gene mutation, inhibiting apoptosis or stimulating angiogenesis and cell proliferation. In particular, NF-κB is emerging as a key factor to provide a mechanistic link between inflammation and CRC [138], and its activation by TLR ligands from intestinal microorganisms has been hypothesized to mediate the intestinal tumor growth under steady state conditions [139].

Reduced temporal stability and increased diversity of the fecal microbiota exists in subjects with established CRC and polyposis [70]. Colonic polyps demonstrate higher bacterial diversity and richness when compared with control patients, with higher abundance of mucosal Proteobacteria and lower abundance of Bacteroidetes [140]. This may in part be explained by the mucosal defensive strategies designed to manage the commensal microbiota. For example, α-defensin expression is significantly increased in adenomas resulting in an increased antibacterial activity compared with normal mucosa [141]. Despite limited human studies, a small number of specific pathobionts have now been linked with adenomas and CRC including *Streptococcus gallolyticus* [142], *Enterococcus faecalis* [143], and *B. fragilis* [144]. *E. coli* is also overexpressed on CRC mucosa; it expresses genes that confer properties relevant to oncological transformation including M-cell translocation, angiogenesis, and genotoxicity [17,145]. Enrichment of *Fusobacterium nucleatum* has also been identified in adenoma versus adjacent normal tissue and is more abundant in stools from CRC and adenoma cases than in healthy controls. *F. nucleatum*'s fadA, a unique adhesin, allows it to adhere to and invade human epithelial cells, eliciting an inflammatory response [146] and stimulating cell proliferation [147].

Research focusing on the metabolic function of the gut microbiome and dietary microbiome interactions in the etiology of CRC involves the long held view of Burkitt's fiber hypothesis. Dietary fiber is critical to this. Metagenomic analyses have consistently identified a reduction of butyrate-producers in patients with CRC [148], a finding replicated in animals [145]. The microbiome also plays an important role in the metabolism of sulfate, through assimilatory sulfate-reduction to produce cysteine and methionine, and dissimilatory sulfate-reduction to produce hydrogen sulfide (H_2S). H_2S is likely to contribute to CRC development, as colonic detoxification of H_2S is also reduced in patients with CRC; it also induces colonic mucosal hyper-proliferation [149].

Furthermore, the impairment of the barrier function of the ageing GM, together with the immunosenescence dependent promotion of bacterial overgrowth on the epithelial cell surface, may enhance the risk to develop CRC, favoring persistent GI colonization by toxigenic bacterial strains [70]. These features are potentially procarcinogenic, and persistent GI infection by these toxigenic microorganisms is viewed as a paradigm for bacteria-induced cancer [137]. Besides toxins, lipopolysaccharides have been associated with metastatic colorectal tumor growth [150].

Gut Microbiome and Pancreatic Cancer

Infectious agents are estimated to be responsible for 10–20% of all cancers globally [151], yet none has been established as causative for pancreatic cancer so far. However, in the last few years several studies have presented tangential evidence that suggests a possible role of GM in the pathogenesis of pancreatic cancer [152]. The most plausible mechanism for the procarcinogenic effect of microbes seems to involve chronic low-grade activation of the immune system and perpetuation of tumor-associated inflammation, rather than direct mutagenic effects [153]. Indirect evidence for GM in relation to pancreatic cancer is related to T2DM, metabolic syndrome, and obesity [120,154].

Interestingly, *Helicobacter pylori*, the ubiquitous bacterium that colonizes the human stomach, has been studied for its possible association with pancreatic carcinoma [155]. There are two plausible hypotheses explaining the role of *H. pylori* in pancreatic carcinogenesis. First, colonization of the antrum by *H. pylori* reduces the number of antral D cells, thus suppressing the production of somatostatin, which results in an increase in the secretion of secretin and pancreatic bicarbonate output [155]. Secretin has been shown to have a positive effect on murine pancreatic growth as well as DNA synthesis in pancreatic ductal cells [156], and it can be inferred that induced ductal epithelial cell proliferation could enhance the carcinogenic effect of known carcinogens, leading to the development of pancreatic cancer [155]. Another hypothesis is that *H. pylori* growth in the gastric corpus, leads to atrophic gastritis and hypochlorhydria, which promotes bacteria overgrowth [155,157].

Oral microbiota overlaps with the GM, providing multiple avenues for dysbiosis [158]. A retrospective case-control study examined the relationship between antibodies to 25 oral bacteria and pancreatic cancer risk. Individuals with high levels of antibodies against *Porphyromonas gingivalis* ATTC 53978, a pathogenic periodontal bacteria, had a twofold higher risk of pancreatic cancer than individuals with lower levels of these antibodies. They also noted increased levels of antibodies against specific commensal oral bacteria, which can inhibit growth of pathogenic bacteria,

might reduce the risk of pancreatic cancer [158]. The possible mechanisms include Toll-like receptor signaling, nitrosamine exposure and increase of systemic inflammation [155,159].

Together these findings point to novel anticancer strategies aimed at promoting GI tract homeostasis.

ALZHEIMER'S DISEASE

GM may contribute to the regulation of multiple neurochemical and neurometabolic pathways through a complex series of highly interactive and symbiotic host-microbiome signaling systems that mechanistically interconnect the gastrointestinal (GI) tract, skin, liver, and other organs with the central nervous system (CNS) [160]. Established pathways of GI-CNS communication currently include the autonomic nervous system, the enteric nervous system (ENS), the neuroendocrine system, and the immune system [160–166]. Studies of the ENS in "germ-free" mice (missing their microbiome) indicate that commensal GM are absolutely essential for passive membrane characteristics, action potentials within the ENS, and the excitability of sensory neurons, providing a potential mechanistic link for the initial exchange of signaling information between the GI tract microbiome and the CNS [165,167,168]. Secretory products of the GM and translocation of these signaling molecules via the lymphatic and systemic circulation throughout the CNS are beginning to be identified. The GIT abundant Gram-positive facultative anaerobic or microaerophilic *Lactobacillus*, and other *Bifidobacterium* species, are capable of metabolizing glutamate to produce gamma-aminobutyric acid (GABA), the major inhibitory neurotransmitter in the CNS; dysfunctions in GABA-signaling are linked to anxiety, depression, defects in synaptogenesis, and cognitive impairment including AD [161,167,169–171].

Microbiome interactions with the N-methyl-D-aspartate (NMDA) glutamate receptor, a prominent CNS device that regulates synaptic plasticity and cognition, have also been shown [172]. The NMDA-, glutamate-targeting, glutathione-depleting and oxidative-stress-inducing neurotoxin β-N-methylamino-L-alanine (BMAA), found elevated in the brains of patients with amyotrophic-lateral sclerosis, Parkinson–dementia (PD) complex of Guam and AD, has been hypothesized to be generated by cyanobacteria of the intestinal microbiome, and may contribute to neurological dysfunction [160,173]. BMAA (neurotoxic amino acid) is not normally incorporated into protein and has been linked with intraneuronal protein misfolding, a hallmark feature of the amyloid peptide-enriched senile plaque lesions, and resultant inflammatory neurodegeneration, that characterize PD, AD, and prion disease [166,167,174,175]. Therefore the microbiome can potentially alter CNS neurochemistry and neurotransmission and secrete molecules that potentially modulate systemic- and CNS-amyloidosis [160].

TABLE 4.6 GM and Autoimmune Disease [176]

Taxa or Bacterial Feature	Background
Bacteroidetes:Firmicutes ratio	Low ratio associated with T1DM in affected children
Lactobacillus	Potentially protective role in T1DM, low abundances in affected children
Clostridium, Bacteroides, Veillonella, Alistipes	High abundances associated with high T1DM incidence and increased acetate and propionate in the distal gut
Bifidobacterium	Low abundance in T1DM affected children
Faecalibacterium and Subdoligranulum	Low abundance associated with low butyrate concentrations in the distal gut and high T1DM incidence. Decreased abundance (F. prausnitzii) in multiple sclerosis
Prevotella and Akkermansia	Low abundance associated with low mucin degradation potential and high T1DM incidence
Prevotella copri and Prevotella related taxa	High abundance in RA
Lachnospiraceae	High abundance and prevalence in affected mice in a SLE model
Clostridiales	Increased in collagen induced model of RA
Parabacteroides and Barnesiella	High abundances in disease resistant females in a collagen induced model of RA
Roseburia, Blautia, Coprococcus, Bilophila	High abundances in susceptible females in a murine T1DM model
Porphyromonadaceae	High abundance in males resistant to T1DM in a murine model
Enterobacteriaceae, Peptostreptococcaceae and segmented filamentous bacteria	High abundance in males resistant to T1DM in a murine model

RA, Rheumatoid arthritis; SLE, systemic lupus erythematosus; T1DM, type 1 diabetes mellitus.

AUTOIMMUNE DISEASES

A growing body of evidence has linked GM to various autoimmune diseases, gastrointestinal and otherwise [176]. The relationship and interactions among commensal microbes, immune system, and epigenetics are an emerging field of interest in autoimmune disease research [177].

Type I diabetes (T1D) is perhaps the most studied disorder for correlations with the GM. Patients with T1D have shown shifts in ratios of the main phyla within the GM exhibiting decreased Bacteroidetes:Firmicutes ratios, lower abundance of potential butyrate producers, and lower bacterial diversity [178]. Another study showed T1D incidence was associated with an increased abundance of specific taxa, such as *Clostridium* and decreased *Bifidobacterium* and *Lactobacillus* compared to healthy subjects [179].

Exploratory analyses in female-biased multiple sclerosis patients showed a decrease in abundance of *F. prausnitzii*, a taxon with known butyrate producing potential, in affected individuals [180].

An association between gut commensals and rheumatoid arthritis (RA) was inferred by the discovery that some microbial DNA, likely of the gut origin, is present in the synovial fluids of patients [181,182]. The involvement of resident commensals in driving RA progression is also supported by a high recurrence of periodontal inflammatory disorders in RA patients, caused by *Porphyromonas gingivalis* [183]. The involvement of GM in the pathogenesis of RA is further supported by increased gut permeability, suggesting that compromised intestinal barrier integrity

may lead to egress of luminal antigens including bacterial fragments and metabolites resulting in enhanced proinflammatory response against specific commensals in RA [184]. The therapeutic effect of some antibiotics (i.e., sulfasalazine and minocycline) for RA patients may be related to the bactericidal activity of these molecules, as they are likely modulating GM [185].

Systemic lupus erythematosus (SLE) has also been shown to be related to the GM [177]. The strongest evidence linking a bacterial antigen and the production of self-antibodies in SLE-related murine models is that injection of LPS in mice-induced production of anti-ds DNA antibodies, associated with an increased formation of immune complexes in kidneys, glomerular dysfunction and chronic kidney dysfunction [178,186,187]. Whether alterations in the GM are pathogenic or an effect of inflammation and autoimmune disease remains an essential question to be addressed [176,188] (Table 4.6).

IBS AND GUT MICROBIOTA

Epidemiological observations have demonstrated that onset of symptoms of IBS is often preceded by a disruption of the individual's normal intestinal microbiota, suggesting that intestinal dysbiosis can be an important etiological factor in IBS [189]. In a healthy individual the small intestine contains a much lower density of bacteria than the large intestine. IBS has been suggested to be associated with SIBO, defined as a bacterial density (colonic bacteria) of

$\geq 10^5$ colony-forming units per milliliter of intestinal fluid [190], measured by the "gold standard" jejunal culture method [191,192]. IBS also sometimes develops following recovery from enteric infections [193] and this may be the strongest predictor for the development of IBS with a six- to sevenfold increased risk and an average incidence of about 10% [194,195]. Absorbable antibiotics can also perturb the composition of the GM and this alteration has also been shown to be associated with symptoms in IBS [196].

Overall diversity of the GM of patients with IBS is reduced compared with the microbiota of healthy individuals [197–199]. The majority of the IBS patients have increase of Firmicutes-associated taxa and a depletion of Bacteroidetes-related taxa [200]. However, data on the abundance of specific species within these two microbial phyla are not consistent [189]. Additional details on this topic are discussed in another chapter.

Nonabsorbable antibiotics, such as neomycin [201] and rifaximin [202–204] are beneficial, providing partial alleviation of symptoms supporting the influence of microbiota on gut well-being [205]. Altering GM composition through the use of probiotics, such as *Bifidobacterium* spp. and *Lactobacillus* spp. [206] has been shown to improve symptoms of IBS [207–209], but not always [205].

DRUG METABOLISM

It is well known that the older adult takes a minimum of five drugs often to control diabetes, hypertension and for cardiovascular or neurologic diseases in addition to a proton pump inhibitor. The interaction of drugs, the consequent change in GM and its effect on the drug metabolism are not often considered in therapy.

GM affects the metabolism of several pharmaceutical agents, with a significant impact on drug efficacy and toxicity [210,211]. The involvement of bacterial enzymatic activities in drug metabolism and assimilation is going to become an important field of interest toward an efficient, personalized medicine, even more significant in the case of the elderly, who are more likely to use multiple drugs and for longer periods than young people [70]. GM secrete a diverse array of enzymes and at least 30 drugs which are, or have been, available commercially, were subsequently shown to be substrates for these bacterial enzymes (Table 4.7) [211]. One example of drug–microbiota interaction of particular interest for older people is simvastatin, a commonly prescribed drug [212]. It has been recently shown that the metabolism of this molecule involves gut microbial processes and has a pathway of colonic microbial metabolism [213]. Part of the variability in the therapeutic response to statin is linked to the variable metabolism related in turn to the high interindividual variability in the GM composition [70]. The determination of the structure–function relationships of the GM, may lead to the design of pharmacological and/or dietary interventions improving

TABLE 4.7 Examples of Drug Substrates for the Gastrointestinal Microbiota [211]

Drug Substrates Which Undergo Bacterial Metabolism		
Prontosil	Phenacetin	Simvastatin
Neoprontosil	Digoxin	Sulindac
Sulfasalazine	Ranitidine	Isosorbide dinitrate
Balsalazide	Nizatidine	Omeprazole
Olsalazine	Insulin	Metronidazole
L-Dopa	Calcitonin	Risperidone
Chloramphenicol	Nitrazepam	Methamphetamine
5-Aminosalicylic acid	Clonazepam	Zonisamide

drug efficacy by altering the phylogenetic functional architecture of the GM [34,211,214].

PROBIOTICS AND PREBIOTICS

The elderly, in particular those in senior care facilities, are prone to develop several age-related diseases and infections with increased morbidity and mortality compared to younger people [215]. Pro-/prebiotics are dietary supplements targeting the homeostasis of the intestinal microbial ecosystem, the composition and functionality of which has been associated with the health maintenance of the elderly [54] and can be hypothesized to impact on the host longevity. The use of pro-/prebiotics as well as a combination of them (synbiotics) in elderly has been recently reviewed in detail by Tiihonen et al. [11] and Toward et al. [216], from both a preventive and therapeutic point of view. Pro-/prebiotics and synbiotics have been shown to alter the composition of the GM, by inducing an increase in the fecal amount of bifidobacteria and lactobacilli [67,217–219]. However, higher *Bifidobacterium* and/or *Lactobacillus* levels do not represent a health benefit themselves; at best they are indicators of good intestinal health [11]. The potential benefits are tabulated [21,70,220–231] (Table 4.8).

TABLE 4.8 Potential Benefits of Pro- and Prebiotic Therapy in the Elderly

Decreases frailty [232]
Reduces the incidence of diarrhea associated with antibiotics and *C. difficile* associated diseases
An immune-modulatory role, prevent and/or limit the effects of immunosenescence
Decreases the duration of common infectious diseases of the airway and GI tract
Alleviates constipation
May induce colon polyps and other cancers

SUMMARY, CONCLUSIONS, PROJECTIONS FOR THE FUTURE

The adult GM consists of 10 to 100 trillion microorganisms almost 10 times the number of cells in the body. They exist in the gut in harmony and symbiotically with the host.

The GM may have beneficial or detrimental influences on the host. Microbiota determine the gut immunity and influence the occurrence of colonic infections, such as *C. difficile* colitis, inflammatory bowel disease, neoplasms in the GIT and elsewhere, irritable bowel syndrome, obesity, lipid disorders, coronary artery disease, insulin resistance, and obesity.

Understanding the determinants of stability the healthy GM of the young adult to old age is important to provide a relatively disease free state up to ripe old age and improve the healthcare of the growing number of elderly. It will also help in decreasing the growing cost of hospitalization of the older adult. Well-designed clinical studies on GM and diseases of the elderly need priority in allotment of research funds in medicine. In this regard the role of life style modifications, diet and nutrition throughout adult life, benefits of prebiotics and probiotics are to be carefully studied while exploitation of the elderly.

There is a window of opportunity to qualitatively improve the GM by diets rich in prebiotics, and oral supplementation of probiotics raising considerable hope in improving the quality of life of the growing number of elderly. Further probiotics reduce the side effects of antibiotics and immunosuppressive medications if needed [34] (Table 4.9).

TABLE 4.9 Key Points

Gut Microbiota in the Elderly

Physiological and pathological changes of the GI tract occur in the elderly (cause and/or effect of GM changes)

Although the intestinal microbiota is relatively stable throughout adult life modifications occur in the composition in the elderly.

The changes are dependent on extrinsic factors, country of living, food, drugs, comorbid conditions, hospitalizations, and other factors.

The Firmicutes/Bacteroidetes ratio undergoes an increase from birth to adulthood and is further altered with advanced age.

A reduction in species diversity, less resistant to environmental injuries

Decline in beneficial microorganisms in GM. The three dominant elderly phyla: Bacteroidetes, Firmicutes, and Actinobacteria

Effect of GM changes include the following:

- low grade chronic systemic inflammation (inflammaging contributes to frailty),
- reduced humoral and cellular immunity and gut immunity,
- susceptibility to intestinal and extraintestinal cancers, weight gain, IBS, IBD, metabolic syndrome.

Therapeutic potential for GM modification by prebiotics and probiotics in the prevention and management of many chronic diseases of the older adult and promotion of overall health.

IBD, Inflammatory bowel disorders.

REFERENCES

[1] Pirkl JJ. Aging, Available from: http://www.transgenerational.org/aging/aging-process.htm

[2] Woodmansey EJ. Intestinal bacteria and ageing. J Appl Microbiol 2007;102(5):1178–86.

[3] Dicarlo AL, Fuldner R, Kaminski J, Hodes R. Aging in the context of immunological architecture, function and disease outcomes. Trends Immunol 2009;30(7):293–4.

[4] Mitsuoka T. Intestinal flora and aging. Nutr Rev 1992;50(12):438–46.

[5] Biagi E, Nylund L, Candela M, et al. Through ageing, and beyond: gut microbiota and inflammatory status in seniors and centenarians. PLoS One 2010;5(5):e10667.

[6] Biagi E, Candela M, Franceschi C, Brigidi P. The aging gut microbiota: new perspectives. Ageing Res Rev 2011;10(4):428–9.

[7] Biagi E, Candela M, Martin F-P, et al. Metabonomics and gut microbial paradigm in healthy aging. In: Kochhar S, Martin F-P, editors. Metabonomics and gut microbiota in nutrition and disease. London: Springer; 2015. p. 169–84.

[8] Maccaferri S, Biagi E, Brigidi P. Metagenomics: key to human gut microbiota. Digest Dis 2011;29(6):525–30.

[9] Martin FP, Collino S, Rezzi S, Kochhar S. Metabolomic applications to decipher gut microbial metabolic influence in health and disease. Front Physiol 2012;3:113.

[10] Woodmansey EJ, McMurdo ME, Macfarlane GT, Macfarlane S. Comparison of compositions and metabolic activities of fecal microbiotas in young adults and in antibiotic-treated and non-antibiotic-treated elderly subjects. Appl Environ Microbiol 2004;70(10):6113–22.

[11] Tiihonen K, Ouwehand AC, Rautonen N. Human intestinal microbiota and healthy ageing. Ageing Res Rev 2010;9(2):107–16.

[12] Jeffery IB, Lynch DB, O'Toole PW. Composition and temporal stability of the gut microbiota in older persons. ISME J 2016;10(1):170–82.

[13] Inglis JE, Ilich JZ. The Microbiome and osteosarcopenic obesity in older individuals in long-term care facilities. Curr Osteoporos Rep 2015;13(5):358–62.

[14] Saraswati S, Sitaraman R. Aging and the human gut microbiota-from correlation to causality. Front Microbiol 2014;5:764.

[15] Brussow H. Microbiota and healthy ageing: observational and nutritional intervention studies. Microb Biotechnol 2013;6(4):326–34.

[16] Patrignani P, Tacconelli S, Bruno A. Gut microbiota, host gene expression, and aging. J Clin Gastroenterol 2014;48(Suppl. 1):S28–31.

[17] Marchesi JR, Adams DH, Fava F, et al. The gut microbiota and host health: a new clinical frontier. Gut 2016;65(2):330–9.

[18] Foxx-Orenstein AE, Chey WD. Manipulation of the gut microbiota as a novel treatment strategy for gastrointestinal disorders. Am J Gastroenterol Suppl 2012;1(1):41–6.

[19] Hooper LV, Gordon JI. Commensal host-bacterial relationships in the gut. Science 2001;292(5519):1115–8.

[20] Dethlefsen L, Eckburg PB, Bik EM, Relman DA. Assembly of the human intestinal microbiota. Trends Ecol Evolut 2006;21(9): 517–23.

[21] Biagi E, Candela M, Fairweather-Tait S, Franceschi C, Brigidi P. Aging of the human metaorganism: the microbial counterpart. Age (Dordr) 2012;34(1):247–67.

[22] Gill SR, Pop M, Deboy RT, et al. Metagenomic analysis of the human distal gut microbiome. Science 2006;312(5778):1355–9.

[23] Rodriguez JM, Murphy K, Stanton C, et al. The composition of the gut microbiota throughout life, with an emphasis on early life. Microb Ecol Health Dis 2015;26:26050.

[24] Saffrey MJ. Aging of the mammalian gastrointestinal tract: a complex organ system. Age (Dordr) 2014;36(3):9603.

[25] Blechman M, Gelb AM. Aging and gastrointestinal physiology. Clin Geriatr Med 1999;15(3):429–38.

[26] D'Souza AL. Ageing and the gut. Postgrad Med J 2007;83(975):44–53.

[27] Wernette CM, White BD, Zizza CA. Signaling proteins that influence energy intake may affect unintentional weight loss in elderly persons. J Am Diet Assoc 2011;111(6):864–73.

[28] Champion A. Anorexia of aging. Ann Long-term Care 2011;19(10).

[29] Atalayer D, Astbury NM. Anorexia of aging and gut hormones. Aging Dis 2013;4(5):264–75.

[30] Katz D, Pitchumoni CS. Management of the hiatal hernia-esophagitis complex in the elderly. Geriatrics 1973;28(10):84–7.

[31] McDonald LC, Killgore GE, Thompson A, et al. An epidemic, toxin gene-variant strain of *Clostridium difficile*. N Engl J Med 2005;353(23):2433–41.

[32] Bartlett JG, Perl TM. The new *Clostridium difficile*—what does it mean? N Engl J Med 2005;353(23):2503–5.

[33] Aronsson B, Mollby R, Nord CE. Antimicrobial agents and *Clostridium difficile* in acute enteric disease: epidemiological data from Sweden, 1980–1982. J Infect Dis 1985;151(3):476–81.

[34] Kinross JM, Darzi AW, Nicholson JK. Gut microbiome-host interactions in health and disease. Genome Med 2011;3(3):14.

[35] McMurdo ME, Roberts H, Parker S, et al. Improving recruitment of older people to research through good practice. Age Ageing 2011;40(6):659–65.

[36] Cusack S, O'Toole PW. Challenges and implications for biomedical research and intervention studies in older populations: insights from the ELDERMET study. Gerontology 2013;59(2):114–21.

[37] Palmer C, Bik EM, DiGiulio DB, Relman DA, Brown PO. Development of the human infant intestinal microbiota. PLoS Biol 2007;5(7):e177.

[38] Bäckhed F, Ley RE, Sonnenburg JL, Peterson DA, Gordon JI. Host-bacterial mutualism in the human intestine. Science 2005;307(5717):1915–20.

[39] Macpherson AJ, Harris NL. Interactions between commensal intestinal bacteria and the immune system. Nat Rev Immunol 2004;4(6):478–85.

[40] Mowat AM. Anatomical basis of tolerance and immunity to intestinal antigens. Nat Rev Immunol 2003;3(4):331–41.

[41] Franks AH, Harmsen JH, Raangs GC, Jansen GJ, Schut F, Welling GW. Variations of bacterial populations in human feces measured by fluorescent in situ hybridization with group-specific 16S rRNA-targeted oligonucleotide probes. Appl Environ Microbiol 1998;64(9):3336–45.

[42] Guarner F, Malagelada JR. Gut flora in health and disease. Lancet 2003;361(9356):512–9.

[43] Beaugerie L, Petit JC. Microbial-gut interactions in health and disease. Antibiotic-associated diarrhoea. Best Pract Res Clin Gastroenterol 2004;18(2):337–52.

[44] Vedantam G, Hecht DW. Antibiotics and anaerobes of gut origin. Curr Opin Microbiol 2003;6(5):457–61.

[45] Hebuterne X. Gut changes attributed to ageing: effects on intestinal microflora. Curr Opin Clin Nutr Metab Care 2003;6(1):49–54.

[46] Hopkins MJ, Macfarlane GT. Changes in predominant bacterial populations in human faeces with age and with *Clostridium difficile* infection. J Med Microbiol 2002;51(5):448–54.

[47] Lay C, Sutren M, Rochet V, Saunier K, Doré J, Rigottier-Gois L. Design and validation of 16S rRNA probes to enumerate members of the *Clostridium leptum* subgroup in human faecal microbiota. Environ Microbiol 2005;7(7):933–46.

[48] Ley RE, Turnbaugh PJ, Klein S, Gordon JI. Microbial ecology: human gut microbes associated with obesity. Nature 2006;444(7122):1022–3.

[49] Mariat D, Firmesse O, Levenez F, et al. The Firmicutes/Bacteroidetes ratio of the human microbiota changes with age. BMC Microbiol 2009;9:123.

[50] Zoetendal EG, Rajilic´-Stojanović M, de Vos WM. High-throughput diversity and functionality analysis of the gastrointestinal tract microbiota. Gut 2008;57(11):1605–15.

[51] Rajilic´-Stojanovic´ M, Heilig HG, Molenaar D, et al. Development and application of the human intestinal tract chip, a phylogenetic microarray: analysis of universally conserved phylotypes in the abundant microbiota of young and elderly adults. Environ Microbiol 2009;11(7):1736–51.

[52] Bartosch S, Fite A, Macfarlane GT, McMurdo ME. Characterization of bacterial communities in feces from healthy elderly volunteers and hospitalized elderly patients by using real-time PCR and effects of antibiotic treatment on the fecal microbiota. Appl Environ Microbiol 2004;70(6):3575–81.

[53] Hopkins MJ, Sharp R, Macfarlane GT. Age and disease related changes in intestinal bacterial populations assessed by cell culture, 16S rRNA abundance, and community cellular fatty acid profiles. Gut 2001;48(2):198–205.

[54] Claesson MJ, Jeffery IB, Conde S, et al. Gut microbiota composition correlates with diet and health in the elderly. Nature 2012;488(7410):178–84.

[55] Guigoz Y, Dore J, Schiffrin EJ. The inflammatory status of old age can be nurtured from the intestinal environment. Curr Opin Clin Nutr Metab Care 2008;11(1):13–20.

[56] Yatsunenko T, Rey FE, Manary MJ, et al. Human gut microbiome viewed across age and geography. Nature 2012;486(7402):222–7.

[57] Mueller S, Saunier K, Hanisch C, et al. Differences in fecal microbiota in different European study populations in relation to age, gender, and country: a cross-sectional study. Appl Environ Microbiol 2006;72(2):1027–33.

[58] Kurokawa K, Itoh T, Kuwahara T, et al. Comparative metagenomics revealed commonly enriched gene sets in human gut microbiomes. DNA Res 2007;14(4):169–81.

[59] Tannock GW. What immunologists should know about bacterial communities of the human bowel. Semin Immunol 2007;19(2): 94–105.

[60] Salazar N, Arboleya S, Valdés L, et al. The human intestinal microbiome at extreme ages of life. Dietary intervention as a way to counteract alterations. Front Genet 2014;5:406.

[61] Makivuokko H, Tiihonen K, Tynkkynen S, Paulin L, Rautonen N. The effect of age and non-steroidal anti-inflammatory drugs on human intestinal microbiota composition. Br J Nutr 2010;103(2):227–34.

[62] Hayashi H, Sakamoto M, Kitahara M, Benno Y. Molecular analysis of fecal microbiota in elderly individuals using 16S rDNA library and T-RFLP. Microbiol Immunol 2003;47(8):557–70.

[63] Claesson MJ, Cusack S, O'Sullivan O, et al. Composition, variability, and temporal stability of the intestinal microbiota of the elderly. Proc Natl Acad Sci USA 2011;108(Suppl. 1):4586–91.

[64] Zwielehner J, Liszt K, Handschur M, Lassl C, Lapin A, Haslberger AG. Combined PCR-DGGE fingerprinting and quantitative-PCR indicates shifts in fecal population sizes and diversity of *Bacteroides*, bifidobacteria and *Clostridium* cluster IV in institutionalized elderly. Exp Gerontol 2009;44(6–7):440–6.

[65] Tiihonen K, Tynkkynen S, Ouwehand A, Ahlroos T, Rautonen N. The effect of ageing with and without non-steroidal anti-inflammatory drugs on gastrointestinal microbiology and immunology. Br J Nutr 2008;100(1):130–7.

[66] van Tongeren SP, Slaets JP, Harmsen JH, Welling GW. Fecal microbiota composition and frailty. Appl Environ Microbiol 2005;71(10):6438–42.

[67] Lahtinen SJ, Tammela L, Korpela J, et al. Probiotics modulate the *Bifidobacterium* microbiota of elderly nursing home residents. Age (Dordr) 2009;31(1):59–66.

[68] Pedron T, Sansonetti P. Commensals, bacterial pathogens and intestinal inflammation: an intriguing menage a trois. Cell Host Microbe 2008;3(6):344–7.

[69] Magrone T, Jirillo E. The interaction between gut microbiota and age-related changes in immune function and inflammation. Immun Ageing 2013;10(1):31.

[70] Biagi E, Candela M, Turroni S, Garagnani P, Franceschi C, Brigidi P. Ageing and gut microbes: perspectives for health maintenance and longevity. Pharmacol Res 2013;69(1):11–20.

[71] Larbi A, Franceschi C, Mazzatti D, Solana R, Wikby A, Pawelec G. Aging of the immune system as a prognostic factor for human longevity. Physiology (Bethesda) 2008;23:64–74.

[72] De Vuyst L, Leroy F. Cross-feeding between bifidobacteria and butyrate-producing colon bacteria explains bifdobacterial competitiveness, butyrate production, and gas production. Int J Food Microbiol 2011;149(1):73–80.

[73] Schiffrin EJ, Morley JE, Donnet-Hughes A, Guigoz Y. The inflammatory status of the elderly: the intestinal contribution. Mutat Res 2010;690(1–2):50–6.

[74] Johansson ME, Phillipson M, Petersson J, Velcich A, Holm L, Hansson GC. The inner of the two Muc2 mucin-dependent mucus layers in colon is devoid of bacteria. Proc Natl Acad Sci USA 2008;105(39):15064–9.

[75] Hooper LV, Macpherson AJ. Immune adaptations that maintain homeostasis with the intestinal microbiota. Nat Rev Immunol 2010;10(3):159–69.

[76] Macdonald TT, Monteleone G. Immunity, inflammation, and allergy in the gut. Science 2005;307(5717):1920–5.

[77] Maynard CL, Weaver CT. Intestinal effector T cells in health and disease. Immunity 2009;31(3):389–400.

[78] Franceschi C, Capri M, Monti D, et al. Inflammaging and anti-inflammaging: a systemic perspective on aging and longevity emerged from studies in humans. Mech Ageing Dev 2007;128(1):92–105.

[79] Round JL, Mazmanian SK. The gut microbiome shapes intestinal immune responses during health and disease. Nat Rev Immunol 2009;9(5):313–23.

[80] Moschen AR, Wieser V, Tilg H. Dietary factors: major regulators of the gut's microbiota. Gut Liver 2012;6(4):411–6.

[81] Turnbaugh PJ, Ridaura VK, Faith JJ, Rey FE, Knight R, Gordon JI. The effect of diet on the human gut microbiome: a metagenomic analysis in humanized gnotobiotic mice. Sci Transl Med 2009;1(6):6ra14.

[82] Hildebrandt MA, Hoffmann C, Sherrill-Mix SA, et al. High-fat diet determines the composition of the murine gut microbiome independently of obesity. Gastroenterology 2009;137(5):1716–24.

[83] Zimmer J, Lange B, Frick JS, et al. A vegan or vegetarian diet substantially alters the human colonic faecal microbiota. Eur J Clin Nutr 2012;66(1):53–60.

[84] Kabeerdoss J, Devi RS, Mary RR, Ramakrishna BS. Faecal microbiota composition in vegetarians: comparison with omnivores in a cohort of young women in southern India. Br J Nutr 2012;108(6):953–7.

[85] Gibson GR, Roberfroid MB. Dietary modulation of the human colonic microbiota: introducing the concept of prebiotics. J Nutr 1995;125(6):1401–12.

[86] Lakshminarayanan B, Stanton C, O'Toole PW, Ross RP. Compositional dynamics of the human intestinal microbiota with aging: implications for health. J Nutr Health Aging 2014;18(9):773–86.

[87] Power SE, O'Toole PW, Stanton C, Ross RP, Fitzgerald GF. Intestinal microbiota, diet and health. Br J Nutr 2014;111(3):387–402.

[88] Cusack S, O'Toole PW. Diet, the gut microbiota and healthy ageing: How dietary modulation of the gut microbiota could transform the health of older populations. Agro FOOD Ind Hi Tech 2013;24(2):54–7.

[89] Wu GD, Chen J, Hoffmann C, et al. Linking long-term dietary patterns with gut microbial enterotypes. Science 2011;334(6052):105–8.

[90] Wu GD, Compher C, Chen EZ, et al. Comparative metabolomics in vegans and omnivores reveal constraints on diet-dependent gut microbiota metabolite production. Gut 2016;65(1):63–72.

[91] Tan J, McKenzie C, Potamitis M, Thorburn AN, Mackay CR, Macia L, et al. The role of short-chain fatty acids in health and disease. Adv Immunol 2014;121:91–119.

[92] Schneider SM, Le Gall P, Girard-Pipau F. Total artificial nutrition is associated with major changes in the fecal flora. Eur J Nutr 2000;39(6):248–55.

[93] Brandt LJ, Reddy SS. Fecal microbiota transplantation for recurrent *Clostridium difficile* infection. J Clin Gastroenterol 2011;45 Suppl. :S159–67.

[94] Yoon SS, Brandt LJ. Treatment of refractory/recurrent *C. difficile*-associated disease by donated stool transplanted via colonoscopy: a case series of 12 patients. J Clin Gastroenterol 2010;44(8):562–6.

[95] Surawicz CM, Brandt LJ, Binion DG, et al. Guidelines for diagnosis, treatment, and prevention of *Clostridium difficile* infections. Am J Gastroenterol 2013;108(4):478–98. quiz 499.

[96] Brandt LJ, Aroniadis OC, Mellow M, et al. Long-term follow-up of colonoscopic fecal microbiota transplant for recurrent *Clostridium difficile* infection. Am J Gastroenterol 2012;107(7):1079–87.

[97] van Nood E, Vrieze A, Nieuwdorp M, et al. Duodenal infusion of donor feces for recurrent *Clostridium difficile*. N Engl J Med 2013;368(5):407–15.

[98] Hookman P, Barkin JS. *Clostridium difficile* associated infection, diarrhea and colitis. World J Gastroenterol 2009;15(13):1554–80.

[99] Ferreira MR, Andreyev HJ. Editorial: Gut microbiota and chemotherapy- or radiation-induced gastrointestinal mucositis. Aliment Pharmacol Ther 2014;40(6):733–4.

[100] Benson AB 3rd, Ajani JA, Catalano RB, et al. Recommended guidelines for the treatment of cancer treatment-induced diarrhea. J Clin Oncol 2004;22(14):2918–26.

[101] Bosset JF, Calais G, Daban A, et al. Preoperative chemoradiotherapy versus preoperative radiotherapy in rectal cancer patients: assessment of acute toxicity and treatment compliance. Report of the 22921 randomised trial conducted by the EORTC Radiotherapy Group. Eur J Cancer 2004;40(2):219–24.

[102] Sasse AD, Clark LG, Sasse EC, Clark OA. Amifostine reduces side effects and improves complete response rate during radiotherapy: results of a meta-analysis. Int J Radiat Oncol Biol Phys 2006;64(3):784–91.

[103] Wada M, Nagata S, Saito M, et al. Effects of the enteral administration of *Bifidobacterium breve* on patients undergoing chemotherapy for pediatric malignancies. Supp Care Cancer 2010;18(6):751–9.

[104] Liong MT. Safety of probiotics: translocation and infection. Nutr Rev 2008;66(4):192–202.

[105] Ciorba MA, Riehl TE, Rao MS, et al. *Lactobacillus* probiotic protects intestinal epithelium from radiation injury in a TLR-2/cyclo-oxygenase-2-dependent manner. Gut 2012;61(6):829–38.

[106] Flegal KM, Carroll MD, Kuczmarski RJ, Johnson CL. Overweight and obesity in the United States: prevalence and trends, 1960–1994. Int J Obes Relat Metab Disord 1998;22(1):39–47.

[107] Tremaroli V, Backhed F. Functional interactions between the gut microbiota and host metabolism. Nature 2012;489(7415):242–9.

[108] Ley RE. Obesity and the human microbiome. Curr Opin Gastroenterol 2010;26(1):5–11.

[109] Hansen TH, Gobel RJ, Hansen T, Pedersen O. The gut microbiome in cardio-metabolic health. Genome Med 2015;7(1):33.

[110] Napolitano A, Miller S, Nicholls AW, et al. Novel gut-based pharmacology of metformin in patients with type 2 diabetes mellitus. PLoS One 2014;9(7):e100778.

[111] Kootte RS, Vrieze A, Holleman F, et al. The therapeutic potential of manipulating gut microbiota in obesity and type 2 diabetes mellitus. Diabetes Obes Metab 2012;14(2):112–20.

[112] DiBaise JK, Zhang H, Crowell MD, Krajmalnik-Brown R, Decker GA, Rittmann BE. Gut microbiota and its possible relationship with obesity. Mayo Clin Proc 2008;83(4):460–9.

[113] Roberfroid MB, Bornet F, Bouley C, Cummings JH. Colonic microflora: nutrition and health. Summary and conclusions of an International Life Sciences Institute (ILSI) [Europe] workshop held in Barcelona, Spain. Nutr Rev 1995;53(5):127–30.

[114] Backhed F, Ding H, Wang T, et al. The gut microbiota as an environmental factor that regulates fat storage. Proc Natl Acad Sci USA. 2004;101(44):15718–23.

[115] Backhed F, Manchester JK, Semenkovich CF, Gordon JI. Mechanisms underlying the resistance to diet-induced obesity in germ-free mice. Proc Natl Acad Sci USA 2007;104(3):979–84.

[116] Turnbaugh PJ, Ley RE, Mahowald MA, Magrini V, Mardis ER, Gordon JI. An obesity-associated gut microbiome with increased capacity for energy harvest. Nature 2006;444(7122):1027–31.

[117] Furet JP, Kong LC, Tap J, et al. Differential adaptation of human gut microbiota to bariatric surgery-induced weight loss: links with metabolic and low-grade inflammation markers. Diabetes 2010;59(12):3049–57.

[118] Armougom F, Henry M, Vialettes V, Raccah D, Raoult D. Monitoring bacterial community of human gut microbiota reveals an increase in *Lactobacillus* in obese patients and Methanogens in anorexic patients. PLoS One 2009;4(9):e7125.

[119] Million M, Maraninchi M, Henry M, et al. Obesity-associated gut microbiota is enriched in *Lactobacillus reuteri* and depleted in *Bifidobacterium animalis* and *Methanobrevibacter smithii*. Int J Obes (Lond) 2012;36(6):817–25.

[120] Larsen N, Vogensen FK, van den Berg FW, et al. Gut microbiota in human adults with type 2 diabetes differs from non-diabetic adults. PLoS One 2010;5(2):e9085.

[121] Ley RE, Bäckhed F, Turnbaugh P, Lozupone CA, Knight RD, Gordon JI. Obesity alters gut microbial ecology. Proc Natl Acad Sci USA 2005;102(31):11070–5.

[122] Turnbaugh PJ, Hamady M, Yatsunenko T, et al. A core gut microbiome in obese and lean twins. Nature 2009;457(7228):480–4.

[123] Tang WH, Wang Z, Levison BS, et al. Intestinal microbial metabolism of phosphatidylcholine and cardiovascular risk. N Engl J Med 2013;368(17):1575–84.

[124] Wang Z, Klipfell E, Bennett BJ, et al. Gut flora metabolism of phosphatidylcholine promotes cardiovascular disease. Nature 2011;472(7341):57–63.

[125] Swidsinski A, Ladhoff A, Pernthaler A, et al. Mucosal flora in inflammatory bowel disease. Gastroenterology 2002;122(1):44–54.

[126] Abraham C, Cho JH. Inflammatory bowel disease. N Engl J Med. 2009;361(21):2066–78.

[127] Sartor RB, Mazmanian SK. Intestinal microbes in inflammatory bowel diseases. Am J Gastroenterol 2012;Suppl. 1(1):15–21.

[128] Rutgeerts P, Goboes K, Peeters M, et al. Effect of faecal stream diversion on recurrence of Crohn's disease in the neoterminal ileum. Lancet 1991;338(8770):771–4.

[129] Koboziev I, Reinoso Webb C, Furr KL, Grisham MB. Role of the enteric microbiota in intestinal homeostasis and inflammation. Free Radic Biol Med 2014;68:122–33.

[130] Maslowski KM, Mackay CR. Diet, gut microbiota and immune responses. Nat Immunol 2011;12(1):5–9.

[131] Neish AS. Microbes in gastrointestinal health and disease. Gastroenterology 2009;136(1):65–80.

[132] Gionchetti P, Rizzello F, Venturi A, et al. Oral bacteriotherapy as maintenance treatment in patients with chronic pouchitis: a double-blind, placebo-controlled trial. Gastroenterology 2000;119(2):305–9.

[133] Naidoo K, Gordon M, Fagbemi AO, Thomas AG, Akobeng AK. Probiotics for maintenance of remission in ulcerative colitis. Cochrane Database Syst Rev 2011;(12):CD007443.

[134] Sang LX, Chang B, Zhang WL, Wu XM, Li XH, Jiang M. Remission induction and maintenance effect of probiotics on ulcerative colitis: a meta-analysis. World J Gastroenterol 2010;16(15):1908–15.

[135] Boyle P, Ferlay J. Mortality and survival in breast and colorectal cancer. Nat Clin Pract Oncol 2005;2(9):424–5.

[136] Ou J, Carbonero F, Zoetendal EG, et al. Diet, microbiota, and microbial metabolites in colon cancer risk in rural Africans and African Americans. Am J Clin Nutr 2013;98(1):111–20.

[137] Candela M, Guidotti M, Fabbri A, Brigidi P, Franceschi C, Fiorentini C. Human intestinal microbiota: cross-talk with the host

and its potential role in colorectal cancer. Crit Rev Microbiol 2011;37(1):1–14.

[138] Kraus S, Arber N. Inflammation and colorectal cancer. Curr Opin Pharmacol 2009;9(4):405–10.

[139] Lee SH, Hu LL, Gonzalez-Navajas J, et al. ERK activation drives intestinal tumorigenesis in Apc(min/+) mice. Nat Med 2010;16(6):665–70.

[140] Shen XJ, Rawls JF, Randall T, et al. Molecular characterization of mucosal adherent bacteria and associations with colorectal adenomas. Gut Microbes 2010;1(3):138–47.

[141] Pagnini C, Corleto VD, Mangoni ML, et al. Alteration of local microflora and alpha-defensins hyper-production in colonic adenoma mucosa. J Clin Gastroenterol 2011;45(7):602–10.

[142] Boleij A, Tjalsma H. The itinerary of *Streptococcus gallolyticus* infection in patients with colonic malignant disease. Lancet Infect Dis 2013;13(8):719–24.

[143] Wang X, Yang Y, Huycke MM. Commensal bacteria drive endogenous transformation and tumour stem cell marker expression through a bystander effect. Gut 2015;64(3):459–68.

[144] Wu S, Rhee JK, Albesiano E, et al. A human colonic commensal promotes colon tumorigenesis via activation of T helper type 17 T cell responses. Nat Med 2009;15(9):1016–22.

[145] Prorok-Hamon M, Friswell MK, Alswied A, et al. Colonic mucosa-associated diffusely adherent afaC+ *Escherichia coli* expressing lpfA and pks are increased in inflammatory bowel disease and colon cancer. Gut 2014;63(5):761–70.

[146] Han YW, Shi W, Huang GT, et al. Interactions between periodontal bacteria and human oral epithelial cells: *Fusobacterium nucleatum* adheres to and invades epithelial cells. Infect Immun 2000;68(6):3140–6.

[147] Rubinstein MR, Wang X, Liu W, Hao Y, Cai G, Han YW. *Fusobacterium nucleatum* promotes colorectal carcinogenesis by modulating E-cadherin/beta-catenin signaling via its FadA adhesin. Cell Host Microbe 2013;14(2):195–206.

[148] Wu N, Yang X, Zhang R, et al. Dysbiosis signature of fecal microbiota in colorectal cancer patients. Microb Ecol 2013;66(2):462–70.

[149] Carbonero F, Benefiel AC, Zlizadeh-Ghamsari AH, Gaskins HR. Microbial pathways in colonic sulfur metabolism and links with health and disease. Front Physiol 2012;3:448.

[150] Killeen SD, Wang JH, Andrews EJ, Redmond HP. Bacterial endotoxin enhances colorectal cancer cell adhesion and invasion through TLR-4 and NF-kappaB-dependent activation of the urokinase plasminogen activator system. Br J Cancer 2009;100(10):1589–602.

[151] Chang AH, Parsonnet J. Role of bacteria in oncogenesis. Clin Microbiol Rev 2010;23(4):837–57.

[152] Zambirinis CP, Pushalkar S, Saxena D, Miller G. Pancreatic cancer, inflammation, and microbiome. Cancer J 2014;20(3):195–202.

[153] Zambirinis CP, Pushalkar S, Saxena D, Miller G. Pancreatic cancer, inflammation and microbiome. Cancer J 2014;20(3):195–202.

[154] Shen J, Obin MS, Zhao L. The gut microbiota, obesity and insulin resistance. Mol Aspects Med 2013;34(1):39–58.

[155] Wang C, Li J. Pathogenic microorganisms and pancreatic cancer. Gastrointest Tumors 2015;2(1):41–7.

[156] Bulajic M, Panic N, Lohr JM. *Helicobacter pylori* and pancreatic diseases. World J Gastrointest Pathophysiol 2014;5(4):380–3.

[157] Kokkinakis DM, Reddy MK, Norgle JR, Baskaran K. Metabolism and activation of pancreas specific nitrosamines by pancreatic ductal cells in culture. Carcinogenesis 1993;14(8):1705–9.

[158] Michaud DS, Izard J, Wilhelm-Benartzi CS, et al. Plasma antibodies to oral bacteria and risk of pancreatic cancer in a large European prospective cohort study. Gut 2013;62(12):1764–70.

[159] Michaud DS, Izard J. Microbiota, oral microbiome, and pancreatic cancer. Cancer J 2014;20(3):203–6.

[160] Bhattacharjee S, Lukiw WJ. Alzheimer's disease and the microbiome. Front Cell Neurosci 2013;7:153.

[161] Aziz Q, Doré J, Emmanuel A, Guarner F, Quigley EM. Gut microbiota and gastrointestinal health: current concepts and future directions. Neurogastroenterol Motil 2013;25(1):4–15.

[162] Camfield DA, Owen L, Scholey AB, Pipingas A, Stough C. Dairy constituents and neurocognitive health in ageing. Br J Nutr 2011;106(2):159–74.

[163] Forsythe P, Kunze WA, Bienenstock J. On communication between gut microbes and the brain. Curr Opin Gastroenterol 2012;28(6):557–62.

[164] Diaz Heijtz R, Wang S, Anuar F, et al. Normal gut microbiota modulates brain development and behavior. Proc Natl Acad Sci USA 2011;108(7):3047–52.

[165] Foster JA, McVey Neufeld KA. Gut-brain axis: how the microbiome influences anxiety and depression. Trends Neurosci 2013;36(5):305–12.

[166] Schwartz K, Boles BR. Microbial amyloids—functions and interactions within the host. Curr Opin Microbiol 2013;16(1):93–9.

[167] Hornig M. The role of microbes and autoimmunity in the pathogenesis of neuropsychiatric illness. Curr Opin Rheumatol 2013;25(4):488–795.

[168] McVey Neufeld KA, Mao YK, Bienenstock J, Foster JA, Kunze WA. The microbiome is essential for normal gut intrinsic primary afferent neuron excitability in the mouse. Neurogastroenterol Motil 2013;25(2):183-e88.

[169] Mitew S, Kirkcaldie MT, Dickson TC, Vickers JC. Altered synapses and gliotransmission in Alzheimer's disease and AD model mice. Neurobiol Aging 2013;34(10):2341–51.

[170] Paula-Lima AC, Brito-Moreira J, Ferreira ST. Deregulation of excitatory neurotransmission underlying synapse failure in Alzheimer's disease. J Neurochem 2013;126(2):191–202.

[171] Saulnier DM, Ringel Y, Heyman MB, et al. The intestinal microbiome, probiotics and prebiotics in neurogastroenterology. Gut Microbes 2013;4(1):17–27.

[172] Lakhan SE, Caro M, Hadzimichalis N. NMDA receptor activity in neuropsychiatric disorders. Front Psychiatry 2013;4:52.

[173] Brenner SR. Blue-green algae or cyanobacteria in the intestinal micro-flora may produce neurotoxins such as Beta-*N*-Methylamino-L-Alanine (BMAA) which may be related to development of amyotrophic lateral sclerosis, Alzheimer's disease and Parkinson-Dementia-Complex in humans and Equine Motor Neuron Disease in horses. Med Hypotheses 2013;80(1):103.

[174] He F, Balling R. The role of regulatory T cells in neurodegenerative diseases. Wiley Interdisc Rev Syst Biol Med 2013;5(2):153–80.

[175] Mulligan VK, Chakrabartty A. Protein misfolding in the late-onset neurodegenerative diseases: common themes and the unique case of amyotrophic lateral sclerosis. Proteins 2013;81(8):1285–303.

[176] Gomez A, Luckey D, Taneja V. The gut microbiome in autoimmunity: sex matters. Clin Immunol 2015;159(2):154–62.

[177] Sanchez B, Hevia A, González S, Margolles A. Interaction of intestinal microorganisms with the human host in the framework of autoimmune diseases. Front Immunol 2015;6:594.

[178] Giongo A, Gano KA, Crabb DB, et al. Toward defining the autoimmune microbiome for type 1 diabetes. ISME J 2011;5(1):82–91.

[179] Murri M, Leiva I, Gomez-Zumaquero JM, et al. Gut microbiota in children with type 1 diabetes differs from that in healthy children: a case-control study. BMC Med 2013;11:46.

[180] Bhargava P, Mowry EM. Gut microbiome and multiple sclerosis. Curr Neurol Neurosci Rep 2014;14(10):492.

[181] Kempsell KE, Cox CJ, Hurle M, et al. Reverse transcriptase-PCR analysis of bacterial rRNA for detection and characterization of bacterial species in arthritis synovial tissue. Infect Immun 2000;68(10):6012–26.

[182] Moen K, Brun JG, Valen M, et al. Synovial inflammation in active rheumatoid arthritis and psoriatic arthritis facilitates trapping of a variety of oral bacterial DNAs. Clin Exp Rheumatol 2006;24(6):656–63.

[183] Scher JU, Abramson SB. Periodontal disease, *Porphyromonas gingivalis*, and rheumatoid arthritis: what triggers autoimmunity and clinical disease? Arthritis Res Ther 2013;15(5):122.

[184] Gomez A, Luckey D, Yeoman CJ, et al. Loss of sex and age driven differences in the gut microbiome characterize arthritis-susceptible 0401 mice but not arthritis-resistant 0402 mice. PLoS One 2012;7(4): pe36095.

[185] Wu H-J, Wu E. The role of gut microbiota in immune homeostasis and autoimmunity. Gut Microbes 2012;3(1):4–14.

[186] Gilkeson GS, Grudier JP, Karounos DG, Pisetsky DS. Induction of anti-double stranded DNA antibodies in normal mice by immunization with bacterial DNA. J Immunol 1989;142(5):1482–6.

[187] Granholm NA, Cavallo T. Long-lasting effects of bacterial lipopolysaccharide promote progression of lupus nephritis in NZB/W mice. Lupus 1994;3(6):507–14.

[188] Barin JG, Tobias LD, Peterson DA. The microbiome and autoimmune disease: report from a Noel R. Rose Colloquium. Clin Immunol 2015;159(2):183–8.

[189] Ringel Y, Ringel-Kulka T. The intestinal microbiota and irritable bowel syndrome. J Clin Gastroenterol 2015;49(Suppl. 1):S56–9.

[190] Bardhan PK, Gyr K, Beglinger C, Vögtlin J, Frey R, Vischer W. Diagnosis of bacterial overgrowth after culturing proximal small-bowel aspirate obtained during routine upper gastrointestinal endoscopy. Scand J Gastroenterol 1992;27(3):253–6.

[191] Quigley EM, Quera R. Small intestinal bacterial overgrowth: roles of antibiotics, prebiotics, and probiotics. Gastroenterology 2006;130(2 Suppl. 1):S78–90.

[192] Posserud I, Stotzer PO, Björnsson ES, Abrahamsson H, Simrén M. Small intestinal bacterial overgrowth in patients with irritable bowel syndrome. Gut 2007;56(6):802–8.

[193] Spiller R, Garsed K. Postinfectious irritable bowel syndrome. Gastroenterology 2009;136(6):1979–88.

[194] Thabane M, Kottachchi DT, Marshall JK. Systematic review and meta-analysis: the incidence and prognosis of post-infectious irritable bowel syndrome. Aliment Pharmacol Ther 2007;26(4):535–44.

[195] Halvorson HA, Schlett CD, Riddle MS. Postinfectious irritable bowel syndrome—a meta-analysis. Am J Gastroenterol 2006;101(8):1894–9. quiz 1942.

[196] Spiller R, Campbell E. Post-infectious irritable bowel syndrome. Curr Opin Gastroenterol 2006;22(1):13–7.

[197] Rajilic´-Stojanovic´ M, Jonkers DM, Salonen A, et al. Intestinal microbiota and diet in IBS: causes, consequences, or epiphenomena? Am J Gastroenterol 2015;110(2):278–87.

[198] Carroll IM, et al. Molecular analysis of the luminal- and mucosal-associated intestinal microbiota in diarrhea-predominant irritable bowel syndrome. Am J Physiol Gastrointest Liver Physiol 2011;301(5):G799–807.

[199] Carroll IM, Ringel-Kulka T, Siddle JP, Ringel Y. Alterations in composition and diversity of the intestinal microbiota in patients with diarrhea-predominant irritable bowel syndrome. Neurogastroenterol Motil 2012;24(6):521–30. e248.

[200] Jeffery IB, O'Toole PW, Öhman L, et al. An irritable bowel syndrome subtype defined by species-specific alterations in faecal microbiota. Gut 2012;61(7):997–1006.

[201] Pimentel M, Chatterjee S, Chow EJ, Park S, Kong Y. Neomycin improves constipation-predominant irritable bowel syndrome in a fashion that is dependent on the presence of methane gas: sub-analysis of a double-blind randomized controlled study. Dig Dis Sci 2006;51(8):1297–301.

[202] Brandt LJ, Chey WD, Foxx-Orenstein AE, et al. An evidence-based position statement on the management of irritable bowel syndrome. Am J Gastroenterol 2009;104(Suppl. 1):S1–S35.

[203] Pimentel M, Morales W, Chua K, et al. Effects of rifaximin treatment and retreatment in nonconstipated IBS subjects. Dig Dis Sci 2011;56(7):2067–72.

[204] Menees SB, Maneerattanaporn M, Kim HM, Chey WD. The efficacy and safety of rifaximin for the irritable bowel syndrome: a systematic review and meta-analysis. Am J Gastroenterol 2012;107(1):28–35. quiz 36.

[205] Bennet SM, Ohman L, Simren M. Gut microbiota as potential orchestrators of irritable bowel syndrome. Gut Liver 2015;9(3): 318–31.

[206] Gomes AMP, Malcata FX. *Bifidobacterium* spp. and *Lactobacillus acidophilus*: biological, biochemical, technological and therapeutical properties relevant for use as probiotics. Trends Food Sci Technol 1999;10(4–5):139–57.

[207] Kruis W, Chrubasik S, Boehm S, Stange C, Schulze J. A double-blind placebo-controlled trial to study therapeutic effects of probiotic *Escherichia coli* Nissle 1917 in subgroups of patients with irritable bowel syndrome. Int J Colorectal Dis 2012;27(4):467–74.

[208] Sisson G, Ayis S, Sherwood RA, Bjarnason I. Randomised clinical trial: a liquid multi-strain probiotic vs. placebo in the irritable bowel syndrome--a 12 week double-blind study. Aliment Pharmacol Ther 2014;40(1):51–62.

[209] Sinn DH, Song JH, Kim HJ, et al. Therapeutic effect of *Lactobacillus acidophilus*-SDC 2012, 2013 in patients with irritable bowel syndrome. Dig Dis Sci 2008;53(10):2714–8.

[210] Wilson ID. Drugs, bugs, and personalized medicine: pharmacometabonomics enters the ring. Proc Natl Acad Sci USA 2009;106(34):14187–8.

[211] Sousa T, Paterson R, Moore V, Carlsson A, Abrahamsson B, Basit AW. The gastrointestinal microbiota as a site for the biotransformation of drugs. Int J Pharm 2008;363(1–2):1–25.

[212] Gotto AM Jr. Statin therapy and the elderly: SAGE advice? Circulation 2007;115(6):681–3.

[213] Aura AM, Mattila I, Hyötyläinen T, et al. Drug metabolome of the simvastatin formed by human intestinal microbiota in vitro. Mol Biosyst 2011;7(2):437–46.

[214] Kaddurah-Daouk R, Baillie RA, Zhu H, et al. Enteric microbiome metabolites correlate with response to simvastatin treatment. PLoS One 2011;6(10):pe25482.

[215] Nagata S, Asahara T, Ohta T, et al. Effect of the continuous intake of probiotic-fermented milk containing *Lactobacillus casei* strain Shirota on fever in a mass outbreak of norovirus gastroenteritis and the faecal microflora in a health service facility for the aged. Br J Nutr 2011;106(4):549–56.

[216] Toward R, Montandon S, Walton G, Gibson GR. Effect of prebiotics on the human gut microbiota of elderly persons. Gut Microbes 2012;3(1):57–60.

[217] Björklund M, Ouwehand AC, Forssten SD, et al. Gut microbiota of healthy elderly NSAID users is selectively modified with the administration of *Lactobacillus acidophilus* NCFM and lactitol. Age (Dordr) 2012;34(4):987–99.

[218] Walton GE, van den Heuvel EG, Kosters H, Rastall RA, Tuohy KM, Gibson GR. A randomised crossover study investigating the effects of galacto-oligosaccharides on the faecal microbiota in men and women over 50 years of age. Br J Nutr 2012;107(10):1466–75.

[219] Lahtinen SJ, Forssten S, Aakko J, et al. Probiotic cheese containing *Lactobacillus rhamnosus* HN001 and *Lactobacillus acidophilus* NCFM(R) modifies subpopulations of fecal lactobacilli and *Clostridium difficile* in the elderly. Age (Dordr) 2012;34(1):133–43.

[220] Hickson M, D'Souza AL, Muthu N, et al. Use of probiotic *Lactobacillus* preparation to prevent diarrhoea associated with antibiotics: randomised double blind placebo controlled trial. BMJ 2007;335(7610):80.

[221] Kee VR. *Clostridium difficile* infection in older adults: a review and update on its management. Am J Geriatr Pharmacother 2012;10(1):14–24.

[222] An HM, Baek EH, Jang S, et al. Efficacy of lactic acid bacteria (LAB) supplement in management of constipation among nursing home residents. Nutr J 2010;9:5.

[223] Carlsson M, Gustafson Y, Haglin L, Eriksson S. The feasibility of serving liquid yogurt supplemented with probiotic bacteria, *Lactobacillus rhamnosus* LB 21, and *Lactococcus lactis* L1A—a pilot study among old people with dementia in a residential care facility. J Nutr Health Aging 2009;13(9):813–9.

[224] Pitkala KH, Strandberg TE, Finne Soveri UH, Ouwehand AC, Poussa T, Salminen S. Fermented cereal with specific bifidobacteria normalizes bowel movements in elderly nursing home residents. A randomized, controlled trial. J Nutr Health Aging 2007;11(4):305–11.

[225] Candore G, Balistreri CR, Colonna-Romano G, et al. Immunosenescence and anti-immunosenescence therapies: the case of probiotics. Rejuvenation Res 2008;11(2):425–32.

[226] Fukushima Y, Miyaguchi S, Yamano T, et al. Improvement of nutritional status and incidence of infection in hospitalised, enterally fed elderly by feeding of fermented milk containing probiotic *Lactobacillus johnsonii* La1 (NCC533). Br J Nutr 2007;98(5):969–77.

[227] Ibrahim F, Ruvio S, Granlund L, Salminen S, Viitanen M, Ouwehand AC. Probiotics and immunosenescence: cheese as a carrier. FEMS Immunol Med Microbiol 2010;59(1):53–9.

[228] Matsumoto M, Benno Y. Anti-inflammatory metabolite production in the gut from the consumption of probiotic yogurt containing *Bifidobacterium animalis* subsp. lactis LKM512. Biosci Biotechnol Biochem 2006;70(6):1287–92.

[229] Moro-Garcia MA, Alonso-Arias R, Baltadjieva M, et al. Oral supplementation with *Lactobacillus delbrueckii* subsp. bulgaricus 8481 enhances systemic immunity in elderly subjects. Age (Dordr) 2013;35(4):1311–26.

[230] Guillemard E, Tondu F, Lacoin F, Schrezenmeir J. Consumption of a fermented dairy product containing the probiotic *Lactobacillus casei* DN-114001 reduces the duration of respiratory infections in the elderly in a randomised controlled trial. Br J Nutr 2010;103(1):58–68.

[231] Cusack S, Claesson MJ, O'Toole PW. How beneficial is the use of probiotic supplements for the aging gut? Aging Health 2011;(7):179–86.

[232] Jackson M, Jeffery IB, Beaumont M, et al. Signatures of early frailty in the gut microbiota. Genome Med 2016;8:8.

Part B

Common Organisms and Probiotics

Chapter 5

Escherichia coli Nissle 1917

M. Schultz* and J.P. Burton**,†

*Department of Medicine, Dunedin School of Medicine, University of Otago, Dunedin, New Zealand; **Department of Surgery, Department of Microbiology & Immunology, Division of Urology, Western University, London, ON, Canada; †Lawson Health Research Institute and Canadian Research and Development Centre for Probiotics, London, ON, Canada

INTRODUCTION—HISTORY

Escherichia coli strain Nissle 1917 (EcN, DSM 6601 in the German Collection for Microorganisms) is probably one of the most exciting probiotic strains, both from a historical as well as a scientific and clinical point of view. It is certainly one of the most extensively studied probiotic bacteria worldwide. EcN is the active component of Mutaflor (Ardeypharm GmbH, Herdecke, Germany), which since 1917 has been used to successfully treat a variety of gastrointestinal disorders. It is named after Professor Alfred Nissle (1874–1965), a German physician and hygienist who observed that various *E. coli* strains differed in their ability to inhibit the growth of typhus pathogens when cultured together in a Petri dish. He isolated an antagonistically very strong *E. coli* strain from the feces of a soldier who, on the battlefields of south-eastern Europe during World War I, was not affected by the then-rampant *Shigella*-induced diarrhea. In early studies at the Institute for Hygiene of the University of Freiburg, Germany, he successfully treated patients suffering from diarrhea [1,2] with this *E. coli* strain. Based on this success, he began filling gelatine capsules with the bacteria, grown on agar plates, and in 1917 applied for a patent for *Mutaflor*. Since then, Mutaflor, unchanged in its composition and containing $2.5–25 \times 10^9$ colony forming units of lyophilized viable *E. coli* Nissle strain 1917 bacteria, has been manufactured and marketed as a treatment for a number of gastrointestinal disorders.

Despite decades of research however, the fundamental mechanisms by which EcN exerts its probiotic benefits are still incompletely understood. The recent publication of the complete genome sequence of EcN will certainly be helpful for further identification of the mechanisms responsible for the strain's probiotic properties [3].

FITNESS FACTORS AND PROBIOTIC MECHANISMS

The FAO/WHO defines probiotics as "live bacteria which when administered in adequate amounts confer benefits to the host" (http://www.who.int/foodsafety/publications/biotech/biotech_en.pdf?ua=1, accessed January 2, 2016). The human gastrointestinal tract harbors approximately 10^{14} microorganisms of up to 1200 distinct species [4] and therefore intestinal colonization is difficult to achieve for any probiotic, even temporarily. We were able to demonstrate that culturable EcN disappeared almost completely from the feces of rats 14 days after oral administration of EcN was ceased [5]. Similarly, in humans it has been shown that 2 weeks after oral administration was stopped, EcN can only be detected in the feces of approximately 50% of volunteers and this detection rate continues to drop rapidly over time [6]. Therefore, in order to colonize the gastrointestinal tract and to exert beneficial effects, probiotics need to exhibit certain so-called "fitness factors," allowing them to outcompete other microorganisms. These fitness factors include microcins, adhesins, and proteases. It should be noted however, that the distinction between fitness factors and probiotic effects is often a matter of interpretation (Table 5.1).

EcN is a gram-negative bacterium classified as serotype O6:K5:H1 and thus related to other bacteria in the common O6 serogroup which include nonpathogenic commensals and also pathogenic strains [38,39]. The H serogroup is determined by the flagellin antigen, a major subunit of the bacterial flagella which serves as a potent activator of the host innate immune system and also leads to β-defensin 2 induction thought to be associated with strengthening of the epithelial barrier to limit adhesion and invasion of other microorganisms.

TABLE 5.1 Fitness Factors and Probiotic Mechanisms of *E. coli* Nissle 1917

Summary: Properties of *E. coli* Nissle 1917	Leading References
Temporary colonization of the murine and human gastrointestinal tract	[5,6]
Several iron uptake mechanisms as a fitness factor to facilitate intestinal survival and to limit growth of pathogens	[7,8]
Microcin H47 and M production to compete against other enterobacteria that utilize catecholate siderophores	[9–11]
Improved mucosal adhesion due to flagellum	[12]
EcN is serum sensitive	[13]
Limitation of invasion or adhesion of *E. coli* strains (mainly EHEC), *Salmonella*, *Yersinia enterocolitica*, *Shigella flexneri*, *Legionella pneumophila* and *Listeria monocytogenes*, *Candida albicans*, *Clostridium perfringens*	[14–22]
Immunomodulatory effects mediated through H1 flagellin and K5 capsule via Toll-like receptors 4 and 5	[23–32]
Strengthening of the intestinal epithelial barrier by ZO-2 and PKCζ redistribution following injury and through biofilm formation	[33–35]
Induction of human β-defensin 2	[36,37]

Due to its fitness factors, EcN is well equipped to survive in the intestine. For example, EcN utilizes an array of outer membrane iron receptors similar to uropathogenic *E. coli* (e.g., enterobactin, salmochelin, aerobactin, yersiniabactin, EfeU, ChuA), to facilitate siderophore and heme import from within the iron-limited gastrointestinal tract [7]. This provides EcN with a potentially significant survival advantage as iron is an important nutrient for microbes and is essential for growth through generation of energy through ATP [40]. These superior iron uptake systems not only provide an advantage regarding colonization but also form the basis of a major probiotic effect in the treatment of *Salmonella*-induced colitis. Deriu et al. provided a mechanistic explanation for this and demonstrated that only the administration of EcN with intact iron uptake systems resulted in reduced *Salmonella typhimurium* colonization whereas the administration of mutated EcN with defective iron metabolism systems did not [8].

Further assisting colonization is the production of microcins—antibacterial peptides, produced by enterobacteria. Microcins are secreted under certain conditions, especially of nutrient depletion, and lead to potent antibacterial activity mainly against closely related species [41]. It is believed that Alfred Nissle first observed these microcins

in action in vitro when he noticed that EcN inhibited the growth of *Salmonella* species and certain other *E. coli* strains when cultured together on agar plates [42,43]. Initially called colicins, EcN produces the microcins H47 and M (for Mutaflor) [9,10]. Recent work shows that again competition for iron plays a role and these microcins, also called siderophore-microcins [44], produced under iron-restricted conditions, could aid the producing strain in competing against enterobacteria that also utilize catecholate siderophores to capture iron [11].

EcN is a safe probiotic strain despite its close relationship to uropathogenic *E. coli*. The lack of defined virulence factors (i.e., alpha-hemolysin, P-fimbrial adhesins, and the semirough lipopolysaccharide phenotype) [45] combined with the expression of fitness factors, such as microcins, different iron uptake systems, adhesins (type 1-, Curli-, F1C-fimbria) [46], and proteases, which may support its survival and successful colonization through improved adhesion to the mucosa of the human gut, most likely contributes to the probiotic character of EcN [47]. Furthermore, while the EcN flagella is the major antigen and responsible for immune system activation, the flagellum itself serves as an important adhesin aiding in the competition for binding sites [12]. In contrast to *Lactobacilli*, EcN is serum-sensitive as a result of shortened O6 side chains which is an important biosafety feature [13].

Not all antibactericidal and host-protective effects can be related to microcin production and a variety of mechanisms seem to be utilized. For example, EcN shows strong antagonistic effects leading to a reduction of invasion, adhesion, or toxicity toward a variety of enteropathogenic microorganisms including different *E. coli* strains (mainly EHEC), *Salmonella*, *Y. enterocolitica*, *S. flexneri*, *L. pneumophila* and *L. monocytogenes*, *C. albicans*, *C. perfringens*, and others [14–21]. Interestingly, at least in the case of *Salmonella* infection, the observed effect was not due to the production of microcins because a microcin-negative mutant was as effective and direct contact with the pathogen was not necessary, implying the presence of other secreted factors [22]. However, while in vitro, EcN showed strong antagonistic effects against EHEC serotype O104:H4 and O157:H7 which caused a significant outbreak in Germany in 2011 [15]; we were unable to demonstrate in vivo efficacy of EcN to contain an outbreak of multidrug resistant *E. coli* in a human trial [48].

These additional antimicrobial and immunomodulatory effects are mediated primarily through the H1 flagellin antigen via Toll-like receptors 4 and 5 (TLR4 and 5) [23,24] and the K5 capsule [25,26] which seem to potentiate each others effect in the activation and modulation of the innate immune system [27]. The K5 capsule-induced TLR5 signaling induces MyD88 and TRIF as well as interleukin-8 (IL-8) secretion via the mitogen-activated protein kinase pathway which in turn acts as a chemoattractant to neutrophils [23].

Furthermore, as shown by Hafez et al. the loss of the K5 capsule significantly reduced the induction of a number of chemokines [monocyte chemoattractant-1, RANTES, macrophage inflammatory protein 2α (MIP-2α), MIP-2β, IL-8, and gamma interferon-inducible protein 10] [23,28]. Further immunomodulatory effects include increased IL-10 and 12 levels and IgA secretion [29], primarily in in vitro experiments using various cell lines [30]. It has further been suggested that EcN can also interact via TLR-2 with certain T-cell subsets by activating γδT cells, important in innate and adaptive immune system response to microbial antigens [31]. EcN stimulation leads to reduced TNF-α but increased IL-6 and CXCL8 secretion. In turn however, following activation, EcN induces apoptosis of these T cells thereby regulating the immune response [32].

The intestinal epithelial barrier, separating the host from the environment, is made up of several components. These include not only the physical cell barrier itself, the mucus layer, but also secreted antimicrobial substances. While the previously mentioned mechanisms indirectly affect barrier function, thereby limiting microbial adhesion and invasion, EcN has been demonstrated to directly strengthen the physical epithelial barrier by zonula occludens-2 [49] and PKCζ redistribution following injury [33,34] and through biofilm formation [35]. EcN flagellin induces β-defensin-2 (HBD-2) secretion via NF-κB and AP-1 [36,37]. In addition, it seems as if EcN is capable of further direct modulation of intestinal epithelial cell differentiation involving different Notch-dependent and Notch-independent pathways [50]. Most interestingly, HBD-2 has a broad antimicrobial spectrum against a variety of gram-positive and gram-negative bacteria, fungi, and viruses and interestingly, inflammatory bowel diseases (IBD) are characterized by low defensin levels [51].

The preceding discussion highlights the fact that EcN is indeed a remarkable probiotic, equipped with a number of fitness factors to allow at least temporary colonization of the gastrointestinal tract, in the complete absence of any virulence factor. This is a prerequisite for any probiotic strain to exert beneficial effects on the host. EcN exhibits a variety of probiotic properties, some of which are also fitness factors leading to immunomodulatory and barrier strengthening effects, and forms the basis of successful clinical application.

CLINICAL INDICATIONS AND APPLICATIONS (TABLE 5.2)

Gastrointestinal Disorders

The therapeutic properties of *E. coli* strain Nissle 1917 have been demonstrated in a number of animal models for chronic and acute colitis as well as intestinal infection but review of this is not part of this chapter. Undoubtedly the therapeutic strength of this probiotic lies in its intestinal immunomodulatory and competitive colonization properties. However, more recently, EcN applications have been extended to use EcN as a delivery vehicle due to its innocuous nature as a probiotic bacterium.

A detailed review of clinical applications of EcN was published in 2008 by the author and others later [66–68] however, several recent contributions to the field mean that it is timely to again review the topic. This chapter is restricted predominantly to the use of EcN in humans.

Inflammatory Bowel Diseases

Mounting evidence links the IBD, ulcerative colitis (UC), and Crohn's disease (CD) to the presence of certain bacteria [69]. Whether these organisms are the cause or the consequence of the resulting inflammation is still the subject of speculation. For instance, various *E. coli* strains have been found in higher percentages in patients with IBD compared to healthy controls [70–72]. Genetic defects found in CD patients make these patients more susceptible to infections with, for example, adherent-invasive *E. coli* [73]. Most interestingly, EcN shares many genetic traits with other, mainly pathogenic, *E. coli* belonging to the phylogenetic group B2 [74–76]. Using cell culture models, EcN affects the adhesion and invasion of various other *E. coli* strains associated with CD [15,18,21,77]. It is therefore conceivable that in the treatment of IBD, EcN outcompetes these pathogenic bacteria and leads to clinical improvement.

Ulcerative Colitis

Interest in the therapeutic use of EcN increased following the demonstration of equivalent efficacy of EcN and mesalazine in the maintenance of remission of UC. This was demonstrated in three trials comprising a total of 116 patients with active UC [53] and 447 patients with inactive colitis [52,54]. The first trial by Rembacken et al. was complex and comprised a heterogeneous group of patients with mild to severe colitis and proctitis to pancolitis [53]. This makes the assessment of the efficacy of the investigational product difficult. Based on a modified Rachmilewitz index [78], patients with active colitis (clinical activity index; CAI ≥ 4) were randomized in a double-blind, double-dummy fashion to receive either mesalazine 800 mg TDS (2.4 g/day) or EcN (Mutaflor) two capsules BD. Treatment was started following 1 week of gentamycin. In addition, patients with mild proctitis were treated with hydrocortisone acetate enemas once daily. Patients with mild-to-moderate at least left-sided disease received prednisolone 30 mg/day orally. Patients not in remission after a maximum of 12 weeks, or showing clinical deterioration, were excluded. Once the patient achieved remission, Mutaflor was reduced to two capsules OD and mesalazine to 1.2 g/day. Enemas and prednisolone were tapered. After 1 year, 75% of patients

TABLE 5.2 Clinical Trials

Diagnosis	Study Design	Number of Patients Per Treatment Group	Duration	Aim	Outcome	Adverse Events	References
IBD							
UC							
Inactive UC (CAI ≤ 4)	Double-blind, placebo-controlled, randomized	EcN ($n = 50$) Mesalazine ($n = 53$)	12 weeks	Equivalence of EcN and mesalazine	No significant difference between the two study arms	EcN 8.6% Mesalazine 13.3%	[52]
Active and inactive UC	Double-blind, placebo-controlled, randomized	EcN ($n = 59$) Mesalazine ($n = 57$)	1 year	Time to relapse, rate of relapse, steroid sparing effect	EcN is as effective in maintaining remission	EcN 18.6% Mesalazine 14%	[53]
Inactive UC (CAI ≤ 4)	Double-blind, placebo-controlled, randomized	EcN ($n = 162$ Mesalazine ($n = 165$)	1 year	Equivalence of EcN and mesalazine	EcN is as effective in maintaining remission (significant equivalence)	EcN 7 serious adverse events, 42% adverse events Mesalazine 6 serious adverse events, 35.2% adverse events	[54]
Inactive UC	Open-label	Children (11–18 years) EcN ($n = 24$) Mesalazine ($n = 10$)	1 year	Relapse rate	Relapse rate: EcN 25% Mesalazine 30%	No serious adverse events	[55]
Active UC (CAI ≥ 6)	Double-blind, placebo-controlled, randomized	Ciprofloxacin followed by EcN ($n = 25$) Ciprofloxacin followed by placebo ($n = 25$) Placebo followed by EcN ($n = 25$) Placebo followed by placebo ($n = 25$)	12 weeks	Remission rate: Effect of add-on treatment to conventional medical therapy	Remission rate: Ciprofloxacin/EcN 66% Ciprofloxacin/ placebo 78% Placebo/EcN 54% Placebo/placebo 89%	26 pts were withdrawn (11/26 from EcN containing regime)	[56]
Acute and active proctitis	Double-blind, placebo-controlled, randomized	Rectal enema EcN 40 mL ($n = 24$) EcN 20 mL ($n = 23$) EcN 10 mL ($n = 23$) Placebo ($n = 20$)	8 weeks	Rate of remission	Remission rates: EcN 40 mL 52.9% EcN 20 mL 44.4% EcN 10 mL 27.3% Placebo 18.2% (not significant in the intention to treat analysis but in the Jonckheere–Terpstra rank-correlation)	EcN 40 mL 43.5% EcN 20 mL 65.2% EcN 10 mL 54.5% Placebo 50%	[57]
Active pouchitis (PDAI 13–15)	Case report	$N = 2$	315 days respectively 56 days	Induction and maintenance of remission	Significant reduction in PDAI score to 6 resp. 3	Not mentioned	[58]

Active and inactive CD (CDAI > 150)	Double-blind, placebo-controlled, randomized	Prednisolone + EcN (n = 16) Prednisolone + placebo (n = 12)	1 year	Steroid sparing effect Maintenance of remission	Nonsignificant trend in favor of EcN (relapse rate EcN 33.3%, placebo 63.6%)	No data given	[59]
IBS							
Constipation	crossover, randomized, double-blind	EcN (n = 35) Placebo (n = 35)	9 weeks (crossover after 4 weeks)	Increase in stool frequency	Significant increase in stool frequency		[60]
IBS (Rome II)	Double-blind, placebo-controlled, randomized	EcN (n = 60) Placebo (n = 60)	12 weeks	Assessment of efficacy and safety of EcN in subgroups	No significant difference between the groups	EcN 50% Placebo 45% (and 1 SAE—abscess)	[61]
IBS (Rome II)	Double-blind, placebo-controlled, randomized	139 in total	12 weeks	Assessment of efficacy and safety of EcN in subgroups	No significant difference between the groups	No data given	[62]
Diarrhea							
Infectious, acute diarrhea (<3 days)	Double-blind, placebo-controlled, randomized	EcN (n = 55) Placebo (n = 58)	10 days	Duration of diarrhea	Reduction in diarrhea: EcN 2.5 days Placebo 4.8 days, p = 0.0007	EcN 3.6% Placebo 3.4%	[63]
Chronic diarrhea (>4 days)	Double-blind, placebo-controlled, randomized	EcN (n = 75) Placebo (n = 76)	21 days	Duration of diarrhea	Response rate at day 21: EcN 98.7% Placebo 71.1%, p < 0.001	EcN 4% Placebo 5.3%	[64]
Collagenous colitis	Open-label	EcN (n = 14)	Max. 18 weeks	Reduction in stool frequency	Significant improvement		[65]

CD, Crohn's disease; IBD, inflammatory bowel diseases; IBS, irritable bowel syndrome; UC, ulcerative colitis.

receiving mesalazine and 68% of patients receiving EcN achieved remission. There was no statistical difference between the two groups, nor was there a difference in the time to remission which was 44, versus 42 days.

In a further trial by Kruis et al., 118/120 patients with inactive colitis (CAI ≤ 4) [78] were randomized in a double-blind, double-dummy fashion to receive either mesalazine 500 mg TDS or Mutaflor, 2 capsules/day [52]. After 12 weeks, there was no difference concerning the CAI or the histological analysis of sigmoidoscopically obtained biopsies between the two groups. The relapse rate in the EcN group was 16% and in the mesalazine group was 11.3%.

In a final study with an increased patient number and longer follow-up period (1 year), Kruis et al. recruited 327 patients with inactive colitis (CAI ≤ 4; [78], endoscopic index ≤4) to receive in a double-blind, double-dummy fashion either mesalazine, 500 mg TDS or Mutaflor, 2 capsules/day. A relapse occurred in 36.4% in the EcN group and in 33.9% in the mesalazine group resulting in statistical significant equivalence. This was also observed concerning endoscopic activity and histological assessment.

These results in adult populations were mirrored in an open-label pilot trial of 34 children and adolescents with UC in remission [55]. Patients received either two capsules of Mutaflor or a median of 1.5 g mesalazine daily. After 1 year, the relapse rate was 25% in the EcN group and 30% in the mesalazine group, not showing a statistical difference.

Topical treatment of proctitis or left-sided colitis is superior to systemic treatment [79]. Matthes et al. explored the effect of rectal EcN administration in acute, moderate proctitis [57]. In this placebo-controlled, randomized trial, 88/90 patients received either a placebo enema or an enema containing 10, 20, or 40 mL of 10^8 EcN/mL daily for 2 weeks. Follow-up period was 8 weeks. Remission rates and time to remission were clearly dose dependant with highest remission rates (40 mL 52.9% vs. placebo 18.2%) achieved with the highest concentration of EcN.

All trials showed very good tolerability of the investigational product with a low number of adverse effects.

An interesting study evaluated the effect of EcN and/or ciprofloxacin in addition to conventional medication in the induction of remission in active UC [56]. It was postulated that combining these therapies would improve the clinical outcome. This study followed on from a pilot trial in healthy volunteers that evaluated the survival of EcN if given in combination with mesalazine [6]. There was no significant difference in the fecal recovery rate of EcN, thereby opening the way for evaluation of a potential improvement in efficacy for a combination of EcN and mesalazine. The study by Petersen et al. was a double-blind, placebo-controlled trial. Hundred patients with active UC (CAI ≥ 6) were randomized to receive either placebo or

ciprofloxacin for 1 week followed by EcN or placebo for 7 weeks in addition to their standard medication. Patients on steroids and/or anti-TNF-α drugs were excluded but those on 5-ASA medication were included. This is important as both EcN and mesalazine are equally effective, and when combined, might work in synergy. Interestingly, more patients on placebo-only achieved remission (89%) compared to all other groups. The combination of ciprofloxacin and EcN achieved a remission rate of 66% while EcN on its own achieved only 54%. Ciprofloxacin treatment on its own led to 78% of patients achieving remission. The high withdrawal rate in the groups receiving either placebo or EcN was notable. This was attributed to the lack of effect rather than to adverse events. Unfortunately, the authors do not comment on the group of patients receiving mesalazine and EcN in combination.

The trials discussed earlier were included in a recently published metaanalysis that evaluated the role of EcN in the treatment of UC [68]. Combined, these trials included 719 patients and the metaanalysis concluded that EcN is equivalent to mesalazine treatment and can be recommended in the maintenance of remission as per current guidelines on the treatment of UC [80]. However, EcN cannot be recommended for induction of remission as the only information available is the result of a single trial comparing EcN to placebo. In contrast, VSL#3, a probiotic mixture of eight strains but not containing EcN, can be recommended in induction and maintenance of remission of UC at least in children [81].

Pouchitis

An ileal pouch and anal anastomosis following proctocolectomy is the ultimate curative attempt in UC when medical therapy fails. However, pouchitis (chronic inflammation of the pouch) is a common complication and is present in varying degrees of severity in up to 50% of patients [82]. Dysbiosis seems to play a role in chronic pouchitis making this disease potentially amenable to probiotic treatment [83]. A case report, published in 2001 and comprising of two patients, showed a reduction in the pouch disease activity index (PDAI) score to levels indicating remission [58].

Crohn's Disease

Despite the in vitro evidence, only one small trial examined the effect of EcN in active CD [59]. In this double-blind, randomized, placebo-controlled pilot study, 28 patients were enrolled. All patients were started on a tapering regime of prednisolone and randomized to receive either Mutaflor with an increase in dose to 100 mg or placebo. After 1 year, the overall rate of remission was comparable in both groups. Within the EcN group, 33.3% of patients had a relapse compared with 63.6% in the placebo group. Of patients who had stopped taking steroids before the relapse, not only 30% of the EcN but also 70% of the placebo patients experienced a relapse. None of the results were statistically significant.

Irritable Bowel Syndrome

Recently, the effect of EcN was studied in patients with irritable bowel syndrome (IBS). The etiology of IBS is incompletely understood and it is now believed that IBS does not represent a single disease but rather the result of different pathologies. IBS is possibly the result of heightened visceral hypersensitivity, intestinal dysmobility, dysbiosis, and an altered gut immune function. Clinically, several subgroups can be distinguished. In the majority of patients, IBS presents with either abdominal pain, diarrhea, or constipation—there is also a group of patients with mixed symptoms [84]. EcN would appear to have the capabilities to affect some of these pathomechanisms, including a modulation of the intestinal immune response and dysbiosis. In addition, using an in vitro system, EcN supernatants were able to modulate (increase) contractility of colonic circular smooth muscle by direct stimulation, independent of the nervous system [85]. This is contradicted by a more recent study in a murine model that demonstrated decreased motility in response to EcN extract [86]. The authors explain this discrepancy with methodological issues in the earlier study. In the context of IBS treatment, a further recent finding is of interest. Nzakizwanayo et al. demonstrated in an in vitro model that EcN increases the bioavailability of serotonin in the gut by interacting with several pathways involved in secretion and clearance [87]. This is important as low levels of serotonin have been implicated in constipation.

A small number of clinical trials have addressed the clinical efficacy of EcN in patients with IBS. In a crossover, randomized, double-blind study, Möllenbrink et al. treated 70 patients suffering from constipation with EcN or placebo. After 4 weeks treatment the number of weekly stools was significantly higher in the EcN treated group compared to the placebo group. This remained significant at week 8 [60].

In a more recent, prospective, double-blind, and placebo-controlled study, 120 patients with IBS according to Rome II criteria [88] were recruited to be treated for 12 weeks with EcN or placebo [61]. Interestingly, after 12 weeks, the study did not yield consistent, statistically significant differences between the groups. Subgroup analysis favored patients with antibiotic induced ($n = 5$) or postinfectious IBS ($n = 15$).

A randomized, placebo-controlled, double-blind study by Faghihi et al. yielded similarly disappointing results [62]. Here, 139 patients with IBS according to the Rome II criteria [88] were enrolled. While some items on the Birmingham IBS Symptom Questionnaire [89] indicated improvement, overall, no statistically significant differences were observed between the two groups.

Diarrhea

Henker and his group evaluated the effect of EcN on infectious diarrhea in infants and toddlers in two studies. For the first double-blind trial, EcN or placebo was administered to 113 infants with acute, nonbloody diarrhea (>3 watery stools/day) lasting for a maximum of 3 days [63]. Depending on the age of the children, EcN at a dose of $1–3 \times 10^8$ CFU was administered daily. Response was defined as a reduction in stool frequency to less than three watery stools in 24 h for 2 consecutive days. Children receiving EcN responded more rapidly than children receiving placebo (2.5 days vs. 4.8 days; $p \leq 0.0007$). EcN was found to be well tolerated and safe. In a second, confirmatory trial, 151 children with nonspecific diarrhea lasting for more than 4 days but less than 14, were given either EcN or placebo [64]. In a Kaplan–Meier analysis, the time to response was 2.4 days in the EcN group compared to 5.7 days in the placebo group and this was significant. Only three patients in the EcN group experienced adverse events and these were not thought to be related to the investigational product.

One small, open-label trial was performed in collagenous colitis [65]. One to six capsules of EcN ($2.5–25 \times 10^8$ CFU/capsule) was administered to 14 patients for at least 4 weeks (4–18 weeks). Stool frequency decreased significantly during the duration of the trial.

Two postmarketing evaluations confirmed the results of the clinical trials. Krammer et al. analyzed the efficacy and safety of EcN in 3807 patients [90]. EcN was given for more than 20 indications but predominantly for diarrhea, IBD, IBS, and constipation. According to prescribers, EcN was effective in 81.4% with very few side effects (2.8%). Similar results were seen in children. Rohrenbach prospectively analyzed treatment data for 668 children [91]. The main indication was for gastrointestinal disorders. Overall efficacy was good in 84.4%, with side effects noted in only 2.9% of patients.

Other Indications

Early findings from a murine trial on the effects of EcN to prevent or treat grass pollen allergy [92] formed the basis for a randomized, double-blind, placebo-controlled trial [93]. To prevent an allergic skin reaction, mice were treated with oral EcN. Not only was the local allergic reaction ameliorated, as seen by a reduced swelling and other signs of dermatitis, but a decrease in infiltrating CD4$^+$ T cells and mast cells, accompanied by a reduced inflammatory cytokine response, was also noted. These findings were mirrored by a reduced systemic allergen-specific response. Unfortunately, when extrapolated to the human system, results were not confirmed [93]. In a double-blind, placebo-controlled fashion, grass pollen-allergic volunteers received EcN or placebo before and during the grass pollen season. While a significant increase in specific IgA was observed, no difference regarding symptoms was seen between the groups.

Use of EcN as Delivery Vehicle

The innocuous nature of EcN and its flexibility for bioengineering makes this bacterium an ideal candidate as a delivery vehicle for vaccines, cytokines, and other substances. For example, genetically engineered EcN can be used to secrete the nematode immunomodulator cystatin in the intestine to treat murine or pig experimental colitis [94]. Others programmed EcN to secrete biologically active IL-10 [95]. In the future, both systems could prove beneficial in the treatment of IBD. EcN has also been engineered to produce omega-3 fatty acids for potential industrial use [96] or pyrroloquinoline quinone to treat alcohol-induced liver damage [97].

Another striking feature of EcN has recently gained scientific attention. It has been noted that certain bacteria preferentially colonize tumor tissue when administered intravenously [98]. Probiotic bacteria accumulating in tumors can be used to deliver targeted cytotoxicity either through bioengineering or by using their natural enzymatic activity to activate chemotherapeutic prodrugs [99]. Engineered bacteria can also be detected using light magnetic resonance imaging or positron emission tomography. Oral administration of EcN has been successfully used to deliver a signal for the presence of liver metastasis which is detectable in urine [100]. Bioengineering, coupled with the safe nature of EcN, offers endless possibilities in the ongoing search for drug discoveries.

CONCLUSIONS

E. coli strain Nissle 1917 is a remarkable probiotic bacterium. Its antagonistic effects toward intestinal pathogens were first noted in 1917 and it is still the active ingredient of Mutaflor, a remedy used for the treatment of various gastrointestinal disorders. In the absence of virulence factors, EcN possesses an interesting profile of fitness factors that are responsible for temporary intestinal colonization and probiotic effects. Besides direct antagonistic effects toward gastrointestinal pathogens, EcN has immunomodulatory effects and directly strengthens the epithelial barrier. Most convincingly demonstrated in large clinical trials, EcN is equivalent to mesalazine in the maintenance of remission of UC and also shortens the duration of chronic and acute diarrhea in toddlers and children. However, less effect was seen in the treatment of IBS.

The whole genome was published recently which will undoubtedly lead to the demonstration of further probiotic mechanisms. Already, biogenetically engineered EcN is being used as a vehicle to deliver beneficial substances to the organ of interest.

REFERENCES

[1] Nissle A. Ueber die Grundlagen einer neuen ursaechlichen Bekaempfung der pathologischen Darmflora (On the fundamentals for new causal control of pathological intestinal microflora). Dt Med Wschr 1916;42(39):1181–4.

[2] Nissle A. Weiteres ueber die Mutaflorbehandlung unter besonderer Beruecksichtigung der chronischen Ruhr. MMW 1919;25:678–81.

[3] Reister M, Hoffmeier K, Krezdorn N, et al. Complete genome sequence of the gram-negative probiotic *Escherichia coli* strain Nissle 1917. J Biotechnol 2014;187:106–7.

[4] Rajilic-Stojanovic M, Smidt H, Vos de W. Diversity of the human gastrointestinal tract microbiota revisited. Environ Microbiol 2007;9(9):2125–36.

[5] Schultz M, Watzl S, Oelschlaeger TA, et al. Green fluorescent protein for detection of the probiotic microorganism *Escherichia coli* strain Nissle 1917 (EcN) in vivo. J Microbiol Methods 2005;61(3):389–98.

[6] Joeres-Nguyen-Xuan HT, Boehm KS, Joeres L, et al. Survival of the probiotic *Escherichia coli* Nissle 1917 (EcN) in the gastrointestinal tract given in combination with oral mesalamine to healthy volunteers. Inflamm Bowel Dis 2010;16(2):256–62.

[7] Garcia CE, Brumbaugh RA, Mobley LH. Redundancy and specificity of *Escherichia coli* iron acquisition systems during urinary tract infection. Infect Immun 2011;79(3):1225–35.

[8] Deriu E, Liu ZJ, Pezeshki M, et al. Probiotic bacteria reduce *Salmonella typhimurium* intestinal colonization by competing for iron. Cell Host Microbe 2013;14(1):26–37.

[9] Papavassiliou J. Production of colicines in Simmons's citrate agar. Nature 1959;184(4965):1339–40.

[10] Papavassiliou J. Biological characteristics of colicine X. Nature 1961;190(4770):110.

[11] Patzer IS, Baquero RM, Bravo D, et al. The colicin G, H and X determinants encode microcins M and H47, which might utilize the catecholate siderophore receptors FepA, Cir, Fiu and IroN. Microbiology 2003;149(Pt. 9):2557–70.

[12] Troge A, Scheppach W, Schroeder OB, et al. More than a marine propeller—the flagellum of the probiotic *Escherichia coli* strain Nissle 1917 is the major adhesin mediating binding to human mucus. Int J Med Microbiol 2012;302(7–8):304–14.

[13] Grozdanov L, Zahringer U, Blum-Oehler G, et al. A single nucleotide exchange in the wzy gene is responsible for the semirough O6 lipopolysaccharide phenotype and serum sensitivity of *Escherichia coli* strain Nissle 1917. J Bacteriol 2002;184(21):5912–25.

[14] Schierack P, Kleta S, Tedin K, et al. *E. coli* Nissle 1917 affects *Salmonella* adhesion to porcine intestinal epithelial cells. PLoS One 2011;6(2):e14712.

[15] Rund AS, Rohde H, Sonnenborn U, et al. Antagonistic effects of probiotic *Escherichia coli* Nissle 1917 on EHEC strains of serotype O104:H4 and O157:H7. Int J Med Microbiol 2013;303(1):1–8.

[16] Jiang Y, Kong Q, Roland LK, et al. Multiple effects of *Escherichia coli* Nissle 1917 on growth, biofilm formation, and inflammation cytokines profile of *Clostridium perfringens* type A strain CP4. Pathog Dis 2014;70(3):390–400.

[17] Kleta S, Nordhoff M, Tedin K, et al. Role of F1C fimbriae, flagella, and secreted bacterial components in the inhibitory effect of probiotic *Escherichia coli* Nissle 1917 on atypical enteropathogenic *E. coli* infection. Infect Immun 2014;82(5):1801–12.

[18] Mohsin M, Guenther S, Schierack P, et al. Probiotic *Escherichia coli* Nissle 1917 reduces growth, Shiga toxin expression, release

and thus cytotoxicity of enterohemorrhagic *Escherichia coli*. Int J Med Microbiol 2015;305(1):20–6.

[19] Mandel L, Trebichavsky I, Splichal I, et al. Stimulation of intestinal immune cells by *E. coli* in gnotobiotic piglets. In: Mestecky J, editor. Advances in mucosal immunology. New York, NY: Plenum Press; 1995. p. 463–4.

[20] Lorenz A, Schulze J. Establishment of *E. coli* Nissle 1917 and its interaction with *Candida albicans* in gnotobiotic rats. Microecol Ther 1996;24:45.

[21] Boudeau J, Glasser LA, Julien S, et al. Inhibitory effect of probiotic *Escherichia coli* strain Nissle 1917 on adhesion to and invasion of intestinal epithelial cells by adherent-invasive *E. coli* strains isolated from patients with Crohn's disease. Aliment Pharmacol Ther 2003;18(1):45.

[22] Altenhoefer A, Oswald S, Sonnenborn U, et al. The probiotic *Escherichia coli* strain Nissle 1917 interferes with invasion of human intestinal epithelial cells by different enteroinvasive bacterial pathogens. FEMS Immunol Med Microbiol 2004;40:223–9.

[23] Hafez M, Hayes K, Goldrick M, et al. The K5 capsule of *Escherichia coli* strain Nissle 1917 is important in stimulating expression of Toll-like receptor 5, CD14, MyD88, and TRIF together with the induction of interleukin-8 expression via the mitogen-activated protein kinase pathway in epithelial cells. Infect Immun 2010;78(5):2153–62.

[24] Miao AE, Andersen-Nissen E, Warren ES, et al. TLR5 and Ipaf: dual sensors of bacterial flagellin in the innate immune system. Semin Immunopathol 2007;29(3):275–88.

[25] Nzakizwanayo J, Kumar S, Ogilvie AL, et al. Disruption of *Escherichia coli* Nissle 1917 K5 capsule biosynthesis, through loss of distinct kfi genes, modulates interaction with intestinal epithelial cells and impact on cell health. PLoS One 2015;10(3):e0120430.

[26] Hafez M, Hayes K, Goldrick M, et al. The K5 capsule of *Escherichia coli* strain Nissle 1917 is important in mediating interactions with intestinal epithelial cells and chemokine induction. Infect Immun 2009;77(7):2995–3003.

[27] Trebichavsky I, Splichal I, Rada V, et al. Modulation of natural immunity in the gut by *Escherichia coli* strain Nissle 1917. Nutr Rev 2010;68(8):459–64.

[28] Ukena NS, Westendorf MA, Hansen W, et al. The host response to the probiotic *Escherichia coli* strain Nissle 1917: specific upregulation of the proinflammatory chemokine MCP-1. BMC Med Genet 2005;6:43.

[29] Cukrowska B, Lodinova-Zadnikova R, Enders C, et al. Specific proliferative and antibody responses of premature infants to intestinal colonization with nonpathogenic probiotic *E. coli* strain Nissle 1917. Scand J Immunol 2002;55(2):204–9.

[30] Sturm A, Rilling K, Baumgart CD, et al. *Escherichia coli* Nissle 1917 distinctively modulates T cell cycling and expansion via Toll-like receptor 2 signaling. Infect Immun 2005;73(3):1452–65.

[31] Wesch D, Peters C, Oberg HH, et al. Modulation of gammadelta T cell responses by TLR ligands. Cell Mol Life Sci 2011;68(14):2357–70.

[32] Guzy C, Paclik D, Schirbel A, et al. The probiotic *Escherichia coli* strain Nissle 1917 induces gammadelta T cell apoptosis via caspase- and FasL-dependent pathways. Int Immunol 2008;20(7):829–40.

[33] Zyrek AA, Cichon C, Helms S, et al. Molecular mechanisms underlying the probiotic effects of *Escherichia coli* Nissle 1917 involve ZO-2 and PKCzeta redistribution resulting in tight junction and epithelial barrier repair. Cell Microbiol 2007;9(3):804–16.

[34] Hering AN, Richter FJ, Fromm A, et al. TcpC protein from *E. coli* Nissle improves epithelial barrier function involving PKC-zeta and ERK1/2 signaling in HT-29/B6 cells. Mucosal Immunol 2014;7(2):369–78.

[35] Hancock V, Dahl M, Klemm P. Probiotic *Escherichia coli* strain Nissle 1917 outcompetes intestinal pathogens during biofilm formation. J Med Microbiol 2010;59(Pt. 4):392–9.

[36] Wehkamp J, Harder J, Wehkamp K, et al. NF-kappaB- and AP-1-mediated induction of human beta defensin-2 in intestinal epithelial cells by *Escherichia coli* Nissle 1917: a novel effect of a probiotic bacterium. Infect Immun 2004;72(10):5750–8.

[37] Schlee M, Wehkamp J, Altenhoefer A, et al. Induction of human beta-defensin 2 by the probiotic *Escherichia coli* Nissle 1917 is mediated through flagellin. Infect Immun 2007;75(5):2399–407.

[38] Valdebenito M, Crumbliss LA, Winkelmann G, et al. Environmental factors influence the production of enterobactin, salmochelin, aerobactin, and yersiniabactin in *Escherichia coli* strain Nissle 1917. Int J Med Microbiol 2006;296(8):513–20.

[39] Blum G, Marre R, Hacker J. Properties of *Escherichia coli* strains of serotype O6. Infection 1995;23(4):234–6.

[40] Weiss G. Intestinal irony: how probiotic bacteria outcompete bad bugs. Cell Host Microbe 2013;14(1):3–4.

[41] Duquesne S, Destoumieux-Garzon D, Peduzzi J, et al. Microcins, gene-encoded antibacterial peptides from enterobacteria. Nat Prod Rep 2007;24(4):708–34.

[42] Nissle A. Die antagonistische Behandlung chronischer Darmstoerungen mit Colibakterien (The antagonistical therapy of chronic intestinal disturbances). Med Klinik 1918;2:29–30.

[43] Nissle A. Weiteres ueber die Grundlagen und Praxis der Mutaflor-behandlung. Dtsch Med Wochenschr 1925;44:1809–13.

[44] Vassiliadis G, Destoumieux-Garzon D, Lombard C, et al. Isolation and characterization of two members of the siderophore-microcin family, microcins M and H47. Antimicrob Agents Chemother 2010;54(1):288–97.

[45] Toloza L, Gimenez R, Fabrega JM, et al. The secreted autotransporter toxin (Sat) does not act as a virulence factor in the probiotic *Escherichia coli* strain Nissle 1917. BMC Microbiol 2015;15:250.

[46] Lasaro AM, Salinger N, Zhang J, et al. F1C fimbriae play an important role in biofilm formation and intestinal colonization by the *Escherichia coli* commensal strain Nissle 1917. Appl Environ Microbiol 2009;75(1):246–51.

[47] Grozdanov L, Raasch C, Schulze J, et al. Analysis of the genome structure of the nonpathogenic probiotic *Escherichia coli* strain Nissle 1917. J Bacteriol 2004;186(16):5432–41.

[48] Tannock WG, Tiong SI, Priest P, et al. Testing probiotic strain *Escherichia coli* Nissle 1917 (Mutaflor) for its ability to reduce carriage of multidrug-resistant *E. coli* by elderly residents in long-term care facilities. J Med Microbiol 2011;60(Pt. 3):366–70.

[49] Sha S, Xu B, Kong X, et al. Preventive effects of *Escherichia coli* strain Nissle 1917 with different courses and different doses on intestinal inflammation in murine model of colitis. Inflamm Res 2014;63(10):873–83.

[50] Becker S, Oelschlaeger AT, Wullaert A, et al. Bacteria regulate intestinal epithelial cell differentiation factors both in vitro and in vivo. PLoS One 2013;8(2):e55620.

[51] Cobo RE, Chadee K. Antimicrobial human beta-defensins in the colon and their role in infectious and non-infectious diseases. Pathogens 2013;2(1):177–92.

[52] Kruis W, Schutz E, Fric P, et al. Double-blind comparison of an oral *Escherichia coli* preparation and mesalazine in maintaining

remission of ulcerative colitis. Aliment Pharmacol Ther 1997;11(5): 853–8.

[53] Rembacken JB, Snelling MA, Hawkey MP, et al. Non-pathogenic *Escherichia coli* versus mesalazine for the treatment of ulcerative colitis: a randomised trial. Lancet 1999;354(9179):635.

[54] Kruis W, Fric P, Pokrotnieks J, et al. Maintaining remission of ulcerative colitis with the probiotic *Escherichia coli* Nissle 1917 is as effective as with standard mesalazine. Gut 2004;53(11):1617–23.

[55] Henker J, Muller S, Laass WM, et al. Probiotic *Escherichia coli* Nissle 1917 (EcN) for successful remission maintenance of ulcerative colitis in children and adolescents: an open-label pilot study. Z Gastroenterol 2008;46(9):874–5.

[56] Petersen MA, Mirsepasi H, Halkjaer IS, et al. Ciprofloxacin and probiotic *Escherichia coli* Nissle add-on treatment in active ulcerative colitis: a double-blind randomized placebo controlled clinical trial. J Crohns Colitis 2014;8(11):1498–505.

[57] Matthes H, Krummenerl T, Giensch M, et al. Clinical trial: probiotic treatment of acute distal ulcerative colitis with rectally administered *Escherichia coli* Nissle 1917 (EcN). BMC Complement Altern Med 2010;10:13.

[58] Kuzela L, Kascak M, Vavrecka A. Induction and maintenance of remission with nonpathogenic *Escherichia coli* in patients with pouchitis. Am J Gastroenterol 2001;96(11):3218–9.

[59] Malchow HA. Crohn's disease and *Escherichia coli*. A new approach in therapy to maintain remission of colonic Crohn's disease? J Clin Gastroenterol 1997;25(4):653–8.

[60] Mollenbrink M, Bruckschen E. Treatment of chronic constipation with physiologic *Escherichia coli* bacteria. Results of a clinical study of the effectiveness and tolerance of microbiological therapy with the *E. coli* Nissle 1917 strain (Mutaflor). Med Klinik 1994;89(11):587.

[61] Kruis W, Chrubasik S, Boehm S, et al. A double-blind placebo-controlled trial to study therapeutic effects of probiotic *Escherichia coli* Nissle 1917 in subgroups of patients with irritable bowel syndrome. Int J Colorectal Dis 2012;27(4):467–74.

[62] Faghihi HA, Agah S, Masoudi M, et al. Efficacy of probiotic *Escherichia coli* Nissle 1917 in patients with irritable bowel syndrome: a double blind placebo-controlled randomized trial. Acta Med Indones 2015;47(3):201–8.

[63] Henker J, Laass M, Blokhin MB, et al. The probiotic *Escherichia coli* strain Nissle 1917 (EcN) stops acute diarrhoea in infants and toddlers. Eur J Pediatr 2007;166(4):311–8.

[64] Henker J, Laass WM, Blokhin MB, et al. Probiotic *Escherichia coli* Nissle 1917 versus placebo for treating diarrhea of greater than 4 days duration in infants and toddlers. Ped Infect Dis J 2008;27(6):494–9.

[65] Tromm A, Niewerth U, Khoury M, et al. The probiotic *E. coli* strain Nissle 1917 for the treatment of collagenous colitis: first results of an open-label trial. Z Gastroenterol 2004;42(5):365–9.

[66] Schultz M. Clinical use of *E. coli* Nissle 1917 in inflammatory bowel disease. Inflamm Bowel Dis 2008;14(7):1012–8.

[67] Jacobi AC, Malfertheiner P. *Escherichia coli* Nissle 1917 (Mutaflor): new insights into an old probiotic bacterium. Dig Dis 2011;29(6):600–7.

[68] Losurdo G, Iannone A, Contaldo A, et al. *Escherichia coli* Nissle 1917 in ulcerative colitis treatment: systematic review and meta-analysis. J Gastrointest Liv Dis 2015;24(4):499–505.

[69] Kaser A, Zeissig S, Blumberg SR. Inflammatory bowel disease. Ann Rev Immunol 2010;28:573–621.

[70] Darfeuille-Michaud A, Neut C, Barnich N, et al. Presence of adherent *Escherichia coli* strains in ileal mucosa of patient with Crohn's disease. Gastroenterology 1998;115:1.

[71] Baumgart M, Dogan B, Rishniw M, et al. Culture independent analysis of ileal mucosa reveals a selective increase in invasive *Escherichia coli* of novel phylogeny relative to depletion of Clostridiales in Crohn's disease involving the ileum. ISME J 2007;1(5):403–18.

[72] Swidsinski A, Ladhoff A, Pernthaler A, et al. Mucosal flora in inflammatory bowel disease. Gastroenterology 2002;122(1):44.

[73] Glasser LA, Darfeuille-Michaud A. Abnormalities in the handling of intracellular bacteria in Crohn's disease: a link between infectious etiology and host genetic susceptibility. Arch Immunol Ther Exp 2008;56(4):237–44.

[74] Kotlowski R, Bernstein NC, Sepehri S, et al. High prevalence of *Escherichia coli* belonging to the B2 + D phylogenetic group in inflammatory bowel disease. Gut 2007;56(5):669–75.

[75] Vejborg MR, Hancock V, Petersen MA, et al. Comparative genomics of *Escherichia coli* isolated from patients with inflammatory bowel disease. BMC Genom 2011;12:316.

[76] Petersen MA, Halkjaer IS, Gluud LL. Intestinal colonization with phylogenetic group B2 *Escherichia coli* related to inflammatory bowel disease: a systematic review and meta-analysis. Scand J Gastroenterol 2015;50(10):1199–207.

[77] Huebner C, Ding Y, Petermann I, et al. The probiotic *Escherichia coli* Nissle 1917 reduces pathogen invasion and modulates cytokine expression in Caco-2 cells infected with Crohn's disease-associated *E. coli* LF82. Appl Environ Microbiol 2011;77(7):2541–4.

[78] Rachmilewitz D. Coated mesalazine (5-aminosalicylic acid) versus sulphasalazine in the treatment of active ulcerative colitis: a randomised trial. BMJ 1989;298(6666):82.

[79] Bressler B, Marshall KJ, Bernstein NC, et al. Clinical practice guidelines for the medical management of nonhospitalized ulcerative colitis: the Toronto consensus. Gastroenterology 2015;148(5). 1035-1058.e3.

[80] Dignass A, Lindsay OJ, Sturm A, et al. Second European evidence-based consensus on the diagnosis and management of ulcerative colitis part 2: current management. J Crohns Colitis 2012;6(10):991–1030.

[81] Miele E, Pascarella F, Giannetti E, et al. Effect of a probiotic preparation (VSL#3) on induction and maintenance of remission in children with ulcerative colitis. Am J Gastroenterol 2009;104(2):437–43.

[82] Zezos P, Saibil F. Inflammatory pouch disease: the spectrum of pouchitis. World J Gastroenterol 2015;21(29):8739–52.

[83] Tannock WG, Lawley B, Munro K, et al. Comprehensive analysis of the bacterial content of stool from patients with chronic pouchitis, normal pouches, or familial adenomatous polyposis pouches. Inflamm Bowel Dis 2012;18(5):925–34.

[84] Chey DW, Kurlander J, Eswaran S. Irritable bowel syndrome: a clinical review. JAMA 2015;313(9):949–58.

[85] Bar F, Koschitzky VH, Roblick U, et al. Cell-free supernatants of *Escherichia coli* Nissle 1917 modulate human colonic motility: evidence from an in vitro organ bath study. Neurogastroenterol Motil 2009;21(5):559–66. e16–e17.

[86] Dalziel EJ, Mohan V, Peters J, et al. The probiotic *Escherichia coli* Nissle 1917 inhibits propagating colonic contractions in the rat isolated large intestine. Food Funct 2015;6(1):257–64.

[87] Nzakizwanayo J, Dedi C, Standen G, et al. *Escherichia coli* Nissle 1917 enhances bioavailability of serotonin in gut tissues through modulation of synthesis and clearance. Sci Rep 2015;5:17324.

[88] Thompson GW, Longstreth FG, Drossman AD, et al. Functional bowel disorders and functional abdominal pain. Gut 1999;45(Suppl. 2): II43–7.

[89] Taub E, Cuevas LJ, Cook WE, et al. Irritable bowel syndrome defined by factor analysis. Gender and race comparisons. Dig Dis Sci 1995;40(12):2647–55.

[90] Krammer JH, Kamper H, Bunau v R, et al. Probiotic drug therapy with *E. coli* strain Nissle 1917 (EcN): results of a prospective study of the records of 3,807 patients. Z Gastroenterol 2006;44(8):651–6.

[91] Rohrenbach J, Matthess A, Maier R, et al. *Escherichia coli* Stamm Nissle 1917 (EcN) bei Kindern (*E. coli* Nissle 1917 in children). Kinder Jugendarzt 2007;38(3):1–5.

[92] Weise C, Zhu Y, Ernst D, et al. Oral administration of *Escherichia coli* Nissle 1917 prevents allergen-induced dermatitis in mice. Exp Dermatol 2011;20(10):805–9.

[93] Dolle S, Berg J, Rasche C, et al. Tolerability and clinical outcome of coseasonal treatment with *Escherichia coli* strain Nissle 1917 in grass pollen-allergic subjects. Int Arch Allergy Immunol 2014;163(1):29–35.

[94] Whelan AR, Rausch S, Ebner F, et al. A transgenic probiotic secreting a parasite immunomodulator for site-directed treatment of gut inflammation. Mol Ther 2014;22(10):1730–40.

[95] Pohlmann C, Thomas M, Forster S, et al. Improving health from the inside: use of engineered intestinal microorganisms as in situ cytokine delivery system. Bioengineered 2013;4(3):172–9.

[96] Amiri-Jami M, Abdelhamid GA, Hazaa M, et al. Recombinant production of omega-3 fatty acids by probiotic *Escherichia coli* Nissle 1917. FEMS Microbiol Lett 2015;362(20).

[97] Singh KA, Pandey KS, Kumar NG. Pyrroloquinoline quinone-secreting probiotic *Escherichia coli* Nissle 1917 ameliorates ethanol-induced oxidative damage and hyperlipidemia in rats. Alcohol Clin Exp Res 2014;38(7):2127–37.

[98] Forbes SN. Engineering the perfect (bacterial) cancer therapy. Nat Rev Cancer 2010;10(11):785–94.

[99] Lehouritis P, Stanton M, McCarthy OF, et al. Activation of multiple chemotherapeutic prodrugs by the natural enzymolome of tumour-localised probiotic bacteria. J Contr Rel 2015;222:9–17.

[100] Danino T, Prindle A, Kwong AG, et al. Programmable probiotics for detection of cancer in urine. Sci Transl Med 2015;7(289):289ra84.

Chapter 6

Probiotics of the Acidophilus Group: *Lactobacillus acidophilus, delbrueckii* subsp. *bulgaricus* and *johnsonii*

A.S. Neish

Department of Pathology, Emory University School of Medicine, Atlanta, GA, United States

INTRODUCTION

Probiotics are defined as "live microorganisms, which when administered in adequate amounts, confer a health benefit upon the host" (FAO/WHO, 2002). Ingested probiotic bacteria have long been accepted to contribute to overall health of the gastrointestinal (GI) tract and vagina, as well beneficially regulate other systemic processes including infection, innate and adaptive immunity, secondary metabolism, hepatic biology, and neurobehavioral systems by multiple postulated mechanisms. For example, probiotic bacteria have been reported to directly antagonize enteric pathogens via competitive inhibition of binding to host cells (colonization resistance) and/or production of antimicrobial metabolites (bacteriocins). Additionally, probiotics have been implicated in multiple mechanisms of regulation of intrinsic defenses such as increased antimicrobial peptide production by Paneth cells, increased mucin production by goblet cells, and degradation of luminal antigens and toxins. An extensive literature describes multiple mechanisms by which probiotics regulate/induce innate immunity/inflammatory signaling, and elicit/modulate mucosal and systemic adaptive immunity. Finally, and perhaps most significantly, emerging evidence indicates probiotics can augment epithelial barrier function via upregulation of tight junction proteins, stimulation of intestinal crypt proliferation and differentiation, and activation of prosurvival pathways that eventuate in enhanced mucosal homeostasis [1,2]. Increasing evidence suggests that the collective effect at the gut-microbe epithelial interface diminishes a "leaky gut" and reduces systemic access of bacterial components Microbial associated molecular pattern (MAMPs) that can stimulate chronic inflammatory reactions and contribute to a range of systemic disease. The interested reader is directed to several excellent reviews available discussing proposed mechanisms of probiotic benefits [3–5].

Most probiotics were first identified in food products or isolated from the intestinal environment. Prominent grouping of commercially important and scientifically validated probiotics are members of the lactic acid bacteria (LAB), a highly diverse group of commensal and environmental microbes historically defined by the ability to ferment hexose sugars to lactate [6]. The LAB can be further subdivided into multiple genera including *Enterococcus, Leuconostoc, Bifidobacterium, Streptococcus, Lactococcus,* and *Lactobacillus.* The lactobacilli are classified as phylum Firmicutes, class Bacilli, order Lactobacillales, and family Lactobacillus. A definitive classification of 123 *Lactobacillus* strains by full genome sequencing was achieved in 2015 [7]. The genus *Lactobacillus* has approximately 125 species [8] and its members are primarily found in environmental niches that provide an abundant energy source, including fruit and dairy ecosystems, sewage and decaying plant material, as well as the metazoan GI tract [9]. It is of interest that they (along with bifidobacteria) are among the first bacteria to colonize the infant mammalian gut [10], and are key members of invertebrate microbiota [11], wherein both diverse host ecosystems the lactobacilli have striking roles in epithelial growth and differentiation [1,2]. These observations suggest that the lactobacilli have an ancient history of coevolution with metazoan hosts, from which mutually beneficial interactions have formed. Thus it is not surprising this bacterial taxon has been a prominent source of probiotic strains. Extensive reviews are available discussing the biology of host lactobacilli interactions [9], the potential mechanisms by which lactobacilli can mediate beneficial effects, and assessment of health claims and supporting clinical trials [4].

Lactobacilli of the acidophilus complex (*acidophilus, bulgaricus,* and *johnsonii,* discussed here, along with *gasseri, helveticus*) have been exploited for millennia in the production of fermented milk products, such as yogurts. Members of acidophilus group are widely known to colonize human mucosa sites, included the oral cavity, GI tract, and the vagina. They have provided a fruitful source of candidate probiotic bacteria. Indeed, Elie Metchnikoff, a professor at the Pasteur Institute in Paris, observed the vigorous old age enjoyed by Eastern European residents who commonly made yogurt a central part of their diet, prompting his recommendation that "Bulgarian milk" be utilized as a tonic for healthy living, a hypothesis put forth in a paper often considered to be the founding document of probiotics [12].

While reports of the clinical utility of probiotics have been marked by often lofty claims, the most recent (2015) critical assessment of clinical efficacy of probiotics concludes sufficiently high quality evidence exists to support recommendation for probiotic use in the following conditions: treatment of infectious diarrhea and prevention of antibiotic associated diarrhea, maintenance of remission in inflammatory bowel disease and pouchitis, modulation of immune responses, and treatment for atopic eczema associated with cow's milk allergy and in hepatic encephalopathy. Less robust evidence suggests roles in treating necrotizing enterocolitis, irritable bowel syndrome, and emerging data may indicate potential roles in treating radiation enteritis, vaginosis and vaginitis, alcoholic liver disease, and nonalcoholic fatty liver diseases [13]. We now turn our attention to individual strains of the acidophilus group used in therapeutic contexts.

LACTOBACILLUS ACIDOPHILUS

The *Lactobacillus* strain *L. acidophilus* was first isolated in 1900 by Moro, from the feces and oral cavity of normal infants [14]. He correctly deduced the organism was a representative of multiple strains that flourished in an acid growth media. Formal description occurred [15], and based on DNA—DNA hybridization studies, an "acidophilus complex" was defined [16], further refined with the development of 16s ribosomal RNA sequencing [17]. The full genome was sequenced in 2005 [18] which was notable for mucus and fibronectin binding proteins proposed to be involved in the ability of *L. acidophilus* to adhere to mammalian mucosa. Recent studies provide genomic-based support to the concept of the acidophilus complex, with *L. acidophilus* and closely related *Lactobacillus johnsonii* and slightly less related *Lactobacillus bulgarius* [7,17]. As mentioned, *L. acidophilus* as long been used in fermented food, such as yogurt, other dairy products, and fermented soy products, such as miso and tempeh, and a range of functional foods

[19]. Its use in probiotics is supported by data showing postgastric survival and persistence in feces (up to 10 days postingestion) in humans [20].

Contemporary Clinical Success

L. acidophilus has been employed as a therapeutic for decades, inadvertently for centuries, in attempts to treat a wide range of clinical conditions in humans (and domestic animals) with varying degrees of success, in vaginal disorders (candidiasis, vaginosis), hypercholesterolemia, *Helicobacter pylori* and other enteric infections, allergic dermatitis and rhinitis, hyperoxaluria, renal and hepatic failure, infectious and antibiotic associated diarrhea, various forms of inflammatory bowel disease, functional bowel disorders, and others. The following discussion will be limited to recent (last 5 years) controlled human studies of *L. acidophilus* used in isolation or as a simple combination. A difficulty in assessing the specific efficacy of *L. acidophilus* is its common utilization in combination with other bacteria. A typical use as an additive to yogurt-like products necessitates inclusion of other strains (e.g., *Lactobacillus bulgaricus*). Furthermore, multiple acidophilus stains exist (as we shall see, *L. johnsonii* was originally classified as *acidophilus*). The commonly used proprietary probiotic preparation VSL#3 (VSL pharmaceuticals Inc., Fort Lauderdale, USA) comprises four strains of lactobacilli (*acidophilus, delbrueckii* subsp. *bulgaricus, casei, plantarum*), three strains of bifidobacteria (*breve, longum, infantis*), and one strain of *Streptococcus salivarius* subsp. *thermophilus*. Thus any evaluation of the efficacy of probiotics must bear in mind the highly complex and variable formulations used in any given study.

CLINICAL TRIALS OF IMMUNOMODULATION

Probiotics are commonly assumed to influence systemic immunity, presumably by interaction with immunoregulatory cells in the gut that are present in the lamina propria and epithelial layer of the mucosa and in organized gut associated lymphoid tissue. *L. acidophilus* has successfully been reported in modulation of systemic immune related conditions. In a double-blind, placebo-controlled study of 326 children treated with both *L. acidophilus* NCFM alone or in combination with *B. animalis* as lyophilized powder twice daily for 6 months showed reduced fever, rhinorrhea, cough, antibiotic prescription incidence, and lost school days [21]. In adults with atopic dermatitis, a double-blind, placebo-controlled randomized trial of *L. acidophilus* L-92 showed highly significant reduction in the clinical assessment of dermatitis severity and alterations of circulating cytokine levels [22]. Interestingly, in that study *L. acidophilus* was utilized in a heat-killed form, suggesting that immunomodulatory effects stem from stimulation of immunocompetent cells via pattern recognition receptors. In a similar study of children with atopic dermatitis, a double-blind randomized controlled study of 40 pediatric patients treated a combination of *L. acidophilus, Bifidobacterium bifidum, Lactobacillus casei,* and *Lactobacillus salivarius* for 8 weeks showed significant reduction in clinical manifestations, as well as reduction of serum IgE [23]. This study utilized live probiotics. The marked differences in probiotic preparations used in these studies underscore the difficulty of assessing a common mechanism of action for identified beneficial effects.

CLINICAL TRIALS IN DIGESTIVE DISEASES

As discussed, the lactobacilli as occupants of the gut can directly affect epithelial biology and immunity, influence the normal microbiota and digestive processes at the biochemical level, and stimulate systemic metabolism, and thus could plausibly affect many aspects of intestinal physiology. For example, *L. acidophilus* containing probiotic preparations have shown promise in treating functional bowel disorders. A synbiotic combination of polydextrose, *L. acidophilus* NCFM, and *Bifidobacterium lactis* HN019 (in a yogurt vehicle) was used in a randomized, double-blinded, controlled study to demonstrate shortened colonic transit time and improvement of clinical assessments of constipation over a 2-week treatment period [24]. Similarly, in a double-blind, placebo-controlled trial of *L. acidophilus* NCFM and *B. lactis* HN019 (2×10^{11} cfu daily) administered over 8 weeks showed improvement in bloating in a 60 patient cohort of nonconstipated functional bowel disorder [25]. These workers followed up that study showing that similar probiotic treatment regimen trended toward improvement of symptoms of chronic mild to moderate abdominal pain. Interestingly, in that study *L. acidophilus* NCFM was shown to selectively induce expression of the mu-opioid receptors, potentially identifying a mechanism of action for the observed effects [26]. In a distinct clinical population of hemodialysis patients, Viramontes-Horner et al. used a symbiotic preparation of *L. acidophilus* and *B. lactis* with inulin in a double-blinded placebo-controlled trial over 2 months and demonstrated reduced severity and frequency GI symptoms, including vomiting, heartburn, and abdominal pain [27]. Similarly, a report of 24 female patients treated with pelvic irradiation for gynecological malignancies, who received *L. acidophilus* orally administered daily (along with 6.5% lactulose), showed reduction of radiotherapy-associated diarrhea [28]. Such benefits may not apply to a healthy population. For example, a randomized, double-blind, crossover study of a yogurt preparation containing *L. acidophilus* La5, *B. lactis* Bb12, and *L. casei* CRL431 and inulin in 65 asymptomatic adults showed no significant differences in GI transit time or GI symptoms, though decreased energy and fat intake were observed [29]. *L. acidophilus* has shown efficacy in antibiotic associated

diarrhea. In a multisite, randomized, placebo-controlled trial of 140 children prescribed antibiotics, a yogurt containing *L. acidophilus, Lactobacillus rhamnosus* GG, and *B. lactis* (Bb12), or control (pasteurized yogurt) for the duration of antibiotic treatment showed highly significant reduction in minor diarrhea [30]. Finally, the probiotic preparation VSL#3 which contains *L. acidophilus* has been shown to be quite effective in the treatment of pouchitis following ileal pouch—anal anastomosis for ulcerative colitis [31,32]. These results suggest *L. acidophilus* may have beneficial effects on range intestinal disorders, though as we have discussed, the use of *L. acidophilus* as a combination therapy with other probiotic organisms, or with a prebiotic substrate complicates assessment of specific activity mediated by *L. acidophilus*.

CONTROL OF INFECTIOUS DISEASE

Beneficial bacteria are well known to complete with pathogens for niche occupancy in the gut. This property, also called colonization resistance, has prompted studies of probiotics in treatment or prevention of enteric infections. Past studies have been mixed in reported efficacy in a variety of enteric infections. More recently, a double-blind, placebo-controlled study of pediatric patients with viral gastroenteritis utilized 10^9 cfu of several strains of probiotic bacteria administered individually twice daily for a week. *L. acidophilus* showed the second greatest effect (*B. longum* had the greatest efficacy) in shortening the duration of diarrhea [33]. Additionally, in a randomized controlled series of 64 cases of acute rotaviral diarrhea, slightly significant reduction of vomiting was observed in children treated with a probiotic mixture containing *L. acidophilus* (though more significant reduction in clinical symptoms were seen with a single agent *Saccharomyces boulardii* regimen) [34]. In 2 separate controlled randomized studies of acute pediatric diarrhea, over 300 children were treated with a synbiotic preparation (probiotical) containing multiple probiotic strains including *L. acidophilus,* showing a slight but significant decrease in diarrhea duration and reduced prescription of medication [35,36]. Similar effects were reported in a metaanalysis of *L. acidophilus* treatment of acute pediatric diarrhea [37]. Interestingly, predosing 20 human subjects with 10^9 cfu of *L. acidophilus* twice daily for 2 weeks did not protect from diarrhea and GI discomfort induced by a challenge with a model enterotoxigenic *Escherichia coli* infection (attenuated live ETEC vaccine) [38]. Additionally, a large well-designed and -controlled series of over 3000 adults over the age of 65 treated with 10^{10} cfu of a multistrain preparation of lactobacilli and bidifobacteria failed to detect a benefit in preventing antibiotic- or *Clostridium difficile*-associated diarrhea [39]. Overall, the evidence is suggestive of a role for probiotics, usually as complex mixture, in the treatment of pediatric diarrhea; however, a specific antiinfective role of *L. acidophilus* is not established.

USE IN NECROTIZING ENTEROCOLITIS

Necrotizing enterocolitis (NEC) is a devastating complication of prematurity, likely associated with the inability of an immature intestine to tolerate extrauterine environmental challenges. Given the ability of lactobacilli to induce epithelial growth and differentiation, a clear rational exists for probiotic use in this condition. To date, clinical trials are mixed. In a prospective, blinded, randomized multicenter controlled trial of seven NICUs in Taiwan, very low birth weight infants received *L. acidophilus* and *B. bifidum* in breast milk or formula twice daily for 6 weeks, and showed a significantly lowered incidence of NEC (4 of 217 vs. 14 of 217) [40]. A second randomized, double-blind study of 150 neonatal patients (<1500 g) utilized a multispecies probiotic of *L. acidophilus, L. rhamnosus, Lactobacillus plantarum, Bifidobacterium infantis,* and *Streptococcus thermophilus*, administered 1 g daily, found no significant risk reduction, though a clear trend was evident [41]. In a 2 years observational study of a cohort of 5351, very low birth weight infants in multiple centers assigned no probiotics or administration of a *L. acidophilus/B. infantis* preparation showed reduced risk of NEC and hospital mortality, and increased weight gain [42]. While these promising results are not conclusive, given the severe nature of NEC, further studies are warranted.

USE IN GYNECOLOGICAL DISORDERS

It is well known that the lactobacilli are the dominant taxa in the female genitourinary tract, providing a clear biological rationale for their use in gynecological disorders. Molecular studies have demonstrated that oral intake of probiotics, including *L. acidophilus*, results in detectable probiotic bacteria in the vagina, even 1 week after cessation of consumption [43]. In a (uncontrolled) clinical trial of recurrent vulvovaginal candidiasis, after treatment with fluconazol, maintenance therapy with *L. acidophilus* LA02 and *Lactobacillus fermentum* LF10 in a slow release vaginal tablet form was effective in reducing recurrence [44]. Other workers report similar but inconclusive trends [45]. Thus, while promising and highly biologically plausible, use of *L. acidophilus* in gynecological disorders is not yet established.

USE IN HEPATIC AND METABOLIC DISORDERS

Past studies using the combination probiotic VSL#3 has shown efficacy in hepatic encephalopathy secondary to cirrhosis, and in nonalcoholic fatty liver disease showing improvement in clinical and serum parameters of hepatic injury [46–48]. In a study of minimal hepatic encephalopathy, a randomized control trial of 90 patients receiving *L. acidophilus*, demonstrated a decrease in the development of overt encephalopathy and improved blood ammonia, as well brain neurometabolites imaged by magnetic resonance spectroscopy [49]. A complex mixture of probiotics (*L. acidophilus, L. rhamnosus, Bifidobacterium bifidum, B. lactis, Bifidobacterium longum,* and *S. thermophilus* were used in a randomized, controlled trial in 53 patients with chronic liver disease. Probiotic treatment was significantly effective in reducing symptoms of small intestinal overgrowth, with positive trends in reducing general GI symptoms and intestinal permeability [50]. Other metabolic effects have been reported. In the context of a clinical population with type 2 diabetes, a probiotic yogurt preparation containing *L. acidophilus* La5 and *B. lactis* Bb12 was used in a randomized, double-blind, controlled trial for 6 weeks and demonstrated significantly decreased total cholesterol and LDL-C, without significantly affecting triglycerides or HDL-C [51]. Additionally, a randomized, crossover study of 14 subjects given yogurt containing *L. acidophilus* and *B. lactis*, with unsupplemented yogurt as control, detected a significant reduction in serum total cholesterol after 6 weeks [52]. Thus, *L. acidophilus* may influence systemic metabolic diseases.

In summary, *L. acidophilus* and *L. acidophilus* containing preparations have been widely used as probiotic therapy, often in synbiotic formulations. Most clinical trials use a combination of probiotic agents, hampering the ability to distinguish effects of a single species. Additionally, as *L. acidophilus* has an origin in dairy science, it is often delivered in a yogurt vehicle, which may have probiotic effects as a stand-alone product. Similarly, administration with a pre- or symbiotic component likely would have independent effects on the native microbiota. Nevertheless, a positive influence of *L. acidophilus* on health and disease is widely reported in clinical trials (Table 6.1).

LACTOBACILLUS BULGARICUS

This strain has an equally venerable history as *L. acidophilus*. *Lactobacillus delbrueckii* subsp. *bulgaricus* was first isolated from a starter culture used in production of fermented milk product called "kiselo mleko" in 1905 by Bulgarian physician Stamen Grigorov who named it *Bacillus bulgaricus* [53]. *L. delbrueckii* subsp. *bulgaricus* was the strain present in the "sour milk" recommended by Metchniov as a means of "detoxifying" the colon and extending life. By the early decades of the 20th century, both *L. delbrueckii* subsp. *bulgaricus* and *L. acidophilus* were used in "sour milk" therapy for constipation, diarrhea, and other GI complaints, the first "modern" use of probiotics [54,55].

Genome sequence was achieved in 2006 [56] demonstrating gene loss consistent with its highly dairy adapted lifestyle. Renamed *L. delbrueckii* subsp. *bulgaricus* in 2014, strain is used with *S. thermophilus* as a starter culture in yogurt manufacture. Indeed, yogurt products are legally defined as containing these two strains (substitution

TABLE 6.1 Clinical Studies Involving *L. acidophilus* as a Major Component

Strains Used	Pathology	Design (References)
L. acidophilus NCFM	Pediatric fever colds, lost school days	Double blind, controlled [21]
L. acidophilus NCFM (heat killed)	Adult dermatitis	Double blind, controlled [22]
L. acidophilus, B. bifidum, L. casei, L. salivarius	Pediatric dermatitis	Double blind, controlled [23]
L. acidophilus, B. bifidum, symbiotic in yogurt; *L. acidophilus, B. bifidum*	IBS	Double blind, controlled [24,25]
L. acidophilus, B. lactis, and symbiotic	GI symptoms in hemodialysis patients	Double blind, controlled [27]
L. acidophilus and synbiotic	Radiation-induced diarrhea	Double blind, controlled [28]
L. acidophilus, B. lactis, L. rhamnosus GG in yogurt	Pediatric antibiotic associated diarrhea	Double blind, controlled [30]
L. acidophilus	Pediatric viral gastroenteritis	Double blind, controlled [33]
L. acidophilus, B. bifidum	Necrotizing enterocolitis	Double blind, controlled [40]
L. acidophilus, B. infantis	Necrotizing enterocolitis	Observational [42]
L. acidophilus LA02, *L. fermentum* as vaginal suppository	Vulvovaginal candidiasis	Uncontrolled clinical trial [44]
L. acidophilus	Hepatic encephalopathy	Double blind, controlled [49]
L. acidophilus, B. lactis in yogurt	Serum cholesterol and lipids	Double blind, controlled [51,52]

GI, Gastrointestinal; IBS, Irritable bowel syndrome.

of *bulgaricus* for another *Lactobacillus* species will earn a "yogurt-like product" designation). This fact complicates our discussion as vast numbers of probiotic trials using other candidate probiotic strains are delivered in a yogurt vehicle that necessarily includes *L. bulgaricus* and *S. thermophilus*, hampering isolation of specific probiotic strains and activities. The point may be moot, as certain combinations of microorganisms may have a synergistic interaction. This is seen in yogurt manufacture, where the two species work in tandem, with *L. delbrueckii* subsp. *bulgaricus* degrading milk proteins to generate amino acids, which are then used by *S. thermophilus*. Both species ferment lactic acid providing the yogurt with flavor and texture, and acting as a preservative. Multiple studies have shown traditional yogurt alone can mediate improved lactose digestion and absorption, so the food product itself may be considered to have probiotic activity [57]. Thus, any clinical probiotic study that utilizes yogurt as a vehicle for delivery of a selected probiotic agent will necessarily include this strain.

Contemporary Clinical Success

In a randomized, double-blinded placebo-controlled study of 135 elderly hospitalized patients receiving antibiotic therapy were treated with a yogurt drink (Actimel) containing *L. casei* DN-114 001 (*L. casei* immunitas) along with *L. delbrueckii* subsp. *bulgaricus*. reduced the incidence of antibiotic-associated and *C. difficile*-associated diarrhea [58]. *L. delbrueckii* subsp. *bulgaricus* 8481 was also used in a multicenter, double-blind, placebo-controlled of 61 elderly volunteers treated with daily with 3×10^7 in capsule form 3 times a day for 6 months. Evaluation of peripheral blood revealed increased percentage of NK cells, undifferentiated T-cell subsets, and decease in circulating IL-8. No actual clinical parameters were characterized [59].

In summary *L. delbrueckii* subsp. *bulgaricus* is only rarely used a single probiotic agent, but is a invariant component of all yogurts used in probiotics research. Thus, whether this strain possesses specific probiotic properties is unclear (Table 6.2).

TABLE 6.2 Clinical Studies Involving *L. delbrueckii* subsp. *bulgaricus* as a Major Component

Strains Used	Pathology	Design (References)
L. bulgaricus, L. casei DN-114 001 in yogurt	Antibiotic and *C. difficile*-associated diarrhea	Double blind, controlled [58]
L. bulgaricus 8481	Immunomodulation in the elderly	Double blind, controlled [59]

LACTOBACILLUS JOHNSONII

L. johnsonii is closely related to *L. acidophilus*, first recognized as distinct from *L. acidophilus* by biochemical and DNA-hybridization studies in 1992 [16]. The strain was originally designated *L. acidophilus* LA1, as part of the Nestec collection in Lausanne Switzerland. In vitro work showed this strain was notably more adherent to cultured human intestinal CaCo2 epithelial cells and could effectively compete with pathogens binding on these on cells, including enterotoxigenic *E. coli*, enteropathogenic *E. coli* (EPEC), and *Salmonella typhimurium* strains [60] and furthermore, showed antibacterial activity in a mouse model of Salmonellosis, a property attributed to a secreted factor present in conditioned media [61]. In human studies *L. johnsonii* La1 was shown to stably adhere to colon mucosa for several days postadministration, and resulted in alterations of local dendritic cell immunophenotypes [62]. Furthermore, the strain was show to modulate the human fecal microbiota while not stably colonizing [63]. Based on these in vitro and in vivo findings, the organism was utilized in a series of probiotic trials aiming to exploit these observed activities.

Contemporary Clinical Success

Control of Infectious Disease and Immunological Disorders

Based on proposed antimicrobial effects, a dietary product containing *L. johnsonii* La1 was utilized in a double-blinded controlled study of 326 asymptomatic children in Santiago, and was shown to mediate a statistically significant decrease in *H. pylori* carriage [64]. Another reported application includes modulation of allergic disorders. *L. johnsonii* EM1 in tandem with levocetirizine, was administered to 63 children to treat perennial allergic rhinitis, the tandem combination was more effective than levocetirizine alone [65]. However, in a negative result, *L. johnsonii* La1 was shown to have no effect on Crohn's disease postoperative recurrence [66,67].

Lactobacillus johnsonii-*Induced Influences on Dermatological Health*

More recently, *L. johnsonii* La1 has been explored for effects on ultraviolet induced skin damage. In 139 patients, 10 weeks of *L. johnsonii* La1 in combination with carotenoid supplementation reduced clinical and laboratory assessment of skin damages including reduction of loss of Langerhans cells [68]. Another group used a randomized, double-blind, placebo-controlled trial of 54 volunteers with simulated solar UV irradiation and demonstrated *L. johnsonii* La1 supplementation protected against UV-induced suppression of contact hypersensitivity, normalized of CD1a expression and increased in IL-10, thus showing

TABLE 6.3 Clinical Studies Involving *L. johnsonii* as a Major Component

Strains Used	Pathology	Design (References)
L. johnsonii La1	*H. pylori* carriage	Double blind, controlled [64]
L. johnsonii Em1	Allergic rhinitis	Double blind, controlled [65]
L. johnsonii La1 and dietary supplement	UV-induced skin damage	Double blind, controlled [68–71]

immunomodulatory capacity [69,70]. Similarly, a randomized, double-blind, placebo-controlled trial of the common photodermatosis, polymorphic light eruption, showed clinical efficacy with supplementation with *L. johnsonii* La1, lycopene, and beta carotene [71].

In summary, *L. johnsonii* has shown occasional used as a single agent probiotic, with several successful claims in the clinical contexts it has been deployed in. This strain was selected for use based on ability to influence proinflammatory signaling in in vitro cell culture based assays. Whether this approach to identify individual strains of lactobacilli with probiotic properties will be useful going forward is not established (Table 6.3).

SUMMARY

Lactobacilli of the acidophilus group have been utilized in human food products for centuries. While experiments in model systems implies that lactobacilli have taxa level properties that allows them to influence mammalian biology, the specific effects of individual strains remains elusive. Overall, a theme that emerges from recent reports is that the diversity of site of action (gut, vagina, systemic immunity, metabolism, etc.) and diversity of delivery and administration (whether via isolate organisms, food products, and heat killed microbes) all imply very different mechanisms of action. Clearly further research is needed to characterize the mechanism of action of this group of probiotics, indeed all probiotics, and clinically, standardization of formulations will be necessary to rationally assess efficacy of probiotics as single agents or synergistically acting consortia of beneficial bacteria.

REFERENCES

[1] Jones RM, Desai C, Darby TM, et al. Lactobacilli modulate epithelial cytoprotection through the Nrf2 pathway. Cell Rep 2015;12(8):1217–25.

[2] Jones RM, Luo L, Ardita CS, et al. Symbiotic lactobacilli stimulate gut epithelial proliferation via Nox-mediated generation of reactive oxygen species. EMBO J 2013;32(23):3017–28.

[3] Bienenstock J, Gibson G, Klaenhammer TR, et al. New insights into probiotic mechanisms: a harvest from functional and metagenomic studies. Gut Microbes 2013;4(2):94–100.

[4] Di Cerbo A, Palmieri B, Aponte M, et al. Mechanisms and therapeutic effectiveness of lactobacilli. J Clin Pathol 2016;69(3):187–203.

[5] Reid G, Sanders ME, Gaskins HR, et al. New scientific paradigms for probiotics and prebiotics. J Clin Gastroenterol 2003;37(2):105–18.

[6] Orla-Jensen S. The lactic acid bacteria. Mem Acad Royal Soc Denmark Ser 1919;5:81–197.

[7] Sun Z, Harris HM, McCann A, et al. Expanding the biotechnology potential of lactobacilli through comparative genomics of 213 strains and associated genera. Nat Commun 2015;6:8322.

[8] Felis GE, Dellaglio F. Taxonomy of Lactobacilli and Bifidobacteria. Curr Issues Intest Microbiol 2007;8(2):44–61.

[9] O'Callaghan J, O'Toole PW. *Lactobacillus*: host-microbe relationships. Curr Top Microbiol Immunol 2013;358:119–54.

[10] Reuter G. The *Lactobacillus* and *Bifidobacterium* microflora of the human intestine: composition and succession. Curr Issues Intest Microbiol 2001;2(2):43–53.

[11] Storelli G, Defaye A, Erkosar B, et al. *Lactobacillus plantarum* promotes Drosophila systemic growth by modulating hormonal signals through TOR-dependent nutrient sensing. Cell Metab 2011;14(3):403–14.

[12] Metchnikoff E. In: Chalmers P, Mitchell GP, editors. The prolongation of life. New York, London: Putnam's Sons; 1908.

[13] Floch MH, Walker WA, Sanders ME, et al. Recommendations for probiotic use—2015 update: proceedings and consensus opinion. J Clin Gastroenterol 2015;49(Suppl. 1):S69–73.

[14] Moro E. Ueber den *Bacillus acidophilus*. Jahrb Kinderh 1900;52: 38–55.

[15] Hanson PA, Mocquot G. *Lactobacillus acidophilus* (Moro) comb. nov. Int J Syst Bacteriol 1970;20:325–7.

[16] Fujisawa T, Benno Y, Yaeshima T, et al. Taxonomic study of the *Lactobacillus acidophilus* group, with recognition of *Lactobacillus gallinarum* sp. nov. and *Lactobacillus johnsonii* sp. nov. and synonymy of *Lactobacillus acidophilus* group A3 (Johnson et al. 1980) with the type strain of *Lactobacillus amylovorus* (Nakamura 1981). Int J Syst Bacteriol 1992;42(3):487–91.

[17] Zhang ZG, Ye ZQ, Yu L, et al. Phylogenomic reconstruction of lactic acid bacteria: an update. BMC Evol Biol 2011;11:1.

[18] Altermann E, Russell WM, Azcarate-Peril MA, et al. Complete genome sequence of the probiotic lactic acid bacterium *Lactobacillus acidophilus* NCFM. Proc Nat Acad Sci USA 2005;102(11):3906–12.

[19] Anjum N, Maqsood S, Masud T, et al. *Lactobacillus acidophilus*: characterization of the species and application in food production. Crit Rev Food Sci Nutr 2014;54(9):1241–51.

[20] Hutt P, Koll P, Stsepetova J, et al. Safety and persistence of orally administered human *Lactobacillus* sp. strains in healthy adults. Benef Microbes 2011;2(1):79–90.

[21] Leyer GJ, Li S, Mubasher ME, et al. Probiotic effects on cold and influenza-like symptom incidence and duration in children. Pediatrics 2009;124(2):e172–9.

[22] Inoue Y, Kambara T, Murata N, et al. Effects of oral administration of *Lactobacillus acidophilus* L-92 on the symptoms and serum cytokines of atopic dermatitis in Japanese adults: a double-blind, randomized, clinical trial. Int Arch Allergy Immunol 2014;165(4):247–54.

[23] Yesilova Y, Calka O, Akdeniz N, et al. Effect of probiotics on the treatment of children with atopic dermatitis. Ann Dermatol 2012;24(2):189–93.

[24] Magro DO, de Oliveira LM, Bernasconi I, et al. Effect of yogurt containing polydextrose, *Lactobacillus acidophilus* NCFM and *Bifidobacterium lactis* HN019: a randomized, double-blind, controlled study in chronic constipation. Nutr J 2014;13:75.

[25] Ringel-Kulka T, Palsson OS, Maier D, et al. Probiotic bacteria Lactobacillus acidophilus NCFM and *Bifidobacterium lactis* Bi-07 versus placebo for the symptoms of bloating in patients with functional bowel disorders: a double-blind study. J Clin Gastroenterol 2011;45(6):518–25.

[26] Ringel-Kulka T, Goldsmith JR, Carroll IM, et al. *Lactobacillus acidophilus* NCFM affects colonic mucosal opioid receptor expression in patients with functional abdominal pain–a randomised clinical study. Aliment Pharmacol Ther 2014;40(2):200–7.

[27] Viramontes-Horner D, Marquez-Sandoval F, Martin-del-Campo F, et al. Effect of a symbiotic gel (*Lactobacillus acidophilus* + *Bifidobacterium lactis* + inulin) on presence and severity of gastrointestinal symptoms in hemodialysis patients. JRen Nutr 2015;25(3):284–91.

[28] Salminen E, Elomaa I, Minkkinen J, et al. Preservation of intestinal integrity during radiotherapy using live *Lactobacillus acidophilus* cultures. Clin Radiol 1988;39(4):435–7.

[29] Tulk HM, Blonski DC, Murch LA, et al. Daily consumption of a synbiotic yogurt decreases energy intake but does not improve gastrointestinal transit time: a double-blind, randomized, crossover study in healthy adults. Nutr J 2013;12:87.

[30] Fox MJ, Ahuja KD, Robertson IK, et al. Can probiotic yogurt prevent diarrhoea in children on antibiotics? A double-blind, randomised, placebo-controlled study. BMJ Open 2015;5(1):e006474.

[31] Gionchetti P, Rizzello F, Helwig U, et al. Prophylaxis of pouchitis onset with probiotic therapy: a double-blind, placebo-controlled trial. Gastroenterology 2003;124(5):1202–9.

[32] Gionchetti P, Rizzello F, Venturi A, et al. Oral bacteriotherapy as maintenance treatment in patients with chronic pouchitis: a double-blind, placebo-controlled trial. Gastroenterology 2000;119(2): 305–9.

[33] Lee do K, Park JE, Kim MJ, et al. Probiotic bacteria, *B. longum* and *L. acidophilus* inhibit infection by rotavirus in vitro and decrease the duration of diarrhea in pediatric patients. Clin Res Hepatol Gastroenterol 2015;39(2):237–44.

[34] Grandy G, Medina M, Soria R, et al. Probiotics in the treatment of acute rotavirus diarrhoea. A randomized, double-blind, controlled trial using two different probiotic preparations in Bolivian children. BMC Infect Dis 2010;10:253.

[35] Vandenplas Y, De Hert SG. group PR-s. Randomised clinical trial: the synbiotic food supplement Probiotical vs. placebo for acute gastroenteritis in children. Aliment Pharmacol Ther 2011;34(8): 862–7.

[36] Dinleyici EC, Dalgic N, Guven S, et al. The effect of a multispecies synbiotic mixture on the duration of diarrhea and length of hospital stay in children with acute diarrhea in Turkey: single blinded randomized study. Eur J Pediatr. 2013;172(4):459–64.

[37] Szajewska H, Ruszczynski M, Kolacek S. Meta-analysis shows limited evidence for using *Lactobacillus acidophilus* LB to treat acute gastroenteritis in children. Acta Paediatr 2014;103(3):249–55.

[38] Ouwehand AC, ten Bruggencate SJ, Schonewille AJ, et al. Lactobacillus acidophilus supplementation in human subjects and their resistance to enterotoxigenic *Escherichia coli* infection. Br J Nutr 2014;111(3):465–73.

[39] Allen SJ, Wareham K, Wang D, et al. Lactobacilli and bifidobacteria in the prevention of antibiotic-associated diarrhoea and

Clostridium difficile diarrhoea in older inpatients (PLACIDE): a randomised, double-blind, placebo-controlled, multicentre trial. Lancet 2013;382(9900):1249–57.

[40] Lin HC, Hsu CH, Chen HL, et al. Oral probiotics prevent necrotizing enterocolitis in very low birth weight preterm infants: a multicenter, randomized, controlled trial. Pediatrics 2008;122(4):693–700.

[41] Fernandez-Carrocera LA, Solis-Herrera A, Cabanillas-Ayon M, et al. Double-blind, randomised clinical assay to evaluate the efficacy of probiotics in preterm newborns weighing less than 1500 g in the prevention of necrotising enterocolitis. Arch Dis Child Fetal Neonatal Ed 2013;98(1):F5–9.

[42] Hartel C, Pagel J, Rupp J, et al. Prophylactic use of *Lactobacillus acidophilus/Bifidobacterium infantis* probiotics and outcome in very low birth weight infants. J Pediatr 2014;165(2):285-289.e1.

[43] De Alberti D, Russo R, Terruzzi F, et al. Lactobacilli vaginal colonisation after oral consumption of Respecta((R)) complex: a randomised controlled pilot study. Arch Gynecol Obstet 2015;292(4):861–7.

[44] Murina F, Graziottin A, Vicariotto F, et al. Can *Lactobacillus fermentum* LF10 and *Lactobacillus acidophilus* LA02 in a slow-release vaginal product be useful for prevention of recurrent vulvovaginal candidiasis? A clinical study. J Clin Gastroenterol 2014;48(Suppl. 1):S102–5.

[45] Homayouni A, Bastani P, Ziyadi S, et al. Effects of probiotics on the recurrence of bacterial vaginosis: a review. J Low Genit Tract Dis 2014;18(1):79–86.

[46] Mittal VV, Sharma BC, Sharma P, et al. A randomized controlled trial comparing lactulose, probiotics, and L-ornithine L-aspartate in treatment of minimal hepatic encephalopathy. Eur J Gastroenterol Hepatol 2011;23(8):725–32.

[47] Agrawal A, Sharma BC, Sharma P, et al. Secondary prophylaxis of hepatic encephalopathy in cirrhosis: an open-label, randomized controlled trial of lactulose, probiotics, and no therapy. Am J Gastroenterol 2012;107(7):1043–50.

[48] Loguercio C, Federico A, Tuccillo C, et al. Beneficial effects of a probiotic VSL#3 on parameters of liver dysfunction in chronic liver diseases. JClin Gastroenterol 2005;39(6):540–3.

[49] Ziada DH, Soliman HH, El Yamany SA, et al. Can *Lactobacillus acidophilus* improve minimal hepatic encephalopathy? A neurometabolite study using magnetic resonance spectroscopy. Arab J GastroenterolV 14 2013;(3):116–22.

[50] Kwak DS, Jun DW, Seo JG, et al. Short-term probiotic therapy alleviates small intestinal bacterial overgrowth, but does not improve intestinal permeability in chronic liver disease. Eur J Gastroenterol Hepatol 2014;26(12):1353–9.

[51] Ejtahed HS, Mohtadi-Nia J, Homayouni-Rad A, et al. Effect of probiotic yogurt containing *Lactobacillus acidophilus* and *Bifidobacterium lactis* on lipid profile in individuals with type 2 diabetes mellitus. J Dairy Sci 2011;94(7):3288–94.

[52] Ataie-Jafari A, Larijani B, Alavi Majd H, et al. Cholesterol-lowering effect of probiotic yogurt in comparison with ordinary yogurt in mildly to moderately hypercholesterolemic subjects. Ann Nutr Metabol 2009;54(1):22–7.

[53] Grigoroff S. Etude sur le lait fermenté comestible: le 'Kissélo-mléko' de Bulgarie. Revue Médicale de la Suisse Romande (in French), Genéve: Libraires-Éditeurs. Librairie de L'Université; 1905.

[54] Kulp WL, Rettger LF. Comparative Study of *Lactobacillus acidophilus* and *Lactobacillus bulgaricus*. J Bacteriol 1924;9(4):357–95.

[55] Cheplin HA, Rettger LF. Studies on the transformation of the intestinal flora, with special reference to the implantation of *Bacillus*

acidophilus: II. Feeding experiments on man. Proc Natl Acad Sci USAV 6 1920;(12):704–5.

[56] van de Guchte M, Penaud S, Grimaldi C, et al. The complete genome sequence of *Lactobacillus bulgaricus* reveals extensive and ongoing reductive evolution. Proc Nat Acad Sci USA 2006;103(24):9274–9.

[57] Guarner F, Perdigon G, Corthier G, et al. Should yoghurt cultures be considered probiotic? Br J Nutr 2005;93(6):783–6.

[58] Hickson M, D'Souza AL, Muthu N, et al. Use of probiotic *Lactobacillus* preparation to prevent diarrhoea associated with antibiotics: randomised double blind placebo controlled trial. BMJ 2007;335(7610):80.

[59] Moro-Garcia MA, Alonso-Arias R, Baltadjieva M, et al. Oral supplementation with *Lactobacillus delbrueckii* subsp. *bulgaricus* 8481 enhances systemic immunity in elderly subjects. Age 2013;35(4):1311–26.

[60] Bernet MF, Brassart D, Neeser JR, et al. *Lactobacillus acidophilus* LA 1 binds to cultured human intestinal cell lines and inhibits cell attachment and cell invasion by enterovirulent bacteria. Gut 1994;35(4):483–9.

[61] Bernet-Camard MF, Lievin V, Brassart D, et al. The human Lactobacillus acidophilus strain LA1 secretes a nonbacteriocin antibacterial substance(s) active in vitro and in vivo. Appl Environ Microbiol 1997;63(7):2747–53.

[62] Gianotti L, Morelli L, Galbiati F, et al. A randomized double-blind trial on perioperative administration of probiotics in colorectal cancer patients. World J Gastroenterol 2010;16(2):167–75.

[63] Garrido D, Suau A, Pochart P, et al. Modulation of the fecal microbiota by the intake of a *Lactobacillus johnsonii* La1-containing product in human volunteers. FEMS Microbiol Lett 2005;248(2):249–56.

[64] Cruchet S, Obregon MC, Salazar G, et al. Effect of the ingestion of a dietary product containing *Lactobacillus johnsonii* La1 on *Helicobacter pylori* colonization in children. Nutrition 2003;19(9):716–21.

[65] Lue KH, Sun HL, Lu KH, et al. A trial of adding *Lactobacillus johnsonii* EM1 to levocetirizine for treatment of perennial allergic rhinitis in children aged 7–12 years. Int J Pediatr Otorhinolaryngol 2012;76(7):994–1001.

[66] Marteau P, Lemann M, Seksik P, et al. Ineffectiveness of *Lactobacillus johnsonii* LA1 for prophylaxis of postoperative recurrence in Crohn's disease: a randomised, double blind, placebo controlled GETAID trial. Gut 2006;55(6):842–7.

[67] Van Gossum A, Dewit O, Louis E, et al. Multicenter randomized-controlled clinical trial of probiotics (*Lactobacillus johnsonii*, LA1) on early endoscopic recurrence of Crohn's disease after Ileo-caecal resection. Inflamm Bowel Dis 2007;13(2):135–42.

[68] Bouilly-Gauthier D, Jeannes C, Maubert Y, et al. Clinical evidence of benefits of a dietary supplement containing probiotic and carotenoids on ultraviolet-induced skin damage. Br J Dermatol 2010;163(3):536–43.

[69] Peguet-Navarro J, Dezutter-Dambuyant C, Buetler T, et al. Supplementation with oral probiotic bacteria protects human cutaneous immune homeostasis after UV exposure-double blind, randomized, placebo controlled clinical trial. Eur J Dermatol 2008;18(5):504–11.

[70] Gueniche A, Benyacoub J, Buetler TM, et al. Supplementation with oral probiotic bacteria maintains cutaneous immune homeostasis after UV exposure. Eur J Dermatol 2006;16(5):511–7.

[71] Marini A, Jaenicke T, Grether-Beck S, et al. Prevention of polymorphic light eruption by oral administration of a nutritional supplement containing lycopene, beta-carotene, and *Lactobacillus johnsonii*: results from a randomized, placebo-controlled, double-blinded study. Photodermatol Photoimmunol Photomed 2014;30(4):189–94.

Chapter 7

Lactobacillus rhamnosus GG

S. Gorbach*,, S. Doron†,‡ and F. Magro†,‡**
*Department of Public Health & Medicine, Tufts University School of Medicine, Boston, MA, United States; **Division of Infectious Diseases, Tufts Medical Center, Boston, MA, United States; †Department of Medicine, Tufts University School of Medicine, Boston, MA, United States; ‡Division of Geographic Medicine and Infectious Diseases, Tufts Medical Center, Boston, MA, United States

HISTORY

Lactobacillus rhamnosus strain GG (LGG), ATCC 53103 was originally isolated from fecal samples of a healthy human adult by Sherwood Gorbach and Barry Goldin; the GG is derived from their names [1]. It was discovered in 1985 as part of an attempt to isolate a strain of lactobacillus which would meet the following necessary characteristics for an ideal probiotic: resistance to stomach acid and bile so it can survive transit to the lower gastrointestinal tract, ability to consistently implant on human intestinal epithelial cells (IECs) and colonize the intestine, production of an antimicrobial substance, rapid growth rate, and beneficial effects on health [2]. It is one of the most widely studied probiotic strains and its effects on human health have been examined in numerous clinical trials [1].

BACTERIOLOGY

LGG is a Gram-positive rod with a distinct buttery odor when cultured and its morphology is that of a creamy white colony. On Gram stain, it forms a palisading structure unlike other lactobacilli, which helps in identifying it. Chemical testing reveals that it ferments cellobiose, fructose, glucose, mannitol, mannose, melezitose, rhamnose, ribose, saliciin, sorbitol, trehalose, and xylose. It does not ferment the sugars lactose, maltose, or sucrose, nor will it ferment amygdalin, arabinose, erythritol, glycogen, inositol, melibiose, or affinose.

LGG produces a compound, also known as a bacteriocin, with antimicrobial activity against anaerobic bacteria, such as Clostridium, Bacteroides, and Bifidobacterium, as well as *Escherichia coli*, Pseudomonas, Staphylococcus, Streptococcus, and Salmonella [3]. The genome sequence of LGG encodes several bacteriocin-related genes [4] but it has been challenging to induce the production of these compounds in vitro. Not all of the various secreted

antimicrobials made by LGG have been definitively identified, though lactic acid is a major antimicrobial substance made by all lactobacilli. Lactic acid may facilitate the activity of other compounds with antimicrobial action [5]. In one study, seven heat-stable peptides with antibacterial activity against enteroaggregative *E. coli* strain EAEC 042, *Salmonella typhi*, and *Staphylococcus aureus* were identified in LGG culture medium [6], but were not further characterized. It has been shown elsewhere that LGG is able to greatly inhibit the growth of *Salmonella typhimurium* specifically. A survey of the antimicrobial compounds produced by *L. rhamnosus* GG in spent culture supernatant of the organism grown in MRS medium found that it contained five important compounds, but based on different experimental approaches, lactic acid was shown to be the important chemical which inhibited growth of *S. typhimurium*. This compound not only chelates important growth factors, such as iron from the nutritional environment, but also increases permeability of Gram-negative membranes and specifically has effects on salmonella virulence gene hilA expression [7]. Lactic acid comes in two stereoisomers, D-lactic acid and L-lactic acid, and different strains of lactobacilli create these lactic acid isomers in idiosyncratic ratios. No strain creates 100% of one isomer alone but there is wide variability in ratio of D- versus L-forms across the lactobacilli. *L. rhamnosus* species notably creates L-lactic acid almost exclusively, but does form a minority in D-lactic acid form (94.2 and 5.8%), whereas other species have the opposite pattern [8] (Fig. 7.1).

ANTIMICROBIAL SUSCEPTIBILITY

L. rhamnosus GG is susceptible to many commonly used antimicrobials, including penicillin, ampicillin, imipenem, and erythromycin. However, when an individual receives concurrent treatment with these antibiotics while being given LGG, the organism still retains the ability to colonize the

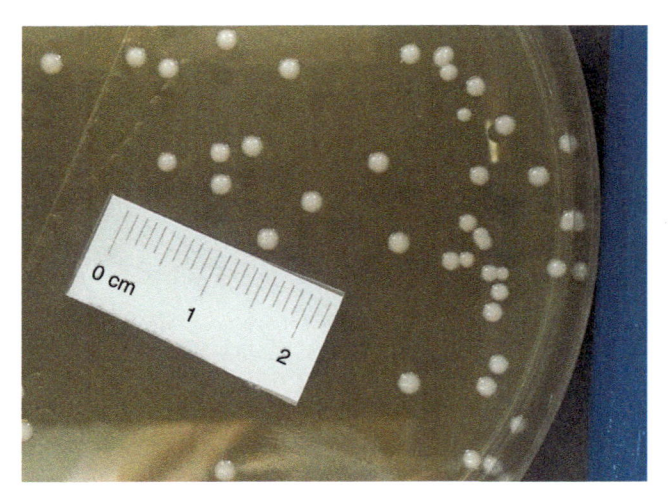

FIGURE 7.1 **Appearance of *L. rhamnosus* GG colonies growing on a plate.**

gut [9]. A more recent study looked at clinical isolates of numerous lactobacillus species including LGG causing bacteremia and found important interspecies differences. All *Lactobacillus* isolates here demonstrated low minimum inhibitory concentrations to imipenem, piperacillin–tazobactam, erythromycin, and clindamycin via E test. In the study, 22 of the 85 isolates were identified as *L. rhamnosus* GG. LGG specifically had reliable susceptibility to ciprofloxacin as well. Conversely, cephalosporin susceptibility varied widely between species, with cefuroxime demonstrating the highest activity and ceftriaxone the lowest, and this remained true in LGG. Minimum inhibitory concentrations of vancomycin were high ($>256\ \mu g/mL$) for most species including LGG, as the organism is, importantly, inherently resistant to glycopeptide antibiotics. Disk diffusion testing correlated well with E testing [10].

MOLECULAR BASIS OF LGG–HOST INTERACTIONS

LGG is known for its adhesive capacity. When LGG binds to enteric epithelium, the adhesion of other pathogens is inhibited. This process is called competitive colonization and is one of the primary factors underlying the effectiveness of LGG in preventing intestinal infection [2]. It is active at adhering to mucus compared to other related strains and this is largely due to its pili. Pili or fimbriae are long and thin proteinaceous protrusions of the cell surface present on specific Gram-positive and Gram-negative bacteria [11]. Importantly, pili are susceptible to shearing. Cells subjected to $8000g$ centrifugal forces are completely devoid of pili. Since much of the probiotic effect of LGG may depend on its unique pili and adhesion abilities, this vulnerability could be important to avoid during industrial production of the probiotic to retain a functional product [12]. There are genes for three secreted LPXTG-like pilins (*spaCBA*). Structurally,

LGG pilus contains three distinct pilin monomers which are covalently linked, the major pilin SpaA and the minor pilins SpaB and SpaC. Mass spectrometry revealed a SpaA/SpaC/SpaB ratio of 5:2:1. SpaA exerts merely a structural function forming the pilus backbone, while the accessory pilins play a functional role [13]. The larger minor pilin SpaC, located on the tip of the pilus and covering the pilus length is thought to play a pivotal role in adhesion to mucus; the smaller minor pilin SpaB is thought to act as a molecular switch responsible for pilus termination and initiation of peptidoglycan binding [1,13]. Adherence of lactobacillus strain GG to human intestinal mucus was blocked by SpaC antiserum and abolished in a mutant carrying an inactivated SpaC gene. Similarly, binding to mucus was demonstrated for the purified SpaC protein, suggesting that SpaC is essential for the mucus interaction of *L. rhamnosus* GG and likely explains its ability to persist in the human intestinal tract longer than LC705, a similar strain, during an intervention trial [4,11].

Using single-molecule atomic force microscopy the binding mechanism of LGG was further elucidated to show broad binding ability of spaC. Although it binds primarily mucin and collagen, which are present throughout host cell epithelial layers, it also shows good "self-adhesion" which allows for aggregation of multiple lactobacilli in the form of homophilic *trans*-interactions. Homophilic and heterophilic interactions were of similar binding strengths. In pulling experiments on live bacteria, there was a zipper-like adhesion function of the numerous SpaC proteins distributed along the pilus, as well as nanospring properties which allowed the pili to resist high forces [12,14].

Potentially owing to its excellent binding abilities, the organism can be cultured in saliva 2 weeks after ingestion [15], cultured from stool for 7 days [9], and cultured from intestinal biopsy specimens for 28 days [16]. Of note, the colonization capacity of the organism is significantly better in newborns, who also have a less established microbiome. In addition to colonizing the small intestine preferentially, it also adheres well to the colon and can also be recovered from the oral cavity, tonsils, and vagina [1].

Exopolysaccharides are another adaptation factor in LGG. A layer of extracellular polysaccharide is often found in lactobacilli and its components are structurally diverse [1]. The surface extracellular polysaccharide layer of LGG has two primary polysaccharides, a long galactose-rich type and a short glucose- and mannose-rich type. Interestingly, genetically knocking out the former type causes the organism to adhere better to intestinal epithelial cells (IECs) at the expense of less defense against complement-mediated lysis and cathelicidins, specific cationic antimicrobial peptides [17].

LGG promotes the survival of IECs by preventing cytokine-induced apoptosis through blocking of p38 MAP kinase. The effect is also present in other probiotic strains, such as *Lactobacillus acidophilus* ATCC 393 and

Lactobacillus casei ATCC 4356, but LGG demonstrates this trait most strongly [18]. The supernatant of LGG was shown to induce heat shock proteins [19]. Consequently, two proteins from the LGG supernatants were found to cause the antiapoptic effect [18] and were named major secreted protein Msp1 and Msp2. Each activates the Akt signaling peptide, inhibits cytokine-induced IEC apoptosis, and reduces tumor necrosis factor (TNF)-induced epithelial damage. The proteins also promoted cell growth in human and mouse colon IECs and cultured mouse colon explants [20]. Moreover, they were shown to protect the intestinal epithelial barrier function from hydrogen peroxide–induced damage through blocking of MAP kinases [21].

Lactobacillus GG also competes for gut monosaccharides and thus slows the growth of other potential pathogens in its vicinity. This has particular relevance to *Clostridium difficile*, which is an infection precipitated by depletion of normal colonic flora. One of the reasons for this phenomenon is that *C. difficile* is not as efficient in utilizing nutrients; therefore presence of other flora normally suppresses its growth by outcompeting it. For example, monosaccharides *N*-acetylglucosamine and *N*-acetylneuraminic acid do not occur in nature as free monosaccharides but are both found in abundance in the side chains of gastrointestinal mucin. *C. difficile* does not contain the necessary enzymes to capture these monosaccharides from their oligosaccharide side chains and so *C. difficile* must compete directly for free monosaccharides, but it does so poorly compared to other colonic microorganisms [22].

GENE EXPRESSION PATTERN IN SMALL BOWEL

The pattern of gene expression in the small bowel was found to be affected in complex ways by the presence of LGG and the genes altered are involved in widely different functions. A small randomized study of duodenal biopsies in six patients with esophagitis treated with a PPI and either LGG or placebo for 1 month revealed that LGG was associated with up- and downregulation of 334 and 92 genes, respectively. Affected genes included those, such as TGF-beta, TNF family members, cytokines, nitric oxide synthase 1, and defensin alpha 1 which are involved in inflammation and immune response; genes related to cell growth, differentiation, and apoptosis, such as cyclins, caspases, and oncogenes; cell–cell signaling ICAMs and integrins; cell adhesion cadherins; and genes for signal transcription and transduction [23].

The impact of LGG on the structure and functional dynamics (gene expression) of the gut microbiota in a study of 12 individuals, 65–80 years old, was undertaken using volunteers recruited from the Boston area for an open-label clinical trial to assess the safety and tolerability of LGG. The major finding was that though overall organism composition was stable (challenging the notion that probiotics must significantly alter the gut microbiota), the transcriptional response of the organisms present was reliably modulated by LGG treatment. The effects noted across all 12 individuals was similar, and had to do with an increased expression of genes related to chemotaxis, flagellar motility, and adhesion from *Bifidobacterium* and butyrate producers *Roseburia* and *Eubacterium* [24]. Of note, the short-chain fatty acid butyrate is a beneficial compound. As the main energy source for colonocytes, it is involved in cellular apoptosis and NF-κB signaling that lend it anticancer and antiinflammatory effects [25,26] while also decreasing epithelial permeability [24]. Therefore, LGG may exert a functional effect and benefit by promoting the interaction of other organisms with host epithelium. The study also found a high correlation between abundance of LGG genome recovery (absolute amount of LGG present in stool) and transcript abundance, suggesting that the relative amount of LGG present in the gut might play a strong role in transcriptional activity ($r = 0.73$, $P = 0.0074$). Additionally, the transcript coverage of the LGG genome was limited to the time period that LGG was being ingested and then subsided after the probiotic was no longer part of the subjects' intake, supporting evidence that the effect of LGG may unfortunately be short-lived if not continuously supplied in the diet [24].

Of note, others have suggested pili as a reason for LGG not seeming to alter overall microbiota populations in its favor. Since the pili of LGG allow firm attachment to intestinal mucosa, it is possible that the majority of microbial abundance change could be mucosal only, and the mucosal surface has different populations than stool, so an abundance of LGG may not be seen in stool microbiota after supplementation [27].

The effect of LGG on the microbiota in cystic fibrosis has also been studied. Given that intestinal inflammation is a feature of cystic fibrosis, it was posited that reduction in intestinal inflammation by LGG may benefit cystic fibrosis patients, potentially even from the standpoint of pulmonary exacerbations. In one trial that sought to compare the microbiota of CF versus normal hosts and markers of inflammation, LGG was shown to improve diversity of intestinal microbiota in 22 children, aged 2–9, who received LGG or placebo for 1 month. The authors argue that intestinal flora in children with CF is altered because of numerous factors including cystic fibrosis transmembrane conductance regulator malfunction, frequent antibiotics, pancreatic enzyme supplementation, and pharmacological suppression of gastric acid. There was a correlation between reduced microbial richness and intestinal inflammation, as measured by fecal calprotectin and rectal nitric oxide markers, and CF children were found to have less microbial diversity and the levels of the organisms *Eubacterium rectale*, *Bacteroides uniformis*, *Bacteroides vulgatus*, *Bifidobacterium adolescentis*, *Bifidobacterium catenulatum*, and *Faecalibacterium prausnitzii* were all reduced compared to non-CF controls [28].

POTENTIAL SAFETY CONCERNS

The Agency for Healthcare Research and Quality (AHRQ) released an NIH-sponsored report in 2011 reviewing the safety of probiotics in general, and lactobacillus was one of the six organisms studied for the report which included data from 622 studies. Of note, LGG was not the specific focus of the report, but the overall conclusion was that there has not been any evidence that probiotics in clinical trial settings have led to any specific increased risks, however, the most important statement in their conclusion was, "the current literature is not well equipped to answer questions on the safety of probiotics in intervention studies with confidence" [29].

The World Health Organization (WHO) and Food and Agriculture Organization (FAO) issued a joint report in 2002 that probiotics may theoretically be responsible for four types of side effects, namely, systemic infections, deleterious metabolic activities, excessive immune stimulation in susceptible individuals, and gene transfer [30]. Regarding the possibility of the first type of effect, systemic infection, the best data regarding LGG come from Finland, where there has been widespread use of the organism as a probiotic. There is Finnish surveillance data from 1990 to 2000 during which there was increasing popularity and use of LGG (from 1 L per person per year in Finland to 6 L per person per year over this time period) however, there was no concomitant increase in incidence of lactobacillus bacteremias during this time. Of all positive blood cultures, lactobacillus represented 0.02%. Out of the 89 strains isolated from blood, 11 appeared identical to LGG by pulse field gel electrophoresis, however, phenotypically, these were different from LGG during pathogenicity testing, specifically regarding in vitro adhesion rates and induction of respiratory burst [31,32].

The second potential side effect noted by WHO/FAO is deleterious metabolic activities. One example of an effect in this category is the potential for adverse events from D-lactate being produced by lactobacilli. There are case reports of D-lactic acidosis in the literature, usually associated with short gut syndromes. The essential process is carbohydrate malabsorption in the small intestine which then leads to a lower pH favoring overgrowth of D-lactic acid–producing bacteria. Acidosis is induced by absorbed D-lactic acid in the small intestine. These risks are not specific to LGG, however, as the condition could theoretically be caused by any D-lactate producing organism either already present in the gut or introduced by probiotic supplementation, but does introduce a potential class of higher risk patients for whom risks and benefits of probiotic use should be carefully weighed. Of note, *L. rhamnosus* has been shown to produce largely only the L-lactate form as noted previously, which is metabolized differently than D-lactate and would therefore not be expected to cause lactic acidosis to the same extent as other strains of lactobacilli [8]. LGG has

not been specifically linked to an increased incidence of the syndrome of metabolic acidosis due to lactic acid [33–35].

Excessive immune stimulation especially in susceptible individuals, such as those with autoimmune diseases is another theoretical side effect of probiotics given cytokine modulation and effects on dendritic cells which are the chief antigen presenting cells within the gut. However, LGG has not been shown to result in autoimmune phenomena thus far. In fact, unlike pathogenic organisms, lactobacilli including LGG have been shown to have a much lower inflammatory response than typical pathogen species, and LGG's interaction with dendritic cells results in only a moderate upregulation of costimulatory molecules and a very low production of TNF-alpha and CCL20 [36].

Gene transfer is another concern with use of lactobacilli as probiotics, as lactic acid bacteria have plasmids with genes that confer resistance to multiple antibiotic classes. As LGG and resistant organisms such as vancomycin-resistant Enterococcus (VRE) colonize the gut simultaneously, transfer of the Van A or Van B resistance genes between VRE and LGG has been raised as a concern. However, there has not been any detection of vancomycin resistance genes, such as vanA or vanB in lactobacilli. While many lactobacilli, including LGG, are resistant to vancomycin inherently, this is not mediated by the same mechanism as in VRE. One study using PCR-based methods with LGG was unable to find amplification using numerous vanA primers, nor did enterococcal vanA, vanB, vanH, vanX, vanZ, vanY, vanS, and vanR genes hybridize with DNA of LGG [37]. Gene transfer thus is more of a theoretical safety concern and there has not been suggestion of this occurring with LGG specifically [30].

A phase I study evaluated safety and tolerability of LGG compared to placebo in cirrhotic patients with minimal hepatic encephalopathy, a population particularly susceptible to perturbed gut flora and whose treatment often relies on intentional gut flora manipulation. Thirty patients completed the study and during follow-up period of 8 weeks, though self-limited diarrhea was more frequent in LGG patients, there was no difference in adverse events and no change in cognition scores [27].

CLINICAL USES OF LGG

Antibiotic-Associated Diarrhea

Antibiotic-associated diarrhea is widespread, occurring with a frequency between 11% and 30% in children on oral antibiotics [38,39]. Numerous clinical trials have looked at the effectiveness of LGG for preventing antibiotic-associated diarrhea [40–49]. Almost all of them have shown benefit over placebo [50].

A 2015 multisite, randomized, double-blind, placebo-controlled clinical trial studied 72 children up to 12 years old who were on antibiotics, and received either a probiotic yogurt [containing a combination of LGG, *Bifidobacterium lactis* (Bb-12), and *L. acidophilus* (La-5)] versus pasteurized yogurt without probiotics for the duration of receiving antibiotics. The two groups were overall similar in terms of baseline characteristics and class of antibiotics being used. The most common reason for antibiotics in these children was for ear infection in 53% of probiotic group and 56% of placebo recipients. Throat infection was next most common, at 26 and 22%, respectively. Class of antibiotics used was most often beta lactams (89 and 94%) followed by macrolides (12 and 3%).There were six episodes of severe diarrhea in the placebo group and none in the probiotic group (Fisher's exact $P = 0.025$). There were 21 episodes of mild diarrhea in the placebo group and none in the probiotic group (Fisher's exact $P < 0.001$). The study was unable to say which component of the three-strain probiotic mixture was responsible for the effect or if the combination was required [48].

Given that the effects of probiotics are strain specific, the effects should ideally be evaluated separately. A meta-analysis in 2015 looked at the use of *L. rhamnosus* GG alone in preventing antibiotic-associated diarrhea in children and adults. Twelve RCTs including 1499 patients were included. The overall conclusion was that LGG versus either no treatment or placebo reduced the risk of antibiotic-associated diarrhea from 22.4% to 12.3% [RR: 0.49, 95% confidence interval (CI): 0.29–0.83] in patients treated with antibiotics for any reason. The number needed to treat to avoid one case of antibiotic associated diarrhea was 9. The quality of evidence was only moderate at best, however, according to this assessment [49]. The analysis also evaluated the effect of LGG on risk of *C. difficile*–associated diarrhea (CDAD) as part of a secondary outcome, and four trials ($n = 891$) were included in this evaluation. The pooled results revealed no significant difference in *C. difficile* risk between LGG and control groups (RR 0.75, 95% CI: 0.31–1.84). Overall, LGG was well tolerated without dissimilar adverse event rates compared to the control groups. In the included trials there was significant heterogeneity and also a variable daily dose of probiotic, follow-up timeframe, and definitions of antibiotic-associated diarrhea, which are limitations [49].

The authors of the previously mentioned metaanalysis point out that the existing literature does not elucidate what the best regimen is for prevention of antibiotic-associated diarrhea. It is unclear whether the probiotic should be started before, or at the same time as antibiotics, though it seems reasonable to start before gut microbiota modification occurs as a result of antibiotics. The ideal duration of probiotic is also poorly understood, whether it should be continued as long as antibiotics or for some longer duration. Additionally, the optimal dose of daily probiotic is unknown, though at least in children, the highest dose studied

(1–2×10^{10} CFU) seems most effective [41], and in another study of children, LGG dose at 10^{10} CFU was as effective as 10^{12} CFU when treating diarrhea for other reasons, such as rotavirus [51]. Thus far adults do not seem to have a noticeable effect size correlation with dose [49].

Infectious Diarrhea

A trial evaluating LGG in Indian children admitted for acute watery diarrhea (due to rotavirus in 57%) compared different doses of LGG and found that the lowest dose used in the study was as effective as higher doses (10^{10} vs. 10^{12} CFU twice daily until diarrhea stopped or for 7 days) [51]. This dose reduced frequency and duration of diarrhea and also reduced hospital length of stay. Frequency of diarrhea became significantly reduced compared to controls starting on the fourth day; duration of diarrhea was reduced by 2 days. Additionally, the duration of hospital stay was reduced by 3 days on average. This study reported a more profound impact of their intervention (LGG) than prior probiotic trials for childhood diarrhea have, though metaanalyses of prior trials included many other types of probiotics in addition to LGG [51–53].

The effect of LGG in Indian children with cryptosporidial or rotavirus diarrhea was studied in another setting. Of 124 children enrolled, 82 and 42 had rotavirus and cryptosporidial diarrhea, respectively. Fewer children with rotavirus diarrhea on LGG had repeated diarrheal episodes (25% vs. 46%; $P = 0.048$) and impaired intestinal function (48% vs. 72%; $P = 0.027$). Among children with rotavirus or cryptosporidial diarrhea, those receiving LGG showed significant improvement in intestinal permeability as measured by a lactulose to mannitol assay. Children in the LGG group had impaired function compared to those in the placebo group (37/57 [65%]; $P = 0.046$) [54].

LGG has been used to prevent recurrent CDAD in a few small trials. No randomized trial has been completed, though a preliminary report of a controlled trial suggested a lower 3-week CDAD relapse rate in subjects given LGG [55]. Prior uncontrolled studies suggested a benefit from LGG in recurrent CDAD as well [56–58], but each of these series included only a handful of patients.

Obesity

In the last 15 years, the influence of gut microbiota on energy efficiency in the host and potential link to obesity [59,60] has become a topic of great interest. Given that the gut microbiome has an important role in maintaining the equilibrium in energy metabolism, and alterations in the microbiome may alter how the body handles energy [61–64], it is plausible that this dynamic interplay may be favorably manipulated with the right probiotics in a way that impacts obesity-related factors. LGG has been used to alter the microbiome of the rodent small intestine and was associated with a potential antiobesity effect [65]. In this particular study LGG and *Lactobacillus sakei* NR28 (from kimchi), alone and in combination, were fed to the animals daily for 3 weeks. There was a consequent significant reduction of fat mass (epididymal) as well as decrease in biomarkers associated with obesity [66,67] including fatty acid synthase, acetyl-CoA carboxylase, and stearoyl-CoA desaturase-1 in the liver in groups receiving either probiotic compared to control. Importantly, the probiotics did not induce weight loss in this study [65].

Perinatal use of LGG has been evaluated for impact on childhood growth patterns including overweight development during 10-year follow up. Luoto et al. theorized that modifying the gut microbial environment early in life with probiotics (by manipulating the mother's microbiome, from which the infant's microbiome would be seeded) would reduce the risk of overweight development in the child's early life. In a double-blinded fashion, 159 women were randomized to receive probiotics or placebo for 4 weeks before delivery and to continue the intervention for 6 months postnatally. Children were followed up at ages 3, 6, 12, and 24 months, as well as 4, 7, and 10 years (capturing 72% of children). Based on anthropometric measurements, excessive weight gain during the first 2 years of life was less in the LGG group, though weight gain in later years was not different between groups. Those who had been overweight in early life continued to be so at every measurement point during follow up. The authors speculated that to affect long term weight gain, modifying the flora had to happen early in life before growth patterns were determined [68].

Respiratory Benefits

With regard to respiratory benefits, a randomized, double-blind, placebo-controlled trial in 281 children who attend day care centers showed that LGG consumption (in the form of milk with LGG added to it at a dose of 10^9 colony-forming units in 100 mL of fermented milk) during a 3-month intervention period reduced the number of upper respiratory infections (but not lower respiratory infections) compared to placebo [69].

A prospective, randomized, blinded, controlled trial in 146 critically ill adults at high risk for ventilator-associated pneumonia (VAP) based on anticipated mechanical ventilation requirements for 72 h or more, showed a benefit in reduction of incidence of VAP based on both clinical and microbiological definitions. Out of patients given LGG, 19.1% developed microbiologically confirmed VAP compared to 40.0% of placebo patients ($P = 0.007$). Importantly, this study assessed for measurable changes in pathogenic flora by sampling oropharyngeal swabs, gastric aspirates, and non-bronchoscopic bronchoalveolar lavages prior to study entry and at 72 h, as well as after a VAP diagnosis. At baseline both groups had similar oral colonization with pathogenic species (41.4% in the placebo group vs. 42.6% in the LGG group, $P = 0.88$) and gastric colonization (31.4% vs. 32.3%, $P = 0.49$); but by 72 h the placebo patients had significantly higher rates of oral pathogens (70.0% in placebo group vs. 38.2% in LGG group, $P < 0.001$) and gastric pathogens (45.7% vs. 32.3%, $P = 0.03$) and the changes in oral colonization were significantly correlated with going on to develop microbiologically confirmed VAP (Pearson correlation coefficient, 0.22; $P = 0.009$). The estimated number needed to treat with LGG to prevent one case of VAP in this highly selected ICU population is approximately 5. The authors note, however, that the study has limited generalizability to ICU patients as a whole because of numerous exclusion criteria. This study excluded 95% of intubated patients (since, e.g., a patient intubated for less than 48 h cannot by definition develop a VAP) and the long enrollment period in order to attain the necessary sample size [70].

Immune Regulation

LGG has also been studied as an adjuvant for live-attenuated flu vaccination. This was based on previous findings that probiotics can improve the immune response to vaccines delivered mucosally. It had been shown that mice given Lactobacillus before an influenza virus challenge had higher levels of influenza specific IgG and protection against illness [71]. Therefore, 42 healthy adults were randomized during the 2007–08 influenza season to a proof-of-concept study to receive the live-attenuated flu vaccine plus either LGG or placebo twice daily for 28 days starting immediately after LAIV administration. Thirty-nine subjects completed the per-protocol analysis for the proof-of-concept investigation, and the intervention was well tolerated. Antibody titers to each of the vaccine components was measured by the serum hemagglutinin inhibition assay. Of note, protection rates against the vaccine H1N1 and B strains were suboptimal in subjects receiving LGG and placebo (overall, 15% seroconverted for the H1N1 strain and 41% for the B strain). For the H3N2 strain, 84% receiving LGG versus 55% receiving placebo had a protective titer 28 days after vaccination (odds of having a protective titer was 1.84, 95% CI: 1.04–3.22, $P = 0.048$). The researchers posited that it might be necessary to administer probiotics ahead of time rather than concurrently with the vaccine, to achieve better protection, and that future studies may show greater benefit in groups, such as the elderly who might stand to gain more from an adjuvant to boost immune response [72].

A mouse model of asthma has also shown benefit from oral LGG. The probiotic inhibited allergen-induced pulmonary inflammation and airway hyperreactivity to methacholine. This animal model was sensitized with ovalbumin intraperitoneally and received intranasal challenge with ovalbumin. Thereafter, airway inflammation and hyperresponsiveness were determined by bronchoalveolar lavage analysis with special stains and enzyme-linked immunosorbent assay. The lungs were also assessed for histopathologic changes with additional stains. In addition, various inflammatory markers and cytokines related to asthma and allergy were measured in serum. LGG suppressed hyperresponsiveness to methacholine and decreased infiltrating inflammatory cells and Th2 cytokines in bronchoalveolar lavage fluid and serum. It also reduced ovalbumin-specific IgE levels in serum. There was also a decrease in matrix metalloproteinase 9 expression in lung tissue in the LGG group. The findings support the potential role of LGG as a therapy for asthma and allergic response in general [73].

Lactobacilli lower the colonic pH which favors less pathogenic organisms by production of short chain fatty acids [2]. Modulation of antigen transport in the gut by LGG may prevent food allergies in children. LGG was shown to close intestinal pores in rats given LGG. As intestinal inflammation allows antigens to cross the intestinal mucosal barrier and sensitizes the immune system, restoring the integrity of this barrier may protect at-risk infants from developing sensitization to antigens which would lead to food allergy [74].

Atopic Diseases

The ability of LGG to reduce the incidence of atopic disease in infants has been studied with varying results. Some studies show no reduction, however, a randomized placebo-controlled trial in at-risk children during the first 2 years of life showed a reduced incidence of atopy in LGG users (the incidence was halved). The LGG was taken daily by pregnant women for 4 weeks before expected delivery. After delivery, it was taken postnatally for 6 months, either by the mother during breast-feeding or by the infant. The same authors also followed up with the children at 4 years and documented that the benefit appeared to be maintained beyond infancy. Atopic disease at the age of 4 years was based on clinical exam and a questionnaire and was found in 14 of 53 children who received LGG compared to 25 of 54 receiving placebo (relative risk 0.57, 95% CI: 0.33–0.97) [75].

The management of atopic eczema was also studied in 27 infants, mean age 4.6 months, who manifested atopic eczema during exclusive breast-feeding, and they were randomized to receive formula either enriched with or without probiotics, with LGG as one of two possible probiotic formulas. The severity of eczema as well as levels of cytokines and soluble cell surface adhesion molecules in serum and methyl-histamine and eosinophilic protein X in urine were determined. Within 2 months, there was significant improvement noted in the infants' skin condition if given LGG compared to no probiotic, with a similar reduction in the concentration of soluble CD4 in serum and eosinophilic protein X in urine of these patients [76].

Another study looked at maturation of humoral immune responses in 39 infants with atopic dermatitis, who received formula supplemented with or without LGG. The proportions of IgA- and IgM-secreting cells decreased significantly in the circulation of the treated group, and the proportions of CD19, CD27 memory B cells among peripheral blood leukocytes increased in the probiotic-treated infants, but not in the untreated, reflecting gut maturation of humoral immune responses [77].

Cow's milk allergic colitis, which occurs in both breast- and formula-fed infants, is an allergic condition causing hematochezia in infants as a reaction to dietary antigens. LGG-supplemented formula has shown some benefit in infants with this condition. The disorder is relatively benign and so is rarely diagnosed by biopsy (which would show rectal eosinophils); rather it is often determined presumptively based on clinical picture: the presence of mucus or blood in the stools, with or without diarrhea. The condition usually subsides after dietary antigen elimination, further supporting the diagnosis. A study assessing LGG in this context also used another marker of intestinal inflammation called calprotectin in conjunction with presence of blood in stools as a marker of treatment response. It found that addition of LGG to formula resulted in a statistically significantly greater decrease in values of fecal calprotectin and presence of fecal blood after 4 weeks of dietary therapy. The study did not determine the mechanism of this response, but supported the idea that LGG may improve the gut mucosal barrier and help direct the immune system away from an atopic response [78].

A randomized, double-blind, controlled trial of LGG included 30 infants with colic. The infants' mothers adhered to an allergen avoidance diet. The findings showed that over a 4-week intervention, although daily crying as objectively recorded by diaries was comparable between groups, in parental interviews mothers reported decrease of 68% (95% CI: 58–78) in daily crying in the probiotic group and decrease of 49% (95% CI: 32–66) in the placebo group ($P = 0.05$). Regurgitation and stool frequency were initially similar between the treatment and placebo groups of infants, but during follow up the treatment group had fewer regurgitant symptoms [27% vs. 73%, odds ratio (OR): 6.19; 95% CI: 1.36–38.04; $P = 0.04$]; stooling remained similar. PCR testing analyzed the fecal gut microbiota at the beginning of the study and again at the end of the study to look for differences between the two groups at baseline and after intervention. Overall there remained no major difference in gut microbiota between groups either before or after intervention, but certain bifidobacterium species were increased in the LGG group and this correlated with colic; the median number of stool *Bifidobacterium* genera were lower in colicky infants (1.19 vs. 4.610, respectively; $P = 0.04$). This study also made sure to utilize many other potentially effective strategies against colic as well, such as behavioral training and the allergen elimination diet, in order to assess if there was additional benefit beyond the usual interventions by adding LGG [79].

Functional Abdominal Pain and IBS

Another important area where LGG has shown benefit is in functional abdominal pain, a disorder that affects 10–15% of school-aged children and a common reason for referral to gastroenterologists. Irritable bowel syndrome (IBS) is also a common disorder in children with a prevalence of about one in five. Symptom-based criteria specific to each diagnosis exist, and recurrent abdominal pain is a characteristic of both disorders, however, in IBS the child also experiences an alteration in defecation, such as diarrhea or constipation. In a double-blind study involving multiple primary care sites, 141 children with IBS or functional abdominal pain were randomized to LGG or placebo for 8 weeks and had follow up for an additional 8 weeks. LGG significantly reduced frequency and severity of abdominal pain, and the reduction remained significant at the end of the follow up. At entry and at the end of the trial, participants had a double-sugar intestinal permeability test; at entry, 59% of the children had abnormal results. LGG led to a significant decrease in the number of patients with abnormal results from the intestinal permeability testing ($P < 0.03$) though these effects mainly were in children with IBS. The baseline number of pain episodes experienced weekly was 3.7 (2.5) in the probiotic and 3.5 (2.4) in the placebo group. At 12 weeks this decreased to 1.1 (0.8) and 2.2 (1.2), respectively ($P < 0.01$). At the end of the follow-up period, episodes of pain decreased to 0.9 (0.5) in the probiotic group and 1.5 (1.0) in the placebo group ($P < 0.02$). This effect may be secondary to improvement in the gut barrier, though that was not definitively established in this study [80].

SUMMARY

L. rhamnosus strain GG has been studied in a variety of contexts as this chapter indicates and is perhaps the most widely studied organism among all probiotics. It has potential beneficial effects in many organ systems and disease states and has been shown to be extremely safe and well tolerated. However, despite such extensive research, there is still much to be elucidated about which conditions probiotics such as LGG can prevent or treat, how these probiotic strains exert their beneficial effects, which ones are most useful for particular conditions, whether they should be used alone or in combination, and the optimal dosage and mode of delivery.

REFERENCES

[1] Segers M, Lebeer S. Towards a better understanding of *Lactobacillus rhamnosus* GG–host interactions. Microb Cell Fact 2014;13:S7.

[2] Doron S, Snydman DR, Gorbach SL. Lactobacillus GG: bacteriology and clinical applications. Gastroenterol Clin North Am 2005;34(3):483–98.

[3] Silva M, Jacobus NV, Deneke C, et al. Antimicrobial substance from a human Lactobacillus strain. Antimicrob Agents Chemother 1987;31(8):1231–3.

[4] Kankainen M, Paulin L, Tynkkynen S, et al. Comparative genomic analysis of *Lactobacillus rhamnosus* GG reveals pili containing a human-mucus binding protein. Proc Natl Acad Sci 2009;106(40):17193–8.

[5] Marianelli C, Cifani N, Pasquali P. Evaluation of antimicrobial activity of probiotic bacteria against *Salmonella enterica* subsp. enterica serovar typhimurium 1344 in a common medium under different environmental conditions. Res Microbiol 2010;161:673–80.

[6] Lu R, Fasano S, Madayiputhiya N, et al. Isolation, identification, and characterization of small bioactive peptides from Lactobacillus GG conditional media that exert both anti-Gram-negative and Gram-positive bactericidal activity. J Pediatr Gastroenterol Nutr 2009;49:23–30.

[7] De Keersmaecker SCJ, Verhoeven TLA, Desair J, et al. Strong antimicrobial activity of *Lactobacillus rhamnosus* GG against *Salmonella typhimurium* is due to accumulation of lactic acid. FEMS Microbiol Lett. 2006;259(1):89–96.

[8] Manome A, Okada S, Uchimura T, et al. The ratio of L-form to D-form of lactic acid as a criteria for the identification of lactic acid bacteria. J Gen Appl Microbiol 1998;44(6):371–4.

[9] Goldin BR, Gorbach SL, Saxelin M, et al. Survival of *Lactobacillus* species (strain GG) in human gastrointestinal tract. Dig Dis Sci 1992;37(1):121–8.

[10] Salminen MK, Rautelin H, Tynkkynen S, et al. Lactobacillus bacteremia, species identification, and antimicrobial susceptibility of 85 blood isolates. Clin Infect Dis 2006;42(5):e35–44.

[11] Krishnan V. Pilins in Gram-positive bacteria: a structural perspective. IUBMB Life 2015;67(7):533–43.

[12] Tripathi P, Dupres V, Beaussart A, et al. Deciphering the nanometer-scale organization and assembly of *Lactobacillus rhamnosus* GG pili using atomic force microscopy. Langmuir 2012;28(4):2211.

[13] Reunanen J, von Ossowski I, Hendrickx APA, et al. Characterization of the SpaCBA pilus fibers in the probiotic *Lactobacillus rhamnosus* GG. Appl Environ Microbiol 2012;78(7):2337–44.

[14] Tripathi P, Beaussart A, Alsteens D, et al. Adhesion and nanomechanics of pili from the probiotic *Lactobacillus rhamnosus* GG. ACS Nano 2013;7(4):3685–97.

[15] Meurman JH, Antila H, Salminen S. Recovery of Lactobacillus strain GG (ATCC 53103) from saliva of healthy volunteers after consumption of yoghurt prepared with the bacterium. Microb Ecol Health Dis 1994;7(6):295–8.

[16] Alander M, Satokari R, Korpela R, et al. Persistence of colonization of human colonic mucosa by a probiotic strain, *Lactobacillus rhamnosus* GG, after oral consumption. Appl Environ Microbiol 1999;65(1):351–4.

[17] Lebeer S, Claes IJ, Verhoeven TL, et al. Exopolysaccharides of *Lactobacillus rhamnosus* GG form a protective shield against innate immune factors in the intestine. Microb Biotechnol 2011;4:368–74.

[18] Yan F, Polk DB. Probiotic bacterium prevents cytokine-induced apoptosis in intestinal epithelial cells. J Biol Chem 2002;277:50959–65.

[19] Tao Y, Drabik KA, Waypa TS, et al. Soluble factors from Lactobacillus GG activate MAPKs and induce cytoprotective heat shock proteins in intestinal epithelial cells. Am J Physiol 2006;290:C1018–30.

[20] Yan F, Cao H, Cover TL, et al. Soluble proteins produced by probiotic bacteria regulate intestinal epithelial cell survival and growth. Gastroenterology 2007;132:562–75.

[21] Seth A, Yan F, Polk DB, et al. Probiotics ameliorate the hydrogen peroxide-induced epithelial barrier disruption by a PKC- and MAP kinase-dependent mechanism. Am J Physiol 2008;294:G1060–9.

[22] Wilson KH, Perini F. Role of competition for nutrients in suppression of *Clostridium difficile* by the colonic microflora. Infect Immun 1988;56(10):2610–4.

[23] Di Caro S, Tao H, Grillo A, et al. Effects of Lactobacillus GG on genes expression pattern in small bowel mucosa. Dig Liver Dis 2005;37:320–9.

[24] Eloe-Fadrosh EA, Brady A, Crabtree J, et al. Functional dynamics of the gut microbiome in elderly people during probiotic consumption. MBio 2015;6(2):1–119.

[25] Hamer HM, Jonkers D, Venema K, et al. Review article: the role of butyrate on colonic function. Aliment Pharmacol Ther 2008;27(2):104–19.

[26] Luhrs H, Kudlich T, Neumann M, et al. Butyrate-enhanced TNFalpha-induced apoptosis is associated with inhibition of NF-kappaB. Anticancer Res 2002;22(3):1561–8.

[27] Bajaj J, Heuman D, Hylemon P, et al. Randomised clinical trial: Lactobacillus GG modulates gut microbiome, metabolome and endotoxemia in patients with cirrhosis. Aliment Pharmacol Ther 2014;39(10):1113–25.

[28] Bruzzese E, Callegari ML, Raia V, et al. Disrupted intestinal microbiota and intestinal inflammation in children with cystic fibrosis and its restoration with Lactobacillus GG: a randomised clinical trial. PLoS One 2014;9(2):e87796.

[29] Hempel S, Newberry S, Ruelaz A, et al. Safety of probiotics used to reduce risk and prevent or treat disease. Evid Rep Technol Assess (Full Rep). 2011;200:1–645.

[30] Joint FAO/WHO. WHO Working Group on Drafting Guidelines for the Evaluation of Probiotics in Food. Guidelines for the evaluation of probiotics in food: report of a Joint FAO/WHO Working Group on Drafting Guidelines for the Evaluation of Probiotics in Food. London (ON, Canada), April 30 and May 1. Available from: ftp://ftp.fao.org/es/esn/food/wgreport2.pdf; 2002.

[31] Ouwehand AC, Saxelin M, Salminen S. Phenotypic differences between commercial *Lactobacillus rhamnosus* GG and *L. rhamnosus* strains recovered from blood. Clin Infect Dis 2004;39(12):1858–60.

[32] Doron S, Snydman DR. Risk and safety of probiotics. Clin Infect Dis 2015;60(Suppl. 2):S129–34.

[33] Munakata S, Arakawa C, Kohira R, et al. A case of D-lactic acid encephalopathy associated with use of probiotics. Brain Dev 2010;32(8):691–4.

[34] Ku W, Lau D, Huen K. Other articles probiotics provoked D-lactic acidosis in short bowel syndrome: case report and literature review. HK J Paediatr 2006;11(3):246–54.

[35] Oh MS, Phelps KR, Traube M, et al. D-lactic acidosis in a man with the short-bowel syndrome. N Engl J Med 1979;301(5):249–52.

[36] Veckman V, Miettinen M, Pirhonen J, et al. *Streptococcus pyogenes* and *Lactobacillus rhamnosus* differentially induce maturation and production of Th1-type cytokines and chemokines in human monocyte-derived dendritic cells. J Leukoc Biol 2004;75(5):764–71.

[37] Tynkkynen S, Singh KV, Varmanen P. Vancomycin resistance factor of *Lactobacillus rhamnosus* GG in relation to enterococcal vancomycin resistance (van) genes. Int J Food Microbiol 1998;41(3):195–204.

[38] Turck D, Bernet J, Marx J, et al. Incidence and risk factors of oral antibiotic-associated diarrhea in an outpatient pediatric population. J Pediatr Gastroenterol Nutr 2003;37(1):22–6.

[39] Kotowska M, Albrecht P, Szajewska H. *Saccharomyces boulardii* in the prevention of antibiotic-associated diarrhoea in children: a randomized double-blind placebo-controlled trial. Aliment Pharmacol Ther 2005;21(5):583–90.

[40] Wenus C, Goll R, Loken E, et al. Prevention of antibiotic-associated diarrhoea by a fermented probiotic milk drink. Eur J Clin Nutr 2008;62(2):299–301.

[41] Vanderhoof JA, Whitney DB, Antonson DL, et al. Lactobacillus GG in the prevention of antibiotic-associated diarrhea in children. J Pediatr 1999;135(5):564–8.

[42] Arvola T, Laiho K, Torkkeli S, et al. Prophylactic Lactobacillus GG reduces antibiotic-associated diarrhea in children with respiratory infections: a randomized study. Pediatrics 1999;104(5):e64.

[43] Szajewska H, Kotowska M, Mrukowicz JZ, et al. Efficacy of Lactobacillus GG in prevention of nosocomial diarrhea in infants. J Pediatr 2001;138(3):361–5.

[44] Thomas M, Litin S, Osmon D, et al. Lack of effect of Lactobacillus GG on antibiotic-associated diarrhea: a randomized, placebo-controlled trial. Mayo Clin Proc 2001;76(9):883–9.

[45] Armuzzi A, Cremonini F, Ojetti V, et al. Effect of Lactobacillus GG supplementation on antibiotic-associated gastrointestinal side effects during *Helicobacter pylori* eradication therapy: a pilot study. Digestion 2001;63(1):1–7.

[46] Armuzzi A, Cremonini F, Bartolozzi F, et al. The effect of oral administration of Lactobacillus GG on antibiotic-associated gastrointestinal side-effects during *Helicobacter pylori* eradication therapy. Aliment Pharmacol Ther 2001;15(2):163–9.

[47] Cremonini F, Di Caro S, Covino M, et al. Effect of different probiotic preparations on anti-*Helicobacter pylori* therapy-related side effects: a parallel group, triple blind, placebo-controlled study. Am J Gastroenterol 2002;97(11):2744–9.

[48] Fox MJ, Ahuja KD, Robertson IK, et al. Can probiotic yogurt prevent diarrhoea in children on antibiotics? A double-blind, randomised, placebo-controlled study. BMJ Open 2015;5(1):e006474.

[49] Szajewska H, Kołodziej M. Systematic review with meta-analysis: *Lactobacillus rhamnosus* GG in the prevention of antibiotic-associated diarrhoea in children and adults. Aliment Pharmacol Ther 2015;42(10):1149–57.

[50] Doron SI, Hibberd PL, Gorbach SL. Probiotics for prevention of antibiotic-associated diarrhea. J Clin Gastroenterol 2008;42(Suppl. 2):S58–63.

[51] Basu S, Paul DK, Ganguly S, et al. Efficacy of high-dose *Lactobacillus rhamnosus* GG in controlling acute watery diarrhea in Indian children: a randomized controlled trial. J Clin Gastroenterol 2009;43(3):208–13.

[52] Szajewska H, Mrukowicz JZ. Probiotics in the treatment and prevention of acute infectious diarrhea in infants and children: a systematic review of published randomized, double-blind, placebo-controlled trials. J Pediatr Gastroenterol Nutr 2001;33:S17–25.

[53] Huang JS, Bousvaros A, Lee JW, et al. Efficacy of probiotic use in acute diarrhea in children: a meta-analysis. Dig Dis Sci 2002;47(11):2625–34.

[54] Sindhu KN, Sowmyanarayanan TV, Paul A, et al. Immune response and intestinal permeability in children with acute gastroenteritis treated with *Lactobacillus rhamnosus* GG: a randomized, double-blind, placebo-controlled trial. Clin Infect Dis 2014;58(8):1107–15.

[55] Pochapin M. The effect of probiotics on *Clostridium difficile* diarrhea. Am J Gastroenterol 2000;95(1):S11–3.

[56] Gorbach S, Chang T, Goldin B. Successful treatment of relapsing *Clostridium difficile* colitis with Lactobacillus GG. Lancet 1987;330(8574):1519.

[57] Biller J, Katz A, Flores A, et al. Treatment of recurrent *Clostridium difficile* colitis with Lactobacillus GG. J Pediatr Gastroenterol Nutr 1995;21(2):224–6.

[58] Bennett RG, Gorbach SL, Goldin BR, et al. Treatment of relapsing *Clostridium difficile* diarrhea with Lactobacillus GG. Nutr Today 1996;31(6):39S.

[59] DiBaise JK, Zhang H, Crowell MD, et al. Gut microbiota and its possible relationship with obesity. Mayo Clin Proc 2008;83(4):460–9.

[60] Gionchetti P, Rizzello F, Venturi A, et al. Oral bacteriotherapy as maintenance treatment in patients with chronic pouchitis: a double-blind, placebo-controlled trial. Gastroenterology 2000;119(2):305–9.

[61] Ley RE, Backhed F, Turnbaugh P, et al. Obesity alters gut microbial ecology. Proc Natl Acad Sci USA 2005;102(31):11070–5.

[62] Ley RE, Turnbaugh PJ, Klein S, et al. Microbial ecology: human gut microbes associated with obesity. Nature 2006;444(7122):1022–3.

[63] Turnbaugh PJ, Bäckhed F, Fulton L, et al. Diet-induced obesity is linked to marked but reversible alterations in the mouse distal gut microbiome. Cell Host Microbe 2008;3(4):213–23.

[64] Turnbaugh PJ, Ley RE, Mahowald MA, et al. An obesity-associated gut microbiome with increased capacity for energy harvest. Nature 2006;444(7122):1027–131.

[65] Ji Y, Kim H, Park H, et al. Modulation of the murine microbiome with a concomitant anti-obesity effect by *Lactobacillus rhamnosus* GG and *Lactobacillus sakei* NR28. Benef Microbes 2012;3(1):13–22.

[66] Horton JD, Goldstein JL, Brown MS. SREBPs: activators of the complete program of cholesterol and fatty acid synthesis in the liver. J Clin Invest 2002;109(9):1125–32.

[67] Tappy L, Le KA. Metabolic effects of fructose and the worldwide increase in obesity. Physiol Rev 2010;90(1):23–46.

[68] Luoto R, Kalliomäki M, Laitinen K, et al. The impact of perinatal probiotic intervention on the development of overweight and obesity: follow-up study from birth to 10 years. Int J Obes 2010;34(10):1531–7.

[69] Hojsak I, Snovak N, Abdovic S, et al. Lactobacillus GG in the prevention of gastrointestinal and respiratory tract infections in children who attend day care centers: a randomized, double-blind, placebo-controlled trial. Clin Nutr 2015;29(3):312–6.

[70] Morrow LE, Kollef MH, Casale TB. Probiotic prophylaxis of ventilator-associated pneumonia: a blinded, randomized, controlled trial. Am J Respir Crit Care Med 2010;182:1058–64.

[71] Yasui H, Kiyoshima J, Hori T. Reduction of influenza virus titer and protection against influenza virus infection in infant mice fed *Lactobacillus casei* Shirota. Clin Diagn Lab Immunol 2004;11(4):675–9.

[72] Davidson LE, Fiorino A, Snydman DR, et al. Lactobacillus GG as an immune adjuvant for live-attenuated influenza vaccine in healthy adults: a randomized double-blind placebo-controlled trial. Eur J Clin Nutr 2011;65(4):501–7.

[73] Wu C, Chen P, Lee Y, et al. Effects of immunomodulatory supplementation with *Lactobacillus rhamnosus* on airway inflammation in a mouse asthma model. J Microbiol Immunol Infect 2014;.

[74] Isolauri E, Majamaa H, Arvola T, et al. *Lactobacillus casei* strain GG reverses increased intestinal permeability induced by cow milk in suckling rats. Gastroenterology 1993;105(6):1643–50.

[75] Kalliomaki M, Salminen S, Poussa T, et al. Probiotics and prevention of atopic disease: 4-year follow-up of a randomised placebo-controlled trial. Lancet 2003;361:1869–71.

[76] Isolauri E, Arvola T, Sutas Y, et al. Probiotics in the management of atopic eczema. Clin Exp Allergy 2000;30(11):1604–10.

[77] Nermes M, Kantele JM, Atosuo TJ, et al. Interaction of orally administered *Lactobacillus rhamnosus* GG with skin and gut microbiota and humoral immunity in infants with atopic dermatitis. Clin Exp Allergy 2011;41(3):370–7.

[78] Baldassarre ME, Laforgia N, Fanelli M, et al. Lactobacillus GG improves recovery in infants with blood in the stools and presumptive allergic colitis compared with extensively hydrolyzed formula alone. J Pediatr 2010;156(3):397–401.

[79] Pärtty A, Lehtonen L, Kalliomäki M, et al. Probiotic *Lactobacillus rhamnosus* GG therapy and microbiological programming in infantile colic: a randomized, controlled trial. Pediatr Res 2015;78(4):470–5.

[80] Francavilla R, Miniello V, Magista AM, et al. A randomized controlled trial of Lactobacillus GG in children with functional abdominal pain. Pediatrics 2010;126(6):e1445–52.

Chapter 8

Lactobacillus reuteri

R.A. Britton

Department of Molecular Virology and Microbiology, Center for Metagenomics and Microbiome Research, Baylor College of Medicine, Houston, TX, United States

HISTORY

Lactobacillus reuteri is named after Gerhard Reuter, a German microbiologist who spent most of his career attempting to understand the human intestinal microbiota with a focus on lactobacilli and bifidobacteria [1]. His lab undertook a challenge that we still face in microbial ecology today—how do we access the small intestine with minimal disruption in order to understand the ecology of the normal small intestine? Using regulated capsule sampling technology and cultivation from sudden death victims, Reuter found that lactobacilli were abundant in the small intestine and present in feces of many of the individuals they sampled, which included children and adults [1]. In addition, they found significant levels of lactobacilli in the stomach and duodenum, which for a long time were considered sterile sites of the GI tract. While it is well appreciated now that these sites are colonized with bacterial communities, this was quite controversial 50 years ago.

From these human studies Reuter identified *L. reuteri* as well as a few other lactobacillus species (e.g., *Lactobacillus gasseri*) as autochthonous to the intestinal microbiome [1]. *L. reuteri* strains were frequently isolated from the jejunum/ileum and feces of human subjects and were occasionally found at other sites of the GI tract. In addition, *L. reuteri* were consistently isolated from the same subjects over time, supporting that *L. reuteri* constitutes a stable member of the human microbiome. As discussed later, the location that provides the ecological niche for *L. reuteri* strains in humans remains to be defined.

Lactobacillus reuteri STRAINS USED IN HUMAN TRIALS, ANIMAL MODELS, AND IN VITRO STUDIES

During the characterization of strains isolated in these studies, an organism originally identified as a *Lactobacillus fermentum* species was found to be distinct from *L. fermentum*

and was renamed *L. reuteri*. The type strain of the species is designated in multiple ways that can result in some confusion when reviewing the literature and genomic databases. The original designation is F275 from a strain isolated in 1962. This strain was deposited with the ATCC as strain ATCC 23272 and later with DSM as DSM 20016. The full genome sequence of two "F275" strains that were contributed from laboratories in Japan (JCM 1112) and New Zealand (DSM 20016) were released in 2008. Although designated as the same F275 strain, there were many differences between the two strains including numerous SNPs and up to 40 kB of DNA that has been deleted in the DSM 20016 derivative [2]. The significant changes in the genomes between the two lab derivatives of the original F275 strains is a stark reminder that genomes are dynamic and care should be taken when passaging nonlab adapted strains in the laboratory environment.

Research investigating the effects of *L. reuteri* in a variety of disease and inflammation models has greatly increased in the past 15 years with 928 papers published between 2006 and 2016 versus 134 papers prior to 2001 that mention *L. reuteri*. As the functionality of bacteria resides at the strain level rather than the species level, it is critical to understand the origins and functions of individual strains. As will become clear later, not all *L. reuteri* strains possess the same genetic and functional potential, which is true for any other bacterial species whether a commensal or pathogen. *L. reuteri* as a probiotic species has been mainly driven by BioGaia AB that has marketed *L. reuteri* ATCC 55730 and more recently the daughter strain *L. reuteri* ATCC 17938 as probiotic strains. These two strains have been utilized the most in human trials to date with the initial human trials conducted with ATCC 55730, which was isolated from the breast milk of a Peruvian female. This strain has been shown to be resistant to high levels of acid and bile that would be encountered in the intestinal tract and has been shown to be viable in the human gut [3–5]. When the genome of ATCC 55730 was sequenced it was revealed that in addition to the single ~2 Mb chromosome the genome was found to contain two plasmids that contained possible antibiotic resistance markers for tetracycline and lincomycin. To remove these plasmids and the unwanted associated antibiotic resistance, ATCC 55730 was subjected to protoplast formation screened for sensitivity to first tetracycline and lincomycin [6]. The resulting strain was indeed sensitive to both antibiotics, devoid of both plasmids and is referred to as ATCC 17938. ATCC 17938 has been designated as the functional equivalent of ATCC 55730 and is now used in most clinical trials that have appeared in the past 5 years.

Additional strains that have been used in clinical trials include the oral isolate *L. reuteri* ATCC PTA 5289, which has been used in several human clinical trials of gingivitis and periodontal disease (with limited success), often with *L. reuteri* 17938 [7,8]. In addition, the vaginal tract isolate *L. reuteri* RC-14 has been tested (often with *Lactobacillus rhamnosus* GR-1) for the ability to improve bacterial vaginosis with some success [9,10]. Interestingly, *L. reuteri* RC-14 does not make the antimicrobial compound reuterin [11], which delineates it away from the human breast milk and gut isolates discussed later. Oral and vaginal *L. reuteri* probiotics will not be further discussed in this chapter.

Finally, three strains that have received considerable attention in in vitro and animal models are the human-derived strain *L. reuteri* ATCC PTA 6475 and the rodent-derived strain *L. reuteri* 100-23. *L. reuteri* 6475 has potent antiinflammatory activity in vitro and in a mouse colitis model (discussed more thoroughly later) [12,13]. The rat isolate *L. reuteri* 100-23 has a fully sequenced closed genome, has been used extensively for studying mechanisms by which *L. reuteri* colonizes the gut, and can promote the formation of T_{reg} cells in lactobacilli free animals [14–17].

PHYLOGENY AND ECOLOGY OF *Lactobacillus reuteri*

In addition to colonizing the human intestinal tract, strains of *L. reuteri* also colonize the intestinal tracts of many other mammalian and vertebrate animals including mice, rats, turkeys, pigs, and chickens. An in-depth analysis of the genomes from dozens of isolates from these animals demonstrated that *L. reuteri* strains have coevolved with their animal hosts [18,19]. Jens Walter and coworkers used a combination of amplified fragment polymorphism analysis, multiple-locus sequence analysis, microarray analysis for DNA content, and genome sequencing to clearly show that *L. reuteri* isolates clustered primarily with the animal hosts from which they were isolated [18,19]. The 165 strains grouped into 6 distinct clades that corresponded with the host animals: clade I (rodents I), clade II (humans), clade III (rodents II), clade IV (pigs I), clade V (pigs II), and clade VI (poultry/humans). For the most part, *L. reuteri* strains

were confined to the clade associated with the animal host with the exception of a cluster of human-derived strains that clustered with poultry (clade VI), suggesting that this particular group may represent a direct transmission between poultry and humans. Importantly, the authors utilized strains isolated from several distinct geographical locations around the world within each animal group to prevent sampling bias. This coevolution with the host had been appreciated for some time for pathogenic bacteria but this was the first demonstration of host adaptation within a commensal or mutualistic species [20].

The ecological niches that *L. reuteri* strains colonize in animal intestines have been well characterized, while the location where *L. reuteri* resides in the human intestine remains a mystery. Animal strains colonize the squamous epithelium of the forestomach (rodents) and poultry crop (chickens and turkeys) [21]. Humans lack this type of squamous epithelium and thus it is difficult to predict the ecological niche in the human intestinal tract based on animal studies. In the mouse genome there exists niche specific factors, such as the large surface protein located on the cell wall of mouse-derived strains, which are required for colonization of the forestomach [16]. This factor is not present in human-derived strains and is likely part of the reason that human strains are unable to compete with rodent strains in lactobacilli free mice [18].

PROBIOTIC PROPERTIES AND MECHANISMS OF ACTION

Reuterin

The key finding that launched strains of *L. reuteri* as probiotics was the discovery of the antimicrobial compound reuterin. Originally described in the Dobrogosz laboratory in the late 1980s, a porcine strain was shown to produce a compound in the presence of glycerol that was able to kill bacteria, fungi, and protozoa and inactivate viruses [22,23]. Reuterin was identified as the compound 3-hydroxypropionaldehye (3-HPA) and is an intermediate in the conversion of glycerol to 1,3-propanediol [23,24]. Glycerol is first converted to 3-HPA by the vitamin B_{12} dependent enzyme glycerol dehydratase, which is further reduced to 1,3-propanediol by 1,3-propanediol oxidoreductase. This latter step utilizes NADH and regenerates NAD, which has been proposed as a possible mechanism by which *L. reuteri* can reset its redox status to favor glycolysis. Indeed it has been shown that addition of glycerol to growth medium increases the ability of *L. reuteri* to improve growth yield despite not being able to utilize glycerol as a carbon source [24,25].

Although many other bacteria also contain genes to convert glycerol into 1,3-PD via 3-HPA, *L. reuteri* appears to be unique in its ability to secrete high levels of 3-HPA out into the extracellular millieu. 3-HPA is a highly reactive aldehyde and can form other molecules in solution including HPA-hydrate, HPA-dimer, or can be dehydrated to acrolein [26]. Thus the mechanism by which 3-HPA is antimicrobial remained a mystery until just recently. Two groups conclusively demonstrated that the mechanism by which *L. reuteri* mediates its effects is via 3-HPA interacting with thiol groups within the cell [27,28]. As several enzymes (e.g., ribonucleotide reductase) and metabolites in the cell have free thiols that are critical to their function, it appears 3-HPA mediates cell killing via interactions with these molecules.

The ability to produce reuterin is contained within a large horizontally transferred DNA island that also encodes genes for the production of vitamin B_{12}, ethanolamine metabolism, and the microcompartment structure in which reuterin conversion takes place [29]. All human isolates analyzed to date have the ability to produce reuterin while most animal isolates lack this ability [18]. Comparison of *L. reuteri* human isolates demonstrated that human clade VI organisms have a higher capacity for reuterin production in vitro [30]. Although there has been a lot of work demonstrating the role of reuterin as an antimicrobial in vitro, it still has not been demonstrated that it serves this function in vivo. One argument against reuterin as an antimicrobial in vivo is the fact that in vitro killing requires concentrations in the low millimolar range [27,28]. Achieving this concentration in vivo would require a large source of glycerol and given the high degree of reactivity of 3-HPA, it seems unlikely that such a concentration could be reached except for very close to *L. reuteri*. The fact that *L. reuteri* strains are much more resistant to the effects of reuterin than most other bacteria suggests that it is indeed produced in the gut and the strains have evolved intrinsic resistance to the compound [27,31]. Thus the role of reuterin in mediating pathogen resistance in the human microbiota remains to be proven and possible other metabolic possibilities need to be explored. Applications for reuterin production in food preservation are being investigated [32–34].

Reutericyclin

Another antimicrobial compound produced by sourdough isolates of *L. reuteri* is reutericyclin, a tetramic acid molecule that has potent antimicrobial activity against Grampositive bacteria [35,36]. Reutericyclin functions as a proton ionophore and prevents bacterial growth by dissipating the transmembrane ΔpH. The minimal inhibitory concentration for reutericyclin against most bacteria is in the high nanomolar to low micromolar range and strains that produce reutericyclin are resistant. Reutericyclin is currently being investigated as a potential food preservative as well as a potential probiotic supplement to reduce the carriage of pathogens in food animals [35]. One recent study demonstrated that reutericyclin producing strains altered the microbiota of piglets, although the direct participation of the antimicrobial compound was not proven [37].

Vitamins

Human-derived *L. reuteri* strains have the ability to produce a number of vitamins including cobalamin (vitamin B_{12}), folic acid, and thiamine [38]. Of these, cobalamin has received the most attention due to its requirement for glycerol dehydratase activity during reuterin production and the role of vitamin B_{12} deficiency in the human population in pernicious anemia [39]. Experiments in a cobalamin-deficient diet mouse model supported the hypothesis that *L. reuteri* could produce bioactive vitamin B_{12} and alleviate symptoms associated with B_{12} deficiency [40]. However, mass spectrometry of vitamin B_{12} produced from *L. reuteri* demonstrated that the produced B_{12} is a pseudo-B_{12} that differs in the alpha ligand present on the cobalamin ring compared to vitamin B_{12} [41,42]. Pseudo-B_{12} produced has 500 times less affinity for intrinsic factor than vitamin B_{12} and thus is unlikely to provide much health benefit to B_{12}-deficient humans or animals [43].

Immunomodulation

The impact of probiotics on the immune system has become a major field of investigation in the past 15 years. While many probiotic strains have the ability to modulate the immune system, the mechanisms by which they regulate immune function have lagged far behind. Recently much has been learned about how human-derived *L. reuteri* strains impact host immunity at the mechanistic level. Initial work with a pig isolate of *L. reuteri* demonstrated that proinflammatory cytokines produced by murine bone marrow–derived dendritic cells could be suppressed by *L. reuteri* DSM 12246 conditioned supernatant [44]. Noteworthy in this study was the ability of several different lactobacilli to promote expression of the proinflammatory cytokine IL-12 from BMDCs, whereas *L. reuteri* was unique in not only failing to strongly induce IL-12 but could suppress IL-12 and other proinflammatory cytokines induced by other lactobacilli. Similarly, *L. reuteri* DSM 20016 promoted DCs to drive T-cell maturation toward a T_{reg} phenotype and stimulated the strong production of IL-10 [44].

Versalovic and coworkers found that a clade II *L. reuteri* strain, *L. reuteri* 6475, was capable of reducing the production of TNF by greater than 90% while two clade VI strains lacked this activity [13]. Suppression was mediated by secretion of a less than 3-kDa metabolite that was resistant to heat and protease treatment. Subsequent biochemical, mass spectrometry, and NMR analysis of supernatant from *L. reuteri* 6475 identified the bioactive compound as histamine and inhibitors of histamine receptors showed that signaling via the histamine H2 receptor was required for the suppression of TNF [45]. Analysis of the genome sequence of *L. reuteri* strains identified a three-gene histidine decarboxylase (*hdc*) operon that was present only in clade II human isolates of *L. reuteri*, supporting a genetic basis for the ability of these strains to produce histamine from histidine. Using newly developed recombineering technology in *L. reuteri* 6475, individual genes of the *hdc* locus were disrupted and the resulting strains partially lost the ability to suppress TNF from activated monocytes [45].

Although histamine is normally thought of in the context of the allergic response as a proinflammatory cytokine, histamine has been shown to play an antiinflammatory role in the intestinal tract. Mice that have a deletion of their histidine decarboxylase gene display spontaneous colonic inflammation and are more prone to develop colon cancer than wild-type mice, highlighting a role for host-derived histamine in reducing levels of intestinal inflammation [46]. A recent study investigated the role of histamine derived from *L. reuteri* in the amelioration of intestinal inflammation in a murine colitis model [12]. The found *L. reuteri* could suppress TNBS-driven colitis in a manner that was dependent on the *hdc* locus and the presence of histidine in the diet. Either disruption of *hdcA* or removing dietary histidine was sufficient to alleviate the protective effects of *L. reuteri* [12]. Interestingly, signaling via the H2 receptor in vivo was required for the antiinflammatory effect, as was shown in vitro with activated monocytes [45]. These studies highlight the ability of diet-derived metabolites produced by metabolism by intestinal microbes to have a large impact on the health of the host.

A recent series of studies of *L. reuteri* 6475 from the Erdman group has highlighted a number of beneficial effects of this strain on the health of mice. These include rapid wound healing, protection from obesity, increased testosterone levels in aging mice, and the ability to more rapidly regrow hair [47–50]. Although the mechanistic connections between *L. reuteri* supplementation and amelioration of these apparently disparate conditions remain to be elucidated, one common theme appears to be the ability of *L. reuteri* to promote a T_{reg} phenotype while reducing expression of the proinflammatory cytokine IL-17. Indeed, blocking IL-17 signaling phenocopies some of the *L. reuteri* beneficial effects including increased testes size and more rapid healing of wounds [48,50]. In wound healing, it was shown that systemic oxytocin levels were increased by 2.5 times in *L. reuteri*–treated animals and was associated with a more rapid wound healing than control animals [48]. Oxytocin knockout mice were unable to respond to *L. reuteri* ingestion and wounds healed at the same rate as control animals, supporting that oxytocin was required for microbial stimulation of wound healing. A mechanistic link between oxytocin and the immune system was established by demonstrating that adoptive transfer of Foxp3+ CD45lo CD25+ T_{reg} cells from *L. reuteri*–treated wild-type mice, but

not from *L. reuteri*–treated oxytocin knockout mice, were sufficient to stimulate wound healing. These studies show that *L. reuteri* 6475 can significantly impact the immune system of mice leading to numerous beneficial effects. The compounds produced by *L. reuteri* that drive these effects, with the exception of histamine, remain to be discovered.

USE OF *Lactobacillus reuteri* AS A PROBIOTIC FOR THE PREVENTION AND TREATMENT OF HUMAN DISEASE

As with most strains of probiotic bacteria that have shown promise in the laboratory and animal models, a number of small-scale clinical trials have been performed for a variety of purposes using *L. reuteri* strains. Many of the human trials conducted to date have been with the strain *L. reuteri* ATCC 55730 or the daughter strain ATCC 17938 (mentioned earlier). Nearly all human trials associated with the testing of probiotics for the treatment and prevention of disease were severely underpowered and thus the positive effects attributed to the *L. reuteri* strains discussed later should be viewed as pilot studies. Nonetheless, several of these trials have demonstrated beneficial effects in a variety of human and animal trials and will be highlighted in subsequent sections.

Colic

Colic is a condition that is marked by infants suffering from uncontrolled crying spells that can last for hours per episode. The clinical definition of colic is when a baby cries for more than 3 h a day, 3 times per week for more than 3 weeks [51]. The etiology of colic remains undefined and there are no satisfactory treatments for alleviating the symptoms of colic. Given the enormous amount of stress for parents that these long-duration crying spells create, there is a need to find solutions for colicky babies.

In 2007, Savino et al. described the effects of *L. reuteri* 55730 on the reduction of crying times in infants with colic [52]. In this trial babies were either assigned to receive *L. reuteri* 55730 or simethicone, a compound that has been routinely used in the treatment of colic despite little evidence of a beneficial effect against colic. At the beginning of the study infants had mean crying times of just under 3 h a day and were given either 10^8 live *L. reuteri* or 60 mg of simethicone, a compound often prescribed but clinically shown to be ineffective against colic. Within 2 weeks the *L. reuteri*–treated infants were displaying significantly decreased crying times and at the end of the 28-day study 95% (39 of 41) of the infants experienced a >50% decrease in crying times versus only 7% (3 of 42) in the simethicone group. A second double-blind, placebo-controlled trial demonstrated that *L. reuteri* 17938 was superior to placebo in reducing colicky symptoms, although in this trial a strong beneficial effect was also observed in the placebo group [53]. Fecal microbiota analysis of a subset of the subjects demonstrated that *L. reuteri* administration did not significantly alter the microbial community structure [54].

Following up on these initial studies there have been five additional double-blind, placebo-controlled trials that have addressed if *L. reuteri* 17938 can prevent or treat infant colic [55–59]. Four of these trials showed that *L. reuteri* provided a beneficial effect while one of them showed no benefit of *L. reuteri* supplementation. The largest trial to date consisted of 468 infants (238 in the treatment group vs. 230 in the control group) [55]. Rather than treating symptoms of colic, this study investigated whether or not *L. reuteri* supplementation beginning at birth would be sufficient to reduce the emergence of colicky symptoms as measured by daily mean crying time. Indeed, at both 1 month and 3 months of age the *L. reuteri*–treated children displayed half the duration of crying episodes when compared to placebo and had reduced levels of regurgitation and evacuations at 3 months of age. These benefits also led to reduced numbers of emergency room and pediatrician visits while also reducing the number of days missed at work for parents. These exciting results are tempered by the study by Sung et al. that showed no improvement in colic in a study of 167 infants displaying active colic [58]. While these results are contrary to the aforementioned successful interventional trials; they suggest that *L. reuteri* 17938 supplementation may not be beneficial for colic in all settings.

Colic is likely a heterogeneous disorder and we understand little with regard to how *L. reuteri* 17938 may be functioning to suppress symptoms associated with colic. To begin to address mechanistic aspects of *L. reuteri* supplementation in premature infants, Indrio and coworkers assessed gut motility in a group of 34-week premature infants that were fed breast milk, formula, or formula plus *L. reuteri* 55730 (*n* = 10 in each group) [60]. Using ultrasound to monitor gastric emptying after 30 days of treatment, they found that premature infants given formula plus *L. reuteri* showed increased motility, less regurgitation, and increased evacuations than with formula alone. While this does not necessarily speak directly to the cause of colic, the results show that *L. reuteri* 55730 can impact intestinal motility and function. A similar conclusion was reached in a study vetting *L. reuteri* 17938 in infants with gastroesophageal reflux [61].

Necrotizing Enterocolitis

Necrotizing enterocolitis (NEC) is a disease that results in the death of intestinal tissue that primarily affects premature infants [62]. The onset of the disease is quite rapid and is fatal in up to 30–50% of patients diagnosed with NEC. As with colic, the etiology of NEC remains a mystery although recent work has implicated the microbiota and bile

acid metabolism as possible drivers of disease. In support of a role for microbial communities in NEC it has been found that premature infants fed human milk have a significantly reduced incidence of NEC than formula-fed infants [63]. Nonetheless, NEC remains a condition that is difficult to predict in infants and once diagnosis has been made disease has generally progressed to a significant degree.

The Rhoads laboratory has investigated the ability of *L. reuteri* strains to reduce the severity of NEC in an infant rat model. *L. reuteri* ATCC 17938 and ATCC 4659 (clade II human strain) were both effective in reducing the severity and incidence of NEC [64]. Both strains, which differ in their antiinflammatory capabilities in vitro (clade II human strains are antiinflammatory in a human monocyte assay), can suppress proinflammatory cytokine production in the intestine due to NEC. Both strains inhibited the expression of the proinflammatory cytokines of TNF and IL-1B while increasing the antiinflammatory cytokine IL-10. *L. reuteri* strains were able to inhibit LPS-induced NF-κB activation by preventing the phosphorylation of IκB, while *Lactobacillus acidophilus* DDS and *Escherichia coli* DH5a were unable to provide this function demonstrating specificity for *L. reuteri* in this pathway. These findings are noteworthy because it highlights the limitations of predicting how in vitro cell culture assays will translate to functional output in an in vivo setting as well as showing strain specificity of *L. reuteri* [64,65]. The mechanism by which *L. reuteri* suppresses NEC in rodents is still unknown but two subsequent studies in rat and mouse models demonstrated that *L. reuteri* 17938 improved NEC while increasing the levels of T_{reg} cells in the ileum and mesenteric lymph nodes [66,67]. Indeed, T_{reg} immunotherapy was sufficient to reduce the severity of NEC in their rat model, suggesting that this is the mechanism by which *L. reuteri* is impacting disease [68].

In 2012, *L. reuteri* 17938 was reported to significantly decrease the amount of NEC in extremely low birth weight infants in North Carolina [69]. In this neonatal intensive care unit the rate of NEC was higher than the national average and motivated the physicians to try new strategies to lower the NEC rate in this high-risk population. Prior to supplementation with *L. reuteri* 17938 in 2009 the rate of NEC was 15.1%, which dropped to 2.5% after *L. reuteri* 17938 was used to treat extremely low birth weight infants. Thus *L. reuteri* was successfully able to reduce NEC such that one case of NEC would be prevented for every eight infants that were treated. Further studies are urgently needed to replicate this study and identify if *L. reuteri* 17938 may serve as a preventative measure for NEC.

Infectious Diseases

Several trials in adults and children have addressed the ability of *L. reuteri* supplementation to prevent and treat diarrheal episodes. Two initial studies in 1997 demonstrated that *L. reuteri* 55730 was effective in reducing the duration and severity of diarrhea in children 6–36 months old, most of which were afflicted with rotavirus infection [70,71]. In a study of infants (4–10 months old) in daycare setting, *L. reuteri* 55730 and *Bifidobacterium* BB-12 were given daily for 12 weeks [72]. During the randomized double-blind, placebo-controlled trial a number of outcomes were measured including the incidence and duration of febrile episodes, respiratory illness, and diarrhea, as well as clinic visits, antibiotic use, and absences from daycare. *L. reuteri* 55730 and BB-12 were successful in reducing the incidence and duration of diarrhea and fever but had no effect on respiratory illnesses. Interestingly, only *L. reuteri* 55730 also reduced absences from childcare, prescribed antibiotics, and clinic visits while BB-12 was not different than the placebo controls. This study is noteworthy in that it compared multiple probiotics and they were able to observe differences between the organisms used. Most human and animal trials normally only test a single probiotic strain and thus the specificity of beneficial effects cannot be addressed.

Since this initial study, a number of trials have been conducted to specifically address if *L. reuteri* 17938 can prevent and reduce the duration in children worldwide [73–79]. In most trials, *L. reuteri* 17938 supplementation was sufficient to improve diarrhea in children with the exception of the prevention of diarrhea in hospitalized children. One recent study of 336 children in daycare settings in Mexico repeated the initial Weizman study and also found that *L. reuteri* 17938 improved diarrheal disease in kids during the 12-week treatment period [77]. They also noted that respiratory illnesses were reduced in the *L. reuteri* treatments group as well. Children were followed up for an additional 12 weeks after *L. reuteri* supplementation was stopped and this showed that the reduction in diarrhea and respiratory infections was still observed a full 3 months after treatment was halted. These successful trials need to be followed up with larger, more rigorous studies such that physicians can make more informed recommendations for probiotic treatment.

Other Inflammatory Diseases

As noted earlier, *L. reuteri* 6475 can ameliorate chemically induced colitis [12]. Several studies have also shown this impact in chemical and genetic mouse models of colitis, although the possible mechanisms by which suppression of inflammation occurs in these models are not well understood [80–84]. Recent work has addressed the ability of *L. reuteri* 23272 (the equivalent of DSM 20016) to ameliorate infectious colitis that is enhanced by psychological stress [81,82]. Using a model in which the colonization of the gut by the murine pathogen *Citrobacter rodentium* and ensuing colitis is enhanced by social stress, Bailey and coworkers have shown that *L. reuteri* 23272 supplementation can

suppress stressor-induced colitis. They further showed that stress-induced colitis is dependent on the chemokine CCL2 and that *L. reuteri* supplementation was able to significantly reduce CCL2 mRNA levels in the colon, suggesting a possible mechanism for how the probiotic effect is being mediated.

Another inflammatory disease that is ameliorated by *L. reuteri* supplementation in animal models is osteoporosis [85–87]. During menopause, bone resorption increases due to the loss of estrogen and several animal studies have shown that inflammation in the bone marrow, notably increased TNF levels, drives osteoporosis [88]. Ovariectomized mice (ovx) show a rapid loss of trabecular bone over 4 weeks that is suppressed by *L. reuteri* 6475. In the ovx model, *L. reuteri* supplementation completely suppressed bone loss in femur and vertebrae trabecular bone [85]. Although the mechanism by which *L. reuteri* impacts bone health is still under investigation, secreted factors are capable of suppressing the formation of bone resorbing osteoclasts in vitro. It will be interesting to find out if histamine suppression of TNF in the gut is important for ameliorating osteoporosis. In addition, *L. reuteri* 6475 also suppresses bone loss in streptozocin-treated mice (Type I diabetes model), but only in males [87]. These studies suggest that probiotic *L. reuteri* may be worthwhile investigating for the ability to ameliorate bone loss in humans.

FUTURE DIRECTIONS AND APPLICATIONS FOR *Lactobacillus reuteri*

The past decade has seen a number of exciting developments in how microbes impact human health. Although probiotic bacterial strains have shown promise in larger trials recently, the fact still remains that translation of probiotics into meaningful therapeutic options are yet to be realized. This is due in part to the fact that many of the strains available on the market were not selected based on a function that will ameliorate a disease but rather on their ability to be grown easily in industrial settings and their ability to be grandfathered into GRAS status due to long-term use in foods. Moving forward we will need to have a more rational selection of probiotic strains for the prevention and treatment of specific diseases which will require a much better understanding of the mechanistic interactions of these strains with the host and indigenous microbiota.

L. reuteri strains are poised to play an important role in this transition. As highlighted earlier, strains that impact histamine metabolism and oxytocin production are beginning to uncover the interactions of *L. reuteri* with the host. Several independent studies have also shown the ability of different *L. reuteri* strains to impact the formation of T_{regs} in the intestinal tract, which may partially explain the antiinflammatory properties of these strains. However, we still have an incomplete understanding of the products produced by *L. reuteri* that impact health and disease. In fully understanding the complex cross talk between *L. reuteri* and the host, it will be necessary to establish powerful genetic tools in strains that will be available for public consumption. Thus, the development of recombineering and CRISPR technology for precision genome engineering of *L. reuteri* 6475 and *Lactococcus lactis* have been, and will continue to be, critical tools in understanding the factors produced by *L. reuteri* that impact host physiology [89–92]. The use of new human enteroid and human intestinal organoid technologies will also be important for understanding how human-derived probiotics interact with the human gut [93,94].

ACKNOWLEDGMENTS

Research in the Britton laboratory on *L. reuteri* has been supported by the National Institutes of Health (1RO1AT007695, R21-RAT005472), the National Cancer Institute (U01 CA170930), and the Gerber Foundation.

REFERENCES

[1] Reuter G. The *Lactobacillus* and *Bifidobacterium* microflora of the human intestine: composition and succession. Curr Issues Intest Microbiol 2001;2(2):43–53.

[2] Morita H, Toh H, Fukuda S, et al. Comparative genome analysis of *Lactobacillus reuteri* and *Lactobacillus fermentum* reveal a genomic island for reuterin and cobalamin production. DNA Res 2008;15(3):151–61.

[3] Valeur N, Engel P, Carbajal N, et al. Colonization and immunomodulation by *Lactobacillus reuteri* ATCC 55730 in the human gastrointestinal tract. Appl Environ Microbiol 2004;70(2):1176–81.

[4] Wall T, Bath K, Britton RA, et al. The early response to acid shock in *Lactobacillus reuteri* involves the ClpL chaperone and a putative cell wall-altering esterase. Appl Environ Microbiol 2007;73(12):3924–35.

[5] Whitehead K, Versalovic J, Roos S, et al. Genomic and genetic characterization of the bile stress response of probiotic *Lactobacillus reuteri* ATCC 55730. Appl Environ Microbiol 2008;74(6):1812–9.

[6] Rosander A, Connolly E, Roos S. Removal of antibiotic resistance gene-carrying plasmids from *Lactobacillus reuteri* ATCC 55730 and characterization of the resulting daughter strain, *L. reuteri* DSM 17938. Appl Environ Microbiol 2008;74(19):6032–40.

[7] Iniesta M, Herrera D, Montero E, et al. Probiotic effects of orally administered *Lactobacillus reuteri*-containing tablets on the subgingival and salivary microbiota in patients with gingivitis. A randomized clinical trial. J Clin Periodontol 2012;39(8):736–44.

[8] Romani Vestman N, Chen T, Lif Holgerson P, et al. Oral microbiota shift after 12-week supplementation with *Lactobacillus reuteri* DSM 17938 and PTA 5289; a randomized control trial. PLoS One 2015;10(5):e0125812.

[9] Anukam K, Osazuwa E, Ahonkhai I, et al. Augmentation of antimicrobial metronidazole therapy of bacterial vaginosis with oral probiotic *Lactobacillus rhamnosus* GR-1 and *Lactobacillus reuteri* RC-14: randomized, double-blind, placebo controlled trial. Microbes Infect 2006;8(6):1450–4.

[10] Martinez RC, Franceschini SA, Patta MC, et al. Improved cure of bacterial vaginosis with single dose of tinidazole (2 g), *Lactobacillus rhamnosus* GR-1, and *Lactobacillus reuteri* RC-14: a randomized, double-blind, placebo-controlled trial. Can J Microbiol 2009;55(2): 133–8.

[11] Cadieux P, Wind A, Sommer P, et al. Evaluation of reuterin production in urogenital probiotic *Lactobacillus reuteri* RC-14. Appl Environ Microbiol 2008;74(15):4645–9.

[12] Gao C, Major A, Rendon D, et al. Histamine H2 receptor-mediated suppression of intestinal inflammation by probiotic *Lactobacillus reuteri*. MBio 2015;6(6):e01358-15.

[13] Lin YP, Thibodeaux CH, Pena JA, et al. Probiotic *Lactobacillus reuteri* suppress proinflammatory cytokines via c-Jun. Inflamm Bowel Dis 2008;14(8):1068–83.

[14] Livingston M, Loach D, Wilson M, et al. Gut commensal *Lactobacillus reuteri* 100-23 stimulates an immunoregulatory response. Immunol Cell Biol 2010;88(1):99–102.

[15] Tannock GW, Ghazally S, Walter J, et al. Ecological behavior of *Lactobacillus reuteri* 100-23 is affected by mutation of the luxS gene. Appl Environ Microbiol 2005;71(12):8419–25.

[16] Walter J, Chagnaud P, Tannock GW, et al. A high-molecular-mass surface protein (Lsp) and methionine sulfoxide reductase B (MsrB) contribute to the ecological performance of *Lactobacillus reuteri* in the murine gut. Appl Environ Microbiol 2005;71(2):979–86.

[17] Walter J, Loach DM, Alqumber M, et al. D-alanyl ester depletion of teichoic acids in *Lactobacillus reuteri* 100-23 results in impaired colonization of the mouse gastrointestinal tract. Environ Microbiol 2007;9(7):1750–60.

[18] Frese SA, Benson AK, Tannock GW, et al. The evolution of host specialization in the vertebrate gut symbiont *Lactobacillus reuteri*. PLoS Genet 2011;7(2):e1001314.

[19] Oh PL, Benson AK, Peterson DA, et al. Diversification of the gut symbiont *Lactobacillus reuteri* as a result of host-driven evolution. ISME J 2010;4(3):377–87.

[20] Walter J, Britton RA, Roos S. Host-microbial symbiosis in the vertebrate gastrointestinal tract and the *Lactobacillus reuteri* paradigm. Proc Natl Acad Sci USA 2011;108(Suppl. 1):4645–52.

[21] Wesney E, Tannock GW. Association of rat, pig, and fowl biotypes of lactobacilli with the stomach of gnotobiotic mice. Microb Ecol 1979;5(1):35–42.

[22] Talarico TL, Casas IA, Chung TC, et al. Production and isolation of reuterin, a growth inhibitor produced by *Lactobacillus reuteri*. Antimicrob Agents Chemother 1988;32(12):1854–8.

[23] Talarico TL, Dobrogosz WJ. Chemical characterization of an antimicrobial substance produced by *Lactobacillus reuteri*. Antimicrob Agents Chemother 1989;33(5):674–9.

[24] Talarico TL, Axelsson LT, Novotny J, et al. Utilization of glycerol as a hydrogen acceptor by *Lactobacillus reuteri*: purification of 1,3-propanediol:NAD oxidoreductase. Appl Environ Microbiol 1990;56(4):943–8.

[25] Luthi-Peng Q, Dileme FB, Puhan Z. Effect of glucose on glycerol bioconversion by *Lactobacillus reuteri*. Appl Microbiol Biotechnol 2002;59(2–3):289–96.

[26] Vollenweider S, Lacroix C. 3-Hydroxypropionaldehyde: applications and perspectives of biotechnological production. Appl Microbiol Biotechnol 2004;64(1):16–27.

[27] Schaefer L, Auchtung TA, Hermans KE, et al. The antimicrobial compound reuterin (3-hydroxypropionaldehyde) induces oxidative stress via interaction with thiol groups. Microbiology 2010;156(Pt. 6): 1589–99.

[28] Vollenweider S, Evers S, Zurbriggen K, et al. Unraveling the hydroxypropionaldehyde (HPA) system: an active antimicrobial

agent against human pathogens. J Agric Food Chem 2010;58(19): 10315–22.

[29] Santos F, Vera JL, van der Heijden R, et al. The complete coenzyme B12 biosynthesis gene cluster of *Lactobacillus reuteri* CRL1098. Microbiology 2008;154(Pt. 1):81–93.

[30] Spinler JK, Taweechotipatr M, Rognerud CL, et al. Human-derived probiotic *Lactobacillus reuteri* demonstrate antimicrobial activities targeting diverse enteric bacterial pathogens. Anaerobe 2008;14(3):166–71.

[31] Cleusix V, Lacroix C, Vollenweider S, et al. Inhibitory activity spectrum of reuterin produced by *Lactobacillus reuteri* against intestinal bacteria. BMC Microbiol 2007;7:101.

[32] Angiolillo L, Conte A, Zambrini AV, et al. Biopreservation of Fior di Latte cheese. J Dairy Sci 2014;97(9):5345–55.

[33] Langa S, Martin-Cabrejas I, Montiel R, et al. Short communication: combined antimicrobial activity of reuterin and diacetyl against foodborne pathogens. J Dairy Sci 2014;97(10):6116–21.

[34] Montiel R, Martin-Cabrejas I, Langa S, et al. Antimicrobial activity of reuterin produced by *Lactobacillus reuteri* on *Listeria monocytogenes* in cold-smoked salmon. Food Microbiol 2014;44:1–5.

[35] Ganzle MG. Reutericyclin: biological activity, mode of action, and potential applications. Appl Microbiol Biotechnol 2004;64(3):326–32.

[36] Holtzel A, Ganzle MG, Nicholson GJ, et al. The first low molecular weight antibiotic from lactic acid bacteria: reutericyclin, a new tetramic acid. Angew Chem Int Ed Engl 2000;39(15):2766–8.

[37] Yang Y, Zhao X, Le MH, et al. Reutericyclin producing *Lactobacillus reuteri* modulates development of fecal microbiota in weanling pigs. Front Microbiol 2015;6:762.

[38] Saulnier DM, Santos F, Roos S, et al. Exploring metabolic pathway reconstruction and genome-wide expression profiling in *Lactobacillus reuteri* to define functional probiotic features. PLoS One 2011;6(4):e18783.

[39] Talarico TL, Dobrogosz WJ. Purification and characterization of glycerol dehydratase from *Lactobacillus reuteri*. Appl Environ Microbiol 1990;56(4):1195–7.

[40] Molina VC, Medici M, Taranto MP, et al. *Lactobacillus reuteri* CRL 1098 prevents side effects produced by a nutritional vitamin B deficiency. J Appl Microbiol 2009;106(2):467–73.

[41] Santos F, Teusink B, de Vos WM, et al. The evidence that pseudovitamin B(12) is biologically active in mammals is still lacking—a comment on Molina et al.'s (2009) experimental design. J Appl Microbiol 2009;107(5):1763. author reply 1764.

[42] Santos F, Vera JL, Lamosa P, et al. Pseudovitamin B(12) is the corrinoid produced by *Lactobacillus reuteri* CRL1098 under anaerobic conditions. FEBS Lett 2007;581(25):4865–70.

[43] Stupperich E, Nexo E. Effect of the cobalt-N coordination on the cobamide recognition by the human vitamin B12 binding proteins intrinsic factor, transcobalamin and haptocorrin. Eur J Biochem 1991;199(2):299–303.

[44] Christensen HR, Frokiaer H, Pestka JJ. Lactobacilli differentially modulate expression of cytokines and maturation surface markers in murine dendritic cells. J Immunol 2002;168(1):171–8.

[45] Thomas CM, Hong T, van Pijkeren JP, et al. Histamine derived from probiotic *Lactobacillus reuteri* suppresses TNF via modulation of PKA and ERK signaling. PLoS One 2012;7(2):e31951.

[46] Yang XD, Ai W, Asfaha S, et al. Histamine deficiency promotes inflammation-associated carcinogenesis through reduced myeloid

maturation and accumulation of CD11b+Ly6G+ immature myeloid cells. Nat Med 2011;17(1):87–95.

[47] Levkovich T, Poutahidis T, Smillie C, et al. Probiotic bacteria induce a 'glow of health'. PLoS One 2013;8(1):e53867.

[48] Poutahidis T, Kearney SM, Levkovich T, et al. Microbial symbionts accelerate wound healing via the neuropeptide hormone oxytocin. PLoS One 2013;8(10):e78898.

[49] Poutahidis T, Kleinewietfeld M, Smillie C, et al. Microbial reprogramming inhibits Western diet-associated obesity. PLoS One 2013;8(7):e68596.

[50] Poutahidis T, Springer A, Levkovich T, et al. Probiotic microbes sustain youthful serum testosterone levels and testicular size in aging mice. PLoS One 2014;9(1):e84877.

[51] Wessel MA, Cobb JC, Jackson EB, et al. Paroxysmal fussing in infancy, sometimes called colic. Pediatrics 1954;14(5):421–35.

[52] Savino F, Pelle E, Palumeri E, et al. *Lactobacillus reuteri* (American Type Culture Collection Strain 55730) versus simethicone in the treatment of infantile colic: a prospective randomized study. Pediatrics 2007;119(1):e124–30.

[53] Savino F, Cordisco L, Tarasco V, et al. *Lactobacillus reuteri* DSM 17938 in infantile colic: a randomized, double-blind, placebo-controlled trial. Pediatrics 2010;126(3):e526–33.

[54] Roos S, Dicksved J, Tarasco V, et al. 454 pyrosequencing analysis on faecal samples from a randomized DBPC trial of colicky infants treated with *Lactobacillus reuteri* DSM 17938. PLoS One 2013;8(2):e56710.

[55] Indrio F, Di Mauro A, Riezzo G, et al. Prophylactic use of a probiotic in the prevention of colic, regurgitation, and functional constipation: a randomized clinical trial. JAMA Pediatr 2014;168(3):228–33.

[56] Mi GL, Zhao L, Qiao DD, et al. Effectiveness of *Lactobacillus reuteri* in infantile colic and colicky induced maternal depression: a prospective single blind randomized trial. Antonie Van Leeuwenhoek 2015;107(6):1547–53.

[57] Savino F, Ceratto S, Poggi E, et al. Preventive effects of oral probiotic on infantile colic: a prospective, randomised, blinded, controlled trial using *Lactobacillus reuteri* DSM 17938. Benef Microbes 2015;6(3):245–51.

[58] Sung V, Hiscock H, Tang ML, et al. Treating infant colic with the probiotic *Lactobacillus reuteri*: double blind, placebo controlled randomised trial. BMJ 2014;348:g2107.

[59] Szajewska H, Gyrczuk E, Horvath A. *Lactobacillus reuteri* DSM 17938 for the management of infantile colic in breastfed infants: a randomized, double-blind, placebo-controlled trial. J Pediatr 2013;162(2):257–62.

[60] Indrio F, Riezzo G, Raimondi F, et al. The effects of probiotics on feeding tolerance, bowel habits, and gastrointestinal motility in preterm newborns. J Pediatr 2008;152(6):801–6.

[61] Indrio F, Riezzo G, Raimondi F, et al. *Lactobacillus reuteri* accelerates gastric emptying and improves regurgitation in infants. Eur J Clin Invest 2011;41(4):417–22.

[62] Gephart SM, Spitzer AR, Effken JA, et al. Discrimination of GutCheck(NEC): a clinical risk index for necrotizing enterocolitis. J Perinatol 2014;34(6):468–75.

[63] Kantorowska A, Wei JC, Cohen RS, et al. Impact of donor milk availability on breast milk use and necrotizing enterocolitis rates. Pediatrics 2016;137(3):1–8.

[64] Liu Y, Fatheree NY, Mangalat N, et al. *Lactobacillus reuteri* strains reduce incidence and severity of experimental necrotizing

enterocolitis via modulation of TLR4 and NF-kappaB signaling in the intestine. Am J Physiol 2012;302(6):G608–17.

[65] Liu Y, Fatheree NY, Mangalat N, et al. Human-derived probiotic *Lactobacillus reuteri* strains differentially reduce intestinal inflammation. Am J Physiol 2010;299(5):G1087–96.

[66] Liu Y, Fatheree NY, Dingle BM, et al. *Lactobacillus reuteri* DSM 17938 changes the frequency of Foxp3+ regulatory T cells in the intestine and mesenteric lymph node in experimental necrotizing enterocolitis. PLoS One 2013;8(2):e56547.

[67] Liu Y, Tran DQ, Fatheree NY, et al. *Lactobacillus reuteri* DSM 17938 differentially modulates effector memory T cells and Foxp3+ regulatory T cells in a mouse model of necrotizing enterocolitis. Am J Physiol 2014;307(2):G177–86.

[68] Dingle BM, Liu Y, Fatheree NY, et al. FoxP3(+) regulatory T cells attenuate experimental necrotizing enterocolitis. PLoS One 2013;8(12):e82963.

[69] Hunter C, Dimaguila MA, Gal P, et al. Effect of routine probiotic, *Lactobacillus reuteri* DSM 17938, use on rates of necrotizing enterocolitis in neonates with birthweight <1000 grams: a sequential analysis. BMC Pediatr 2012;12:142.

[70] Shornikova AV, Casas IA, Isolauri E, et al. *Lactobacillus reuteri* as a therapeutic agent in acute diarrhea in young children. J Pediatr Gastroenterol Nutr 1997;24(4):399–404.

[71] Shornikova AV, Casas IA, Mykkanen H, et al. Bacteriotherapy with *Lactobacillus reuteri* in rotavirus gastroenteritis. Pediatr Infect Dis J 1997;16(12):1103–7.

[72] Weizman Z, Asli G, Alsheikh A. Effect of a probiotic infant formula on infections in child care centers: comparison of two probiotic agents. Pediatrics 2005;115(1):5–9.

[73] Agustina R, Kok FJ, van de Rest O, et al. Randomized trial of probiotics and calcium on diarrhea and respiratory tract infections in Indonesian children. Pediatrics 2012;129(5):e1155–64.

[74] Cimperman L, Bayless G, Best K, et al. A randomized, double-blind, placebo-controlled pilot study of *Lactobacillus reuteri* ATCC 55730 for the prevention of antibiotic-associated diarrhea in hospitalized adults. J Clin Gastroenterol 2011;45(9):785–9.

[75] Dinleyici EC, Dalgic N, Guven S, et al. *Lactobacillus reuteri* DSM 17938 shortens acute infectious diarrhea in a pediatric outpatient setting. J Pediatr 2015;91(4):392–6.

[76] Dinleyici EC, Group PS, Vandenplas Y. *Lactobacillus reuteri* DSM 17938 effectively reduces the duration of acute diarrhoea in hospitalised children. Acta Paediatr 2014;103(7):e300–5.

[77] Gutierrez-Castrellon P, Lopez-Velazquez G, Diaz-Garcia L, et al. Diarrhea in preschool children and *Lactobacillus reuteri*: a randomized controlled trial. Pediatrics 2014;133(4):e904–9.

[78] Urbanska M, Gieruszczak-Bialek D, Szymanski H, et al. Effectiveness of *Lactobacillus reuteri* DSM 17938 for the prevention of nosocomial diarrhea in children: a randomized, double-blind, placebo-controlled trial. Pediatr Infect Dis J 2016;35(2):142–5.

[79] Wanke M, Szajewska H. Lack of an effect of *Lactobacillus reuteri* DSM 17938 in preventing nosocomial diarrhea in children: a randomized, double-blind, placebo-controlled trial. J Pediatr 2012;161(1):40-3.e.1.

[80] Dicksved J, Schreiber O, Willing B, et al. *Lactobacillus reuteri* maintains a functional mucosal barrier during DSS treatment despite mucus layer dysfunction. PLoS One 2012;7(9):e46399.

[81] Mackos AR, Eubank TD, Parry NM, et al. Probiotic *Lactobacillus reuteri* attenuates the stressor-enhanced severity of *Citrobacter rodentium* infection. Infect Immun 2013;81(9):3253–63.

[82] Mackos AR, Galley JD, Eubank TD, et al. Social stress-enhanced severity of *Citrobacter rodentium*-induced colitis is CCL2-dependent and attenuated by probiotic *Lactobacillus reuteri*. Mucosal Immunol 2015;9:515–26.

[83] Madsen KL, Doyle JS, Jewell LD, et al. *Lactobacillus* species prevents colitis in interleukin 10 gene-deficient mice. Gastroenterology 1999;116(5):1107–14.

[84] Schreiber O, Petersson J, Phillipson M, et al. *Lactobacillus reuteri* prevents colitis by reducing P-selectin-associated leukocyte- and platelet-endothelial cell interactions. Am J Physiol 2009;296(3):G534–42.

[85] Britton RA, Irwin R, Quach D, et al. Probiotic *L. reuteri* treatment prevents bone loss in a menopausal ovariectomized mouse model. J Cell Physiol 2014;229(11):1822–30.

[86] McCabe LR, Irwin R, Schaefer L, et al. Probiotic use decreases intestinal inflammation and increases bone density in healthy male but not female mice. J Cell Physiol 2013;228(8):1793–8.

[87] Zhang J, Motyl KJ, Irwin R, et al. Loss of bone and Wnt10b expression in male type 1 diabetic mice is blocked by the probiotic *Lactobacillus reuteri*. Endocrinology 2015;156(9):3169–82.

[88] Weitzmann MN, Pacifici R. Estrogen regulation of immune cell bone interactions. Ann NY Acad Sci 2006;1068:256–74.

[89] Oh JH, van Pijkeren JP. CRISPR-Cas9-assisted recombineering in *Lactobacillus reuteri*. Nucleic Acids Res 2014;42(17):e131.

[90] van Pijkeren JP, Britton RA. High efficiency recombineering in lactic acid bacteria. Nucleic Acids Res 2012;40(10):e76.

[91] van Pijkeren JP, Britton RA. Precision genome engineering in lactic acid bacteria. Microb Cell Fact 2014;13(Suppl. 1):S10.

[92] Van Pijkeren JP, Neoh KM, Sirias D, et al. Exploring optimization parameters to increase ssDNA recombineering in *Lactococcus lactis* and *Lactobacillus reuteri*. Bioengineered 2012;3(4):209–17.

[93] Dedhia PH, Bertaux-Skeirik N, Zavros Y, et al. Organoid models of human jastrointestinal development and disease. Gastroenterology 2016;150:1098–112.

[94] Zachos NC, Kovbasnjuk O, Foulke-Abel J, et al. Human enteroids/colonoids and intestinal organoids functionally recapitulate normal intestinal physiology and pathophysiology. J Biol Chem 2016;291(8):3759–66.

Chapter 9

The Use of *Lactobacillus casei* and *Lactobacillus paracasei* in Clinical Trials for the Improvement of Human Health

R.M. Jones

Department of Pediatrics, Emory University School of Medicine, Atlanta, GA, United States

INTRODUCTION

The genus *Lactobacillus* is a part of the lactic acid bacteria group of bacteria of which *Lactobacillus casei, Lactobacillus paracasei,* and *Lactobacillus rhamnosus* are well-defined representative species [1,2]. Together, these three species are taxonomically classified within a narrow clade (*L. casei* group) of facultative heterofermentative lactobacilli, although discourse over this classification continues [2–6]. Before 1989, the *L. casei* group consisted of five subspecies, with the type strain being ATCC 393 [7]. Thereafter, the group was further classified into the current three species within this group, namely *L. casei* (including the ATCC 393 strain), *L. paracasei* that was further divided into two subspecies (subsp. *tolerans* and subsp. *paracasei*), and *L. rhamnosus* [8].

In subsequent years, appeals were put forward to the International Committee on Systematic Bacteriology (ICSB) to modify the *L. casei* group, most notably to reject the name *L. paracasei*, and to reclassify *L. casei* ATCC 393 [9,10]. In addition, the status of the nomenclature in this group has been somewhat contested since 1989, with classification wavering between *L. casei* and *L. paracasei*. This is because the strains in the earlier *L. casei* subsp. *casei* clade that are remaining in the new *L. casei* group, only consist of strains that are similar to ATCC 393. Thus, the key bone of contention is the extent to which strains formerly classified as *L. casei* actually possess a phylogenetic association with ATCC 393. However these proposals were rejected [11] and the present taxonomy of the *L. casei* group is still based on the proposition by Collins et al. [8].

Isolation of the majority of the *L. casei* group of bacteria has occurred from fermented foodstuffs such as wine, kimchi, or pickle, as well as fermented or raw dairy, such as cheese as the source. Many of the isolated *L. casei* and *L. paracasei* strains have been used as single probiotics, or as part of a synbiotic consortium within formulations. Of these formulations, it has been proposed that about a third of commercial strains intended for probiotic use were misidentified [12]. This may be due to the fact that the classification of the *L. casei* group strains is a challenging task because ribotyping or 16S rRNA gene sequences do not adequately discriminate between closely related taxa. Thankfully, other improved genetic approaches have been developed to classify members of this group, based on partial hsp60 gene sequences [13], the comparative sequence analysis of a recA gene [4], and the identification of lactobacilli by pheS and rpoA gene sequence analyses [14]. Together, a combination of these approaches holds promise to establish a veritable classification of the strains in the *L. casei* group, which is crucial for culture gathering and curating in both academic and in industrial settings.

L. casei and *L. paracasei* strains have been extensively used in trials to assess the extent to which they influence human health. At least in studies undertaken in mice, 20–40% of administered *L. casei* and *L. paracasei* can survive and populate within the physiological conditions of the stomach and the duodenum after oral administration [15]. However, whereas animal models are useful for the initial identification of novel probiotics, and for the elucidation of the underlying molecular mechanisms of probiotic action, they do not substitute for clinical trials to study the safety and efficacy of a probiotic intervention in humans. Here, we highlight a number of examples of the use of *L. casei* and *L. paracasei* in clinical trials since 2010. A number of other reviews have adumbrated findings in clinical trials involving *L. casei* and *L. paracasei* up to 2010 [16–19].

CLINICAL TRIALS EMPLOYING *L. casei* AND *L. paracasei* TO TREAT DIGESTIVES DISEASES

It is essential to establish the safety and efficacy of a probiotic intervention in humans, in human clinical trials. For example, a study that assessed *L. casei* strain Shirota in 28 critically ill children revealed no evidence of either colonization or bacteremia with *L. casei* strain Shirota in bacterial cultures obtained from these patients. *L. casei* strain Shirota administration was tolerated with no observable side effects, suggesting that the use of *L. casei* strain Shirota as a probiotic in enterally fed, and even in critically ill children is safe [20]. In addition, a study examining the influence of *L. casei* LBC80R on antibiotic-associated diarrhea (AAD) found that probiotics reduced the risk of AAD and, in particular, *Clostridium difficile* diarrhea in hospitalized patients on antibiotics with no reported adverse effects [21].

As well as *C. difficile* diarrhea, probiotic interventions have identified effectiveness in the treatment of Norovirus-induced diarrhea in humans. An open-case controlled study that enrolled 77 adults with an average age of 84 years was conducted to assay the efficacy of *L. casei* strain Shirota against norovirus in the elderly. The study reported that while daily consumption of milk fermented with the probiotic reduced the mean duration of fever, importantly, no adverse observable side effects were observed [22]. A nutritional supplement program based on zinc, prebiotics, probiotics, and vitamins to prevent radiation-related gastrointestinal disorders was studied in a phase II trial. The aim was to evaluate the tolerance and safety of Dixentil, a nutritional supplement based on zinc with the addition of prebiotics (galactooligosaccharides) and *L. casei* that is given as prophylaxis to patients undergoing pelvic radiotherapy. The study concluded that the use of Dixentil is an easy, safe, and feasible approach to protect patients against the risk of radiation-induced diarrhea [23].

Two recent randomized control trials assessed the efficacy of synbiotics for the management of acute gastroenteritis. Trial parameters included the duration of diarrhea, stool amount, and stool frequency. The first trial compared the efficacy of five probiotic strains (*L. rhamnosus*, *Lactobacillus acidophilus*, *Streptococcus thermophilus*, *Bifidobacterium infantis*, and *Bifidobacterium lactis*) in 111 children with acute diarrhea (median age: 40 months). Compared with the placebo group, the median duration of diarrhea was significantly shorter in the synbiotic group

where also fewer medications, such as antibiotics were required to be administered [24]. In the second clinical trial, it was reported that a synbiotic combination of *L. paracasei* B21060 plus arabinogalactan and xilooligosaccharides protects against acute gastroenteritis. In this study of 107 children aged between 3 and 36 months, the rate of resolution of diarrhea at 72 h postsupplementation was significantly higher in children who received the synbiotic combination. In addition, children in the synbiotic group had significant reduced diarrhea duration and number of stool outputs [25]. In another important randomized control trial assessing *L. paracasei* ST11 (10^{10} CFU/day for 5 days) supplementation found reduced duration of illness in children with moderate nonrotavirus diarrhea [26].

Dysbiosis in peridiverticular bacterial flora has been postulated as a key step in the pathogenesis of diverticulitis, thus making the administration of probiotics an attractive interventional therapy for this disease. Indeed, many studies have investigated the efficacy of symbiotic preparations containing *L. paracasei*, *L. acidophilus*, *Lactobacillus helveticus*, and *Bifidobacterium* on ameliorating abdominal pain in patients with uncomplicated diverticular disease. Of these, *L. paracasei* may be particularly suited as a probiotic treatment for diverticulitis since it is able to survive the passage through the gastrointestinal tract, to persist in stool after administration is discontinued, and to temporarily associate throughout different sites of the entire human colon [27]. Indeed, the benefit of *L. paracasei* subsp. *paracasei* *F19* administration with a high-fiber diet on abdominal pain and bloating in symptomatic uncomplicated diverticular disease has been documented [28].

A randomized clinical trial assessing the effectiveness of mesalazine and *L. casei* subsp. DG in maintaining remission in symptomatic uncomplicated diverticular disease (SUDD) found that both cyclic mesalazine and *L. casei* DG treatments, particularly when given in combination, appear to be better than placebo for maintaining remission of SUDD [29]. Another study demonstrated the positive effects of a synergistic combination of *L. paracasei* B21060 and arabinogalactan/xylooligosaccharides on abdominal symptoms in patients with uncomplicated diverticulitis [30]. This synbiotic combination has been shown to be effective in relieving symptoms associated with irritable bowel syndrome (IBS). For example, a study involving a clinical trial to determine the efficacy of the symbiotic formulation, marketed as Flortec, in patients with IBS was carried out. In IBS-predominant diarrhea, Flortec significantly reduced bowel movements, pain, and IBS scores [31]. In a follow-up study, *L. paracasei* B21060 (Flortec) was shown to be effective in the treatment of acute diarrhea in adults treated at a primary care setting [32].

A number of clinical trials have reported on the positive influences of *L. casei* and *L. paracasei* in various intestinal diseases. For example, probiotic drinks containing *L. casei*

DN-114001 was shown to reduce AAD. In an observational study, the frequency of bowel movements from 258 patients with infections in a primary care hospital in western Germany were assessed and showed that *L. casei* DN 114001 cost-efficiently reduced the prevalence of AAD during antibiotic treatment [33]. Also, a double-blind comparative study to evaluate the effects of a fermented milk beverage containing *L. casei* strain Shirota in patients showed that continuous consumption of *L. casei*-fermented milk relieves irregular bowel movement in gastrectomized patients [34].

A clinical trial to evaluate the effect of probiotic treatment on the small bowel injuries induced by chronic low-dose aspirin use showed that coadministration of *L. casei* is effective for the treatment of aspirin-associated small bowel injury [35]. Also, a study to evaluate the orocecal intestinal transit time (ITT) before and after administration of a dairy product containing *L. casei* CRL 431and fiber supplemet in healthy women, showed that ITT decreased significantly after consumption of the synbiotics [36]. The effect of the continuous intake of *L. casei* strain Shirota on fever in a mass outbreak of norovirus gastroenteritis and on the fecal microflora in a health service facility for the aged in Japan showed that *L. casei* strain Shirota positively contributed to the alleviation of fever caused by norovirus gastroenteritis by correcting the imbalance of the intestinal microflora peculiar to the elderly, although such consumption could not protect them from norovirus gastroenteritis itself [37].

A study examining the effect of a fermented dairy product containing *L. casei* DN-114 on the incidence of respiratory and gastrointestinal common infectious diseases and on immune functions in healthy shift workers concluded that daily consumption of a fermented dairy product containing *L. casei* DN-114 could reduce the risk of common infections in stressed shift workers [38]. Characterization of the effect of the *L. casei* DG on colonic-associated microbiota, mucosal cytokine balance, and toll-like receptor expression in 26 patients with mild left-sided UC showed that a modulation of the diversity of the mucosal microbiota by *L. casei* DG occurs during the beneficial activities of probiotics in UC patients [39]. Furthermore, the role of *L. casei* strain Shirota in preventing acute diarrhea in children in a community-based field trial in an urban slum in India showed that daily intake of a probiotic drink can play a role in prevention of acute diarrhea in young children in a community setting of a developing country [40].

A randomized, double-blind clinical trial undertaken to assess the effectiveness of probiotics in the prevention of necrotizing enterocolitis (NEC) in newborns weighing <1500 g supplementing with a multispecies probiotic (*L. acidophilus*, *L. rhamnosus*, *L. casei*, *L. plantarum*, *B. infantis*, and *S. thermophillus*) at a rate of 1 g/day plus their regular feedings reported that probiotics may offer potential benefits for premature infants and are a promising strategy in the reduction of the risk of NEC in preterm newborns

[41]. In another study, the combined use of *L. casei* and *Bifidobacterium breve* on the prevention of the occurrence of NEC stage ≥2 by the criteria of Bell in very-low-birth-weight preterm infants showed that oral supplementation of these two probiotics reduced the occurrence of NEC by the improvement of intestinal motility [42].

Finally, a metaanalysis of randomized controlled trials on the effect of probiotics, including *L. casei* strain Shirota, on functional constipation in adults revealed that probiotics may improve whole gut transit time, stool frequency, and stool consistency, with subgroup analysis indicating beneficial effects of *L. casei* strain Shirota in particular [43]. However, caution is needed with the interpretation of these data due to their high heterogeneity and risk of bias. Adequately powered clinical trials are required to better determine the species or strains, doses, and duration of use of probiotics that are most efficacious. Nevertheless, the substantial body of literature generated from clinical trials measuring the beneficial influences of *L. casei* or *L. paracasei* supplementation is irrefutable, and importantly, none of the mentioned studies described adverse effects of probiotic supplementation (Table 9.1)

CONTROL OF BACTERIAL INFECTIONS

The effectiveness of *L. casei* strain Shirota in controlling infections among the residents of an aged care facility was assessed in a Randomized Placebo-Controlled Double-Blind Trial among residents (average age, 85) and staff members (average age, 37). The participants randomly received either *L. casei* strain Shirota-fermented milk or a placebo beverage once daily for 6 months. Clinical data and enteric health were compared and concluded that long-term consumption of *L. casei* strain Shirota may be useful for decreasing the daily risk of infection and improving the quality of life among the residents and staff of facilities for the elderly [44]. An in vivo microbiological study assessing the effect of *L. casei* strain Shirota on salivary *Streptococcus mutans* count revealed that daily consumption of milk containing this probiotic can reduce the numbers of *S. mutans* and may contribute to the prevention of dental caries [45]. In a study to determine the effectiveness of two different *L. casei* strain Shirota doses in decreasing urea concentrations in patients with chronic kidney disease, the study showed more than a 10% decrease in the serum urea concentrations following *L. casei* strain Shirota supplementation in patients with stage 3 and 4 chronic kidney disease [46].

A number of trials have been undertaken using *L. casei* strain Shirota on upper respiratory tract infections (URTIs), which present clinically as the common cold. These include a multicenter study involving 1072 volunteers (median age, 76) investigated the influence of *L. casei* on the prevention of viral respiratory tract infections. Here, randomized trials for the consumption of either dairy products containing the

TABLE 9.1 Clinical Trials Employing *L. casei* or *L. paracasei* for the Treatment of Digestives Diseases

Strain Used	Clinical Pathology	References
L. casei LBC80R	Improve symptoms of *C. difficile* diarrhea patients	[21]
L. casei strain Shirota	Improve symptoms of norovirus in the elderly	[22]
L. casei (unspecified)	Reduce radiation-related gastrointestinal disorders	[23]
L. paracasei B21060	Improve symptoms of acute gastroenteritis	[25]
L. paracasei ST11	Improve symptoms of moderate nonrotavirus diarrhea	[26]
L. paracasei F19	Reduce abdominal pain in symptomatic diverticulitis	[27,28]
L. casei subsp. DG	Maintaining remission following diverticulitis	[29]
L. paracasei B21060	Improve symptoms of uncomplicated diverticulitis	[30]
L. paracasei B21060	Improved diarrhea symptoms	[31,32]
L. casei DN-114001	Reduce incidences of AAD	[33]
L. casei strain Shirota	Relieves irregular bowel movement in gastrectomized patients	[34]
L. casei	Prevention of aspirin-associated small bowel injury	[35]
L. casei CRL 431	Slowing of intestinal transit time	[36]
L. casei strain Shirota	Alleviation of fever caused by norovirus gastroenteritis	[37]
L. casei DN-114	Prevent common infections in stressed shift workers	[38]
L. casei DG	Dampen mucosal immune system in UC patients	[39]
L. casei strain Shirota	Preventing acute diarrhea in children	[40]
L. casei (unspecified)	Prevention of NEC	[41,42]
L. casei Shirota	Improve whole gut transit time, stool frequency, and stool consistency	[43]

AAD, Antibiotic-associated diarrhea; NEC, necrotizing enterocolitis.

L. casei for 3 months recorded reduced episodes and cumulative durations of URTIs, and rhinopharyngitis [47]. Another multicenter study that registered 154 senior citizens investigated the influence of fermented milk containing *L. casei* strain Shirota (and placebo drinks as control) concluded that *L. casei* strain Shirota probably reduces the duration of acute URTIs [48].

In another study, consumption of *L. paracasei and L. plantarum* reduced the occurrence of contracting one or more common cold episode from 67% in the control group to 55%. Furthermore, during the 12-week period, the number of days with common cold symptoms was significantly reduced from 8.6 days in the control group to 6.2 days in the probiotic group [49]. In addition, the daily intake of fermented milk with *L. casei* strain Shirota was shown to reduce the incidence and duration of URTIs in healthy middle-aged office workers. Here, a randomized controlled trial of 96 male workers aged 30–49 years consumed *L. casei* strain Shirota 1.0×10^{11} CFU's for 12 weeks during the winter season. URTI episodes were evaluated by a questionnaire of URTI symptoms and data concluded that the daily intake of fermented milk with *L. casei* strain Shirota may reduce the risk of URTIs in healthy middle-aged office workers [50].

However, the general effect of probiotics on respiratory tract infections may be bacterial strain-specific. Two studies failed to find significant protective effects of *L. casei* and *L. paracasei* on URTIs. These include a study to determine the influence of *L. paracasei* subsp. *paracasei, L. casei* 431 on the immune response to influenza vaccination and URTIs in healthy adult volunteers by a randomized, double-blind, placebo-controlled, and parallel-group study in 1104 healthy subjects aged 18–60 years at 2 centers in Germany and Denmark. Subjects randomly assigned to receive an acidified milk drink containing 10^9 CFU *L. casei* 431 ($n = 553$) or placebo ($n = 551$) for 42 days showed that daily consumption of this strain resulted in no observable effect on the components of the immune response to influenza vaccination but did reduce the duration of upper respiratory symptoms [51]. Other studies investigating the efficacy of daily intake of *L. casei* strain Shirota on respiratory symptoms and influenza vaccination immune response concluded that daily consumption of a fermented milk drink that contains *L. casei* strain Shirota has no statistically or clinically significant effect on the protection against respiratory symptoms [52].

Probiotics have been used in association with antibiotics for the treatment of *Helicobacter pylori* infections. *H. pylori* can colonize the gastric mucosa, and often leads to chronic gastritis, peptic ulcer disease, gastric mucosa-associated lymphoid tissue lymphoma, and gastric adenocarcinoma. Clinical trials in children investigated the effects of probiotics on *H. pylori* eradication rates as an adjuvant to an eliminating regimen. The earliest study on the inclusion of *L. casei* DN-114001 to a standard triple therapy against *H. pylori* observed that *L. casei* DN-114001 confers an enhanced therapeutic benefit on *H. pylori* eradication in children with gastritis [53]. However, another study assessing the effect of a probiotic food as an adjuvant to triple therapy for eradication of *H. pylori* infection found no significant differences in *H. pylori* eradication rates at 1 and 3 months [54]. A metaanalysis of the effect of probiotics supplementation on *H. pylori* eradication rates and side effects during eradication therapy concluded that supplementation with specific strains of probiotics compared with eradication therapy may be considered an option for increasing eradication rates, particularly when antibiotic therapies are relatively ineffective [55]. Thus, at this moment, the impact on side effects remains unclear and more high-quality trials on specific probiotic strains are thus needed to clearly assess the efficacy of probiotics in *H. pylori*-related diseases.

THE EFFECTS OF PROBIOTICS ON IMMUNITY

A probiotic yogurt containing *L. casei* DN001 was shown to modulate gene expression of a specific subset of transcription factors and cytokines in peripheral blood mononuclear cells (PBMC) of obese and overweight people. In a randomized doubled-blind controlled clinical trial that enrolled 75 individuals with BMI 25–35, participants received 200 g/day yogurt containing *L. casei* DN001, *L. acidophilus* La5, and *Bifidobacterium* Bb12. The gene expression of the FOXP3, T-bet, GATA3, TNF-α, IFN-γ, TGF-β, and ROR-γt in PBMC genes were measured, before and after intervention. In probiotic-treated groups, IFN-γ and ROR-γt expression were reduced, and FOXP3 was increased showing that probiotic yogurt had synergistic effects on T-cell specific gene expression in PBMCs among overweight and obese individuals [56]. In addition, a study aimed at investigating the effects of *L. casei* 01 supplementation on symptoms and inflammatory biomarkers of rheumatoid arthritis in women, revealed that probiotic supplementation may be an appropriate adjunct therapy for rheumatoid arthritis patients [57].

A daily intake of *L. casei* strain Shirota was shown to increase natural killer cell (NK) activity in a double-blind, placebo-controlled, randomized study conducted on 72 healthy male smokers. Specifically, the probiotic prevented the expected smoke-dependent NK activity reduction in comparison to the placebo intake group [58]. Probiotics have also been shown to enhance immunlization. Evaluation of the immune benefits of two probiotic strains, namely *Bifidobacterium animalis* ssp. *lactis*, BB-12 and *L. casei* 431 in an influenza vaccination model by a randomized, double-blind, placebo-controlled study showed that supplementation with BB-12 or *L. casei* 431 may be an effective means to improve immune function by priming systemic and mucosal immune responses to challenge [59].

Probiotic supplements have been investigated by a few research groups as a therapy for established food allergies. These include two recent similar studies that yielded conflicting results, both investigating the effects of *L. rhamnosus* GG and *L. casei* CRL431 on cow's milk allergy in infants. First, a randomized controlled trial investigated the effect of feeding extensively hydrolyzed casein formula in combination with *L. rhamnosus* GG (*n* = 28 in treatment, *n* = 27 in control group). It was found that the children in the treatment group had a significantly higher chance of gaining tolerance at 6 and 12 months of treatment [60]. Another study investigated the influence *L. casei* 431 and *B. lactis* BB-12 to accelerate tolerance in infants with cow's milk allergy. Here, 119 infants were randomized to a combination of *L. casei* 431 and *B. lactis* BB-12 in hydrolyzed formula or control for 12 months. However, in this study neither probiotic bacteria induced any detectable increased tolerance at 6 and 12 months [61], suggesting that the probiotic effects are likely to be highly strain-specific, and that the combination of *L. rhamnosus* GG and *L. casei* 431 employed in the earlier study had a higher efficacy. Moving forward, there is an urgent need to identify the specific efficacious strains that dampen proinflammatory responses such that they may become routine supplements to boost immune function and thereby promote efficient resolution of disease (Table 9.2).

PROBIOTICS IN THE ENHANCEMENT OF LIPID METABOLISM

Probiotics offer an attractive strategy for optimizing lipid metabolism in populations that have increasing prevalence of obesity, and the chronic diseases associated with obesity. Intense investigations have focused on the capacity of lactobacilli to facilitate cellular lipid metabolism. For example, a study on cholesterol to coprostanol conversion by strains of lactobacilli, including *L. casei* 393, reported intracellular and extracellular cholesterol reductase activity in all tested bacterial strains. The study showed that cholesterol was converted to coprostanol both intracellularly and extracellularly and that fermentation reactions performed by probiotic strains reduced cholesterol concentration in the medium [62]. Furthermore, *L. paracasei* F19 was shown to reduce fat storage in association with sharp changes in the activity of peroxisome proliferator-activated receptors, which are a group of nuclear receptor proteins that function as transcription factors regulating the expression of genes involved in metabolism [63].

In addition, *L. casei* strain Shirota supplementation was also shown to be effective in the prevention of high-fat diet-induced insulin resistance in human subjects [64], and a 4 week supplementation of *L. paracasei* subsp. *paracasei* W8 was reported to induce modest effect on triacylglycerol levels in healthy young adults [65]. A study investigating whether probiotics had an effect on proinflammatory markers in overweight and obese individuals, and assessing whether they could have synergistic effects with weight-loss

TABLE 9.2 Clinical Trials Employing *Lactobacillus casei* or *Lactobacillus paracasei* for the Treatment of Bacterial Infectious Diseases and for the Enhancement of Immune Functions

Strain Used	Clinical Pathology	References
Control of Bacterial Infections		
L. casei strain Shirota	Lowering risk of infection	[44]
L. casei strain Shirota	Lower salivary *Streptococcus mutans* count	[45]
L. casei strain Shirota	Improve symptoms chronic kidney disease	[46]
L. casei strain Shirota	URTIs	[47,48,50]
L. casei 431	Augmentation of immune response to flu vaccination	[51]
L. casei strain Shirota	Augmentation of immune response to flu vaccination	[52]
L. casei DN-114001	*H. pylori* infection in children	[53–55]
Effects on Immunity		
L. casei DN001	Enhance T-cell gene expression	[56]
L. casei 01	Improve symptoms of rheumatoid arthritis	[57]
L. casei strain Shirota	Increase NK activity	[58]
L. paracasei	Priming systemic and mucosal immune responses	[59]
L. casei CRL431	Accelerate tolerance in cow's milk allergy	[60]

NK, Natural killer cell; URTIs, upper respiratory tract infections.

diets in improving health showed that probiotic yogurt that included *L. casei* DN001 lowered proinflammatory markers, lowered fat percentages, and lowered body weight [66].

The effect of probiotic fermented milk (kefir) on glycemic control and lipid profile in type 2 diabetic patients was measured in a randomized double-blind placebo-controlled clinical trial of 60 diabetic patients aged 35–65 years. Patients were randomly and equally assigned to consume either probiotic fermented milk or conventional fermented milk for 8 weeks. The probiotic group consumed 600 mL/day probiotic fermented milk containing *L. casei*, *L. acidophilus*, and *Bifidobacteria*. The study concluded that probiotic fermented milk can be useful as an adjuvant therapy in the treatment of diabetes [67]. Finally, a study designed to determine the effects of multispecies probiotic supplements on metabolic profiles and oxidative stress in diabetic patients showed that ingestion of a supplement that included *L. casei* for 8 weeks in diabetic patients prevented a rise in fasting plasma glucose and resulted in a decrease serum levels of C-reactive protein, and caused an increase in plasma total glutathione levels [68]. Together, the effects of probiotic lactobacilli on lipid metabolism are well developed. The future challenge will be to incorporate these supplements into general use by overweight individuals who do not visit clinical settings. Indeed, routine use of lipid metabolism promoting probiotics should be incorporated into official recommendations that define a healthy diet.

PROBIOTICS AND CANCER PREVENTION

Many studies have reported the beneficial effect of probiotics against breast cancer. A recent study that enrolled a group of Japanese women postulated that regular consumption of beverages containing *L. casei* strain Shirota and soy isoflavones was inversely associated with the incidence of breast cancer. Specifically, in a population-based case-control study, 360 cases of breast cancer and 662 healthy controls aged 40–55 were enrolled. Odds ratios analysis revealed an association between soy consumption and breast cancer incidence, where the higher the isoflavone consumption, the odds of breast cancer becomes lower [69].

The preventative effects exerted by the soymilk and *L. casei* strain Shirota was recapitulated in a Sprague–Dawley rat carcinogenic model where the multiplicity and volume of mammary tumors were reduced by soymilk in combination with *L. casei* strain Shirota. Immunohistochemical analysis revealed that soymilk in combination with *L. casei* strain Shirota reduced the numbers of Estrogen Receptorα-positive and Ki-67-positive cells (marker for cell proliferation) in tumors leading the investigators to postulate that the habitual consumption of *L. casei* strain Shirota in combination with soymilk might be a beneficial dietary style for breast cancer prevention [70].

In Western diets, fermented dairy foods are more common than soy-based products. In an early investigation, dairy products fermented by different lactobacilli, including *L. paracasei* were evaluated in vitro, and were shown to inhibit the growth of a breast cancer cell line [71]. In addition, perioperative synbiotics were shown to decrease postoperative complications in periampullary neoplasms (PNs), which are progressive tumors with a poor prognosis and high mortality rates. Evaluation of the use of probiotics, including *L. casei* in a double-blind study of patients undergoing surgery for PNs revealed that perioperative administration of probiotics reduces postoperative mortality and complication rates [72]. Together, these are compelling new data, all published between 2013 and 2015 for an ameliorating effects of *L. casei* and *L. paracasei* on cancer. Due to the severe morbidity associated with cancer, any type of positive treatment is considered valuable. Thus, the assessment of the effects of probiotics on multiple types of cancers must be a high priority in developing inexpensive and efficacious adjuvant treatments.

INTERVENTIONS FOR DEPRESSIVE DISORDERS

Clinical and metabolic responses to probiotic administration in patients with major depressive disorder (MDD) was assessed in a randomized, double-blind, placebo-controlled trial. Patients were randomly allocated into two groups to receive either *L. casei* (2×10^9 CFU/g) supplements ($n = 20$) or placebo ($n = 20$) for 8 weeks. Data revealed that *L. casei* administration to patients with MDD had beneficial effects on Beck Depression Inventory, on insulin resistance, and on serum glutathione concentrations [73]. *L. casei* strain Shirota was also shown to prevent the onset of physical symptoms in medical students under academic examination stress. In a double-blind, placebo-controlled trial, 24 healthy medical students consumed fermented milk containing *L. casei* strain Shirota once a day for 8 weeks until the day before the examination whereupon psychophysical state was assessed. Results suggest that the daily consumption of *L. casei* strain Shirota prevented the onset of outward physical symptoms in healthy subjects exposed to stressful situations [74].

A randomized controlled trial tested the effect of multispecies probiotics on cognitive reactivity to sad mood. The study tested if a multispecies probiotic containing *L. casei* W56 could reduce cognitive reactivity in nondepressed individuals. In a triple-blind, placebo-controlled, randomized, pre- and postintervention assessment design, 20 healthy participants without current mood disorder received a 4-week probiotic food-supplement intervention, while 20 control participants received an inert placebo for the same period. Here, it was shown that the intake of probiotics may help reduce negative thoughts associated with sad mood [75]. Thus, probiotic supplementation warrants further research as a potential preventive strategy for depression. Clearly, standardization of defined parameters such as the Beck Depression Inventory is required such that transstudy metaanalysis assessments can be made to corroborate and substantiate conclusions. In addition, heterogeneity in lifestyles is certainly a confounding factor in studies examining depression. Future studies, must focus more on clearly defined parameters of depression, and must identify tight cohorts of homogenous patients for trials (Table 9.3).

TABLE 9.3 Clinical Trials Employing *Lactobacillus casei* or *Lactobacillus paracasei* for Therapeutic Interventions for Obesity, for Cancer Prevention, and for Depression Disorders

Strain Used	Clinical Pathology	References
Enhancement of Lipid Metabolism		
L. casei strain Shirota	Prevention of high-fat diet-induced insulin resistance	[64]
L. paracasei W8	Lowering of triacylglycerol levels	[65]
L. casei DN001	Lowering of fat percentages and body weight	[66]
L. casei	Adjuvant therapy in the treatment of diabetes	[67]
L. casei	Prevention of increased fasting plasma glucose in diabetic patients	[68]
Cancer Prevention		
L. casei strain Shirota	Lower chances of developing breast cancer	[69,70]
L. casei	Decrease postoperative complications in PNs	[72]
Interventions for Depressive Disorders		
L. casei	Improve outcomes in patients with MDD	[73]
L. casei strain Shirota	Beneficial effects preventing the onset of physical symptoms of stress	[74]
L. casei W56	Improve cognitive reactivity to sad mood	[75]

MDD, Major depressive disorder; PNs, periampullary neoplasms.

CONCLUSIONS

In summary, *L. casei* and *L. paracasei* have been widely used as a single probiotic therapy or in synbiotic formulations. The positive influence of *L. casei* and *L. paracasei* on health and disease is widely reported in clinical trials excreting beneficial influences on digestives diseases, infectious diseases, immunity, and on depression disorders. The vast majority of the clinical trials found in the literature report significant beneficial influences of *L. casei* and *L. paracasei*, whereas cases where no beneficial effect were found make up a small fraction of all studies. Indeed many of the cases where no significant beneficial influences were detected often involve the treatment of diseases with complex immunological etiology, such as rheumatoid arthritis [76], or following severe trauma or injury. In the future, isolating further strains of *L. casei* and *L. paracasei* with probiotic activities is necessary to advance the use of probiotics and synbiotics in therapeutic interventions to promote health and recovery from disease.

REFERENCES

[1] Makarova K, Slesarev A, Wolf Y, et al. Comparative genomics of the lactic acid bacteria. Proc Natl Acad Sci USA 2006;103(42): 15611–6.

[2] Sun Z, Harris HM, McCann A, et al. Expanding the biotechnology potential of lactobacilli through comparative genomics of 213 strains and associated genera. Nat Commun 2015;6:8322.

[3] Diancourt L, Passet V, Chervaux C, et al. Multilocus sequence typing of Lactobacillus casei reveals a clonal population structure with low levels of homologous recombination. Appl Environ Microbiol 2007;73(20):6601–11.

[4] Felis GE, Dellaglio F, Mizzi L, et al. Comparative sequence analysis of a recA gene fragment brings new evidence for a change in the taxonomy of the Lactobacillus casei group. Int J Syst Evol Microbiol 2001;51(Pt 6):2113–7.

[5] Sato H, Torimura M, Kitahara M, et al. Characterization of the Lactobacillus casei group based on the profiling of ribosomal proteins coded in S10-spc-alpha operons as observed by MALDI-TOF MS. Syst Appl Microbiol 2012;35(7):447–54.

[6] Toh H, Oshima K, Nakano A, et al. Genomic adaptation of the Lactobacillus casei group. PLoS One 2013;8(10).

[7] Skerman VBD, Mcgowan V, Sneath PHA. Approved lists of bacterial names. Int J Syst Bacteriol 1980;30(1):225–420.

[8] Collins MD, Phillips BA, Zanoni P. Deoxyribonucleic-acid homology studies of Lactobacillus-Casei, Lactobacillus-Paracasei Sp-Nov, Subsp Paracasei and Subsp Tolerans, and Lactobacillus-Rhamnosus Sp-Nov, Comb-Nov. Int J Syst Bacteriol 1989;39(2):105–8.

[9] Dellaglio F, Dicks LMT, Dutoit M, et al. Designation of Atcc 334 in place of Atcc 393 (Ncdo 161) as the Neotype Strain of Lactobacillus-Casei Subsp. Casei and rejection of the name Lactobacillus-Paracasei. Int J Syst Bacteriol 1991;41(2):340–2.

[10] Dicks LMT, DuPlessis EM, Dellaglio F, et al. Reclassification of Lactobacillus casei subsp. casei ATCC 393 and Lactobacillus rhamnosus ATCC 15820 as Lactobacillus zeae nom rev, designation of ATCC 334 as the neotype of L-casei subsp. casei, and rejection of the Lactobacillus paracasei. Int J Syst Bacteriol 1996;46(1):337–40.

[11] Tindall BJ, Syste JCIC. The type strain of Lactobacillus casei is ATCC 393, ATCC 334 cannot serve as the type because it represents a different taxon, the name Lactobacillus paracasei and its subspecies names are not rejected and the revival of the name "Lactobacillus zeae" contravenes Rules 51 b (1) and (2) of the International Code of Nomenclature of Bacteria. Opinion 82. Int J Syst Evol Microbiol 2008;58:1764–5.

[12] Huys G, Vancanneyt M, D'Haene K, et al. Accuracy of species identity of commercial bacterial cultures intended for probiotic or nutritional use. Res Microbiol 2006;157(9):803–10.

[13] Blaiotta G, Fusco V, Ercolini D, et al. Lactobacillus strain diversity based on partial hsp60 gene sequences and design of PCR-restriction fragment length polymorphism assays for species identification and differentiation. Appl Environ Microbiol 2008;74(1):208–15.

[14] Naser SM, Dawyndt P, Hoste B, et al. Identification of lactobacilli by pheS and rpoA gene sequence analyses. Int J Syst Evol Microbiol 2007;57(Pt 12):2777–89.

[15] Lee YK, Ho PS, Low CS, et al. Permanent colonization by Lactobacillus casei is hindered by the low rate of cell division in mouse gut. Appl Environ Microbiol 2004;70(2):670–4.

[16] Gao XW, Mubasher M, Fang CY, et al. Dose-response efficacy of a proprietary probiotic formula of Lactobacillus acidophilus CL1285 and Lactobacillus casei LBC80R for antibiotic-associated diarrhea and Clostridium difficile-associated diarrhea prophylaxis in adult patients. Am J Gastroenterol 2010;105(7):1636–41.

[17] Guandalini S. Probiotics for prevention and treatment of diarrhea. J Clin Gastroenterol 2011;45:S149–53.

[18] Venuto C, Butler M, Ashley ED, et al. Alternative therapies for Clostridium difficile infections. Pharmacotherapy 2010;30(12):1266–78.

[19] Chmielewska A, Szajewska H. Systematic review of randomised controlled trials: probiotics for functional constipation. World J Gastroenterol 2010;16(1):69–75.

[20] Srinivasan R, Meyer R, Padmanabhan R, et al. Clinical safety of Lactobacillus casei shirota as a probiotic in critically ill children. J Pediatr Gastr Nutr 2006;42(2):171–3.

[21] Gao XW, Mubasher M, Fang CY, et al. Dose-response efficacy of a proprietary probiotic formula of Lactobacillus acidophilus CL1285 and Lactobacillus casei LBC80R for antibiotic-associated diarrhea and Clostridium difficile-associated diarrhea prophylaxis in adult patients. Am J Gastroenterol 2010;105(7):1636–41.

[22] Nagata S, Asahara T, Ohta T, et al. Effect of the continuous intake of probiotic-fermented milk containing Lactobacillus casei strain Shirota on fever in a mass outbreak of norovirus gastroenteritis and the faecal microflora in a health service facility for the aged. Br J Nutr 2011;106(4):549–56.

[23] Scartoni D, Desideri I, Giacomelli I, et al. Nutritional supplement based on zinc, prebiotics, probiotics, and vitamins to prevent radiation-related gastrointestinal disorders. Anticancer Res 2015;35(10):5687–92.

[24] Vandenplas Y, De Hert SG. group PR-s. Randomised clinical trial: the synbiotic food supplement probiotical vs. placebo for acute gastroenteritis in children. Aliment Pharmacol Therapeut 2011;34(8):862–7.

[25] Passariello A, Terrin G, Cecere G, et al. Randomised clinical trial: efficacy of a new synbiotic formulation containing Lactobacillus paracasei B21060 plus arabinogalactan and xilooligosaccharides in children with acute diarrhoea. Aliment Pharmacol Therapeut 2012;35(7):782–8.

[26] Sarker SA, Sultana S, Fuchs GJ, et al. Lactobacillus paracasei strain ST11 has no effect on rotavirus but ameliorates the outcome of nonrotavirus diarrhea in children from Bangladesh. Pediatrics 2005;116(2):E221–8.

[27] Morelli L, Garbagna N, Rizzello F, et al. In vivo association to human colon of Lactobacillus paracasei B21060: map from biopsies. Dig Liver Dis 2006;38(12):894–8.

[28] Annibale B, Maconi G, Lahner E, et al. Efficacy of Lactobacillus paracasei sub. paracasei F19 on abdominal symptoms in patients with symptomatic uncomplicated diverticular disease: a pilot study. Minerva Gastroenterol Dietol 2011;57(1):13–22.

[29] Tursi A, Brandimarte G, Elisei W, et al. Randomised clinical trial: mesalazine and/or probiotics in maintaining remission of symptomatic uncomplicated diverticular disease—a double-blind, randomised, placebo-controlled study. Alim Pharmacol Therapeut 2013;38(7):741–51.

[30] Lahner E, Esposito G, Zullo A, et al. High-fibre diet and Lactobacillus paracasei B21060 in symptomatic uncomplicated diverticular disease. World J Gastroenterol 2012;18(41):5918–24.

[31] Andriulli A, Neri M, Loguercio C, et al. Clinical trial on the efficacy of a new symbiotic formulation, Flortec, in patients with irritable bowel syndrome—a multicenter, randomized study. J Clin Gastroenterol 2008;42(8):S218–23.

[32] Grossi E, Buresta R, Abbiati R, et al. Clinical trial on the efficacy of a new symbiotic formulation, Flortec, in patients with acute diarrhea: a multicenter, randomized study in primary care. J Clin Gastroenterol 2010;44(Suppl. 1):S35–41.

[33] Dietrich CG, Kottmann T, Alavi M. Commercially available probiotic drinks containing *Lactobacillus casei* DN-114001 reduce antibiotic-associated diarrhea. World J Gastroenterol 2014;20(42): 15837–44.

[34] Aoki T, Asahara T, Matsumoto K, et al. Effects of the continuous intake of a milk drink containing *Lactobacillus casei* strain Shirota on abdominal symptoms, fecal microbiota, and metabolites in gastrectomized subjects. Scand J Gastroenterol 2014;49(5):552–63.

[35] Endo H, Higurashi T, Hosono K, et al. Efficacy of *Lactobacillus casei* treatment on small bowel injury in chronic low-dose aspirin users: a pilot randomized controlled study. J Gastroenterol 2011;46(7): 894–905.

[36] Malpeli A, Gonzalez S, Vicentin D, et al. Randomised, double-blind and placebo-controlled study of the effect of a synbiotic dairy product on orocecal transit time in healthy adult women. Nutr Hosp 2012;27(4):1314–9.

[37] Nagata S, Asahara T, Ohta T, et al. Effect of the continuous intake of probiotic-fermented milk containing *Lactobacillus casei* strain Shirota on fever in a mass outbreak of norovirus gastroenteritis and the faecal microflora in a health service facility for the aged. Br J Nutr 2011;106(4):549–56.

[38] Guillemard E, Tanguy J, Flavigny A, et al. Effects of consumption of a fermented dairy product containing the probiotic *Lactobacillus casei* DN-114 001 on common respiratory and gastrointestinal infections in shift workers in a randomized controlled trial. J Am Coll Nutr 2010;29(5):455–68.

[39] D'Inca R, Barollo M, Scarpa M, et al. Rectal administration of *Lactobacillus casei* DG modifies flora composition and toll-like receptor expression in colonic mucosa of patients with mild ulcerative colitis. Digest Dis Sci 2011;56(4):1178–87.

[40] Sur D, Manna B, Niyogi SK, et al. Role of probiotic in preventing acute diarrhoea in children: acommunity-based, randomized, double-blind placebo-controlled field trial in an urban slum. Epidemiol Infect 2011;139(6):919–26.

[41] Fernandez-Carrocera LA, Solis-Herrera A, Cabanillas-Ayon M, et al. Double-blind, randomised clinical assay to evaluate the efficacy of probiotics in preterm newborns weighing less than 1500 g in the prevention of necrotising enterocolitis. Arch Dis Child Fetal Neonatal Ed 2013;98(1):F5–9.

[42] Braga TD, da Silva GAP, de Lira PIC, et al. Efficacy of *Bifidobacterium breve* and *Lactobacillus casei* oral supplementation on necrotizing enterocolitis in very-low-birth-weight preterm infants: a double-blind, randomized, controlled trial. Am J Clin Nutr 2011;93(1):81–6.

[43] Dimidi E, Christodoulides S, Fragkos KC, et al. The effect of probiotics on functional constipation in adults: a systematic review and meta-analysis of randomized controlled trials. Am J Clin Nutr 2014;100(4):1075–84.

[44] Nagata S, Asahara T, Wang C, et al. The effectiveness of *Lactobacillus* beverages in controlling infections among the residents of an aged care facility: a randomized placebo-controlled double-blind trial. Ann Nutr Metab 2016;68(1):51–9.

[45] Yadav M, Poornima P, Roshan NM, et al. Evaluation of probiotic milk on salivary mutans streptococci count: an in vivo microbiological study. J Clin Pediatr Dent 2014;39(1):23–6.

[46] Miranda Alatriste PV, Urbina Arronte R, Gomez Espinosa CO, et al. Effect of probiotics on human blood urea levels in patients with chronic renal failure. Nutr Hosp 2014;29(3):582–90.

[47] Guillemard E, Tondu F, Lacoin F, et al. Consumption of a fermented dairy product containing the probiotic *Lactobacillus casei* DN-114001 reduces the duration of respiratory infections in the elderly in a randomised controlled trial. Br J Nutr 2010;103(1):58–68.

[48] Fujita R, Iimuro S, Shinozaki T, et al. Decreased duration of acute upper respiratory tract infections with daily intake of fermented milk: a multicenter, double-blinded, randomized comparative study in users of day care facilities for the elderly population. Am J Infect Control 2013;41(12):1231–5.

[49] Berggren A, Ahren IL, Larsson N, et al. Randomised, double-blind and placebo-controlled study using new probiotic lactobacilli for strengthening the body immune defence against viral infections. Eur J Nutr 2011;50(3):203–10.

[50] Shida K, Sato T, Iizuka R, et al. Daily intake of fermented milk with *Lactobacillus casei* strain Shirota reduces the incidence and duration of upper respiratory tract infections in healthy middle-aged office workers. Eur J Nutr 2015;. Epub ahead of print.

[51] Jespersen L, Tarnow I, Eskesen D, et al. Effect of *Lactobacillus paracasei* subsp. *paracasei*, L. casei 431 on immune response to influenza vaccination and upper respiratory tract infections in healthy adult volunteers: a randomized, double-blind, placebo-controlled, parallel-group study. Am J Clin Nutr 2015;101(6):1188–96.

[52] Van Puyenbroeck K, Hens N, Coenen S, et al. Efficacy of daily intake of *Lactobacillus casei* Shirota on respiratory symptoms and influenza vaccination immune response: a randomized, double-blind, placebo-controlled trial in healthy elderly nursing home residents. Am J Clin Nutr 2012;95(5):1165–71.

[53] Sykora J, Valeckova K, Amlerova J, et al. Effects of a specially designed fermented milk product containing probiotic *Lactobacillus casei* DN-114 001 and the eradication of H. pylori in children: a prospective randomized double-blind study. J Clin Gastroenterol 2005;39(8):692–8.

[54] Goldman CG, Barrado DA, Balcarce N, et al. Effect of a probiotic food as an adjuvant to triple therapy for eradication of *Helicobacter pylori* infection in children. Nutrition 2006;22(10):984–8.

[55] Dang Y, Reinhardt JD, Zhou X, et al. The effect of probiotics supplementation on *Helicobacter pylori* eradication rates and side effects during eradication therapy: a meta-analysis. PLoS One 2014;9(11):e111030.

[56] Zarrati M, Shidfar F, Nourijelyani K, et al. *Lactobacillus acidophilus* La5, Bifidobacterium BB12, and *Lactobacillus casei* DN001 modulate gene expression of subset specific transcription factors and cytokines in peripheral blood mononuclear cells of obese and overweight people. BioFactors 2013;39(6):633–43.

[57] Alipour B, Homayouni-Rad A, Vaghef-Mehrabany E, et al. Effects of *Lactobacillus casei* supplementation on disease activity and inflammatory cytokines in rheumatoid arthritis patients: a randomized double-blind clinical trial. Int J Rheum Dis 2014;17(5):519–27.

[58] Reale M, Boscolo P, Bellante V, et al. Daily intake of *Lactobacillus casei* Shirota increases natural killer cell activity in smokers. Br J Nutr 2012;108(2):308–14.

[59] Rizzardini G, Eskesen D, Calder PC, et al. Evaluation of the immune benefits of two probiotic strains *Bifidobacterium animalis* ssp. *lactis*,

BB-12 (R) and *Lactobacillus paracasei* ssp. *paracasei, L. casei* 431 (R) in an influenza vaccination model: a randomised, double-blind, placebo-controlled study. Br J Nutr 2012;107(6):876–84.

[60] Berni Canani R, Nocerino R, Terrin G, et al. Effect of Lactobacillus GG on tolerance acquisition in infants with cow's milk allergy: a randomized trial. J Allergy Clin Immunol 2012;129(2). 580-2, 2 e1–e5.

[61] Hol J, van Leer EH, Elink Schuurman BE, et al. The acquisition of tolerance toward cow's milk through probiotic supplementation: a randomized, controlled trial. J Allergy Clin Immunol 2008;121(6):1448–54.

[62] Lye HS, Rusul G, Liong MT. Removal of cholesterol by lactobacilli via incorporation and conversion to coprostanol. J Dairy Sci 2010;93(4):1383–92.

[63] Aronsson L, Huang Y, Parini P, et al. Decreased fat storage by *Lactobacillus paracasei* is associated with increased levels of angiopoietin-like 4 protein (ANGPTL4). PLoS One 2010;5(9):e13087.

[64] Hulston CJ, Churnside AA, Venables MC. Probiotic supplementation prevents high-fat, overfeeding-induced insulin resistance in human subjects. Br J Nutr 2015;113(4):596–602.

[65] Bjerg AT, Kristensen M, Ritz C, et al. Four weeks supplementation with *Lactobacillus paracasei* subsp. paracasei L. casei W8(R) shows modest effect on triacylglycerol in young healthy adults. Benef Microbes 2015;6(1):29–39.

[66] Zarrati M, Salehi E, Nourijelyani K, et al. Effects of probiotic yogurt on fat distribution and gene expression of proinflammatory factors in peripheral blood mononuclear cells in overweight and obese people with or without weight-loss diet. J Am Coll Nutr 2014;33(6):417–25.

[67] Ostadrahimi A, Taghizadeh A, Mobasseri M, et al. Effect of probiotic fermented milk (kefir) on glycemic control and lipid profile in type 2 diabetic patients: a randomized double-blind placebo-controlled clinical trial. Iran J Public Health 2015;44(2):228–37.

[68] Asemi Z, Zare Z, Shakeri H, et al. Effect of multispecies probiotic supplements on metabolic profiles, hs-CRP, and oxidative stress in patients with type 2 diabetes. Ann Nutr Metab 2013;63(1–2):1–9.

[69] Toi M, Hirota S, Tomotaki A, et al. Probiotic beverage with soy isoflavone consumption for breast cancer prevention: a case-control study. Curr Nutr Food Sci 2013;9(3):194–200.

[70] Kaga C, Takagi A, Kano M, et al. *Lactobacillus casei* Shirota enhances the preventive efficacy of soymilk in chemically induced breast cancer. Cancer Sci 2013;104(11):1508–14.

[71] Biffi A, Coradini D, Larsen R, et al. Antiproliferative effect of fermented milk on the growth of a human breast cancer cell line. Nutr Cancer 1997;28(1):93–9.

[72] Sommacal HM, Bersch VP, Vitola SP, et al. Perioperative synbiotics decrease postoperative complications in periampullary neoplasms: a randomized, double-blind clinical trial. Nutr Cancer 2015;67(3): 457–62.

[73] Akkasheh G, Kashani-Poor Z, Tajabadi-Ebrahimi M, et al. Clinical and metabolic response to probiotic administration in patients with major depressive disorder: a randomized, double-blind, placebo-controlled trial. Nutrition 2016;32(3):315–20.

[74] Kato-Kataoka A, Nishida K, Takada M, et al. Fermented milk containing *Lactobacillus casei* strain Shirota prevents the onset of physical symptoms in medical students under academic examination stress. Benef Microbes 2016;7(2):153–6.

[75] Steenbergen L, Sellaro R, van Hemert S, et al. A randomized controlled trial to test the effect of multispecies probiotics on cognitive reactivity to sad mood. Brain Behav Immun 2015;48:258–64.

[76] Vaghef-Mehrabany E, Homayouni-Rad A, Alipour B, et al. Effects of probiotic supplementation on oxidative stress indices in women with rheumatoid arthritis: a randomized double-blind clinical trial. J Am Coll Nutr 2015;1–9.

Chapter 10

Beneficial Influences of *Lactobacillus plantarum* on Human Health and Disease

T.M. Darby and R.M. Jones
Department of Pediatrics, Emory University School of Medicine, Atlanta, GA, United States

INTRODUCTION

The naming of the species *Streptobacterium plantarum* by Orla Jensen in 1919 was originally part of characterizing a series of bacterial strains isolated from butter, milk, cheese, fermenting potatoes, beets, cabbage, and dough. One of these strains, now known as *Lactobacillus plantarum* [1] is a versatile and flexible species of lactic acid bacterium that is found in a wide range of different ecological niches including vegetable [2], meat and dairy substrates [2,3], and the gastrointestinal tract of humans and metazoans [4–8]. The diversity of niches is probably due to the fact that this bacterium is able to ferment a broad range of sugars [9]. This flexibility of *L. plantarum* is reflected by its relatively large genome size, with a large number of proteins involved in regulation and transport functions, and a high metabolic potential [10]. At the genetic level, *L. plantarum* has a relatively large genome compared to other *Lactobacillus* spp. It has a 3.3 Mb circular chromosome consisting of 3052 protein-encoding genes and only 39 of these genes are pseudogenes. The genome also contains a total of 62 tRNA encoding genes as well as the two classes of transposase mobile genetic elements [10].

The *Lactobacillus* species is a phylogenetically homogeneous group that comprises six species/subspecies including *L. plantarum*, *Lactobacillus paraplantarum*, *Lactobacillus pentosus*, *Lactobacillus fabifermentans*, and *Lactobacillus xiangfangensis*. Combined, they form a closely related taxa known as the "*L. plantarum* group." Within *L. plantarum*, a number of subspecies have been identified. These include *L. plantarum* subsp. *argentoratensis* and *L. plantarum* subsp. *plantarum* [11]. All of these species are facultatively heterofermentative (Group II), produce sodium D,L-lactate, and contain meso-diaminopimelic acid in their cell wall [12,13]. *L. pentosus*, formerly considered similar to *L. plantarum* [14], is generally positive for xylose fermentation, whereas *Lactobacillus arizonensis*

and *L. paraplantarum* are unable to ferment this carbohydrate [15–17]. Unlike *L. plantarum* and *L. paraplantarum*, *L. pentosus* ferments glycerol but not melezitose [13,16,18]. *L. fabifermentans* [19] and *L. xiangfangensis* [20] are much more recent species, isolated from cocoa fermentations and Chinese pickles.

L. plantarum is also a constituent of the normal bacterial flora of the gastrointestinal tract. It is frequently isolated from the human intestinal lumen and noted for being a species that is capable of surviving the low pH of the stomach and duodenum, adept at resisting the effect of bile acids in the small intestine, and transiently occupying the gastrointestinal tract by binding to the intestinal and colonic mucosa. In addition, enteral ingestion of *L. plantarum* decreases bacterial groups with gas-producing ability, such as *Veillonella* spp. and *Clostridia* spp. Evidence has now been accrued to substantiate the efficacy of certain *L. plantarum*, in particular strain Lp299v, which has been shown to be capable of bringing about a significant positive impacts on health and disease as will now be discussed in subsequent sections.

Not all lactic acid bacteria possess the ability to confer health benefits for the host. Thus, it becomes necessary to screen and characterize numerous strains to obtain ideal probiotics. In vitro and in vivo tests to demonstrate the capacities of lactic acid bacteria as probiotics are the focus of intense interest in academic and industrial settings. A typical example of such a study on the characterization of the probiotic potential of *L. plantarum* collected 98 isolates from Italian and Argentinean cheeses, which were evaluated for probiotic potential. Considerable heterogeneity was found among a number of *L. plantarum* strains screened in this study, leading to the design of multiple cultures to cooperatively link strains showing the widest range of useful traits [21]. In a more directed study, the probiotic properties of *L. plantarum* CECT 7315 and CECT 7316 isolated from feces of healthy children was assessed. Both strains, due to

The Microbiota in Gastrointestinal Pathophysiology

109

their high ability to survive at gastrointestinal tract conditions and to adhere to intestinal epithelial cells, as well as inhibitory activity against a wide range of enteropathogens, and ability to induce the production of antiinflammatory cytokine IL-10, highlight their potential as excellent candidates for being tested in clinical trials aimed to demonstrate beneficial effects on human health [22]. Typically, the safety of potential probiotics is also rigorously assessed. For example, for *L. plantarum* 423, safety was determined in trials with Wistar rats. Results commonly used are the health of the spleen and liver, and whether blood counts are normal, thus showing that the strain is not pathogenic. In addition, evidence that strains produce antimicrobial peptides active against pathogens are often assessed such that the strain and may be considered as potent probiotic [23].

Lactic acid fermentation is a simple way of preserving food and has probably been used by humans for hundreds, if not thousands of years. *Lactobacillus paracasei* and *Lactobacillus rhamnosus* are frequently associated with dairy products, whereas by contrast *L. plantarum* is used in fermented plant foods. Of note, a probiotic in lactic acid fermented oatmeal gruel that is mixed in a fruit drink was launched in Sweden in 1994 that contained 5×10^{10} CFU of *L. plantarum* 299v/L. The strain, *L. plantarum* 299 originates from the human intestinal mucosa [24], and is by far the most well-characterized strain of *L. plantarum* in clinical settings. This strain has been shown in humans to increase the concentration of carboxylic acids in feces and decrease abdominal bloating in patients with irritable bowel disease, as well as decreasing fibrinogen concentrations in blood. As will be described later, *L. plantarum* 299v also affects the bacterial flora within the intestine, and modulates the host's immunologic defense [25].

Colonization and endurance are considered as key elements in the success of probiotic supplements. Preliminary studies are often done in rodents, however, human studies are necessary to corroborate findings. The considerable variances in endurance in the gastrointestinal tract and ability to influence cytokine production after passage through the stomach and small intestine are well demonstrated by an early study in 1996, comparing four different probiotic species in *L. plantarum*, *L. paracasei*, *L. rhamnosus*, and *Bifidobacterium animalis*. Of the administered 10^8 CFU/mL of each probiotic, it was estimated that 10^7 *L. plantarum* bacterial cells remained after passage through the stomach and small intestine. This was in sharp contrast to the other probiotics tested and showed evidence of the potent capacity of *L. plantarum* to influence cytokine production after passage through the stomach and small intestine [26].

In another study, the long-term colonization of a *L. plantarum* synbiotic preparation was assessed in the neonatal gut. Healthy newborns >35 weeks of gestational age and >1800 g birth weight were randomized between 1 and 3 days after birth to receive an oral synbiotic preparation of *L. plantarum* and fructooligosaccharides. The study found that the synbiotic preparation colonized quickly after 3 days of administration and the infants stayed colonized for several months after therapy was stopped. There was also a reported increase in bacterial diversity of gram-positive organisms and a reduction of gram-negative bacterial load in the treatment group [27]. In addition to colonization, the potential fate of ingested *L. plantarum* as a probiotic must be considered. One such study examined 61 enrolled subjects receiving daily doses of fermented milk containing 2×10^{11} CFU of *L. plantarum* LP115. At 15 and 45 days after discontinuing supplementation, the number of lactobacilli was reduced back to the baseline levels (similar to those at time zero), showing that this particular probiotic is a transient, rather than a permeant colonizer of the gut [28]. Once the safety and efficacy of the probiotic has been established, the next step is to undertake clinical trials to examine the efficacy of the probiotic in clinical environments. In this chapter, we focus on the clinical trials and uses of *L. plantarum*, many of which have focused on the positive influences on health and disease of *L. plantarum* 299v.

TRIALS THAT REPORT THE SAFETY OF *Lactobacillus plantarum* AS A PROBIOTIC

Many clinical trials have reported on the safety of using *L. plantarum* as probiotic or symbiotic treatment for acute and chronic diseases with varying degrees of success. These include studies showing that *L. plantarum* 299v reduces colonization of *Clostridium difficile* in critically ill patients treated with antibiotics [29]. It was also shown in a controlled trial in patients with major abdominal surgery that early enteral supply of fiber and 2×10^{10} CFUs/day of *L. plantarum* 299V versus conventional nutrition had measured beneficial influences and few adverse effects [30]. A controlled trial in liver transplant recipients showed that early enteral supply of 2×10^{10} CFUs/day of *L. plantarum* 299V and fiber was well tolerated. *L. plantarum* 299V also markedly decreased the rate of postoperative infections both in comparison with inactivated *L. plantarum* 299 and a standard enteral nutrition formula [31]. The effect of *L. plantarum* enteral feeding on the gut permeability and septic complications in patients with acute pancreatitis in a surgical care unit has also been assessed, where *L. plantarum* was found to attenuate disease severity, improve intestinal permeability, and improve clinical outcomes with no adverse effects [32].

In addition, the adhesion of the *L. plantarum* 299v onto the gut mucosa in critically ill patients was measured in a randomized open trial. The study concluded that *L. plantarum* 299v could survive the passage from the stomach to the rectum and was able to adhere onto the rectal mucosa also in critically ill, antibiotic-treated patients. Again, these studies report no adverse effects of probiotic treatment, although bowel distension was reported in two patients [33]. Another study involved the prospective and randomized assessment of *L. plantarum* 299V on indices of bacterial translocation, gastric colonization, and septic complications in elective surgical patients. A total of 129 patients completed the study with 64 patients given *L. plantarum* 299V. This study concluded that administration of *L. plantarum* 299v in elective surgical patients does not influence the rate of bacterial translocation, gastric colonization, or incidence of postoperative septic morbidity [34].

To support these investigations into the safe use of probiotics, studies analyzing *L. plantarum* passage through an orogastrointestinal tract simulator showed a carrier matrix effect on transcriptional activity of genes associated with probiosis. This study proposed that the food matrix used to deliver beneficial bacteria may contribute to their probiotic action and may be variable in enhancing survival and gut colonization. Specifically, the investigators assessed the survival of *L. plantarum* WCFS1 in a human orogastrointestinal in vitro system, using different carrier matrices to compare protective and buffering properties. The study found higher survival of *L. plantarum* WCFS1 when in complex and/or nutrient-rich matrices [35]. Together, these studies report that the administration of *L. plantarum* to severely compromised patients have no measurable adverse effects on health, and indeed, in many cases have a positive influence.

Randomized Trials in Critically Ill Patients

A double-center and double-blind randomized clinical trial assessed the effects of perioperative probiotic treatment on serum Zonulin, a newly discovered protein that has an important role in the regulation of intestinal permeability, on subsequent postoperative infectious complications after colorectal cancer surgery. A total of 150 patients with colorectal carcinoma were randomly assigned to the control group ($n = 75$), which received placebo, or the probiotics group ($n = 75$). Patients in the probiotics group received a mixture of three probiotics bacteria composed of *L. plantarum* (CGMCC no. 1258), *L. acidophilus*-11, and *B. longum*-88. Both the probiotics and placebo were given orally for 6 days preoperatively and 10 days postoperatively. The study concluded that probiotic treatment can reduce the rate of postoperative septicemia which is associated with reduced serum Zonulin concentrations in patients [36].

A prospective randomized trial of probiotics in critically ill patients aimed to study the effect of the probiotic *L. plantarum* 299v on gut barrier function and the systemic inflammatory response in critically ill patients. One hundred and three critically ill patients were randomized to receive an oral preparation containing *L. plantarum* 299v (ProViva) in addition to conventional therapy. This study concluded that enteral administration of *L. plantarum* 299v to critically ill patients was associated with a late attenuation of the systemic inflammatory response, but was not accompanied by any significant changes in the intestinal microflora, intestinal permeability, endotoxin exposure, septic morbidity, or mortality [37]. These data are important since it is speculated that supplementation of any type of bacteria to patients with a compromised gut barrier, for example, after colorectal cancer surgery, would result in increased bacterial antigen translocation into the subepithelial compartments. These antigens, regardless of the fact that they are from a recognized probiotic bacteria, may still be indiscriminantly sensed by Toll-like receptors and evoke an inflammatory response.

Lactobacillus plantarum–RELATED TRIALS FOR IRRITABLE BOWEL SYNDROME

A controlled, double-blind, randomized study was conducted to assess the efficacy of *L. plantarum* 299V in patients with irritable bowel syndrome (IBS). Forty patients were randomized to receive either *L. plantarum* 299V in liquid suspension (20 patients) or placebo (20 patients) over a period of 4 weeks. All patients treated with *L. plantarum* 299V reported resolution of their abdominal pain as compared to 11 patients from the placebo group. A trend toward normalization of stool frequency in constipated patients was also reported in 6 out of 10 patients treated with *L. plantarum* 299V, together concluding that *L. plantarum* 299V has beneficial effect in patients with IBS [38]. In another IBS focused clinical trial for the relief of abdominal symptoms in a large subset of patients fulfilling the Rome III criteria, *L. plantarum* 299v (DSM 9843) was shown to improve symptoms of IBS. Here, a double-blind, placebo-controlled, parallel-designed study was set up, where subjects were randomized to receive daily either one capsule of *L. plantarum* 299v (DSM 9843) or placebo for 4 weeks. Assessment of the frequency and intensity of abdominal pain, bloating, and feeling of incomplete rectal emptying in 214 IBS enrolled patients concluded that 4-week treatment with *L. plantarum* 299v (DSM 9843) provided effective symptom relief, particularly on abdominal pain and bloating, in IBS patients fulfilling the Rome III criteria [39].

However, in another study that involved a randomized clinical trial looking at the effects of *L. plantarum* 299v over an 8-week period on symptoms of IBS concluded that *L. plantarum* 299v did not provide symptomatic relief, particularly of abdominal pain and bloating, in patients fulfilling the Rome II criteria [40]. The effect of *L. plantarum* 299v on colonic fermentation and symptoms of IBS were also assessed in a double-blind, placebo-controlled, crossover, 4-week trial of *L. plantarum* 299V in 12 previously untreated patients with IBS. Symptoms were assessed daily by a validated composite score and fermentation by calorimetry techniques. However, in this study, *L. plantarum*

299V did significantly alter colonic fermentation or improve symptoms in patients with IBS [41]. Thus it appears that some studies report amelioration of symptoms following *L. plantarum* 299V supplementation, whereas other report no effects. Importantly, in studies where no significant beneficial effects were discovered, at least no adverse effects were reported. Moreover, these differing reports underscore the need to undertake a metaanalysis of the group of similar studies to ascertain significance across clinical studies involving probiotics.

METAANALYSIS OF CLINICAL TRIALS

It is always important to critically assess clinical trials for the efficacy of probiotics on the basis of methodologic bias. Sufficient trials on the use of lactobacilli for digestive diseases have been undertaken to carry out metaanalysis on these studies. These include a metaanalysis to determine the overall efficacy of lactobacilli in major abdominal surgery assessing nine controlled trials with 733 patients enrolled. Despite these trials having used different probiotic formulations, in general, patients who were administered lactobacilli had shorter length of hospital stay, shorter period of antimicrobial therapy, and decreased incidence of infection [42]. Another metaanalysis on the effect of probiotic species on IBS symptoms concluded that some probiotics are an effective therapeutic option for IBS patients and that the effects on each IBS symptom are likely bacterial species specific. The authors added that future studies must focus on the role of probiotics in modulating intestinal microbiota and the immune system while considering individual patient symptom profiles [43,44].

A further metaanalysis, this time focusing on the role of *Lactobacillus* in the prevention of *C. difficile*–associated diarrhea in randomized controlled trials, concluded that there is sufficient evidence to recommend *L. acidophilus* and *Lactobacillus casei*, but not *L. plantarum* as a prevention therapy for treatment of this disease [45]. Finally, another assessment that compared 10 trials of critically ill patients administered probiotics, concluded that probiotics did not appear to influence mortality or duration of hospitalization. Importantly, the authors did note that recipients of the probiotics had fewer infectious episodes. Indeed, *L. plantarum* was one probiotic agent in which beneficial effects were significantly associated across many trials. This analysis affirmed that probiotics also reduced the incidence of antibiotic-associated diarrhea in hospitalized patients, although they added that trials did not specifically focus only on those who were critically ill. In addition, the authors of these analyses urge caution since, in general, methodological shortcomings introduce biases into the trials [46]. These analysis of the effectiveness of clinical trials together conclude that it is still not clear that probiotics are beneficial in the critically ill patient group.

CARDIOVASCULAR DISEASES, PANCREATIC DISEASES, AND RESPIRATORY TRACT INFECTIONS

The effect of *L. plantarum* 299v on cardiovascular disease risk factors in smokers has also been assessed. In this trial, 36 healthy volunteers (18 women and 18 men) aged 35–45 years were enrolled. The experimental group drank 400 mL/day of a rose-hip drink containing *L. plantarum* 299v and the control group consumed the same volume of product without bacteria over a 6-week period. Significant decreases in systolic blood pressure, leptin levels, fibrinogen, and interleukin 6 levels were recorded in the experimental group, whereas no such changes were observed in the control group. This trial concluded that *L. plantarum* administration leads to a reduction in cardiovascular disease risk factors and identified *L. plantarum* 299v as a useful protective agent in the primary prevention of atherosclerosis in smokers [47].

For patients with acute pancreatitis, a randomized clinical trial of *L. plantarum* 299 and fiber supplement by early enteral nutrition was carried out. Here, the aim was to determine the extent to which *L. plantarum* 299 could prevent colonization of gut pathogens in patients, thus reducing acute pancreatitis-associated endotoxemia. A total of 45 patients were enrolled, of which 50% were treated with live *L. plantarum* 299 and 50% with heat-killed *L. plantarum* 299 for 7 days. Examination of pancreatic necrosis and abscesses, as well as mean length of stay concluded that supplementary live *L. plantarum* 299 was significantly effective in reducing pancreatic sepsis and the number of necessary surgical interventions [48]. For respiratory tract infections, the consumption of *L. plantarum* was shown to lower the occurrence rate of acquiring rhinopharyngitis episodes from 67% in the control group to 55% in the probiotic group. Furthermore, during the 12-week period, the duration of the cold symptoms was significantly reduced from 8.6 days in the control group to 6.2 days in the probiotic group [49]. These data show that *L. plantarum* 299 supplementation has a significantly salutary benefit on patients with cardiovascular diseases, pancreatic diseases, or respiratory tract infections.

GYNECOLOGICAL AND IRON ABSORPTION INFLUENCES

The vaginal microbiota is predominantly constituted by *Lactobacillus* spp. [50]. Some probiotic bacteria, such as *L. plantarum* P17630 can suppress vulvovaginal candidiasis (VVC), caused by *Candida albicans*. Here, the *L. plantarum* P17630 is able to attach to vaginal epithelial cells and significantly reduce the adhesion of *C. albicans* [51]. Indeed, a major study reported in 2007 that *L. plantarum* P17630 enteral ingestion (after treatment with oral fluconazole 150 mg) significantly increased the proportion of asymptomatic VVC patients [52].

A retrospective comparative study published in 2014, again using *L. plantarum* P17630 for preventing *Candida vaginitis* recurrence was undertaken to further evaluate the effect of the application of *L. plantarum* P17630 in restoring the vaginal microbiota and the prevention of relapses among women with acute VVC. Here, 89 women were enrolled each with a diagnosis of VVC. They were placed into two groups on the basis of reported treatment. The control group was treated with a daily dose of azole vaginal cream and placebo, whereas the experimental group was treated with the same azole-based protocol but followed by vaginal application of a capsule containing *L. plantarum* P17630 (about 10^8 CFU total) once a day for 6 days and then once a week for another 4 weeks. The results of the study confirmed *L. plantarum* P17630 (now sold as Gyno-Canesflor by Bayer) is a preventive agent for reducing vaginal discomfort after conventional treatment for acute VVC, plausibly by a mechanism that improves the vaginal pH by shifting the milieu toward a predominance of lactobacilli [53].

In addition, in a double-isotope crossover single-blind study in women of reproductive age, *L. plantarum* 299v was found to facilitate iron absorption into the body from an iron-supplemented fruit drink. The aim of the study was to see if nonheme iron absorption from a fruit drink is improved following supplementation of *L. plantarum* 299v in healthy women of reproductive age. In two clinical trials that enrolled 55 and 59 females, respectively, mean iron absorption from the drink containing 10^9 CFU of *L. plantarum* 299v was significantly higher than from the control drink [54]. Thus from these three studies, clear evidence has emerged of tangible influences of *L. plantarum* on gynecological-related health. In addition, due to the direct vaginal application of gynecological probiotics, it is anticipated that this approach will hold significant promise as an inexpensive and efficacious treatment for conditions like VVC.

Lactobacillus plantarum–INDUCED INFLUENCES ON INFLAMMATION

The effect of *L. plantarum* IS-10506 and zinc supplementation on the humoral immune response and zinc status of Indonesian preschool children was examined where a 90-day randomized, double-blind, placebo-controlled, pre–post trial was conducted in four groups of Indonesian children aged 12–24 months: placebo, probiotic, zinc, and a combination of probiotic and zinc ($n = 12$ per group). *L. plantarum* IS-10506 was supplemented at a dose of 10^{10} CFU/day as a probiotic and zinc was supplemented as 20 mg zinc sulfate monohydrate (8 mg zinc elemental). Analysis of blood and stool samples for fecal sIgA and serum zinc concentrations showed that supplementation with the probiotic *L. plantarum* IS-10506 and zinc resulted in a significantly increased humoral immune response, as well as improved zinc status, in young children [55].

A study investigating the effect of *L. plantarum* 299v on intestinal permeability and tumor necrosis factor (TNF) p55 receptor concentrations in patients with obstructive jaundice undergoing biliary drainage was undertaken. Patients undergoing biliary drainage were recruited and randomized into three groups to receive either *L. plantarum* 299v, inactivated *L. plantarum* 299v (placebo), or water. These were administered daily at noon for 7 days after biliary drainage. Intestinal permeability was measured using the lactulose/mannitol (L/M) dual sugar absorption test on admission, on the day before biliary drainage, and on days 1 and 7 after biliary drainage. Blood and urine were collected to determine the L/M ratio and the TNF p55 receptor levels at each time point. The study concluded that pretreatment with probiotic *L. plantarum* 299v improves intestinal permeability after biliary drainage and attenuates parameters of inflammatory response [56].

An analysis of NF-κB pathway induction by *L. plantarum* in healthy humans found correlations with immune tolerance in the duodenum. This study focused on human responses to *L. plantarum* in a randomized double-blind placebo-controlled crossover study where healthy adults ingested preparations of living and heat-killed *L. plantarum*. Analysis of biopsies of the duodenal mucosa by whole-genome microarrays showed that expression profiles of human mucosa had marked differences in NF-κB pathway activity after consumption of living *L. plantarum* bacteria. This study thus identified mucosal gene expression patterns that correlated with the establishment of immune tolerance in healthy adults in the human duodenum environment [57]. These studies highlight the potential use of *L. plantarum* in the dampening of proinflammatory signaling pathways and the resolution of inflammation.

Lactobacillus plantarum–INDUCED INFLUENCES ON METABOLISM

The influence of *L. plantarum* TENSIA (DSM 21380) in different dairy products on anthropometric and blood biochemical indices of healthy adults was assessed in a two double-blinded randomized placebo-controlled trial of healthy adults over a 3-week period. *L. plantarum* TENSIA is a novel microorganism with antimicrobial and antihypertensive functional properties. Its safety in probiotic cheese was recently assessed according to a variety of health indices in different age groups [58].

L. plantarum TENSIA was administered in the trial by a daily dose of 1×10^{10} CFU in probiotic cheese or a daily dose of 6×10^9 CFU total in yogurt containing different content of carbohydrates, proteins, and lipids did not significantly change the body mass index (BMI), plasma glucose and lipid levels, or inflammatory markers in the blood. The trial concluded that consumption of the probiotic *L. plantarum* TENSIA either in cheese or yogurt lowered diastolic and systolic blood pressure regardless of food matrix and baseline values of blood pressure and BMI, implicating the positive functional activity of this probiotic strain [59].

In another trial, the triglyceride-lowering effect of supplementation with the dual probiotic strains *L. plantarum* KY1032 and *Lactobacillus curvatus* HY7601 was assessed in relation to the reduction of fasting plasma lysophosphatidylcholines in nondiabetic and hypertriglyceridemic subjects. A randomized, double-blind, placebo-controlled study was conducted on 92 participants with nondiabetic hypertriglyceridemia. Over 12 weeks the probiotic group consumed daily, 5×10^9 CFU total each of *L. plantarum* KY1032 and *L. curvatus* HY7601. After 12 weeks of treatment, the trial detected significant reduced palmitoleamide, palmitic amide, oleamide, and lysophosphatidyl choline (lysoPC) in the probiotic supplementation group [60], together showing the promise of the use of *L. plantarum* as a

healthy supplement to control metabolism and lipid levels within the body (Table 10.1).

Lactobacillus plantarum–INDUCED INFLUENCES ON DERMATOLOGICAL HEALTH

It was previously shown that *L. plantarum* HY7714 improves skin hydration and has antiphotoaging effects. Clinical evidence of effects of *L. plantarum* HY7714 on skin aging was shown in a randomized, double-blind, placebo-controlled study. The trial included 110 volunteers aged 41 and 59 years who have dry skin and wrinkles. Participants took 1×10^{10} CFU/day of *L. plantarum* HY7714 or a placebo for 12 weeks. There were significant increases in the skin water content, and a significant reduction in wrinkle depth in the probiotic group at week 12, confirming the antiaging benefit of *L. plantarum* HY7714 to the skin, and

TABLE 10.1 Clinical Studies Involving *L. plantarum* Supplementation

Strain Used	Clinical Pathology	References
Clinical trials reporting no adverse effects following the use of *L. plantarum*		
L. plantarum 299v	*C. difficile* infection patients	[29]
L. plantarum 299v	Recovery following abdominal surgery	[30]
L. plantarum 299v	Liver transplant recipients	[31]
L. plantarum 299v	Acute pancreatitis in a surgical care	[32]
L. plantarum 299v	Random critically ill patients	[33]
L. plantarum 299v	Elective surgical patients	[34]
Clinical studies reporting the positive influences of *L. plantarum*		
L. plantarum (CGMCC no. 1258)	Reduces colorectal carcinoma-associated septicemia	[36]
L. plantarum 299v	Attenuation of inflammatory response in the critically ill	[37]
L. plantarum 299v	Normalization of digestive functions in patients with IBS	[38–41]
L. plantarum 299v	Reduces cardiovascular disease risk in smokers	[47]
L. plantarum 299v	Reduction of acute pancreatitis sepsis	[48]
L. plantarum 299v	Reduces rates of respiratory tract infections	[49]
L. plantarum P17630	Suppression of VVC	[51–53]
L. plantarum 299v	Facilitates iron absorption into the body	[54]
L. plantarum IS-10506	Promotes humoral immune response	[55]
L. plantarum 299v	Suppression of inflammation	[56]
L. plantarum (unspecified)	Suppression of NF-κB pathway	[57]
L. plantarum TENSIA (DSM 21380)	Antimicrobial and antihypertensive properties	[58]
L. plantarum TENSIA (DSM 21380)	Lowering of BMI and lipid levels	[59]
L. plantarum KY1032	Triglyceride-lowering effect	[60]
L. plantarum HY7714	Improves skin hydration and has antiphotoaging effects	[61]
L. plantarum (unspecified)	Improves mild-to-moderate AD	[62]

AD, atopic dermatitis; BMI, Body mass index; IBS, irritable bowel syndrome; VVC, vulvovaginal candidiasis.

its potential use as a nutricosmetic agent [61]. Of note is another trial performed using a probiotic mixture (*L. plantarum*, *L. rhamnosus*, *L. casei*, and *Bifidobacterium lactis*) to evaluate a therapeutic efficacy in the treatment of children with mild-to-moderate atopic dermatitis (AD). However, this study showed that while probiotics successfully colonized in the intestine after 6 weeks' intervention; nevertheless, they could not find an additional therapeutic or immunomodulatory effect on the treatment of AD [62]. Thus, although these are only two studies, what we may conclude is that *L. plantarum* may have positive influences on enhancing the appearance of healthy skin; it did not, at least in this trial, have any detectable influences on a more aggressive skin condition like AD.

Lactobacillus plantarum in Synbiotic Formulations

Many studies have specified that administration of a mixture of probiotic organisms is more beneficial than administration of a single probiotic strain; although the optimal number of different probiotic species in a symbiotic formulation is unclear. In addition, many studies involving synbiotic therapy have used the *Lactobacillus* species combined with an oat fiber substrate. There are numerous instances of synbiotics reported in the scientific literature from in vitro studies (e.g., *L. plantarum* or *L. acidophilus* + xylo- and fructooligosaccharides) as well as numerous observations in vivo [63].

Other examples of the use of *L. plantarum* in association with other bacteria is the synbiotically formulated VSL#3 preparation. *L. plantarum* is one of the eight probiotic strains in VSL#3. The efficacy of VSL#3 has been shown in patients with ulcerative colitis (UC) where its positive efficacy in 34 adult patients with mild–moderate UC was observed [64,65]. VSL#3's positive effect on 20 UC patients that were intolerant or allergic to 5-aminosalicylic acid (an antiinflammatory drug used to treat inflammatory bowel disease) has also been described [66]. In addition, a decrease in frequency of liquid stool in enterally fed critically ill patients given VSL#3 doses was observed in a pilot trial [67].

As well as VSL#3, *L. plantarum* is a component of Synbiotic 2000 FORTE. This formulation was employed in a randomized study of 65 critically ill, mechanically ventilated, polytrauma patients, receiving daily supplementation of the formula. Here, treatment was associated with a significant reduction in rate of infection, sepsis, and mortality, as well as with a significantly shortened period of time on mechanical ventilation [68]. However, another study showed no difference in incidence of ventilator-associated pneumonia in probiotic-treated patients. This trial also investigated the effect of Synbiotic 2000 FORTE (which contains *L. plantarum* and other probiotics, plus 2.5 g of inulin, oat bran, pectin, and resistant starch) on ventilator-associated pneumonia in critically ill patients. This trial enrolled 259 critically ill patients on mechanical ventilation and found that oropharyngeal microbial flora and colonization rates were unaffected by enteral feeding of Synbiotic 2000 FORTE twice per day for 28 days [69]. In addition, a randomized pilot trial of a synbiotic dietary supplement in chronic HIV-1 infection over 4 weeks where HIV-infected women on antiretroviral therapy were fed daily the Synbiotic 2000 FORTE formulation concluded that markers of systemic immune activation appear largely unchanged between treated and control patients [70]. Thus in synbiotics, *L. plantarum* contributes a critical role to the beneficial influence of these formulations and it is apparent that a consortium of probiotic lactobacilli may have synergistic positive influence whose affect are greater than the sum of the lactobacilli administered independently.

CONCLUSIONS

In summary, *L. plantarum* has been widely used as a single probiotic, as in the case of *L. plantarum* 299v, or in synbiotic formulations. The positive influence of *L. plantarum* in health and disease extends across diverse physiological processes, including IBS, cardiovascular disease, pancreatic, respiratory tract infections, gynecological influences, modulation of immunity in the gastrointestinal tract, as well as metabolic and dermatological influences. Thorough clinical trial analysis have been undertaken to show the efficacy of *L. plantarum* in disease situations, with no adverse effects reported. Going forward, because of the capacity of *L. plantarum* to efficiently survive transit through the stomach and duodenum, it makes this *L. plantarum* a very attractive probiotic candidate for clinical use. Identification of further strains of *L. plantarum* with potent probiotic capacities must be a priority to develop inexpensive therapeutic interventions to promote health in acute and chronically ill patients.

REFERENCES

[1] Bergey DH. Society of American Bacteriologists. Bergey's manual of determinative bacteriology: a key for the identification of organisms of the class Schizomycetes. Baltimore, MD: Williams & Wilkins company; 1923. xi, 1, 442 p.

[2] Aymerich T, Martin B, Garriga M, et al. Microbial quality and direct PCR identification of lactic acid bacteria and nonpathogenic *Staphylococci* from artisanal low-acid sausages. Appl Environ Microbiol 2003;69(8):4583–94.

[3] Ercolini D, Hill PJ, Dodd CE. Bacterial community structure and location in Stilton cheese. Appl Environ Microbiol 2003;69(6): 3540–8.

[4] Siezen RJ, Tzeneva VA, Castioni A, et al. Phenotypic and genomic diversity of *Lactobacillus plantarum* strains isolated from various environmental niches. Environ Microbiol 2010;12(3):758–73.

[5] Ahrne S, Nobaek S, Jeppsson B, et al. The normal *Lactobacillus* flora of healthy human rectal and oral mucosa. J Appl Microbiol 1998;85(1):88–94.

[6] Jones RM, Desai C, Darby TM, et al. Lactobacilli modulate epithelial cytoprotection through the Nrf2 pathway. Cell Rep 2015;12(8): 1217–25.

[7] Jones RM, Luo L, Ardita CS, et al. Symbiotic lactobacilli stimulate gut epithelial proliferation via Nox-mediated generation of reactive oxygen species. EMBO J 2013;32(23):3017–28.

[8] Neish AS, Jones RM. Redox signaling mediates symbiosis between the gut microbiota and the intestine. Gut Microbes 2014;5(2):250–3.

[9] Bringel F, Quenee P, Tailliez P. Polyphasic investigation of the diversity within *Lactobacillus plantarum* related strains revealed two *L. plantarum* subgroups. Syst Appl Microbiol 2001;24(4):561–71.

[10] Kleerebezem M, Boekhorst J, van Kranenburg R, et al. Complete genome sequence of *Lactobacillus plantarum* WCFS1. Proc Natl Acad Sci USA 2003;100(4):1990–5.

[11] Siezen RJ, van Hylckama Vlieg JE. Genomic diversity and versatility of *Lactobacillus plantarum*, a natural metabolic engineer. Microb Cell Fact 2011;10(Suppl. 1):S3.

[12] Hammes WP, Hertel C. 3rd ed. The genera *Lactobacillus* and *Carnobacterium*. Prokaryotes: a handbook on the biology of bacteria, vol. 4. New York, NY: Springer; 2006. p. 320–403.

[13] Curk MC, Hubert JC, Bringel F. *Lactobacillus paraplantarum* sp. now., a new species related to *Lactobacillus plantarum*. Int J Syst Bacteriol 1996;46(2):595–8.

[14] Dellaglio F, Bottazzi V, Vescovo M. Deoxyribonucleic-acid homology among *Lactobacillus* species of subgenus *Streptobacterium* Orla-Jensen. Int J Syst Bacteriol 1975;25(2):160–72.

[15] Swezey JL, Nakamura LK, Abbott TP, et al. *Lactobacillus arizonensis* sp. nov., isolated from jojoba meal. Int J Syst Evol Micr 2000;50:1803–9.

[16] Zanoni P, Farrow JAE, Phillips BA, et al. *Lactobacillus pentosus* (Fred, Peterson, and Anderson) sp. nov., nom. rev. Int J Syst Bacteriol 1987;37(4):339–41.

[17] Kandler O, Weiss N. Bergey's manual of systematic bacteriology. In: Sneath PHE, Mair NS, Sharpe ME, Holt JG, editors. Baltimore, MD: Williams & Wilkins; 1986.

[18] Bringel F, Curk MC, Hubert JC. Characterization of lactobacilli by Southern-type hybridization with a *Lactobacillus plantarum* pyrDFE probe. Int J Syst Bacteriol 1996;46(2):588–94.

[19] De Bruyne K, Camu N, De Vuyst L, et al. *Lactobacillus fabifermentans* sp. nov. and *Lactobacillus cacaonum* sp. nov., isolated from Ghanaian cocoa fermentations. Int J Syst Evol Microbiol 2009;59(Pt. 1):7–12.

[20] Gu CT, Wang F, Li CY, et al. *Lactobacillus xiangfangensis* sp. nov., isolated from Chinese pickle. Int J Syst Evol Microbiol 2012;62(Pt. 4): 860–3.

[21] Zago M, Fornasari ME, Carminati D, et al. Characterization and probiotic potential of *Lactobacillus plantarum* strains isolated from cheeses. Food Microbiol 2011;28(5):1033–40.

[22] Bosch M, Rodriguez M, Garcia F, et al. Probiotic properties of *Lactobacillus plantarum* CECT 7315 and CECT 7316 isolated from faeces of healthy children. Lett Appl Microbiol 2012;54(3): 240–6.

[23] Ramiah K, Ten Doeschate K, Smith R, et al. Safety assessment of *Lactobacillus plantarum* 423 and *Enterococcus mundtii* ST4SA determined in trials with Wistar rats. Probiotics Antimicrob Proteins 2009;1(1):15–23.

[24] Johansson ML, Molin G, Jeppsson B, et al. Administration of different *Lactobacillus* strains in fermented oatmeal soup: in vivo colonization of human intestinal mucosa and effect on the indigenous flora. Appl Environ Microbiol 1993;59(1):15–20.

[25] Molin G. Probiotics in foods not containing milk or milk constituents, with special reference to *Lactobacillus plantarum* 299v. Am J Clin Nutr 2001;73(2):380s–5s.

[26] Miettinen M, VuopioVarkila J, Varkila K. Production of human tumor necrosis factor alpha, interleukin-6, and interleukin-10 is induced by lactic acid bacteria. Infect Immun 1996;64(12):5403–5.

[27] Panigrahi P, Parida S, Pradhan L, et al. Long-term colonization of a *Lactobacillus plantarum* synbiotic preparation in the neonatal gut. J Pediatr Gastroenterol Nutr 2008;47(1):45–53.

[28] Costa GN, Marcelino-Guimaraes FC, Vilas-Boas GT, et al. Potential fate of ingested *Lactobacillus plantarum* and its occurrence in human feces. Appl Environ Microbiol 2014;80(3):1013–9.

[29] Klarin B, Wullt M, Palmquist I, et al. *Lactobacillus plantarum* 299v reduces colonisation of *Clostridium difficile* in critically ill patients treated with antibiotics. Acta Anaesthesiol Scand 2008;52(8): 1096–102.

[30] Rayes N, Hansen S, Seehofer D, et al. Early enteral supply of fiber and Lactobacilli versus conventional nutrition: a controlled trial in patients with major abdominal surgery. Nutrition 2002;18(7–8): 609–15.

[31] Rayes N, Seehofer D, Hansen S, et al. Early enteral supply of lactobacillus and fiber versus selective bowel decontamination: a controlled trial in liver transplant recipients. Transplantation 2002;74(1): 123–7.

[32] Qin HL, Zheng JJ, Tong DN, et al. Effect of *Lactobacillus plantarum* enteral feeding on the gut permeability and septic complications in the patients with acute pancreatitis. Eur J Clin Nutr 2008;62(7): 923–30.

[33] Klarin B, Johansson ML, Molin G, et al. Adhesion of the probiotic bacterium *Lactobacillus plantarum* 299v onto the gut mucosa in critically ill patients: a randomised open trial. Crit Care 2005;9(3): R285–93.

[34] McNaught CE, Woodcock NP, MacFie J, et al. A prospective randomised study of the probiotic *Lactobacillus plantarum* 299V on indices of gut barrier function in elective surgical patients. Gut 2002;51(6):827–31.

[35] Bove P, Russo P, Capozzi V, et al. *Lactobacillus plantarum* passage through an oro-gastro-intestinal tract simulator: carrier matrix effect and transcriptional analysis of genes associated to stress and probiosis. Microbiol Res 2013;168(6):351–9.

[36] Liu ZH, Huang MJ, Zhang XW, et al. The effects of perioperative probiotic treatment on serum zonulin concentration and subsequent postoperative infectious complications after colorectal cancer surgery: a double-center and double-blind randomized clinical trial. Am J Clin Nutr 2013;97(1):117–26.

[37] McNaught CE, Woodcock NP, Anderson AD, et al. A prospective randomised trial of probiotics in critically ill patients. Clin Nutr 2005;24(2):211–9.

[38] Niedzielin K, Kordecki H, Birkenfeld B. A controlled, double-blind, randomized study on the efficacy of *Lactobacillus plantarum* 299V in patients with irritable bowel syndrome. Eur J Gastroenterol Hepatol 2001;13(10):1143–7.

[39] Ducrotte P, Sawant P, Jayanthi V. Clinical trial: *Lactobacillus plantarum* 299v (DSM 9843) improves symptoms of irritable bowel syndrome. World J Gastroenterol 2012;18(30):4012–8.

[40] Stevenson C, Blaauw R, Fredericks E, et al. Randomized clinical trial: effect of *Lactobacillus plantarum* 299 v on symptoms of irritable bowel syndrome. Nutrition 2014;30(10):1151–7.

[41] Sen S, Mullan MM, Parker TJ, et al. Effect of *Lactobacillus plantarum* 299v on colonic fermentation and symptoms of irritable bowel syndrome. Dig Dis Sci 2002;47(11):2615–20.

[42] Pitsouni E, Alexiou V, Saridakis V, et al. Does the use of probiotics/synbiotics prevent postoperative infections in patients undergoing abdominal surgery? A meta-analysis of randomized controlled trials. Eur J Clin Pharmacol 2009;65(6):561–70.

[43] Ortiz-Lucas M, Tobias A, Saz P, et al. Effect of probiotic species on irritable bowel syndrome symptoms: a bring up to date meta-analysis. Rev Esp Enferm Dig 2013;105(1):19–36.

[44] Sanchez AA, Rey E. Probiotics for irritable bowel syndrome: should we give them full names? Rev Esp Enferm Dig 2013;105(1):1–2.

[45] Wu ZJ, Du X, Zheng J. Role of *Lactobacillus* in the prevention of *Clostridium difficile*-associated diarrhea: a meta-analysis of randomized controlled trials. Chin Med J 2013;126(21):4154–61.

[46] Koretz RL. Probiotics, critical illness, and methodologic bias. Nutr Clin Pract 2009;24(1):45–9.

[47] Naruszewicz M, Johansson ML, Zapolska-Downar D, et al. Effect of *Lactobacillus plantarum* 299v on cardiovascular disease risk factors in smokers. Am J Clin Nutr 2002;76(6):1249–55.

[48] Olah A, Belagyi T, Issekutz A, et al. Randomized clinical trial of specific lactobacillus and fibre supplement to early enteral nutrition in patients with acute pancreatitis. Br J Surg 2002;89(9):1103–7.

[49] Berggren A, Ahren IL, Larsson N, et al. Randomised, double-blind and placebo-controlled study using new probiotic lactobacilli for strengthening the body immune defence against viral infections. Eur J Nutr 2011;50(3):203–10.

[50] Cribby S, Taylor M, Reid G. Vaginal microbiota and the use of probiotics. Interdiscip Perspect Infect Dis 2008;2008:256490.

[51] Bonetti A, Morelli L, Campominosi E, et al. Adherence of *Lactobacillus plantarum* 17630 in soft-gel capsule formulation versus Doderlein's bacillus in tablet formulation to vaginal epithelial cells. Minerva Ginecol 2003;55(3):279–84. 284–287.

[52] Carriero C, Lezzi V, Mancini T, Selvaggi L. Vaginal capsules of *Lactobacillus plantarum* P17630 for prevention of relapse of Candida vulvovaginitis: an Italian Multicentre Observational Study. Int J Probiotics Prebiotics 2007;2:155–62.

[53] De Seta F, Parazzini F, De Leo R, et al. *Lactobacillus plantarum* P17630 for preventing Candida vaginitis recurrence: a retrospective comparative study. Eur J Obstet Gynecol Reprod Biol 2014;182:136–9.

[54] Hoppe M, Onning G, Berggren A, et al. Probiotic strain *Lactobacillus plantarum* 299v increases iron absorption from an iron-supplemented fruit drink: a double-isotope cross-over single-blind study in women of reproductive age. Br J Nutr 2015;114(8):1195–202.

[55] Surono IS, Martono PD, Kameo S, et al. Effect of probiotic *L. plantarum* IS-10506 and zinc supplementation on humoral immune response and zinc status of Indonesian pre-school children. J Trace Elem Med Biol 2014;28(4):465–9.

[56] Jones C, Badger SA, Regan M, et al. Modulation of gut barrier function in patients with obstructive jaundice using probiotic LP299v. Eur J Gastroenterol Hepatol 2013;25(12):1424–30.

[57] van Baarlen P, Troost FJ, van Hemert S, et al. Differential NF-kappaB pathways induction by *Lactobacillus plantarum* in the duodenum of healthy humans correlating with immune tolerance. Proc Natl Acad Sci USA 2009;106(7):2371–6.

[58] Songisepp E, Hutt P, Ratsep M, et al. Safety of a probiotic cheese containing *Lactobacillus plantarum* Tensia according to a variety of health indices in different age groups. J Dairy Sci 2012;95(10):5495–509.

[59] Hutt P, Songisepp E, Ratsep M, et al. Impact of probiotic *Lactobacillus plantarum* TENSIA in different dairy products on anthropometric and blood biochemical indices of healthy adults. Benef Microbes 2015;6(3):233–43.

[60] Ahn HY, Kim M, Ahn YT, et al. The triglyceride-lowering effect of supplementation with dual probiotic strains, *Lactobacillus curvatus* HY7601 and *Lactobacillus plantarum* KY1032: reduction of fasting plasma lysophosphatidylcholines in nondiabetic and hypertriglyceridemic subjects. Nutr Metab Cardiovasc Dis 2015;25(8):724–33.

[61] Lee DE, Huh CS, Ra J, et al. Clinical evidence of effects of *Lactobacillus plantarum* HY7714 on skin aging: a randomized, double blind, placebo-controlled study. J Microbiol Biotechnol 2015;25(12):2160–8.

[62] Yang HJ, Min TK, Lee HW, et al. Efficacy of probiotic therapy on atopic dermatitis in children: a randomized, double-blind, placebo-controlled trial. Allergy Asthma Immunol Res 2014;6(3):208–15.

[63] Iqbal S, Nguyen TH, Nguyen TT, et al. Beta-galactosidase from *Lactobacillus plantarum* WCFS1: biochemical characterization and formation of prebiotic galacto-oligosaccharides. Carbohydr Res 2010;345(10):1408–16.

[64] Bibiloni R, Fedorak RN, Tannock GW, et al. VSL#3 probiotic-mixture induces remission in patients with active ulcerative colitis. Am J Gastroenterol 2005;100(7):1539–46.

[65] Chapman TM, Plosker GL, Figgitt DP. Spotlight on VSL#3 probiotic mixture in chronic inflammatory bowel diseases. BioDrugs 2007;21(1):61–3.

[66] Venturi A, Gionchetti P, Rizzello F, et al. Impact on the composition of the faecal flora by a new probiotic preparation: preliminary data on maintenance treatment of patients with ulcerative colitis. Aliment Pharmacol Ther 1999;13(8):1103–8.

[67] Frohmader TJ, Chaboyer WP, Robertson IK, et al. Decrease in frequency of liquid stool in enterally fed critically ill patients given the multispecies probiotic VSL#3: a pilot trial. Am J Crit Care 2010;19(3):e1–e11.

[68] Kotzampassi K, Giamarellos-Bourboulis EJ, Voudouris A, et al. Benefits of a synbiotic formula (Synbiotic 2000Forte) in critically Ill trauma patients: early results of a randomized controlled trial. World J Surg 2006;30(10):1848–55.

[69] Knight DJ, Gardiner D, Banks A, et al. Effect of synbiotic therapy on the incidence of ventilator associated pneumonia in critically ill patients: a randomised, double-blind, placebo-controlled trial. Intensive Care Med 2009;35(5):854–61.

[70] Schunter M, Chu HT, Hayes TL, et al. Randomized pilot trial of a synbiotic dietary supplement in chronic HIV-1 infection. BMC Complement Altern Med 2012;12:84.

Chapter 11

Use of *Bacillus* in Human Intestinal Probiotic Applications

M. Schultz*, J.P. Burton**,†,§ and R.M. Chanyi‡,§

*Department of Medicine, Dunedin School of Medicine, University of Otago, Dunedin, New Zealand; **Division of Urology, Department of Surgery/ Department of Microbiology & Immunology, Western University, London, ON, Canada; †Lawson Health Research Institute and Canadian Research and Development Centre for Probiotics, London, ON, Canada; ‡Department of Microbiology & Immunology, Western University, London, ON, Canada; §Lawson Health Research Institute, Canadian Centre for Human Microbiome and Probiotics Research, London, ON, Canada

INTRODUCTION

Bacillus species are ubiquitous in nature but found in higher concentrations in soil, water, and food products that have a plant origin. Strains of *Bacillus* are very good potential candidates to be used as probiotics. Metabolically, *Bacillus* species are very active and previous research has identified a number of useful enzymes and numerous antibiotics they produce. In addition to these secreted products, *Bacillus* remains stable in probiotic products much longer than conventional probiotics due to their ability to form endospores [1]. Most survive the rigors of food processing, including those designed to deplete microorganisms such as pasteurization. They have a history of use in fermented foods largely in Africa and Asia, but are becoming more prominent in global probiotics relatively recently. While there are pathogenic species of *Bacillus*, including *B. cereus* and *B. anthracis*, the more benign members have a good record of safety and their appearance in randomized controlled studies in humans is increasing. This chapter demonstrates the evolution of the use of *Bacillus* from environmental and food microorganism to that of a clinically used probiotic type and its potential future uses.

USE OF *Bacillus* IN FOOD

Bacillus has shown probiotic benefits since the 1950s [2] despite that it has previously only been associated with its use in the production of fermented foods. Typically, fermentation by most other food additives involve an end product that typically acidifies the food product to some degree; however, fermentation by *Bacillus* actually increases the pH due to the liberation of ammonia (alkaline fermented). Examples of these foods come from across the world, but largely Southeast Asia and African countries: including Japanese fermented soy in the form of natto, where cooked soya beans are purposely inoculated with *B. subtilis* var. *natto* and fermented for 18–20 h at 40–45°C [3]. There are other soy products, such as Kinema (India, Nepal) and Thua-nao from Thailand that use a similar process. Condiments, such as Dawadawa from West and Central Africa are made from the fermentation of locust beans by *B. subtilis* and *B. licheniformis*. Ogiri is produced by fermenting melon seeds by *B. subtilis*, *B. megaterium*, and *B. firmu*. Ugba is made from oilbean seeds fermented by *B. subtilis* and Pidan is *Bacillus*-fermented duck eggs. This alkaline fermentation process serves to increase the nutritional value of foods, provide foods with different taste properties, help preservation, and decrease toxicity [3].

ADVANTAGES OF FORMING ENDOSPORES

Bacillus are fairly unique bacteria that are currently used as probiotic organisms, perhaps with the exception of probiotic yeasts, as they have an ability to sporulate. Sporulation is a complex process that results in a modified cell that can survive very harsh environments without any nutrient consumption, including dessication, ultraviolet radiation, high to extreme cold temperatures, hydrated environments, and even many chemical disinfectants for months to years. The spore of *Bacillus* is termed an endospore as it is not a true spore by definition because offspring are not produced by these endogenously contained seeds. Instead, many receptors on the surface detect when the endospore enters favorable growth conditions and the process of sporulation is reversed and the mother cell can replicate by binary fission once again. Typically, sporulation is employed when nutrients become depleted and growth is limited [4]. The process of sporulation involves an unequal division of the cytoplasm, giving rise to both small and large progeny each with a complete genome. The smaller "forespore" becomes the mature spore, while the larger portion engulfs the forespore resulting in a cell within a mother cell. Eventually, the mother cell will lyse and by doing so release the mature spore into the environment. The mature spores are considered dormant as they have no detectable metabolism and therefore can survive for extremely long periods of time.

The longer survival of *Bacillus* strains is quite different to other bacterial types that are used as probiotics (lactobacilli, streptococci, and bifidobacteria). These organisms, when in their vegetative states, have short lives. Most are only able to survive for a few weeks when refrigerated in dairy products. Many probiotic supplements; however, rely on lyophilisation to extend the lives of these cultures. But as a comparison, the freeze-dried products produced under optimal manufacturing and storage conditions can survive in very dry products for up to several years. Due to the requirement of these to remain dry and be protected from even atmospheric moisture, and locked in protective packaging, limited the use of these probiotics to a limited number of products. Endospores of *Bacillus* can easily attain this viability in nonideal conditions and therefore show great promise for probiotic application in a much wider array of applications.

THE RISE IN RESPECTABILITY OF *Bacillus* PROBIOTICS

The main species of *Bacillus* used as probiotics include *B. subtilis, B. coagulans, B. clausii, B. pumilus, B. licheniformis,* and *B. cereus* [5]. Although, the latter has been associated with foodborne outbreaks of disease due to its potential toxin production. There are also a number of *Bacillus* probiotic products that contain nonrecognized species, such as *B. laterosporus, B. polyfermenticus,* and *B. polymyxa* [5].

The awareness of *Bacillus*-containing probiotics has risen dramatically, as has their use and scientific credibility. However, many of the *Bacillus*-containing products were often incorrectly labeled as *Lactobacillus sporogenes* in dietary supplements in North America when in fact they contained *B. coagulans* and sometimes even *B. cereus*. This comes from a misclassification dating back to the 1915 when *B. coagulans* was assigned to the genus *Lactobacillus* [6]. However, even though the correction has been made, many commercial products and medical literature retain the *Lactobacillus* name, although it has no official standing. There was contention and damage to the reputation of the probiotic industry when this organism was blatantly passed off as *Lactobacillus*, though clearly not. Some saw it as a marketing ploy to make the product more appealing to consumers by relating the organism with an established record of safety and numerous studies relating to

efficacy for various conditions while others thought of it as blatantly fraudulent. Some papers published in reputable journals still use this incorrect terminology and there are many dietary supplements that still list the active ingredient as *L. sporogenes*, rather than *B. coagulans* [7–11].

TRANSIENT OR INHABITANT?

B. subtilis is found mainly in soil and also in the intestinal tract of humans and animals. They can survive in a wide range of environments and conditions, adverse in pH and temperature. Some prefer growing at optimal temperatures of around 50°C and conditions of pH quite different to those found in the intestinal tract, while some *B. subtilis* were routinely detected in feces. There was some debate whether this was part of the human microbiota or merely a transient organism picked up that had found its way to the gastrointestinal tract. Additionally, while they were detected in feces, *B. subtilis* did not seem to be a predominant member of the normal microflora. More recent studies have added further insight on the composition of *Bacillus* species in the human intestinal tract [12]. Hoyles et al. characterized 124 isolates from the feces of 10 healthy adult donors by 16S rRNA gene sequence analyses [12]. They showed that the majority of strains belonged to *B. clausii, B. fordii, B. licheniformis, B. pumilus, B. simplex, B. sonorensis, B. thermoamylovorans,* and potentially three other species yet to be characterized. Hong et al. [13] analyzed the fecal material from 29 healthy volunteers and the ileal sample from 6 patients undergoing an endoscopy for dyspeptic symptoms. They mainly recovered *B. subtilis,* but *B. pumilus, B. licheniformis, B. amyloliquefaciens, B. cereus, B. megaterium,* and *B. flexus* were also found [14]. They estimated that the numbers of *Bacillus* were too large to be accounted for by simply the ingestion of food and that *Bacillus* species were able to form biofilm structures for survival at the mucosal wall.

CAN AN ENDOSPORE FORMER BE METABOLICALLY ACTIVE AND BE AN EFFECTIVE PROBIOTIC?

Members of the *Bacillus* genus are now widely used as a probiotic, whose applications even extend to aquaculture and beyond. Despite being a promoting factor for its use as a probiotic, endospore formation also has its drawbacks as it had been questioned whether or not the spores became active within the gastrointestinal tract for probiotic benefit and to be an efficacious probiotic in humans [15]. It is possible that nongerminated spores still provide an immunologic benefit to the host but it is likely that the spore must germinate and grow within the gut to be fully active. Hong et al. showed that the robustness of spores enables them to survive transit through the stomach, after which the spore can germinate and grow, proliferate and then resporulate before excretion in the feces [13,16]. Without coprophagy, one would expect that the cycle is not complete without the spore able to reinoculate the host. Even with the advancements in sanitation, perhaps only a few spores are required for human reinoculation or enough are consumed through food and water or via other environmental means.

Tam et al. [17] found that natural isolates of *Bacillus* that were able to form biofilms were able to survive in the intestine of mice longer than a laboratory strain. RT-PCR detection of mRNA specific to vegetative genes and sporulation genes determined that *Bacillus* was actively germinating and sporulating. Interestingly, Tam et al. observed that the *Bacillus* strains were able to grow, germinate, and sporulate significantly faster in the intestine of the mice than what was possible in culture conditions in the laboratory. These results demonstrated that *Bacillus* is not just a transient organism but an inhabitant and that they were actively germinating and interacting within the host [16,17]. Casula and Cutting also showed using a murine model and RT-PCR detection of RNA expression of the *ftsH* gene that spores were germinating as early as the jejunum and ileum of the small intestine after dosing with *B. subtilis* [18]. It was also shown that the numbers of *B. subtilis* in some cases exceeded the numbers originally input into the model suggesting strong growth.

Bacillus species have been widely used in different countries with various regulatory approvals. *B. coagulans* was added by the European Food Safety Authority (EFSA) to their Qualified Presumption of Safety (QPS) list while the *B. coagulans* BC30 strain has undertaken the generally recognized as safe (GRAS) affirmed status. Given its prior use in foods and dietary supplements, it is becoming an acceptably safe probiotic for adults and even children, even though its efficacy is not always established [19,20]. Table 11.1 summarizes the various clinical trials involving humans and different species and strains of *Bacillus* that have been conducted. There are many studies to suggest the potential intestinal benefit in taking *Bacillus* probiotics including double-blind placebo-controlled studies. However, the evidence for *Bacillus* as a probiotic is not yet as comprehensive as for the typically used bacterial species.

THERAPEUTIC OUTCOMES ASSOCIATED WITH *Bacillus* PROBIOTICS

Several human studies have shown that feeding *Bacillus* probiotics, sometimes in combination with other bacterial types, has the propensity to positively change the intestinal environment (Table 11.1). Consumption of *Bacillus mesentericus* increased the amount of bifidobacteria and lactobacilli in the intestinal tract while also increasing serum IL-10 and decreasing TNF-α levels [20]. Overall this was associated with decreased severity of diarrhea, shorter hospital

stay, reconstitution of a healthy microbiota, decreased proinflammatory markers and increased antiinflammatory markers. This was similar to a porcine study where probiotic *B. subtilis* in the diet decreased the copy number and percentage of Bacteroidetes, while increasing the percentage of health-associated Firmicutes in the cecal contents [21]. It was concluded that the addition of *B. subtilis* improves growth performance and upregulates lipid metabolism in subcutaneous fat of pigs by lipid metabolism through regulation of the proportion of Bacteroidetes and Firmicutes in the gut. Therefore, *Bacillus* may have a further role in the metabolic regulation of host metabolism [22]. Other studies have shown clinical improvements associated with bloating, intestinal discomfort, gas, and diarrhea associated with conditions including IBS, HIV, drug therapies, acute diarrhea, and constipation (Table 11.1).

FUTURE INTESTINAL USES

Given the symphony of metabolic capabilities by different strains of *Bacillus*, it seems that the number of applications using these probiotics will increase. Some potential uses include detoxification of food toxins and contaminants, such as ochratoxins to prevent human poisoning [23]. *Bacillus* also has the potential to be involved in the bioremediation of heavy metals. Heavy metals can be sometimes contaminate food and drinking water but are more prominent in an industrial setting and as such these bacteria may have a larger impact in these applications [24].

A more conventional use of *Bacillus*-based probiotics may be for the treatment of *Clostridium difficile* infection. In murine studies, *B. coagulans* BC30 improved some parameters of *C. difficile*-induced colitis in mice [25,26] while increasing the survival of *C. difficile* infected mice. More specifically, a strain of *Bacillus thuringiensis* isolated from human feces has been shown to produce a posttranslationally modified bacteriocin (thuricin) with a narrow spectrum of activity against *C. difficile* [27]. In reality, *Bacillus* species produce a huge variety of antimicrobial peptides (AMPs) that have been shown to have a broad spectrum of activity against pathogenic microbes, hence their long use as food preservatives by direct addition of the bacteria to foods or by use of their metabolic products. *Bacillus*-derived AMPs can be synthesized both ribosomally and nonribosomally and vary in their biosynthesis, structure, and molecular weight. However, the mechanism of action for most of these AMP's has not yet been elucidated. Given the relative benign intentions of *Bacillus* in humans there is great potential to develop targeted therapeutics based upon these organisms [1].

Due to the previous mislabeling controversies of *Bacillus*-based probiotics and the uniqueness of utilizing a spore-forming bacteria for human applications, the potential of these organisms have not yet been realized. With studies

TABLE 11.1 Examples of Human Studies With *Bacillus* Probiotics With Intestinal Benefit

Author and Study	Methods	Outcome
B. subtilis CU1 supplementation to stimulate immune responses in elderly subjects [28]	100 subjects aged 60–74 were included in this randomized, double-blind, placebo-controlled, parallel-arms study. Subjects consumed either the placebo or 2×10^9 *B. subtilis* daily for 4 months	Did not decrease primary measure. However, of 44 subjects providing biologic samples, showed that *B. subtilis* significantly increased fecal and salivary secretory IgA
B. coagulans BC30 6086 increased beneficial groups of bacteria in the human gut and cytokines [29,30]	36 volunteers aged 65–80 received either *B. coagulans* (1×10^9) colony-forming units) or placebo/day, followed by a 21-day washout period before switching to the other treatment	*B. coagulans* BC30 increased populations of *Faecalibacterium prausnitzii* LPS-stimulated PBMCs showed a 0.2 ng/mL increase in the antiinflammatory cytokine IL-10 28 day after consumption
B. subtilis combination therapy with mosapride is effective for relief of symptoms in patients with nondiarrheal-type IBS [31]	258 IBS patients were randomly assigned to either a combination of probiotics, *B. subtilis* and *Streptococcus faecium* and mosapride at one of four different doses or placebo	The proportion of AR at week 4 was significantly higher in all treatment groups compared to the placebo group
Possible benefit of *B. coagulans* GBI-30 6086 for residual inflammation in treated HIV-1 infection [32]	17 HIV-1-infected persons with suppressed viremia on stable antiretroviral therapy in a 3-month double-blind placebo-controlled trial	Improved chronic gastrointestinal symptoms.
B. coagulans improves abdominal pain and diarrhea in IBS patients [33]	85 adult IBS patients randomized to receive synbiotic containing *B. coagulans* or placebo for 12 weeks	More reduction in abdominal pain frequency was observed with synbiotic compared with placebo (score reduction 4.2 ± 1.8 vs. 1.9 ± 1.5, $P < 0.001$). Diarrhea frequency was decreased in the synbiotic group, but not on placebo
B. mesentericus reduced the severity of diarrhea and length of hospital stay in children with acute diarrhea [20]	304 children aged 3 months to 6 years hospitalized for acute diarrhea were randomized to receive Bio-three (a mixture of *B mesentericus*, *E. faecalis*, and *Clostridium butyricum*) or placebo orally 3 times daily for 7 days	Duration of diarrhea after start of therapy was for 60.1 h in the probiotics group versus 86.3 h in the placebo group ($P = 0.003$). Hospital stay was shorter in the probiotics group than in the placebo ($P = 0.009$). *Bifidobacterium* and *Lactobacillus* were elevated with probiotics
B. coagulans GBI-30, 6086 improved the quality of life and reduced gastrointestinal symptoms in adults with postprandial intestinal gas-related symptoms [34]	61 randomized to either *B. coagulans* or placebo. Study subjects were evaluated every 2 weeks over a 4-week period using validated questionnaires	Subjects in the probiotic group achieved significant improvements in GSRS abdominal pain subscore ($P = 0.046$) and the GSRS total score ($P = 0.048$)
B. subtilis and *S. faecium* as part of bowel preparation significantly improves colonic mucosa visualization during colonoscopy in constipated patients [35]	104 patients with constipation were prospectively randomized to receive a 2-week of a mixture of *B. subtilis* and *S. faecium* or placebo	Probiotics pretreatment was more effective at bowel cleansing for colonoscopy compared with placebo in constipated patients (54.9 vs. 20.8%; $P < 0.001$)
B coagulans GBI-30, 6086 for relief of abdominal pain and bloating for IBS patients [36]	44 randomized, double-blind, parallel-group, placebo-controlled study received either placebo or *B. coagulans* once a day for 8 weeks	Improvements from baseline abdominal pain and bloating scores in the *B. coagulans* group were statistically significant for all 7 weekly comparisons ($P < 0.01$)
Bacillus clausii reduces the side-effects related to anti-*H. pylori* antibiotic therapy [37]	120 *H. pylori*-positive patients randomly received either a 7-day triple therapy with or without and *B. clausii* 2×10^9 for 14 days starting from the first day of treatment	The incidences of nausea, diarrhea, and epigastric pain in patients treated with *B. clausii* were significantly lower than in placebo group

showing their metabolic activity in the gut, increasing numbers of positive clinical trials, superior stability to conventional probiotics, the vast metabolic toolbox of products that this species offers with a history of safe use as probiotics, and historically in food, it is likely that many more *Bacillus* strains will be developed for clinically targeted applications.

REFERENCES

[1] Sumi CD, Yang BW, Yeo I-C, Hahm YT. Antimicrobial peptides of the genus *Bacillus*: a new era for antibiotics. Can J Microbiol 2015;61(2):93–103.

[2] Henry R, Mielle F, Mohr H. *Bacillus subtilis* in intestinal therapy. Gaz Med Fr 1950;57:537–41.

[3] Wang J, Fung DY. Alkaline-fermented foods: a review with emphasis on pidan fermentation. Crit Rev Microbiol 1996;22(2):101–38.

[4] Setlow P. Spores of *Bacillus subtilis*: their resistance to and killing by radiation, heat, and chemicals. J Appl Microbiol 2006;514–25.

[5] Cutting SM. *Bacillus* probiotics. Food Microbiol 2011;28(2): 214–20.

[6] Hammer B. Bacteriological studies on the coagulation of evaporated milk. Iowa Agric Exp Stn Res Bull 1915;19:119–31.

[7] De Vecchi E, Drago L. *Lactobacillus* sporogenes or *Bacillus* coagulans: misidentification or mislabelling? Int J Probiotics Prebiotics 2006;1(1):3–10.

[8] Drago L, De Vecchi E. Should *Lactobacillus* sporogenes and *Bacillus* coagulans have a future? J. Chemother 2009;21(4):371–7.

[9] Sari FN, Dizdar EA, Oguz S, Erdeve O, Uras N, Dilmen U. Oral probiotics: *Lactobacillus* sporogenes for prevention of necrotizing enterocolitis in very low-birth weight infants: a randomized, controlled trial. Eur J Clin Nutr 2011;65(4):434–9.

[10] Colacurci N, De Franciscis P, Atlante M, et al. Endometrial, breast and liver safety of soy isoflavones plus *Lactobacillus* sporogenes in post-menopausal women. Gynecol Endocrinol 2013;29(3):209–12.

[11] Saneian H, Tavakkol K, Adhamian P, Gholamrezaei A. Comparison of *Lactobacillus sporogenes* plus mineral oil and mineral oil alone in the treatment of childhood functional constipation. J Res Med Sci 2013;18(2):85–8.

[12] Hoyles L, Honda H, Logan NA, Halket G, La Ragione RM, McCartney AL. Recognition of greater diversity of *Bacillus* species and related bacteria in human faeces. Res Microbiol 2012;163(1):3–13.

[13] Hong HA, Khaneja R, Tam NMK, et al. *Bacillus subtilis* isolated from the human gastrointestinal tract. Res Microbiol 2009;160(2):134–43.

[14] Hong HA, Khaneja R, Tam NMK, et al. *Bacillus subtilis* isolated from the human gastrointestinal tract. Res Microbiol 2009;160(2):134–43.

[15] Spinosa MR, Braccini T, Ricca E, et al. On the fate of ingested *Bacillus* spores. Res Microbiol 2000;151(5):361–8.

[16] Le Duc H, Hong HA, Barbosa TM, Henriques AO, Cutting SM. Characterization of *Bacillus* probiotics available for human use characterization of *Bacillus* probiotics available for human use. Appl Environ Microbiol 2004;70(4):2161–71.

[17] Tam NKM, Uyen NQ, Hong HA, et al. The intestinal life cycle of *Bacillus subtilis* and close relatives. J Bacteriol 2006;188(7): 2692–700.

[18] Casula G, Cutting SM. Probiotics: spore germination in the gastrointestinal tract. Society 2002;68(5):2344–52.

[19] Dutta P, Mitra U, Dutta S, Rajendran K, Saha TK, Chatterjee MK. Randomised controlled clinical trial of *Lactobacillus sporogenes* (*Bacillus coagulans*), used as probiotic in clinical practice, on acute watery diarrhoea in children. Trop Med Int Health 2011;16(5): 555–61.

[20] Chen C-C, Kong M-S, Lai M-W, et al. Probiotics have clinical, microbiologic, and immunologic efficacy in acute infectious diarrhea. Pediatr Infect Dis J 2010;29(2):135–8.

[21] Cui C, Shen CJ, Jia G, Wang KN. Effect of dietary *Bacillus subtilis* on proportion of Bacteroidetes and Firmicutes in swine intestine and lipid metabolism. Genet Mol Res 2013;12(2):1766–76.

[22] Ley R, Turnbaugh P, Klein S, Gordon J. Microbial ecology: human gut microbes associated with obesity. Nature 2006;444(7122):1022–3.

[23] Shi L, Liang Z, Li J, et al. Ochratoxin A biocontrol and biodegradation by *Bacillus subtilis* CW 14. J Sci Food Agric 2014;94(9): 1879–85.

[24] Monachese M, Burton JP, Reid G. Bioremediation and tolerance of humans to heavy metals through microbial processes: a potential role for probiotics? Appl Environ Microbiol 2012;78(18):6397–404.

[25] Fitzpatrick LR, Small JS, Greene WH, Karpa KD, Keller D. *Bacillus coagulans* GBI-30 (BC30) improves indices of *Clostridium difficile*-induced colitis in mice. Gut Pathog 2011;3(1):16.

[26] Fitzpatrick LR, Small JS, Greene WH, Karpa KD, Farmer S, Keller D. *Bacillus coagulans* GBI-30, 6086 limits the recurrence of *Clostridium difficile*-induced colitis following vancomycin withdrawal in mice. Gut Pathog 2012;4:13.

[27] Rea MC, Dobson A, O'Sullivan O, et al. Effect of broad- and narrow-spectrum antimicrobials on *Clostridium difficile* and microbial diversity in a model of the distal colon. Proc Natl Acad Sci USA 2011;108(Suppl.):4639–44.

[28] Lefevre M, Racedo SM, Ripert G, et al. Probiotic strain *Bacillus subtilis* CU1 stimulates immune system of elderly during common infectious disease period: a randomized, double-blind placebo-controlled study. Immun Ageing 2015;12(1):24.

[29] Nyangale EP, Farmer S, Cash HA, Keller D, Chernoff D, Gibson GR. *Bacillus coagulans* GBI-30, 6086 modulates *Faecalibacterium prausnitzii* in older men and women. J Nutr 2015;145(7):1446–52.

[30] Nyangale EP, Farmer S, Keller D, Chernoff D, Gibson GR. Effect of prebiotics on the fecal microbiota of elderly volunteers after dietary supplementation of *Bacillus coagulans* GBI-30, 6086. Anaerobe 2014;30:75–81.

[31] Choi CH, Kwon JG, Kim SK, et al. Efficacy of combination therapy with probiotics and mosapride in patients with IBS without diarrhea: a randomized, double-blind, placebo-controlled, multicenter, phase II trial. Neurogastroenterol Motil 2015;27(5):705–16.

[32] Yang OO, Kelesidis T, Cordova R, Khanlou H. Immunomodulation of antiretroviral drug-suppressed chronic HIV-1 infection in an oral probiotic double-blind placebo-controlled trial. AIDS Res Hum Retroviruses 2014;30(10):988–95.

[33] Rogha M, Esfahani MZ, Zargarzadeh AH. The efficacy of a synbiotic containing *Bacillus coagulans* in treatment of irritable bowel syndrome: a randomized placebo-controlled trial. Gastroenterol Hepatol Bed Bench 2014;7(3):156–63.

[34] Kalman DS, Schwartz HI, Alvarez P, Feldman S, Pezzullo JC, Krieger DR. A prospective, randomized, double-blind, placebo-controlled parallel-group dual site trial to evaluate the effects of a *Bacillus coagulans*-based product on functional intestinal gas symptoms. BMC Gastroenterol 2009;9:85.

[35] Lee H, Kim YH, Kim JH, et al. A feasibility study of probiotics pretreatment as a bowel preparation for colonoscopy in constipated patients. Dig Dis Sci 2010;55(8):2344–51.

[36] Hun L. *Bacillus coagulans* significantly improved abdominal pain and bloating in patients with IBS. Postgrad Med 2009;121(2): 119–24.

[37] Nista EC, Candelli M, Cremonini F, et al. *Bacillus clausii* therapy to reduce side-effects of anti-*Helicobacter pylori* treatment: randomized, double-blind, placebo controlled trial. Aliment Pharmacol Ther 2004;20(10):1181–8.

Chapter 12

Bifidobacteria as Probiotic Organisms: An Introduction

E.M.M. Quigley

Lynda K and David M Underwood Center for Digestive Disorders, Houston Methodist Hospital and Weill Cornell Medical College, Houston, TX, United States

The genus *Bifidobacterium* is one of five genera within the family Bifidobacteriaceae, in turn, one of two families in the order Bifidobacteriales. Bifidobacteriales are one of six orders within the phylum Actinobacteria, one of the largest phyla in the domain Bacteria. Members of this phylum are Gram-positive and, with few exceptions, characterized by a high guanine and cytosine content (high-G + C) in their DNA.

Bifidobacteria are anaerobic, nonmotile, nonspore forming, and nongas producing. They are important members of the commensal bacterial populations of the gastrointestinal tract, vagina, and oral cavity. Given their anaerobic metabolism and sensitivity to oxygen *Bifidobacteria*, in the gastrointestinal tract, are predominantly located in the colon.

The discovery of *Bifidobacteria* is attributed to the French pediatrician Henry Tissier who, in 1899, isolated a Y-shaped or branched bacterium (thus named bifidus) from the gastrointestinal tract of breast-fed infants [1,2]. Over the decades since then the taxonomy of this genus has undergone several revisions with associated name changes which can be confusing [3]. This process continues as the complete genome sequences of individual strains of *Bifidobacteria* are described and precise relationships with other strains defined [4]. This has and will, undoubtedly, continue to lead to changes in the actual names of individual strains, including those described in this chapter. A name change does not, of course imply a change of properties or function; it merely provides a more precise taxonomy. It is important, therefore, that the reader remain vigilant for name changes among *Bifidobacteria*.

The potential for *Bifidobacteria* to act as probiotics was recognized from the very outset; their relative profusion in the gut of breast-fed infants being held to account for the lower rates of diarrhea in these infants as compared to those who were bottle fed [5]. Since then the prebiotic properties of oligosaccharides in human breast milk in promoting

the growth of *Bifidobacteria* have been demonstrated [6]. Indeed, the genus *Bifidobacterium* possesses a unique fructose-6-phosphate phosphoketolase pathway which leads to the fermentation of carbohydrates. Homeostatic and beneficial effects that have been demonstrated for *Bifidobacteria*, in general, include protection against pathogens, enhancement of the gut barrier, synthesis of water-soluble vitamins, digestion of plant oligo- and polysaccharides, suppressing the production of potentially toxic and carcinogenic metabolites and, through modulating the host immune response, promoting an antiinflammatory environment [3,6–10].

To date, at least 31 species have been identified within the genus *Bifidobacteria*; of these, 9 (*adolescentis, angulatum, bifidum, breve, catenulatum, dentium, gallicum, longum, and pseudocatenulatum*) have been recovered from either the feces or oral cavity of man. Furthermore, subspecies have been defined among some of these (e.g., *animalis* and *longum*) [3]. This and subsequent chapters will address the microbiological properties, in vivo effects, and clinical benefits of one very specific strain in each of the following Bifidobacterial species or subspecies: *animalis* spp. *animalis, animalis* spp. *lactis, breve, longum* spp. *infantis,* and *longum* spp. *longum.* These have been selected based on their use as a probiotic in man, as well as on the availability of scientific and clinical data. It must be emphasized that the laboratory data and clinical evidence presented for a given strain relates to that particular strain and that strain alone and should not be extrapolated to other members of the same species or to *Bifidobacteria*, in general.

While the literature on probiotic safety is far from complete or satisfactory, critical assessments to date have suggested a very favorable safety profile for probiotic organisms, in general and for *Bifidobacteria*, in particular, and with perhaps some notable exceptions, even among critically ill and high-risk individuals [11–13].

REFERENCES

[1] Tissier H. Le bacterium coli et la reaction chromophile d'Escherich. Crit Rev Soc Biol 1899;51:943–5.

[2] Tissier H. Recherches sur la flore intestinale des nourissons (etat normal et pathologique). M.D. Thesis. University of Paris; 1900.

[3] Lee J-H, O'Sullivan DJ. Genomic insights into *Bifidobacteria*. Microbiol Mol Biol Rev 2010;74:378–416.

[4] Lewis ZT, Shani G, Masarweh CF, et al. Validating bifidobacterial species and subspecies identity in commercial probiotic products. Pediatr Res 2016;79:445–52.

[5] Tissier H. Traitement des infections intestinales par la method de la flore bacterienne de l'intestin. Crit Rev Soc Biol 1906;60:359–61.

[6] Underwood MA, German JB, Lebrilla CB, Mills DA. *Bifidobacterium longum* subspecies *infantis*: champion colonizer of the infant gut. Pediatr Res 2015;77:229–35.

[7] Dunne C, Murphy L, Flynn S, et al. Probiotics: from myth to reality. Demonstration of functionality in animal models of disease and in human clinical trials. Antonie Van Leeuwenhoek 1999;76:279–92.

[8] Konieczna P, Akdis CA, Quigley EM, Shanahan F, O'Mahony L. Portrait of an immunoregulatory *Bifidobacterium*. Gut Microbes 2012;3:261–6.

[9] Ventura M, Turroni F, Lugli GA, van Sinderen D. *Bifidobacteria* and humans: our special friends, from ecological to genomics perspectives. J Sci Food Agric 2014;94:163–8.

[10] Turroni F, Milani C, Duranti S, et al. Deciphering bifidobacterial-mediated metabolic interactions and their impact on gut microbiota by a multi-omics approach. ISME J 2016;10:1656–68.

[11] Sempel S, Newberry S, Ruelaz A, et al. Safety of probiotics used to reduce risk and prevent or treat disease. Evid Rep Technol Assess 2011;200:1–645.

[12] Sanders ME, Akkermans LM, Haller D, et al. Safety assessment of probiotics for human use. Gut Microbes 2010;1:164–85.

[13] Didari T, Solki S, Mozaffari S, Nikfar S, Abdollahi M. A systematic review of the safety of probiotics. Expert Opin Drug Saf 2014;13:227–39.

Chapter 13

Bifidobacterium animalis spp. *lactis*

E.M.M. Quigley

Lynda K and David M Underwood Center for Digestive Disorders, Houston Methodist Hospital and Weill Cornell Medical College, Houston, TX, United States

In the past, *Bifidobacterium animalis* and *Bifidobacterium lactis* were considered to be distinct species but are now regarded as members of the same species *B. animalis* that contains two subspecies: *animalis* and *lactis* [1]. Nowadays, the differentiation of different strains within *B. animalis* spp. *lactis* can be rapidly and effectively achieved employing molecular microbiological methods [2]. Two commercial preparations contain strains of *B. animalis* spp. *lactis*, one *B. animalis* spp. *lactis* DN-173 010, often referred to a *B. lactis* DN-173 010, is commonly available in a fermented milk product combined with two yogurt starter cultures: *Streptococcus thermophilus* and *Lactobacillus bulgaricus* and the other contains *B. animalis* spp. *lactis* BB-12 is available either on its own or in combination with other probiotics or a prebiotic (as a synbiotic) in a variety of preparations (dietary supplements, infant formula, and fermented milk products). *B. animalis* is a normal inhabitant of the mammalian colon and has been isolated from dairy cultures. *B. animalis* spp. *lactis* is very resistant to acidity and oxidative stress [3,4], adheres to intestinal mucin [5], and grows in milk-based media; important characteristics given its frequent formulation as a yogurt and its need to survive transit through the gut. Detailed studies have identified and characterized an endopeptidase in this bacterium, thereby, conferring an ability to utilize milk proteins and milk-derived peptides [6].

The complete genome of *B. animalis* spp. *lactis* BB-12 has been sequenced [7] and a study of its extracellular proteome identified proteins relevant to clinical benefits [8].

In an animal model *B. animalis* spp. *lactis* DN-173 010 has been show to exert antiinflammatory effects in a knockout model of colitis. These effects were associated with certain metabolic changes in the colon including a lowering of luminal pH and increased concentrations of the short chain fatty acids (SCFAs) acetate, propionate, and butyrate creating an environment which was hostile to Enterobacteriaceae, known drivers of colitis in this particular model [9].

This same strain has been shown to reduce the incidence of aberrant crypts [10], fecal mutagenicity, as well as levels of beta glucuronidase and UDP-glucuronyl-transferase activity in the feces while simultaneously altering bile acid metabolism and promoting immunoglobulin A secretion; effects which could prevent the development of colon cancer (results summarized in Ref. [11]). This bacterium demonstrates bile salt hydrolyase activity in vivo, adheres to epithelial cells and promotes the integrity of the gut barrier [11,12]. *B. animalis* spp. *lactis* has been shown to accelerate colonic transit in man [11]; in an interesting mouse study this same subspecies was shown to hydrolyze sennosides to their active moieties from a herbal preparation, thereby, accentuating their laxative effects [13]. In an animal model of irritable bowel syndrome, *B. animalis* spp. *lactis* BB-12 activated the stress response but downregulated some of the inflammatory changes associated with IBS [14].

HUMAN STUDIES

B. animalis spp. *lactis* DN-173 010 has been shown to survive transit though the gastrointestinal tract in man in either it's more usual fermented milk format or in a lyophilized formulation [15]. *B. animalis* spp. *lactis* BB-12 has also been shown to successfully transit though the gastrointestinal tract [16–18].

Initial studies involving *B. animalis* spp. *lactis* DN-173 010 demonstrated its ability to accelerate whole gut and colonic transit in elderly individuals with normal and prolonged transit times [11,19–21] and in female healthy volunteers [11,22]. Among subjects with constipation-predominant irritable bowel syndrome these effects have translated into clinical benefits [23,24]. In one of these studies a reduction in the subjective symptom of bloating and in objective measures of abdominal distention were also evident [24]. This strain has also been shown to reduce minor gastrointestinal symptoms in the general population

[25,26]. This strain also appears to benefit chronic constipation in adults [27,28]; though a randomized placebo-controlled trial in children with constipation did not show a benefit over placebo [29].

In a single study, B. animalis spp. lactis BB-12 was shown to increase defecation frequency among otherwise healthy adults who had constipation and abdominal discomfort [30].

Given its antiinflammatory properties and the importance of Bifidobacteria in the infant gut, B. animalis strains have also been studied in preterm infants for their potential to prevent or reduce the impact of necrotizing enterocolitis; however, available data (based on underpowered studies) has not shown a benefit [31].

Many of the clinical studies of B. animalis spp. lactis BB-12 have focused, in contrast, on immunity. Based on the demonstration of a variety of immunomodulatory effects in healthy and diseased human subjects [32–35], the strain has been assessed for its potential to prevent common infections, as well as manage inflammatory disorders. B. animalis spp. lactis BB-12 has also been administered to healthy adults in conjunction with Lactobacillus paracasei spp. paracasei (CRL-431) and a number of immunological parameters assayed. Overall, no significant changes were detected in response to the probiotic preparation [36]. However, the same combination did boost the immune response to influenza vaccination, again without any changes in plasma cytokines or other immune parameters [37]. In another formulation (Gut Balance) these two bacteria were combined with Lactobacillus acidophilus (LA-5), Lactobacillus rhamnosus (LGG), the prebiotics raftiline and raftilose, bovine whey derived lactoferrin and immunoglobulins, and compared with acacia gum in terms of effects on the fecal microbiota, SCFAs, gut permeability, salivary lactoferrin, and serum cytokines in healthy individuals undergoing a period of exercise training [38]. Though the symbiotic product increased the recovery of L. paracasei in feces, neither the symbiotic compound nor the acacia gum had any significant effect on fecal SCFA concentrations, measures of mucosal immunity, or gut permeability [38].

The clinical impact of the aforementioned immune effects of B. animalis spp. lactis BB-12 has been addressed in a number of studies. In a study among 182 healthy children in the United States, B. animalis spp. lactis BB-12, failed to reduce instances of school absences due to illness [39]. However, when combined with L. rhamnosus LGG, a significant reduction in the duration and severity of upper respiratory infections was noted among college students randomized to this preparation in comparison to a placebo; this led to fewer days being missed from classes [40]. The same combination, while not impacting on the overall morbidity related to respiratory and gastrointestinal infections among a group of military conscripts in Finland, did reduce specific respiratory infection symptoms in a subgroup of the population [41]. Other positive impacts on the occurrence of respiratory infections in adults [32] and children [42,43] have been reported; it should be noted that some studies failed to demonstrate an effect [44]. In a further study the probiotic combination of B. animalis spp. lactis BB-12 and L. rhamnosus LGG was shown to decrease the prevalence of picornaviruses but had little impact on other viruses implicated in common respiratory and enteric infections [45].

The administration of B. animalis spp. lactis BB-12 and L. rhamnosus LGG to pregnant women from the 36th week of gestation reduced the incidence of allergic disorders in their newborns and up to 2 years of age [46]; a follow up study indicated that the initial benefits persisted up to 6 years of age [46] though at neither point of time was there any impact on the incidence of asthma and atopic sensitization [46,47]. No clinical benefits were seen in studies of ulcerative [48] or collagenous [49] colitis.

A considerable body of work has been performed on the molecular microbiology of B. animalis spp. lactis BB-12 and has revealed considerable detail about its metabolic machinery [50–52]. Initial animal [53] and human [34] studies have, indeed, shown promise in glycemic control.

REFERENCES

[1] Masco L, Ventura M, Zink R, et al. Polyphasic taxonomic analysis of Bifidobacterium animalis and Bifidobacterium lactis reveals relatedness at the subspecies level: reclassification of Bifidobacterium animalis as Bifidobacterium animalis subsp. animalis subsp. nov. and Bifidobacterium lactis as Bifidobacterium animalis subsp. lactis subsp. nov. Int J Syst Evol Microbiol 2004;54:1137–43.

[2] Lomonaco S, Furumoto EJ, Loquasto JR, et al. Development of a rapid SNP-typing assay to differentiate Bifidobacterium animalis ssp. lactis strains used in probiotic-supplemented dairy products. J Dairy Sci 2015;98:804–12.

[3] Jayamanne VS, Adams MR. Determination of survival, identity and stress resistance of probiotic bifidobacteria in bio-yoghurts. Lett Appl Microbiol 2006;42:189–94.

[4] Matsumoto M, Ohishi H, Benno Y. H+-ATPase activity in Bifidobacterium with special reference to acid tolerance. Int J Food Microbiol 2004;93:109–13.

[5] Matsumoto M, Tani H, Ono H, et al. Adhesive property of Bifidobacterium lactis LKM512 and predominant bacteria of intestinal microflora to human intestinal mucin. Curr Microbiol 2002;44:212–5.

[6] Janer C, Arigoni F, Lee BH, et al. Enzymatic ability of Bifidobacterium animalis subsp. lactis to hydrolyze milk proteins: identification and characterization of endopeptidase O. Appl Environ Microbiol 2005;71:8460–5.

[7] Garrigues C, Johansen E, Pedersen MB. Complete genome sequence of Bifidobacterium animalis subsp. lactis BB-12, a widely consume probiotic. J Bacteriol 2010;192(9):2467–8.

[8] Gilad O, Svensson B, Viborg AH, et al. The extracellular proteome of Bifidobacterium animalis subsp. lactis BB-12 reveals proteins with putative roles in probiotic effects. Proteomics 2011;11:2503–14.

[9] Veiga P, Gallini CA, Beal C, et al. Bifidobacterium animalis subsp. lactis fermented milk product reduces inflammation by altering a niche for colitogenic microbes. Proc Natl Acad Sci USA. 2010;107:18132–7.

[10] Tavan E, Cayuela C, Antoine JM, et al. Effects of dairy products on heteocyclic aromatic amine-induced rat colon carcinogenesis. Carcinogenesis 2002;23:477–83.

[11] Picard C, Fioramonti J, Francois A, et al. Review article: bifidobacteria as probiotic agents—physiological effects and clinical benefits. Aliment Pharmacol Therapeut 2005;22:495–512.

[12] Lepercq P, Relano P, Cayuela C, Juste C. *Bifidobacterium animalis* strain DN-173 010 hydrolyses bile salts in the gastrointestinal tract of pigs. Scand J Gastroenterol 2004;39:1266–71.

[13] Matsumoto M, Ishige A, Yazawa Y, et al. Promotion of intestinal peristalsis by *Bifidobacterium* spp. capable of hydrolysing sennosides in mice. PLoS One 2012;7:e31700.

[14] Barouie J, Moussavi M, Hodgson DM. Effect of maternal probiotic intervention on HPA axis, immunity and gut microbiota in a rat model of irritable bowel syndrome. PLoS One 2012;7:e46051.

[15] Rochet V, Rigottier-Gois L, Ledaire A, et al. Survival of *Bifidobacterium animalis* DN-173 010 in the faecal microbiota after administration in lyophilized form or in a fermented product—a randomized study in health adults. J Mol Microbiol Biotechnol 2008;14:128–36.

[16] Larsen CN, Nielsen S, Kaestel P, et al. Dose-response study of probiotic bacteria *Bifidobacterium animalis* subsp. *lactis* BB-12 and *Lactobacillus paracasei* subsp *paracasei* CRL-341 in healthy young adults. Eur J Clin Nutr 2006;60:1284–93.

[17] Savard P, Lamarche B, Paradis ME, et al. Impact of *Bifidobacterium animalis* subsp. *lactis* BB-12 and, *Lactobacillus* LA-5-containing yoghurt, on fecal bacterial counts of healthy adults. Int J Food Microbiol 2011;149:59–67.

[18] Dotterud CK, Avershina E, Sekelja M, et al. Does maternal perinatal probiotic supplementation alter the intestinal microbiota of mother and child? J Pediatr Gastroenterol Nutr 2015;61:200–7.

[19] Meance S, Cayuela C, Turchet P, et al. A fermented milk with a *Bifidobacterium* probiotic strain DN-173 010 shortened oro-fecal gut transit time in elderly. Microb Ecol Health Dis 2001;13:217–22.

[20] Meance S, Cayuela C, Raimondi A, et al. Recent advances in the use of functional foods: effects of the commercial fermented milk with *Bifidobacterium animalis* strain DN-173 010 and yoghurt strains on gut transit time in the elderly. Microb Ecol Health Dis 2003;15:15–22.

[21] Miller LE, Ouwehand AC. Probiotic supplementation decreases intestinal transit time: meta-analysis of randomized controlled trials. World J Gastroenterol 2013;19:4718–25.

[22] Marteau P, Cuillerier E, Meance S, et al. *Bifidobacterium animalis* strain DN-173 010 shortens the colonic transit time in healthy women: a double-blind, randomised, controlled study. Aliment Pharmacol Therapeut 2002;16:587–93.

[23] Guyonnet D, Chassany O, Ducrotte P, et al. Effect of a fermented milk containing *Bifidobacterium animalis* DN-173 010 on the health-related quality of life and symptoms in irritable bowel syndrome in adults in primary care: a multicentre, randomised, double-blind, controlled trial. Aliment Pharmacol Therapeut 2007;26:475–86.

[24] Agrawal A, Houghton LA, Morris J, et al. Clinical trial: the effects of a fermented milk product containing *Bifidobacterium lactis* DN-173 010 on abdominal distension and gastrointestinal transit in irritable bowel syndrome with constipation. Aliment Pharmacol Therapeut 2009;29:104–14.

[25] Guyonnet D, Woodcock A, Stefani B, et al. Fermented milk containing *Bifidobacterium lactis* DN-173 010 improved self-reported digestive comfort amongst a general population of adults. A randomized, open-label, controlled, pilot study. J Dig Dis 2009;10:61–70.

[26] Guyonnet D, Schlumberger A, Mhamdi L, et al. Fermented milk containing *Bifidobacterium lactis* DN-173 010 improves gastrointestinal well-being and digestive symptoms in women reporting minor digestive symptoms: a randomised, double-blind, parallel, controlled study. Br J Nutr 2009;102:1654–62.

[27] Chmielewska A, Szajewska H. Systematic review of randomised controlled trials: probiotics for functional constipation. World J Gastroenterol 2010;16:69–75.

[28] De Paula JA, Carmuega E, Weill R. Effect of the ingestion of a symbiotic yogurt on the bowel habits of women with functional constipation. Acta Gastroenterol Latinoam 2008;38:16–25.

[29] Tabbers MM, Chmielewska A, Roseboom MG, et al. Fermented milk containing *Bifidobacterium lactis* DN-173 010 in childhood constipation: a randomized, double-blind, controlled trial. Pediatrics 2011;127:e1392–9.

[30] Eskesen D, Jespersen L, Michelsen B, et al. Effect of the probiotic strain *Bifidobacterium animalis* subsp. *lactis* BB-12®, on defecation frequency in health adults with low defecation frequency and abdominal discomfort: a randomized, double-blind, placebo-controlled, parallel-group trial. Br J Nutr 2015;114:1638–46.

[31] Szajewska H, Guandalini S, Morelli L, et al. Effect of *Bifidobacterium animalis* subsp *lactis* supplementation in preterm infants: a systematic review of randomized controlled trials. J Pediatr Gastroenterol Nutr 2010;51:203–9.

[32] Meng H, Lee Y, Ba Z, et al. Consumption of *Bifidobacterium animalis* subsp. *lactis* BB-12 impacts upper respiratory tract infection and the function of NK and T cells in health adults. Mol Nutr Food Res 2016;60(5):1161–71.

[33] Meng H, Ba Z, Lee Y, et al. Consumption of *Bifidobacterium animalis* subsp. *lactis* BB-12 in yogurt reduced expression of TLR-2 on peripheral blood-derived monocytes and pro-inflammatory cytokine secretion in young adults. Eur J Nutr 2015. [epub ahead of print].

[34] Tonucci LB, Olbrich Dos Santos KM, Licursi de Oliveira L, et al. Clinical application of probiotics in type 2 diabetes mellitus: a randomized double-blind, placebo-controlled study. Clin Nutr 2015. [epub ahead of print].

[35] Sheikhi A, Shakerian M, Baghaeifar M, et al. Probiotic yogurt culture *Bifidobacterium animalis* subsp. *lactis* BB-12 modulate the cytokine secretion by peripheral blood mononuclear cells from patients with ulcerative colitis. Drug Res 2016;66(6):300–5.

[36] Christensen HR, Larsen CN, Kaestel P. Immunomodulating potential of supplementation with probiotics: a dose-response study in healthy young adults. FEMS Immunol Med Microbiol 2006;47:380–90.

[37] Rizzardini G, Eskesen D, Calder PC, et al. Evaluation of the immune benefits of two probiotic strains *Bifidobacterium animalis* ssp. *lactis*, BB-12® and *Lactobacillus paracasei* ssp. i, *L. casei* 431® in an influenza vaccination model: a randomised, double-blind, placebo-controlled study. Br J Nutr 2012;107:876–84.

[38] West NP, Pyne DB, Cripps AW, et al. Gut Balance, a synbiotic supplement, increases fecal *Lactobacillus paracasei* but has little effect on immunity in healthy physically active individuals. Gut Microbes 2012;3:221–7.

[39] Merenstein DJ, Smith KH, Scriven M, et al. The study to investigate the potential benefits of probiotics in yogurt, a patient-oriented, double-blind, cluster-randomised, placebo-controlled, clinical trial. Eur J Clin Nutr 2010;64:685–91.

[40] Smith TJ, Rigassio-Radler D, Denmark R, et al. Effect of *Lactobacillus rhamnosus* LGG® and *Bifidobacterium animalis* ssp. *lactis* BB-12® on health-related quality of life in college students affected by upper respiratory infections. Br J Nutr 2013;109:1999–2007.

[41] Kalima K, Lehtoranta L, He L, et al. Probiotics and respiratory and gastrointestinal tract infections in Finnish military conscripts—a randomized placebo-controlled double-blinded study. Benef Microbes 2016. [Epub ahead of print].

[42] Taipale T, Pienihakkinen K, Isolauri E, et al. *Bifidobacterium animalis* subsp. *lactis* BB-12 in reducing the risk of infections in infancy. Br J Nutr 2011;105:409–16.

[43] Taipale TJ, Pienihakkinen K, Isolauri E, et al. *Bifidobacterium animalis* subsp. *lactis* BB-12 in reducing the risk of infections in early childhood. Pediatr Res 2016;79:65–9.

[44] Hojsak I, Mocic Pavic A, Kos T, et al. *Bifidobacterium animalis* subsp. *lactis* in prevention of common infections in healthy children attending day care centers—randomized, double-blind, placebo-controlled study. Clin Nutr 2015;35(3):587–91.

[45] Lehtoranta L, Kalima K, He L, et al. Specific probiotics and virological findings in symptomatic conscripts attending military service in Finland. J Clin Virol 2014;60:276–81.

[46] Dotterud CK, Storro O, Johnsen R, et al. Probiotics in pregnant women to prevent allergic disease: a randomized, double-blind trial. Br J Dermatol 2010;163:616–23.

[47] Simpson MR, Dotterud CK, Storro O, et al. Perinatal probiotic supplementation in the prevention of allergy related disease: 6 year follow up of a randomized controlled trial. BMC Dermatol 2015;15:13.

[48] Wildt S, Nordgaard I, Hansen U, et al. A randomized double-blind placebo controlled trial with *Lactobacillus acidophilus* La-5 and *Bifidobacterium animalis* subsp. *lactis* BB-12 for maintenance of remission in ulcerative colitis. J Crohns Colitis 2011;5:115–21.

[49] Wildt S, Munck LK, Vinter-Jensen L, et al. Probiotic treatment of collagenous colitis: a randomized, double-blind controlled trial with *Lactobacillus acidophilus* and *Bifidobacterium animalis* subsp. *lactis*. Inflamm Bowel Dis 2006;12:395–401.

[50] Gonzalez-Rodriguez I, Gaspar P, Sanchez B, et al. Catabolism of glucose and lactose in *Bifidobacterium animalis* spp. *lactis*, studied by ^{13}C nuclear magnetic resonance. Appl Environ Microbiol 2013;79:7628–38.

[51] Viborg AH, Sorenson KI, Gilad O, et al. Biochemical and kinetic characterization of a novel xyloologosaccheride-upregulated GH43 β-d-xylosidase/α-1-arabinofuranosidase (BXA43) from the probiotic *Bifidobacterium animalis* spp. *lactis* BB-12. AMB Express 2013;3:65.

[52] Zhu DQ, Liu F, Sun Y, et al. Genome-wide identification of small RNAs in *Bifidobacterium animalis* subsp. *lactis* KLDS 2.0603 and their regulation role in the adaptation to gastrointestinal environment. PLoS One 2015;10:e0117373.

[53] Bomhof MR, Saha DC, Reid DT, et al. Combined effects of oligofructose and *Bifidobacterium animalis* on gut microbiota and glycemia in rats. Obesity 2014;22:763–71.

Chapter 14

Bifidobacterium bifidum

E.M.M. Quigley

Lynda K and David M Underwood Center for Digestive Disorders, Houston Methodist Hospital and Weill Cornell Medical College, Houston, TX, United States

Bifidobacterium bifidum was first isolated from the infant intestine and characterized in some detail over 60 years ago as an anaerobic, Gram-positive nonmotile, nonspore-forming bacterium [1–4]. *B. bifidum* is an important constituent of the colonic microbime and is especially prevalent in the colon of breast-fed infants which, along with *Bifidobacterium longum* and *Bifidobacterium breve*, constitutes one of the dominant commensal bacterial species [5]. Since then a considerable volume of laboratory and clinical studies have defined its microbiological and molecular characteristics, revealed its biological actions, and identified clinical benefits. The probiotic potential of number of strains have been evaluated in the laboratory (BB-06, mimbb-75, YIT 4007, YIT-10347, G9-1, R0071, NCFB 1454, BbVK3, BGN4, and PRL 2010) and some have been formulated as probiotics and studied in man. As is the case with several other Bifidobacterial strains, *B. bifidum* strains have been studied in isolation, in conjunction with another putative probiotic organism, with a prebiotic, or in a probiotic cocktail, such as the formulation that includes *Lactobacillus acidophilus*, *Lactobacillus casei*, and *B. bifidum* [6] or the probiotic Dahi containing *L. acidophilus* LaVK2 and *B. bifidum* BbVK3 [7]. As the relative contributions of the individual strains to the effects of these cocktails are impossible to discern, these formulations will not be discussed any further.

CHARACTERIZATION AND LABORATORY STUDIES

Sequencing of the genome of several *B. bifidum* strains, together with proteomics, has provided insights into their metabolic pathways, including their ability to digest mucus [8–16]. Laboratory studies have also defined the properties of the bacterium that allow it to survive transit through the gastrointestinal tract [17,18] and establish itself among the other constituents of the microbiome [19]; the possession of an exopolysaccharide coat appears to be critical to the latter characteristic [19]. Survival of an administered *B. bifidum* probiotic can be enhanced by microencapsulation [20]. *B. bifidum* strains produce bacteriocins [21–24] and their antibacterial activity is also enhanced by the aforementioned exopolysaccharide coat [19,25]. Colonization of host tissues, evasion of immune responses, and the production of biofilms in *B. bifidum* is orchestrated by the presence of pilus-like structures as appendages; these have been well characterized [26–28]. In vitro studies have also identified the enzymatic repertoire of this species [29,30] as well as its immunological [31,32] and antiviral (rotavirus, to be precise) [33] effects. Laboratory studies have also revealed effects that enhance the gut barrier [34] and suppress carcinogenesis [7,35]. Potentially clinically important impacts on necrotizing enterocolitis [36,37], aging and immune senescence [38,39], pollenosis [40], *Helicobacter pylori* infection [41,42], inflammatory bowel disease [43,44], and mucositis related to cancer chemotherapy [45] have been demonstrated in appropriate animal models.

CLINICAL STUDIES

Some of these laboratory findings have translated into clinical effects. Thus potentially beneficial immunological effects have been demonstrated in the elderly [46] and a probiotic preparation containing *B. bifidum* has been shown to reduce the occurrence of upper respiratory infections in otherwise healthy but "academically stressed" students [47]. Benefits in eczema [48] and *Clostridium difficile* infection [49], in the management of radiation-related diarrhea [50], and in the prevention of rotavirus-related diarrhea in children [51] have also been demonstrated. Metabolic and neuromodulatory effects have translated into benefits in hyperlipidemia [52] and gastrointestinal responses to stress [53]. A preparation including *B. bifidum* has also been shown to accelerate the establishment of enteral feeding in very low birth weight infants after birth [54]. Despite its in vitro effect clinical benefits against *H. pylori* have been less consistent [55,56]. Strains of *B. bifidum* have demonstrated clinical efficacy in functional disorders affecting both the upper [57,58] and lower (i.e., irritable bowel syndrome) [59] gastrointestinal tract.

SAFETY

Generally regarded as safe, *B. bifidum* has had a good safety record even among very low birth weight infants [60–62].

REFERENCES

[1] Malyoth G, Bauer A. Observations on *Bacterium bifidum*. Z Kinderheilkd 1950;68:358–67.

[2] Malyoth G, Bauer A. Attempt at a classification of the *Bacterium bifidum*. Klin Wochenschr 1950;28:451–2.

[3] Frisell E. Studies on *Bacterium bifidum* in healthy infants: a clinical bacteriological investigation. Acta Paediatr 1951;40(Suppl. 80): 1–120.

[4] Bamberger P. Morphology and biochemistry of *Bacterium bifidum*. Bibl Paediatr 1957;64:13–23.

[5] Barrett E, Deshpandey AK, Ryan CA, et al. The neonatal gut harbours distinct bifidobacterial strains. Arch Dis Child Fetal Neonatal Ed 2015;100:F405–10.

[6] Karamali M, Dadkhah F, Sadrkhanlou M, et al. Effects of probiotic supplementation on glycaemic control and lipid profiles in gestational diabetes: a randomized, double-blind, placebo-controlled trial. Diabetes Metab 2016.

[7] Mohania D, Kansal VK, Kruzliak P, et al. Probiotic Dahi containing *Lactobacillus acidophilus* and *Bifidobacterium bifidum* modulates the formation of aberrant crypt foci, mucin-depleted foci, and cell proliferation on 1,2-dimethylhydrazine-induced colorectal carcinogenesis in Wistar rats. Rejuvenation Res 2014;17:325–33.

[8] Wei X, Wang S, Zhao X, et al. Proteomic profiling of *Bifidobacterium bifidum* S17 cultivated under in vitro conditions. Front Microbiol 2016;7:97.

[9] Morita H, Toh H, Oshima K, et al. Complete genome sequence of *Bifidobacterium bifidum* JCM 1255(T) isolated from feces of a breast-fed infant. J Biotechnol 2015;210:66–7.

[10] Duranti S, Milani C, Lugli GA, et al. Insights from genomes of representatives of the human gut commensal *Bifidobacterium bifidum*. Environ Microbiol 2015;17:2515–31.

[11] Gueimonde M, Ventura M, Margolles A, et al. Genome sequence of the immunomodulatory strain *Bifidobacterium bifidum* LMG 13195. J Bacteriol 2012;194:6997.

[12] Yu DS, Jeong H, Lee DH, et al. Complete genome sequence of the probiotic bacterium *Bifidobacterium bifidum* strain BGN4. J Bacteriol 2012;194:4757–8.

[13] Turroni F, Foroni E, Montanini B, et al. Global genome transcription profiling of *Bifidobacterium bifidum* PRL2010 under in vitro conditions and identification of reference genes for quantitative real-time PCR. Appl Environ Microbiol 2011;77:8578–87.

[14] Turroni F, Milani C, van Sinderen D, et al. Genetic strategies for mucin metabolism in *Bifidobacterium bifidum* PRL2010: an example of possible human-microbe co-evolution. Gut Microbes 2011;2:183–9.

[15] Zhurina D, Zomer A, Gleinser M, et al. Complete genome sequence of *Bifidobacterium bifidum* S17. J Bacteriol 2011;193:301–2.

[16] Turroni F, Bottacini F, Foroni E, et al. Genome analysis of *Bifidobacterium bifidum* PRL2010 reveals metabolic pathways for host-derived glycan foraging. Proc Natl Acad Sci 2010;107:19514–9.

[17] Kim GB, Miyamoto CM, Meighen EA, et al. Cloning and characterization of the bile salt hydrolase genes (bsh) from *Bifidobacterium bifidum* strains. Appl Environ Microbiol 2004;70:5603–12.

[18] Andriantsoanirina V, Allano S, Butel MJ, et al. Tolerance of *Bifidobacterium* human isolates to bile, acid and oxygen. Anaerobe 2013;21:39–42.

[19] Li S, Chen T, Xu F, Dong S, et al. The beneficial effect of exopolysaccharides from *Bifidobacterium bifidum* WBIN03 on microbial diversity in mouse intestine. J Sci Food Agric 2014;94:256–64.

[20] Zhang F, Li XY, Park HJ, et al. Effect of microencapsulation methods on the survival of freeze-dried *Bifidobacterium bifidum*. J Microencapsul 2013;30:511–8.

[21] Yildirim Z, Winters DK, Johnson MG. Purification, amino acid sequence and mode of action of bifidocin B produced by *Bifidobacterium bifidum* NCFB 1454. J Appl Microbiol 1999;86:45–54.

[22] Silva AM, Bambirra EA, Oliveira AL, et al. Protective effect of bifidus milk on the experimental infection with *Salmonella enteritidis* subsp. *typhimurium* in conventional and gnotobiotic mice. J Appl Microbiol 1999;86:331–6.

[23] Yildirim Z, Johnson MG. Characterization and antimicrobial spectrum of bifidocin B, a bacteriocin produced by *Bifidobacterium bifidum* NCFB 1454. J Food Prot 1998;61:47–51.

[24] Bayoumi MA, Griffiths MW. In vitro inhibition of expression of virulence genes responsible for colonization and systemic spread of enteric pathogens using *Bifidobacterium bifidum* secreted molecules. Int J Food Microbiol 2012;156:255–63.

[25] Li S, Huang R, Shah NP, et al. Antioxidant and antibacterial activities of exopolysaccharides from *Bifidobacterium bifidum* WBIN03 and *Lactobacillus plantarum* R315. J Dairy Sci 2014;97:7334–43.

[26] Foroni E, Serafini F, Amidani D, et al. Genetic analysis and morphological identification of pilus-like structures in members of the genus *Bifidobacterium*. Microb Cell Fact 2011;10(Suppl. 1):S16.

[27] Turroni F, Serafini F, Foroni E, et al. Role of sortase-dependent pili of *Bifidobacterium bifidum* PRL2010 in modulating bacterium-host interactions. Proc Natl Acad Sci USA 2013;110:11151–6.

[28] Turroni F, Serafini F, Mangifesta M, et al. Expression of sortase-dependent pili of *Bifidobacterium bifidum* PRL2010 in response to environmental gut conditions. FEMS Microbiol Lett 2014;357:23–33.

[29] Katayama T, Sakuma A, Kimura T, et al. Molecular cloning and characterization of *Bifidobacterium bifidum* 1,2-alpha-L-fucosidase (AfcA), a novel inverting glycosidase (glycoside hydrolase family 95). J Bacteriol 2004;186:4885–93.

[30] Turroni F, Strati F, Foroni E, et al. Analysis of predicted carbohydrate transport systems encoded by *Bifidobacterium bifidum* PRL2010. Appl Environ Microbiol 2012;78:5002–12.

[31] Ohno H, Tsunemine S, Isa Y, et al. Oral administration of *Bifidobacterium bifidum* G9-1 suppresses total and antigen specific immunoglobulin E production in mice. Biol Pharm Bull 2005;28:1462–6.

[32] Turroni F, Taverniti V, Ruas-Madiedo P, et al. *Bifidobacterium bifidum* PRL2010 modulates the host innate immune response. Appl Environ Microbiol 2014;80:730–40.

[33] Duffy LC, Zielezny MA, Riepenhoff-Talty M, et al. Effectiveness of *Bifidobacterium bifidum* in mediating the clinical course of murine rotavirus diarrhea. Pediatr Res 1994;35:690–5.

[34] Hsieh CY, Osaka T, Moriyama E, et al. Strengthening of the intestinal epithelial tight junction by *Bifidobacterium bifidum*. Physiol Rep 2015;3:e12327.

[35] You HJ, Oh DK, Ji GE. Anticancerogenic effect of a novel chiroinositol-containing polysaccharide from *Bifidobacterium bifidum* BGN4. FEMS Microbiol Lett 2004;240:131–6.

[36] Khailova L, Mount Patrick SK, et al. *Bifidobacterium bifidum* reduces apoptosis in the intestinal epithelium in necrotizing enterocolitis. Am J Physiol 2010;299:G1118–27.

[37] Khailova L, Dvorak K, Arganbright KM, et al. *Bifidobacterium bifidum* improves intestinal integrity in a rat model of necrotizing enterocolitis. Am J Physiol 2009;297:G940–9.

[38] Kaushal D, Kansal VK. Probiotic Dahi containing *Lactobacillus acidophilus* and *Bifidobacterium bifidum* alleviates age-inflicted oxidative stress and improves expression of biomarkers of ageing in mice. Mol Biol Rep 2012;39:1791–9.

[39] Fu YR, Yi ZJ, Pei JL, et al. Effects of *Bifidobacterium bifidum* on adaptive immune senescence in aging mice. Microbiol Immunol 2010;54:578–83.

[40] Tsunemine S, Isa Y, Ohno H, et al. Longitudinal study of effects of oral dosage of *Bifidobacterium bifidum* G9-1 on Japanese cedar pollen-induced allergic nasal symptoms in guinea pigs. Microbiol Immunol 2015;59:690–9.

[41] Chenoll E, Casinos B, Bataller E, et al. Novel probiotic *Bifidobacterium bifidum* CECT 7366 strain active against the pathogenic bacterium *Helicobacter pylori*. Appl Environ Microbiol 2011;77:1335–43.

[42] Shirasawa Y, Shibahara-Sone H, Iino T, et al. *Bifidobacterium bifidum* BF-1 suppresses *Helicobacter pylori*-induced genes in human epithelial cells. J Dairy Sci 2010;93:4526–34.

[43] Kim N, Kunisawa J, Kweon MN, et al. Oral feeding of *Bifidobacterium bifidum* (BGN4) prevents CD4(+) CD45RB(high) T cell-mediated inflammatory bowel disease by inhibition of disordered T cell activation. Clin Immunol 2007;123:30–9.

[44] Jadhav SR, Shandilya UK, Kansal VK. Exploring the ameliorative potential of probiotic Dahi containing *Lactobacillus acidophilus* and *Bifidobacterium bifidum* on dextran sodium sulphate induced colitis in mice. J Dairy Res 2013;80:21–7.

[45] Yeung CY, Chan WT, Jiang CB, et al. Amelioration of chemotherapy-induced intestinal mucositis by orally administered probiotics in a mouse model. PLoS One 2015;10:e0138746.

[46] Spaiser SJ, Culpepper T, Nieves C Jr, et al. *Lactobacillus gasseri* KS-13, *Bifidobacterium bifidum* G9-1, and *Bifidobacterium longum* MM-2 ingestion induces a less inflammatory cytokine profile and a potentially beneficial shift in gut microbiota in older adults: a randomized, double-blind, placebo-controlled, crossover study. J Am Coll Nutr 2015;34:459–69.

[47] Langkamp-Henken B, Rowe CC, Ford AL, et al. *Bifidobacterium bifidum* R0071 results in a greater proportion of healthy days and a lower percentage of academically stressed students reporting a day of cold/flu: a randomised, double-blind, placebo-controlled study. Br J Nutr 2015;113:426–34.

[48] Lin RJ, Qiu LH, Guan RZ, et al. Protective effect of probiotics in the treatment of infantile eczema. Exp Ther Med 2015;9:1593–6.

[49] McFarland LV. Probiotics for the primary and secondary prevention of *C. difficile* infections: a meta-analysis and systematic review. Antibiotics 2015;4:160–78.

[50] Chitapanarux I, Chitapanarux T, Traisathit P, et al. Randomized controlled trial of live *Lactobacillus acidophilus* plus *Bifidobacterium bifidum* in prophylaxis of diarrhea during radiotherapy in cervical cancer patients. Radiat Oncol 2010;5:31.

[51] Saavedra JM, Bauman NA, Oung I, et al. Feeding of *Bifidobacterium bifidum* and *Streptococcus thermophilus* to infants in hospital for prevention of diarrhoea and shedding of rotavirus. Lancet 1994;344:1046–9.

[52] Rerksuppaphol S, Rerksuppaphol L. A Randomized double-blind controlled trial of *Lactobacillus acidophilus* plus *Bifidobacterium bifidum* versus placebo in patients with hypercholesterolemia. Clin Diagn Res 2015;9:KC01–4.

[53] Culpepper T, Christman MC, Nieves C Jr, et al. *Bifidobacterium bifidum* R0071 decreases stress-associated diarrhoea-related symptoms and self-reported stress: a secondary analysis of a randomised trial. Benef Microbes 2016;7:327–36.

[54] Totsu S, Yamasaki C, Terahara M, et al. *Bifidobacterium* and enteral feeding in preterm infants: cluster-randomized trial. Pediatr Int 2014;56:714–9.

[55] Miki K, Urita Y, Ishikawa F, et al. Effect of *Bifidobacterium bifidum* fermented milk on *Helicobacter pylori* and serum pepsinogen levels in humans. J Dairy Sci 2007;90:2630–40.

[56] Navarro-Rodriguez T, Silva FM, Barbuti RC, et al. Association of a probiotic to a *Helicobacter pylori* eradication regimen does not increase efficacy or decreases the adverse effects of the treatment: a prospective, randomized, double-blind, placebo-controlled study. BMC Gastroenterol 2013;13:56.

[57] Urita Y, Goto M, Watanabe T, et al. Continuous consumption of fermented milk containing *Bifidobacterium bifidum* YIT 10347 improves gastrointestinal and psychological symptoms in patients with functional gastrointestinal disorders. Biosci Microbiota Food Health 2015;34:37–44.

[58] Gomi A, Iino T, Nonaka C, et al. Health benefits of fermented milk containing *Bifidobacterium bifidum* YIT 10347 on gastric symptoms in adults. J Dairy Sci 2015;98:2277–83.

[59] Guglielmetti S, Mora D, Gschwender M, et al. Randomised clinical trial: *Bifidobacterium bifidum* MIMBb75 significantly alleviates irritable bowel syndrome and improves quality of life—a double-blind, placebo-controlled study. Aliment Pharmacol Ther 2011;33:1123–32.

[60] Cruchet S, Furnes R, Maruy A, et al. The use of probiotics in pediatric gastroenterology: a review of the literature and recommendations by Latin-American experts. Paediatr Drugs 2015;17:199–216.

[61] Allen SJ, Jordan S, Storey M, et al. Dietary supplementation with lactobacilli and bifidobacteria is well tolerated and not associated with adverse events during late pregnancy and early infancy. J Nutr 2010;140:483–8.

[62] Yamasaki C, Totsu S, Uchiyama A, et al. Effect of *Bifidobacterium* administration on very-low-birthweight infants. Pediatr Int 2012;54:651–6.

Chapter 15

Bifidobacterium breve

E.M.M. Quigley

Lynda K and David M Underwood Center for Digestive Disorders, Houston Methodist Hospital and Weill Cornell Medical College, Houston, TX, United States

As has been the case with other Bifidobacterial species, strains within the species *Bifidobacterium breve* were first isolated from the feces of healthy infants; indeed, *B. breve* is regarded as one of the first colonizers of the infant gut. A number of *B. breve* strains have been evaluated in man: BBG-001, BR-03, B632, M-16V, BB536, CNCM I-4035, C-50, and strain Yakult, either as a single organism, in combination with another probiotic (typically a *Lactobacillus* or a *Streptococcus*) or as a synbiotic in combination with a prebiotic (such as a galactooligosaccharide). *B. breve* has also been a component of multiorganism probiotics, such as VSL#3 and a commercially available preparation that includes *Lactobacillus rhamnosus* GG, *L. rhamnosus* LC705, *B. breve* Bb99, and *Propionibacterium freudenreichii* ssp. *shermanii* JS. However, as the contribution of *B. breve* to clinical benefits in ulcerative colitis, pouchitis, [1] and irritable bowel syndrome [2], for example, which have been demonstrated with these cocktails, is difficult to distinguish from those of the other probiotics included, these preparations will not be considered further in this chapter.

CHARACTERIZATION AND LABORATORY STUDIES

Strains within the species *B. breve* have been extensively studied in the laboratory and the complete genome sequence of several defined [3–8]; information that has proven invaluable in predicting function. The molecular basis for the ability of this species to survive transit through the gut and in particular to resist the effects of bile has also been studied in some detail [9–12]. The enzymatic repertoire possessed by *B. breve* strains facilitates their digestion of plant polysaccharides and mucin [13–16]; *B breve* strains can also, in the right circumstances, fundamentally influence host fatty acid metabolism [17,18]. In the laboratory, *B. breve* strains have also been shown to be antipathogenic [19], antiinflammatory, and immune-modulating [20–24]; the latter effects predicting an ability to modulate allergic responses in man. As with other Bifidobacteria, the elaboration of an exopolysaccharide coat may play an important role in some of these immune effects [23]. Efficacy for specific strains was demonstrated in animal models of allergic asthma [25], necrotizing enterocolitis [26], and ulcerative colitis [27,28].

CLINICAL STUDIES

Various strains and preparations of *B. breve* have been evaluated in clinical studies. Their ability to colonize the gastrointestinal tract of infants (including very low birth weight infants) [29–31], children, [32] and adults [33,34] has been demonstrated and beneficial effects on the infant microbiota reported [30,32]. Microencapsulation has been shown to increase survival as the strain traverses the gastrointestinal tract [33]. Positive immunological effects, including an enhancement of transforming growth factor (TGF) β [35] as well as increases in secretory IgA and the antiinflammatory cytokines interleukins (ILs) 10 and 4 in combination with a reduction in the proinflammatory cytokine IL-12 [36,37], have been demonstrated. However, these immune-modulatory effects did not translate into enhanced immunogenicity to an oral inactivated cholera vaccine [38].

Not surprisingly, given findings from the laboratory, considerable effort has been exerted on assessing the effect of *B. breve* strains and formulations in allergic disorders, and especially, in infants and children. Thus, benefits have been demonstrated in cow's milk allergy [39,40], in preventing allergic disorders in children through its administration to their mothers during pregnancy [41], in atopic dermatitis, [42] and asthma [43]. However, it is important to note that other studies have shown no benefits in atopic dermatitis [44,45].

The Microbiota in Gastrointestinal Pathophysiology

135

Results of studies in animal models together with favorable immunological effects of probiotics in very low birth weight infants [46] and positive results in initial studies [47,48] suggested considerable promise for *B. breve* strains in the prevention of a devastating complication of extreme prematurity: necrotizing enterocolitis; however, a large multicenter study involving over 1300 infants whose gestational age was between 23 and 30 weeks failed to show any benefit from supplementation with *B. breve*-001 [49].

Beneficial effects with formulations containing *B. breve* strains have been demonstrated, in single studies, in acute diarrhea in children [50], constipation in adults [51], antibiotic-associated diarrhea [52], immune-compromised, [53] or severely stressed [54] individuals, celiac disease, [55] and ulcerative colitis [56]. A formulation incorporating *B. breve*, *Lactobacillus casei* and the prebiotic galactooligosaccharide was shown to reduce the occurrence of infectious complications following liver transplantation [57].

SAFETY

As for other *Bifidobacteria* species that have been utilized as probiotics in man, strains within the species, *B. breve* have generally been regarded as safe. Isolated cases of septicemia [58] and meningitis [59] have, however, been reported in neonates.

REFERENCES

[1] Holubar SD, Cima RR, Sandborn WJ, et al. Treatment and prevention of pouchitis after ileal pouch-anal anastomosis for chronic ulcerative colitis. Cochrane Database Syst Rev 2010;6. CD001176.

[2] Kajander K, Hatakka K, Poussa T, et al. A probiotic mixture alleviates symptoms in irritable bowel syndrome patients: a controlled 6-month intervention. Aliment Pharmacol Therapeut 2005;22:387–94.

[3] Kwak MJ, Yoon JK, Kwon SK, et al. Complete genome sequence of the probiotic bacterium *Bifidobacterium breve* KCTC 12201BP isolated from a healthy infant. J Biotechnol 2015;214:156–7.

[4] Morita H, Toh H, Oshima K, et al. Complete genome sequence of *Bifidobacterium breve* JCM 1192(T) isolated from infant feces. J Biotechnol 2015;210:81–2.

[5] Bottacini F, O'Connell Motherway M, Kuczynski J, et al. Comparative genomics of the *Bifidobacterium breve* taxon. BMC Genom 2014;15:170.

[6] Jiménez E, Villar-Tajadura MA, Marín M, et al. Complete genome sequence of *Bifidobacterium breve* CECT 7263, a strain isolated from human milk. J Bacteriol 2012;194:3762–3.

[7] Guinane CM, Barrett E, Fitzgerald GF, et al. Genome sequence of *Bifidobacterium breve* DPC 6330, a strain isolated from the human intestine. J Bacteriol 2011;193:6799–800.

[8] O'Connell Motherway M, Zomer A, Leahy SC, et al. Functional genome analysis of *Bifidobacterium breve* UCC2003 reveals type IVb tight adherence (Tad) pili as an essential and conserved host-colonization factor. Proc Natl Acad Sci USA 2011;108:11217–22.

[9] Andriantsoanirina V, Allano S, Butel MJ, et al. Tolerance of *Bifidobacterium* human isolates to bile, acid and oxygen. Anaerobe 2013;21:39–42.

[10] Ruiz L, O'Connell-Motherway M, Zomer A, et al. A bile-inducible membrane protein mediates bifidobacterial bile resistance. Microb Biotechnol 2012;5:523–35.

[11] Ruiz L, Zomer A, O'Connell-Motherway M, et al. Discovering novel bile protection systems in *Bifidobacterium breve* UCC2003 through functional genomics. Appl Environ Microbiol 2012;78:1123–31.

[12] Watson D, Sleator RD, Hill C, et al. Enhancing bile tolerance improves survival and persistence of *Bifidobacterium* and *Lactococcus* in the murine gastrointestinal tract. BMC Microbiol 2008;8:176.

[13] Kelly ED, Bottacini F, O'Callaghan J, et al. Glycoside hydrolase family 13 α-glucosidases encoded by *Bifidobacterium breve* UCC2003; A comparative analysis of function, structure, and phylogeny. Int J Food Microbiol 2016;224:55–65.

[14] Turroni F, Milani C, Duranti S, et al. Deciphering bifidobacterial-mediated metabolic interactions and their impact on gut microbiota by a multi-omics approach. ISME J 2016;10(7):1656–8.

[15] Egan M, O'Connell Motherway M, Ventura M, et al. Metabolism of sialic acid by *Bifidobacterium breve* UCC2003. Appl Environ Microbiol 2014;80:4414–26.

[16] O'Connell Motherway M, Kinsella M, Fitzgerald GF, et al. Transcriptional and functional characterization of genetic elements involved in galacto-oligosaccharide utilization by *Bifidobacterium breve* UCC2003. Microb Biotechnol 2013;6:67–79.

[17] Barrett E, Fitzgerald P, Dinan TG, et al. *Bifidobacterium breve* with α-linolenic acid and linoleic acid alters fatty acid metabolism in the maternal separation model of irritable bowel syndrome. PLoS One 2012;7:e48159.

[18] Wall R, Ross RP, Shanahan F, et al. Impact of administered *Bifidobacterium* on murine host fatty acid composition. Lipids 2010;45:429–36.

[19] Delcaru C, Alexandru I, Podgoreanu P, et al. Antagonistic activities of some *Bifidobacterium* sp. strains isolated from resident infant gastrointestinal microbiota on Gram-negative enteric pathogens. Anaerobe 2016;39:39–44.

[20] Wickramasinghe S, Pacheco AR, Lemay DG, et al. Bifidobacteria grown on human milk oligosaccharides downregulate the expression of inflammation-related genes in Caco-2 cells. BMC Microbiol 2015;15:172.

[21] Verheijden KA, Willemsen LE, Braber S, et al. The development of allergic inflammation in a murine house dust mite asthma model is suppressed by synbiotic mixtures of non-digestible oligosaccharides and *Bifidobacterium breve* M-16V. Eur J Nutr 2016;55:1141–51.

[22] Bermudez-Brito M, Muñoz-Quezada S, Gomez-Llorente C, et al. Cell-free culture supernatant of *Bifidobacterium breve* CNCM I-4035 decreases pro-inflammatory cytokines in human dendritic cells challenged with *Salmonella typhi* through TLR activation. PLoS One 2013;8:e59370.

[23] Fanning S, Hall LJ, van Sinderen D. *Bifidobacterium breve* UCC2003 surface exopolysaccharide production is a beneficial trait mediating commensal-host interaction through immune modulation and pathogen protection. Gut Microbes 2012;3:420–5.

[24] Ohtsuka Y, Ikegami T, Izumi H, et al. Effects of *Bifidobacterium breve* on inflammatory gene expression in neonatal and weaning rat intestine. Pediatr Res 2012;71:46–53.

[25] Sagar S, Vos AP, Morgan ME, et al. The combination of *Bifidobacterium breve* with non-digestible oligosaccharides suppresses airway inflammation in a murine model for chronic asthma. Biochim Biophys Acta 2014;1842:573–83.

[26] Satoh T, Izumi H, Iwabuchi N, et al. *Bifidobacterium breve* prevents necrotising enterocolitis by suppressing inflammatory responses in a preterm rat model. Benef Microbes 2015;. [Epub ahead of print].

[27] Zheng B, van Bergenhenegouwen J, Overbeek S, et al. *Bifidobacterium breve* attenuates murine dextran sodium sulfate-induced colitis and increases regulatory T cell responses. PLoS One 2014;9: e95441.

[28] Hayes CL, Natividad JM, Jury J, et al. Efficacy of *Bifidobacterium breve* NCC2950 against DSS-induced colitis is dependent on bacterial preparation and timing of administration. Benef Microbes 2014;5:79–88.

[29] Patole S, Keil AD, Chang A, et al. Effect of *Bifidobacterium breve* M-16V supplementation on fecal bifidobacteria in preterm neonates—a randomised double blind placebo controlled trial. PLoS One 2014;9:e89511.

[30] Li Y, Shimizu T, Hosaka A, et al. Effects of *Bifidobacterium breve* supplementation on intestinal flora of low birthweight infants. Pediatr Int 2004;46:509–15.

[31] Kitajima H, Sumida Y, Tanaka R, et al. Early administration of *Bifidobacterium breve* to preterm infants: randomized controlled trial. Arch Dis Child Fetal Neonatal Ed 1997;76:F101–7.

[32] Mogna L, Del Piano M, Mogna G. Capability of the two microorganisms *Bifidobacterium breve* B632 and *Bifidobacterium breve* BR03 to colonize the intestinal microbiota of children. J Clin Gastroenterol 2014;48(Suppl. 1):S37–9.

[33] Del Piano M, Carmagnola S, Andorno S, et al. Evaluation of the intestinal colonization by microencapsulated probiotic bacteria in comparison with the same uncoated strains. J Clin Gastroenterol 2010;44(Suppl. 1):S42–6.

[34] Shimakawa Y, Matsubara S, Yuki N, et al. Evaluation of *Bifidobacterium breve* strain Yakult-fermented soymilk as a probiotic food. Int J Food Microbiol 2003;81:131–6.

[35] Fujii T, Ohtsuka Y, Lee T, et al. *Bifidobacterium breve* enhances transforming growth factor beta1 signaling by regulating Smad7 expression in preterm infants. J Pediatr Gastroenterol Nutr 2006;43: 83–8.

[36] Plaza-Diaz J, Gomez-Llorente C, Campaña-Martin L, et al. Safety and immunomodulatory effects of three probiotic strains isolated from the feces of breast-fed infants in healthy adults: SETOPROB study. PLoS One 2013;8:e78111.

[37] Campeotto F, Suau A, Kapel N, et al. A fermented formula in preterm infants: clinical tolerance, gut microbiota, down-regulation of faecal calprotectin and up-regulation of faecal secretory IgA. Br J Nutr 2011;105:1843–51.

[38] Matsuda F, Chowdhury MI, Saha A, et al. Evaluation of a probiotics, *Bifidobacterium breve* BBG-01, for enhancement of immunogenicity of an oral inactivated cholera vaccine and safety: a randomized, double-blind, placebo-controlled trial in Bangladeshi children under 5 years of age. Vaccine 2011;29:1855–8.

[39] Morisset M, Aubert-Jacquin C, Soulaines P, et al. A non-hydrolyzed, fermented milk formula reduces digestive and respiratory events in infants at high risk of allergy. Eur J Clin Nutr 2011;65:175–83.

[40] Burks AW, Harthoorn LF, Van Ampting MT, et al. Synbiotics-supplemented amino acid-based formula supports adequate growth in cow's milk allergic infants. Pediatr Allergy Immunol 2015;26:316–22.

[41] Enomoto T, Sowa M, Nishimori K, et al. Effects of bifidobacterial supplementation to pregnant women and infants in the prevention of allergy development in infants and on fecal microbiota. Allergol Int 2014;63:575–85.

[42] Iemoli E, Trabattoni D, Parisotto S, et al. Probiotics reduce gut microbial translocation and improve adult atopic dermatitis. J Clin Gastroenterol 2012;46(Suppl.):S33–40.

[43] van der Aa LB, van Aalderen WM, Heymans HS, et al. Synbiotics prevent asthma-like symptoms in infants with atopic dermatitis. Allergy 2011;66:170–7.

[44] van der Aa LB, Lutter R, Heymans HS, et al. No detectable beneficial systemic immunomodulatory effects of a specific synbiotic mixture in infants with atopic dermatitis. Clin Exp Allergy 2012;42:531–9.

[45] van der Aa LB, Heymans HS, van Aalderen WM, et al. Effect of a new synbiotic mixture on atopic dermatitis in infants: a randomized-controlled trial. Clin Exp Allergy 2010;40:795–804.

[46] Campeotto F, Suau A, Kapel N, et al. A fermented formula in preterm infants: clinical tolerance, gut microbiota, down-regulation of faecal calprotectin and up-regulation of faecal secretory IgA. Br J Nutr 2011;105:1843–51.

[47] Janvier A, Malo J, Barrington KJ. Cohort study of probiotics in a North American neonatal intensive care unit. J Pediatr 2014;164: 980–5.

[48] Braga TD, da Silva GA, de Lira PI, et al. Efficacy of *Bifidobacterium breve* and *Lactobacillus casei* oral supplementation on necrotizing enterocolitis in very-low-birth-weight preterm infants: a double-blind, randomized, controlled trial. Am J Clin Nutr 2011;93:81–6.

[49] Costeloe K, Hardy P, Juszczak E, et al. *Bifidobacterium breve* BBG-001 in very preterm infants: a randomised controlled phase 3 trial. Lancet 2016;387:649–60.

[50] Thibault H, Aubert-Jacquin C, Goulet O. Effects of long-term consumption of a fermented infant formula (with *Bifidobacterium breve* c50 and *Streptococcus thermophilus* 065) on acute diarrhea in healthy infants. J Pediatr Gastroenterol Nutr 2004;39:147–52.

[51] Del Piano M, Carmagnola S, Anderloni A, et al. The use of probiotics in healthy volunteers with evacuation disorders and hard stools: a double-blind, randomized, placebo-controlled study. J Clin Gastroenterol 2010;44(Suppl. 1):S30–4.

[52] Souza DN, Jorge MT. The effect of Lactobacillus casei and *Bifidobacterium breve* on antibiotic-associated diarrhea treatment: randomized double-blind clinical trial. Rev Soc Bras Med Trop 2012;45:112–6.

[53] Wada M, Nagata S, Saito M, et al. Effects of the enteral administration of *Bifidobacterium breve* on patients undergoing chemotherapy for pediatric malignancies. Support Care Cancer 2010;18:751–9.

[54] Shimizu K, Ogura H, Goto M, et al. Synbiotics decrease the incidence of septic complications in patients with severe SIRS: a preliminary report. Dig Dis Sci 2009;54:1071–8.

[55] Klemenak M, Dolinšek J, Langerholc T, et al. Administration of *Bifidobacterium breve* decreases the production of TNF-α in children with celiac disease. Dig Dis Sci 2015;60:3386–92.

[56] Ishikawa H, Matsumoto S, Ohashi Y, et al. Beneficial effects of probiotic *Bifidobacterium* and galacto-oligosaccharide in patients with ulcerative colitis: a randomized controlled study. Digestion 2011;84:128–33.

[57] Eguchi S, Takatsuki M, Hidaka M, et al. Perioperative synbiotic treatment to prevent infectious complications in patients after elective living donor liver transplantation: a prospective randomized study. Am J Surg 2011;201:498–502.

[58] Ohishi A, Takahashi S, Ito Y, et al. *Bifidobacterium septicemia* associated with postoperative probiotic therapy in a neonate with omphalocele. J Pediatr 2010;156:679–81.

[59] Nakazawa T, Kaneko K, Takahashi H, et al. Neonatal meningitis caused by *Bifidobacterium breve*. Brain Dev 1996;18:160–2.

Chapter 16

Bifidobacterium longum

E.M.M. Quigley

Lynda K and David M Underwood Center for Digestive Disorders, Houston Methodist Hospital and Weill Cornell Medical College, Houston, TX, United States

The species of Bifidobacteria now referred to as *Bifidobacterium longum* contains three subspecies: *longum, infantis,* and *suis* which were originally categorized as separate species. Subsequent analyses of their sequences revealed a high degree of homology leading to their inclusion under one species, *B. longum* [1,2]. As *B. longum* spp. *suis* is found in the porcine intestine and has not been explored for probiotic properties in man, it will not be considered further. *B. longum* spp. *infantis* and *longum* have been isolated from the infant and adult intestine and several strains from both subspecies have been studied in the laboratory, as well as in clinical trials as probiotics [1–4]. Indeed, *B. longum* strains are typically dominant in the breast-fed infant gut [4]. One such strain, *B. longum* spp. *infantis* 35624, has already been the subject of a separate chapter and will not be discussed further here. Several other *longum* strains will be discussed: *B. longum* bb536; *B. longum* es 1; *B. longum* w11; *B. longum* NCC 3001; *B. longum* 1714; *B. longum* KACC 91563; *B. longum* spp. *longum* SPM 1205, 1206, 1207; *B. longum* subsp. *infantis* ATCC 15697. As has been emphasized earlier and in the introductory section on Bifidobacteria as probiotics, there is considerable homology between these strains and, as their complete genomes are described, it is likely that some strains well be reclassified and renamed [5,6]. A name change does not, of course imply a change of properties or function; it merely provides a more precise taxonomy. It is important, therefore that the reader remain vigilant for name changes among *Bifidobacteria*, especially in the species *B. longum*.

CHARACTERIZATION AND LABORATORY STUDIES

The entire genome sequence and plasmids of a number of *B. longum* strains have been described [7–9]. Such studies are valuable, not only in predicting likely biological actions, but also safety. Thus, Wei and coworkers, in sequencing *B. longum* JDM301, identified 36 genes associated with antibiotic resistance, 5 enzymes related to harmful metabolites, and 162 nonspecific virulence factors mainly associated with transcriptional regulation, adhesion, and sugar and amino acid transport. Of greatest concern was a gene conferring resistance to tetracycline that was capable of transfer [10]. Their work highlights the importance of knowing your probiotic at genome level; a strategy increasingly adopted by those who work in this field and strive to develop high quality products. Additional information on the potential effects of strains has also been gained from studies involving proteomics and transcriptomics [11,12]. Supporting their ability to survive transit through the intestine, strains of *B. longum* are bile resistant and many produce a well-defined exopolysaccharide coat that may be relevant to biological effects [13,14]. Interestingly, some differences in cell surface properties have been identified between strains isolated from preterm and term infants suggesting an adaptation to differing environments [15].

In vitro and in vivo animal studies have identified a number of properties of potential clinical importance among *B. longum* strains. These have included antiviral [16,17], metabolic [18], antioxidant [19], immunomodulatory [20–26], and neuromodulatory [27–29] effects, as well as potentially beneficial impacts on the systemic circulation [30], intestinal transport [31], and the gut barrier [32–34].

CLINICAL STUDIES

Given that some of the laboratory studies described earlier were performed in animal models of human disease, such as colitis [33], celiac disease [22], food and systemic allergy [23,24], necrotizing enterocolitis [26], and *Salmonella* infection [25], it should come as no surprise that efficacy has also been evaluated and defined in a number of clinical situations and disease states. First, *B. longum* strains have been shown to exert antiinflammatory effects in the elderly (MM-2)

[35,36], patients on peritoneal dialysis (A101) [37], and those with celiac disease (CECT 7347 and ES1) [38,39]. Strain BB536 improved several parameters of disease activity among patients with mild-to-moderate ulcerative colitis [40] and also normalized bowel function in elderly patients receiving enteral nutrition via a nasojejunal catheter [41]. Similarly, *B. longum* strains have been among those that, either alone or in combination with other probiotic organisms, ameliorated symptoms in children and adults with constipation [42,43]. Though the quality of the evidence was adjudged to be low, *B. longum* strains were also among those probiotic organisms whose use lowered the prevalence of ventilator-related pneumonia [44]. The clinical impact of a *B. longum* strain on impaired gut barrier function was assessed in a group of individuals with liver disease; though the prevalence of small intestinal bacterial overgrowth (a common occurrence in cirrhosis) was reduced, no impact on permeability was detected [45].

SAFETY

Though *B. longum* strains are generally regarded as safe [46,47], instances of bacteremia have been reported in preterm infants [48,49] and a case of peritonitis documented in a 42-year-old man with perforated diverticulitis [50]. As in many such instances the precise impact of the probiotic strain on clinical outcome is difficult to discern.

REFERENCES

[1] Lee J-H, O'Sullivan DJ. Genomic insights into *Bifidobacteria*. Microbiol Mol Biol Rev 2010;74:378–416.

[2] Leahy SC, Higgins DG, Fitzgerald GF, et al. Getting better with bifidobacteria. J Appl Microbiol 2005;98:1303–15.

[3] Picard C, Fioramonti J, Francois A, et al. Review article: bifidobacteria as probiotic agents—physiological effects and clinical benefits. Aliment Pharmacol Ther 2005;22:495–512.

[4] Underwood MA, German JB, Lebrilla CB, et al. *Bifidobacterium longum* subspecies *infantis*: champion colonizer of the infant gut. Pediatr Res 2015;77:229–35.

[5] Lewis ZT, Shani G, Masarweh CF, et al. Validating bifidobacterial species and subspecies identity in commercial probiotic products. Pediatr Res 2016;79:445–52.

[6] Srůtková D, Spanova A, Spano M, et al. Efficiency of PCR-based methods in discriminating *Bifidobacterium longum* ssp. *longum* and *Bifidobacterium longum* ssp. *infantis* strains of human origin. J Microbiol Methods 2011;87:10–6.

[7] Ham J-H, Lee T, Byun M-J, et al. Complete genome sequence of *Bifidobacterium longum* subsp. *longum* KACC 91563. J Bacteriol 2011;193:5044.

[8] Yu H, Liu L, Chang Z, Wang S, et al. Genome sequence of the bacterium *Bifidobacterium longum* strain CMCC P0001, a probiotic strain used for treating gastrointestinal disease. Genome Announc 2013;1(5):e00716-13.

[9] Zakharevich NV, Averina OV, Klimina KM, et al. Complete genome sequence of *Bifidobacterium longum* GT15: identification and characterization of unique and global regulatory genes. Microb Ecol 2015;70:819–34.

[10] Wei YX, Zhang ZY, Liu C, et al. Safety assessment of *Bifidobacterium longum* JDM301 based on complete genome sequences. World J Gastroenterol 2012;18:479–88.

[11] Mozzetti V, Grattepanche F, Moine D, et al. Transcriptome analysis and physiology of *Bifidobacterium longum* NCC2705 cells under continuous culture conditions. Benef Microbes 2012;3:261–72.

[12] An H, Douillard FP, Wang G, et al. Integrated transcriptomic and proteomic analysis of the bile stress response in a centenarian-originated probiotic *Bifidobacterium longum* BBMN68. Mol Cell Proteomics 2014;13:2558–72.

[13] Salazar N, Ruas-Madiedo P, Prieto A, et al. Characterization of exopolysaccharides produced by *Bifidobacterium longum* NB667 and its cholate-resistant derivative strain IPLA B667dCo. J Agric Food Chem 2012;60:1028–35.

[14] Salazar N, López P, Garrido P, et al. Immune modulating capability of two exopolysaccharide-producing *Bifidobacterium* strains in a Wistar rat model. Biomed Res Int 2014;2014:106290.

[15] Andriantsoanirina V, Teolis AC, Xin LX, et al. *Bifidobacterium longum* and *Bifidobacterium breve* isolates from preterm and full term neonates: comparison of cell surface properties. Anaerobe 2014;28:212–5.

[16] Kawahara T, Takahashi T, Oishi K, et al. Consecutive oral administration of *Bifidobacterium longum* MM-2 improves the defense system against influenza virus infection by enhancing natural killer cell activity in a murine model. Microbiol Immunol 2015;59:1–12.

[17] Muñoz JA, Chenoll E, Casinos B, et al. Novel probiotic *Bifidobacterium longum* subsp. *infantis* CECT 7210 strain active against rotavirus infections. Appl Environ Microbiol 2011;77:8775–83.

[18] An HM, Park SY, Lee DK, et al. Antiobesity and lipid-lowering effects of *Bifidobacterium* spp. in high fat diet-induced obese rats. Lipids Health Dis 2011;10:116.

[19] Gagnon M, Savard P, Rivière A, et al. Bioaccessible antioxidants in milk fermented by *Bifidobacterium longum* subsp. *longum* strains. Biomed Res Int 2015;2015:169381.

[20] MacSharry J, O'Mahony C, Shalaby KH, et al. Immunomodulatory effects of feeding with *Bifidobacterium longum* on allergen-induced lung inflammation in the mouse. Pulm Pharmacol Ther 2012;25:325–34.

[21] You J, Yaqoob P. Evidence of immunomodulatory effects of a novel probiotic, *Bifidobacterium longum* bv. *infantis* CCUG 52486. FEMS Immunol Med Microbiol 2012;66:353–62.

[22] Olivares M, Laparra M, Sanz Y. Oral administration of *Bifidobacterium longum* CECT 7347 modulates jejunal proteome in an in vivo gliadin-induced enteropathy animal model. J Proteomics 2012;77:310–20.

[23] Kim JH, Jeun EJ, Hong CP, et al. Extracellular vesicle-derived protein from *Bifidobacterium longum* alleviates food allergy through mast cell suppression. J Allergy Clin Immunol 2016;137:507.e8–16.e8.

[24] Schabussova I, Hufnagl K, Wild C, et al. Distinctive anti-allergy properties of two probiotic bacterial strains in a mouse model of allergic poly-sensitization. Vaccine 2011;29:1981–90.

[25] Symonds EL, O'Mahony C, Lapthorne S, et al. *Bifidobacterium infantis* 35624 protects against salmonella-induced reductions in digestive enzyme activity in mice by attenuation of the host inflammatory response. Clin Transl Gastroenterol 2012;3:e15.

[26] Underwood MA, Arriola J, Gerber CW, et al. *Bifidobacterium longum* subsp. *infantis* in experimental necrotizing enterocolitis: alterations in inflammation, innate immune response, and the microbiota. Pediatr Res 2014;76:326–33.

[27] Khoshdel A, Verdu EF, Kunze W, et al. *Bifidobacterium longum* NCC3001 inhibits AH neuron excitability. Neurogastroenterol Motil 2013;25:e478–84.

[28] Savignac HM, Tramullas M, Kiely B, et al. Bifidobacteria modulate cognitive processes in an anxious mouse strain. Behav Brain Res 2015;287:59–72.

[29] Savignac HM, Kiely B, Dinan TG, et al. Bifidobacteria exert strain-specific effects on stress-related behavior and physiology in BALB/c mice. Neurogastroenterol Motil 2014;26:1615–27.

[30] Ha GE, Chang OK, Jo S-M, et al. Identification of antihypertensive peptides derived from low molecular weight casein hydrolysates generated during fermentation by *Bifidobacterium longum* KACC 91563. Korean J Food Sci Anim Resour 2015;35:738–47.

[31] Lomasney KW, Houston A, Shanahan F, et al. Selective influence of host microbiota on cAMP-mediated ion transport in mouse colon. Neurogastroenterol Motil 2014;26:887–90.

[32] Sultana R, McBain AJ, O'Neill CA. Strain-dependent augmentation of tight-junction barrier function in human primary epidermal keratinocytes by *Lactobacillus* and *Bifidobacterium* lysates. Appl Environ Microbiol 2013;79:4887–94.

[33] Srutkova D, Schwarzer M, Hudcovic T, et al. *Bifidobacterium longum* CCM 7952 promotes epithelial barrier function and prevents acute DSS-induced colitis in strictly strain-specific manner. PLoS One 2015;10:e0134050.

[34] Rodes L, Saha S, Tomaro-Duchesneau C, et al. Microencapsulated *Bifidobacterium longum* subsp. *infantis* ATCC 15697 favorably modulates gut microbiota and reduces circulating endotoxins in F344 rats. Biomed Res Int 2014;2014:602832.

[35] Spaiser SJ, Culpepper T, Nieves C Jr, et al. *Lactobacillus gasseri* KS-13, *Bifidobacterium bifidum* G9-1, and *Bifidobacterium longum* MM-2 ingestion induces a less inflammatory cytokine profile and a potentially beneficial shift in gut microbiota in older adults: a randomized, double-blind, placebo-controlled, crossover study. J Am Coll Nutr 2015;34:459–69.

[36] Akatsu H, Iwabuchi N, Xiao JZ, et al. Clinical effects of probiotic *Bifidobacterium longum* BB536 on immune function and intestinal microbiota in elderly patients receiving enteral tube feeding. JPEN 2013;37:631–40.

[37] Wang IK, Wu YY, Yang YF, et al. The effect of probiotics on serum levels of cytokine and endotoxin in peritoneal dialysis patients: a randomised, double-blind, placebo-controlled trial. Benef Microbes 2015;6:423–30.

[38] Medina M, De Palma G, Ribes-Koninckx C, et al. *Bifidobacterium* strains suppress in vitro the pro-inflammatory milieu triggered by the large intestinal microbiota of coeliac patients. J Inflamm 2008;5:19.

[39] Olivares M, Castillejo G, Varea V, et al. Double-blind, randomised, placebo-controlled intervention trial to evaluate the effects of *Bifidobacterium longum* CECT 7347 in children with newly diagnosed coeliac disease. Br J Nutr 2014;112:30–40.

[40] Tamaki H, Nakase H, Inoue S, et al. Efficacy of probiotic treatment with *Bifidobacterium longum* 536 for induction of remission in active ulcerative colitis: a randomized, double-blinded, placebo-controlled multicenter trial. Dig Endosc 2016;28:67–74.

[41] Kondo J, Xiao JZ, Shirahata A, et al. Modulatory effects of *Bifidobacterium longum* BB536 on defecation in elderly patients receiving enteral feeding. World J Gastroenterol 2013;19:2162–70.

[42] Guerra PV, Lima LN, Souza TC, et al. Pediatric functional constipation treatment with *Bifidobacterium*-containing yogurt: a crossover, double-blind, controlled trial. World J Gastroenterol 2011;17:3916–21.

[43] Chmielewska A, Szajewska H. Systematic review of randomised controlled trials: probiotics for functional constipation. World J Gastroenterol 2010;16:69–75.

[44] Bo L, Li J, Tao T, et al. Probiotics for preventing ventilator-associated pneumonia. Cochrane Database Syst Rev 2014;10: CD009066.

[45] Kwak DS, Jun DW, Seo JG, et al. Short-term probiotic therapy alleviates small intestinal bacterial overgrowth, but does not improve intestinal permeability in chronic liver disease. Eur J Gastroenterol Hepatol 2014;26:1353–9.

[46] Simakachorn N, Bibiloni R, Yimyaem P, et al. Tolerance, safety, and effect on the faecal microbiota of an enteral formula supplemented with pre- and probiotics in critically ill children. J Pediatr Gastroenterol Nutr 2011;53:174–81.

[47] Choi SS, Kang BY, Chung MJ, et al. Safety assessment of potential lactic acid bacteria *Bifidobacterium longum* SPM1205 isolated from healthy Koreans. J Microbiol 2005;43:493–8.

[48] Bertelli C, Pillonel T, Torregrossa A, et al. *Bifidobacterium longum* bacteremia in preterm infants receiving probiotics. Clin Infect Dis 2015;60:924–7.

[49] Zbinden A, Zbinden R, Berger C, et al. Case series of *Bifidobacterium longum* bacteremia in three preterm infants on probiotic therapy. Neonatology 2015;107:56–9.

[50] Tena D, Losa C, Medina MJ, et al. Peritonitis caused by *Bifidobacterium longum*: case report and literature review. Anaerobe 2014;27:27–30.

Chapter 17

Bifidobacterium longum spp. *infantis*

E.M.M. Quigley

Lynda K and David M Underwood Center for Digestive Disorders, Houston Methodist Hospital and Weill Cornell Medical College, Houston, TX, United States

Bifidobacterium longum spp. *infantis* 35624 was originally isolated over 20 years ago from the mucosal surface of a segment of normal human distal ileum which had been harvested as part of surgical procedure to reconstruct or replace the urinary bladder. Its isolation was part of a larger project which set out to identify lactic acid bacteria (LAB) from man that might have probiotic properties. In a rigorous selection process which demonstrated the ability of putative probiotic strains to survive transit through the gut and, thus, resist the effects of gastric acid and bile, *B. longum* spp. *infantis* 35624, along with a number of other strains, emerged as a successful survivor in the enteric microenvironment and progressed to further assessments of its in vitro and in vivo properties [1].

IMMUNOLOGICAL AND PHYSIOLOGICAL EFFECTS

From the outset, the immunological effects of this bacterial strain have attracted considerable attention. Initial experiments in an animal model of inflammatory bowel disease demonstrated a potent antiinflammatory effect associated with a suppression of pro- and preservation of antiinflammatory cytokines [2,3]. Subsequent work confirmed its ability, in contrast to pathogens [4,5] and other probiotic strains [6], to generate an antiinflammatory response. This effect has been demonstrated in normal human volunteers through the ability of the orally administered organism to lead to the elevation, in serum, of the important antiinflammatory cytokine IL-10 [7]. The immunological and molecular basis for these effects have been elaborated in considerable detail and appear to be based on preferential engagement with dendritic cells (as antigen-presenting cells) leading to the induction of regulatory T-cells rather than the activation on an inflammatory cascade [5–9]. These effects have been shown to blunt the inflammatory response to a common pathogen, *Salmonella typhimurium* [10].

Potentially beneficial effects on gastrointestinal physiology have also been demonstrated including an acceleration of small intestinal transit [9] and enhancement of the gut barrier [11,12]. Of particular relevance to its role in irritable bowel syndrome (IBS), *B. longum* spp. *infantis* 35624 has been shown to display visceral antinociceptive effects in the rat [13] and, of great interest in relation to postinfection IBS, to reduce visceral hypersensitivity in a postinflammatory model [14]. Of further relevance to IBS, this same organism has been shown to normalize the immune response, reverse behavioral deficits and restore basal concentrations of norepinephrine in the brainstem in an animal model of depression and IBS [15].

HUMAN STUDIES

B. longum spp. *infantis* 35624 has been shown to successfully transit through the gastrointestinal tract in man and to be recoverable in fecal samples in numbers equivalent to those that have been associated with clinical benefits [16]. Antiinflammatory effects have been demonstrated in normal volunteers [7] and this organism has also been shown to be capable of suppressing important biomarkers of systemic inflammation, such as levels of C-reactive protein (CRP) and tumor necrosis factor alpha (TNFα) in common disorders associated with a proinflammatory phenotype, such as psoriasis, ulcerative colitis, and chronic fatigue syndrome [17]; it must be conceded that associated clinical benefits have yet to be demonstrated in any of these disorders.

Clinical benefits in comparison to placebo [18,19] and a *Lactobacillus* [18] have, however, been demonstrated in two randomized, placebo-controlled trials in IBS [18–20]. In a large dose-ranging study performed in primary care the organism demonstrated efficacy over placebo for pain, global response, and all of the other cardinal symptoms of IBS when administered in a dose of 10^8 colony forming units per milliliter but not in a dose of 10^6 [19]. Unlike some pharmacological agents or even other probiotics, this organism was similarly effective among all IBS subjects regardless of subtype (i.e., diarrhea-predominant, constipation-predominant, or mixed) [19]. Furthermore, it was well tolerated and not associated with any major adverse events. Safety and tolerance have also been reported in a study of this same strain among children with diarrhea [21].

In the only other study of a gastrointestinal disorder reported in full to date, *B. longum* spp. *infantis* 35624 failed to augment the impact of mesalamine on symptoms or recurrence rates among subjects who had experienced an episode of uncomplicated diverticultis [22].

CONCLUSIONS

B. longum spp. *infantis* 35624 has been extensively studied in the laboratory and shown to exert important, and reproducible antiinflammatory effects whose immunological basis has been described in some detail. While these antiinflammatory actions effects have been reproduced in man, efficacy in clinical disorders has to date been confined to irritable bowel syndrome. Other potentially relevant homeostatic effects have also been demonstrated in animal models and may be relevant to effects in man.

REFERENCES

[1] Symonds EL, O'Mahony C, Lapthorne S, et al. *Bifidobacterium infantis* 35624 protects against salmonella-induced reductions in digestive enzyme activity in mice by attenuation of the host inflammatory response. Clin Transl Gastroenterol 2012;3:e15.

[2] McCarthy J, O'Mahony L, O'Callaghan L, et al. Double blind, placebo controlled trial of two probiotic strains in interleukin 10 knockout mice and mechanistic link with cytokine balance. Gut. 2003;52: 975–80.

[3] Konieczna P, Akdis CA, Quigley EM, et al. Portrait of an immuncregulatory *Bifidobacterium*. Gut Microbes 2012;3:261–6.

[4] O'Hara AM, O'Regan P, Fanning A, et al. Functional modulation of human intestinal epithelial cell responses by *Bifidobacterium infantis* and *Lactobacillus salivarius*. Immunology 2006;118:202–15.

[5] O'Mahony L, O'Callaghan L, McCarthy J, et al. Differential cytokine response from dendritic cells to commensal and pathogenic bacteria in different lymphoid compartments in humans. Am J Physiol Gastrointest Liver Physiol 2006;290:G839–45.

[6] Gad M, Ravn P, Søborg DA, et al. Regulation of the IL-10/IL-12 axis in human dendritic cells with probiotic bacteria. FEMS Immunol Med Microbiol 2011;63:93–107.

[7] Konieczna P, Groeger D, Ziegler M, et al. *Bifidobacterium infantis* 35624 administration induces Foxp3 T regulatory cells in human

[8] O'Mahony C, Scully P, O'Mahony D, et al. Commensal-induced regulatory T cells mediate protection against pathogen-stimulated NF-kappaB activation. PLoS Pathog 2008;4:e1000112.

[9] Konieczna P, Ferstl R, Ziegler M, et al. Immunomodulation by *Bifidobacterium infantis* 35624 in the murine lamina propria requires retinoic acid-dependent and independent mechanisms. PLoS One 2013;8:e62617.

[10] Scully P, Macsharry J, O'Mahony D, et al. *Bifidobacterium infantis* suppression of Peyer's patch MIP-1α and MIP-1β secretion during Salmonella infection correlates with increased local CD4 + CD25+ T cell numbers. Cell Immunol 2013;281:134–40.

[11] Lomasney KW, Cryan JF, Hyland NP. Converging effects of a *Bifidobacterium* and *Lactobacillus* probiotic strain on mouse intestinal physiology. Am J Physiol Gastrointest Liver Physiol 2014;307. G241-G247.

[12] Lomasney KW, Houston A, Shanahan F, et al. Selective influence of host microbiota on cAMP-mediated ion transport in mouse colon. Neurogastroenterol Motil 2014;26:887–90.

[13] McKernan DP, Fitzgerald P, Dinan TG, et al. The probiotic *Bifidobacterium infantis* 35624 displays visceral antinociceptive effects in the rat. Neurogastroenterol Motil 2010;22:1029–35.

[14] Johnson AC, Greenwood-Van Meerveld B, McRorie J. Effects of *Bifidobacterium infantis* 35624 on post-inflammatory visceral hypersensitivity in the rat. Dig Dis Sci 2011;56:3179–86.

[15] Desbonnet L, Garrett L, Clarke G, et al. Effects of the probiotic *Bifidobacterium infantis* in the maternal separation model of depression. Neuroscience 2010;170:1179–88.

[16] Charbonneau D, Gibb RD, Quigley EM. Fecal excretion of *Bifidobacterium infantis* 35624 and changes in fecal microbiota after eight weeks of oral supplementation with encapsulated probiotic. Gut Microbes 2013;4:201–11.

[17] Groeger D, O'Mahony L, Murphy EF, et al. *Bifidobacterium infantis* 35624 modulates host inflammatory processes beyond the gut. Gut Microbes 2013;4:325–39.

[18] O'Mahony L, McCarthy J, Kelly P, et al. *Lactobacillus* and *Bifidobacterium* in irritable bowel syndrome: symptom responses and relationship to cytokine profiles. Gastroenterology 2005;128:541–51.

[19] Whorwell PJ, Altringer L, Morel J, et al. Efficacy of an encapsulated probiotic *Bifidobacterium infantis* 35624 in women with irritable bowel syndrome. Am J Gastroenterol 2006;101:1581–90.

[20] Brenner DM, Moeller MJ, Chey WD, et al. The utility of probiotics in the treatment of irritable bowel syndrome: a systematic review. Am J Gastroenterol 2009;104:1033–49.

[21] Hoy-Schulz YE, Jannat K, Roberts T, et al. Safety and acceptability of *Lactobacillus reuteri* DSM 17938 and *Bifidobacterium longum* subspecies *infantis* 35624 in Bangladeshi infants: a phase I randomized clinical trial. BMC Complement Altern Med 2016;16:44.

[22] Stollman N, Magowan S, Shanahan F, et al. A randomized controlled study of mesalamine after acute diverticulitis: results of the DIVA trial. J Clin Gastroenterol 2013;47:621–9.

Chapter 18

Common Organisms and Probiotics: *Saccharomyces boulardii*

L.V. McFarland
Department of Medicinal Chemistry, School of Pharmacy, University of Washington, Seattle, WA, United States

INTRODUCTION

Probiotics have become increasing popular in the United States and are rapidly reaching levels of use seen in Europe and Asia, where there is a longer history of use of probiotics. Probiotics are defined as "live microorganisms, which when administered in adequate amounts, confer a health benefit on the host" [1]. Probiotics are generally recommended to help strengthen host systems and assist in recovery from certain diseases. However, the general public and many health-care providers are still confused about the how best to utilize probiotics and which types of probiotics may be effective for specific diseases. There are several challenges in choosing the appropriate probiotics including the wide diversity of probiotic strains, quality control of commercially available probiotic products, and the degree of evidence-based trials for each disease and type of probiotic. The ability of an organism to be an effective probiotic has been found to be strain specific and microbial organisms are defined by their genus, species, and strain. Probiotics products may contain a single strain or be a mixture of multistrain types. To add to this confusion, probiotic products are available in diverse forms: capsules of freeze-dried or lyophilized cultures, heat-dried culture supernatants, mixed in dairy foods (such as yogurts, cheese, milks, or ice cream), or other foods (kefir, chocolate, wafers). The variety of probiotic products is also regulated under different guidelines according to their use (food, dietary supplement, over-the-counter use, or prescription). Differing standards of quality assurance also varies by the type of product, indication for use, and country. Despite these challenges, the awareness of probiotics as a therapeutic modality has increased dramatically and the frequency of peer-reviewed randomized clinical trials has kept pace with the global interest in this innovative method of therapy. Numerous probiotic strains have been investigated for clinical efficacy and this chapter examines evidence for *Saccharomyces* probiotics focusing on *Saccharomyces boulardii* for the efficacy and safety of different diseases.

HISTORY

Saccharomyces is a genus of fungus that encompasses numerous species and strains used for a wide diversity of purposes. *Saccharomyces cerevisiae* strains have a long history of use in baking and brewing preparations, but have only infrequently been investigated for probiotic properties [2,3]. Another closely related strain, *S. boulardii*, as shown in Fig. 18.1, was discovered by a French microbiologist, Henri Boulard in 1920 when he was in IndoChina searching for new strains of yeast that could be used in fermenting processes. He was visiting during a cholera outbreak and noticed that some people who did not develop cholera were drinking a special tea. This tea was made by taking the outer skin from a tropical fruit (lychee and mangosteens) and cooking them down to make tea. He succeeded in isolating the agent responsible. It was a special strain of yeast he named "*S. boulardii*." The patent for this yeast was bought by Laboratories Biocodex in 1947, which began research and manufacturing protocols. As shown in Fig. 18.2, this yeast probiotic has been extensively studied with 359 published studies (from January 1975 to October 2015) on a wide variety of topics: randomized controlled trials for efficacy, safety and kinetic studies, mechanism of action, reviews, safety reports, and quality control reports.

Currently, this yeast probiotic is used in over 80 countries spanning Europe, North and South America, the Middle East and Asia [4]. However, there are many different *Saccharomyces* products available commercially sold as probiotics as either as lyophilized or heat-dried powders in capsules, or as one of several strains in a probiotic mixture in capsules, or in liquid beverages. It is important to recognize that the choice of which product to choose needs to be based on several factors: (1) if the probiotic product is backed by evidence-based clinical trials, (2) if the probiotic product is manufactured by a reliable company with a good history of producing a quality product, (3) the probiotic strain in the

The Microbiota in Gastrointestinal Pathophysiology

145

FIGURE 18.1 **Photograph of *S. boulardii*.** *Photograph used with permission. S. boulardii Biocodex in cell culture of human epithelium.* Thierry Meylheuc-Harry, Sokol-Philippe Langella, Mima2-INRA Micalis.

product is the identical strain that is backed by clinical evidence of efficacy and safety, and (4) the evidence for the probiotic is for the disease or condition under consideration. Studies of other probiotics have found a wide diversity in both quality and contamination in products available on the Internet [5]. Marcobal et al. tested 14 commercial probiotics in the United States and found 93% were incorrectly labeled (57% had contaminants and 36% were missing strains listed on the label) [6]. Quality of different *S. boulardii* products has also varied. Although most products state they contain at least 1×10^9 *S. boulardii*/mg, independent assays have determined 50% of the products contained a dose less than on the label. In one study comparing six *S. boulardii* products,

all had identical PCR typing profiles, but only 50% [Floratil (Merck), Flomicin (NeoChemical), and Florazin (Herald's)], had the same concentration identified on their label. One product [Lactipan (Sigma Pharm)] had 2×10^4 fewer *S. boulardii* than stated on its label. One product [Floratil (Merck)], had the highest concentration (1×10^9/100 mL) and maintained high levels (9.5×10^8) 6 months later [7]. Even if the label states it contains *S. boulardii*, a variation in efficacy may occur due to lower than stated dose or inaccurate strain composition. Four *S. boulardii* products were tested (along with one *S. cerevisiae* product) in Brazil. Only two (50%) of the *S. boulardii* products were protective in a *Salmonella typhimurum* mouse model (two *S. boulardii* products were ineffective, as well as the one *S. cerevisiae* product) [8]. A more recent of 15 probiotic products found in Belgium that were labeled as containing *S. boulardii*, the concentration was found to be within one log of that listed on the label for 13 (3 products did not list the concentration on their label) [9]. One method of selecting a probiotic product is to find a product in which the manufacturing company has sponsored original clinical trials, as this indicates a degree of commitment that may not be present in companies that do not sponsor original research.

TAXONOMY

Probiotics may be either bacterial or fungal (yeast) microbes. Yeast probiotics, such as *S. boulardii*, are different from bacterial probiotics (different physiologic structures, larger in size, do not acquire antibiotic-resistant genes as easily and are not affected by antibiotics). Not only is there a confusing array of probiotic products on

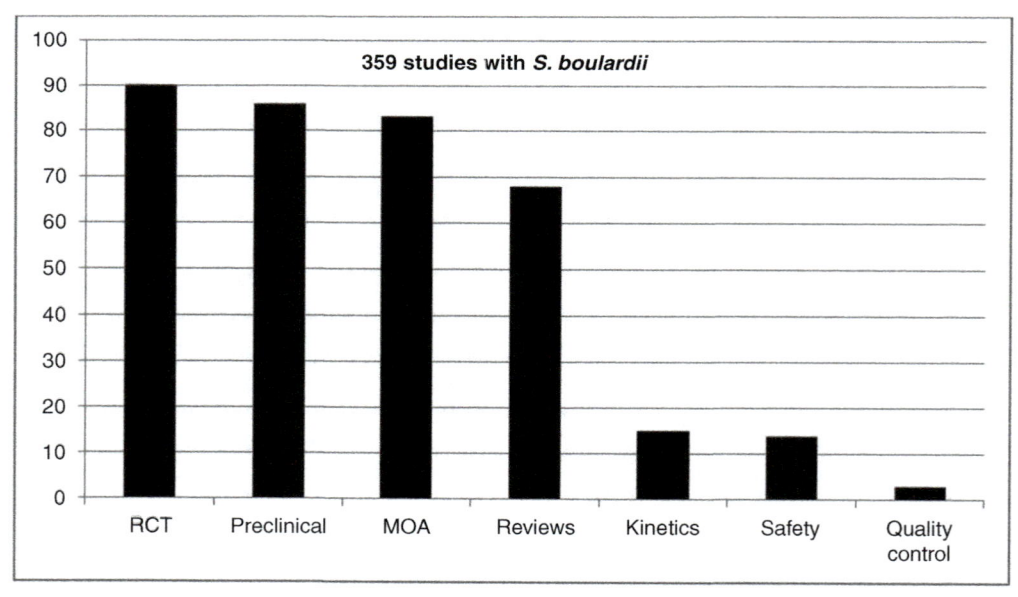

FIGURE 18.2 **The types of published studies (1977–2015) for the *S. boulardii* CNCM I-745 strain.** *MOA*, mechanism of action; *RCT*, randomized controlled trials.

the global market, but also the taxonomy of *Saccharomyces* strains has been debated [10–12]. Originally named *S. boulardii* in the 1950s, advances in typing methods opened a debate as to whether this strain should be reclassified as a variant of *S. cerevisiae* or remain as a separate species. *S. boulardii* and *S. cerevisiae* strains were found to be similar, but *S. boulardii* has different behavior and growth profiles. *S. boulardii* is different physiologically and metabolically in that its optimum growth temperature is 37°C, and it is resistant to low pH and is tolerant of bile acids; whereas other strains of *S. cerevisiae* prefer cooler temperatures (30–33°C) and do not survive well in acid pH ranges [13]. Microsatellite polymorphism analysis and retrotransposon hybridization analyses showed that *S. boulardii* has a unique and specific microsatellite allele that differs from *S. cerevisiae* isolates [14].

As shown in Table 18.1, there are several probiotic products within the *S. cerevisiae–S. boulardii* realm, but only one strain (manufactured by Laboratoires Biocodex, France) has been supported by over 88 randomized controlled trials. Over time, the designation of this strain has shifted from *S. cerevisiae* Hansen Collection of Biologic Specimens (CBS) 5926, to *S. boulardii* or *S. boulardii* lyo, *S. cerevisiae* var. *boulardii*. This strain was deposited by Laboratoires Biocodex with the Pasteur Institut in France in 2014 with the designation of *S. boulardii* CNCM [Collection Nationale de Cultures de microorganismes (Pasteur Institute)] I-745 [15]. *S. boulardii* CNCM 1079 and *S. cerevisiae* CNCM

I-3856 have recently started their investigational studies into their efficacy for disease and have been determined to be separate strains from *S. boulardii* CNCM I-745 [16–18]. European experts have ruled that clinical efficacy from *S. boulardii* CNCM I-745 (also known as *S. cerevisiae* Hansen CBS 5926) cannot be extrapolated to other *Saccharomyces* strains [19]. Other probiotic products on the market labeled as "*S. boulardii*" have not been supported by their own independent clinical trials, and it is unclear if the strain in their product is a strain of *S. boulardii* supported by clinical evidence. The remaining of the discussion of this chapter will focus on the *S. boulardii* CNCM I-745 strain.

MECHANISMS OF ACTION

An advantage of probiotics is that they are living organisms incorporating a delivery system that survives to the host destination (typically skin, intestine, vagina, etc.) bringing an arsenal of antipathogenic strategies to the target site. *S. boulardii* has several different types of mechanisms of action (Fig. 18.3) which may be classified into several main areas: (1) antitoxin effects, (2) physiologic protection, (3) modulation of the normal microbiota, (4) metabolic regulation, (5) trophic or nutritional effects, (6) immune system regulation, and (7) direct pathogen inhibition [14]. Within the intestinal lumen, *S. boulardii* may interfere with pathogenic toxins, preserve cellular physiology, interfere with pathogen attachment, interact with normal microbiota, or assist in reestablishing short-chain fatty acid (SCFA) levels.

TABLE 18.1 Description of Examples of Different Types of Probiotic Products Listed as *S. boulardii* on the European and USA Market

Strains of Sb	Manufactured by	Dose (cfu/Capsule)	Evidence-Based Health Claims	Published References
S. boulardii CNCM I-745 (other names: *S. cerevisiae* Hansen CBS 5926, *S. boulardii* 17, *S. boulardii* ATCC 74012, *S. boulardii* lyo)	Laboratoires Biocodex, France	250 mg; 5×10^9	Treatment pediatric diarrhea, AAD, *H. pylori* infections, IBD, adult acute diarrhea, enteral feed diarrhea, CDI, giardiasis	>88 RCTs; Dinleyici et al. [15], McFarland [14]
S. boulardii + MOS	Jarrow Labs Los Angeles CA	mg nr; 1.5×10^9	"Keep refrigerated"	Case report only; Ooi et al. [20]
S. boulardii	Klaire Lab Reno NV	150 mg; 3×10^9	No claims	No original studies
S. boulardii CNCM I-1079	Pure Encapsulations, Inc. Sudbury, MA and Lallemand, Canada	mg nr; 1.4×10^{10}	Reduction of cholesterol	Open, uncontrolled study only; Ryan et al. [16]
Non-*S. boulardii* strains under development				
S. cerevisiae CNCM I-3856	Lesaffre Intl. France	500 mg, 8×10^9	New probiotic being developed in preclinical, one RCT for IBS	Cordonnier et al. [17], De Chambrun et al. [18]

AAD, Antibiotic-associated diarrhea; ATCC, American type culture collection; CA, California; CDI, *Clostridium difficile* infections; CNCM, Collection Nationale de Cultures de microorganismes (Pasteur Institute); IBD, inflammatory bowel disease; IBS, irritable bowel syndrome; MA, Massachusetts; mg, milligram; MOS, prebiotic; nr, not reported; NV, Nevada.

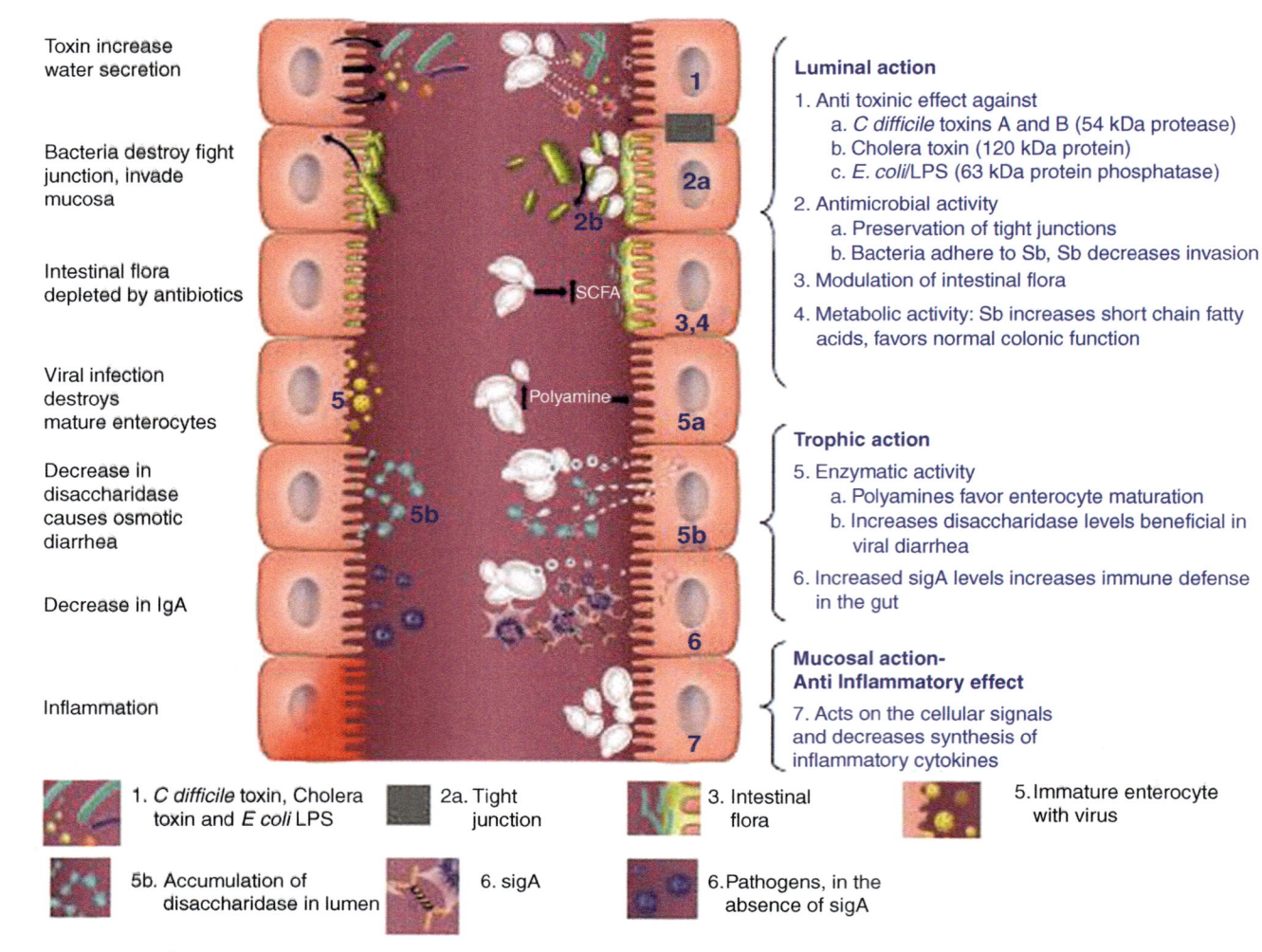

Toxin increase water secretion

Bacteria destroy fight junction, invade mucosa

Intestinal flora depleted by antibiotics

Viral infection destroys mature enterocytes

Decrease in disaccharidase causes osmotic diarrhea

Decrease in IgA

Inflammation

Luminal action

1. Anti toxinic effect against
   a. *C difficile* toxins A and B (54 kDa protease)
   b. Cholera toxin (120 kDa protein)
   c. *E. coli*/LPS (63 kDa protein phosphatase)
2. Antimicrobial activity
   a. Preservation of tight junctions
   b. Bacteria adhere to Sb, Sb decreases invasion
3. Modulation of intestinal flora
4. Metabolic activity: Sb increases short chain fatty acids, favors normal colonic function

Trophic action

5. Enzymatic activity
   a. Polyamines favor enterocyte maturation
   b. Increases disaccharidase levels beneficial in viral diarrhea
6. Increased sigA levels increases immune defense in the gut

Mucosal action-Anti Inflammatory effect

7. Acts on the cellular signals and decreases synthesis of inflammatory cytokines

1. *C difficile* toxin, Cholera toxin and *E coli* LPS

2a. Tight junction

3. Intestinal flora

5. Immature enterocyte with virus

5b. Accumulation of disaccharidase in lumen

6. sigA

6. Pathogens, in the absence of sigA

FIGURE 18.3 **The different mechanisms-of-action for the probiotic *S. boulardii* CNCM I-745 strain.**

S. boulardii also may act as an immune regulator, both within the lumen and systemically.

Antitoxin effects. S. boulardii may act by blocking pathogen toxin receptor sites [21] or act as a decoy receptor for pathogenic toxins [22], or act by direct destruction of the pathogenic toxin. Castagliuolo et al. found a 54 kDa serine protease produced by *S. boulardii* that directly degrades *C. difficile* toxin A and B [23]. The efficacy of other strains of *Saccharomyces* has also been investigated. Only *S. boulardii* produces a protease capable of degrading *C. difficile* toxins and receptors sites on the enterocyte cell surface, other strains of *S. cerevisiae* do show exhibit this ability [21,24]. Buts et al. found a 63 kDa phosphatase produced by *S. boulardii* that destroys the endotoxin of pathogenic *E. coli* [25]. Several investigators showed *S. boulardii* could reduce the effects of cholera toxin and this may be due to a 120 kDa protein produced by *S. boulardii* [26,27].

Physiologic protection. S. boulardii has been shown preserve the tight junctions between enterocytes in the intestine, reducing fluid loss (diarrhea) [28]. Garcina-Vilela et al. showed decreased intestinal permeability when patients with Crohn's disease were given *S. boulardii* (1.6×10^9 per day for 4 months) compared to placebo [29]. *Modulation of normal microbiota.* Newer techniques, including metagenomics and PCR probes have documented that a typical human may carry over 40,000 bacterial species in the collective intestinal microbiome [30,31]. The normal intestinal flora has many functions, including digestion of food, but the one that is most germane for this discussion is called "colonization resistance" [32,33]. This involves the interaction of many bacterial microflora and results in a barrier effect against colonization of pathogenic organisms. Normal microflora may act by competitive exclusion of nutrients or attachment sites, produce bacteriocins, or produce enzymes detrimental to pathogenic growth. Factors that disrupt this protective barrier, for example, antibiotic use or surgery, result in host susceptibility to pathogen colonization until such time as the normal microflora can become reestablished. Typically, it takes 6–8 weeks for normal microbiota to recover after antibiotic exposure or disease resolution [34]. Probiotics are uniquely qualified to fit into this window of susceptibility and may act as surrogate normal microflora until recovery is achieved. *S. boulardii* has been shown to help restore the disrupted normal microflora in patients receiving antibiotics or having surgery [35].

Restoration of metabolic activities. S. boulardii has been shown to be able to increase short chain fatty acids (SCFA), which are depressed during disease indicating altered colonic fermentation [36,37].

Trophic effects. S. boulardii can reduce mucositis [38], restore fluid transport pathways [27,39], stimulate protein and energy production [40], or act through a trophic effect by releasing spermine and spermidine, or other brush border enzymes that aid in the maturation of enterocytes [41,42].

Immune response. S. boulardii may also regulate immune responses, either acting as an immune stimulant or by reducing proinflammatory responses. *S. boulardii* may cause an increase in secretory IgA levels in the intestine [8,43]. *S. boulardii* has also been found associated with higher levels of serum IgG to *C. difficile* toxins A and B [44]. *S. boulardii* may also interfere in NF-kappa B-mediated signal transduction pathways, which stimulate proinflammatory cytokine production [45,46]. Chen et al. found that *S. boulardii* blocks activation of ERK1/2 and MAP kinases, which typically stimulate IL-8 production and cell necrosis in mice ileal loop models and in in vitro models [47]. *S. boulardii* has also been shown to cause the trapping of T helper cells into mesenteric lymph nodes, thereby reducing inflammation [48].

Antimicrobial activity. S. boulardii is capable of directly or indirectly interfering with intestinal pathogens. *S. boulardii* may directly inhibit the growth of pathogens (such as *Candida albicans, S. typhimurum, Yersinia enterocolitium, Aeromonas hemolysin*) [49–51].

PHARMACOKINETICS

S. boulardii, when given orally to humans, achieves steady-state concentrations within 3 days and is cleared within 3–5 days after it is discontinued [52,53]. Blehaut et al. gave eight healthy human volunteers *S. boulardii* (oral dose of 5×10^9) for 6 days and followed them for time-to-clearance [53]. They determined *S. boulardii* has half-life of 6 h, fecal steady-state concentrations (2×10^7 per g) were reached by day 3 and the yeast was cleared after 4 days postadministration. Klein et al. confirmed these findings in their studies of human volunteers and also found levels of *S. boulardii* were 23 times higher in volunteers with disturbed intestinal microbiota (due to ampicillin exposure) than volunteers who were not given ampicillin [54].

CLINICAL EFFICACY OF *Saccharomyces boulardii*

The number of randomized, controlled clinical trials testing *S. boulardii* has increased exponentially from 1995 to 2015, as shown in Fig. 18.4.

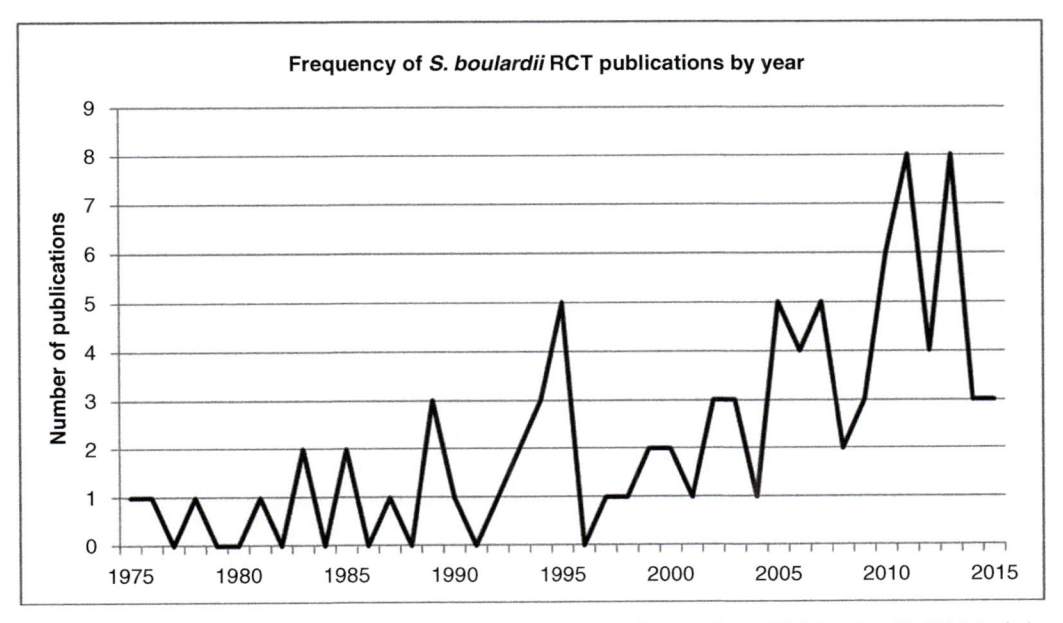

FIGURE 18.4   **Frequency of peer-reviewed publications on *S. boulardii* CNCM I-745 strain from 1976-October 28, 2015.** Includes studies in pediatric and adult populations, "randomized controlled trials" includes only those with control and treatment groups of good–excellent quality.

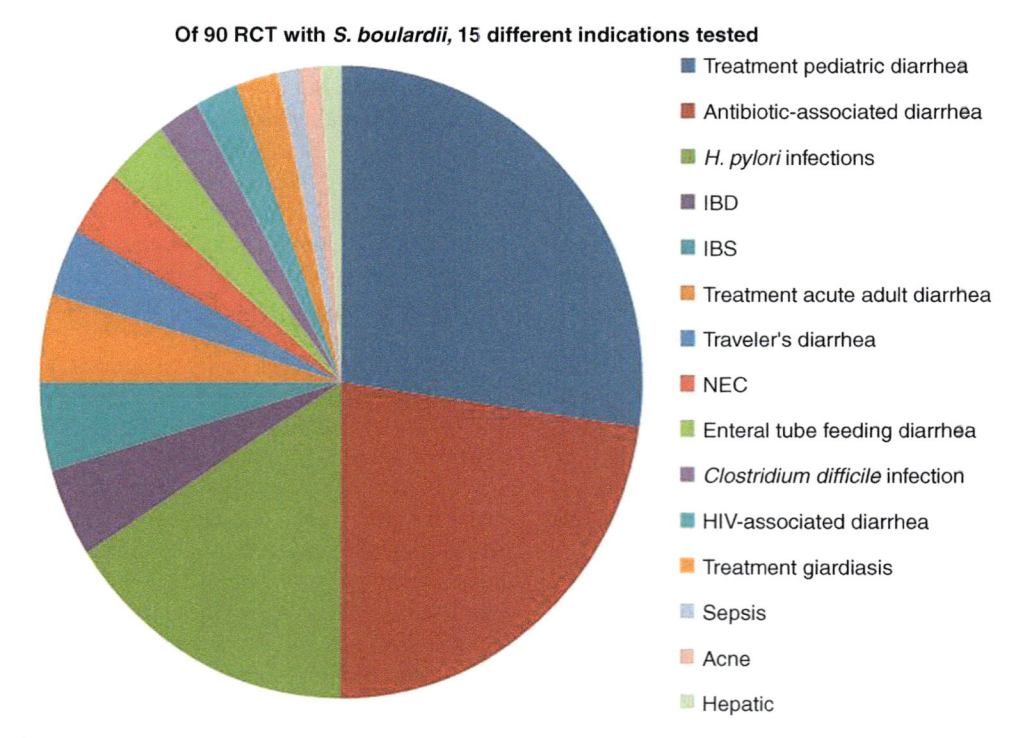

Of 90 RCT with *S. boulardii*, 15 different indications tested

- Treatment pediatric diarrhea
- Antibiotic-associated diarrhea
- *H. pylori* infections
- IBD
- IBS
- Treatment acute adult diarrhea
- Traveler's diarrhea
- NEC
- Enteral tube feeding diarrhea
- *Clostridium difficile* infection
- HIV-associated diarrhea
- Treatment giardiasis
- Sepsis
- Acne
- Hepatic

FIGURE 18.5 **The distribution by 15 different disease indications for 90 randomized controlled clinical trials testing the *S. boulardii* CNCM I-745 strain (from 1977 to 2015). *IBD* inflammatory bowel disease, *IBS* irritable bowel syndrome, *NEC* necrotizing enterocolitis, *HIV* human immunodeficiency virus.**

Not only has the number of trials increased over time, but *S. boulardii* has been studied for 15 different types of diseases, as shown in Fig. 18.5. Most of these trials ($n = 88$) were done with the strain *S. boulardii* CNCM I-745.

Of 90 randomized, controlled trials, the most commonly studied has been for the treatment of acute pediatric diarrhea (27% of all trials), for the prevention of antibiotic-associated diarrhea (AAD) (22% of all trials), and for *Helicobacter pylori* infections (16% of all trials).

TREATMENT OF ACUTE PEDIATRIC DIARRHEA

Epidemiology of acute pediatric diarrhea. Acute pediatric diarrhea occurs in 10–15% of children due to a variety of causes ranging from infection with pathogenic organisms (rotaviruses, *Escherichia coli, Shigella,* and others), or diarrhea may result from a shift in nutrition or life-style changes [55]. Consequences of acute pediatric diarrhea may include severe dehydration in younger children, increased hospitalizations, sepsis, and death. An increasing number of children developing *C. difficile* infections are being reported (covered in subsequent section in this chapter) [56].

Clinical efficacy for acute pediatric diarrhea. For these trials, children with acute diarrhea are enrolled and randomized to either *S. boulardii* or a control group. Then they are followed to observe how quickly they respond.

Efficacy is measured by several methods: the number of children cured of diarrhea by a specific day, or by the improvement of diarrheal symptoms or by the reduction of the mean days of diarrhea. Most of the trials compare one group of children randomized to *S. boulardii* to another group randomized to a control (either placebo or no treatment). There have been 24 randomized, controlled trials published from 1985 to 2015 using *S. boulardii* to treat children with acute diarrhea. One trial had two treatment arms (one with just *S. boulardii* and another testing a mixture of probiotic strains, one of which was *S. boulardii* [57]).

Table 18.2 lists a description of the clinical trials testing *S. boulardii* for the treatment of acute pediatric diarrhea. These trials enrolled children 3 months to 18 years old who had acute diarrhea. Oral rehydration therapy was typically given to both the probiotic and control group children. The daily dose of *S. boulardii* varied and the duration given was usually less than 1 week. Of the 24 randomized trials, 83% found *S. boulardii* was significantly effective in treating pediatric acute diarrhea and no serious adverse events were reported in any of these trials.

One method to summarize the results of many studies is to use a statistical tool called metaanalysis. Metaanalysis pools the results of many studies into a single outcome, allowing easier interpretation of a treatment's ability to improve health status. However,

TABLE 18.2 Randomized Controlled Trials for *S. boulardii* and the Treatment of Acute Pediatric Acute Diarrhea

Number Enrolled	Daily Dose (cfu/d)	Duration Given (d)	Outcome in *S. boulardii*	Outcome in Control Group	References
38	1×10^{10}	5	95%[a] cured	79%	Chapoy [58]
130	2×10^{10}	4	85%[a] cured	40%	Cetina-Sauri and Sierra Basto [59]
18	Varied	30	59%[a] cured	24%	Chouraqui et al. [60]
100	5×10^{9}	3	84%[a] cured	64%	Urganci et al. [61]
109	1×10^{10}	6	Mean stools (4.3/d)[a] on day 3	Mean stools (5.7/d) on day 3	Hafeez et al. [62]
89	1.75×10^{12}	5	Mean stools (2.0 ± 2/d)[a]	Mean stools (5.2 ± 3.0/d)	Gaon et al. [63]
200	5×10^{9}	5	Duration 4.7 d[a]	Duration 5.5	Kurugol and Koturoglu [64]
100	1×10^{10}	5	Duration 3.6 d[a]	Duration 4.8 d	Billoo et al. [65]
100	5×10^{9} if <1-year old, or 1×10^{10} if >1-year old	6	Duration 4.7 d[a], 93%[a] cured	Duration 6.2 d, 73% cured	Villarruelet al. [66]
183	5×10^{9}	5	Duration 4.4 d	Duration 4.8 d	Canani et al. [67]
27	500 mg	7	Mean stools (0.4 ± 0.2/d)[a] on day 4	Mean stools (1.8 ± 0.4/d) on day 4	Ozkan et al. [68]
100	1×10^{10}	5	Duration 3.1 d[a]	Duration 4.7 d	Htwe et al. [69]
50	1×10^{10}	7	Duration 42.2 ± 17.4 h[a]	Duration 72. ± 28.5 h	Dinleyici et al. [70]
85	5×10^{6}	10	Duration 4.5 d	Duration 5 d	Savas-Erdeve et al. [71]
67	250 mg if ≤2 years or 500 mg if ≥2 years	5	46.4%[a] cured	22.2%	Eren et al. [72]
40	500 mg Sb only	5	Duration 58 h[a]	Duration 84.5 h	Grandy et al. [57]
40	500 mg in mixture with Sb	5	Duration 60 h	Duration 84.5 h	Grandy et al. [57]
70	200 mg	6	Duration 35.4 ± 3.7 h[a]	Duration 67.1 ± 5 h	Le Luyer et al. [73]
33	500 mg	10	77.7%[a]	40%	Dinleyici et al. [74]
480	250 mg	5	Duration 3.1 ± 1.8 d[a]	Duration 5.3 ± 1.8 d	Dalgic et al. [75]
176	8×10^{9}	5	69.5%[a] cured	44%	Correa et al. [76]
144	15–20 mg/kg/d	365	Sign improved Symptoms[a]	No change	Jozefczuket al. [77]
108	500 mg	5	Duration 52.1 ± 24.6 h[a]	Duration 64.0 ± 30.4 h	Riaz et al. [78]
50	282.5 mg	5–10	Duration 6.6 ± 1.7 d	7.0 ± 1.6 d	Erdogan et al. [79]
363	500 mg	5	Duration 75.4 ± 33.1 h[a]	Duration 99.8 ± 32.5 h	Dinleyiciet al. [80]

[a]*p < 0 0.5 compared to controls.*
cfu, Colony-forming units; d, day; h, hours; mg, milligram; nr, not reported.

it must be cautioned that metaanalysis using probiotics should always be done pooling probiotics of the same strain, as the efficacy varies widely with the type strain used. A metaanalysis of 22 randomized trials including 2440 children found *S. boulardii* was able to reduce the duration of diarrhea significantly by a mean of 1 day [81]. Another metaanalysis, which pooled the results from 13 trials (11 pediatric and 2 in adults) four 1,266 subjects also found *S. boulardii* reduced acute diarrhea by a mean of 1 day [82]. Although reducing diarrhea by 1 day may not sound impressive, it can be clinically important, especially for young or undernourished children.

ANTIBIOTIC-ASSOCIATED DIARRHEA

Epidemiology of AAD. The reported incidence of AAD ranges from 12/100,000 to 34/100 depending upon the type of antibiotic, host factors (age, health status, etc.), etiology, hospitalization status, and presence of a nosocomial outbreak [83]. The highest frequency of AAD is found (26–60%) during healthcare-associated outbreaks, when susceptible patients are clustered by time, exposure, and proximity [84]. Healthcare-associated (hospital, long-term care facilities, nursing homes) outbreaks of AAD are to be expected because inciting agents (antibiotics), infectious agents, and a susceptible patient population are intermixed.

Clinical efficacy for AAD. In clinical trials for the prevention of AAD, patients without diarrhea who are beginning a course of antibiotics are randomized to either *S. boulardii* or a control group (typically given a placebo). The investigated treatment is usually given for the duration of the antibiotic course and may be given for an additional period, then they are followed to observe if they develop diarrhea associated with their antibiotic exposure. To detect delayed-onset AAD (occurs after antibiotics), they should also be followed for an additional 4–8 weeks. There have been 22 randomized controlled trials in adults using *S. boulardii* for the prevention of AAD (Table 18.3).

One trial combined *S. boulardii* or placebo with one of two different antibiotics (sulbactam-ampicillin or azthromycin), resulting in two treatment arms [92]. These studies have used a variety of daily doses, durations of treatment, length of follow-up, and types of patient population. Of the 23 treatment arms, 65% showed significant efficacy for the prevention of AAD compared to controls. One of the earliest trials, Adams et al. randomized 388 hospitalized patients to *S. boulardii* [4×10^9 cfu (colony forming units)/day] or placebo for 7 days and found a significant reduction in AAD in the *S. boulardii* group (4.5%) compared to the controls (17.5%, $P < 0.05$) [85]. The protective effect of *S. boulardii* was confirmed in later studies. McFarland et al. enrolled 193 hospitalized inpatients receiving beta-lactam antibiotics and randomized them to either *S. boulardii* or placebo for the duration of the antibiotic plus 3 additional days [88]. Significantly fewer developed AAD while on the beta-lactam antibiotics or in the 7 week follow-up period when given *S. boulardii* (7.2%, $P < 0.05$) compared to 14.6% of those given placebo.

It is important to have sufficiently long follow-up times (usually 4–6 weeks postantibiotic cessation) to capture delayed-onset AAD. Several studies have shown that AAD may occur while on antibiotics, but AAD may also be delayed for up to 2 months (in up to 38% of patients) after antibiotics are discontinued [88,107]. Most studies of AAD follow patients 4–6 weeks postantibiotics to document delayed-onset AAD [94–96]. In contrast, Lewis et al. did not find a significant reduction in AAD in a study of 69 elderly patients randomized to *S. boulardii* (4.5×10^9 cfu/day) or placebo for 14 days

[89]. This failure may have been due to a flawed study design, in that patients were only followed while on antibiotics and no follow-up was done after antibiotics were discontinued. The study by Lewis et al. therefore may have missed a significant number of AAD cases due to too short a follow-up period. Short or nonexistent follow-up times postantibiotic exposure may explain why some studies did not find a significant protective effect of *S. boulardii* [89,97,108]. No adverse reactions associated with *S. boulardii* were noted in any of these studies. No other strains of *Saccharomyces* have been found to be effective to prevent AAD.

While most metaanalyses have concluded that probiotics are effective for preventing AAD [84,109], it is inappropriate to conclude that all probiotic strains are effective [110]. The effectiveness of probiotics is strain specific and disease specific, so each probiotic strain must be linked to the disease. Szajewska et al. limited her metaanalysis to randomized controlled trials of one type of probiotic (*S. boulardii*) in adults and children receiving antibiotics [111]. Pooling the results from 21 trials (involving 4,780 subjects), a significantly protective effect of *S. boulardii* use was found [pooled relative risk (RR) = 0.47, 95% confidence interval (CI) 0.38, 0.57]. Some metaanalyses have done sensitivity analysis, separating out subgroups by patient type (e.g., just adults or just pediatrics) or by type of probiotic strain [84,112]. In one metaanalysis, only two probiotic strains had sufficient evidence to show significant efficacy for the prevention of AAD, namely *S. boulardii* and *Lactobacillus rhamnosus* GG. However, *L. rhamnosus* GG probiotics may be contraindicated when the patient is prescribed a type of antibiotic that this strain is susceptible [84]. Another metaanalysis found *S. boulardii* effectively prevented AAD in pediatric patients [113], and another metaanalysis found *S. boulardii* was also effective in adults [14]. Although many metaanalyses have been done, pooling studies must be performed with these limitations in mind. Another metaanalysis looked at patients with *H. pylori* infections who were treated with antibiotics to eradicate *H. pylori* and found that *S. boulardii* was effective in preventing AAD associated with this treatment [114].

HELICOBACTER PYLORI INFECTIONS

Epidemiology of H. pylori. *H. pylori* colonizes the gastric mucosa and usually induces a chronic, asymptomatic carrier state, but some people may develop gastroduodenal ulcers as a result. Carriage of *H. pylori* is a risk factor for developing gastric lymphoma or adenocarcinoma in later life. The prevalence of *H. pylori* carriage is 50–80% worldwide [115]. Standard treatment for *H. pylori* involves a 2-week course of triple or quadruple

TABLE 18.3 Randomized Controlled Trials for *S. boulardii* and the Prevention of Antibiotic-Associated Diarrhea

Population	*S. boulardii* Dose/Day (cfu/d)	*S. boulardii* Duration and Follow-up	AAD in *S. boulardii*	AAD in Controls	References
388 hospitalized adults, 25 different sites in France	4×10^9	7 d, f/up: none	4.5%[a]	17.5%	Adam et al. [85]
240 adults on oral antibiotics, Portugal	4 caps	6 d, f/up: none	15.7%[a]	27.7%	Monteiro [86]
180 hospitalized adults, Seattle WA	2×10^{10}	Duration abx + 2 wks, 23% f/up: 2–3 weeks	9.5%[a]	21.8%	Surawicz et al. [87]
193 hospitalized adults (1 or more beta-lactam antibiotics 4 USA hospitals	2×10^{10}	Duration abx + 3 d f/up: 7 wks	7.2%[a]	14.6%	McFarland et al. [88]
72 enrolled, 69 done; elderly patients (>65 yrs old) UK	4.5×10^9	Duration of abx (mean = 6 d) No f/up	21%	13.9%	Lewis et al. [89]
779 enrolled on abx, (1–5 yrs)	4.5×10^9	Abx duration (mean = 8 ± 2 d) f/up: none	7.6%	5.5%	Benhamouet al. [90]
43 adults asymptomatic *H. pylori* + on std triple therapy	5×10^9	14 d f/up: 2 wks	5%[a]	30%	Cremoniniet al. [91]
234 children enrolled, 1–15 yrs old on sulfbactam-ampicillin	5×10^9	During abx f/up: 2 wks	5.7%[a]	25.6%	Erdeve et al. [92]
232 children enrolled, 1–15 yrs old on azthromycin (AZT)	5×10^9	During abx f/up: 2 wks	5.5%	11.4%	Erdeve et al. [92]
269 enrolled, in and outpatient 6 months–14 yrs Poland	1×10^{10}	During abx 7–13 d f/up: 2 wks	8%[a]	23%	Kotowskaet al. [93]
389 adults enrolled, with *H. pylori* + peptic ulcers	2×10^{10}	2 wks f/up: 4 wks	6.9%[a]	15.6%	Duman et al. [94]
151 adult inpatients, 25–50 yrs old, Turkey	1×10^{10}	Duration abx f/up: 4 wks	1.4%[a]	9.0%	Can et al. [95]
124 adults with *H. pylori* + dyspepsia	2×10^{10}	2 wks f/up: 6 wks	14.5%[a]	30.6%	Cindoruk et al. [96]
86 adult outpatients on amoxicillin	1×10^{10}	12 d f/up: 9 d	7.3%	11.1%	Bravo et al. [97]
991 adult outpatients *H. pylori* +, Korea	3×10^{10}	4 wks f/up: 4 wks	3%	6%	Song et al. [98]
Hospitalized adults (>50 yrs) at Lucco Hosp, Italy	1×10^{10}	During abx and 7 d f/up: 12 wks	15.1%	13.3%	Pozzoni et al. [99]
100 adults with *H. pylori* + peptic ulcers China	500 mg	2 wks f/up: 1 yr	6%	16%	Chu et al. [100]
333 inpatient children (6 months–14 yrs old) with respiratory infections China	1×10^{10}	2 wks f/up: 2 wks	7.9%[a]	29.2%	Shan et al. [101]
140 children (6 months–18 yrs, in or outpatients, IV or oral abx)	500 mg	Duration abx f/up: not reported	16.7%	23.5%	Casem [102]
70 inpatient adults with dyspepsia, *H. pylori* + Greece	6×10^6	2 wks f/up: none	2.8%[a]	20.6%	Kyriakos et al. [103]
160 *H. pylori* + adults with gastritis or ulcers Iran	1×10^{10}	2 wks f/up: 8 wks	12.5%[a]	26%	Zojaji et al. [104]
240 children *H. pylori* + (5–11 yrs old) with gastritis, ulcers or inflammation China	1×10^{10}	14 d f/up: 4 wks	22.5%[a]	39.1%	Zhao et al. [105]
205 *H. pylori* + children (22 months–16 yrs) China	500 mg	14 d f/up: 4 wks	11.8	28.3	Bin et al. [106]

[a]$p < 0 0.5$ compared to controls.
cfu, Colony-forming units; d, days; f/up, follow-up period; h, hours; mg, milligram; wks, weeks; yrs, years.

therapy (antibiotics, typically clarithromycin or amoxicillin, or metronidazole, and a proton-pump inhibitor, typically lansoprazole or omeprazole). However, these eradication therapies have a high rate of side effects (adverse events like gastrointestinal symptoms), which cause patients to prematurely stop the therapy before *H. pylori* is completely eradicated.

Clinical Efficacy for H. pylori. Clinical trials enroll patients infected with *H. pylori* and typically all patients receive a standard eradication therapy aimed at clearing *H. pylori*. The patients are then randomized to *S. boulardii* or a control group and followed for adverse reactions, and tested 6–8 weeks later to see if they have cleared the *H. pylori* bacterium. These eradication therapies include at least one–two antibiotics and a proton-pump inhibitor. Three outcomes have been assessed when using probiotics as an addition to standard therapies focused on eradicating the *H. pylori* pathogen: (1) increasing the eradication rate for *H. pylori*, (2) preventing adverse events associated with the antibiotic-containing standard therapies, and/or (3) preventing diarrhea associated with the antibiotic-containing standard therapies. There have been 13 trials testing *S. boulardii* in *H. pylori* in-

fected patients (Table 18.4), with one trial having two treatment arms [98].

Of the 13 treatment groups who tested for *H. pylori* eradication, *S. boulardii* was able to significantly improve the eradication rates in only 31% of the trials. However, in the 14 treatment arms evaluating the prevention of any adverse reaction or the prevention of AAD, *S. boulardii* significantly prevented these types of reactions in 93% of the treatment groups.

An example of this type of study was one by Cindoruk et al. who assessed *S. boulardii* for both eradication of *H. pylori* and the reduction of side effects of the standard triple eradication treatment [96]. This study enrolled 124 adults with *H. pylori* dyspepsia in Turkey who were receiving the triple antibiotic therapy and then randomized them to either *S. boulardii* (2×10^{10} per day for 2 weeks) or placebo. Patients were followed for 6 weeks for side effects and *H. pylori* clearance. Although there was no significant difference in *H. pylori* eradication (71% in *S. boulardii* vs. 60% in placebo), significantly fewer patients randomized to *S. boulardii* reported epigastric distress (14.5%, $P < 0.05$) compared to placebo (43.5%), as well as lower global symptom

TABLE 18.4 Strength of Clinical Evidence for Various Other Disease Conditions for the Probiotic *S. boulardii*

	Strength of Clinical Evidence[a]		
	Strong	Moderate	Weak
Treatment trials			
Pediatric acute diarrhea	83% of 24 RCT		
H. pylori eradication		31% of 13 RCT	
IBD	75% of 4 RCT		
IBS		50% of 4 RCT	
Treatment of adult acute diarrhea	75% of 4 RCT		
C. difficile infection		100% of 2 RCT	
Giardia lamblia diarrhea		100% of 2 RCT	
HIV diarrhea			50% of 2 RCT
Acne			100% of 1 RCT
Hepatic disease			0% of 1 RCT
Prevention trials			
Antibiotic-associated diarrhea	65% of 23 RCT		
Side effects of *H. pylori* treatments[b]	93% of 14 arms		
Traveler's diarrhea		67% of 3 RCT	
Enteral feed diarrhea		100% of 3 RCT	
Necrotizing enterocolitis			33% of 3 RCT
Sepsis			100% of 1 RCT

[a]*Strong, 3 or more RCTs and >60% of all trials showing significant efficacy; moderate, between 40–60% showing significant efficacy in ≥4 RCTs or only 2–3 RCTs done; weak, <40% showing significant efficacy of ≥ 3 RCTs or only 1 RCT done.*
[b]*Secondary outcomes from H. pylori trials with either prevention of any adverse event or prevention of antibiotic-associated diarrhea.*
IBD, Inflammatory bowel disease; IBS, irritable bowel syndrome; HIV, human immunodeficiency virus; RCT, randomized controlled trial.

scores. The frequency of AAD was also significantly reduced in those receiving *S. boulardii* (14.5%) compared to the placebo group (30.6%). A recent metaanalysis of probiotics tested in randomized, controlled trials to treat patients with *H. pylori* infections found that in the 10 studies using *S. boulardii*, this probiotic was effective in eradicating the pathogen in an average of 82% of those treated, reduced the risk of developing side effects from the antibiotics given to eradicate *H. pylori* by 58%, and reduced the incidence of AAD by 53% [114]. Another metaanalysis of 11 trials confirmed *S. boulardii* was effective in reducing the risk of developing adverse events by 56% and prevented the development of AAD by 49% [111]. These studies the value of adding a probiotic to a standard therapy as an adjunct. While *S. boulardii* may be effective in eradicating the *H. pylori*, its value may be to reduce the development of side effects of the eradication therapy, improving patient compliance, which may improve the eradication rates even further.

OTHER DISEASES

Not all disease conditions have as many supporting clinical trials as the three described previously, but they may offer promising leads as to potential indications for future trials.

INFLAMMATORY BOWEL DISEASE

Epidemiology of inflammatory bowel diseases (IBD). IBD are chronic, immune-related inflammatory diarrheas, which include ulcerative colitis, pouchitis, and Crohn's disease. The incidence of Crohn's disease varies by country, with the highest incidences (8–66/100,000) found in Wales, New Zealand, Canada, Scotland, France, Netherlands, intermediate rates (4–7/100,000) found in United Kingdom., USA, Europe, Australia, and the lowest rates (0.2–3/100,000) found in South America, China, Korea, Japan [116]. In Crohn's disease, typically the lower small intestine and colon are the most affected organs. Symptoms include diarrhea, abdominal pain or cramping, and loss of appetite. Unlike ulcerative colitis, lesions of Crohn's disease are deep and patchy, and often involve thickened areas resulting in intestinal obstruction, which can be life threatening. Disruption of the intestinal wall also allows intestinal bacteria to translocate and incite the immune system. Crohn's disease is a challenging condition for clinical trials in that it is sporadic, has no defined etiology, has multiple measurable disease outcome measures and requires lengthy treatment and follow-up times. Complications of Crohn's disease are common, in that 40% of Crohn's patients report lower GI bleeding, intestinal obstruction, or perforation [117]. Lifetime risk for surgery in Crohn's patients is extremely high (70–80%) [118]. Unfortunately,

the cause of Crohn's disease is not known, so current therapies are directed at symptom relief. But 10–60% suffer recurrences of Crohn's disease after treatment is completed [119].

Clinical efficacy for IBD. These types of trials enroll patients with existing chronic gastrointestinal conditions and randomized them to either *S. boulardii* or placebo/control groups. Subjects are then followed to observe for improvement in disease symptoms or remission of disease. There have been four randomized trials using *S. boulardii* for IBD, one trial with patients with UC or Crohn's disease [120], and three with just patients having Crohn's disease [121–123]. Despite promising results from two small trials with Crohn's disease [121,122] a larger trial failed to confirm these earlier results [123], as shown in Table 18.4. Plein et al. randomized 20 patients with Crohn's disease to either *S. boulardii* (1.5×10^{10} per day) or placebo for 7 weeks [121]. All patients continued their maintenance medications during the trial. At the end of 7 weeks, patients treated with *S. boulardii* were significantly improved (mean = 3.3 ± 1.2 stools/day, $P < 0.05$) compared to the placebo group (mean = 4.6 ± 1.9 stools/day). Guslandi et al. studied the effect of *S. boulardii* on the rate of Crohn's disease relapses in a study of 32 patients with Crohn's disease in Italy [122]. Adults (23–49 years old) who were in remission were randomized to either *S. boulardii* (2×10^{10} per day) and mesalamine (2 g/day), or mesalamine alone (3 g/day) for 6 months. Significantly fewer treated with *S. boulardii* (6%, $P = 0.04$) relapsed compared to the control group (38%). No adverse reactions associated with *S. boulardii* were reported in any of these trials. A large multisite trial randomizing 159 adults with Crohn's disease in remission to either *S. boulardii* (2×10^{10} per day) or placebo for 1 year failed to find a significant improvement in remission rates (47.5% and 53.3%, respectively) [123]. Another study tested a different strain (*S. cerevisiae* or Baker's yeast) in 19 adults with Crohn's disease, but found the disease was worse in those receiving this strain compared to controls [124].

IRRITABLE BOWEL SYNDROME

Epidemiology of irritable bowel syndrome (IBS). IBS is a frequent disorder characterized by a triad of symptoms (bloating, abdominal pain, and intestinal transit disturbances). The prevalence of IBS is high (3–20%) in the general population and IBS may account for 25–50% of gastroenterologists practice [125]. Consequences of IBS include worse quality of life, higher health-care costs ($1.7 billion in health-care costs), and higher incidences of depression [126,127].

Clinical efficacy for IBS. Standard treatments target symptom relief, but have not been more effective than placebo. Four randomized trials have been done with *S. boulardii* and 50% found significant improvement in IBS symptoms (Table 18.4) [128–131]. One of those trials randomized 34 patients with IBS to *S. boulardii* (9×10^9 day^{-1}) or placebo for 4 weeks [131]. A significant decrease in the daily number of stools was found in the *S. boulardii* group and 87.5% of this group reported their IBS had improved compared to 72% of the placebo group ($P < 0.05$). No adverse reactions were noted in this trial. Another study enrolled 72 adults with IBS with diarrhea in Pakistan and treated them with either *S. boulardii* or placebo for 6 weeks [130]. Although there was no significant improvement in diarrheal symptoms, patients reported a significant improvement in inflammatory markers and quality of life. Another study also reported this improvement in quality of life when treated with *S. boulardii* [128]. The challenge of determining whether a treatment is effective for IBS rests on the multiple measures for this condition. A metaanalysis of 20 randomized, controlled trials (with 23 treatment arms) including 1404 subjects found a pooled RR for improvement in global IBS symptoms in 14 probiotic treatment arms (RR = 0.77, 95% CI 0.62, 0.94) [132].

ACUTE ADULT DIARRHEAS

Epidemiology of acute adult diarrhea. Acute adult diarrhea is a broad classification for diarrhea that includes illnesses that may develop quickly, but are typically short-lived and are sporadic. The etiologies of acute adult diarrhea may include infectious agents (*Entamoeba histolytica*, *E. coli*, or *Salmonella*) or may be idiopathic. Diarrhea in adults occurring in outbreaks or due to other situations (AAD, *C. difficile*-associated diarrhea, traveler's diarrhea (TD), IBD or irritable bowel disease) is not included in this classification.

Clinical efficacy for acute adult diarrhea. *S. boulardii* was found to be effective in 75% of the four randomized trials published so far [133–136]. Hochter et al. enrolled 92 German outpatient adults with acute diarrhea and randomized them to *S. boulardii* (8×10^9 day^{-1}) or placebo for 8 days [133]. By the 3rd day, the 43 given *S. boulardii* had a significantly lower diarrhea severity score (5.5 ± 6.8, $P = 0.04$) compared to the 49 given placebo (6.7 ± 8.7). Mansour-Ghanaei et al. enrolled 57 patients with acute *Entamoeba histolytica* dysentery and treated them with *S. boulardii* (1.5×10^{10} per day for 10 days) or nothing (control) [135]. Both groups also received metronidazole and iodoquinol for 10 days. At the end of 4 weeks, 100% given *S. boulardii* were cured compared to fewer (19%) of those not given the yeast. Unfortunately, the number of trials in this area is small

and the etiologies were different in the four trials, so conclusions are limited.

TRAVELER'S DIARRHEA

Epidemiology of TD. TD is a common health complaint among travelers; globally 12 million cases are reported every year [137]. Rates of TD range from 5% to 50%, depending upon the destination, with equatorial countries having the highest rates of TD. TD usually occurs as sporadic cases, but outbreaks of TD may occur in large groups traveling or located together (tourists on cruise ships, group tours, military personnel, disaster-relief groups). The etiologies of TD vary widely, but commonly include enterotoxigenic and entroaggregative *E. coli*, *Campylobacter jejuni*, *S. typhimurum*, viruses (Norwalk or Rotavirus), or parasites (*E. histolytica*, *Giardia lamblia*). Traditional preventive measures including frequent handwashing, avoiding high-risk foods and water, and ingestion of bismuth subsalicylate show only modest protection [138]. A metaanalysis of 12 randomized, controlled trials of various probiotics (including *S. boulardii*, lactobacilli, and mixtures of different probiotic strains) for the prevention of TD found a significant reduction in the risk of TD if probiotics were used, but did not separate out the effect by the different types of probiotics (RR = 0.85, 95% CI 0.79, 0.91) [139].

Clinical efficacy for TD. Three randomized controlled trials have been done with *S. boulardii* [140–142]. Two trials included a total of four different probiotic treatment arms for the prevention of TD and one study tested *S. boulardii* for the treatment of TD. Kollaritsch et al. enrolled 1231 Austrian tourists traveling to hot climates and randomized travelers to either of 2 doses of *S. boulardii* (250 or 500 mg/day) or placebo for 3 weeks [140]. The treatment was started 5 days prior to the trip and continued through the duration of the trip. TD developed in 43% given placebo and significantly fewer in those given the low dose *S. boulardii* (34%) and the higher dose *S. boulardii* (32%). A second study by Kollaritsch et al. was done with 3000 Austrian tourists traveling to northern African, the Middle East, and the Far East [141]. Travelers were randomized to either 250 mg/day (5×10^9) *S. boulardii*, 1000 mg/day (2×10^{10}) *S. boulardii* or placebo, starting 5 days before leaving and throughout the duration of their trip (median of 3 weeks). Only 1016 (34%) completed the study, but *S. boulardii* significantly reduced the incidence of TD in a dose-dependent manner. Of the travelers given placebo, 39% developed diarrhea whereas only 34% of those given the lower dose and 29% of those given the higher dose of *S. boulardii* developed TD ($P < 0.05$). No adverse reactions were noted by the travelers in either of these studies. Bruns et al. tested *S. cerevisiae* Hansen

CBS 5926 (now designated *S. boulardii* CNCM I-745) to treat travelers who had developed TD in Tunisia [142]. Travelers with TD were randomized to *S. cerevisiae* (1×10^{10} per day) for 5 days and the duration of their disease tracked. Of 60 enrolled, 43 completed the trial and the duration of diarrhea not significantly different for those given *S. boulardii* (2.1 days) compared to those given ethacridine-lactate and tannalbuminate (1.4 days). No randomized controlled trials have been done for the treatment of TD using *S. boulardii*. These limited number of studies indicate that probiotics may be more effective in preventing TD, rather than treating the diarrhea once it becomes symptomatic.

ENTERAL NUTRITION-RELATED DIARRHEA

Epidemiology of enteral nutrition-related diarrhea. Diarrhea is a common complication associated with enteral tube feeding and may result in a loss of nutrition in an already serious ill patient. The frequency of diarrhea in enteral tube fed patients has been reported as high as 50–60% and complications may include life-threatening acidosis, increased morbidity and mortality, and increased health-care costs [143].

Clinical efficacy for enteral nutrition-related diarrhea. All three (100%) of the randomized, controlled trials found *S. boulardii* reduced diarrhea in patients receiving this type of nutritional intake. Tempe et al. compared 40 enteral-fed patients randomized to either *S. boulardii* (1×10^{10} per day) or placebo for 11–21 days and found significantly fewer (8.7%) given *S. boulardii* developed diarrhea compared to placebo (16.9%) [144]. Schlotter-er M et al. randomized 18 patients with burns who were receiving enteral nutrition to *S. boulardii* (4×10^{10} per day) or placebo for 8–28 days [145]. Those patients given *S. boulardii* suffered fewer diarrhea days (3/204 days, 1.5%) compared to those given placebo (19/208 days, 9.1%, $P < 0.001$). The efficacy of *S. boulardii* was confirmed in a later study of 128 intensive care unit (ICU) patients randomized to either *S. boulardii* (4×10^{10} per day) or placebo for 21 days [146]. Those given *S. boulardii* reported significantly fewer diarrheal days (7.7%) compared to those given placebo (12.7%). No adverse reactions associated with *S. boulardii* were reported in any of these three trials.

NECROTIZING ENTEROCOLITIS

Epidemiology of necrotizing enterocolitis (NEC). NEC is the most common gastrointestinal infection in very low birthweight neonates, with an incidence of 9% and a mortality of 20% [147].

Clinical efficacy for NEC. Only one of the three randomized trials (Table 18.4) showed any significant effect of *S. boulardii* in low birthweight neonates (<32 weeks old) at risk for developing NEC [148–150]. In this study, there was no difference in the rates of NEC (4% in *S. boulardii* vs. 5% in controls), but those neonates given *S. boulardii* had significantly less episodes of sepsis (35%) compared to controls (48%) [148].

Clostridium difficile INFECTIONS

Epidemiology of C. difficile disease. For over three decades, *C. difficile* infections (CDI) continue to persist as a leading cause of nosocomial gastrointestinal illness [151]. In 2011, there were an estimated 453,000 CDI cases in the USA., 83,000 recurrences of the disease, and 29,300 CDI-associated deaths [152]. Consequences of CDI include longer hospital stays, increased mortality, higher health-care costs, and increased rates of colectomy surgeries. CDI typically occurs in hospitalized adults (66% of CDI cases), but more cases of pediatric CDI and community-onset cases (not associated with hospitals stays) are being reported [56,152]. Outbreaks of an emergent strain, BI/NAP1/027, caused large outbreaks of severe CDI with high rates of mortality in Canada during 2003–05) [153]. Studies have documented that CDI extends hospital stays for hospitalized patients from 4 to 36 days [154–156]. In a study of 1034 CDI cases in Massachusetts during 2000, the average cost ranged from $10,212 to $13,675/patient, projecting a national cost of CDI of $3.2 billion/year [157]. There are only three standard antibiotic treatments for CDI (vancomycin, metronidazole, and fidaxomicin) and the response rate of metronidazole has been declining [156]. In addition, 20–60% of patients may develop recurrent episodes of CDI despite additional antibiotic treatment. Although other investigational antibiotics are under development, no new antibiotics are superior to the two standard antibiotics.

Clinical efficacy for C. difficile disease. Probiotics may offer promise as an adjunctive therapy (given along with standard antibiotics vancomycin or metronidazole) for CDI. There have been two randomized, double blinded, controlled trials for the use of *S. boulardii* both showed significant efficacy for reducing the recurrence rates of CDI episodes [158,159]. A randomized, controlled double-blinded study had also found efficacy of the combination treatment using a standard antibiotic (vancomycin or metronidazole) and *S. boulardii* for patients with *C. difficile* disease [158]. This study did not control the dose or duration of either vancomycin or metronidazole (it was under the physician's discretion), but randomized patients to either *S. boulardii* (1 g/day) or placebo for 28 days. All patients were followed for 2 months for subsequent recurrences. Approximately half of the enrolled patients were having their first episode and half had recurrent *C. difficile* disease. Significantly fewer (26%) of the 57 given *S. boulardii* had a recurrence of *C. difficile* disease compared to 45% of the 67 given placebo ($P < 0.05$). The strongest effect was in the 60 patients with recurrent *C. difficile* disease; significantly fewer patients (35%) given *S. boulardii* had a recurrence compared with 65% of patients given placebo ($P = 0.04$). Two adverse reactions were associated with *S. boulardii* (thirst and constipation). In another multisite, randomized, double-blinded, placebo-controlled trial of 168 patients with recurrent *C. difficile* disease, standard antibiotics were combined with *S. boulardii* or placebo [159]. Three antibiotic regimens were used for 10 days: either high-dose vancomycin (2 g/day), low-dose vancomycin (500 mg/day) or metronidazole (1 g/day). Then either *S. boulardii* or placebo (1 g/day for 28 days) was added to the antibiotic treatment. The patients were followed for 2 months for subsequent *C. difficile* recurrences. A significant decrease in recurrences was observed only in patients treated with the high-dose vancomycin and *S. boulardii* treatment (16.7%) compared with patients who received high-dose vancomycin and placebo (50%, $P = 0.05$). There were no significant reductions in recurrence rates in either the low-dose vancomycin or metronidazole treatment group, regardless if *S. boulardii* was used. No serious adverse reactions were noted in any of these patients. By the end of the antibiotic treatment, only high-dose vancomycin completely cleared *C. difficile* toxin from the colon, low-dose vancomycin and metronidazole could not clear the toxin. A subsequent investigation of 163 patients with recurrent CDI confirmed this finding [160]. In this study, stool samples were assayed at the end of antibiotic treatment to determine if there was a difference in toxin clearance by the type or dose of antibiotic. Treatment with high dose (2 g/day) of vancomycin completely cleared *C. difficile* toxin, but 11% of those treated with lower doses of vancomycin and 41% treated with metronidazole were positive for *C. difficile* at the end of antibiotic treatment. *S. boulardii* may be more effective if complete *C. difficile* toxin clearance is achieved before sole reliance upon the yeast probiotic is required.

GIARDIASIS

Epidemiology of giardiasis. This condition is characterized by long lasting diarrhea with symptoms ranging from mild to severe diarrhea, weight loss, abdominal pain, and weakness. The incidence may be high in developing countries, such as Colombia (13%) and lower in developed countries, such as Germany (12/100,000) [161]. It is also found in people enjoying outdoor hiking and camping who drink contaminated untreated water that appears clean.

Clinical efficacy for giardiasis. Only two trials have been done with *S. boulardii* to treat people with giardiasis and both found a significant effect [136,162]. Besirbellioglu et al. randomized 65 adults in giardiasis in Turkey to either *S. boulardii* (1×10^{10} per day) or placebo for 10 days with both groups receiving metronidazole for the same duration [136]. Two weeks later, both groups reported a resolution of their diarrhea, but none of those on *S. boulardii* had detectable Giardia cysts, while significantly more (17%) on placebo still carried Giardia cysts. Another trial enrolled children in Cuba with chronic diarrhea (88% with Giardia and 10% with Shigella infections). Treatment with *S. boulardii* for 1 month resulted in 70% cure compared to only 10% cured if given placebo [162].

HUMAN IMMUNODEFICIENCY VIRUS DIARRHEA

Epidemiology of human immunodeficiency virus (HIV)-related diarrhea. Patients infected with the HIV are susceptible to a variety of diseases and may also develop chronic life-threatening diarrhea. The prevalence of diarrhea in HIV patients ranges from 50% to 60% in developed countries and nearly 100% in developing countries [163].

Clinical efficacy for HIV-related diarrhea. Several small, uncontrolled trials have been done showing promising results for *S. boulardii* to prevent HIV-related diarrhea, but only one of two randomized, controlled study showed efficacy [164,165]. Saint-Marc et al. randomized 35 Stage IV AIDS adult patients with chronic diarrhea to either *S. boulardii* (3 g/day or 6×10^{10} per day) or placebo for 1 week [164]. Significantly more of those given *S. boulardii* (61%, $P = 0.002$) had their diarrhea resolution compared to placebo (12%) patients. Even with those HIV patients given high doses of *S. boulardii* (3 g/day) did not develop any adverse reactions and this probiotic was well tolerated by all patients [165].

SEPSIS

Epidemiology of sepsis. Sepsis is a complication of infections, often in immunocompromised hospitalized patients and can be due to a wide variety of pathogens. High mortality rates are associated with sepsis.

Clinical efficacy for sepsis. One randomized clinical trial showed a significant reduction in the rates of sepsis [166]. Premature infants with low birthweight were randomized to either *S. boulardii* or nystatin (control) for the duration of their hospitalization. Although fewer infants developed *C. albicans* colonization (5.6% in *S. boulardii* vs. 11.2% in controls), the only significant effect was in less sepsis rates in the *S. boulardii* group (32%) compared to the controls (54%) [166]. *S. boulardii* has also been shown to reduce the incidence of sepsis in neonates at risk for NEC, as discussed previously.

ACNE

Epidemiology of acne. The intestinal tract is not the only target organ for probiotics. Probiotics may be helpful in other sites that involve the disruption of normal microflora, which may include the skin. Acne is common in younger adults and may have complications, not the least is the psychological stress of having this condition. *Clinical efficacy for acne.* Only one randomized controlled trial has been done with *S. boulardii*, but it showed promising results. In Germany, 80% of the 70 adults treated with *S. boulardii* (Hansen CBS 5926) for 5 months reported an improvement or clearance of their acne compared to only 20% of the placebo group [167]. No other trials were found.

HEPATIC FUNCTION

Epidemiology of hepatic conditions. Hepatic diseases and hyperbilirubinemia in infants may result in increased intestinal permeability and translocation of intestinal microbes to other sites in the body.
Clinical efficacy for hepatic conditions. Although several early studies showed promise, a randomized controlled trial of 199 neonates did not show any improvement in liver function in those treated with *S. boulardii* compared with those treated with placebo [168].

SAFETY OF *Saccharomyces boulardii*

The use of a living organism as therapy raises the potential for risk in four general areas: (1) transfer of antibiotic-resistance genes, (2) translocation of the living organism from the intestine to other areas of the body, (3) persistence in the intestines, and (4) development of adverse reactions. The first three theoretical concerns are of minimal impact for *S. boulardii*. Unlike other bacterial strains of probiotics, such as *Enterococcus faecium* and *L. rhamnosus*, that have been shown to acquire antibiotic-resistance genes, *S. boulardii* has not developed any antibiotic or antifungal resistance [169,170]. Data from animal models shows reduced translocation with *S. boulardii* treatment [171,172], unlike other strains of *S. cerevisiae* [173]. As pharmacokinetic studies in human volunteers indicate, *S. boulardii* does not persist past 3–5 days after oral ingestion is discontinued, so persistence is not a concern for this probiotic [54].

S. boulardii has been used as a probiotic since 1950s in Europe and has been investigated in clinical trials worldwide. Safety and adverse event data collected during clinical trials, when patients are closely monitored for problems associated with the investigational treatment, has documented a remarkable safety profile of *S. boulardii*. In the 90 randomized, controlled trials, none reported any serious adverse reactions associated with *S. boulardii* and few reported mild–moderate reactions. One trial reported the only adverse reactions associated with *S. boulardii* were thirst (in five patients) and constipation (in eight patients) in a trial of patients with CDI [158]. No cases of *S. boulardii* fungemia have been reported in these diverse patient populations enrolled in clinical trials.

Infrequent cases of fungemia have been reported in case reports or case series in the literature. Most of the adult cases of *S. boulardii* fungemia are in patients with serious comorbidities and have central venous catheters. Most cases responded well to treatment with fluconazole or amphotericin B [174]. In some cases, patients developed fungemia, not from the direct ingestion of *S. boulardii* probiotics, but acquired the yeast from contaminated environmental fomites [175]. Fungemia from *S. cerevisiae* (nonboulardii strains) have also been reported and are similar to *S. boulardii* cases, but have poorer prognosis [176,177]. A challenge to determine valid incidence of fungemia in older reports is the lack of available advanced yeast identification assays which can distinguish *S. cerevisiae* from *S. boulardii*.

CONCLUSIONS

The use of *S. boulardii* as a therapeutic probiotic is supported by its mechanisms of action, pharmacokinetics, and efficacy from animal models, and clinical trials. The overall safety profile for *S. boulardii* is beneficial. *S. boulardii* can be recommended for several diseases. Other strains of *Saccharomyces* may have probiotic properties, but clinical efficacy evidence for these other strains is currently lacking. Guidelines for the most effective use of *S. boulardii* is generally, giving a daily dose of $> 10^9$ per day. The duration of treatment can vary from 7 days to 6 months and it may be given alone or as an adjunctive treatment, depending upon the disease indication. *S. boulardii* is effective for the treatment of acute pediatric diarrhea, for the prevention of AAD, for the reduction of side effects of *H. pylori* treatment. Although less well studied, *S. boulardii* shows promise for the treatment of IBD and for the prevention of enteral nutrition-related diarrhea and prevention of TD. More clinical trials are encouraged for the treatment of chronic diseases (HIV-related diarrhea and IBS) and the prevention of *C. difficile* disease recurrences and other diseases.

ACKNOWLEDGMENTS

The author has been a paid speaker/consultant for the following companies: Biocodex, Lallemand, and BioK+. The author is not currently employed nor owns stock or equity in any of these companies.

REFERENCES

[1] Guarner F, Khan AG, Garisch J, et al. World Gastroenterology Organisation Global Guidelines: probiotics and prebiotics October 2011. J Clin Gastroenterol 2012;46(6):468–81.

[2] Moyad MA, Robinson LE, Kittelsrud JM, et al. Immunogenic yeast-based fermentation product reduces allergic rhinitis-induced nasal congestion: a randomized, double-blind, placebo-controlled trial. Adv Ther 2009;26(8):795–804.

[3] Pennacchia C, Blaiotta G, Pepe O, Villani F. Isolation of Saccharomyces cerevisiae strains from different food matrices and their preliminary selection for a potential use as probiotics. J Appl Microbiol 2008;105(6):1919–28.

[4] McFarland LV. From yaks to yogurt: the history, development and current use of probiotics. Clin Infect Dis 2015;60(S2):S85–90.

[5] Weese JS. Evaluation of deficiencies in labeling of commercial probiotics. Can Vet J 2003;44(12):982–3.

[6] Marcobal A, Underwood MA, Mills DA. Rapid determination of the bacterial composition of commercial probiotic products by terminal restriction fragment length polymorphism analysis. J Pediatr Gastroenterol Nutr 2008;46(5):608–11.

[7] Martins FS, Nardi RM, Arantes RM, et al. Screening of yeasts as probiotic based on capacities to colonize the gastrointestinal tract and to protect against enteropathogen challenge in mice. J Gen Appl Microbiol 2005;51(2):83–92.

[8] Martins FS, Silva AA, Vieira AT, et al. Comparative study of Bifidobacterium animalis, Escherichia coli, Lactobacillus casei and Saccharomyces boulardii probiotic properties. Arch Microbiol 2009;191(8):623–30.

[9] Vanhee LM, Goemé F, Nelis HJ, Coenye T. Quality control of fifteen probiotic products containing Saccharomyces boulardii. J Appl Microbiol 2010;109(5):1745–52.

[10] Sankoff D. Reconstructing the history of yeast genomes. PLoS Genet 2009;5(5):e1000483.

[11] McFarland LV. Saccharomyces boulardii is not Saccharomyces cerevisiae. Clin Infect Dis 1996;22(1):200–1.

[12] McCullough MJ, Clemons KV, McCusker JH, Stevens DA. Species identification and virulence attributes of Saccharomyces boulardii (nom. inval.). J Clin Microbiol 1998;36(9):2613–7.

[13] Fietto JL, Araújo RS, Valadão FN, et al. Molecular and physiological comparisons between Saccharomyces cerevisiae and Saccharomyces boulardii. Can J Microbiol 2004;50(8):615–21.

[14] McFarland LV. Systematic review and meta-analysis of Saccharomyces boulardii in adult patients. World J Gastroenterol 2010;16(18):2202–22.

[15] Dinleyici EC, Kara A, Ozen M, Vandenplas Y. Saccharomyces boulardii CNCM I-745 in different clinical conditions. Expert Opin Biol Ther 2014;14(11):1593–609.

[16] Ryan JJ, Hanes DA, Schafer B, Mikolai J, Zwickey H. Effect of the probiotic S. boulardii on cholesterol and lipoprotein particles in hypercholesterolemic adults: a single-arm, open-label pilot study. J Altern Complement Med 2015;21(5):288–93.

[17] Cordonnier C, Thevenot J, Etienne-Mesmin L, et al. Dynamic in vitro models of the human gastointestinal tract as relevant tools to assess the survival of probiotic strains and their interactions with gut microbiota. Microorganisms 2015;3:725–45.

[18] De Chambrun NC, Chau A, Cazaubiel M, et al. A randomized clinical trial of Saccharomyces cerevisiae versus placebo in the irritable bowel syndrome. Dig Liv Dis 2015;47:119–24.

[19] European Food Safety Authority. Scientific opinion on the substantiation of health claims related to S. cerevisiae var boulardii CNCM I-1079 and defence against pathogenic gastrointestinal microorganisms. EFSA J 2010;10(6):2717.

[20] Ooi CY, Dilley AV, Day AS. Saccharomyces boulardii in a child with recurrent Clostridium difficile. Pediatr Int 2009;51(1):156–8.

[21] Pothoulakis C, Kelly CP, Joshi MA, et al. Saccharomyces boulardii inhibits Clostridium difficile toxin A binding and enterotoxicity in rat ileum. Gastroenterol 1993;104(4):1108–15.

[22] Brandão RL, Castro IM, Bambirra EA, et al. Intracellular signal triggered by cholera toxin in Saccharomyces boulardii and Saccharomyces cerevisiae. Appl Environ Microbiol 1998;64(2):564–8.

[23] Castagliuolo I, LaMont JT, Nikulasson ST, Pothoulakis C. Saccharomyces boulardii protease inhibits Clostridium difficile toxin A effects in the rat ileum. Infect Immun 1996;64:5225–32.

[24] Castagliuolo I, Riegler MF, Valenick L, LaMont JT, Pothoulakis C. Saccharomyces boulardii protease inhibits the effects of Clostridium difficile toxins A and B in human colonic mucosa. Infect Immun 1999;67(1):302–7.

[25] Buts JP, Dekeyser N, Stilmant C, et al. Saccharomyces boulardii produces in rat small intestine a novel protein phosphatase that inhibits Escherichia coli endotoxin by dephosphorylation. Pediatr Res 2006;60(1):24–9.

[26] Vidon N, Huchet B, Rambaud JC. Effect of S. boulardii on water and sodium secretions induced by cholera toxin. Gastroenterol Clin Biol 1986;10:1–4.

[27] Czerucka D, Rampal P. Effect of Saccharomyces boulardii on cAMP- and Ca^{2+}-dependent Cl^- secretion in T84 cells. Dig Dis Sci 1999;44(11):2359–68.

[28] Murzyn A, Krasowska A, Augustyniak D, et al. The effect of Saccharomyces boulardii on Candida albicans-infected human intestinal cell lines Caco-2 and Intestin 407. FEMS Microbiol Lett 2010;310(1):17–23.

[29] Garcia-Vilela E, De Lourdes De Abreu Ferrari M, Oswaldo Da Gama Torres H, et al. Influence of Saccharomyces boulardii on the intestinal permeability of patients with Crohn's disease in remission. Scand J Gastroenterol 2008;43(7):842–8.

[30] Frank DN, Pace NR. Gastrointestinal microbiology enters the metagenomics era. Curr Opin Gastroenterol 2008;24(1):4–10.

[31] Preidis GA, Versalovic J. Targeting the human microbiome with antibiotics, probiotics, and prebiotics: gastroenterology enters the metagenomics era. Gastroenterol 2009;136(6):2015–31.

[32] McFarland LV. Normal flora: diversity and functions. Microbial Ecol Health Dis 2000;12:193–207.

[33] Ng SC, Harta AL, Kamm MA, Stagg AJ, Knight SC. Mechanisms of action of probiotics: recent advances. Inflam Bowel Dis 2009;15(2):300–10.

[34] Dethlefsen L, Huse S, Sogin ML, Relman DA. The pervasive effects of an antibiotic on the human gut microbiota, as revealed by deep 16S rRNA sequencing. PLoS Biol 2008;18(11):e280.

[35] McFarland LV. Use of probiotics to correct dysbiosis of normal microbiota following disease or disruptive events: a systematic review. BMJ Open 2014;4:e005047.

[36] Girard-Pipau F, Pompei A, Schneider S, et al. Intestinal microflora, short chain and cellular fatty acids, influence of a probiotic S. boulardii. Microb Ecology Health Dis 2002;14:220–7.

[37] Schneider SM, Girard-Pipau F, Filippi J, et al. Effects of Saccharomyces boulardii on fecal short-chain fatty acids and microflora in

patients on long-term total enteral nutrition. World J Gastroenterol 2005;11(39):6165–9.

[38] Sezer A, Usta U, Cicin I. The effect of *Saccharomyces boulardii* on reducing irinotecan-induced intestinal mucositis and diarrhea. Med Oncol 2009;26(3):350–7.

[39] Schroeder B, Winckler C, Failing K, Breves G. Studies on the time course of the effects of the probiotic yeast *Saccharomyces boulardii* on electrolyte transport in pig jejunum. Dig Dis Sci 2004;49(7–8):1311–7.

[40] Buts JP, Stilmant C, Bernasconi P, Neirinck C, De Keyser N. Characterization of alpha, alpha-trehalase released in the intestinal lumen by the probiotic *Saccharomyces boulardii*. Scand J Gastroenterol 2008;43(12):1489–96.

[41] Buts JP, De Keyser N, De Raedemaeker L. *Saccharomyces boulardii* enhances rat intestinal enzyme expression of endoluminal release of polyamines. Pediatr Res 1994;36:522–7.

[42] Jahn HU, Ullrich R, Schneider T, et al. Immunological and trophical effects of *Saccharomyces boulardii* on the small intestine in healthy human volunteers. Digestion 1996;57(2):95–104.

[43] Buts JP. Twenty-five years of research on *Saccharomyces boulardii* trophic effects: updates and perspectives. Dig Dis Sci 2009;54(1):15–8.

[44] Kyne L, Warny M, Qamar A, Kelly CP. Association between antibody response to toxin A and protection against recurrent *Clostridium difficile* diarrhea. Lancet 2001;357:189–93.

[45] Pothoulakis C. Review article: anti-inflammatory mechanisms of action of *Saccharomyces boulardii*. Aliment Pharmacol Ther 2009;30(8):826–33.

[46] Fidan I, Kalkanci A, Yesilyurt E, et al. Effects of *Saccharomyces boulardii* on cytokine secretion from intraepithelial lymphocytes infected by *Escherichia coli* and *Candida albicans*. Mycoses 2009;52(1):29–34.

[47] Chen X, Kokkotou EG, Mustafa N, Bhaskar KR, Sougioultzis S. *Saccharomyces boulardii* inhibits ERK1/2 mitogen-activated protein kinase activation both in vitro and in vivo and protects against *Clostridium difficile* toxin A-induced enteritis. J Biol Chem 2006;281(34):24449–54.

[48] Dalmasso G, Cottrez F, Imbert V, et al. *Saccharomyces boulardii* inhibits inflammatory bowel disease by trapping T cells in mesenteric lymph nodes. Gastroenterol 2006;131:1812–25.

[49] Ducluzeau R, Bensaada M. Comparative effect of a single or continuous administration of *S. boulardii* on the establishment of various strains of Candida in the digestive tract of gnotobiotic mice. Ann Microbiol (Inst Pasteur) 1982;1338:491–501.

[50] Zbinden R, Bonczi E, Altwegg M. Inhibition of *S. boulardii* (nom. inval.) on cell invasion of *Salmonella typhimurium* and *Yersinia enterocolitica*. Micro Ecol Health Dis 1999;11:158–62.

[51] Altwegg M, Schnack J, Zbinden R. Influence of *Saccharomyces boulardii* on *Aeromonas hemolysin*. Med Microbiol Lett 1995;4:417–25.

[52] Elmer GW, McFarland LV, Surawicz CM, Danko L, Greenberg RN. Behaviour of *Saccharomyces boulardii* in recurrent *Clostridium difficile* disease patients. Aliment Pharmacol Ther 1999;13(12):1663–8.

[53] Blehaut H, Massot J, Elmer GW, Levy RH. Disposition kinetics of *Saccharomyces boulardii* in man and rat. Biopharm Drug Dispos 1989;10(4):353–64.

[54] Klein SM, Elmer GW, McFarland LV, Surawicz CM, Levy RH. Recovery and elimination of the biotherapeutic agent, *Saccharomyces boulardii*, in healthy human volunteers. Pharm Res 1993;10(11):1615–9.

[55] Levine MM, Kotloff KL, Nataro JP, Muhsen K. The Global Enteric Multicenter Study (GEMS): impetus, rationale, and genesis. Clin Infect Dis 2012;55(Suppl. 4):S215–24.

[56] McFarland LV, Brandmarker SA, Guandalini S. Pediatric *Clostridium difficile*: a phantom menace or clinical reality? J Pediatr Gastroenterol Nutr 2000;31(3):220–31.

[57] Grandy G, Medina M, Soria R, Terán CG, Araya M. Probiotics in the treatment of acute rotavirus diarrhoea. A randomized, double-blind, controlled trial using two different probiotic preparations in Bolivian children. BMC Infect Dis 2010;10:253.

[58] Chapoy P. Treatment of acute diarrhea in infants: a controlled trial of *Saccharomyces boulardii*. Ann Pediatr (Paris) 1985;32:561–3.

[59] Cetina-Sauri G, Sierra Basto G. Evaluation of *Saccharomyces boulardii* for the treatment of acute diarrhea in pediatric patients. Ann Pediatr (Paris) 1994;41:397–400.

[60] Chouraqui JP, Dietsch C, Musial H, Blehaut H. *Saccharomyces boulardii* in the management of toddler diarrhea: a double blind-placebo controlled study. J Pediatr Gastroenterol Nutr 1995;20:463. Meeting abstract #71.

[61] Urganci N, Polat T, Uysalol M, Cetinkaya F. Evalution of the efficacy of *Saccharomyces boulardii* in children with acute diarrhea. Arch Gastroenterol 2001;20(3–4):1–7.

[62] Hafeez A, Tariq P, Ali S, et al. The efficacy of *Saccharomyces boulardii* in the treatment of acute watery diarrhea in children: a multicentre randomized controlled trial. J Coll Physic Surg Pakistan 2002;12:432–4.

[63] Gaón D, García H, Winter L, et al. Effect of *Lactobacillus* strains and *Saccharomyces boulardii* on persistent diarrhea in children. Medicina (B Aires) 2003;63(4):293–8.

[64] Kurugöl Z, Koturoglu G. Effects of *Saccharomyces boulardii* in children with acute diarrhoea. Acta Paediatr 2005;94(1):44–7.

[65] Billoo AG, Memon MA, Khaskheli SA, et al. Role of probiotic *Saccharomyces boulardii* in management and prevention of diarrhea. World J Gastroenterol 2006;12:4557–60.

[66] Villarruel G, Rubio DM, Lopez F, et al. *Saccharomyces boulardii* in acute childhood diarrhoea: a randomized, placebo-controlled study. Acta Paediatr 2007;96(4):538–41.

[67] Canani RB, Cirillo P, Terrin G, et al. Probiotics for treatment of acute diarrhoea in children: randomised clinical trial of five different preparations. BMJ 2007;335(7615):340.

[68] Ozkan TB, Sahin E, Erdemir G, Budak F. Effect of *Saccharomyces boulardii* in children with acute gastroenteritis and its relationship to the immune response. J Int Med Res 2007;35(2):201–12.

[69] Htwe K, Yee KS, Tin M, Vandenplas Y. Effect of *Saccharomyces boulardii* in the treatment of acute watery diarrhea in Myanmar children: a randomized controlled study. Am J Trop Med Hyg 2008;78(2):214–6.

[70] Dinleyici EC, Eren M, Yargic ZA, Dogan N, Vandenplas Y. Clinical efficacy of *Saccharomyces boulardii* and metronidazole compared to metronidazole alone in children with acute bloody diarrhea caused by amebiasis: a prospective, randomized, open label study. Am J Trop Med Hyg 2009;80(6):953–5.

[71] Savaş-Erdeve S, Gökay S, Dallar Y. Efficacy and safety of *Saccharomyces boulardii* in amebiasis-associated diarrhea in children. Turk J Pediatr 2009;51(3):220–4.

[72] Eren M, Dinleyici EC, Vandenplas Y. Clinical efficacy comparison of *Saccharomyces boulardii* and yogurt fluid in acute non-bloody diarrhea in children: a randomized, controlled, open label study. Am J Trop Med Hyg 2010;82(3):488–91.

[73] Le Luyer B, Makhoul G, Duhamel JF. A multicentric study of a lactose free formula supplemented with *Saccharomyces boulardii* in children with acute diarrhea. [Article in French]. Arch Pediatr 2010;17(5):459–65.

[74] Dinleyici EC, Eren M, Dogan N, et al. Clinical efficacy of *Saccharomyces boulardii* or metronidazole in symptomatic children with *Blastocystis hominis* infection. Parasitol Res 2011;108(3):541–5.

[75] Dalgic N, Sancar M, Bayraktar B, Pullu M, Hasim O. Probiotic, zinc and lactose-free formula in children with rotavirus diarrhea: are they effective? Pediatr Internl 2011;53(5):677–82.

[76] Corrêa NB, Penna FJ, Lima FM, Nicoli JR, Filho LAT. Treatment of acute diarrhea with *Saccharomyces boulardii* in infants. J Pediatr Gastroenterol Nutr 2011;53(5):497–501.

[77] Józefczuk J, Wo niewicz BM. Diagnosis and therapy of microscopic colitis with presence of foamy macrophages in children. ISRN Gastroenterol 2011;2011:756292.

[78] Riaz M, Alam S, Malik A, Ali SM. Efficacy and safety of *Saccharomyces boulardii* in acute childhood diarrhea: a double blind randomised controlled trial. Indian J Pediatr 2012;79(4):478–82.

[79] Erdog an O, Tanyeri B, Torun E, et al. The comparition of the efficacy of two different probiotics in rotavirus gastroenteritis in children. J Trop Med 2012;2012:787240.

[80] Dinleyici EC, Kara A, Dalgic N, et al. *Saccharomyces boulardii* CNCM I-745 reduces the duration of diarrhoea, length of emergency care and hospital stay in children with acute diarrhoea. Benef Microbes 2015;6(4):415–21.

[81] Feizizadeh S, Salehi-Abargouei S, Akbari V. Efficacy and safety of *Saccharomyces boulardii* for acute diarrhea. Pediatrics 2014;134: e176–91.

[82] Dinleyici EC, Eren M, Ozen M, Yargic ZA, Vandenplas Y. Effectiveness and safety of *Saccharomyces boulardii* for acute infectious diarrhea. Expert Opin Biol Ther 2012;12(4):395–410.

[83] McFarland LV. Antibiotic-associated diarrhea: epidemiology, trends and treatment. Future Microbiol 2008;3:563–78.

[84] McFarland LV. Meta-analysis of probiotics for prevention of antibiotic associated diarrhea and treatment of *Clostridium difficile disease*. Am J Gastroenterol 2006;101:812–22.

[85] Adam J, Barret C, Barret-Bellet A, et al. Controlled double-blind clinical trials of Ultra-Levure: multicentre study by 25 physicians in 388 cases. Gaz Med Fr 1977;84:2072–8.

[86] Monteiro E, Fernandes JP, Vieira MR, et al. Double blind clinica trial on the use of ultra-levure in the prophylaxis of antibiotic induced gastro-intestinal and mucocutaneous disorders. Acta Med Port 1981;3(2):143–5.

[87] Surawicz CM, Elmer GW, Speelman P, et al. Prevention of antibiotic-associated diarrhea by *Saccharomyces boulardii*: a prospective study. Gastroenterol 1989;96:981–8.

[88] McFarland LV, Surawicz CM, Greenberg RN, et al. Prevention of β-lactam-associated diarrhea by *Saccharomyces boulardii* compared to placebo. Am J Gastroenterol 1995;90:439–48.

[89] Lewis SJ, Potts LF, Barry RE. The lack of therapeutic effect of *Saccharomyces boulardii* in the prevention of antibiotic-related diarrhoea in elderly patients. J Infect 1998;36:171–4.

[90] Benhamou PH, Berlier P, Danjou G, et al. The antibiotic-associated diarrhea in children: a computerized study double-blind ambulatory patients comparing a preservative and a probiotic agent. Digest Med Surg 1999;28(4):163–8.

[91] Cremonini F, Di Caro S, Covino M, et al. Effect of different probiotic preparations on anti-*Helicobacter pylori* therapy-related side effects: a parallel group, triple blind, placebo-controlled study. Am J Gastroenterol 2002;97(11):2744–9.

[92] Erdeve O, Tiras U, Dallar Y. The probiotic effect of *Saccharomyces boulardii* in a pediatric age group. J Trop Pediatr 2004;50(4):234–6.

[93] Kotowska M, Albrecht P, Szajewska H. *Saccharomyces boulardii* in the prevention of antibiotic-associated diarrhoea in children: a randomized double-blind placebo-controlled trial. Aliment Pharmacol Ther 2005;21:583–90.

[94] Duman DG, Bor S, Ozütemiz O, et al. Efficacy and safety of *Saccharomyces boulardii* in prevention of antibiotic-associated diarrhoea due to *Helicobacter pylori* eradication. Am J Eur J Gastroenterol Hepatol 2005;17(12):1357–61.

[95] Can M, Bes irbellioglu BA, Avci IY, Beker CM, Pahsa A. Prophylactic *Saccharomyces boulardii* in the prevention of antibiotic-associated diarrhea: a prospective study. Med Sci Monit 2006;12(4):PI19–22.

[96] Cindoruk M, Erkan G, Karakan T, Dursun A, Unal S. Efficacy and safety of *Saccharomyces boulardii* in the 14-day triple anti-*Helicobacter pylori* therapy: a prospective randomized placebo-controlled double-blind study. Helicobacter 2007;12(4):309–16.

[97] Bravo MV, Bunout D, Leiva L, et al. Effect of probiotic *Saccharomyces boulardii* on prevention of AAD in adult outpatients with amoxicillin treatment. Rev Med Chile 2008;136:981–8.

[98] Song MJ, Park DI, Park JH, et al. The effect of probiotics and mucoprotective agents on PPI-based triple therapy for eradication of *Helicobacter pylori*. Helicobacter 2010;15:206–13.

[99] Pozzoni P, Riva A, Bellatorre AG, et al. *Saccharomyces boulardii* for the prevention of antibiotic-associated diarrhea in adult hospitalized patients: a single-center, randomized, double-blind, placebo-controlled trial. Am J Gasteroenterol 2012;107:922–31.

[100] Chu Y, Zhu H, Zhou Y, Lv L, Huo J. Intervention study on *Saccharomyces boulardii* with proton pump inhibitor (PPI)-based triple therapy for *Helicobacter pylori* related peptic ulcer. African J Pharmacy Pharmacol 2012;6(41):2900–4.

[101] Shan L, Hou P, Wang Z, et al. Prevention and treatment of diarrhea with *Saccharomyces boulardii* in children with acute lower respiratory tract infections. Benef Microbes 2013;4(4):329–34.

[102] Casem RA. *S. boulardii* in the prevention of antibiotic-associated diarrhea in children: a randomized controlled trial. Phili Infect Dis Soc Proceed J 2013;13(2):70.

[103] Kyriakos N, Papmichael K, Roussos A, et al. Lyophilized Form of *Saccharomyces boulardii* enhances the *Helicobacter pylori* eradication rates of omeprazole-triple therapy in patients with peptic ulcer disease or functional dyspepsia. Hosp Chronicles 2013;8(3):127–33.

[104] Zojaji H, Ghobakhlou M, Rajabalinia H, et al. The efficacy and safety of adding the probiotic *Saccharomyces boulardii* to standard triple therapy for eradication of *H. pylori*: a randomized controlled trial. Gastroenterol Hepatol From Bed to Bench 2013;6(Suppl. 1):S99–S104.

[105] Zhao HM, Ou-Yang HJ, Duan BP, et al. Clinical effect of triple therapy combined with *Saccharomyces boulardii* in the treatment of *Helicobacter pylori* infection in children. [in Chinese]. Zhongguo Dang Dai Er Ke Za Zhi 2014;16(3):230–3.

[106] Bin Z, Ya-Zheng X, Zhao-Hui D, Bo C, Li-Rong J, Vandenplas Y. The Efficacy of *Saccharomyces boulardii* CNCM I-745 in Addition to Standard *Helicobacter pylori* Eradication Treatment in Children. Pediatr Gastroenterol Hepatol Nutr 2015;18(1):17–22.

[107] Hickson M, D'Souza AL, Muthu N, et al. Use of probiotic *Lactobacillus* preparation to prevent diarrhoea associated with

antibiotics: randomised double blind placebo controlled trial. BMJ 2007;335(7610):80–3.

[108] McFarland LV. Unraveling the causes of negative studies: a case of *S boulardii* for the prevention of antibiotic-associated diarrhea. Rev Med Chil 2009;137(5):719–20.

[109] Meerpohl JJ, Timmer A. News from the Cochrane Library: probiotics for the prevention of paediatric antibiotic-associated diarrhoea. Z Gastroenterol 2007;45(8):715–7.

[110] McFarland LV. Deciphering meta-analytic results: a mini-review of probiotics for the prevention of pediatric AAD and CDI. Benef Microbes 2015;6(2):189–94.

[111] Szajewska H, Horvath A, Kołodziej M. Systematic review with meta-analysis: *Saccharomyces boulardii* supplementation and eradication of *Helicobacter pylori* infection. Aliment Pharmacol Ther 2015;41(12):1237–45.

[112] Johnston BC, Supina AL, Vohra S. Probiotics for pediatric antibiotic-associated diarrhea: a meta-analysis of randomized placebo-controlled trials. CMAJ 2006;175(4):377–83.

[113] McFarland LV, Goh S. Preventing pediatric antibiotic-associated Diarrhea and *Clostridium difficile* Infections with Probiotics: a meta-analysis. World J Meta-Anal 2013;1(3):102–20.

[114] McFarland LV, Malfertheiner P, Huang Y, Wang L. Meta-analysis of single strain probiotics for the eradication of *Helicobacter pylori* and prevention of adverse events. World J Meta-Anal 2015;3(2):97–117.

[115] Jafri W, Yakoob J, Abid S, et al. *Helicobacter pylori* infection in children: population-based age-specific prevalence and risk factors in a developing country. Acta Paediatr 2009;99(2):279–82.

[116] Economou M, Pappas G. New global map of Crohn's disease: genetic, environmental, and socioeconomic correlations. Inflamm Bowel Dis 2008;14(5):709–20.

[117] Lok KH, Hung HG, Ng CH, et al. The epidemiology and clinical characteristics of Crohn's disease in the Hong Kong Chinese population: experiences from a regional hospital. Hong Kong Med J 2007;13(6):436–41.

[118] Roberts SE, Williams JG, Yeates D, Goldacre MJ. Mortality in patients with and without colectomy admitted to hospital for ulcerative colitis and Crohn's disease: record linkage studies. BMJ 2007;335(7628):1033.

[119] Elmer GW, McFarland LV, McFarland M. Inflammatory bowel disease, irritable bowel syndrome and digestive problems. The power of probiotics: improving your health with beneficial microbes. Binghamton, NY: Haworth Press; 2007. p. 111–6. [Chapter 6].

[120] Avalueva EB, Uspenski IP, Tkachenko EI, Sitkin SI. Use of *Saccharomyces boulardii* in treating patients inflammatory bowel diseases (clinical trial). Eksp Klin Gastroenterol 2010;7:103–11. in Russian.

[121] Plein K, Hotz J. Therapeutic effects of *Saccharomyces boulardii* on mild residual symptoms in a stable phase of Crohn's disease with special respect to chronic diarrhea—a pilot study. Z Gastroenterol 1993;31:129–34.

[122] Guslandi M, Mezzi G, Sorghi M, Testoni PA. *Saccharomyces boulardii* in maintenance treatment of Crohn's disease. Dig Dis Sci 2000;45:1462–4.

[123] Bourreille A, Cadiot G, Le Dreau G, et al. FLORABEST Study Group. *Saccharomyces boulardii* does not prevent relapse of Crohn's disease. Clin Gastroenterol Hepatol 2013;11(8):982–7.

[124] Barclay GR, McKenzie H, Pennington J, Parratt D, Pennington CR. The effect of dietary yeast on the activity of stable chronic Crohn's disease. Scand J Gastroenterol 1992;27:196–200.

[125] Park KS, Ahn SH, Hwang JS, et al. A survey about irritable bowel syndrome in South Korea: prevalence and observable organic abnormalities in IBS patients. Dig Dis Sci 2008;53(3):704–11.

[126] Foxx-Orenstein A. IBS—review and what's new. Med Gen Med 2006;8(3):20.

[127] Ladep NG, Okeke EN, Samaila AA, et al. Irritable bowel syndrome among patients attending general outpatients' clinics in Jos. Nigeria. Eur J Gastroenterol Hepatol 2007;19(9):795–9.

[128] Choi CH, Jo SY, Park HJ, et al. A randomized, double-blind, placebo-controlled multicenter trial *of Saccharomyces boulardii* in irritable bowel syndrome: effect on quality of life. J Clin Gastroenterol 2011;45(8):679–83.

[129] Kabir MA, Ishaque SM, Ali MS, Mahmuduzzaman M, Hasan M. Role of *Saccharomyces boulardii* in diarrhea predominant irritable bowel syndrome. Mymensingh Med J 2011;20(3):397–401.

[130] Abbas Z, Yakoob J, Jafri W, et al. Cytokine and clinical response to *Saccharomyces boulardii* therapy in diarrhea-dominant irritable bowel syndrome: a randomized trial. Eur J Gastroenterol Hepatol 2014;26(6):630–9.

[131] Maupas JL, Champemont P, Delforge M. Treatment of irritable bowel syndrome. Double blind trial of *Saccharomyces boulardii*. Medecine Chirurgie Digestives 1983;12(1):77–9.

[132] McFarland LV, Dublin S. Meta-analysis of probiotics for the treatment of irritable bowel syndrome. World J Gastroenterol 2008;14(17):2650–61.

[133] Hochter W, Chase D, Hagenhoff G. *Saccharomyces boulardii* in acute adult diarrhea: efficacy and tolerability of treatment. Munch Med Wschr 1990;132:188–92.

[134] Attar A, Flourié B, Rambaud JC, et al. Antibiotic efficacy in small intestinal bacterial overgrowth-related chronic diarrhea: a cross-over, randomized trial. Gastroenterol 1999;117(4):794–7.

[135] Mansour-Ghanael F, Dehbashi N, Yazdanparast K, Shafaghi A. Efficacy of *Saccharomyces boulardii* with antibiotics in acute amoebiasis. World J Gastroenterol 2003;9:1832–3.

[136] Besirbellioglu BA, Ulcay A, Can M, et al. *Saccharomyces boulardii* and infection due to *Giardia lamblia*. Scand J Infect Dis 2006;38(6–7):479–81.

[137] Cheng AC, Thielman NM. Update on traveler's diarrhea. Curr Infect Dis Rep 2002;4:70–7.

[138] Centers for Disease Control and Prevention. CDC Health information for International Travel 2016. New York: Oxford University Press; 2016.

[139] McFarland LV. Meta-analysis of probiotics for the prevention of traveler's diarrhea. Travel Med Infect Dis 2007;5(2):97–105.

[140] Kollaritsch H, Kremsner P, Wiedermann G, Scheiner O. Prevention of traveller's diarrhea: comparison of different non-antibiotic preparations. Travel Med Internl 1989;7:9–18.

[141] Kollaritsch H, Holst H, Grobara P, Wiedermann G. Prophylaxe der reisediarrhoe mit *Saccharomyces boulardii*. [Prevention of traveler's diarrhea with *Saccharomyces boulardii*. Results of a placebo controlled double-blind study]. Fortschr Med 1993;111(9):152–6.

[142] Bruns R, Raedsch R. Therapy of traveller's diarrhea. Medizinische Welt 1995;46:591–6.

[143] Whelan K, Judd PA, Tuohy K, et al. Fecal microbiota in patients receiving enteral feeding are highly variable and may be altered in those who develop diarrhea. Am J Clin Nutr 2009;89(1):240–7.

[144] Tempe JD, Steidel AL, Blehaut H, et al. Use of *Saccharomyces boulardii* for the prevention of diarrhea during continuous enteral feeding. Sem Hop Paris 1983;59(18):1409–12.

[145] Schlotterer M, Bernasconi P, Lebreton F, Wassermann D. Value of *Saccharomyces boulardii* in the digestive acceptability of continuous-flow enteral nutrition in burnt patients. Nutr Clin Metabol 1987;1:31–4.

[146] Bleichner G, Blehaut H, Mentec H, Moyse D. *Saccharomyces boulardii* prevents diarrhea in critically ill tube-fed patients: a multicenter, randomized, double-blind placebo-controlled trial. Intensive Care Med 1997;23:517–23.

[147] Stoll BJ, Hansen NI, Bell EF, et al. Trends in care practices, morbidity, and mortality of extremely preterm neonates, 1993–2012. JAMA 2015;314(10):1039–51.

[148] Demirel G, Erdeve O, Celik IH, Dilmen U. *Saccharomyces boulardii* for prevention of necrotizing enterocolitis in preterm infants: a randomized, controlled study. Acta Paediatr 2013;102(12):e560–5.

[149] Costalos C, Skouteri V, Gounaris A, et al. Enteral feeding of premature infants with *Saccharomyces boulardii*. Early Hum Dev 2003;74(2):89–96.

[150] Serce O, Benzer D, Gursoy T, Karatekin G, Ovali F. Efficacy of *Saccharomyces boulardii* on necrotizing enterocolitis or sepsis in very low birth weight infants: a randomised controlled trial. Early Hum Dev 2013;89(12):1033–6.

[151] McFarland LV, Mulligan ME, Kwok RYY, Stamm WE. Nosocomial acquisition of *Clostridium difficile* infection. N Engl J Med 1989;320:204–10.

[152] Lessa FC, Winston LG, McDonald LC. Emerging Infections Program *C. difficile* Surveillance Team. Burden of *Clostridium difficile* infection in the United States. N Engl J Med 2015;372(24):2369–70.

[153] Pepin J, Valiquette L, Cossette B. Mortality attributable to nosocomial *Clostridium difficile*-associated disease during an epidemic caused by a hypervirulent strain in Quebec. Can Med Asso J 2005;173(9):1037–42.

[154] McFarland LV, Surawicz CM, Stamm WE. Risk factors for *Clostridium difficile* carriage and *C. difficile*-associated diarrhea in a cohort of hospitalized patients. J Infect Dis 1990;162:678–84.

[155] Al Eidan FA, McElnay JC, Scott MG, Kearney MP. *Clostridium difficile*-associated diarrhoea in hospitalised patients. J Clin PharmTher 2000;25(2):101–9.

[156] McFarland LV, Beneda HW, Clarridge JE, Raugi GJ. Implications of the changing face of *C. difficile* disease for health care practitioners. Amer J Infection Control 2007;35(4):237–53.

[157] O'Brien JA, Lahue BJ, Caro JJ, Davidson DM. The emerging infectious challenge of *Clostridium difficile*-associated disease in Massachusetts hospitals: clinical and economic consequences. Infect Control Hosp Epidemiol 2007;28(11):1219–27.

[158] McFarland LV, Surawicz CM, Greenberg RN, et al. A randomized placebo-controlled trial of *Saccharomyces boulardii* in combination with standard antibiotics for *Clostridium difficile* disease. J Amer Med Asso 1994;271:1913–8.

[159] Surawicz CM, McFarland LV, Greenberg RN, et al. The search for a better treatment for recurrent *Clostridium difficile* disease: use of high-dose vancomycin combined with *Saccharomyces boulardii*. Clin Infect Dis 2000;31(4):1012–7.

[160] McFarland LV, Elmer GW, Surawicz CM. Breaking the cycle: treatment strategies for 163 cases of recurrent *Clostridium difficile* disease. Am J Gastroenterol 2002;97(7):1769–75.

[161] Londoño AL, Mejía S, Gómez-Marín JE. Prevalence and risk factors associated with intestinal parasitism in preschool children from the urban area of Calarcá, Colombia]. Rev Salud Publica (Bogota) 2009;11(1):72–81.

[162] Guillot CC, Bacallao EG, Dominguez MSC, Garcia MF, Gutlerrez PM. Effects of *Saccharomyces boulardii* in children with chronic diarrhea, especially cases due to Giardiasis. Rev Mex de Puericultura Ypediatria 1995;2:1–11.

[163] Rossit AR, Gonçalves AC, Franco C, Machado RL. Etiological agents of diarrhea in patients infected by the human immunodeficiency virus-1: a review. Rev Inst Med Trop Sao Paulo 2009;51(2):59–65.

[164] Saint-Marc T, Blehaut H, Musial C, Touraine JL. AIDS-related diarrhea: a double-blind trial of *Saccharomyces boulardii*. Sem Hôp Paris 1995;71:735–41.

[165] Elmer GW, Moyer KA, Vega R, et al. Evaluation of *Saccharomyces boulardii* for patients with HIV-related chronic diarrhoea and in healthy volunteers receiving antifungals. Microecol Ther 1995;25:23–31.

[166] Demirel G, Celik IH, Erdeve O, et al. *Prophylactic Saccharomyces boulardii* versus nystatin for the prevention of fungal colonization and invasive fungal infection in premature infants. Eur J Pediatr 2013;172(10):1321–6.

[167] Weber G, Adamczyk A, Freytag S. Treatment of acne with a yeast preparation]. Fortschr Med 1989;107(26):563–6.

[168] Serce O, Gursoy T, Ovali F, Karatekin G. Effects of *Saccharomyces boulardii* on neonatal hyperbilirubinemia: a randomized controlled trial. Am J Perinatol 2015;30(2):137–42.

[169] Temmerman R, Pot B, Huys G, Swings J. A quality analysis of commercial probiotic products. Meded Rijksuniv Gent Fak Landbouwkd Toegep Biol Wet 2001;66(3b):535. 537–542.

[170] Salminen MK, Rautelin H, Tynkkynen S, et al. *Lactobacillus* bacteremia, species identification, and antimicrobial susceptibility of 85 blood isolates. Clin Infect Dis 2006;42(5):e35–44.

[171] Karen M, Yuksel O, Akyürek N, et al. Probiotic agent *Saccharomyces boulardii* reduces the incidence of lung injury in acute necrotizing pancreatitis induced rats. J Surg Res 2010;160(1):139–44.

[172] Lessard M, Dupuis M, Gagnon N, et al. Administration of *Pediococcus acidilactici* or *Saccharomyces cerevisiae boulardii* modulates development of porcine mucosal immunity and reduces intestinal bacterial translocation after *Escherichia coli* challenge. J Anim Sci 2009;87(3):922–34.

[173] Byron JK, Clemons KV, McCusker JH, Davis RW, Stevens DA. Pathogenicity of *Saccharomyces cerevisiae* in complement factor five-deficient mice. Infect Immun 1995;63(2):478–85.

[174] Boyle RJ, Robins-Browne RM, Tang ML. Probiotic use in clinical practice: what are the risks? Amer Clinical Nutr 2006;83:1256–64.

[175] Hennequin C, Kauffmann-Lacroix C, Jobert A, et al. Possible role of catheters in *Saccharomyces boulardii* fungemia. Eur J Clin Microbiol Infect Dis 2000;19(1):16–20.

[176] Enache-Angoulvant A, Hennequin C. Invasive *Saccharomyces* infection: a comprehensive review. Clin Infect Dis 2005;41(11):1559–68.

[177] Montineri A, Iacobello C, Larocca L, et al. *Saccharomyces cerevisiae* fungemia associated with multifocal pneumonia in a patient with alcohol-related hepatic cirrhosis. Infez Med 2008;16(4):227–9.

Chapter 19

Common Organisms and Probiotics: *Streptococcus thermophilus* (*Streptococcus salivarius* subsp. *thermophilus*)

J.P. Burton[*,**,†], R.M. Chanyi[†,‡] and M. Schultz[§]

*Division of Urology, Department of Surgery/Department of Microbiology & Immunology, Western University, London, ON, Canada;

**Lawson Health Research Institute and Canadian Research and Development Centre for Probiotics, London, ON, Canada; †Department of Microbiology & Immunology, Western University, London, ON, Canada; ‡Lawson Health Research Institute, Canadian Centre for Human Microbiome and Probiotics Research, London, ON, Canada; §Department of Medicine, Dunedin School of Medicine, University of Otago, Dunedin, New Zealand

INTRODUCTION

While *Streptococcus thermophilus* is likely to be one of the leading bacteria consumed by humans, it is less often utilized as a single strain treatment with regards to probiotics. In fact, there are relatively few studies relating to its use as a single strain probiotic. Its importance in fermentation and its stabilization of other probiotic bacteria less adapted to grow well and survival in milk should not be under recognized. In addition, the yogurt coagulum formed by *S. thermophilus* serves to improve the survival of other transiting bacteria through the gastric region of the intestinal tract and beyond. In probiotic use, *S. thermophilus* is typically found in multicombinational species/strain formulations and has often been included in studies. Given its excellent record of safety and similarity to less desirable streptococcal bacteria, it is perhaps an organism that has been underutilized in the area of probiotics. Typically, people have considered it only as the mainstay of fermentation for the dairy industry.

S. thermophilus has thought to have been used since the domestication of animals and the more consistent availability of milk (for some 10,000 years). Including safety, it has a number of attributes that make it a useful probiotic for humans. *S. thermophilus* has the distinction of being one of the most consumed organisms in fermented food and probiotics, with it being present in the millions of tons of yogurt and cheese produced each year. Given that yogurt containing a live culture typically has 1×10^8 bacteria present, it is likely that it contains *S. thermophilus*. A study by Lang et al. investigating bacterial numbers present in the average adult American diet found that only up to 6×10^6 bacteria were typically consumed per day [1]. Comparatively, North Americans consume relatively little fermented or bacterially containing food but, if they do, they are likely to be ingesting *S. thermophilus* as among the most predominant bacterial types.

Due to its characteristic textural polysaccharide capsule "ropiness" and pleasant aroma characteristics *S. thermophilus* is widely used in yogurt products of the West. Strains of *S. thermophilus* are traditionally paired with *Lactobacillus delbrueckii* and *Lactobacillus helveticus*, or *L. delbrueckii* ssp. *bulgaricus*. During fermentation by *S. thermophilus* formate is produced as a by-product, which synergistically enhances the growth of lactobacilli [2]. In these dairy products, *S. thermophilus* rapidly produces acid and subsequently a coagulum of proteins by utilizing the abundant lactose for metabolism with its highly beta-galactase production [3]. This is a benefit for people with low levels of intestinal lactase who are lactose intolerance.

TAXONOMY

Taxonomic placement of *S. thermophilus* is representative of its recent deviation from a related ancestor. Originally *S. thermophilus* was considered a species in its own right by Orla-Jensen in 1919 [4]. *S. thermophilus* and *Streptococcus salivarius* are highly related (99% at 16S rRNA gene level). Therefore, there is still contention as to whether the two

The Microbiota in Gastrointestinal Pathophysiology

165

should be included as the same species or not. The taxonomic status of *S. thermophilus* had been in question for several years and some investigators proposed that it should be a subspecies of *S. salivarius*. In 1984, DNA–DNA hybridization experiments placed *S. thermophilus* under the umbrella of the *salivarius* species, as *S. salivarius* subspecies *thermophilus* [5]. Schleifer et al. conducted further DNA–DNA reassociation experiments under stringent conditions and determined that these strains probably deserved separate full species status and suggested that the name should be shifted back to its former one, though it is still widely reported as *S. salivarius* ssp. *thermophilus* [4]. This has not been fully ratified by taxonomic committees.

Among the streptococci, *S. thermophilus* sits within the "Viridians" grouping, informally regarded as the alpha hemolytic and generally commensal streptococci, which also includes two other species, *S. salivarius* and *Streptococcus vestibularis*. These two species are both commensal bacteria of the human oral cavity and gut, whereas, the reservoir of *S. thermophilus* has not been identified but it is often detected in milk. This is probably the major reason why *S. thermophilus* is rarely used as a solo probiotic strain, as it is rarely demonstrated to be naturally present in the gut. *S. salivarius* also has some history of use in food from various parts of the world, with literature documenting *S. salivarius* as a member of the natural starter culture bacteria in traditional fermented milks of Europe and Africa [6–11], it is not as well equipped for this role as *S. thermophilus*.

THE SHAPING OF A SPECIES BY LIFE IN MILK

The full genome sequences for at least half a dozen *S. thermophilus* strains have been published at this time. Similarly, there have been another 47 strains that have been analyzed using a comparative genomics approach [12–14]. Substantial evidence has demonstrated that significant horizontal gene acquisition has occurred and is still occurring suggesting that *S. thermophilus* is continuing to adapt to its environment [12,14]. While *S. thermophilus* still appears to be acquiring new material, only 58% of genes have been identified to make up the core genome [12]. This is probably reflective of its relatively new emergence as its own species from a commensal ancestor of the salivarius group [12]. Shaping this bacterial change is its specific adaptation to grow and survive within milk, which has a very limited nutrient composition with regards to bacterial growth requirements. The main source to aid growth is fermentable carbohydrates, which are largely in the form of lactose. One of the most interesting characteristics noticed in the initial genome sequence analysis was that *S. thermophilus* has approximately 1500 genes (80%) that are orthologous to genes found in other streptococcal species, which indicates that *S. thermophilus* and its pathogenic, and commensal

relatives still share a substantial part of their overall physiology and metabolism [14]. Streptococcal virulence-related genes that are not involved in basic cellular processes were either inactivated or absent in *S. thermophilus* [13,14]. Approximately 10% of open reading frames in *S. thermophilus* are pseudogenes, their original functions being unnecessary for growth in milk or have been adapted. For example, genes found in the *eps* cluster of *S. thermophilus* required for exopolysaccharide (EPS) synthesis are related to those involved in capsule synthesis in *Streptococcus pneumoniae* and *Streptococcus agalactiae* [14]. Horizontally acquired genes add some minor confusion to taxonomic placement but those acquired are largely for metabolic processes, bacteriocin production, and extracellular polysaccharides.

THE ANCESTORS OF *Streptococcus thermophilus*

It is important to understand the relationship between the ancestors of *S. thermophilus* and their human host to illuminate its leverage for further probiotic use. *S. salivarius* and *Streptococcu vestibulatus* populations are usually established in the mouth, nasopharynx, and intestinal tract within hours of birth and remain for life [15–19]. At certain times of a human's life, *S. salivarius* can be detected as the predominant microbiota in the intestinal tract [19]. In fact, as early as the 1940s it was shown that the general occurrence of *S. salivarius* in the human intestine was in appreciable numbers by its cultivation from feces [20]. Human milk is an important factor in the initiation, development, and composition of the neonatal gut microflora. Breast-feeding infants consume *S. salivarius*, as it is commonly present in the mothers' breast milk [21]. While the origin of the bacteria found in human milk is debated, it has been suggested that at least some species may be endogenously delivered from the maternal gut to the mammary gland [22,23].

S. salivarius populations are abundant in human saliva in numbers up to 1×10^8 bacteria per mL. These are subsequently ingested and passed through to the intestinal tract [17]. Therefore, most people consume large quantities of these bacteria on a daily basis. *S. salivarius* is the most dominant human intestinal streptococcal species [15]. However, it may be more prevalent in certain parts of the intestinal tract, such as the jejunum [24]. More recent studies have shown the beneficial effects *S. salivarius* has on the gut through in vitro studies mimicking intestinal conditions, such as *S. salivarius* exhibits antiinflammatory effects on intestinal epithelial cells and monocytes [25,26]. It is quite possible that *S. thermophilus* produces the same active molecules as it has also been shown to be antiinflammatory [27]. Interestingly, *S. salivarius* in the feces, like *S. thermophilus*, also appears to be a common producer of potentially beneficial antimicrobial compounds, such as bacteriocins [28]. *S. salivarius* is the numerically predominant nonpathogenic bacterial streptococcal species in the human oral cavity and gut, not found elsewhere in nature and thought to have a key role in mucosal homeostasis and protection [29]. In short, while *S. thermophilus* has lost some of the phenotypical traits of its commensal and pathogenic relatives, it has a number of attributes that remain which are recognized by the host and could be harnessed for probiotic potential.

PROBIOTIC BENEFITS

The first, and unintentional, use of *S. thermophilus* cultures in a probiotic application was in the alleviation of lactose intolerance [30]. This functional fermented food enabled people to consume milk products without intestinal discomfort. The *S. thermophilus* cultures used to make yogurt are particularly rich in beta-galactosidase and are able to rapidly hydrolyze the lactose in milk [3]. Therefore, during the production of the yogurt, the amount of lactose from the milk is reduced due to the fermentation by the starter cultures. The beta-galactosidase from *S. thermophilus* substitutes for human lactase and the slower intestinal transit time compared to milk is due to product viscosity. This also reduces the symptoms that allow sufferers to access the nutrients found in dairy sources. The milk products that are best tolerated by lactose intolerant subjects are yogurts containing viable starter culture bacteria [31,30]. However, some have argued whether this is actually a probiotic or a food process effect, as the bacteria have undertaken this process largely before ingestion, but nevertheless it is still a benefit to the host [32].

There are numerous probiotic studies using *S. thermophilus* as part of a consortium with many other probiotic candidates [33]. However, it is difficult to evaluate the contribution of each component to the health outcome or the mechanistic actions involved and therefore these have not been rigorously reviewed here. It appears that *S. thermophilus* have a number of probiotic attributes, but most notably being able to produce biologically active molecules, a natural ability to downregulate the inflammatory response and immune stimulation of host production of proinflammatory and antiinflammatory cytokines [27]. For example, the VSL#3 probiotic combination is one of the most widely used formulations where the *S. thermophilus* is one of several species used in the mixture. This product has been used in successful clinical studies for Crohn's disease, decreased intestinal permeability and pouchitis, atopic dermatitis, and inhibition of specific gut pathogens, among others [34–40].

THE PROMISE OF IN VITRO AND ANIMAL STUDIES

Due to its benign properties with regards to the safety of the host, *S. thermophilus* has been identified as an organism to be used in future probiotics and potentially as a transgenic platform. *S. thermophilus* has shown promise with in vitro and animal models. One study used only *S. thermophilus* as a probiotic in a mouse model of chronic gastritis with a decrease in inflammation and an increase in mucus [41]. Also, *S. thermophilus* TH-4 was used to treat rats with mucositis caused by chemotherapy drugs [42]. The study showed that rats responded to the treatment by showing a normalization of healthy cell function in the affected areas and a significant reduction of distress to the tissue of the intestines. The relatedness of *S. thermophilus* to other streptococci has probably played a role for its antiinflammatory action which may be similar to the mechanism that *S. salivarius* has shown to downregulate responses in the oral and intestinal tracts [29]. Patients with chronic *Clostridium difficile* infections are key targets for *S. thermophilus* probiotic treatment as their last resort is bowel removal. It has been shown in mice that even simple lactic acid production by *S. thermophilus* alters *C. difficile* infection and in vitro toxin A production [43]. Probiotic treatment of these patients may significantly reduce symptom severity and increase overall patient well-being.

S. thermophilus lends itself as a good vehicle to produce and deliver genetically inserted products as it is naturally transformable. Recombinant strains can be quickly constructed using the plethora of genetic tools already developed [44], including those with markerless selection (no antibiotic selection) for use in food products [45]. *S. thermophilus* has the ability to produce beneficial molecules or to cleave bioactive peptides from simple culture media, such as milk proteins like casein. Finally, its ability to survive the intestinal passage and to be metabolically active in the gastrointestinal tract allows for the consideration of *S. thermophilus* as a potential tool for delivering various biological molecules to the intestinal tract. *S. thermophilus* CRL807 has been proposed as a delivery vehicle for cytokines in the case of inflammatory bowel diseases where it can deliver antiinflammatory IL-10 directly to the site [46]. Similarly, other strains have been devised to deliver antioxidant enzymes to the intestinal tract. Such is the case for *S. thermophilus* CRL 807 that contains transformed genes encoding catalase or superoxide dismutase. Along with its natural antiinflammatory effects, the addition of these enzymes into *S. thermophilus* showed promise on a mouse model of colitis [27].

S. thermophilus is and will continue to be an incredibly useful bacterium in the dairy industry in the production of fermented foods. However, we have undervalued its potential contribution to human health due to its commonality in our everyday products. It is a species that can be used for probiotic benefit in its own right. Given its widespread use and great potential, we should look further into the merits of using *S. thermophilus* as both a probiotic and a delivery platform of natural and engineered products.

REFERENCES

[1] Lang JM, Eisen JA, Zivkovic AM. The microbes we eat: abundance and taxonomy of microbes consumed in a day's worth of meals for three diet types. PeerJ 2014;2:e659.

[2] Perez PF, de Antoni GL, Añon MC. Formate production by *Streptococcus thermophilus* cultures. J Dairy Sci 1991;74:2850–4.

[3] Burton JP, Tannock GW. Properties of porcine and yogurt lactobacilli in relation to lactose intolerance. J Dairy Sci 1997;80:2318–24.

[4] Schleifer KH, Ehrmann M, Krusch U, Neve H. Revival of the species *Streptococcus thermophilus* (ex Orla-Jensen, 1919) nom. rev. Syst Appl Microbiol 1991;14:386–8.

[5] Farrow JA, Collins MD. DNA base composition, DNA–DNA homology and long-chain fatty acid studies on *Streptococcus thermophilus and Streptococcus salivarius*. J Gen Microbiol 1984;130:357–62.

[6] Abdelgadir WS, Hamad SH,LMP, Jakobsen M. Characterisation of the dominant microbiota of Sudanese fermented milk Rob. Int Dairy J 2001;11:63–70.

[7] Callon C, Millet L, Montel MC. Diversity of lactic acid bacteria isolated from AOC Salers cheese. J Dairy Res 2004;71:231–44.

[8] Pesic-Mikulec D. Microbiological study of fresh white cheese. Appl Ecol Environ Res 2005;4:129–34.

[9] Van Hoorde K, Verstraete T, Vandamme P, Huys G. Diversity of lactic acid bacteria in two Flemish artisan raw milk Gouda-type cheeses. Food Microbiol 2008;25:929–35.

[10] Obodai M, Dodd CE. Characterization of dominant microbiota of a Ghanaian fermented milk product, nyarmie, by culture- and nonculture-based methods. J Appl Microbiol 2006;100:1355–63.

[11] Ongol MP, Asano K. Main microorganisms involved in the fermentation of Ugandan ghee. Int J Food Microbiol 2009;133:286–91.

[12] Rasmussen TB, Danielsen M, Valina O, et al. *Streptococcus thermophilus* core genome: comparative genome hybridization study of 47 strains. Appl Environ Microbiol 2008;74:4703–10.

[13] Hols P, Hancy F, Fontaine L, et al. New insights in the molecular biology and physiology of *Streptococcus thermophilus* revealed by comparative genomics. FEMS Microbiol Rev 2005;29:435–63.

[14] Bolotin A, Quinquis B, Renault P, et al. Complete sequence and comparative genome analysis of the dairy bacterium *Streptococcus thermophilus*. Nat Biotechnol 2004;22:1554–8.

[15] Kubota H, Tsuji H, Matsuda K, et al. Non-cultural detection of human intestinal catalase-negative, Gram-positive cocci by rRNA-targeted reverse transcription-PCR. Appl Env Microbiol 2010;76:5440–51.

[16] Birri DJ, Brede DA, Tessema GT, Nes IF. Bacteriocin production, antibiotic susceptibility and prevalence of haemolytic and gelatinase activity in faecal lactic acid bacteria isolated from healthy Ethiopian infants. Microb Ecol 2013;65:504–16.

[17] Burton JP, Wescombe PA, Macklaim JM, et al. Persistence of the oral probiotic *Streptococcus salivarius* M18 is dose dependent and megaplasmid transfer can augment their bacteriocin production and adhesion characteristics. PLoS One 2013;8:e65991.

[18] Favier CF, Vaughan EE, De Vos WM, Akkermans AD. Molecular monitoring of succession of bacterial communities in human neonates. Appl Environ Microbiol 2002;68:219–26.

[19] Park HK, Shim SS, Kim SY, et al. Molecular analysis of colonized bacteria in a human newborn infant gut. J Microbiol 2005;43:345–53.

[20] Sherman JM, Niven CF, Smiley KL. *Streptococcus salivarius* and other non-hemolytic streptococci of the human throat. J Bacteriol 1943;45:249–63.

[21] Urbaniak C, Burton JP, Reid G. Breast, milk and microbes: a complex relationship that does not end with lactation. Womens Health 2012;8:385–98.

[22] Martin R, Heilig HG, Zoetendal EG, et al. Cultivation-independent assessment of the bacterial diversity of breast milk among healthy women. Res Microbiol 2007;158:31–7.

[23] Martín R, Langa S, Reviriego C, et al. The commensal microflora of human milk: new perspectives for food bacteriotherapy and probiotics. Trends Food Sci Technol 2004;15:121–7.

[24] van den Bogert B, Erkus O, Boekhorst J, et al. Diversity of human small intestinal *Streptococcus* and Veillonella populations. FEMS Microbiol Ecol 2013;85:376–88.

[25] Kaci G, Lakhdari O, Dore J, et al. Inhibition of the NF-kappaB pathway in human intestinal epithelial cells by commensal *Streptococcus salivarius*. Appl Environ Microbiol 2011;77:4681–4.

[26] Couvigny B, de Wouters T, Kaci G, et al. Commensal *Streptococcus salivarius* modulates PPARγ transcriptional activity in human intestinal epithelial cells. PLoS One 2015;10:e0125371.

[27] Del Carmen S, de Moreno de LeBlanc A, Martin R, et al. Genetically engineered immunomodulatory *Streptococcus thermophilus* strains producing antioxidant enzymes exhibit enhanced anti-inflammatory activities. Appl Environ Microbiol 2014;80:869–77.

[28] O'Shea EF, Gardiner GE, O'Connor PM, et al. Characterization of enterocin- and salivaricin-producing lactic acid bacteria from the mammalian gastrointestinal tract. FEMS Microbiol Lett 2008;291:24–34.

[29] Cosseau C, Devine DA, Dullaghan E, et al. The commensal *Streptococcus salivarius* K12 downregulates the innate immune responses of human epithelial cells and promotes host-microbe homeostasis. Infect Immun 2008;76:4163–75.

[30] Savaiano Da. Lactose digestion from yogurt: mechanism and relevance. Am J Clin Nutr 2014;99:1251S–5S.

[31] Martini MC, Bollweg GL, Levitt MD, Savaiano Da. Lactose digestion by yogurt beta-galactosidase: influence of pH and microbial cell integrity. Am J Clin Nutr 1987;45:432–6.

[32] Guarner F, Perdigon G, Corthier G, et al. Should yoghurt cultures be considered probiotic? Br J Nutr 2005;93:783–6.

[33] Chapman TM, Plosker GL, Figgitt DP. VSL#3 probiotic mixture: a review of its use in chronic inflammatory bowel diseases. Drugs 2006;66:1371–87.

[34] Chapman TM, Plosker GL, Figgitt DP. Spotlight on VSL#3 probiotic mixture in chronic inflammatory bowel diseases. BioDrugs 2007;21:61–3.

[35] Mennigen R, Nolte K, Rijcken E, et al. Probiotic mixture VSL#3 protects the epithelial barrier by maintaining tight junction protein expression and preventing apoptosis in a murine model of colitis. Am J Physiol Gastrointest Liver Physiol 2009;296:G1140–9.

[36] Gionchetti P, Rizzello F, Venturi A, et al. Oral bacteriotherapy as maintenance treatment in patients with chronic pouchitis: a double-blind, placebo-controlled trial. Gastroenterology 2000;119:305–9.

[37] Chapman TM, Plosker GL, Figgitt DP. VSL#3 probiotic mixture: a review of its use in chronic inflammatory bowel diseases. Drugs 2006;66:1371–87.

[38] Di Marzio L, Centi C, Cinque B, et al. Effect of the lactic acid bacterium *Streptococcus thermophilus* on stratum corneum ceramide levels and signs and symptoms of atopic dermatitis patients. Exp Dermatol 2003;12:615–20.

[39] Chang B, Sang L, Wang Y, et al. The protective effect of VSL#3 on intestinal permeability in a rat model of alcoholic intestinal injury. BMC Gastroenterol 2013;13:151.

[40] Fedorak RN, Feagan BG, Hotte N, et al. The probiotic VSL#3 has anti-inflammatory effects and could reduce endoscopic recurrence after surgery for Crohn's disease. Clin Gastroenterol Hepatol 2014;13:928–35. e2.

[41] Rodríguez C, Medici M, Mozzi F, de Valdez GF. Therapeutic effect of *Streptococcus thermophilus* CRL 1190-fermented milk on chronic gastritis. World J Gastroenterol 2010;16:1622–30.

[42] Whitford EJ, Cummins AG, Butler RN, et al. Effects of *Streptococcus thermophilus* TH-4 on intestinal mucositis induced by the chemotherapeutic agent, 5-Fluorouracil (5-FU). Cancer Biol Ther 2009;8:505–11.

[43] Kolling GL, Wu M, Warren Ca, et al. Lactic acid production by *Streptococcus thermophilus* alters *Clostridium difficile* infection and in vitro toxin a production. Gut Microbes 2012;3: 523–9.

[44] Lecomte X, Gagnaire V, Lortal S, et al. *Streptococcus thermophilus*, an emerging and promising tool for heterologous expression: Advantages and future trends. Food Microbiol 2015;53:2–9.

[45] Blomqvist T, Steinmoen H, Håvarstein LS. A food-grade site-directed mutagenesis system for *Streptococcus thermophilus* LMG 18311. Lett Appl Microbiol 2010;50:314–9.

[46] Del Carmen S, Miyoshi A, Azevedo V, et al. Evaluation of a *Streptococcus thermophilus* strain with innate anti-inflammatory properties as a vehicle for IL-10 cDNA delivery in an acute colitis model. Cytokine 2015;73:177–83.

Chapter 20

Complexities and Pitfalls in the Production of Multispecies Probiotics: The Paradigmatic Case of VSL#3 Formulation and Visbiome

M.G. Cifone*, B. Cinque*, C. La Torre*, F. Lombardi*, P. Palumbo*, M.E. van den Rest**, C. Vuotto[†] and G. Donelli[†]

*Department of Life, Health & Environmental Sciences, University of L'Aquila, L'Aquila, Italy; **Biovisible, Groningen, The Netherlands; [†]Microbial Biofilm Laboratory, IRCCS Fondazione Santa Lucia, Rome, Italy

VSL#3 is a probiotic preparation containing a specific combination of eight different strains of lactic acid bacteria and bifidobacteria which was originally formulated over 20 years ago by Claudio De Simone, an Italian professor of infectious diseases. The original combination of strains, currently marketed under the names of Visbiome in North America and Vivomixx in Europe and hereafter defined as *original De Simone (DS) formulation*, has been taken into consideration from many years in the guidelines of international gastroenterological societies, including the American Gastroenterology Association, and the European Crohn's and Colitis Organization, for the management of chronic pouchitis and the prevention of pouchitis after ileal pouch anastomosis, as well as for maintenance treatment of ulcerative colitis. In 2015 the participants to a consensus session in the occasion of the 4th Triennial Yale/Harvard Workshop on Probiotic Recommendations recognized for the first time level B and C evidence for this probiotic preparation in the management of nonalcoholic fatty liver disease in adults and children and in alcoholic liver diseases [1].

CLAIMS FOR PROBIOTICS ARE PRODUCT SPECIFIC

The accepted definition of probiotics is that they are "live microorganisms that when administered in adequate amounts confer a health benefit on the host" [2]. This definition underlines two important issues: first, the microbes in a probiotic formulation must be living and second, there is a dose–effect relationship with the health benefits depending on adequate doses. Thus, dose regimens and specifications for use must be clearly defined when probiotics are used in clinical practice.

Probiotics are regulated in Europe under the European Regulation on Health Claims for Foods [Regulation (EC) No. 1924/2006]. The aim of these regulations is to offer confidence to consumers and physicians that a marketed probiotic product produces the claimed effect in the defined conditions of use. Integral to this is that the product must be consistent in its composition and properties, and that these must be properly described. The following features are essential when describing a probiotic product: [2] (1) the specific strain(s) must be characterized using appropriate phenotypic and genotypic techniques [3]; (2) the matrix in which the strains are delivered must be defined; and (3) the number of live microorganisms provided in a given dose of the product must be specified (and should apply until the end of the product shelf-life).

In addition, to make any health claims, probiotics are required to demonstrate a beneficial effect through appropriate human clinical trials in the target patient group. The results of such studies are only applicable to the specific product that was used. All claims must be related to a specific strain/combination of strains, given in a defined amount in a properly characterized product, and to a defined cohort of individuals [2]. From a regulatory standpoint, the general minimum dose to be declared in any given probiotic has been indicated as 10^9. Any modification in the formulation or in other characteristics means that the changed product is no longer the entity that generated the evidence supporting a claim and may not produce the same effects.

COMPOSITION OF PROBIOTIC PRODUCTS AND CHARACTERIZATION OF THEIR BENEFICIAL EFFECTS ARE CRUCIAL

There is a large body of literature exploring the mechanisms of action of probiotic strains and their effects on human health. For the majority of the probiotic strains, most of the available evidence deals with their safety and ability to colonize the gut, rather than proving clinical efficacy. On the contrary, for the original VSL#3 preparation, a large body of in vitro results and clinical evidences on its effects has been reported by De Simone and coworkers [4,5].

In other terms, for the reasons described earlier, probiotic products are not interchangeable. Even within the same bacterial species, it is accepted that different strains may have different effects. For example, differences in the produced antimicrobial metabolites are reported both among species of lactobacilli and different strains of the same species [6].

A careful selection of the probiotic agent, standardization of the dose, and detailed characterization of the beneficial effects are essential when considering use of a probiotic for a particular health condition [6]. The probiotic strains should have well-defined and measurable metabolic and functional properties that are relevant to the nature of the condition in question [6].

Different strains of the same or closely related species may possess antagonistic properties, that is, they may produce metabolites with antagonistic effects. Changing the mixture of strains within a multistrains product may thus lead to changes in functionality and efficacy. Thus, any reformulation will require an extensive reevaluation to ensure that the properties documented for the original formulation are retained. A clear example is offered by De Angelis et al. in their experiment on gluten digestion by different probiotic preparations; they observed that no other similar probiotic preparation containing eight or nine strains was able to digest gluten in the way VSL#3 did [7].

As already mentioned, probiotic effects are dose dependent. Dose is particularly important for strains whose mechanism of action is based on release of a soluble factor that produces a concentration-dependent response in gut epithelial cells; an example of this type of effect is probiotic inhibition of inflammatory signals induced by gut pathogens, such as *Escherichia coli* serotypes [6].

As probiotics are live microorganisms, there is a theoretical risk of infection, primarily in immunosuppressed individuals. There have been anecdotal cases of such infection, although it has not been widely documented. The risks of probiotic therapy should be weighed against the benefits in individual cases, both in terms of the clinical condition and the choice of product [6]. Nevertheless, it has been suggested that infection rates arising from the use of probiotics in different conditions should be assessed in randomized controlled trials [8].

BIOSIMILARS—HOW SIMILAR IS SIMILAR?

As biopharmaceuticals are made of living cells, during the manufacturing process they are sensitive to changes in formulation, growth conditions, purification steps, or storage conditions [9,10]. The same is true for probiotic preparations and a change in manufacturing has the potential to influence efficacy and even safety. In their position statement on use of biosimilars in pediatric inflammatory bowel disease, the European Society for Paediatric Gastroenterology Hepatology and Nutrition (ESPGHAN) stress that "Post-marketing surveillance programs for efficacy, safety and immunogenicity should become mandatory in children with IBD using biosimilars, as for all biological drugs" [10]. Even within the highly regulated area of biosimilar biopharmaceuticals, questions remain as to how the efficacy and safety of approved biosimilars should be monitored in clinical practice, and how biosimilars should be labeled to ensure accurate prescribing [11]. Such questions apply equally to probiotic formulations when the evidence base has been generated using one manufacturing process and the process is subsequently changed. In the analyses reported later, we show that such changes do indeed affect the product's characteristics and performance.

When biosimilar drugs are developed, they are required to undergo in vitro characterization studies followed by clinical testing to evaluate pharmacokinetics, pharmacodynamics, efficacy, and safety [10]. However, most probiotics are currently regulated as food supplements and these equivalence tests are not required when changes are made to a probiotic product. Nevertheless, it is important that physicians and patients should be fully informed about any changes to a product, so that they can make informed decisions when choosing treatments.

MANUFACTURING CHANGES CAN ALTER PRODUCT CHARACTERISTICS

Sanders et al. have analyzed thoroughly the possible effects of genetic and processing changes on efficacy and safety of probiotics, highlighting some key issues to determine the need to reconfirm efficacy and safety aspects of probiotics, in particular after the change in manufacturing process, both from a scientific and regulatory standpoint [12]. It is well accepted that growth conditions, growth substrates, cryoprotectants, food formulation, and storage conditions may affect the properties of the probiotic, thereby generating changes in gene expression and possibly in metabolic output. Changes can have an impact on the numerical recovery of the probiotic and may consequently influence the final health outcome. The authors note that the regulatory framework still needs fine-tuning in terms of identifying possible criteria for substantial equivalence, and they propose a decision framework to decide when efficacy evaluation is needed.

The FDA addressed this issue back in 2011 with a guidance document requesting new dietary ingredient notification in such cases. These aspects are only addressed for food/dietary supplements; however, the case with VSL#3 is all the more complex since it is a medical food in the United States and a food supplement in Europe, used mainly in the dietary management of IBD patients. Sanders rightly refers to mouse models that have shown cell surface properties of probiotic microbes directing major immunological responses in the GI tract, reducing inflammation and even polyp formation. This is a critical aspect to be checked in the specific case of VSL#3.

In the United States, the *original DS formulation* is a biologic medicine regulated as a medical food. With any biologic product (drug, food, or medical food), changes to the manufacturing process can and will result in changes to the final product, and these are often extremely difficult to characterize. In the case of the *original DS formulation*, even minor changes to the strains, strain ratios, methods of fermentation, and/or methods of lyophilization can have a material impact on the end product which will be impossible to fully detect without head-to-head comparative clinical trials, in particular in ulcerative colitis, IBS, and pouchitis where the Medical Food status applies.

While the *original DS formulation* is not regulated as a drug, the guidance of the FDA with respect to biologic products is nevertheless relevant. As the FDA notes repeatedly in numerous guidance documents: "In contrast to chemically synthesized small molecular weight drugs, which have a well-defined structure and can be thoroughly characterized, biological products are generally derived from living material-human, animal, or microorganism- are complex in structure, and thus are usually not fully characterized" [13]. The FDA further notes: "Because, in many cases, there is limited ability to identify the identity of the clinically active component(s) of a complex biological product, such products *are often defined by their manufacturing processes* [emphasis added]. Changes in the manufacturing process, equipment or facilities could result in changes in the biological product itself and sometimes require additional clinical studies to demonstrate the product's safety, identity, purity and potency …" [13]. The mere fact of moving to a new manufacture automatically creates an assumption of difference between the two products.

In the case of the *original DS formulation*, the difficulty of "copying" the product is even further amplified by the fact that the active principle is constituted by eight unique living bacteria. Additionally, two of the key strains in the formulation are proprietary to Prof. De Simone (*Streptococcus thermophilus* and *Bifidobacterium breve*, owned by DuPont but exclusive to De Simone in the concentrations required to manufacture the product). While alternative bacteria of the same genus and species are available from other suppliers (i.e., other strains of *S. thermophilus* and *B. breve*), these specific strains are not available from the original manufacturer in concentrations required to manufacture the *original DS formulation*. Fermenting offspring

of the same "parent" cells in different ways can yield final bacteria with vastly different properties, which could have significant clinical implications for patients.

As an example, in one study *Bifidobacterium animalis* subsp. *lactis* INL1 was grown on a particular culture broth (MRS-de Man, Rogosa & Sharpe) at two different pH levels (6.5 and 5.0). When the final cultures were subjected to gastric resistance testing, the viability loss for the cells grown at pH 6.5 was a dramatic *4 logs higher* than for the cells grown at pH 5.0. Additionally, further analysis by electron microscope found that the culture grown at pH 6.5 showed production of extracellular compounds (exopolysaccharide type) not found in the same strain grown at pH 5.0. Authors found that these extracellular compounds could exert an in vivo response not found with the strains grown at lower pH [14].

In another study it was found that pH sensitivity of a given probiotic strain is not only influenced by the pH level in culture production but also by the type of cryoprotectant used for freeze-drying of strains. In Saarela et al. resistance to gastric acidity and bile salts by a strain of *Lactobacillus rhamnosus* was variable depending not only on the fermentation pH level but also on the use of either polydextrose versus sucrose as a cryoprotectant [15].

Kimoto-Nira et al. [16] reported changes to the immunomodulatory activity of a strain of *Bifidobacterium lactis* dependent on the type of fermentation broth used to grow the strain. Specifically, the cytokine interleukin-12 (a proinflammatory "chemical messenger" found in the gut immune system) was found to be enhanced in cell lines by *B. lactis* when grown on M17 fermentation media versus MRS media. In addition, the strains were found to have different cell wall compositions, at the sugar and fatty acid level. Both the cell wall and immunomodulatory differences indicate that "sibling" strains of *B. lactis* grown using different technology will yield bacteria with different clinical functionality.

The previously mentioned examples merely hint at the variables that can impact the functionality of a given probiotic. In the context of biologic production, even the most mundane input variable can impact the end product, often in ways that are difficult to measure. It is the reason why FDA requires, for biosimilar "generic" biologics, drugs comparative human clinical data in addition to in vitro comparative assays. In the case of the *original DS formulation* the complexity is further compounded by the fact that the product contains eight strains which are all individually cultivated and lyophilized.

THE *LATELY MARKETED PRODUCT* VERSUS THE *ORIGINAL DS FORMULATION* OF VSL#3—A COMPARISON OF SELECTED PARAMETERS

In subsequent sections we summarize some of the analyses performed by us showing clear distinctions between the *original DS formulation* (Fig. 20.1A)—with which the

(A)

VSL#3®
Dispensing Pack
Probiotic Food Supplement

Store in a refrigerator at 2 - 8°C
Can be stored at room temperature (up to 25 °C)
for up to 7 days without adversely affecting potency
Best before: See base of pack

Ingredients: Bacteria blend *, maltose, anti-caking agent : silicon dioxide.
* Product information.
Each 4.4g sachet provides a blend of 450 billion bacteria, containing: *Streptococcus thermophilus* DSM 24731,
bifidobacteria (*B. longum* DSM 24736, *B. breve* DSM 24732, *B. infantis* DSM 24737), lactobacilli (*L. acidophilus*
DSM 24735, *L. plantarum* DSM 24730, *L. paracasei* DSM 24733, *L. delbrueckii* subsp. *bulgaricus* DSM 24734).

Directions: For adults and teenagers, take 1 to 4 sachets daily.
Open the sachet and stir the contents into cold water or any cold non-fizzy drink or food and
consume immediately.
Do not exceed the recommended daily intake.
Food supplements should not be used as substitute for a balanced and varied diet and a healthy lifestyle.
Do not use if the sachet is broken or damaged.
Store out of reach of young children.

> **This product does NOT contain soy, gluten, lactose or milk products.**

Distributed in the UK by:
FERRING Pharmaceuticals Ltd.
Drayton Hall, Church Road, West Drayton UB7 7PS.

Distributed in the Republic of Ireland by:
FERRING Ireland Limited, United Drug House,
Magna Business Park, Citywest, Dublin 24

VSL#3® is a trademark of Actial Farmaceutica Lda

Manufacturer:
S.I.I.T srl
Via Ariosto - 50/60
20090 Trezzano sul Naviglio, (MI) Italy

5 015919 450094

(B)

VSL#3®
Dispensing Pack
Food Supplement

Store in a refrigerator at 2 - 8°C
Can be stored at room temperature (up to 25 °C)
for up to 7 days without any alteration.
Best before end: See base of pack

Ingredients: Bacteria blend *, maltose, anti-caking agent: silicon dioxide.
* Average contents.
Each 4.4g sachet provides a blend* of 450 billion bacteria, namely: *Streptococcus thermophilus* BT01,
bifidobacteria (*B. breve* BB02, *B. longum*** BL03, *B. infantis*** BI04), lactobacilli (*L. acidophilus* BA05,
L. plantarum BP06, *L. paracasei* BP07, *L. delbrueckii* subsp. *bulgaricus**** BD08).
**Recently reclassified as *B. animalis* subsp. *lactis*
***Recently reclassified as *L. helveticus*

Directions: For adults and teenagers, take 1 to 2 sachets daily.
Open the sachet, pour the content into cold water or any cold non-fizzy liquid or food, stir and consume
immediately. Do not exceed the recommended daily dose.

Warnings: Food supplements should not be used as substitute for a balanced and varied diet and a
healthy lifestyle.
Do not use if the sachet is broken or damaged. Store out of reach of young children.

> **This product does NOT contain soy, gluten, lactose or milk products**

Distributed in the UK by:
FERRING Pharmaceuticals Ltd.
Drayton Hall, Church Road, West Drayton UB7 7PS

Distributed in the Republic of Ireland by:
FERRING Ireland Limited, United Drug House,
Magna Business Park, Citywest, Dublin 24

VSL#3® is a registered trademark of Actial Farmacêutica Lda

5 015919 450094

FIGURE 20.1 **Packaging of VSL#3.** (A) *Original DS formulation*; (B) *lately marketed product.*

TABLE 20.1 Percentage of Live and Dead Cells From Batches of Lately Marketed and Original DS Formulations of VSL#3

Measurements	Lately Marketed[a] Live (%)	Lately Marketed[a] Dead (%)	Original DS[b] Live (%)	Original DS[b] Dead (%)
1	40.0	60.0	69.1	30.9
2	46.2	53.8	59.7	40.3
3	38.5	61.5	64.7	35.3
4	40.2	59.8	67.0	33.0
5	47.2	52.8	58.2	41.8
6	42.0	58.0	78.6	21.4
7	47.3	52.7	71.9	28.1
8	45.7	54.3	62.0	38.0
9	49.1	50.9	62.1	37.9
10	48.3	51.7	57.6	42.4
Mean	44.46[§]	55.54*	65.08[§]	34.92*
SD	3.88	3.88	6.66	6.66

[§]Mean values significantly different ($P < 0.0001$). *Mean values significantly different ($P < 0.0001$).
[a]VSL#3 labeled lot 507132 (lately marketed formulation).
[b]VSL#3 labeled lot TM091 (original DS formulation).

efficacy of VSL#3 was established—and the *lately marketed product* (Fig. 20.1B), both of them distributed at the beginning of 2016 under the "VSL#3" brand name in the United Kingdom and the Republic of Ireland. Even though the two formulations are claimed to be the same in terms of species identity and colony forming units counting, the strains belonging to each species appear different. Furthermore, according to our results, the two formulations are definitely different on the basis of their effects on the cell cycle and programmed cell death.

Live Versus Dead Bacteria

The proportion of live to dead cells in a bacterial formulation is a key element and has a material effect on the formulation's efficacy. Both the *original DS* and the *lately marketed formulations* are labeled on their packaging as "Live freeze-dried bacteria, 450 billion bacteria per sachet." Our analysis checked the percentage of "live" bacteria versus "dead" bacteria in a sachet of each formulation. Live cells in probiotic products will inevitably lose viability, and the actual products will contain varying amounts of dead cells. In practice, it is not possible to administer only live bacteria to a subject. Since the "dead" bacteria affect the responses of the individual, controlling the quality of the product simply by counting the "live" bacteria is not appropriate. The percentage of "live" and "dead" bacteria in a sachet is an important parameter since: (1) the capability of the product to colonize the gut properly is related to the number of "live" bacteria per sachet; (2) the quantity of "dead" is a good predictor of the quality of the product and its stability, and (3) "dead" bacteria are not an inert material.

The percentage of live and dead cells is reported in Table 20.1. The *lately marketed formulation* contained a considerably lower percentage of live cells ($p < 0.001$) and a markedly higher percentage of dead cells ($p < 0.0001$) than the *original DS formulation*. The difference between the two batches can be clearly visualized from the staining pictures (Figs. 20.2 and 20.3).

It is assumed from the package labeling that each sachet of either product contains 450 billion bacteria. Accordingly, the *original DS formulation* contains 241 billion (34.92%) of dead bacteria in addition to the 450 billion live bacteria

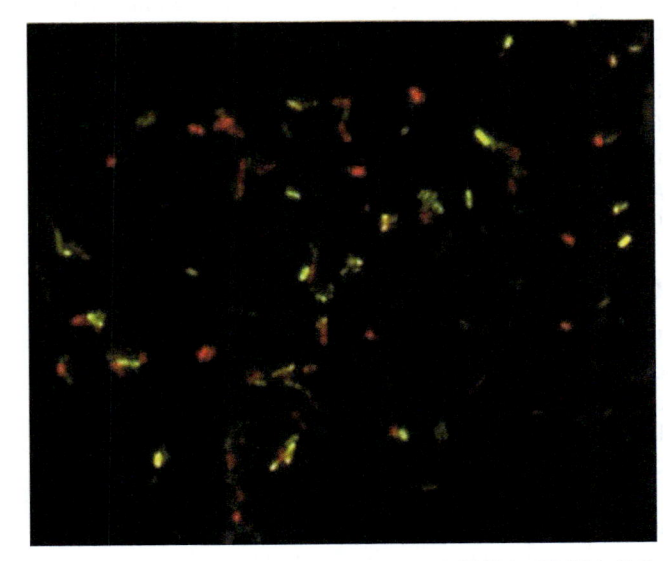

FIGURE 20.2 *"Original DS Formulation"*—**VSL#3 lot TM091.** Cells stained with SYTO 13 and propidium iodide. Magnification 1000×.

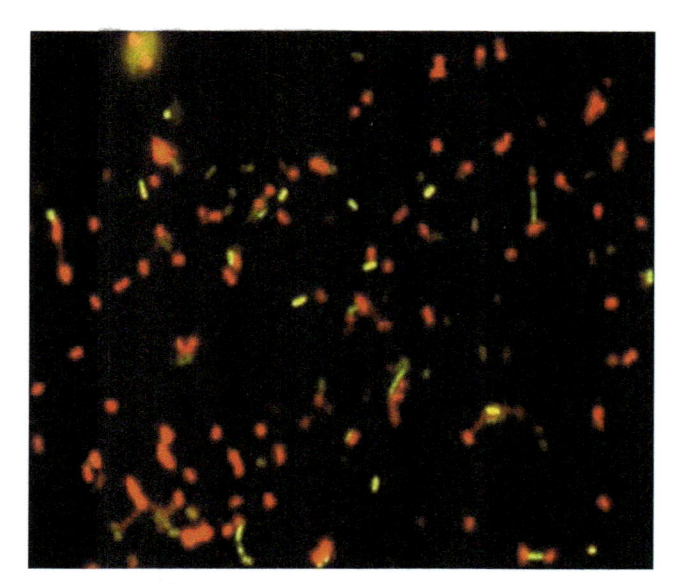

FIGURE 20.3 *Lately marketed formulation—*VSL#3 lot 507132. Cells stained with SYTO 13 and propidium iodide. Magnification 1000×.

(65.08%). The *lately marketed product* contained, on top of the 450 billion live bacteria (44.46%) as per label, also 562 billion dead bacteria (55.54%). Thus, the contents of the two sachets were significantly different ($p < 0.0001$).

Since it is common practice during the manufacturing process to "overfill" the sachet with bacteria (in other words, to maintain the target number of 450 billion live bacteria per sachet at the expiry date of the product, the sachet are "over-filled" with more bacteria) the number of "dead" bacteria ingested by a patient using this new product would be even larger. The number of the dead bacteria ingested will increase proportionally to the increased "overfilling" and to the rate at which the "live" bacteria die during the time interval between the manufacturing and the expiry date of each lot.

The dead bacteria cannot be considered a biologically inert material since they act as biological response modifiers. The single sachet of the *lately marketed product* has more than 130% more dead bacteria than the *original DS product* and this may significantly affect the response and health status of the person ingesting it.

Cell Division and Apoptosis Analysis

Probiotic bacteria are an accepted therapeutic strategy for chronic intestinal inflammation. One of the mechanisms of action by which probiotics exert their effects is the modulation of apoptosis in intestinal immune and/or epithelial cells.

About the cell division cycle, there are several checkpoints to ensure that damaged or incomplete DNA is not passed on to daughter cells. Three main checkpoints exist: the G1/S checkpoint, the G2/M checkpoint, and the

metaphase (mitotic) checkpoint. Dysregulation of the cell cycle components may lead to tumor formation. When the cell is unable to pass the checkpoints, it undergoes apoptosis (programmed cell death) which is a highly regulated and controlled process that confers advantages during an organism's life cycle. Defective apoptotic processes have been implicated in a wide variety of diseases.

According to some experiments carried out by us on the effects of the two probiotic blends (*original DS* and *lately marketed formulations*) on the cell cycle and programmed cell death, the *original DS formulation* showed a statistically significant difference with respect to the *lately marketed product* in terms of its capability to arrest the cancerous cells in G0/G1 and inducing the apoptotic death of cancer cells. In addition, consistent with the increased apoptosis, a statistically significant, reduced number of cancer cell lines (i.e., Jurkat, Caco2, HT1080) treated with the *original DS formulation* were observed in the G2 phase, compared with the *lately marketed formulation*. This means that fewer cancer cells will proliferate when treated with the *original DS formulation* compared to the *lately marketed formulation* of VSL#3. This is a pivotal difference, since it is the final outcome of a very complex cascade of a number of immunological and biochemical pathways activated by the product [17].

The results from one representative experiment on Jurkat cells from three are shown in Fig. 20.4 and the mean values of three experiments are summarized in Table 20.2.

CONCLUSIONS

In this chapter, we highlighted the different parameters by which probiotic preparations are characterized and showed how changes to formulations and/or manufacturing processes can seriously affect the specific properties on which the product's efficacy is based.

As a paradigmatic example, we have examined the case of the multispecies probiotic product VSL#3, which is currently manufactured also under a process different from the original one developed by De Simone 20 years ago. The analytical data reported here strongly support the hypothesis that the manufacturing changes affecting the *lately marketed formulation* result in differences from the *original DS formulation*, so that the efficacy studies carried out on the latter are not generalizable to the *lately marketed product*.

Previously, Grzeskowiak and coworkers, by characterizing phenotypically and genotypically 15 different *L. rhamnosus* strains isolated from different products, demonstrated that *L. rhamnosus* isolates, even if genotypically identical, varied significantly in pathogen exclusion by inhibition and competition. The original probiotic properties resulted influenced by different industrial sources and manufacturing processes, with a possible consequence on efficacy [18].

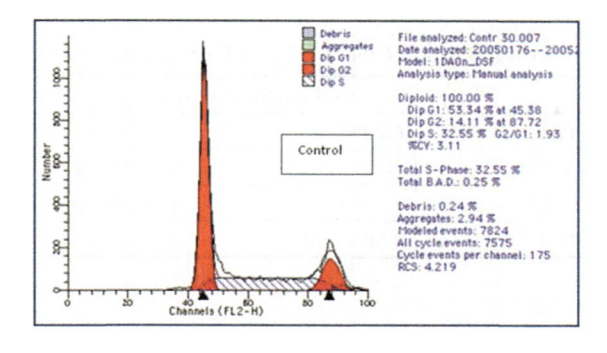

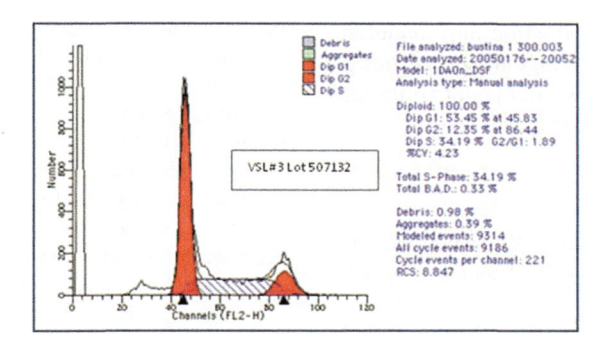

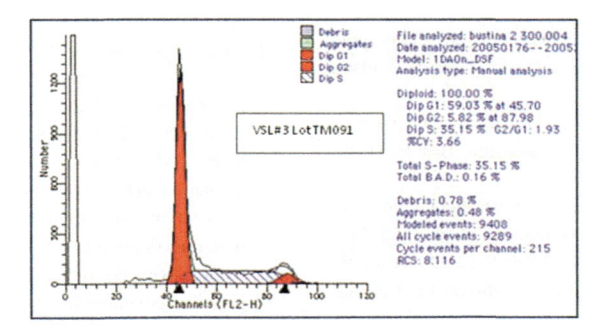

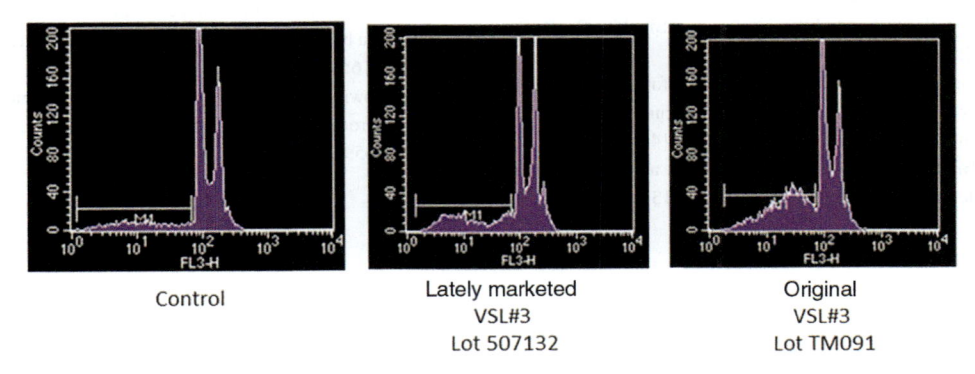

Control

Lately marketed
VSL#3
Lot 507132

Original
VSL#3
Lot TM091

FIGURE 20.4   **Cell cycle profile and apoptosis analysis.**

TABLE 20.2 Effects of Probiotic Blends on the Jurkat Cell Cycle and Programmed Cell Death

Sample	Cell Cycle and Apoptosis Analysis (%)			
	G0/G1 Phase	*G2 Phase*	*S Phase*	*Apoptosis*
Control	55.14 ± 2.55	11.33 ± 3.92	33.52 ± 1.37	10.79 ± 3.45
Lately marketed formulation	54.12 ± 0.82	11.37 ± 1.37	34.50 ± 0.43	18.04 ± 5.31^
Original DS formulation	58.4 ± 0.89*,**	7.58 ± 2.58^,**	34.02 ± 1.60	24.72 ± 8.11^,**

$N = 3$ experiments in duplicate; VSL#3 lot 507132 (*lately marketed* product); VSL#3 lot TM091 (original DS product); *$p < 0.05$ versus control; ^$p < 0.01$ versus control; **$p < 0.01$ versus "*lately marketed* product" (lot 507132).

In conclusion, differences in production and manufacturing methods could significantly change the properties of probiotic strains, and thus have a potential negative impact on the outcome of clinical interventions.

REFERENCES

[1] Floch MH, Walker WA, Sanders ME, et al. Recommendations for probiotic use—2015 update: proceedings and consensus opinion. J Clin Gastroenterol 2015;49:S69–73.

[2] Bertazzoni E, Donelli G, Midtvedt T, Nicoli J, Sanz Y. Probiotics and clinical effects: is the number what counts? J Chemother 2013;25:193–212.

[3] Donelli G, Vuotto C, Mastromarino P. Phenotyping and genotyping are both essential to identify and classify a probiotic microorganism. Microb Ecol Health Dis 2013;11:24.

[4] Lichtenstein L, Avni-Biron I, Ben-Bassat O. The current place of probiotics and prebiotics in the treatment of pouchitis. Best Pract Res Clin Gastroenterol 2016;30:73–80.

[5] Derikx LAAP, Dieleman LA, Hoentjen F. Probiotics and prebiotics in ulcerative colitis. Best Pract Res Clin Gastroenterol 2016;30:55–71.

[6] Mikelsaar M, Lazar V, Onderdonk AB, Donelli G. Do probiotic preparations for humans really have efficacy? Microb Ecol Health Dis 2011;22:10128.

[7] De Angelis M, Rizzello CG, Fasano A, et al. VSL#3 probiotic preparation has the capacity to hydrolyze gliadin polypeptides responsible for Celiac Sprue probiotics and gluten intolerance. Biochim Biophys Acta 2006;1762:80–93.

[8] Rijkers GT, Bengmark S, Enck P, et al. Guidance for substantiating the evidence for beneficial effects of probiotics: current status and recommendations for future research. J Nutr 2010;140:671s–6s.

[9] Declerck PJ. Biosimilar monoclonal antibodies: a science-based regulatory challenge. Expert Opin Biol Ther 2013;13:153–6.

[10] de Ridder L, Waterman M, Turner D, et al. Use of biosimilars in paediatric inflammatory bowel disease: a position statement of the ESPGHAN Paediatric IBD Porto Group. J Pediatr Gastroenterol Nutr 2015;61:503–8.

[11] Isaacs JD, Cutolo M, Keystone EC, Park W, Braun J. Biosimilars in immune-mediated inflammatory diseases: initial lessons from the first approved biosimilar anti-tumour necrosis factor monoclonal antibody. J Intern Med 2016;279:41–59.

[12] Sanders ME, Klaenhammer TR, Ouwehand AC, et al. Effects of genetic, processing, or product formulation changes on efficacy and safety of probiotics. Ann NY Acad Sci 2014;1309:1–18.

[13] Food and Drug Administration. Scientific Considerations in Demonstrating Biosimilarity to a Reference Product: Guidance for Industry; 2015.

[14] Zacarías MF, Binetti A, Laco M, Reinheimer J, Vinderola G. Preliminary technological and potential probiotic characterisation of bifidobacteria isolated from breast milk for use in dairy products. Int Dairy J 2011;21:548–55.

[15] Saarela MH, Alakomi HL, Puhakka A, Matto J. Effect of the fermentation pH on the storage stability of *Lactobacillus rhamnosus* preparations and suitability of in vitro analyses of cell physiological functions to predict it. J Appl Microbiol 2009;106:1204–12.

[16] Kimoto-Nira H, Suzuki C, Kobayashi M, Mizumachi K. Different growth media alter the induction of interleukin 12 by a *Lactococcus lactis* strain. J Food Prot 2008;71:2124–8.

[17] Cinque B, La Torre C, Lombardi F, Palumbo P, Van der Rest ME, Cifone MG. Production conditions affect the in vitro anti-tumoral effects of a high concentration multi-strain probiotic preparation. PLoS One 2016; (accepted for publication).

[18] Grześkowiak Ł, Isolauri E, Salminen S, Gueimonde M. Manufacturing process influences properties of probiotic bacteria. Br J Nutr 2011;105:887–94.

Chapter 21

The Viruses of the Gut Microbiota

M. Lecuit*,**,† and M. Eloit*,‡

*Institut Pasteur, Biology of Infection Unit, Inserm U1117, Laboratory of Pathogen Discovery, Paris, France; **Paris Descartes University, Sorbonne Paris Cité, Paris, France; †Necker-Enfants Malades University Hospital, Division of Infectious Diseases and Tropical Medicine, Paris, France; ‡Ecole Nationale Vétérinaire d'Alfort, Maisons-Alfort, France

The very rapid increase in untargeted Next Generation Sequencing usage in the field of microbiology has allowed to increase the knowledge of resident and pathogenic viruses, and characterize previously unknown or variant viruses. Technological pipelines uncover viruses referenced in databases with a threshold of sensitivity equivalent to that of quantitative PCRs, and are now becoming even more sensitive than PCR with the increasing depth of sequencing [1–4]. This type of pipeline also allows acquisition and assembly of de novo full-length genomes from biological samples, and thereby the discovery of new viruses, even when very distant from known viruses (for a recent review see Ref. [5]). These tools have begun to reveal the existence and composition of the human gut virome and unravel its intrinsic complexity, and interindividual variability in healthy individuals. Yet, the precise composition and potential impact on health of the human virome remains to be determined.

THE EUKARYOTIC VIROME, A COMPONENT OF THE GUT MICROBIOME

The gut virome is defined as the viral component of the gut microbiome, defined itself as the microbial communities of the gut. The gut virome is composed of eukaryotic viruses able to replicate in human cells, as well as bacteriophages that replicate in gut bacteria, which are the most abundant. Eukaryotic viruses also come from food intake, like plant [6] and animal [7] viruses. Systematic longitudinal studies are still lacking and it is therefore difficult to distinguish with certainty viruses that establish long-lasting infection and can be considered as members of a standard flora, from those responsible for acute infections, in particular for human eukaryotic viruses.

Regarding viruses of bacteria (bacteriophages), their presence is indeed modulated by the presence of their host bacteria and they also might regulate the bacterial contents. Recent reviews [8] have covered the human gut phage communities, showing that the population is highly individual and dominated by DNA phages exhibiting a temperate lifestyle. Nevertheless, phage can also lyze bacteria and impact relative bacterial counts [9,10]. The viral load is roughly similar to that of bacteria [11]. Phage may vector transduction (gene transfer) between strains and even bacterial species, and therefore deliver genes encoding toxins, virulence factors, or alternate metabolisms. We have focused this chapter on human eukaryotic viruses.

The history of human gut eukaryote viruses has until recently been dominated by the discovery of pathogenic viruses (among which *Enterovirus*, *Rotavirus*, *Norovirus* genus) that generally lead to transient and symptomatic infections, but some eukaryotic viruses seem resident of the human gut [12]. Most gut viruses are not cultivable and unbiased metagenomics studies have therefore contributed recently to their better characterization. The eukaryotic virome seems to be acquired progressively with age, in contrast to bacteriophage richness, which seems greatest early in life and then decreases [13]. Acquisition of these resident viruses is associated in healthy infants to no apparent underlying acute or chronic disorders [13]. Enteroviruses, parechoviruses, and sapoviruses were mostly detected. Comparison of the sequences of enterovirus and parechovirus strains from cotwins showed high identity, suggesting that twins harbor similar virome, in part linked to common exposure [13].

Later in age, persistent or intermittent shedding of resident enteric viruses from healthy people is well established. For example, human enterovirus (EV) [14] and parechovirus (HPeV) [15] are excreted by a large fraction of healthy children under the age of five. A 1-year NGS longitudinal study of the stool of two healthy infant siblings in samples taken at 1-week intervals demonstrated that viruses were continuously excreted [16]. The most frequently observed viruses, in decreasing order, were anellovirus (*TTV* but also *TTMV*), picobirnavirus, and *HPeV types 1* and *6*. Bocavirus (*HBoV-1*), Adenovirus C and F, *Aichi virus*, astroviruses, and rotaviruses were less frequently detected. Surprisingly, other enteric viruses, such as noroviruses, coronaviruses, cardioviruses, cosaviruses, saliviruses, and sapoviruses, were not detected, although they are frequently detected in stool, indicating that the results of this survey only provide a first indication of the composition of the gut virome, which would benefit from the study of larger samples. Some viruses, including adenoviruses, anelloviruses, picobirnaviruses, parechoviruses, and *Human bocavirus*, were shed for months. These viruses are more likely to represent a significant portion of the normal human virome, owing to their ability to establish persistent infections. Among these viruses, another study showed that the presence of anelloviruses and circoviruses discriminate twin pairs that include one child with severe acute malnutrition from concordant healthy pairs, but not diseased versus healthy children within a given twin pair. So, it remains unclear if anelloviruses and circoviruses are markers of, or are responsible for, disease pathogenesis [17].

INTERPLAY BETWEEN GUT VIROME AND IMMUNE SYSTEM

The interplay between the virome and the immune system is far from being fully understood. Phage and eukaryotic viral particles are translocated, and independently of replication can activate innate immunity. Physical abundance of phages over eukaryotic human viruses, nevertheless, suggests that such activation is mainly driven by phages. Detection of corresponding viral nucleic acids within cells through several pattern-recognition sensors for RNA (RIG-I and Toll-like receptors TLR7 and TLR8) and DNA (TLR9, cyclic-GMP-AMP) should lead to expression of type interferon (IFN)-α and -β with pleiotropic activities, and inflammatory cytokines, such as interleukin 1 and 6. It has been suggested that such activation of innate immunity could have a positive effect against pathogenic infections [18], as already demonstrated for systemic persistent cytomegalovirus infection [19].

On the other side, it has been shown recently that, quite surprisingly, viruses could also play a beneficial role in the control of gut inflammation. Resident viruses recognized by TLR3 and TLR7 favor intestinal homeostasis through antiinflammatory cytokines as IFN-β secreted mainly by plasmacytoid dendritic cells [20].

Immune impairment has been shown to modify the enteric virome [21,22]. The enteric eukaryotic virome expands during pathogenic SIV infection of rhesus macaques [21]. In HIV-infected patients, there is a relationship between low peripheral CD4 T cell counts and alteration of the virome [23]. Interestingly, an association was evidenced between enteric adenovirus and both advanced HIV/AIDS [23], and SIV-infected macaques [21,22]. This suggests a mechanism where adenovirus replication promotes mucosal lesions and enteropathy leading to bacterial and bacteriophage translocation that promotes chronic immune stimulation.

The extent by which immunosuppressed patients harbor pathogenic and opportunistic viruses in the gut is currently unknown. It is likely, however, that a large number of viral species coexist and persist, and may be transiently cleared to reappear later, in the absence of protective immunity. Prospective longitudinal studies aimed at characterizing at steady state in healthy and immunosuppressed individuals the dynamics of the gut virome are warranted, not only to better characterize the driving forces that shape the gut microbiome, but also to monitor and ideally predict pathogenesis associated with these viruses, and ideally prevent their transmission to other susceptible hosts.

PATHOGENIC VIRUSES OF THE GUT VIROME: ENTEROPATHOGENIC VIRUSES ASSOCIATED WITH SYSTEMIC INFECTIONS

Foodborne viral infections are associated with the presence of infectious viral particles in the gut, and enteropathogenic viruses are therefore either transiently, and in a prolonged manner, present in the gut and part of the gut virome. As for foodborne bacterial infections, which may lead to local intestinal infection, or systemic disease, viruses whose portal of entry is the gut may be associated with either local or systemic disease. Their transmissibility depends on the amount of infectious viral particles released in the feces, and the duration of the viral release. Rotavirus and noroviruses, which replicate in intestinal epithelial cells, lead to local replication in the digestive tract, and very high-level fecal shedding, which is favored by the associated diarrhea, which mechanism can result, as for bacteria, from the action of a toxin [24] or a direct enteropathogenicity, and accounts for their high transmissibility. In contrast, enteroviruses, the leading cause of meningitis, translocate at the gut level and disseminate systemically. While their cell tropism is broader, their replication in mucosal tissue and their release in the intestinal lumen lead to fecal shedding, release in the environment and transmission. The infective potential of enteric viruses depends on the species itself, microbiota, and many host factors, such as age, nutritional status, and immune functions, which are themselves dependent on the microbiota. This underlines the complex interplay between the host, and the bacterial and viral parts of the microbiome, which only begins to be deciphered.

GUT AS A MAJOR SOURCE OF NEUROTROPIC VIRUSES, WHEN PATHOGENESIS AND SHEDDING IS FAVORED BY HUMORAL IMMUNE DEFICIENCY

The composition of the gut virome may vary according to multiple factors, such as exposure to viral species, the composition of the microbiota, and host factors, such as immune functions. It is well known by clinicians in charge of immunodeficient patients that diarrhea is a frequent symptom associated with opportunistic viral infections. Patients under immunosuppressive therapy for organ transplantation may present with chronic diarrhea associated with noroviruses and sapoviruses, leading to chronic fecal shedding, and nosocomial transmission [25]. Patients with congenital agammaglobulinema or profound hypogammaglobulinemia are prone to chronic and recurrent severe enterovirus infection, associated with encephalitis. These patients are frequently persistently colonized with enteroviruses, which can persist for years in the gut, and be associated with multiple episodes of CNS disease. In these patients, the occurrence of encephalitis in the absence of detectable enterovirus by RT-PCR methods in the CSF does not rule out their presence. NGS is indeed able to detect enterovirus genome when RT-PCR is negative, because of its intrinsic sensitivity, and its ability to detect variants that may not be amplifiable by PCR (our unpublished results). Interestingly, astroviruses, which are known to induce enteritis in human have also recently been associated with encephalitis in the context of profound humoral deficiency, underlining the intestine as a reservoir for neurotropic viruses [25a,25b,25c]. Another example of virus associated with gastrointestinal infections leading to meningitis and encephalitis are parechoviruses, and in particular human *Parechovirus 3* [26].

Yet the large number of enteric viruses species, their broad host tropism, the quasispecies nature of *Picornaviridae* within a given host, their capacity to recombine, their high transmissibility, and their interactions with the host microbiota and immune functions illustrate not only the fascinating evolutionary success of enteroviruses, but also the complexity of understanding all the factors to take into account to make sense of the contribution of a single viral species to the gut virome.

AN UNCERTAIN STATUS FOR DIET-DERIVED ANIMAL VIRUSES

The human gut virome is also composed of animal viruses transmitted by the oral route by consumption of contaminated food. *Hepatitis E virus* (HEV) is typically responsible for acute hepatitis in humans. Genotypes 1 and 2 are human specific but type 3 and type 4 have a reservoir in pigs [27]. The virus can be seen as a typical zoonotic virus in the context of acute infections. Nevertheless, long-term chronic infections have been described in immunocompromised patients, highlighting the importance of the host-immune status on virome composition and its potentially pathogenic properties [28,29]. Animal genotypes 3 and 4 frequently infect humans from the animal reservoir (up to 50% of the population is antibody positive in some areas [30]), and might be adapting to humans. The situation is puzzling for gyroviruses. The first gyrovirus in humans (HGyV), which is part of the *Circoviridae* family, was initially found by NGS at the surface of the skin of healthy people [31]. It was later shown that different gyroviruses, including a very similar virus, AGV2, are very prevalent in chickens and in human stools [32,33]. This owns to cross-species transmission of gyroviruses and their replication in humans, or alternatively to passive transit of animal viruses via food intake, as has been observed for plant viruses [6]. Within the *Circoviridae* family, the cycloviruses constitute a new genus and are found in the feces of humans and other animal species, pose similar questions as gyroviruses. [34]. It is noteworthy that gyroviruses harbor an apoptin gene, which encodes a protein that has the rare property of being specifically cytotoxic for cancer cells. Indeed, natural infection could be of benefit in controlling the development of tumor cells [35], in particular colon cancer (Table 21.1).

TABLE 21.1 Eukaryotic Viruses of the Human Gut

Family	Genus	References
Picornaviridae	Enterovirus	[36]
	Parechovirus	[36]
	Cardiovirus	[37]
	Salivirus	[38]
Picobirnaviridae	Picobirnavirus	[16]
Astroviridae	Astrovirus	[39]
Reoviridae	Rotavirus	
Caliciviridae	Norovirus	[40]
	Sapovirus	[41]
Adenoviridae	Mastadenovirus C and F, and others	[42]
Anelloviridae	Anellovirus	[16]
Cycloviridae	Circovirus	[43]
	Cyclovirus	[43]
Parvoviridae	Bocavirus	[44,45]

Recent reviews or informative papers are cited when available.

REFERENCES

[1] Mee ET, Preston MD, Minor PD, Schepelmann S. CS533 Study Participants. Development of a candidate reference material for adventitious virus detection in vaccine and biologicals manufacturing by deep sequencing. Vaccine 2016;34:2035–43.

[2] Cheval J, Sauvage V, Frangeul L, et al. Evaluation of high-throughput sequencing for identifying known and unknown viruses in biological samples. J Clin Microbiol 2011;49:3268–75.

[3] Wylie KM, Mihindukulasuriya KA, Sodergren E, Weinstock GM, Storch GA. Sequence analysis of the human virome in febrile and afebrile children. PLoS ONE 2012;7:e27735.

[4] Greninger AL, Chen EC, Sittler T, et al. A metagenomic analysis of pandemic influenza A (2009 H1N1) infection in patients from North America. PLoS ONE 2010;5:e13381.

[5] Chiu CY. Viral pathogen discovery. Curr Opin Microbiol 2013;16(4):468–78.

[6] Zhang T, Breitbart M, Lee WH, et al. RNA viral community in human feces: prevalence of plant pathogenic viruses. PLoS Biol 2006;4:e3.

[7] Gia Phan T, Phung Vo N, Sdiri-Loulizi K, et al. Divergent gyroviruses in the feces of Tunisian children. Virology 2013;446:346–8.

[8] Ogilvie LA, Jones BV. The human gut virome: a multifaceted majority. Front Microbiol 2015;6:918.

[9] Abeles SR, Robles-Sikisaka R, Ly M, et al. Human oral viruses are personal, persistent, gender-consistent. ISMEJ 2014;8:1753–67.

[10] Duerkop BA, Clements CV, Rollins D, Rodrigues JLM, Hooper LV. A composite bacteriophage alters colonization by an intestinal commensal bacterium. Proc Natl Acad Sci USA 2012;109:17621–6.

[11] Minot S, Sinha R, Chen J, et al. The human gut virome: interindividual variation and dynamic response to diet. Genome Res 2011;21:1616–25.

[12] Reyes A, Haynes M, Hanson N, et al. Viruses in the faecal microbiota of monozygotic twins and their mothers. Nature 2010;466:334–8.

[13] Lim ES, Zhou Y, Zhao G, et al. Early life dynamics of the human gut virome and bacterial microbiome in infants. Nat Med 2015;21:1228–34.

[14] Witsø E, Palacios G, Cinek O, et al. High prevalence of human enterovirus a infections in natural circulation of human enteroviruses. J Clin Microbiol 2006;44:4095–100.

[15] Kolehmainen P, Oikarinen S, Koskiniemi M, et al. Human parechoviruses are frequently detected in stool of healthy Finnish children. J Clin Virol 2012;54:156–61.

[16] Kapusinszky B, Minor P, Delwart E. Nearly constant shedding of diverse enteric viruses by two healthy infants. J Clin Microbiol 2012;50:3427–34.

[17] Reyes A, Blanton LV, Cao S, et al. Gut DNA viromes of Malawian twins discordant for severe acute malnutrition. Proc Natl Acad Sci USA 2015;112:11941–6.

[18] Duerkop BA, Hooper LV. Resident viruses and their interactions with the immune system. Nat Immunol 2013;14:654–9.

[19] Barton ES, White DW, Cathelyn JS, et al. Herpesvirus latency confers symbiotic protection from bacterial infection. Nature 2007;447:326–9.

[20] Yang JY, Kim MS, Kim E, et al. Enteric viruses ameliorate gut inflammation via Toll-like receptor 3 and Toll-like receptor 7-mediated interferon-β production. Immunity 2016;44:889–900.

[21] Handley SA, Thackray LB, Zhao G, et al. Pathogenic simian immunodeficiency virus infection is associated with expansion of the enteric virome. Cell 2012;151:253–66.

[22] Handley SA, Desai C, Zhao G, et al. SIV infection-mediated changes in gastrointestinal bacterial microbiome and virome are associated with immunodeficiency and prevented by vaccination. Cell Host Microbe 2016;19:323–35.

[23] Monaco CL, Gootenberg DV, Zhao G, et al. Altered virome and bacterial microbiome in human immunodeficiency virus-associated acquired immunodeficiency syndrome. Cell Host Microbe 2016;19:311–22.

[24] Lorrot M, Vasseur M. How do the rotavirus NSP4 and bacterial enterotoxins lead differently to diarrhea? Virol J 2007;4:31.

[25] Roos-Weil D, Ambert-Balay K, Lantemier F, et al. Impact of *Norovirus/Sapovirus*-related diarrhea in renal transplant recipients hospitalized for diarrhea. Transplantation 2011;92:61–9.

[25a] Frémond ML, Pérot P, Muth E, Cros G, Dumarest M, Mahlaoui N, Seilhean D, Desguerre I, Hébert C, Corre-Catelin N, Neven B, Lecuit M, Blanche S, Picard C, Eloit M. Next-generation sequencing for diagnosis and tailored therapy: a case report of astrovirus-associated progressive encephalitis. J Pediatric Infect Dis Soc 2015;4(3):12.

[25b] Brown JR, Morfopoulou S, Hubb J, Emmett WA, Ip W, Shah D, Brooks T, Paine SM, Anderson G, Virasami A, Tong CY, Clark DA, Plagnol V, Jacques TS, Qasim W, Hubank M, Breuer J. Astrovirus VA1/HMO-C: an increasingly recognized neurotropic pathogen in immunocompromised patients. Clin Infect Dis 2015;60(6):881–8.

[25c] Naccache SN, Peggs KS, Mattes FM, Phadke R, Garson JA, Grant P, Samayoa E, Federman S, Miller S, Lunn MP, Gant V, Chiu CY. Diagnosis of neuroinvasive astrovirus infection in an immunocompromised adult with encephalitis by unbiased next-generation sequencing. Clin Infect Dis 2015;60(6):919–23.

[26] Renaud C, Harrison CJ. Human *Parechovirus* 3: the most common viral cause of meningoencephalitis in young infants. Infect Dis Clin North Am 2015;29:415–28.

[27] Smith DB, Purdy MA, Simmonds P. Genetic variability and the classification of hepatitis E virus. J Virol 2013;87(8):4161–9.

[28] Lhomme S, Abravanel F, Dubois M, et al. Hepatitis E virus quasispecies and the outcome of acute hepatitis E in solid-organ transplant patients. J Virol 2012;86:10006–14.

[29] Koning L, Pas SD, de Man RA, et al. Clinical implications of chronic hepatitis E virus infection in heart transplant recipients. J. Heart Lung Transplant 2013;32:78–85.

[30] Kamar N, Bendall R, Legrand-Abravanel F, et al. Hepatitis E. Lancet 2012;379:2477–88.

[31] Sauvage V, Cheval J, Foulongne V, et al. Identification of the first human gyrovirus, a virus related to chicken anemia virus. J Virol 2011;85:7948–50.

[32] Phan TG, Li L, O'Ryan MG, et al. A third gyrovirus species in human faeces. J Gen Virol 2012;93:1356–61.

[33] Chu DK, Poon LL, Chiu SS, et al. Characterization of a novel gyrovirus in human stool and chicken meat. J Clin Virol 2012;55:209–13.

[34] Delwart E, Li L. Rapidly expanding genetic diversity and host range of the Circoviridae viral family and other Rep encoding small circular ssDNA genomes. Virus Res 2012;164:114–21.

[35] Los M, Panigrahi, Rashedi I, et al. Apoptin, a tumor-selective killer. Biochim Biophys Acta 2009;1793:1335–42.

[36] de Crom SCM, Rossen JWA, van Furth AM, Obihara CC. Enterovirus and parechovirus infection in children: a brief overview. Eur J Pediatr 2016;175(8):1023–9.

[37] Himeda T, Ohara Y. Saffold virus, a novel human Cardiovirus with unknown pathogenicity. J Virol 2012;86:1292–6.

[38] Yip CC, Lo KL, Que TL, et al. Epidemiology of human parechovirus, Aichi virus and salivirus in fecal samples from hospitalized children with gastroenteritis in Hong Kong. Virol J 2014;11:182.

[39] Kapoor A, Li L, Victoria J, et al. Multiple novel astrovirus species in human stool. J Gen Virol 2009;90:2965–72.

[40] Ahmed SM, Hall AJ, Robinson AE, et al. Global prevalence of *Norovirus* in cases of gastroenteritis: a systematic review and meta-analysis. Lancet Infect Dis 2014;14:725–30.

[41] Koopmans M, Vinj J, Duizer E, de Wit M, van Duijnhoven Y. Molecular epidemiology of human enteric caliciviruses in The Netherlands. Novartis Found Symp 2001;238:197–214.

[42] Kosulin K, Geiger E, Vécsei A, et al. Persistence and reactivation of human adenoviruses in the gastrointestinal tract. Clin Microbiol Infect 2016;22. 381.e1-388.e1.

[43] Li L, Kapoor A, Slikas B, et al. Multiple diverse circoviruses infect farm animals and are commonly found in human and chimpanzee feces. J Virol 2012;84:1674–82.

[44] Ong DSY, Schuurman R, Heikens E. Human bocavirus in stool: a true pathogen or an innocent bystander? J Clin Virol 2016;74:45–9.

[45] Paloniemi M, Lappalainen S, Salminen M, et al. Human bocaviruses are commonly found in stools of hospitalized children without causal association to acute gastroenteritis. Eur J Pediatr 2014;173:1051–7.

Food Substrates Important to the Microbiota

Chapter 22

Dietary Fiber, Soluble and Insoluble, Carbohydrates, Fructose, and Lipids

J.M.W. Wong*,**, E.M. Comelli[†], C.W.C. Kendall[‡,§,¶], J.L. Sievenpiper[††,‡‡], J.C. Noronha[§§,***] and D.J.A. Jenkins[†††,‡‡‡]

*Clinical Nutrition & Risk Factor Modification Center and Li Ka Shing Knowledge Institute, St. Michael's Hospital, Toronto, ON, Canada; **New Balance Obesity Foundation Obesity Prevention Center, Boston Children's Hospital, Boston, MA, United States; [†]Department of Nutritional Sciences and Center for Child Nutrition and Health, University of Toronto, Toronto, ON, Canada; [‡]Department of Nutritional Sciences, University of Toronto, Toronto, ON, Canada; [§]Clinical Nutrition & Risk Factor Modification Center, St. Michael's Hospital, Toronto, ON, Canada; [¶]College of Pharmacy and Nutrition, University of Saskatchewan, Saskatoon, SK, Canada; [††]Department of Nutritional Sciences, University of Toronto, Toronto, ON, Canada; [‡‡]Clinical Nutrition & Risk Factor Modification Center, Li Ka Shing Knowledge Institute, and Division of Endocrinology and Metabolism, St. Michael's Hospital, Toronto, ON, Canada; [§§]Department of Nutritional Sciences, University of Toronto, Toronto, ON, Canada; [***]Clinical Nutrition & Risk Factor Modification Center, St. Michael's Hospital, Toronto, ON, Canada; [†††]Departments of Nutritional Sciences and Medicine, Faculty of Medicine, University of Toronto, Toronto, ON, Canada; [‡‡‡]Clinical Nutrition & Risk Factor Modification Center, Li Ka Shing Knowledge Institute, and Division of Endocrinology and Metabolism, St. Michael's Hospital, Toronto, ON, Canada

INTRODUCTION

Diets that emphasize plant-based foods that are naturally high in fiber, such as fruits, vegetables, and whole grains significantly reduce the risk of chronic diseases including coronary heart disease (CHD), type 2 diabetes (T2DM), obesity, and certain types of cancer [1–6]. Moreover, such diets supply a large energy source, as fermentable substrate, to the gut microbiota which not only is necessary for its survival, but more importantly, can influence the health of the host through various end products and modulation of the gut microbial profile. Recent advances in sequencing techniques for characterizing the human gut microbiota have provided insights into how dietary components, such as dietary fiber, might influence the trillions of microbes in the human gut and consequently, chronic disease risk. The purpose of this chapter is to provide an overview of the link between high fiber diets, human health, and the gut microbiota.

HIGH FIBER PLANT-BASED DIETS AND CHRONIC DISEASE RISK

Many dietary approaches have been associated with a reduction in the risk of chronic diseases. A commonality between these dietary approaches is the emphasis on plant-based foods that are naturally high in fiber with minimal processing, such as fruits, vegetables, legumes, and whole grains. A "prudent" diet containing higher intakes of these foods was associated with reduced risk of CHD [1–3], T2DM [4,5], and all-cause mortality [3]. Whereas, a "Western" diet with lower intakes was associated with increased risk of chronic diseases and mortality [1–5]. Furthermore, in an analysis of three large prospective cohort studies, increased consumption of foods, such as vegetables, nuts, whole grains, and fruits, were inversely associated with weight gain [6].

Cohort studies of vegetarians and vegans provide further evidence of the benefits of diets that emphasize plant-based foods naturally high in fiber. A pooled analysis of five prospective cohort studies of vegetarians and nonvegetarians observed a 24% reduction in mortality from CHD in vegetarians compared to nonvegetarians (death rate ratio: 0.76; 95% CI: 0.62, 0.94) [7]. Furthermore, vegetarian diets were associated with an increase in longevity and decreases in all-cause mortality, T2DM, and obesity [8–10]. Longer-term clinical trials of individuals on vegan or vegetarian diets support the findings of cohort studies where such diets have been shown to reduce the progression of CHD [11] and improve glycemic control [12]. Shorter-term clinical trials focused on the reduction of CHD risk factors also support the results from epidemiological and longer-term clinical studies. Diets rich in fruits and vegetables, such as the DASH and OMNIHeart diets, have been shown to improve blood pressure [13] and blood lipids [14,15]. A metabolically controlled feeding study of a dietary portfolio approach (based on plant-based foods, viscous fiber foods, oats, barley, plant sterols, soy, and nuts) showed reductions in low-density lipoprotein cholesterol (LDL-C) and C-reactive protein (CRP) of 29 and 28%, respectively [16]. The effect was equivalent to a first generation statin (the first line pharmacological agent for cholesterol reduction) provided in conjunction with a very low saturated fat diet where LDL-C reduced by 31% and CRP by 33% [16]. The observed beneficial effects are likely attributed to the higher fiber content of the plant-based foods, as well as the presence of an array of bioactive components.

Strong evidence exists from both epidemiological and clinical studies supporting diets that are comprised of plant-based foods naturally higher in fiber for reducing chronic disease burden. Despite the evidence, national survey data from NHANES showed that current intake levels of fruits, vegetables, and legumes do not meet recommended servings based on national dietary guidelines [17]. Although more than 50% of Americans meet the recommended servings of total grains, almost all do not meet the whole grain serving recommendations, which indicates that the intake of grains is primarily from refined grains that are low in dietary fiber. Indeed, the estimated average intake of dietary fiber is ~16 g/day [18], far below the recommendation of 25–38 g/day or 14 g/1000 kcal [19]. This low intake level of dietary fiber significantly reduces the amount of fermentable substrate available for the gut microbiota for use as metabolic fuel, and in turn, influencing chronic disease risk.

DIETARY FIBER AND UNDIGESTED CARBOHYDRATES

High fiber diets provide an abundant supply of fermentable substrate for the gut microbiota that, through direct and indirect effects, can modulate the systemic health of its host. In addition to dietary fiber, some of the "available carbohydrate" (i.e., "available" for small intestinal digestion and absorption) in many foods may also escape digestion and contribute to the supply of fermentable substrate [20]. Carbohydrate losses were assessed with different types of foods that vary in fiber and available carbohydrate in ileostomates. The authors found that consumption of foods in which the whole grain structure was preserved resulted in greater carbohydrate losses (e.g., lentils and other legumes, β-glucan-containing cereals—oat bran and barley, and pumpernickel bread) [21,22]. Other studies have also shown that available carbohydrate of many starchy foods are incompletely digested and absorbed in the small bowel [23,24]. It is suggested that the dietary fiber content of foods tends to determine the amount of available carbohydrate that may be available as fermentable substrate for the gut microbiota. This is based on a positive association

between carbohydrate malabsorption and the fiber content of the food [22,25]. Furthermore, glycemic index of foods, a quantitative classification of carbohydrate foods based on the rate of carbohydrate absorption as reflected in the glycemic response [26,27], was inversely associated with losses in the ileal effluent reflective of the slower rate of absorption of low glycemic index foods [22]. Many other factors exist that may influence the degree of carbohydrate losses including the rate of digestion [26,28], the food form (physical form, particle size) [29], type of preparation (cooking method and processing) [29–32], type of starch (amylose and amylopectin) [29,33], presence of antinutrients (such as α-amylase inhibitors) [34,35], and transit time [36].

Resistant starches are another major source of fermentable substrate [37,38] where it is estimated that up to 20% of dietary starch is not absorbed in the small bowel [21–23,39,40]. Dietary fiber, unabsorbed sugars, raffinose, along with polydextrose and modified cellulose also represent significant sources of fermentable substrate in the colon [41]. Insoluble fibers (e.g., lignins, cellulose, and some hemicelluloses) are important dietary components that play a significant role in fecal bulking; although these fibers are resistant to fermentation by the gut microbiota, they may carry with them fermentable carbohydrate substrate, including starches and sugars [42]. Soluble fibers (e.g., pectins, gums, mucilages, and some hemicelluloses) are more completely fermented by the gut microbiota with a lesser effect on increasing fecal bulk. Most fiber-containing foods contain about one-third soluble and two-thirds insoluble fiber [37]. Quantitatively, studies have demonstrated that pectins, hemicelluloses, and gums are fermented, whereas dietary fiber components like cellulose or wheat bran are fermented to a far lesser degree [43,44].

A prebiotic is defined as a nonviable food component that confers a health benefit on the host associated with modulation of the microbiota [45]. Polyfructans, such as inulin-type fructans and oligofructose, escape digestion and directly or indirectly (through cross-feeding mechanisms) stimulate the growth of specific gut microbiota. These species include, but not limited to [46], *Bifidobacterium* species [47], and in some cases *Lactobacillus* and *Faecalibacterium prausnitzii* [48], where each has been shown to have metabolic health benefits [49,50]. Inulin-type fructans are oligo- or polymers of D-fructose joined by β-(2-1) bonds with an α-(1-2) linked D-glucose at the terminal end. Oligofructose are sugar chains with degrees of polymerization from 3 to 10, whereas inulin has degrees of polymerization from 10 to 65 [51]. Studies of patients with ileostomies have shown that 88% of inulin and 89% of oligofructose are recovered intact and unhydrolyzed in the effluent [52,53] and are not recovered in the feces, suggesting that they are completely fermented in the colon [54,55]. Increased breath hydrogen excretion (an indirect marker of colonic fermentation) with intake of oligofructose and inulin supports this observation [54,56,57]. Both studies using in vitro fermentation systems of human fecal microbiota [58–60] and clinical studies (varying in dose, substrate, duration, and subject population) [56,61–67] have confirmed the findings that support increased fermentation and selective stimulation of microbial growth, specifically *Bifidobacterium* species, with intake of prebiotics. It has been suggested that prebiotic intake of 5–20 g/day is sufficient to induce a significant increase in colonic microbiota [43,68,69]. These observations have led to growing interest and research on their potential effects on colonic and systemic health. In this respect, it should be mentioned that with advancing age, the proportion of Bifidobacteria decline. It may be that dietary or supplemental use of prebiotics can be used to rejuvenate the colonic microbiome.

DIETARY FIBER AND CARBOHYDRATE FERMENTATION

Anaerobic breakdown of organic matter by gut microbial fermentation involves a number of metabolic reactions that result in metabolizable energy for microbial growth and maintenance, as well as other metabolic end products available for use by its host [70]. Major end products of gut microbial fermentation of undigested or nondigestible carbohydrates are short chain fatty acids (SCFAs) including formic, acetic, butyric, and propionic acids; gases (CO_2, CH_4, and H_2); organic acids (succinic, lactic, etc.); heat; and bacterial cell mass [71]. Data from population surveys showed that fecal SCFA production is generally in the order of acetate > propionate ≥ butyrate [72] in a molar ratio of approximately 60:20:20, respectively [73]. The ratio seems to remain fairly constant [37], although changes in dietary intake may lead to alterations in the production and absorption. The production of SCFAs not only influence colonic function through their uptake and metabolism by colonocytes, but more importantly, they may significantly affect metabolic processes of other tissues of the host. Increases in SCFA production have been associated with decreased pH which may reduce potential pathogenic bacteria, decreased solubility of bile acids, increased absorption of minerals (indirectly), and reduced ammonia absorption by the protonic dissociation of ammonia to ammonium (NH_4) and other amines [37,43,74–76]. Moreover, increased butyrate has been suggested to reduce the risk of colon cancer and improve inflammatory bowel disease [77–79]. Butyrate is the preferred fuel of colonic epithelial cells where up to 70–90% of butyrate is metabolized by colonocytes [80], and butyrate plays a role in the regulation of epithelial cell proliferation and differentiation [43,72,77,81]. Studies have shown that *Ruminococcus bromii* of the Firmicutes phylum is the key primary degrader (keystone species) of resistant starch in the human colon [82,83]. Furthermore, *F. prausnitzii*, an abundance key component of the gut microbiota of the Firmicutes phylum, is also a major butyrate producer [84].

Colonic fermentation reactions are highly dependent on the chemical composition of the substrate source. For example, fermentation of starches primarily yields acetate and butyrate, whereas fermentation of pectin and xylan mainly yields acetate [85]. Resistant starches have consistently been shown to raise fecal butyrate levels [42,86–88]. Intake of wheat bran compared to vegetable fiber resulted in greater fecal excretion of butyrate and propionate [89]. Partially hydrolyzed guar gum yielded greater fecal excretion of all three major SCFAs, but did not change the concentration of propionate and butyrate, or decrease their relative contribution [90]. Studies of acute intake of a nondigestible monosaccharide, L-rhamnose (25 g), selectively increased serum propionate without increasing acetate [91], and this affect did not diminish after 28 days [92]. Polyfructans that selectively stimulate the growth of *Bifidobacterium* species have also been associated with a greater production of propionate, whereby decreasing the acetate:propionate ratio and reducing serum cholesterol levels [93–95]. The propionate-raising properties of polyfructans may offset the acetate-increasing effect of certain viscous fibers. Therefore, their combined intake may result in a synergistic effect, favoring a greater reduction in serum lipids and cardiovascular disease risk. The acute effect of rectal infusion of propionate is to reduce the acetate-induced rise in LDL-C [96], but L-rhamnose ingestion to raise propionate was not associated with a fall in LDL-C in the longer term [92].

GUT MICROBIOTA AND DIET LINK

The evolution of our ancestors likely evolved in parallel with the diet to form the current gut microbiota profile of modern humans. While we do not know for certain the diets of our remote ancestors, evidence from our contemporary great ape cousins may provide some insight into human dietary evolution. The diets of great apes comprise large volumes of plant foods, likely reflective of the diets of early ancestors of man [97]. Such diets consist of large amounts of leafy vegetables, stems and shoots, seeds, nuts, and fruits [97–100], providing very high levels of nondigestible fiber and undigested carbohydrates that serve as fermentable substrate for the gut microbiota. The genetic difference of great apes from humans is approximately 2–3%, suggesting that little has changed in our genome since the time it was adapted to plant-based foods [101,102] and our gut microbiota are likely best suited for such diets. As such, inclusion of plant-based foods, naturally high in fiber, plays a crucial nutritional role in human health and its associated gut microbiota. However, modern human diets are low in plant-based foods that are rich in fiber making them radically different from the diets of our ancestors, which in turn, result in low levels of fermentable substrate. The mismatch of our archaic genome and gut microbiota with our modern diets may provide some explanation to the rise in chronic diseases in Western nations over time.

Clinical studies have been conducted to compare the lipid-lowering effect of diets of the early phases of human evolution, from 5 to 7 million years ago, made up of leafy vegetables, nuts, and fruits (Simian) and an early agricultural diet of 10,000 years ago made up of starchy foods (Neolithic), particularly from cereals and legumes. LDL-C was reduced by 33% on the Simian diet and 23% on the Neolithic diet, in contrast to a reduction of 7% on a modern therapeutic diet for cholesterol reduction [103]. The cholesterol reductions on the Simian diet matched those of first generation statins. The Simian diet consisted of 55 g/1000 kcal of dietary fiber, 1 g of plant sterols, and 123 g of vegetable protein derived from leafy vegetables, fruits, and nuts (~67 g of nuts) per day. The high fiber content of the Simian diet is significantly higher than the currently recommended intake level of 14–18 g/1000 kcal per day [19] and even much higher than those found on a typical Western diet, thereby providing a large pool of fermentable substrate (metabolic fuel) for the gut microbiota. Key components of the Simian diet incorporated into a modern therapeutic diet for cholesterol reduction demonstrated a ~30% LDL-C reduction in individuals with hyperlipidemia [16], as well as a significant reduction in CRP (marker of inflammation). Such components taken together form a dietary portfolio that included viscous fibers in the form of oats, barley, and psyllium; vegetable proteins as soy and other legumes, nuts (e.g., almonds); and plant sterols as an enriched margarine, all of which currently bear a United States Food and Drug Administration health claim for CHD risk reduction through cholesterol lowering [104–108]. Therefore, diets largely based on plant-based foods, naturally high in fiber, and in turn, fermentable substrate, are effective dietary approaches for reducing CHD risk and altering colonic microbiota.

PLANT-BASED DIETS AND GUT MICROBIOTA

Analyses of fecal microbial communities of 106 individual mammals showed a clustering by diet pattern—herbivore, omnivore, and carnivore [109]. Samples from "healthy" humans showed a clustering with other omnivores with respect to microbial composition [109]. The microbial diversity of the gut microbiota is defined as the number and abundance distribution of distinct types of organisms [110]. The phylogenic diversity of diets followed a stepwise pattern of herbivore > omnivore > carnivore [109]. Lower diversity has not only been observed in those on meat-based diets, but it has also been observed in obesity and inflammatory bowel disease [109,111,112]. In contrast, evidence has shown that there is a lesser degree of functional diversity than microbial diversity, suggesting that functional redundancy exists (i.e., a number of microbes exist that perform similar functions)

[110,113,114]. It has been hypothesized that greater microbial diversity ensures that key metabolic functions are unaffected by changes in microbial composition under conditions of stress or change. In the presence of greater microbial diversity, different microbes with similar functions can substitute for one another potentially leading to a host that is more resilient to insults to the gut microbiota. This suggests that the metabolic functions carried out by the gut microbiota play a more significant role in the health of its host than the presence of the specific microbes. As such, a plant-based diet, high in fiber and fermentable substrate, leads to a more resilient or adaptable gut microbiota and may be more advantageous for preserving the health of the host.

The clustering of the gut microbiota has led to the concept of classifying individuals by enterotypes, which is based on the notion that the gut microbiota congregates in a relatively stable composition over time [115]. An individual's enterotype may then be used to predict the response to a specific intervention and by extension, personalized nutrition tailored to one's gut microbial profile. Indeed, analyses of the gut microbiota across different populations demonstrated the emergence of distinct enterotypes based on the relative abundance of the specific genera: *Bacteroides*, *Prevotella*, and *Ruminococcus* [115]; whereas, other studies have shown that the enterotypes are driven primarily by the ratio of the two dominant genera: *Bacteroides* and *Prevotella* [113,116,117]. The ratio of *Bacteroides* and *Prevotella* appear to be influenced by diet patterns [113,116–118], where the *Prevotella* enterotype was associated with higher carbohydrate and simple sugar and the *Bacteroides* enterotype with animal protein, a variety of amino acids, and saturated fats [116]. Results from other studies also support the association between these two enterotypes and diet patterns [118,119]. However, the existence of distinct enterotypes has been challenged. It appears that a continuum or gradient of species functionality (gradient concept) exists rather than distinct groups [120]. Further prospective studies are needed to clarify the relationship between gut microbiota variations and health outcomes.

Gut microbial communities were characterized in 531 individuals (151 families) of various populations from diverse geographical regions and cultural practices and found that one of the major factors influencing the gut microbiota was geographical or cultural traditions (Table 22.1) [117]. A distinct separation in the phylogenic composition of the fecal microbiota was found between the gut communities of those from Malawi and Amerindians in comparison to those from the United States [117]. Those from Malawi and Amerindians had a great abundance of *Prevotella*, characteristic of diets high in plant-based foods (corn and cassava). In contrast, those from the United States had a greater abundance of *Bacteroides*, characteristic of diets that are richer in protein (mostly animal protein) [117]. Similarly, analysis of the fecal microbiota of 249 self-reported vegetarians or

TABLE 22.1 Diet Patterns and Fecal Microbial Profiles

Population	Gut Microbial Profile	Diets
Malawi and Amerindians[a]	Greater abundance of *Prevotella*	High in plant-based foods (e.g., corn and cassava)
Rural Africa[b]	Greater abundance of *Prevotella* and *Xylanibacter*	High fiber-rich plant-based foods mainly from cereals, legumes, and vegetables
United States[a]	Greater abundance of *Bacteroides*	Rich in protein (mostly animal protein)
Western Europe[b]	Greater abundance of *Bacteroides*	Low in fiber and high in animal protein, sugar, and fat

[a]Yatsunenko T, Rey FE, Manary MJ, et al. Human gut microbiome viewed across age and geography. Nature 2012;486(7402):222–7.
[b]De Filippo C, Cavalieri D, Di Paola M, et al. Impact of diet in shaping gut microbiota revealed by a comparative study in children from Europe and rural Africa. Proc Natl Acad Sci USA 2010;107(33):14691–6.

vegans showed a significantly lower presence of *Bacteroides* compared to matched controls following an omnivore diet [119]. Moreover, the traditional diets of children living in rural Africa had a higher abundance of *Prevotella* and *Xylanibacter*, whereas a typical Western diet of children living in Western Europe had a higher abundance of *Bacteroides* [118]. The traditional rural African diet emphasized high fiber plant-based foods mainly from cereals, legumes, and vegetables, where higher presence of substrates, such as cellulose and xylan can be metabolized by the increased abundance of the genera *Prevotella* and *Xylanibacter* (both of the Bacteroidetes phylum). By contrast, the typical Western diet is low in fiber and high in animal protein, sugar, and fat [118]. These differences suggest that microbial profiles, to a certain degree, are reflective of the substrate availability (food environment). Fecal analyses revealed that children in rural Africa had significantly higher levels of SCFAs, primarily from propionic and butyric acids, as well as increased microbial diversity suggesting the influence of a rich supply of fermentable substrate from their traditional diet compared to children in Western Europe. It is clear that diets emphasizing plant-based foods, naturally high in fiber, play a significant role in forming the gut microbial communities. However, further investigation is needed in understanding whether these enterotypes can predict and influence the health outcomes of its host (i.e., causal relationship or a marker of diet pattern and disease).

GUT MICROBIOTA AND DIET-RELATED CHRONIC DISEASES

Recent interest in the modulation of the gut microbiota has gained momentum in part due to the observation that the human gut microbial communities may play a much greater role in human health than previously thought, especially their link to obesity. Animal and human data indicate that obesity is associated with a shift in the relative abundance of the two main phyla: Bacteroidetes and Firmicutes. Relative to lean mice, genetically obese mice have significantly less Bacteroidetes and greater Firmicutes [121] and have a

greater capacity to harvest energy from the diet (assessed by bomb calorimetry of fecal matter) [122]. In contrast, germ-free mice are protected against diet-induced obesity. Proposed mechanisms of action include AMP-activated protein kinase and fasting-induced adipose factor–associated pathways that result in increased fatty acid metabolism [123,124]. Studies of cecal microbiota transplanted from mice with either lean or obese phenotypical traits into germ-free mice showed that these traits were transmissible [122]. The gut microbiota from a mouse model of gastric bypass surgery was transferred to gnotobiotic mice and resulted in weight loss and decreased fat mass [125]. The authors suggested that the observed response was a consequence of altered microbial SCFA production. Successful transplantation of human fresh and frozen fecal microbial communities into germ-free mice has also been demonstrated. The resulting microbial profile in the humanized gnotobiotic mice closely resembled that of the host [126]. Such studies that utilize the gnotobiotic mouse model need to be confirmed by human studies. Also, the stability of the microbiota over time requires demonstration since obesity reduction and body weight stability are long-term problems. It is also not always clear whether changes in the microbiota determine body weight or whether it is the changes in caloric intake and associated change in body weight that determine nature of the microbiota.

Human studies support the findings of differences in the abundance of Bacteroidetes and Firmicutes observed in animal studies. In a year-long weight loss study of 12 obese individuals who were randomly assigned to either a fat- or carbohydrate-restricted low-calorie diet showed that both groups were associated with an increase in the abundance of Bacteroidetes and a decrease in the abundance of Firmicutes, irrespective of diet [127]. The increase in the abundance of Bacteroidetes was correlated with the percentage loss of body weight and not with the calorie content of the diet. The difference in the abundance of Bacteroidetes and Firmicutes was also observed in children of rural Africa following a traditional higher fiber plant-based diet compared to children in Western Europe following a lower fiber typical Western diet [118]. A Western-type diet fed to gnotobiotic mice has been shown to modulate the gut microbial profile by inducing the obesity phenotype in comparison to a low-fat plant polysaccharide-rich diet where there were no significant differences in chow consumption or initial body weight or body fat [126,128]. It would be of interest to determine if the effects would be reversed in a crossover design where the animals were fed both diets. Wolever et al. found that overweight and obese individuals not only had a higher relative abundance of Firmicutes and Firmicutes:Bacteroidetes ratio, but the differences in their microbial profile resulted in a greater efficiency in fermenting substrates and production of colonic SCFAs (not SCFA absorption) than lean individuals [129]. Obese individuals also have decreased phylogenic diversity compared to lean individuals [111], which may reduce their resilience to metabolic stress or change with respect to preserving functional diversity. Furthermore, if a higher abundance of Firmicutes relative to Bacteroidetes is the reason for a lower body weight and fiber determines this microbial profile, then a diet based on plant foods that are naturally high in fiber may be a therapeutic strategy for preventing and treating obesity and its associated metabolic dysfunctions.

TMAO

Recent studies have hypothesized that the gut microbiota may play a role in the pathogenesis of atherosclerosis. Trimethylamine (TMA)-containing compounds (e.g., choline, phosphatidylcholine, and carnitine) are metabolized by the gut microbiota to TMA, which is then converted by hepatic flavin-containing monooxygenases to trimethylamine-N-oxide (TMAO), a proatherogenic metabolite [130,131]. During a 3-year follow-up period in a cohort of over 400 individuals who underwent elective diagnostic cardiac catheterization, those who had major cardiovascular events (i.e., death, myocardial infarction, stroke) had higher baseline TMAO levels than those who had no cardiovascular events, even after adjusting for traditional risk factors [132]. Furthermore, those with the highest quartile of TMAO levels were more than twice as likely to have a major cardiovascular event compared to those in the lowest quartile (HR: 2.54; 95% CI: 1.96, 3.28). The proposed mechanisms of action of TMAO and the pathogenesis of atherosclerosis include the suppression of reverse cholesterol transport [131] and upregulation of proatherogenic scavenger receptors [130]. Key dietary sources of phosphatidylcholine include many animal-based foods, such as liver, eggs, and red meat [130,132]. Evidence from both animal and human studies support the formation of TMAO by gut microbial biotransformation [130–132]. Long-term diet patterns appear to influence the production of TMAO, likely a reflection of the gut microbial profile. In comparison to omnivores, individuals consuming vegetarian or vegan diets had lower fasting levels of

TMAO and produced significantly less TMAO after a carnitine challenge [131]. Upon further classification based on enterotypes, those with a relative abundance of *Prevotella* had higher plasma TMAO than those with a relative abundance of *Bacteroides*. This observation contrasts previous studies showing that diets rich in animal protein have higher relative abundance of *Bacteroides* [116,118,119]. The observed difference cautions against classifications at the genus level which may not reflect differences in species within these genera which may be responsible for the formation of TMAO. Furthermore, some Archaea from the Methanomassiliicoccales can metabolize TMA to methane and in turn, deplete the gut TMA concentration for conversion to TMAO [133]. Their presence or absence also needs to be assessed when examining the association between the gut microbial profile and TMAO production.

Type 2 Diabetes

The gut microbiota have also been suggested to play a role in T2DM. In a small study, those with T2DM had a lower abundance of Firmicutes and class Clostridia, and a nonsignificant increase in Bacteroidetes and Proteobacteria compared to those without T2DM [134]. Furthermore, the Bacteroidetes to Firmicutes ratio was positively associated with plasma glucose concentrations. A larger metagenome-wide association study of Chinese individuals with T2DM reported a decrease in a number of butyrate-producing bacteria and an increase in a number of opportunistic pathogens relative to controls [112]. A similar observation was seen in a European cohort with T2DM [135]. Concentrations of *F. prausnitzii*, a major butyrate producer, have been observed to be lower in those with T2DM [48]. Bariatric surgery has been shown to result in shifts in the gut microbiota that may contribute to the observed remission of diabetes and reduction in adiposity [136–138]. Fecal transplantation of intestinal microbiota from lean donors to recipients with metabolic syndrome resulted in an increase in insulin sensitivity, as well as butyrate-producing bacteria [139]. Obese individuals with CHD who were randomized to follow either a Mediterranean diet or a low-fat, high-complex carbohydrate diet had changes in their gut microbial profiles that although different, both favored improvements in insulin sensitivity [140]. The association between diets higher in plant-based foods (rich in dietary fiber and fermentable substrate) and the reduction in T2DM risk may be partially explained by the shifts in the gut microbiota through dietary intake that favor disease risk reduction.

ANTIBIOTICS, BLOOD LIPIDS, AND GUT MICROBIOTA

Over 40 years ago in the prestatin era antibiotics were shown by Samuel and others to lower serum cholesterol and for a time neomycin treatment was advocated to control serum lipids [141,142]. The neomycin effect was attributed to the reduction in the activity of microbially synthesized 7α-hydroxylase in the colon, presumably related to the antibiotic inhibition of colonic microbial synthetic activity. It has been suggested that the resulting lack of conversion of primary to secondary bile acids (e.g., cholate to deoxycholate) [141] may have resulted in the suppression of hepatic cholesterol synthesis by primary bile acids or removal of the stimulus to cholesterol synthesis related to the absence of secondary bile acids. Overall, one can conclude that neomycin influences serum lipids by modifying colonic microbial metabolism. Since then other antibiotics have been tested and metronidazole was also found to lower LDL-C as doses of 750–225 mg/d in studies of 1–3 week duration [143–145], in contrast, for example, to ciprofloxacin [95]. Again the mechanism is not clear. In a study where ciprofloxacin was compared to metronidazole, aerobes and anaerobe counts were reduced on ciprofloxacin, but surprisingly were increased on metronidazole. When all the data were pooled a significant negative correlation was shown between the change in both aerobes and Bifidobacteria and the change in LDL-C concentration. This unexpected finding of increased rather than reduced aerobe and Bifidobacteria on metronidazole may have been due to the loss of more antibiotic-sensitive microbiota leaving more fecal nutrients for the growth of the less antibiotic-sensitive Bifidobacteria [146]. Indeed neomycin containing agar is used to isolate and enumerate Bifidobacteria [147]. The mechanism by which Bifidobacteria may lower cholesterol is not certain. We have mentioned the effect on conversion of primary to secondary bile acids and the possible effects on hepatic synthesis. Equally, changes in SCFA metabolism may be responsible but have not been observed [95].

Overall, altered colonic metabolism may influence blood lipids. However, due to the danger of antibiotic resistance, pre- and probiotic approaches are more appropriate if they can be identified, since lipid-lowering interventions are by necessity lifelong interventions, and must be without the risk of creating antibiotic-resistant organisms.

FRUCTOSE AND THE GUT

Fructose exists in foods either as free fructose (i.e., fruits, honey, high fructose corn syrup) or fructose bound to glucose (i.e., sucrose). Fructose has long been known to have a low glycemic index [26]. The ability of the small intestine to absorb free fructose varies among healthy adults and is dose dependent. In three separate studies, when the fructose dose was increased from 25 to 50 g in 10% solution, the prevalence of fructose malabsorption increased from 0% to 37.5%, 11% to 58%, and 50% to 80%, respectively [148–150]. Fructose malabsorption has also been shown to be present in children. In one study, ingestion of 2 g/kg body weight of fructose led to incomplete absorption in 71% of

children [151]. It has been suggested that the absorptive capacity for free fructose in healthy individuals ranges from less than 5 g to more than 50 g [152]. This malabsorption can be largely overcome by coingestion with glucose, such as fructose in the form of sucrose or high fructose corn syrup [148–151]. In the classical model of absorption, fructose enters the brush border membrane of the enterocyte through a specific fructose transporter, GLUT5, and diffuses into blood vessels through the GLUT2 transporter located on the basolateral membrane [153]. However, there is evidence that the presence of both glucose and fructose in the small intestine recruits GLUT2 transporters to the brush border membrane which can also facilitate diffusion of fructose [154,155]. This additional recruitment of GLUT2 to the brush border membrane could potentially explain why fructose malabsorption does not occur with coingestion of glucose [156–158]. After absorption, fructose enters the hepatic portal vein where it is rapidly and efficiently extracted by the liver [153].

Any fructose that is not absorbed by the small intestine enters the colon where it interacts with the gut microbiota. Colonic fermentation of unabsorbed fructose produces SCFAs and gases, such as hydrogen and carbon dioxide which may cause flatulence, abdominal discomfort, and/or diarrhea in some individuals [152,156]. There is little research on the impact of high fructose consumption on the gut microbiome and its metabolic activity. One animal study demonstrated that extreme fructose feeding (30% energy) induced a metabolic syndrome phenotype through favoring two bacterial genera, *Coprococcus* and *Ruminococcus*, the reduction of which with antibiotics or fecal transplant appeared to rescue the phenotype [159]. It has also been suggested that colonic fermentation of unabsorbed fructose to SCFAs promotes interaction with its G protein-coupled receptor Gpr41 which are expressed by a subset of enteroendocrine cells in the colonic mucosa [160,161]. When mice that expressed Gpr41 were compared with Gpr41 knockout mice, signal transduction of SCFA bound to Gpr41 increased leptin expression, enhanced appetite, and increased adiposity, despite similar levels of chow consumption [162]. However, no studies to confirm these findings have been undertaken in humans.

It has been hypothesized that colonic metabolism of fructose may contribute to the development of obesity, metabolic syndrome, and their downstream cardiometabolic complications [159,160,162]. Certain human data from prospective cohort studies that have looked at consumption of sugar-sweetened beverages in excess of caloric requirements have indicated an adverse association with metabolic syndrome and T2DM [163]. However, a series of systematic reviews and metaanalyses of >50 controlled feeding trials involving >1000 participants showed a lack of adverse effects of fructose in isocaloric substitution for other sources of carbohydrates (mainly starch) on body weight, established lipid targets, postprandial triglycerides, glycemic control, blood pressure, uric acid, and nonalcoholic fatty liver disease. Advantages were even seen for glycemic control and blood pressure [164–172]. On the other hand, supplementing diets with fructose as excess calories compared to the same diets alone without the excess calories showed adverse effects for cardiometabolic risk factors, suggesting that excess calories mediate the adverse effects [164–172]. We conclude that the data show no clear evidence for a microbiota-induced metabolic derangement caused by fructose, in the absence of excess calories.

HYPOSUCRASIA

This genetic abnormality is rare and is due to an inability of the affected individuals' enterocyte brush border to produce sucrase and thus absorb glucose and fructose. These individuals produce an increase in breath H_2 after sucrose ingestion [173], as seen with lactose in hypolactasia [174], and are troubled often by unexplained bloating, flatulence, and diarrhea that may be especially severe in children and require rehydration. The treatment is reduction in sugar intake. There are no reports of the effect of hyposucrasia on the colonic microbiota.

CONCLUSIONS

Diets that emphasize plant-based foods that are high in fiber favorably alter the gut microbiota and may have significant implications for health. Such diets have evolved in parallel with our ancestors and are likely best suited for our gut microbiota. Moving forward, future studies are needed to differentiate changes in the gut microbial profile as a cause or an effect and how the associated end products affect the health of the host. Results of longitudinal studies may pave the way for targeted dietary approaches that modulate the gut microbiota for optimizing the prevention and management of diet-related chronic diseases.

FUNDING STATEMENT

E.M. Comelli holds the Lawson Family Chair in Microbiome Nutrition Research in the Centre for Child Nutrition and Health, Faculty of Medicine, University of Toronto. J.L. Sievenpiper was funded by a PSI Graham Farquharson Knowledge Translation Fellowship, Canadian Diabetes Association (CDA) Clinician Scientist award, CIHR INMD/CNS New Investigator Partnership Prize, and Banting & Best Diabetes Centre Sun Life Financial New Investigator Award. J.C. Noronha was funded by a CIHR Frederick Banting and Charles Best Canada Graduate Scholarship and Banting and Best Diabetes Centre Novo Nordisk Studentship. D.J.A. Jenkins was funded by the Government of Canada through the Canada Research Chair Endowment.

None of the sponsors had a role in any aspect of the present study, including design and conduct of the study; collection, management, analysis, and interpretation of the data; and preparation, review, approval of the manuscript or decision to publish.

COMPETING INTERESTS

J.M.M. Wong and J.C. Noronha declare no competing interests relevant to this manuscript. E.M. Comelli has received research support from the National Sciences and Engineering Research Council (NSERC) of Canada. C.W.C. Kendall has received research support from the Advanced Foods and Material Network, Agrifoods and Agriculture Canada, the Almond Board of California, the American Pistachio Growers, Barilla, the California Strawberry Commission, the Calorie Control Council, CIHR, the Canola Council of Canada, the Coca-Cola Company (investigator initiated, unrestricted grant), Hain Celestial, the International Tree Nut Council Nutrition Research and Education Foundation, Kellogg, Kraft, Loblaw Companies Ltd., Orafti, Pulse Canada, Saskatchewan Pulse Growers, Solae and Unilever. He has received travel funding, consultant fees, and/or honoraria from Abbott Laboratories, the Almond Board of California, the American Peanut Council, the American Pistachio Growers, Barilla, Bayer, the Canola Council of Canada, the Coca-Cola Company, Danone, General Mills, the International Tree Nut Council Nutrition Research and Education Foundation, Kellogg, Loblaw Companies Ltd., the Nutrition Foundation of Italy, Oldways Preservation Trust, Orafti, Paramount Farms, the Peanut Institute, PepsiCo, Pulse Canada, Sabra Dipping Co., Saskatchewan Pulse Growers, Solae, Sun-Maid, Tate and Lyle, and Unilever. He has served on the scientific advisory board for the Almond Board of California, the International Tree Nut Council, Oldways Preservation Trust, Paramount Farms and Pulse Canada. He is a member of the International Carbohydrate Quality Consortium (ICQC), Executive Board Member of the Diabetes and Nutrition Study Group (DNSG) of the European Association for the Study of Diabetes (EASD), is on the Clinical Practice Guidelines Expert Committee for Nutrition Therapy of the EASD and is a Director of the Toronto 3D Knowledge Synthesis and Clinical Trials foundation. J.L. Sievenpiper has received research support from the Canadian Institutes of health Research (CIHR), Canadian Diabetes Association (CDA), PSI Foundation, Calorie Control Council, Banting and Best Diabetes Centre (BBDC), American Society for Nutrition (ASN), Dr. Pepper Snapple Group (investigator initiated, unrestricted donation), INC International Nut and Dried Fruit Council, and University of Toronto (through an unrestricted donation from Tate & Lyle). He has received speaker fees and/or honoraria from the Canadian Diabetes Association (CDA), Canadian Nutrition Society (CNS), University of Alabama at Birmingham, Abbott Laboratories, Canadian Sugar Institute, Dr. Pepper Snapple Group, The Coca-Cola Company, Dairy Farmers of Canada, C3 Collaborating for Health, White Wave Foods, Rippe Lifestyle, mdBriefcase, Alberta Milk, Food Minds, PepsiCo, and Pulse Canada. He has ad hoc consulting arrangements with Winston & Strawn LLP, Perkins Coie LLP, and Tate & Lyle. He is a member of the European Fruit Juice Association Scientific Expert Panel. He is on the Clinical Practice Guidelines Expert Committees of the Canadian Diabetes Association (CDA), European Association for the study of Diabetes (EASD), and Canadian Cardiovascular Society (CCS), as well as an expert writing panel of the American Society for Nutrition (ASN). He serves as an unpaid scientific advisor for the Food, Nutrition, and Safety Program (FNSP) and the Technical Committee on Carbohydrates of the International Life Science Institute (ILSI) North America. He is a member of the International Carbohydrate Quality Consortium (ICQC), Executive Board Member of the Diabetes and Nutrition Study Group (DNSG) of the EASD, and Director of the Toronto 3D Knowledge Synthesis and Clinical Trials foundation. His wife is an employee of Unilever Canada. *D.J.A. Jenkins* has received research grants from Saskatchewan Pulse Growers, the Agricultural Bioproducts Innovation Program through the Pulse Research Network, the Advanced Foods and Material Network, Loblaw Companies Ltd., Unilever, Barilla, the Almond Board of California, Agriculture and Agri-food Canada, Pulse Canada, Kellogg's Company, Canada, Quaker Oats, Canada, Procter & Gamble Technical Centre Ltd., Bayer Consumer Care, Springfield, NJ, Pepsi/Quaker, International Nut & Dried Fruit (INC), Soy Foods Association of North America, the Coca-Cola Company (investigator initiated, unrestricted grant), Solae, Haine Celestial, the Sanitarium Company, Orafti, the International Tree Nut Council Nutrition Research and Education Foundation, the Peanut Institute, the Canola and Flax Councils of Canada, the Calorie Control Council, the CIHR, the Canada Foundation for Innovation and the Ontario Research Fund. He has been on the speaker's panel, served on the scientific advisory board, and/or received travel support and/or honoraria from the Almond Board of California, Canadian Agriculture Policy Institute, Loblaw Companies Ltd, the Griffin Hospital (for the development of the NuVal scoring system), the Coca-Cola Company, EPICURE, Danone, Saskatchewan Pulse Growers, Sanitarium Company, Orafti, Almond Board of California, the American Peanut Council, the International Tree Nut Council Nutrition Research and Education Foundation, the Peanut Institute, Herbalife International, Pacific Health Laboratories, Nutritional Fundamental for Health, Barilla, Metagenics, Bayer Consumer Care, Unilever Canada and Netherlands, Solae, Kellogg, Quaker Oats, Procter & Gamble, the Coca-Cola Company, the Griffin Hospital, Abbott Laboratories, the Canola Council of Canada, Dean Foods, the California Strawberry Commission, Haine Celestial, PepsiCo, the Alpro Foundation, Pioneer Hi-Bred International, DuPont Nutrition and Health, Spherix Consulting and WhiteWave Foods, the Advanced Foods and Material Network, the Canola and Flax Councils of Canada, the Nutritional Fundamentals for Health, Agri-Culture and Agri-Food Canada, the Canadian Agri-Food Policy Institute, Pulse Canada, the Saskatchewan Pulse Growers, the Soy Foods Association of North America, the Nutrition Foundation of Italy (NFI), NutraSource Diagnostics, the McDougall Program, the Toronto Knowledge Translation Group (St. Michael's Hospital), the Canadian College of Naturopathic Medicine, The Hospital for Sick Children, the Canadian Nutrition Society (CNS), the American Society of Nutrition (ASN), Arizona State University, Paolo Sorbini Foundation and the Institute of Nutrition, Metabolism and Diabetes. He received an honorarium from the United States Department of Agriculture to present the 2013 W.O. Atwater Memorial Lecture. He received the 2013 Award for Excellence in Research from the International Nut and Dried Fruit Council. He received funding and travel support from the Canadian Society of Endocrinology and Metabolism to produce mini cases for the Canadian Diabetes Association. He is a member of the International Carbohydrate Quality Consortium (ICQC). His wife is a director and partner of Glycemic Index Laboratories, and his sister received funding through a grant from the St. Michael's Hospital Foundation to develop a cookbook for one of his studies.

REFERENCES

[1] Fung TT, Willett WC, Stampfer MJ, Manson JE, Hu FB. Dietary patterns and the risk of coronary heart disease in women. Arch Intern Med 2001;161(15):1857–62.

[2] Hu FB, Rimm EB, Stampfer MJ, Ascherio A, Spiegelman D, Willett WC. Prospective study of major dietary patterns and risk of coronary heart disease in men. Am J Clin Nutr 2000;72(4):912–21.

[3] Heidemann C, Schulze MB, Franco OH, van Dam RM, Mantzoros CS, Hu FB. Dietary patterns and risk of mortality from cardiovascular disease, cancer, and all causes in a prospective cohort of women. Circulation 2008;118(3):230–7.

[4] van Dam RM, Rimm EB, Willett WC, Stampfer MJ, Hu FB. Dietary patterns and risk for type 2 diabetes mellitus in U.S. men. Ann Intern Med 2002;136(3):201–9.

[5] Fung TT, Schulze M, Manson JE, Willett WC, Hu FB. Dietary patterns, meat intake, and the risk of type 2 diabetes in women. Arch Intern Med 2004;164(20):2235–40.

[6] Mozaffarian D, Hao T, Rimm EB, Willett WC, Hu FB. Changes in diet and lifestyle and long-term weight gain in women and men. N Engl J Med 2011;364(25):2392–404.

[7] Key TJ, Fraser GE, Thorogood M, et al. Mortality in vegetarians and nonvegetarians: detailed findings from a collaborative analysis of 5 prospective studies. Am J Clin Nutr 1999;70(3 Suppl.):516S–24S.

[8] Fraser GE. Vegetarian diets: what do we know of their effects on common chronic diseases? Am J Clin Nutr 2009;89(5):1607S–12S.

[9] Fraser GE. Diet, life expectancy, and chronic disease: studies of Seventh-day Adventists and other vegetarians. New York, NY: Oxford University Press; 2003.

[10] Key TJ, Appleby PN, Rosell MS. Health effects of vegetarian and vegan diets. Proc Nutr Soc 2006;65(1):35–41.

[11] Ornish D, Brown SE, Scherwitz LW, et al. Can lifestyle changes reverse coronary heart disease? The Lifestyle Heart Trial. Lancet 1990;336(8708):129–33.

[12] Barnard ND, Cohen J, Jenkins DJ, et al. A low-fat vegan diet and a conventional diabetes diet in the treatment of type 2 diabetes: a randomized, controlled 74-wk clinical trial. Am J Clin Nutr 2009;89(5):1588S–96S.

[13] Appel LJ, Moore TJ, Obarzanek E, et al. A clinical trial of the effects of dietary patterns on blood pressure. DASH Collaborative Research Group. N Engl J Med 1997;336(16):1117–24.

[14] Appel LJ, Sacks FM, Carey VJ, et al. Effects of protein, monounsaturated fat, and carbohydrate intake on blood pressure and serum lipids: results of the OmniHeart randomized trial. JAMA 2005;294(19):2455–64.

[15] Obarzanek E, Sacks FM, Vollmer WM, et al. Effects on blood lipids of a blood pressure-lowering diet: the Dietary Approaches to Stop Hypertension (DASH) Trial. Am J Clin Nutr 2001;74(1):80–9.

[16] Jenkins DJ, Kendall CW, Marchie A, et al. Effects of a dietary portfolio of cholesterol-lowering foods vs lovastatin on serum lipids and C-reactive protein. JAMA 2003;290(4):502–10.

[17] Krebs-Smith SM, Guenther PM, Subar AF, Kirkpatrick SI, Dodd KW. Americans do not meet federal dietary recommendations. J Nutr 2010;140(10):1832–8.

[18] King DE, Mainous AG III, Lambourne CA. Trends in dietary fiber intake in the United States, 1999–2008. J Acad Nutr Diet 2012;112(5):642–8.

[19] A Report of the Panel on Macronutrients, Subcommittees on Upper Reference Levels of Nutrients and Interpretation and Uses of Dietary Reference Intakes, and the Standing Committee on the Scientific Evaluation of Dietary Reference Intakes. Dietary reference intakes for energy, carbohydrate, fiber, fat, fatty acids, cholesterol, protein, and amino acids (macronutrients). Washington, DC: The National Academies Press; 2005.

[20] Cummings JH, Branch WJ. Fermentation and the production of short chain fatty acids in the human large intestine. In: Vahonny GV, Kritchevsky D, editors. Dietary fiber: basic and clinical aspects. New York, NY: Plenum Press; 1986. p. 131–49.

[21] Wolever TM, Cohen Z, Thompson LU, et al. Ileal loss of available carbohydrate in man: comparison of a breath hydrogen method with direct measurement using a human ileostomy model. Am J Gastroenterol 1986;81(2):115–22.

[22] Jenkins DJ, Cuff D, Wolever TM, et al. Digestibility of carbohydrate foods in an ileostomate: relationship to dietary fiber, in vitro digestibility, and glycemic response. Am J Gastroenterol 1987;82(8):709–17.

[23] Anderson IH, Levine AS, Levitt MD. Incomplete absorption of the carbohydrate in all-purpose wheat flour. N Engl J Med 1981;304(15):891–2.

[24] Englyst HN, Cummings JH. Digestion of the carbohydrates of banana (*Musa paradisiaca sapientum*) in the human small intestine. Am J Clin Nutr 1986;44(1):42–50.

[25] Steinhart AH, Jenkins DJ, Mitchell S, Cuff D, Prokipchuk EJ. Effect of dietary fiber on total carbohydrate losses in ileostomy effluent. Am J Gastroenterol 1992;87(1):48–54.

[26] Jenkins DJ, Wolever TM, Taylor RH, et al. Glycemic index of foods: a physiological basis for carbohydrate exchange. Am J Clin Nutr 1981;34(3):362–6.

[27] Jenkins DJ, Wolever TM, Jenkins AL, Josse RG, Wong GS. The glycaemic response to carbohydrate foods. Lancet 1984;2(8399):388–91.

[28] Englyst KN, Englyst HN, Hudson GJ, Cole TJ, Cummings JH. Rapidly available glucose in foods: an in vitro measurement that reflects the glycemic response. Am J Clin Nutr 1999;69(3):448–54.

[29] Sheard NF, Clark NG, Brand-Miller JC, et al. Dietary carbohydrate (amount and type) in the prevention and management of diabetes: a statement by the American Diabetes Association. Diabetes Care 2004;27(9):2266–71.

[30] Haber GB, Heaton KW, Murphy D, Burroughs LF. Depletion and disruption of dietary fibre. Effects on satiety, plasma-glucose, and serum-insulin. Lancet 1977;2(8040):679–82.

[31] O'Dea K, Nestel PJ, Antonoff L. Physical factors influencing postprandial glucose and insulin responses to starch. Am J Clin Nutr 1980;33(4):760–5.

[32] Jenkins DJ, Thorne MJ, Camelon K, et al. Effect of processing on digestibility and the blood glucose response: a study of lentils. Am J Clin Nutr 1982;36(6):1093–101.

[33] Wursch P, Del Vedovo S, Koellreutter B. Cell structure and starch nature as key determinants of the digestion rate of starch in legume. Am J Clin Nutr 1986;43(1):25–9.

[34] Isaksson G, Lundquist I, Ihse I. Effect of dietary fiber on pancreatic enzyme activity in vitro. Gastroenterology 1982;82 (5 Pt. 1):918–24.

[35] Yoon JH, Thompson LU, Jenkins DJ. The effect of phytic acid on in vitro rate of starch digestibility and blood glucose response. Am J Clin Nutr 1983;38(6):835–42.

[36] Englyst HN, Kingman SM, Cummings JH. Classification and measurement of nutritionally important starch fractions. Eur J Clin Nutr 1992;46(Suppl. 2):S33–50.

[37] Cummings JH. Short chain fatty acids in the human colon. Gut 1981;22(9):763–79.

[38] Cummings JH, Englyst HN, Wiggins HS. The role of carbohydrates in lower gut function. Nutr Rev 1986;44(2):50–4.

[39] Stephen AM, Haddad AC, Phillips SF. Passage of carbohydrate into the colon. Direct measurements in humans. Gastroenterology 1983;85(3):589–95.

[40] Englyst HN, Cummings JH. Non-starch polysaccharides (dietary fiber) and resistant starch. Adv Exp Med Biol 1990;270:205–25.

[41] Cummings JH, Englyst HN. Fermentation in the human large intestine and the available substrates. Am J Clin Nutr 1987;45(5 Suppl.):1243–55.

[42] Jenkins DJ, Vuksan V, Kendall CW, et al. Physiological effects of resistant starches on fecal bulk, short chain fatty acids, blood lipids and glycemic index. J Am Coll Nutr 1998;17(6):609–16.

[43] Roberfroid MB. Inulin-type fructans: functional food ingredients. Boca Raton, FL: CRC Press; 2005.

[44] Cummings JH, Macfarlane GT. The control and consequences of bacterial fermentation in the human colon. J Appl Bacteriol 1991;70(6):443–59.

[45] FAO Technical Meeting on Prebiotics. Food Quality and Standards Service (AGNS) and Food and Agriculture Organization of the United Nations (FAO); 2007.

[46] Scott KP, Martin JC, Chassard C, et al. Substrate-driven gene expression in *Roseburia inulinivorans*: importance of inducible

enzymes in the utilization of inulin and starch. Proc Natl Acad Sci USA 2011;108(Suppl. 1):4672–9.

[47] Gibson GR, Roberfroid MB. Dietary modulation of the human colonic microbiota: introducing the concept of prebiotics. J Nutr 1995;125(6):1401–12.

[48] Tilg H, Moschen AR. Microbiota and diabetes: an evolving relationship. Gut 2014;63(9):1513–21.

[49] Macfarlane S, Macfarlane GT, Cummings JH. Review article: prebiotics in the gastrointestinal tract. Aliment Pharmacol Ther 2006;24(5):701–14.

[50] Dewulf EM, Cani PD, Claus SP, et al. Insight into the prebiotic concept: lessons from an exploratory, double blind intervention study with inulin-type fructans in obese women. Gut 2013;62(8):1112–21.

[51] Kolida S, Gibson GR. Prebiotic capacity of inulin-type fructans. J Nutr 2007;137(11 Suppl.):2503S–6S.

[52] Andersson HB, Ellegard LH, Bosaeus IG. Nondigestibility characteristics of inulin and oligofructose in humans. J Nutr 1999;129 (7 Suppl.):1428S–30S.

[53] Ellegard L, Andersson H, Bosaeus I. Inulin and oligofructose do not influence the absorption of cholesterol, or the excretion of cholesterol, Ca, Mg, Zn, Fe, or bile acids but increases energy excretion in ileostomy subjects. Eur J Clin Nutr 1997;51(1):1–5.

[54] Alles MS, Hautvast JG, Nagengast FM, Hartemink R, Van Laere KM, Jansen JB. Fate of fructo-oligosaccharides in the human intestine. Br J Nutr 1996;76(2):211–21.

[55] Molis C, Flourie B, Ouarne F, et al. Digestion, excretion, and energy value of fructooligosaccharides in healthy humans. Am J Clin Nutr 1996;64(3):324–8.

[56] Gibson GR, Beatty ER, Wang X, Cummings JH. Selective stimulation of bifidobacteria in the human colon by oligofructose and inulin. Gastroenterology 1995;108:975–82.

[57] Brighenti F, Casiraghi MC, Canzi E, Ferrari A. Effect of consumption of a ready-to-eat breakfast cereal containing inulin on the intestinal milieu and blood lipids in healthy male volunteers. Eur J Clin Nutr 1999;53(9):726–33.

[58] Wang X, Gibson GR. Effects of the in vitro fermentation of oligofructose and inulin by bacteria growing in the human large intestine. J Appl Bacteriol 1993;75(4):373–80.

[59] Gibson GR, Wang X. Bifidogenic properties of different types of fructo-oligosaccharides. Food Microbiol 1994;11(6):491–8.

[60] Gibson GR, Wang X. Enrichment of bifidobacteria from human gut contents by oligofructose using continuous culture. FEMS Microbiol Lett 1994;118(1–2):121–7.

[61] Bouhnik Y, Flourie B, Riottot M, et al. Effects of fructo-oligosaccharides ingestion on fecal bifidobacteria and selected metabolic indexes of colon carcinogenesis in healthy humans. Nutr Cancer 1996;26(1):21–9.

[62] Buddington RK, Williams CH, Chen SC, Witherly SA. Dietary supplement of neosugar alters the fecal flora and decreases activities of some reductive enzymes in human subjects. Am J Clin Nutr 1996;63(5):709–16.

[63] Kleessen B, Sykura B, Zunft HJ, Blaut M. Effects of inulin and lactose on fecal microflora, microbial activity, and bowel habit in elderly constipated persons. Am J Clin Nutr 1997;65(5):1397–402.

[64] Kruse HP, Kleessen B, Blaut M. Effects of inulin on faecal bifidobacteria in human subjects. Br J Nutr 1999;82(5):375–82.

[65] Bouhnik Y, Vahedi K, Achour L, et al. Short-chain fructo-oligosaccharide administration dose-dependently increases fecal bifidobacteria in healthy humans. J Nutr 1999;129(1):113–6.

[66] Rao AV. The prebiotic properties of oligofructose at low intake levels. Nutr Res 2001;21(6):843–8.

[67] Tuohy KM, Kolida S, Lustenberger AM, Gibson GR. The prebiotic effects of biscuits containing partially hydrolysed guar gum and fructo-oligosaccharides—a human volunteer study. Br J Nutr 2001;86(3):341–8.

[68] Tuohy KM, Probert HM, Smejkal CW, Gibson GR. Using probiotics and prebiotics to improve gut health. Drug Discov Today 2003;8(15):692–700.

[69] Gibson GR, Probert HM, Loo JV, Rastall RA, Roberfroid MB. Dietary modulation of the human colonic microbiota: updating the concept of prebiotics. Nutr Res Rev 2004;17(2):259–75.

[70] Macfarlane GT, Gibson GR. Microbiological aspects of production of short-chain fatty acids in the large bowel. In: Cummings JH, Rombeau JL, Sakata T, editors. Physiological and clinical aspects of short-chain fatty acids. Cambridge (England), New York, NY: Cambridge University Press; 1995. p. 87–105.

[71] Macfarlane GT, Macfarlane S. Bacteria, colonic fermentation, and gastrointestinal health. J AOAC Int 2012;95(1):50–60.

[72] Topping DL, Clifton PM. Short-chain fatty acids and human colonic function: roles of resistant starch and nonstarch polysaccharides. Physiol Rev 2001;81(3):1031–64.

[73] Cummings JH, Hill MJ, Bone ES, Branch WJ, Jenkins DJ. The effect of meat protein and dietary fiber on colonic function and metabolism. II. Bacterial metabolites in feces and urine. Am J Clin Nutr 1979;32(10):2094–101.

[74] Vince A, Killingley M, Wrong OM. Effect of lactulose on ammonia production in a fecal incubation system. Gastroenterology 1978;74(3):544–9.

[75] Jackson AA. Aminoacids: essential and non-essential? Lancet 1983;1(8332):1034–7.

[76] Jenkins DJ, Wolever TM, Collier GR, et al. Metabolic effects of a low-glycemic-index diet. Am J Clin Nutr 1987;46(6):968–75.

[77] Roediger WE. Role of anaerobic bacteria in the metabolic welfare of the colonic mucosa in man. Gut 1980;21(9):793–8.

[78] Jenkins DJ, Kendall CW, Vuksan V. Inulin, oligofructose and intestinal function. J Nutr 1999;129(7 Suppl.):1431S–3S.

[79] Floch MH, Hong-Curtiss J. Probiotics and functional foods in gastrointestinal disorders. Curr Treat Options Gastroenterol 2002;5(4):311–21.

[80] Cook SI, Sellin JH. Review article: short chain fatty acids in health and disease. Aliment Pharmacol Ther 1998;12(6):499–507.

[81] Roediger WE. Utilization of nutrients by isolated epithelial cells of the rat colon. Gastroenterology 1982;83(2):424–9.

[82] Walker AW, Ince J, Duncan SH, et al. Dominant and diet-responsive groups of bacteria within the human colonic microbiota. ISME J 2011;5(2):220–30.

[83] Ze X, Duncan SH, Louis P, Flint HJ. *Ruminococcus bromii* is a keystone species for the degradation of resistant starch in the human colon. ISME J 2012;6(8):1535–43.

[84] Louis P, Flint HJ. Diversity, metabolism and microbial ecology of butyrate-producing bacteria from the human large intestine. FEMS Microbiol Lett 2009;294(1):1–8.

[85] Englyst HN, Hay S, Macfarlane GT. Polysaccharide breakdown by mixed populations of human fecal bacteria. FEMS Microbiol Lett 1987;45:163–71.

[86] Van Munster IP, Tangerman A, Nagengast FM. Effect of resistant starch on colonic fermentation, bile acid metabolism, and mucosal proliferation. Dig Dis Sci 1994;39:834–42.

[87] Noakes M, Clifton PM, Nestel PJ, Le Leu R, McIntosh G. Effect of high-amylose starch and oat bran on metabolic variables and bowel function in subjects with hypertriglyceridemia. Am J Clin Nutr 1996;64(6):944–51.

[88] Phillips J, Muir JG, Birkett A, et al. Effect of resistant starch on fecal bulk and fermentation-dependent events in humans. Am J Clin Nutr 1995;62(1):121–30.

[89] Fredstrom SB, Lampe JW, Jung HJ, Slavin JL. Apparent fiber digestibility and fecal short-chain fatty acid concentrations with ingestion of two types of dietary fiber. JPEN 1994;18(1):14–9.

[90] Takahashi H, Yang SI, Hayashi C, Kim M, Yamanaka J, Yamamoto T. Effect of partially hydrolyzed guar gum on fecal output in human volunteers. Nutr Res 1993;13:649–57.

[91] Vogt JA, Pencharz PB, Wolever TM. l-rhamnose increases serum propionate in humans. Am J Clin Nutr 2004;80(1):89–94.

[92] Vogt JA, Ishii-Schrade KB, Pencharz PB, Wolever TM. L-rhamnose increases serum propionate after long-term supplementation, but lactulose does not raise serum acetate. Am J Clin Nutr 2004;80(5):1254–61.

[93] Thacker PA, Salamons MO, Aherne FX, Milligan LP, Bowland JP. Influence of propionic acid on the cholesterol metabolism of pigs fed hypercholesterolemic diets. Can J Anim Sci 1981;61:969–75.

[94] Amaral L, Hoppel C, Stephen AM. Effect of propionate on lipid metabolism in healthy human subjects. Falk Symposium 1993;73:E2.

[95] Jenkins DJ, Kendall CW, Hamidi M, et al. Effect of antibiotics as cholesterol-lowering agents. Metabolism 2005;54(1):103–12.

[96] Wolever TM, Brighenti F, Royall D, Jenkins AL, Jenkins DJ. Effect of rectal infusion of short chain fatty acids in human subjects. Am J Gastroenterol 1989;84(9):1027–33.

[97] Popovich DG, Jenkins DJ, Kendall CW, et al. The western lowland gorilla diet has implications for the health of humans and other hominoids. J Nutr 1997;127(10):2000–5.

[98] Milton K. Nutritional characteristics of wild primate foods: do the diets of our closest living relatives have lessons for us? Nutrition 1999;15(6):488–98.

[99] Cousins D. A review of the diets of captive gorillas (*Gorilla gorilla*). Acta Zool Pathol Antverp 1976;66:91–100.

[100] Lucas PW, Peters CR, Arrandale SR. Seed-breaking forces exerted by orang-utans with their teeth in captivity and a new technique for estimating forces produced in the wild. Am J Phys Anthropol 1994;94(3):365–78.

[101] Kay RF. Diets of early Miocene African hominoids. Nature 1977;268:628–30.

[102] Milton K. Primate diets and gut morphology. Implications for hominid evolution. In: Harris W, Boss EB, editors. Food and nutrition: toward a theory of human and food habits. Philadelphia, PA: Temple University Press; 1987. p. 96–116.

[103] Jenkins DJ, Kendall CW, Popovich DG, et al. Effect of a very-high-fiber vegetable, fruit, and nut diet on serum lipids and colonic function. Metabolism 2001;50(4):494–503.

[104] US Food and Drug Administration. Food labeling: health claims: soluble fiber from certain foods and coronary heart disease. Washington, DC: US Food and Drug Administration; 1998. Docket 96P-0338.

[105] US Food and Drug Administration. FDA final rule for food labeling: health claims: soy protein and coronary heart disease. 64 Federal Register 1999:57699-733.

[106] US Food and Drug Administration. FDA authorizes new coronary heart disease health claim for plant sterol and plant stanol esters. Washington, DC: US Food and Drug Administration; 2000. Docket 001-1275, OOP-6.

[107] US Food and Drug Administration. Food labeling: health claims: soluble fiber from whole oats and risk of coronary heart disease. Washington, DC: US Food and Drug Administration; 2001. p. 15343–4. Docket 95P-0197.

[108] US Food and Drug Administration. Food labeling: health claims: nuts & heart disease. Washington, DC: US Food and Drug Administration; 2003. Docket No. 02P-0505.

[109] Ley RE, Hamady M, Lozupone C, et al. Evolution of mammals and their gut microbes. Science 2008;320(5883):1647–51.

[110] Human Microbiome Project Consortium. Structure, function and diversity of the healthy human microbiome. Nature 2012;486(7402):207–14.

[111] Turnbaugh PJ, Hamady M, Yatsunenko T, et al. A core gut microbiome in obese and lean twins. Nature 2009;457(7228):480–4.

[112] Qin J, Li R, Raes J, et al. A human gut microbial gene catalogue established by metagenomic sequencing. Nature 2010;464(7285): 59–65.

[113] Lozupone CA, Stombaugh JI, Gordon JI, Jansson JK, Knight R. Diversity, stability and resilience of the human gut microbiota. Nature 2012;489(7415):220–30.

[114] Muegge BD, Kuczynski J, Knights D, et al. Diet drives convergence in gut microbiome functions across mammalian phylogeny and within humans. Science 2011;332(6032):970–4.

[115] Arumugam M, Raes J, Pelletier E, et al. Enterotypes of the human gut microbiome. Nature 2011;473(7346):174–80.

[116] Wu GD, Chen J, Hoffmann C, et al. Linking long-term dietary patterns with gut microbial enterotypes. Science 2011;334(6052): 105–8.

[117] Yatsunenko T, Rey FE, Manary MJ, et al. Human gut microbiome viewed across age and geography. Nature 2012;486(7402):222–7.

[118] De Filippo C, Cavalieri D, Di Paola M, et al. Impact of diet in shaping gut microbiota revealed by a comparative study in children from Europe and rural Africa. Proc Natl Acad Sci USA 2010;107(33):14691–6.

[119] Zimmer J, Lange B, Frick JS, et al. A vegan or vegetarian diet substantially alters the human colonic faecal microbiota. Eur J Clin Nutr 2012;66(1):53–60.

[120] Jeffery IB, Claesson MJ, O'Toole PW, Shanahan F. Categorization of the gut microbiota: enterotypes or gradients? Nat Rev Microbiol 2012;10(9):591–2.

[121] Ley RE, Backhed F, Turnbaugh P, Lozupone CA, Knight RD, Gordon JI. Obesity alters gut microbial ecology. Proc Natl Acad Sci USA 2005;102(31):11070–5.

[122] Turnbaugh PJ, Ley RE, Mahowald MA, Magrini V, Mardis ER, Gordon JI. An obesity-associated gut microbiome with increased capacity for energy harvest. Nature 2006;444(7122):1027–31.

[123] Backhed F, Manchester JK, Semenkovich CF, Gordon JI. Mechanisms underlying the resistance to diet-induced obesity in germ-free mice. Proc Natl Acad Sci USA 2007;104(3):979–84.

[124] Mandard S, Zandbergen F, van Straten E, et al. The fasting-induced adipose factor/angiopoietin-like protein 4 is physically associated with lipoproteins and governs plasma lipid levels and adiposity. J Biol Chem 2006;281(2):934–44.

[125] Liou AP, Paziuk M, Luevano JM Jr, Machineni S, Turnbaugh PJ, Kaplan LM. Conserved shifts in the gut microbiota due to gastric bypass reduce host weight and adiposity. Sci Transl Med 2013;5(178). 178ra41.

[126] Turnbaugh PJ, Ridaura VK, Faith JJ, Rey FE, Knight R, Gordon JI. The effect of diet on the human gut microbiome: a metagenomic analysis in humanized gnotobiotic mice. Science Transl Med 2009;1(6). 6ra14.

[127] Ley RE, Turnbaugh PJ, Klein S, Gordon JI. Microbial ecology: human gut microbes associated with obesity. Nature 2006;444(7122):1022–3.

[128] Turnbaugh PJ, Backhed F, Fulton L, Gordon JI. Diet-induced obesity is linked to marked but reversible alterations in the mouse distal gut microbiome. Cell Host Microbe 2008;3(4):213–23.

[129] Rahat-Rozenbloom S, Fernandes J, Gloor GB, Wolever TM. Evidence for greater production of colonic short-chain fatty acids in overweight than lean humans. Int J Obes 2014;38(12):1525–31.

[130] Wang Z, Klipfell E, Bennett BJ, et al. Gut flora metabolism of phosphatidylcholine promotes cardiovascular disease. Nature 2011;472(7341):57–63.

[131] Koeth RA, Wang Z, Levison BS, et al. Intestinal microbiota metabolism of L-carnitine, a nutrient in red meat, promotes atherosclerosis. Nat Med 2013;19(5):576–85.

[132] Tang WH, Wang Z, Levison BS, et al. Intestinal microbial metabolism of phosphatidylcholine and cardiovascular risk. N Engl J Med 2013;368(17):1575–84.

[133] Gaci N, Borrel G, Tottey W, O'Toole PW, Brugere JF. Archaea and the human gut: new beginning of an old story. World J Gastroenterol 2014;20(43):16062–78.

[134] Larsen N, Vogensen FK, van den Berg FW, et al. Gut microbiota in human adults with type 2 diabetes differs from non-diabetic adults. PLoS One 2010;5(2):e9085.

[135] Karlsson FH, Tremaroli V, Nookaew I, et al. Gut metagenome in European women with normal, impaired and diabetic glucose control. Nature 2013;498(7452):99–103.

[136] Furet JP, Kong LC, Tap J, et al. Differential adaptation of human gut microbiota to bariatric surgery-induced weight loss: links with metabolic and low-grade inflammation markers. Diabetes 2010;59(12):3049–57.

[137] Tremaroli V, Karlsson F, Werling M, et al. Roux-en-Y gastric bypass and vertical banded gastroplasty induce long-term changes on the human gut microbiome contributing to fat mass regulation. Cell Metab 2015;22(2):228–38.

[138] Cho YM. A gut feeling to cure diabetes: potential mechanisms of diabetes remission after bariatric surgery. Diabetes Metab J 2014;38(6):406–15.

[139] Vrieze A, Van Nood E, Holleman F, et al. Transfer of intestinal microbiota from lean donors increases insulin sensitivity in individuals with metabolic syndrome. Gastroenterology 2012;143(4): 913–6. e7.

[140] Haro C, Montes-Borrego M, Rangel-Zuniga OA, et al. Two healthy diets modulate gut microbial community improving insulin sensitivity in a human obese population. J Clin Endocrinol Metab 2016;101(1):233–42.

[141] Samuel P. Treatment of hypercholesterolemia with neomycin—a time for reappraisal. N Engl J Med 1979;301(11):595–7.

[142] Samuel P, Holtzman CM, Meilman E, Sekowski I. Effect of neomycin and other antibiotics on serum cholesterol levels and on 7alpha-dehydroxylation of bile acids by the fecal bacterial flora in man. Circ Res 1973;33(4):393–402.

[143] von Bergmann K, Streicher U, Leiss O, Jensen C, Gugler R. Serum-cholesterol-lowering effect of metronidazole and possible mechanisms of action. Klin Wochenschr 1985;63(6):279–81.

[144] Betti R, Palvarini M, Marmini A, Gendarini P, Cattaneo M. Metronidazole in hyperlipidemic subjects. Panminerva Med 1987;29(1):79–81.

[145] Davis JL, Schultz TA, Mosley CA. Metronidazole lowers serum lipids. Ann Intern Med 1983;99(1):43–4.

[146] Lim KS, Huh CS, Baek YJ. Antimicrobial susceptibility of bifidobacteria. J Dairy Sci 1993;76(8):2168–74.

[147] Murray BE. Impact of fluoroquinolones on the gastrointestinal flora. Rev Infect Dis 1989;11(Suppl. 5):S1372–8.

[148] Ravich WJ, Bayless TM, Thomas M. Fructose: incomplete intestinal absorption in humans. Gastroenterology 1983;84(1):26–9.

[149] Truswell AS, Seach JM, Thorburn AW. Incomplete absorption of pure fructose in healthy subjects and the facilitating effect of glucose. Am J Clin Nutr 1988;48(6):1424–30.

[150] Rumessen JJ, Gudmand-Hoyer E. Absorption capacity of fructose in healthy adults. Comparison with sucrose and its constituent monosaccharides. Gut 1986;27(10):1161–8.

[151] Kneepkens CM, Vonk RJ, Fernandes J. Incomplete intestinal absorption of fructose. Arch Dis Child 1984;59(8):735–8.

[152] Skoog SM, Bharucha AE. Dietary fructose and gastrointestinal symptoms: a review. Am J Gastroenterol 2004;99(10):2046–50.

[153] Tappy L, Le KA. Metabolic effects of fructose and the worldwide increase in obesity. Physiol Rev 2010;90(1):23–46.

[154] Helliwell PA, Richardson M, Affleck J, Kellett GL. Regulation of GLUT5, GLUT2 and intestinal brush-border fructose absorption by the extracellular signal-regulated kinase, p38 mitogen-activated kinase and phosphatidylinositol 3-kinase intracellular signalling pathways: implications for adaptation to diabetes. Biochem J 2000;350(Pt. 1):163–9.

[155] Helliwell PA, Richardson M, Affleck J, Kellett GL. Stimulation of fructose transport across the intestinal brush-border membrane by PMA is mediated by GLUT2 and dynamically regulated by protein kinase C. Biochem J 2000;350(Pt. 1):149–54.

[156] Gibson PR, Newnham E, Barrett JS, Shepherd SJ, Muir JG. Review article: fructose malabsorption and the bigger picture. Aliment Pharmacol Ther 2007;25(4):349–63.

[157] Kellett GL, Brot-Laroche E. Apical GLUT2: a major pathway of intestinal sugar absorption. Diabetes 2005;54(10):3056–62.

[158] Kellett GL, Brot-Laroche E, Mace OJ, Leturque A. Sugar absorption in the intestine: the role of GLUT2. Annu Rev Nutr 2008;28:35–54.

[159] Di Luccia B, Crescenzo R, Mazzoli A, et al. Rescue of fructose-induced metabolic syndrome by antibiotics or faecal transplantation in a rat model of obesity. PLoS One 2015;10(8):e0134893.

[160] Payne AN, Chassard C, Lacroix C. Gut microbial adaptation to dietary consumption of fructose, artificial sweeteners and sugar alcohols: implications for host-microbe interactions contributing to obesity. Obes Rev 2012;13(9):799–809.

[161] Tazoe H, Otomo Y, Karaki S, et al. Expression of short-chain fatty acid receptor GPR41 in the human colon. Biomed Res 2009;30(3):149–56.

[162] Samuel BS, Shaito A, Motoike T, et al. Effects of the gut microbiota on host adiposity are modulated by the short-chain fatty-acid binding G protein-coupled receptor, Gpr41. Proc Natl Acad Sci USA 2008;105(43):16767–72.

[163] Malik VS, Popkin BM, Bray GA, Despres JP, Willett WC, Hu FB. Sugar-sweetened beverages and risk of metabolic syndrome and type 2 diabetes: a meta-analysis. Diabetes Care 2010;33(11):2477–83.

[164] Sievenpiper JL, de Souza RJ, Mirrahimi A, et al. Effect of fructose on body weight in controlled feeding trials: a systematic review and meta-analysis. Ann Intern Med 2012;156(4):291–304.

[165] Sievenpiper JL, Carleton AJ, Chatha S, et al. Heterogeneous effects of fructose on blood lipids in individuals with type 2 diabetes: systematic review and meta-analysis of experimental trials in humans. Diabetes Care 2009;32(10):1930–7.

[166] Chiavaroli L, de Souza RJ, Ha V, et al. Effect of fructose on established lipid targets: a systematic review and meta-analysis of controlled feeding trials. J Am Heart Assoc 2015;4(9):e001700.

[167] Wang D, Sievenpiper JL, de Souza RJ, et al. Effect of fructose on postprandial triglycerides: a systematic review and meta-analysis of controlled feeding trials. Atherosclerosis 2014;232(1):125–33.

[168] Cozma AI, Sievenpiper JL, de Souza RJ, et al. Effect of fructose on glycemic control in diabetes: a systematic review and meta-analysis of controlled feeding trials. Diabetes Care 2012;35(7):1611–20.

[169] Ha V, Sievenpiper JL, de Souza RJ, et al. Effect of fructose on blood pressure: a systematic review and meta-analysis of controlled feeding trials. Hypertension 2012;59(4):787–95.

[170] Wang DD, Sievenpiper JL, de Souza RJ, et al. The effects of fructose intake on serum uric acid vary among controlled dietary trials. J Nutr 2012;142(5):916–23.

[171] Chiu S, Sievenpiper JL, de Souza RJ, et al. Effect of fructose on markers of non-alcoholic fatty liver disease (NAFLD): a systematic review and meta-analysis of controlled feeding trials. Eur J Clin Nutr 2014;68(4):416–23.

[172] Sievenpiper JL, Chiavaroli L, de Souza RJ, et al. 'Catalytic' doses of fructose may benefit glycaemic control without harming cardiometabolic risk factors: a small meta-analysis of randomised controlled feeding trials. Br J Nutr 2012;108(3):418–23.

[173] Metz G, Jenkins DJ, Newman A, Blends LM. Breath hydrogen in hyposucrasia. Lancet 1976;1(7951):119–20.

[174] Metz G, Jenkins DJ, Peters TJ, Newman A, Blendis LM. Breath hydrogen as a diagnostic method for hypolactasia. Lancet 1975;1(7917):1155–7.

Chapter 23

Prebiotics: Inulin and Other Oligosaccharides

S. Mitmesser and M. Combs
Department of Nutrition & Scientific Affairs, NBTY, Inc., Ronkonkoma, NY, United States

INTRODUCTION

Interest in the potential of prebiotics for gastrointestinal health and therapeutic benefit began as early as 1986, before the term "prebiotic" was formally introduced. As improved analytical methods have dramatically increased our knowledge on the subject of the human gastrointestinal microbiota in recent years, attention to these nondigestible compounds has also intensified. This emerging area of research is uncovering associations with systems and conditions from allergy and immunology, to inflammatory bowel disease, the gut–brain axis and metabolic syndrome, with mechanisms of action originating in the gut. Therefore, it is important to understand prebiotics—the criteria that define them, their sources and structures, and their actions.

DEFINITIONS

Prebiotics were first defined in 1995 by Gibson and Roberfroid as a "non-digestible food ingredient that beneficially affects the host by selectively stimulating the growth and/or activity of one or a limited number of bacteria in the colon, and thus improves host health" [1,2]. As our grasp of the gut microbiota changed over the next two decades, so too did the definition of prebiotic. Following their inaugural meeting, The International Scientific Association for Probiotics and Prebiotics (ISAPP) proposed a new definition in 2003, which broadened the scope of prebiotic activity beyond the colon to other colonized body sites, and allowed for physiological effects as markers of benefit in addition to health measures [3]. In 2004, Gibson expanded the scope of the original definition to include compositional changes within the whole of the GI microbiota from any fermentable substrate [2,4]. The most precise definition of prebiotics is still the subject of debate. While the early definitions, and indeed much of the prebiotic literature, focus on the increase in abundance and/or activity of a select few

beneficial species to improve host health, some experts feel these definitions are limiting and misaligned with the evolution of our understanding of the microbiome [5]. A more comprehensive definition was recommended in 2007 by the Food and Agricultural Organization (FAO) Technical Meeting of experts, which aimed to release the prebiotic concept from the confines of selectivity, fermentability, and causality. "A prebiotic is a non-viable food component that confers a health benefit on the host associated with modulation of the microbiota" [2,5]. An ISAPP working group of 2008, however, did not take into account the FAO proposal, but maintained these constraints [5–7].

At the 2014 FAO/WHO International Conference on Nutrition, Barbara Burlingame, Deputy Director, Nutrition Division, FAO, acknowledged a lack of follow-up to the FAO Technical Meeting experts' recommendation that a full Expert Consultation on prebiotics be considered. She encouraged attendees who felt this important subject was being ignored to issue a call to action to the heads of the four world food agencies in Rome [8]. Bindels et al. urged further expansion of the prebiotic concept to include "a nondigestible compound that, through its metabolization by microorganisms in the gut, modulates composition and/or activity of the gut microbiota, thus conferring a beneficial physiological effect on the host," in hopes to foster greater development of microbiota-modulating therapies [5].

MEASUREMENTS

Measurement of the growth and/or modulation of commensal gut bacteria associated with their metabolism of prebiotic fibers has primarily been performed by cultural enumeration methods. This practice limits results to only culturable species, most frequently bifidobacteria and lactobacilli. Strict use of culture methods overlooks the many genera within the gut microbial community that cannot be cultured, but that may also prove beneficial to the host, such as *Eubacterium*, *Faecalibacterium*, *Roseburia*, and select species of clostridia [9,10].

Now that advanced molecular methods with broader application are used, their respective shortcomings must also be considered; reverse transcription quantitative polymerase chain reaction (qRT-PCR) and fluorescence in situ hybridization (FISH) focus on selected groups that depend on the choice of probes, while the denaturing gradient gel electrophoresis (DGGE) and terminal-restriction fragment length polymorphism methods only measure the major species representative of the microbial community and thus may not detect all changes with physiological impact [9].

It has been suggested that measures of the activity of the metabolome may better capture changes with potential for health impact [11]. Difficulties inherent in measuring activity and changes within the intestinal lumen have been recognized. In their section on substantiating a prebiotic, the FAO Technical Meeting of experts indicated that fecal analysis is acceptable pending development of more precise, real-time, in situ methodology [2].

TYPES

The nondigestible compounds with the most robust evidence supporting their role as prebiotics include the fructans, inulin and fructooligosaccharides (FOS), as well as galactooligosaccharides (GOS) [2,5,12] (Table 23.1). Here we explore the evidence that these compounds are nondigestible, metabolized by microorganisms in the gut, and modulate the composition and/or activity of the gut microbiota, with brief mention of results for physiological benefits in some cases. A more thorough discussion of the beneficial physiological effects of these compounds on the host will be covered in subsequent chapters.

TABLE 23.1 Accepted Prebiotics Have the Most Robust Evidence Supporting this Classification, While Candidate Prebiotics Have Some Prebiotic Properties, But With Varying Levels of Less Convincing Evidence

Accepted prebiotics
Inulin
Fructooligosaccharides
Galactooligosaccharides
Candidate prebiotics
Xylooligosaccharides
Isomaltooligosaccharides
β-Glucans
Polyphenols

Fructans

Fructans are a group of oligo- and polysaccharides composed of fructose units connected with β-$(2\rightarrow1)$ linkages, and frequently terminating in a glucosyl moiety [13]. The shortest members of this structural classification are called oligofructose or FOS, and consist of 2–9 units, while fructans with 10 or more monomeric units are categorized as inulin [13–17]. The number of units in a polysaccharide chain is also frequently referred to as degrees of polymerization (DP).

Many plants store carbohydrates in the form of inulin. Globe and Jerusalem artichokes, chicory, and agave are plants used for the commercial extraction of inulin, but other foods, such as wheat, bananas, onions, and garlic also contain inulin [16,18–20]. FOS can be produced by the enzymatic hydrolysis of longer chain fructans using endoglycosidase enzymes [13,21]. Fructans can also be enzymatically synthesized from sucrose via transfructosylation [12,22].

Humans lack digestive enzymes to break down fructans' β-$(2\rightarrow1)$ linkages, making them indigestible fibers [16,19,21]. However, bifidobacteria and *Lactobacillus* species produce the 2,1-β-D-fructan-fructanohydrolase enzyme to hydrolyze these fibers [23].

The clinical research on the prebiotic effects of fructans is rich and varied with studies dating back 30 years. FOS are among the first fibers studied for their effects on the intestinal microbiota and human health, even before the prebiotic concept was officially recognized [12]. Japanese researchers hypothesized that their Neosugar, a FOS mixture, would modulate the gastrointestinal microbiota in humans based on preclinical experiments. Indeed, bifidobacteria were increased in healthy humans. Additionally, observations in specific cases indicated that baseline bifidobacteria levels could predict the magnitude of modulation, and hinted at secondary metabolism of lactic acid from bifidobacteria by *Veillonella*, which also increased [22].

That was just the beginning of prebiotic research, which has continued and developed over the years. In 1996 Bouhnik et al. reported that 12.5 g/day FOS was well tolerated in healthy adults, but failed to show a benefit in risk factors for colon cancer [24]. In 1999 Bouhnik confirmed the bifidogenicity of FOS in a dose ranging study, concluding that 10 g/day was optimum for tolerance and modulation. Total anaerobes were not increased, however, as measured using culture methods [25]. Bouhnik et al. later found a dose response relationship for the bifidogenic effect of FOS from 2.5 to 10 g/day FOS in healthy adults [26].

Inulin from Globe artichoke increased both bifidobacteria and lactobacilli, along with the genus *Atopobium*, in healthy adults, while it decreased *Bacteroides–Prevotella*. This expanded view of modulation may have been due in part to the use of the more broadly applicable fluorescent in situ hybridization method [19]. An FOS-enriched inulin, Synergy1, also increased bifidobacteria in healthy middle-aged adults at 8 g/day, but did not alter the immune responses examined. This was attributed to a lack of immune challenge in the study design [27]. In a healthy elderly population, Bouhnik et al. found a significant increase in fecal bifidobacteria after 8 g/d FOS for 4 weeks, along with positive changes in cholesterol metabolism [28].

Agave is another source of mixed fructans with more diverse and highly branched structures that vary depending on the species used. These complex mixtures contain both β-(2→1) and β-(2→6) linkages, as well as internal and external glucose units [18]. In two different trials, agave inulin was shown to modulate the fecal microbiota. In one study at 5 g/day both bifidobacteria and lactobacilli increased significantly, while in the second study 5 and 7.5 g/day increased total *Actinobacteria* and specifically the bifidobacteria, while decreasing *Desulfovibrio*. These seeming disparities may have been due to the different methods of analysis used in these two studies [18,29].

In healthy young males, chicory inulin at 20 g/day increased total anaerobes and lactobacilli, but did not increase bifidobacteria. There were concomitant changes in markers of precarcinogen production in the colon [23]. Another trial in males showed positive effects of an inulin-enriched pasta on markers of intestinal barrier function, including urinary lactulose recovery, and circulating zonulin and GLP-2 [30]. Other studies in healthy male subjects by the same authors showed metabolic benefits as well as improvements in gastric emptying time [31–33].

An equal mixture of inulin and FOS at a dose of 16 g/day administered to obese females resulted in modulation of the microbiota. Bifidobacteria and *Faecalibacterium prausnitzii* were increased, and there was a negative association with serum lipopolysaccharide (LPS). *Bacteroides intestinalis*, *Bacteroides vulgatus*, and *Propionibacterium* were also decreased, an effect that was associated with a slight decrease in fat mass and other markers of obesity and diabetes [34]. This increase in bifidobacteria was confirmed in another study in obese women also taking 16 g/day of inulin. In this study *Bifidobacterium longum*, *Bifidobacterium catenulatum,* and *Bifidobacterium adolescentis* were all increased, the increase in *B. longum* was negatively correlated with serum LPS, and other metabolic risk factors were improved. This study did not look at any other species [35]. Yacon syrup is also a good source of FOS. When obese premenopausal women with insulin resistance took 140 mg/kg for 4 months, their body weight, waist circumference, body mass index, fasting serum insulin, and Homeostasis Model Assessment index were improved, along with defecation frequency and satiety [36].

A combination of inulin and partially hydrolyzed guar gum at a dose of 15 g/day in constipated female health-care workers decreased *Clostridium* genera, but surprisingly there were no increases in bifidobacteria and *Lactobacillus*, nor decreases in *Bacteroides* and *Escherichia*. There was also no difference in the improvement of bowel habits compared to placebo. The authors hypothesized that other species not measured in their study may have increased, causing the decrease in clostridia [37].

In a crossover study, 20 g of inulin per day showed prebiotic activity in women of childbearing age with low iron status, increasing the levels of fecal bifidobacteria and lactate, and decreasing fecal pH. While the authors theorized that iron status could be improved based on preclinical studies and early human cases, no improvement in iron absorption from standardized test meals was detected in this study [38].

Prebiotics seem to work differently in some critically and chronically ill patient populations. Levels of *F. prausnitzii*, along with *Bacteroides–Prevotella*, were reduced in critically ill patients who received an additional 7 g/day inulin and FOS in their fiber containing enteral nutrition. This reduction in *F. prausnitzii* was surprising, as was the lack of increase in bifidobacteria. The authors attributed these results to the ailing status of the subjects and other potential variables [39]. Similarly, FOS given to Crohn's disease patients impacted intestinal dendritic cell function but did not impact *F. prausnitzii* or bifidobacteria levels, or clinical symptoms [17]. In contrast, a prebiotic mixture of inulin and FOS helped gynecological cancer patients recover levels of both bifidobacteria and *Lactobacillus*, which were reduced by radiotherapy [40].

Galactooligosaccharides

GOS are another type of nondigestible fiber with prebiotic activity. β-GOS, also referred to as *trans*-GOS, are β-(1→6) linked galactosyl residues that terminate in a β-(1→4) linked glucose unit, with a mixture of other bond types present [12,41]. β-Linked GOS are made by enzymatic synthesis from lactose using bacterial galactosidases [42,43]. α-GOS are similar, but with α–(1→6) linkages and come primarily from soybeans, and their prebiotic effects are less well defined [12].

Around the same time prebiotic research on the fructans began, investigations with GOS began. Japanese researchers administered an oligosaccharide mixture, Oligomate-50, providing 2.5, 5, or 10 g/day GOS in a single-blind crossover trial to healthy male volunteers. They found a dose response relationship for the increases in both bifidobacteria and lactobacilli. The highest dose was well tolerated, so it was concluded at that time that 10 g/day was the optimal dose [44]. In contrast, a later study gave 7.5 or 15 g/day to healthy males and females, and found that these doses did not increase bifidobacteria compared to placebo, because the placebo group also experienced an increase. Interestingly, there were no effects on bowel habits, or multiple measures of fecal composition, including short-chain fatty acids, even though breath tests indicated that the GOS was completely fermented. The authors speculated that the subjects' baseline bifidobacteria counts might have been too high for the GOS to elicit a difference [45].

Davis et al. compared the bifidogenic effects of a commercial GOS and a novel GOS produced using an enzyme from a specific strain of *Bifidobacteria bifidum*. There was a significant increase in the bifidobacteria counts compared to both placebo and the commercial GOS, as well as a dose response effect up to 7 g/day in the second phase of the study. These results suggest that production of oligosaccharides using enzymes from probiotic bacterial strains, may increase the prebiotic activity of the fiber [46]. A reduction in traveler's diarrhea in healthy volunteers was also attributed to the prebiotic effects of this same novel GOS mixture. It was assumed that the preventative effect occurred due to the known increases in beneficial species, such as bifidobacteria, which can contribute to the resistance of intestinal colonization by pathogenic bacteria and viruses [47].

A novel analytical method, high-throughput multiple community sequencing of 16s rDNA tags, was used to measure changes in the microbiota of healthy subjects after intakes of 2.5, 5, or 10 g/day GOS for 12 weeks. The intervention increased *Actinobacteria*, specifically bifidobacteria, and at the expense of *Bacteroides*. This broader method revealed that GOS was remarkably specific in its enrichment of certain species of bifidobacteria. This was unexpected because a preceding in vitro investigation indicated that GOS could be utilized by other genera in bacterial monocultures. The fact that growth was restricted to bifidobacteria in vivo suggests competition for nutrients [9].

Studies in elderly subjects, a population with known decreases in bifidobacteria, confirm a bifidogenic effect of GOS. Men and women over 50 years of age consumed 4 g/day GOS and placebo in a crossover study for 3 weeks each. Bifidobacteria levels were increased with GOS compared to placebo [48]. Men and women around 70 years of age taking 5.5 g/day of GOS also experienced a beneficial modulation of their gastrointestinal microbiota. *Bifidobacterium* spp., *Lactobacillus–Enterococcus* spp., and the *Clostridium coccoides–Eubacterium rectale* increased while *Bacteroides*, *Clostridium histolyticum*, *Escherichia coli*, and *Desulfovibrio* spp. decreased [49]. Some of these results were confirmed in a second group of volunteers ranging in age from 65 to 80 years old. GOS at a dose of 5.5 g/day reversed the age related decline in bifidobacteria, but in this case also increased *Bacteroides* [50].

There is a suspicion is the scientific community that irritable bowel syndrome (IBS) is associated with the activity of the gut microbiota, and this has led researchers to test the effects of prebiotics in this patient population. In subjects meeting the Rome II criteria for IBS, as with healthy subjects, bifidobacteria were increased with 3.5 and 7 g/day

GOS. Other bacterial groups were modulated including increases in *E. rectale-C. coccoides* at 3.5 g, and reductions in *Clostridium perfringens* subgroup *histolyticum* and *Bacteroides–Prevotella* at both doses. This was accompanied by an improvement in IBS, as well as psychological symptoms [51]. A dose of 5.5 g of the same GOS mixture was given daily to an overweight sample population, who also experienced an increase in bifidobacteria. Similarly, the subjects had decreases in *C. histolyticum* and *Bacteroides*, but there was no change in the *E. rectale-C. coccoides* genera in this particular study. Markers of immune function and metabolic syndrome were also improved [52].

CANDIDATE PREBIOTICS

There are many nondigestible compounds with potential prebiotic properties and with varying levels of scientific evidence that is not quite as convincing as that for the fructans and GOS. These candidate prebiotics are covered briefly here [6] (Table 23.1).

Other Oligosaccharides

The xylooligosaccharides (XOS) are made up of xylose units linked by β-(1→4) bonds with DP ranging from 2 to 10, that can reportedly be found naturally in some foods, including bamboo shoots, fruits, vegetables, milk, and honey [53–55]. XOS are produced commercially from xylan containing lignocellulosic materials by chemical and/or enzymatic methods [54]. Doses ranging from 1 to 5 g/day in healthy young, middle-aged, or elderly adults have been shown to be bifidogenic [15,55,56]. There were varying effects on markers of intestinal function, but further research is needed to determine clinical benefits associated with these changes [15].

Isomaltooligosaccharides (IMO) are another type of nondigestible fiber metabolized by colonic microbiota, and made up of α-(1→6) glucosidic linkages [57,58]. Miso, soy sauce, sake, and honey are some foods that contain IMO [59]. IMO are prepared enzymatically from cornstarch and used commercially as a sweetener in Japan [57,60]. A handful of small studies in limited populations have suggested prebiotic functions of IMO. A mixture of IMO given to healthy adult males in doses from 10 to 13.5 g/day was bifidogenic, especially in those subjects with low baseline levels [58–60]. IMO has also been shown to be bifidogenic in senile and constipated elderly males and females at doses of 10 or 13.5 g/day, respectively [58,60]. Increases in fecal acetate and propionate, and reductions in clostridia were also measured in the constipated elderly. The authors suggested that the higher dose of IMO required to elicit a bifidogenic effect, as compared to FOS which is bifidogenic at doses as low as 2.5 g/day, may be due to the partial digestibility of the IMO in the jejunum [58].

β-Glucans

β-Glucan is comprised of glucose residues which are linked by β-(1–4) and β-(1–3) glycosidic bonds. Single β-(1–3) linkages are generally separated by 2 or 3 β-(1–4) linkages but the ratio between β-(1–4) and β-(1–3) linkages differs between cereal species [61,62]. The linkage ratio can vary between tissues and species with longer cellulose-like regions of continuous β-(1-4) linkages in wheat bran β-glucan [63], as an example, with β-glucan accounting for about 20% of the total cell wall polysaccharides in the starchy endosperm of wheat [64,65]. Food sources of β-glucan include edible mushrooms, yeast, and grains, such as oat, barley, and wheat.

Numerous systematic reviews and meta-analyses indicate the use of β-glucan as a prebiotic fiber [66–72]. The difference in molecular weight (MW) may be the rate limiting factor as to whether or not β-glucan clinical trials result in a clinical benefit. Many factors can affect the MW of β-glucan, such as food processing and the source of β-glucan. Heating, in particular, can decrease the MW of β-glucan and therefore decrease its viscosity inside the GI tract [73]. Different sources of β-glucan may also have different MW and viscosity. It has been noted that the MW among oat varieties differ, which leads to the assumption that the MW can also vary among cereal grain types [74]. β-Glucan from oat and barley also differ in their solubility which can directly affect intestinal viscosity [74,75], with a higher proportion of the total fraction being soluble in oats and barley than in wheat which is likely due to the chain length (shorter chains being more soluble) [69,76]. As mentioned previously, the basic structure (linkage pattern) may account for the differences that occur in the properties of β-glucan, specifically solubility and viscosity.

The possible mechanisms in which β-glucan works within the human body and acts as a prebiotic is most likely through its viscous nature and the production of short-chain fatty acids. The viscosity of β-glucan in the GI tract is the most probable mechanism behind the clinical benefits observed in humans, such as decreases in serum cholesterol levels [65,77,78] and improvements in postprandial glucose metabolism [76,79]. The gelation property may increase intestinal viscosity and thereby decrease bile acid absorption and increase bile acid excretion, resulting in a higher hepatic cholesterol synthesis because of the higher need for bile acid synthesis [80]. The viscosity may also delay glucose absorption into the blood improving postprandial glucose and insulin levels [79].

β-Glucan's ability to produce short-chain fatty acids is also a likely mechanism behind its observed metabolic effects. Larger amounts of propionate are produced from the fermentation of oat β-glucan [81,82]. Propionate has been shown to significantly inhibit cholesterol synthesis in humans and is thought to be due to the inhibition of the rate-limiting enzyme HMG CoA reductase [83,84].

Polyphenols

Polyphenols are a diverse and complex group of phytochemicals metabolized to varying degrees by the gut microbiota, possibly resulting in health benefits, a set of interactions that has been termed "polyphenols–microbiota–host triangle" [85] The metabolites resulting from these interactions may have positive benefits for humans [86]. Polyphenols from a variety of food sources have been shown to increase, decrease, or have no effect on different sets of human colonic microbes, as measured using a variety of analytical methods [85]. Studies published in 2013 and 2014 showed polyphenols from red wine and from wild blueberries increased bifidobacteria in humans [87,88], while a novel polyphenol-based concentrate, Aliva, was helpful in acute gastroenteritis in children and adults when diluted with oral rehydration solution [89]. Additional preclinical and clinical studies are needed to identify the beneficial compounds, determine their efficacious levels, and delineate their specific health benefits in humans [90].

CONCLUSIONS

The study of prebiotics, and their role in the health of gut microbiota and host, is in a stage of rapid progression. There is currently no clear consensus on the precise definition of prebiotics, although a broad scope will certainly open the door for further exploration and benefit to public health. Evolution in this field includes expeditious development of analytical techniques, bringing a flood of information. Established prebiotics that fit the proposed definitions include the fructans, inulin and fructooligosaccharides, as well as galactooligosaccharides. Candidate prebiotics, such as β-glucans, other oligosaccharides, and polyphenols, are scientifically supported for modulating gut microbiota, but further research on associated benefit to host physiology is needed before they are fully embraced.

REFERENCES

[1] Gibson GR, Roberfroid MB. Dietary modulation of the human colonic microbiota: introducing the concept of prebiotics. J Nutr 1995;125(6):1401–12.

[2] Pineiro M, Asp NG, Reid G, et al. FAO technical meeting on prebiotics. J Clin Gastroenterol 2008;42(Suppl. 3 Pt 2):S156–9.

[3] Reid G, Sanders ME, Gaskins HR, et al. New scientific paradigms for probiotics and prebiotics. J Clin Gastroenterol 2003;37(2):105–18.

[4] Gibson GR, Probert HM, Loo JV, Rastall RA, Roberfroid MB. Dietary modulation of the human colonic microbiota: updating the concept of prebiotics. Nutr Res Rev 2004;17(2):259–75.

[5] Bindels LB, Delzenne NM, Cani PD, Walter J. Towards a more comprehensive concept for prebiotics. Nat Rev Gastroenterol Hepatol 2015;12(5):303–10.

[6] Gibson GR, Scott KP, Rastall RA, et al. Dietary prebiotics: current status and new definition. Food Sci Technol Bull Funct Food 2010;7(1):1–19.

[7] Brownawell AM, Caers W, Gibson GR, et al. Prebiotics and the health benefits of fiber: current regulatory status, future research, and goals. J Nutr 2012;142(5):962–74.

[8] Burlingame B. Prebiotics and probiotics and the specialized UN agencies. J Clin Gastroenterol 2014;48(Suppl. 1):S1.

[9] Davis LM, Martinez I, Walter J, Goin C, Hutkins RW. Barcoded pyrosequencing reveals that consumption of galactooligosaccharides results in a highly specific bifidogenic response in humans. PLoS One 2011;6(9):e25200.

[10] Verbeke KA, Boobis AR, Chiodini A, et al. Towards microbial fermentation metabolites as markers for health benefits of prebiotics. Nutr Res Rev 2015;28(1):42–66.

[11] Damen B, Cloetens L, Broekaert WF, et al. Consumption of breads containing in situ-produced arabinoxylan oligosaccharides alters gastrointestinal effects in healthy volunteers. J Nutr 2012;142(3):470–7.

[12] Meyer D. Health benefits of prebiotic fibers. Adv Food Nutr Res 2015;74:47–91.

[13] Roberfroid MB, Van Loo JA, Gibson GR. The bifidogenic nature of chicory inulin and its hydrolysis products. J Nutr 1998;128(1):11–9.

[14] Fedewa A, Rao SSC. Dietary fructose intolerance, fructan intolerance and FODMAPs. Curr Gastroenterol Rep 2014;16(1):370.

[15] Lecerf JM, Depeint F, Clerc E, et al. Xylo-oligosaccharide (XOS) in combination with inulin modulates both the intestinal environment and immune status in healthy subjects, while XOS alone only shows prebiotic properties. Br J Nutr 2012;108(10):1847–58.

[16] Niness KR. Inulin and oligofructose: what are they? J Nutr 1999;129(7 Suppl.):1402s–6s.

[17] Benjamin JL, Hedin CR, Koutsoumpas A, et al. Randomised, double-blind, placebo-controlled trial of fructo-oligosaccharides in active Crohn's disease. Gut 2011;60(7):923–9.

[18] Ramnani P, Costabile A, Bustillo AG, Gibson GR. A randomised, double- blind, cross-over study investigating the prebiotic effect of agave fructans in healthy human subjects. J Nutr Sci 2015;4:e10.

[19] Costabile A, Kolida S, Klinder A, et al. A double-blind, placebo-controlled, cross-over study to establish the bifidogenic effect of a very-long-chain inulin extracted from globe artichoke (*Cynara scolymus*) in healthy human subjects. Br J Nutr 2010;104(7):1007–17.

[20] Srinameb BO, Nuchadomrong S, Jogloy S, Patanothai A, Srijaranai S. Preparation of inulin powder from Jerusalem artichoke (*Helianthus tuberosus* L.) tuber. Plant Foods Hum Nutr 2015;70(2):221–6.

[21] van de Wiele T, Boon N, Possemiers S, Jacobs H, Verstraete W. Inulin-type fructans of longer degree of polymerization exert more pronounced in vitro prebiotic effects. J Appl Microbiol 2007;102(2):452–60.

[22] Hidaka H, Eida T, Takizawa T, Tokunaga T, Tashiro Y. Effects of fructooligosaccharides on intestinal flora and human health. Bifidobacteria Microflora 1986;5(1):37–50.

[23] Slavin J, Feirtag J. Chicory inulin does not increase stool weight or speed up intestinal transit time in healthy male subjects. Food Funct 2011;2(1):72–7.

[24] Bouhnik Y, Flourie B, Riottot M, et al. Effects of fructo-oligosaccharides ingestion on fecal bifidobacteria and selected metabolic indexes of colon carcinogenesis in healthy humans. Nutr Cancer 1996;26(1):21–9.

[25] Bouhnik Y, Vahedi K, Achour L, et al. Short-chain fructo-oligosaccharide administration dose-dependently increases fecal bifidobacteria in healthy humans. J Nutr 1999;129(1):113–6.

[26] Bouhnik Y, Raskine L, Simoneau G, Paineau D, Bornet F. The capacity of short-chain fructo-oligosaccharides to stimulate faecal bifidobacteria: a dose-response relationship study in healthy humans. Nutr J 2006;5:8.

[27] Lomax AR, Cheung LV, Tuohy KM, Noakes PS, Miles EA, Calder PC. beta2-1 Fructans have a bifidogenic effect in healthy middle-aged human subjects but do not alter immune responses examined in the absence of an in vivo immune challenge: results from a randomised controlled trial. Br J Nutr 2012;108(10):1818–28.

[28] Bouhnik Y, Achour L, Paineau D, Riottot M, Attar A, Bornet F. Four-week short chain fructo-oligosaccharides ingestion leads to increasing fecal bifidobacteria and cholesterol excretion in healthy elderly volunteers. Nutr J 2007;6:42.

[29] Holscher HD, Bauer LL, Gourineni V, Pelkman CL, Fahey GC Jr, Swanson KS. Agave inulin supplementation affects the fecal microbiota of healthy adults participating in a randomized, double-blind, placebo-controlled. Crossover Trial J Nutr 2015;145(9):2025–32.

[30] Russo F, Linsalata M, Clemente C, et al. Inulin-enriched pasta improves intestinal permeability and modifies the circulating levels of zonulin and glucagon-like peptide 2 in healthy young volunteers. Nutr Res 2012;32(12):940–6.

[31] Russo F, Chimienti G, Riezzo G, et al. Inulin-enriched pasta affects lipid profile and Lp(a) concentrations in Italian young healthy male volunteers. Eur J Nutr 2008;47(8):453–9.

[32] Russo F, Clemente C, Linsalata M, et al. Effects of a diet with inulin-enriched pasta on gut peptides and gastric emptying rates in healthy young volunteers. Eur J Nutr 2011;50(4):271–7.

[33] Russo F, Riezzo G, Chiloiro M, et al. Metabolic effects of a diet with inulin-enriched pasta in healthy young volunteers. Curr Pharm Des 2010;16(7):825–31.

[34] Dewulf EM, Cani PD, Claus SP, et al. Insight into the prebiotic concept: lessons from an exploratory, double blind intervention study with inulin-type fructans in obese women. Gut 2013;62(8):1112–21.

[35] Salazar N, Dewulf EM, Neyrinck AM, et al. Inulin-type fructans modulate intestinal *Bifidobacterium* species populations and decrease fecal short-chain fatty acids in obese women. Clin Nutr 2015;34(3): 501–7.

[36] Genta S, Cabrera W, Habib N, et al. Yacon syrup: beneficial effects on obesity and insulin resistance in humans. Clin Nutr 2009;28(2):182–7.

[37] Linetzky Waitzberg D, Alves Pereira CC, Logullo L, et al. Microbiota benefits after inulin and partially hydrolyzed guar gum supplementation: a randomized clinical trial in constipated women. Nutr Hosp 2012;27(1):123–9.

[38] Petry N, Egli I, Chassard C, Lacroix C, Hurrell R. Inulin modifies the bifidobacteria population, fecal lactate concentration, and fecal pH but does not influence iron absorption in women with low iron status. Am J Clin Nutr 2012;96(2):325–31.

[39] Majid HA, Cole J, Emery PW, Whelan K. Additional oligofructose/inulin does not increase faecal bifidobacteria in critically ill patients receiving enteral nutrition: a randomised controlled trial. Clin Nutr 2014;33(6):966–72.

[40] Garcia-Peris P, Velasco C, Lozano MA, et al. Effect of a mixture of inulin and fructo-oligosaccharide on *Lactobacillus* and *Bifidobacterium* intestinal microbiota of patients receiving radiotherapy: a randomised, double-blind, placebo-controlled trial. Nutr Hosp 2012;27(6):1908–15.

[41] Coulier L, Timmermans J, Bas R, et al. In-depth characterization of prebiotic galacto-oligosaccharides by a combination of analytical techniques. J Agric Food Chem 2009;57(18):8488–95.

[42] Goulas A, Tzortzis G, Gibson GR. Development of a process for the production and purification of α- and β-galactooligosaccharides from

Bifidobacterium bifidum NCIMB 41171. Int Dairy J 2007;17(6): 648–6456.

[43] Tzortzis G. Development and functional properties of Bimuno®: a second-generation prebiotic mixture. Food Sci Technol Bull: Functl Food 2010;6(7):81–9.

[44] Ito M, Deguchi Y, Miyamori A, et al. Effects of administration of galactooligosaccharides on the human faecal microflora, stool weight and abdominal sensation. Microb Ecol Health Dis 1990;3(6).

[45] Alles MS, Hartemink R, Meyboom S, et al. Effect of transgalactooligosaccharides on the composition of the human intestinal microflora and on putative risk markers for colon cancer. Am J Clin Nutr 1999;69(5):980–91.

[46] Depeint F, Tzortzis G, Vulevic J, I'Anson K, Gibson GR. Prebiotic evaluation of a novel galactooligosaccharide mixture produced by the enzymatic activity of *Bifidobacterium bifidum* NCIMB 41171, in healthy humans: a randomized, double-blind, crossover, placebo-controlled intervention study. Am J Clin Nutr 2008;87(3):785–91.

[47] Drakoularakou A, Tzortzis G, Rastall RA, Gibson GR. A double-blind, placebo-controlled, randomized human study assessing the capacity of a novel galacto-oligosaccharide mixture in reducing travellers' diarrhoea. Eur J Clin Nutr 2010;64(2):146–52.

[48] Walton GE, van den Heuvel EG, Kosters MH, Rastall RA, Tuohy KM, Gibson GR. A randomised crossover study investigating the effects of galacto-oligosaccharides on the faecal microbiota in men and women over 50 years of age. Br J Nutr 2012;107(10): 1466–75.

[49] Vulevic J, Drakoularakou A, Yaqoob P, Tzortzis G, Gibson GR. Modulation of the fecal microflora profile and immune function by a novel trans-galactooligosaccharide mixture (B-GOS) in healthy elderly volunteers. Am J Clin Nutr 2008;88(5):1438–46.

[50] Vulevic J, Juric A, Walton GE, et al. Influence of galacto-oligosaccharide mixture (B-GOS) on gut microbiota, immune parameters and metabonomics in elderly persons. Br J Nutr 2015;114(4): 586–95.

[51] Silk DB, Davis A, Vulevic J, Tzortzis G, Gibson GR. Clinical trial: the effects of a trans-galactooligosaccharide prebiotic on faecal microbiota and symptoms in irritable bowel syndrome. Aliment Pharmacol Ther 2009;29(5):508–18.

[52] Vulevic J, Juric A, Tzortzis G, Gibson GR. A mixture of trans-galactooligosaccharides reduces markers of metabolic syndrome and modulates the fecal microbiota and immune function of overweight adults. J Nutr 2013;143(3):324–31.

[53] Carvalho AFA, Neto PdO, da Silva DF, Pastore GM. Xylo-oligosaccharides from lignocellulosic materials: chemical structure, health benefits and production by chemical and enzymatic hydrolysis. Food Res Int 2013;51(1):75–85.

[54] Aachary AA, Prapulla SG. Xylooligosaccharides (XOS) as an emerging prebiotic: microbial synthesis, utilization, structural characterization, bioactive properties, and applications. Compr Rev Food Sci Food Saf 2011;10(1):2–16.

[55] Chung Y-C, Hsu C-K, Ko C-Y, Chan Y-C. Dietary intake of xylooligosaccharides improves the intestinal microbiota, fecal moisture, and pH value in the elderly. Nutr Res 2007;27(12):756–61.

[56] Finegold SM, Li Z, Summanen PH, et al. Xylooligosaccharide increases bifidobacteria but not lactobacilli in human gut microbiota. Food Funct 2014;5(3):436–45.

[57] Chen HL, Lu YH, Lin JJ, Ko LY. Effects of isomalto-oligosaccharides on bowel functions and indicators of nutritional status in constipated elderly men. J Am Coll Nutr 2001;20(1):44–9.

[58] Kohmoto T, Fukui F, Takaku H, Mitsuoka T. Dose-response test of isomaltooligosaccharides for increasing fecal Bifidobacteria. J Agric Food Chem 1991;55(8):2157–9.

[59] Kohmoto T, Fukui F, Takaku H, Machida Y, Arai M, Mitsuoka T. Effect of isomalto-oligosaccharides on human fecal flora. Bifidobacteria Microflora 1988;7(2):61–9.

[60] Yen CH, Tseng YH, Kuo YW, Lee MC, Chen HL. Long-term supplementation of isomalto-oligosaccharides improved colonic microflora profile, bowel function, and blood cholesterol levels in constipated elderly people—a placebo-controlled, diet-controlled trial. Nutrition 2011;27(4):445–50.

[61] Lazaridou A, Biliaderis C. Molecular aspects of cereal B-glucan functionality: physical properties, technological applications and physiological effects. J Cereal Sci 2007;46:101–18.

[62] Lazaridou A, Biliaderis C, Micha-Scretta M, Steele B. A comparative study on structure-function relations of mixed linkage (1-3), (1-4) linear B-D-glucans. Food Hydrocoll 2004;18:837–55.

[63] Li W, Cui S, Kakuda Y. Extraction, fractionation, structural and physical characterization of wheat B-glucans. Carbohydr Polym 2006;63:408–16.

[64] Mares D, Stone B. Studies on wheat endosperm. Chemical composition and ultrastructure of the cell walls. Aust J Biol Sci. 1973;26: 793–812.

[65] Davidson M, McDonald A. Fiber: forms and functions. Nutr Res 1998;18:617–24.

[66] Lattimer JM, Haub MD. Effects of dietary fiber and its components on metabolic health. Nutrients 2010;2(12):1266–89.

[67] Pastor-Villaescusa B, Rangel-Huerta OD, Aguilera CM, Gil A. A systematic review of the efficacy of bioactive compounds in cardiovascular disease: carbohydrates, active lipids and nitrogen compounds. Ann Nutrition Metab 2015;66(2–3):168–81.

[68] Clemens R, van Klinken BJ. The future of oats in the food and health continuum. Br J Nutr 2014;112(Suppl. 2):S75–9.

[69] Fardet A. New hypotheses for the health-protective mechanisms of whole-grain cereals: what is beyond fibre? Nutr Res Rev 2010;23(1):65–134.

[70] Gangopadhyay N, Hossain MB, Rai DK, Brunton NP. A review of extraction and analysis of bioactives in oat and barley and scope for use of novel food processing technologies. Molecules (Basel, Switzerland) 2015;20(6):10884–1909.

[71] Hou Q, Li Y, Li L, et al. The metabolic effects of oats intake in patients with type 2 diabetes: a systematic review and meta-analysis. Nutrients 2015;7(12):10369–87.

[72] Lafiandra D, Riccardi G, Shewry PR. Improving cereal grain carbohydrates for diet and health. J Cereal Sci 2014;59(3):312–26.

[73] Suortti T, Johansson L, Autio K. Effect of heating and freezing on molecular weight of oat B-glucan. AACC Annual Meeting, November 5–9: Kansas City, MO; 2000.

[74] Gaidosova A, Petruldkova Z, Havrlentova M, et al. The content of water-soluble and water-insoluble beta-D-glucans in selected oats and barley varieties. Carbhohydr Polym 2007;70:46–52.

[75] Clark A, Stone B. Enzymic hydrolysis of barely and other beta-glucans by a beta-(1,4)-glucan hydrolase. J Biochem 1966;99:582–8.

[76] El-Khoury D, Cuda C, Luhovyy B, Anderson G. Beta glucan: health benefits in obesity and metabolic syndrome. J Nutr Metab 2012;28. ID 851362.

[77] Naumann E, van Rees AB, Onning G, Oste R, Wydra M, Mensink RP. Beta-glucan incorporated into a fruit drink effectively lowers serum LDL-cholesterol concentrations. Am J Clin Nutr 2006;83(3):601–5.

[78] Theuwissen E, Mensink RP. Simultaneous intake of beta-glucan and plant stanol esters affects lipid metabolism in slightly hypercholesterolemic subjects. J Nutr 2007;137(3):583–8.

[79] Nazare JA, Normand S, Oste Triantafyllou A, Brac de la Perriere A, Desage M, Laville M. Modulation of the postprandial phase by beta-glucan in overweight subjects: effects on glucose and insulin kinetics. Mol Nutr Food Res 2009;53(3):361–9.

[80] Lia A, Hallmans G, Sandberg AS, Sundberg B, Aman P, Andersson H. Oat beta-glucan increases bile acid excretion and a fiber-rich barley fraction increases cholesterol excretion in ileostomy subjects. Am J Clin Nutr 1995;62(6):1245–51.

[81] Kim H, White P. In vitro bile-acid binding and fermentation of high, medium, and low molecular weight beta-glucan. J Agric Food Chem 2010;58:628–34.

[82] Kim H, White P. In vitro fermentation of oat flours from typical and high beta-glucan oat lines. J Agric Food Chem 2009;57:7529–36.

[83] Amaral L, Morgan D, Stephen A, Whiting S. Effect of propionate on lipid-metabolism in healthy-human subjects. FASEB J 1992;6:A1655.

[84] Ide T, Okamatsu H, Sugano M. Regulation by dietary fats of 3-hydroxy-3-methylglutaryl-Coenzyme A reductase in rat liver. J Nutr 1978;108(4):601–12.

[85] Duenas M, Munoz-Gonzalez I, Cueva C, et al. A survey of modulation of gut microbiota by dietary polyphenols. Biomed Res Int 2015;2015:850902.

[86] Duda-Chodak A, Tarko T, Satora P, Sroka P. Interaction of dietary compounds, especially polyphenols, with the intestinal microbiota: a review. Eur J Nutr 2015;54(3):325–41.

[87] Boto-Ordonez M, Urpi-Sarda M, Queipo-Ortuno MI, Tulipani S, Tinahones FJ, Andres-Lacueva C. High levels of Bifidobacteria are associated with increased levels of anthocyanin microbial metabolites: a randomized clinical trial. Food Funct 2014;5(8):1932–8.

[88] Guglielmetti S, Fracassetti D, Taverniti V, et al. Differential modulation of human intestinal *Bifidobacterium* populations after consumption of a wild blueberry (*Vaccinium angustifolium*) drink. J Agric Food Chem 2013;61(34):8134–40.

[89] Noguera T, Wotring R, Melville CR, Hargraves K, Kumm J, Morton JM. Resolution of acute gastroenteritis symptoms in children and adults treated with a novel polyphenol-based prebiotic. World J Gastroenterol 2014;20(34):12301–7.

[90] Etxeberria U, Fernandez-Quintela A, Milagro FI, Aguirre L, Martinez JA, Portillo MP. Impact of polyphenols and polyphenol-rich dietary sources on gut microbiota composition. J Agric Food Chem 2013;61(40):9517–33.

Chapter 24

The Benefits of Yogurt, Cultures, and Fermentation

M. Freitas

Health Affairs, The Dannon Company Inc., White Plains, NY, United States

GENERAL CONSIDERATIONS ABOUT YOGURT, FERMENTED DAIRY PRODUCTS, AND PROBIOTICS

Definitions of Yogurt, Fermented Dairy Products, Cultures, and Probiotics

Yogurt is defined by the Food and Drug Administration (FDA) as a fermented dairy product derived from the fermentation of milk by two species of bacterial cultures, *Streptococcus thermophilus* (*S. thermophilus*) and *Lactobacillus bulgaricus* (*L. bulgaricus*). There are additional requirements regarding the composition of the product with respect to fat content, acidity, and amounts of nonfat milk solids, mainly protein content. The codex standard for fermented milks (codex standard 243-2003) stipulates that the total content of these two cultures must be 10^7 CFU/g. *L. bulgaricus* and *S. thermophilus* act as "thermophilic" starter in the manufacturing of yogurt.

Yogurts can also contain additional cultures, such as *Lactococcus lactis*, *Lactobacillus casei*, or different species of *Bifidobacterium*, among others. However, in this case, the product is regarded as yogurt with added bacteria. These additional cultures are usually included in the fermentation process either for technological, taste or texture reasons, or for their additional probiotic properties. The Joint Food and Agriculture Organization/World Health Organization Working Group (FAO/WHO) defines probiotics as live microorganisms which, when administered in adequate amounts, confer a health benefit in the host [1].

The majority of the cultures present in yogurt belong to the group of microorganisms known as lactic acid bacteria (LAB) because they use lactose, the naturally occurring milk sugar, and transform it into lactic acid. LAB include, but is not limited to, species of *Lactobacillus* (e.g., *L. casei* and *Lactobacillus acidophilus*), *Lactococcus*, and *S. thermophilus*. The genus *Bifidobacterium* is distinct from the other LAB, but many *Bifidobacterium* species are able to ferment milk. Historically, LAB have been used for preservation of food by fermentation for thousands of years. This was also the case for yogurt, which was initially created as a means to preserve milk and extend its shelf life. Yogurts that are heat treated after fermentation do not contain viable cultures and must be labeled "heat treated." In addition to not having viable microorganisms, the nutrition profile of "heat-treated" yogurt is different from conventional yogurt because the active cultures continue to transform the milk during shelf life.

Apart from its nutritional properties, the most recognized benefit of conventional yogurt is to help with lactose digestion in lactose maldigesters because of the presence and fermentation of milk by *S. thermophilus* and *L. bulgaricus*. More recently, several investigators have hypothesized that milk fermentation by LAB is also associated with benefits related to weight management, diabetes risk management, and heart health.

Safety and Survival of LAB Through the Gastrointestinal Tract

Fermented foods, such as yogurt have a long history of safe use and it is generally recognized that commercial LAB and probiotics strains are safe for the healthy individual. Usually, cultures used in dairy food products belong to the groups of *Lactobacillus*, *Bifidobacterium*, and *S. thermophilus* which have been historically used in fermented foods, such as yogurt, cheese, and milk. These properties are stable during the refrigeration process.

Yogurt cultures are known to suffer from exposure to gastric acid conditions and have a moderate ability to adhere to intestinal epithelial cells [2,3]. Survival during passage through the gastrointestinal tract is generally considered a key feature for probiotics to preserve their expected health-promoting effects [4]. Although most probiotics cultures have been shown to survive passage through the gastrointestinal tract, there have been conflicting studies concerning the recovery of *L. bulgaricus* and *S. thermophilus* from fecal samples after daily yogurt ingestion. Some authors reported that *L. bulgaricus* and *S. thermophilus* were not recovered from the feces of young [5] and elderly [6] subjects, while others showed that yogurt cultures were recovered after healthy volunteers were fed yogurt containing strains of *S. thermophilus* and *L. bulgaricus* [7]. Despite the fact that there are still controversial results regarding the survival of yogurt cultures in the gastrointestinal tract, consumption of yogurt may result in beneficial changes to the balance and metabolic activities of the indigenous microbiota as recently shown by some authors [8,9].

BENEFITS OF YOGURT AND FERMENTED DAIRY PRODUCTS

This chapter focuses on the benefits of yogurts and fermented dairy products, and not on individual probiotic strains, which are described to a large extent in other chapters of this book. The potential underlying mechanisms behind many of the benefits described in this chapter have been linked to the nutrient density of yogurt (e.g., calcium, vitamin D) and to the presence of LAB in yogurt. Although gaps exist in our knowledge regarding the mechanisms by which yogurt cultures and its nutrients modulate the microbiota and various physiological functions, emerging evidence suggests that gut microbiota play an important role in modulating host physiology and that both LAB and certain nutrients can have an impact in the gut microbiota and in other gut physiological functions. The following sections describe a selection of studies evaluating the benefits of yogurt related to nutrient density, balance diet, lactose intolerance, weight management, heart health, diabetes, immune function, and digestive function.

YOGURT AND NUTRIENT DENSITY

Nonfat and low-fat yogurts are nutrient dense dairy products that often contain several nutrients lacking in the American diet, such as calcium, vitamin D (when added), and potassium. Yogurt can also be an excellent source of high-quality protein, which promotes satiety, helps in maintaining a healthy body weight, and helps muscle and bone growth [10]. In addition, yogurt is low in sodium and contributes only 1.0% or less of added sugars to the diets of most individuals in the United States. As defined by the Food and Drug Administration (FDA), a yogurt or food that contains 10% of the required daily intake (RDI) of a specific nutrient (e.g., calcium) is considered a good source of that nutrient, while if it contains 20% or more of the RDI is considered to be an excellent source.

The nutrient composition of yogurt and milk are similar, however, yogurt is a more concentrated source of riboflavin, vitamin B12, calcium, magnesium, and potassium, as well as other nutrients [11]. Low-fat yogurt contains approximately 25% more potassium, calcium, and magnesium per 8-oz serving compared with an equal serving of low-fat milk [10,12]. The nutrient profile of yogurt is therefore unique and is a result of both the original nutrition profile of milk and also the fermentation process. Other nutrients not inherently present in yogurt can also be added either before or after fermentation (e.g., additional vitamins, antioxidants, fiber).

The 2015 Dietary Guidelines for Americans (DGA) recommends that individuals choose foods that provide more calcium, vitamin D, potassium, and dietary fiber; and increase intake of nonfat or low-fat dairy products, such as milk, yogurt, cheese, or fortified soy beverages. In addition, the DGA also recommends that Americans consume 3 servings of nonfat or low-fat dairy products per day. Yet, according to the USDA Center for Nutrition Policy and Promotion and the DGA, less than one-third of the population meets the dietary recommendations for servings of dairy (milk, cheese, yogurt) [13] averaging only 52% of the recommended intake with young adults having the greatest tendency for insufficient intakes [14,15].

Consuming one 8-oz serving of nonfat or low-fat yogurt every day provides, in many cases, 30% of the daily value for calcium, helping to close the calcium deficiency gap. Low-fat yogurt is not high in saturated fat and is lower in sodium than most cheeses. Additionally, choosing a vitamin D–fortified yogurt can make a significant contribution to vitamin D intake.

Potassium

Increasing potassium intake can decrease the blood pressure–raising effects of excess sodium. Inadequate potassium intake may also increase the risk of kidney stones and osteoporosis. The DGAC recommended that Americans increase their intake of potassium, ideally up to 4700 mg a day for adults. Men consume, on average, less than 3200 mg/day and women less than 2400 mg/day [14]. Low-fat and nonfat yogurts are low in sodium and are, for the most part, a good source of potassium. A single serving of yogurt provides 6–14% of the RDI for potassium; however, only about 1% of the current potassium intake of the US population comes from yogurt [16].

Calcium and Vitamin D

Adequate calcium and vitamin D intake may reduce the risk of osteoporosis. Calcium intake increases the rate of bone acquisition and it occupies about one-third of bone mineral content. Increasing calcium intake increases the amount of absorbed calcium and decreases bone resorption [17]. Vitamin D supports calcium absorption and thereby maintains level of calcium and phosphorus in the blood. Vitamin D has also been shown to increase insulin-like growth factor 2 (IGF-2) production and upregulate insulin-like growth factor binding protein (IGFBP)5 mRNA which is important in bone growth [18].

The recommended calcium intake ranges from 800 to 1300 mg/day, depending on the age and gender of the individual [19]. The average total intake of calcium from all sources in the US population ranges from 918 to 1296 mg/day with approximately 72% coming from milk, cheese, yogurt, and foods to which dairy products have been added (pizza, lasagna, dairy desserts) [20]. One 8-oz serving of nonfat or low-fat yogurt provides up to 80% more calcium than an equal serving of milk [10] and three to 4 servings of dairy would ensure adequate intakes of calcium for everyone over the age of 9 years and increase intake of other nutrients in foods within the dairy group, including potassium, magnesium, phosphorus, and vitamins D, A, B12, riboflavin, and niacin [21].

Results from the Framingham Heart Study Offspring Cohort showed that consuming more than 4 servings of yogurt a week, but not milk or cheese, was associated with a protective trend against hip fracture and greater bone mineral density as compared to consuming fewer servings of yogurt a week [22]. In this study, no other dairy food showed a similar protective effect on reducing the rate of hip fractures. In pregnant women at 28 weeks gestation, consumption of fermented dairy products was significantly related to the bone mineral content of their offspring at the age of 6 years [23].

Calcium can also be consumed in the form of supplements. Calcium supplements are not nutritionally equivalent substitutes for dairy foods, such as milk or yogurt, which provide a unique combination of nine essential nutrients and may be more beneficial for bone health [24,25]. If calcium supplements are used, care must be taken to ensure that other essential nutrients provided by dairy foods are met from other food sources or supplements [25]. One study compared the effects of dietary calcium versus supplemental calcium on estrogen levels among healthy postmenopausal women and suggested that, compared to supplemental calcium, dairy sources of calcium had a positive effect on estrogen levels, and, in turn, may have contributed to greater bone densities. Subjects consuming dietary calcium had greater bone mass densities despite a lower average total daily calcium intake than the calcium supplement group [24].

Protein

Yogurt is a source of high-quality protein that can help Americans diversify their protein sources as recommended by the 2010 DGA. The nutritional value of proteins differs substantially depending on their essential amino acid composition and digestibility. The Food and Agriculture Organization (FAO) recently recommended DIAAS (digestible indispensable amino acid score) as the preferred method for assessing protein quality, replacing the PDCAAS rating system, which has been used as the gold standard since 1993 [26]. While the PDCAAS truncates values at 1, the DIASS method differentiates protein sources by their ability to supply amino acids to the body. The DIAAS score for dairy protein, which provides all nine essential amino acids is 1.22, compared to 0.64 for peas and 0.40 for wheat.

The recommended dietary allowance (RDA) of protein for both women and men of 19 years and older is 0.80 g/kg of body weight per day [27]. Protein recommendations for children are higher on a gram-per-body-weight basis (ages 1–3 years, 1.05 g/kg/day; ages 4–13 years, 0.95 g/kg/day; ages 14–18 years, 0.85 g/kg/day). RDAs for protein also increase for women who are pregnant (1.1 g/kg/day) and lactating (1.3 g/kg/day).

High quality protein, such as that found in yogurt, is important for bone health and building and maintaining muscle mass [18]. In addition to calcium and vitamin D, protein is also associated with optimal peak bone mass during growth and reducing bone loss in later life since protein comprises half of the volume of bone. Increasing dietary protein increases the serum IGF-1 levels both during growth and in the elderly. During these life stages, timing is key because a response to increase IGF-1 can increase calcium use, but won't have an effect when bones are stable to turnover [17].

Several active components of milk could have an important role in maintaining bone health, such as lactose, galactooligosaccharides (GOS), casein phosphopeptides, whey protein, and amino acids. These bioactives can be involved in several mechanisms associated to bone strength including enhancing calcium absorption (lactose, casein, and whey protein), increasing the proportion of *Bifidobacterium* which may then mediate increases in calcium absorption, and activating calcium sending receptors (amino acids) [17].

Dietary protein has also been shown to contribute to satiety, which may help in weight management [28]. Whey protein and casein, the two main types of proteins in yogurt, may increase satiety more than other proteins [29–31]. Increases in appetite-regulating hormones, such as pancreatic polypeptide, that may result from increased intake of protein [32] and dairy [33] have been suggested as the mechanism for the satiety stimulating effects.

YOGURT AND A BALANCED DIET

The Healthy Eating Index, a measure of healthful eating developed by the USDA's Center for Nutrition Policy and Promotion to assess the US diet, has shown that approximately 74% of the population needs to improve their diets

[34]. Only 10% of the population reportedly have good diets and 16% were characterized as having poor diets [13].

Consumption of dairy products plays a role in building a nutrient-dense diet. Not only does dairy consumption increase calcium intake and improve bone and muscle function, 3 daily servings of low-fat or nonfat dairy products, such as milk, yogurt, and cheese help improve overall nutrient intake, which can contribute to a reduction in the risk of several conditions [35].

In a recent large prospective study, consumption of yogurt was associated with a more balanced diet. According to this study, yogurt consumers are not only likely to have higher potassium intakes (+120 mg/day), but are less likely to have inadequate intakes of vitamins B2 and B12, calcium, magnesium, and zinc [36]. Another study showed that among adult men in the military, frequent consumption of low-fat yogurt was associated with decreased intake of both saturated and total fat [37].

American children are also consuming diets too high in calories and lacking in important nutrients. In particular, dietary intakes of calcium, vitamin D, and potassium, as well as fiber, are of public health concern in children of all ages [38]. Poor eating habits started at a young age may continue into later life; therefore early interventions are needed to help children adopt healthy nutrition. A recent review examined data from both the 2009–10 National Health and Nutrition Examination Survey (NHANES) and the 2004–05 and 2009–10 School Nutrition Dietary Assessment Study (SNDA). This review found that snacks contribute to 37% of children's caloric intake, but only 15–30% of vital micronutrients and nearly 40% of the added sugar in children's diets [39]. The authors of the review also found that introducing vitamin D–fortified yogurt to children's snack times may help increase dietary intake of nutrients currently lacking in children's diets [35,39]. Adding one 6-oz serving of vitamin D–fortified yogurt each day would help children move closer to almost all nutrients of concern, including calcium, vitamin D, and potassium. Combining yogurt with fruit or vegetables as a snack would also increase dietary intake of all nutrients of concern, since this snack would also provide fiber [39]. Finally, swapping yogurt for salty snacks helps meet DGA recommendations to reduce daily sodium intake and to consume fewer foods with sodium [38].

Another recent prospective study in children aged 2–18 years showed that frequent yogurt consumers had better diet quality than infrequent consumers as indicated by the higher Healthy Eating Index-2005 (HEI-2005) total score. Frequent yogurt consumption was also significantly associated with a lower fasting insulin level and a higher quantitative insulin sensitivity check index [40]. If substituted for more energy dense foods, yogurt could lead to more nutrient-dense diets and provide more of several of the shortfall nutrients identified by the 2010 DGAC in both children and adults [14].

YOGURT AND LACTOSE INTOLERANCE

About 70% of the world's population exhibits various degrees of lactose malabsorption as a result of the physiologic decline in intestinal lactase activity [41]. In the United States, as much as 75% of African Americans, 60% of Hispanics, and 20% of the white population currently suffer from lactose maldigestion [42]. Lactose maldigestion occurs when the dose of lactose eaten exceeds the capacity of the enzyme lactase found in the brush border membrane of the small intestine.

Lactose intolerance is a symptomatic diagnosis, including flatus, bloating, cramps, abdominal pain, belching, and diarrhea in response to undigested lactose [43]. Lactose intolerance occurs only when lactose malabsorption is associated with these clinical manifestations. Bloating and abdominal pain are typically caused by the colonic fermentation of unabsorbed lactose by the gut microbiota generating short-chain fatty acids (SCFA) and different gases, such as hydrogen, methane, and carbon dioxide. The consequence is an increase in abdominal pressure and acceleration of intestinal transit, which can generate diarrhea due to the higher secretion of electrolytes and fluids in the colon, a consequence of the metabolism of the unabsorbed lactose in the ileum [44]. It has been suggested that the difference in gut microbiota between individuals can contribute to variations in the amounts of lactose tolerated [43].

According to the National Institute of Health, self-restriction of dairy foods, associated with self-diagnosis of lactose intolerance is a public health problem. Additionally, a certain percentage of persons who classified themselves as lactose intolerant, when tested with the hydrogen breath test, are actually proved to be lactose digesters. These individuals are usually called perceived lactose intolerant [43]. For those who are lactose intolerant or suffer from perceived lactose intolerance, milk avoidance is a major obstacle for obtaining adequate calcium, vitamin D and for some a high quality protein source in the diet, which may predispose them to, decreased bone accrual, osteoporosis, and other adverse health outcomes. These individuals are also often advised to remove or restrict dairy products in their diet which increases the risk for calcium and vitamin D deficiencies.

For those who suffer from lactose intolerance, yogurt offers a nutrient-dense, more easily digestible alternative to milk or cheese. While cheese provides many of the same nutrients as milk and is low in lactose, it is also a source of sodium and saturated fat. Several studies in different populations and age groups have shown that yogurt consumption improves lactose digestion. Studies have shown better lactose digestion and absorption, as well as a reduction in gastrointestinal symptoms, in subjects who consumed yogurt with live and active cultures. These beneficial effects have not been fully shown for heat-treated yogurt of fermented milks [45]. Heat treatment of yogurt reduced the lactase activity of yogurt by 10-fold [46]. Yogurt is a more easily digestible alternative to milk because on average it contains less lactose per serving and because of the presence of the lactase-producing bacteria. The enhanced lactose digestion in yogurt is the result of the inherent β-galactosidase activity of both yogurt cultures, *S. thermophilus* and *L. bulgaricus* [47]. These two cultures exhibit lactase activity during milk fermentation, as well as in the intestinal tract after consumption, which allow lactose-intolerant individuals to enjoy dairy products with fewer associated symptoms [47,48]. The β-galactosidase in yogurt can pass through the stomach protected by the cell wall and membrane of the yogurt bacteria. The enzyme can have better access to lactose due to the effect of bile salts on the cell wall and membrane of the yogurt cultures. Bile salts likely disrupt the cell wall to increase β-galactosidase/lactose interaction, thus supporting lactose digestion [49].

YOGURT AND WEIGHT MANAGEMENT

The poor quality of the US diet is also evidenced by the prevalence of overweight and obesity among all age groups. Recent data show that 68% of adults in the United States are either overweight or obese, almost 36% are obese, and almost 6% are extremely obese [50,51]. Among children aged 2–5 years, more than 12% are overweight; among children aged 6–11 years, 17% are overweight; and almost 18% of adolescents between the ages of 12 and 19 years are overweight. Among all children and adolescents aged 2–19 years, almost 17% are obese [50].

The association between dairy products and weight change is inspired by evidence from human and animal studies suggesting beneficial effects of yogurt and calcium intake on adiposity and fat loss. In a clinical study with 34 obese subjects placed on a weight loss diet and randomized to a control diet or a yogurt diet, the subjects in the yogurt diet had significantly increased fat loss and trunk fat mass loss while showing a reduction of waist circumference [52].

Several prospective studies follow these initial findings. The Coronary Artery Risk Development in Young Adults (CARDIA) study was among the first cohorts investigated [53]. In this study, the authors reported that total dairy consumption was associated with a lower risk of developing metabolic syndrome among 3157 US adults aged 18–30 years who were overweight at baseline. Among lean participants, this association was null. The authors also found that yogurt was more strongly associated with a lower risk of obesity. More recently, the results from additional prospective studies are consistent with those initially reported in the CARDIA study. A prospective study of three separate cohorts of more than 120,000 women and men in the United States, followed every 4 years for 20 years, showed that frequent yogurt consumption as part

of a healthy dietary pattern was associated with less weight gain over time [54]. This study compared consumption of different foods, including yogurt, fruits, vegetables, and whole grains and showed that consumption of these foods was associated with less weight gain over time, with yogurt showing the best results. Other forms of dairy, including low-fat or nonfat milk, had no measurable association with less weight gain, while whole milk and cheese consumption was associated with weight gain over the two decades. This association was independent of a wide array of lifestyle factors. In a different cohort of 3285 subjects followed over 17 years (Framingham Heart Study Offspring Cohort 1991–2008), yogurt consumption was also associated with less change in annualized weight and waist circumference over time and with healthier levels of circulating glucose [11]. Similar findings have been reported in children and young adolescents. A recent examination of NHANES data (2005–08) has found that higher yogurt consumption was associated with lower measures of adiposity in US children (ages 8–18), such as lower BMI-for-age, lower waist circumference, and smaller subscapular skinfold [35].

The mechanisms associating yogurt consumption with weight management are not entirely explained, though it has been suggested that changes in the microbiota can influence weight gain [55]. The role of gut microbiota and fermentation products in modulating intestinal and adipose tissue physiology, energy balance, and insulin resistance have been reported in studies with specific probiotics cultures, prebiotics, and regular yogurt [56–60]. The role of calcium intake in weight control has been also discussed as a possible underlying mechanism [61,62]. Dietary calcium suppresses parathyroid hormone and 1,25-dihydroxyvitamin D. These hormones can increase adipocyte intracellular Ca^{2+} levels, which lead to fat accumulation in adipocytes [52,63]. In comparison to other dairy products, yogurt contains on average higher amounts of calcium, which also shows to have a higher physiological bioavailability [64]. Among other nutrients present in yogurt, the protein can also be a major factor in the mechanism behind these findings because protein can have an effect on satiety and subsequent energy intake [10].

YOGURT, DIABETES, AND METABOLIC HEALTH

Type 2 diabetes (T2D) is a chronic disease associated with dysfunctional glucose insulin balance. This disease occurs when the body does not efficiently use the insulin it makes (also known as insulin resistance) and when the pancreas fails to produce enough insulin to compensate for this insulin resistance. T2D is characterized by dyslipidemic, proinflammatory, and prothrombotic states that contribute to the increased risk of cardiovascular diseases (CVD) [65] and other complications including microvascular and macrovascular damages (e.g., retinopathy, nephropathy, ischemic heart disease, and stroke) [66].

The American Diabetes Association and the Center for Disease Control estimates that in 2012, 29.1 million Americans, or 9.3% of the population, had diabetes. Approximately 1.25 million American children and adults have type 1 diabetes. Of the 29.1 million, 21.0 million were diagnosed and 8.1 million were undiagnosed. In 2012, 86 million Americans aged 20 and older had prediabetes; this is up from 79 million in 2010. Currently, it is estimated that one out of three people will develop diabetes in their lifetime [67]. Dietary modifications are proposed as a primary target for the prevention and treatment of T2D and studies have suggested that the consumption of a diet rich in low-fat and nonfat dairy, fruits, vegetables, and whole grains protects against insulin-resistant phenotypes among adults without diabetes [68].

A selection of studies performed on the effects of dairy and yogurt on T2D and diabetic markers are described next. A systematic review and metaanalysis of seven prospective studies found that increasing dairy intake by 1 serving/day was associated with a 5% reduced risk for T2D. The risk reduction was higher for low-fat dairy product [69]. Choi and coworkers showed that total dairy consumption was significantly associated with a lower risk of T2D after a 12-year follow-up study in middle-aged and elderly men [70]. The magnitude of the association for yogurt was similar to that of total dairy consumption, although it did not reach statistical significance: comparing ≥2 servings/week to 1 serving/month, the relative risk (RR) [95% confidence interval (CI)] was 0.83 (0.66, 1.06; p for trend 5 .11). Another study found, in women aged 45 years or older, that a decrease of 18% of T2D risk was observed for consumption of more than two portions of yogurt per week compare to less than one portion per month (the RR was 0.82 at 95% CI, 0.70–0.97). This association was not attenuated by adjustments for confounding factors [71]. Other dairy foods were not significantly associated with T2D reduced risk in this study. In another prospective investigation among 82,076 women who participated in the Women's Health Initiative Observational Study, yogurt intake was more strongly associated with a lower T2D risk. The RR (95% CI) was 0.46 (0.31, 0.68; p for trend 5 .004) comparing ≥2 servings/week to 1 serving/month of yogurt intake. Other individual dairy foods did not seem to be associated with T2D risk [72].

The T2D risk was evaluated in a Japanese cohort (Japan Public Health Center-based Prospective Study; 59,796 middle-aged and older men and women; 5-year follow up, 1,114 cases of T2D documented) in relation to calcium, vitamin D, and dairy intake. In women only, the intake of dairy foods was significantly inversely associated with the risk of T2D. In models with adjustment of age and area only, the odds ratios for the highest versus lowest intake category were 0.72 (p = 0.04) for yogurt. However, in multivariable

analyses, these associations were attenuated. Among participants with a higher vitamin D intake, calcium intake was inversely associated with T2D risk [73].

The EPIC-Inter Act study, involving eight European countries of the European Prospective Investigation into Cancer and Nutrition (n = 340,234), investigated the association of total dairy products and different dairy subtypes with incidence of T2D in populations with marked variation of consumption of these food. They found no association between total dairy product intake and T2D. However, an inverse association with T2D was found for cheese intake and for combined fermented dairy product intake (cheese, yogurt, and thick fermented milk; hazard ratio (HR) 0.88; 95% CI, 0.78, 0.99; p = 0.02) in adjusted analyses that compared extreme consumption quintiles (≥122.1 g/day vs. ≤18.8 g/day) [74].

A systematic review and dose–response metaanalysis of studies evaluated the associations between different types of dairy products (including yogurt) and T2D risk. Of all individual dairy products, yogurt was most strongly associated with a lower T2D risk. For every 200 g/day of yogurt consumption, the risk of developing T2D was lowered by 22% (95% CI, 22, 40), although the dose–response relationship may not be linear. When comparing high- to low-yogurt intake levels, the pooled RR (95% CI) was 0.86 (0.75, 0.98), with a p value for heterogeneity of 0.02 [75].

A recent observation analysis on 214,480 men and women in the United States revealed that total dairy consumption was not associated with T2D risk and the pooled HR (95% CI) of T2D for 1 serving/day increase in total dairy was 0.99 (0.98, 1.01). However, a focus on yogurt showed consistent inverse relationship associated with T2D risk (HR 0.83) (CI, 0.75–0.92) for 1 serving/day increment (p for trend <0.001). Similar trend was observed when the authors performed a metaanalysis on 14 prospective cohorts with 459,790 participants. The pooled RRs (95% CIs) were 0.98 (0.96, 1.01) and 0.82 (0.70, 0.96) for 1 serving of total dairy per day and 1 serving of yogurt per day, respectively [76].

Another recent observational study followed 3454 nondiabetic individuals in Europe. Dairy consumption was monitored at baseline and every year using food frequency questionnaires and categorized into total, low-fat, whole-fat, and milk, yogurt, cheeses, fermented dairy, concentrated full fat, and processed dairy. After multivariate adjustment, total yogurt consumption was associated with a lower T2D risk [HR 0.60 (0.42–0.86); p trend = 0.002]. The authors concluded that substituting 1 serving/day of a combination of biscuits and chocolate and whole grain biscuits and homemade pastries for 1 serving/day of yogurt was associated with a 40 and 45% lower risk of T2D, respectively. No significant associations were found for the other dairy subgroups (cheese, concentrated full fat, and processed dairy products) [77].

Very recently, an analysis of the association between diet and metabolic profile in 5124 children aging between 2- and

18-year-old pointed out that high yogurt intake was associated with significant lower fasting insulin. Homeostatic model assessment score was also decreased whereas insulin sensitivity index was increased. The study supported the positive role of yogurt consumption in insulin metabolism [40].

In other cohorts the effect of yogurt did not always show a significant association with a reduced risk of T2D, although for most studies yogurt consumption was very low when compared to the total dairy products [53,78]. In two smaller studies conducted in Australia (the Australian Diabetes Obesity and Lifestyle Study) and in the United Kingdom (the Whitehall II Study), yogurt intake was also not associated with T2D risk regardless of gender [79,80].

Several observational studies were also identified that reported associations between yogurt and some markers of diabetes (e.g., hemoglobin A1c) but not the condition itself. Two of these studies reported beneficial associations with one or more of these parameters [36,81] while two did not report such findings [78,82]. The majority of the evidence associating consumption of yogurt and reduction of the risk of T2D originates from observational prospective studies. Some intervention studies have compared the effect of standard yogurt with other fermented dairy products that included additional beneficial cultures but none employed a control diet that did not contain yogurt. Therefore, these studies may provide useful information on the effect of yogurt compared to baseline, but no such comparisons to a placebo is possible. Some of these studies reported no significant changes in fasting blood glucose versus baseline [83–85] while others report a significant reduction [86,87].

YOGURT, HEART HEALTH, BLOOD PRESSURE, AND HYPERTENSION

Heart diseases are a major cause of death and the WHO has predicted that by 2030, CVD will remain the leading causes of death worldwide. Smoking, overweight, hypertension, high plasma cholesterol levels, and diabetes are risk factors for CVD [88]. Blood/serum low-density lipoprotein (LDL) cholesterol concentration and blood pressure are also recognized by the European Food Safety Agency (EFSA) to be important risk factors for CVD and the maintenance of normal HDL-cholesterol concentrations, triglycerides, and homocysteine levels has been proposed by EFSA to represent beneficial health effect [89].

Consuming a healthy diet is recommended for the prevention of CVD. Importantly, dietary recommendations to lower total and LDL cholesterol with regards to dairy foods are to preferentially consume nonfat milk and yogurt; to moderate intake of low-fat milk, low-fat cheese, and other milk products; and to limit the consumption of regular cheese, cream, whole-fat milk, and whole-fat yogurt. Elevated blood pressure is one of the major independent risk factors for CVD. To prevent hypertension and its adverse outcomes (e.g., stroke, heart, and renal failure), national authorities recommend regular physical activity, moderate alcohol intake, weight control, reduced sodium intake, and increased potassium intake [90,91]. A diet rich in fruit, vegetables, and low-fat dairy products and low in saturated and total fat has been reported to lower blood pressure in the Dietary Approaches to Stop Hypertension (DASH) Trial [92]. The DASH diet plan includes 2–3 servings of low-fat or nonfat dairy foods per day.

Several studies have been conducted to evaluate the association between yogurt intake and CVD risk. Tavani et al. found that higher yogurt intake was significantly associated with lower odds of developing nonfatal myocardial infarction while other dairy foods were not associated with the risk [93]. In an observation study of over 85,000 women to evaluate the hypothesis that mineral intake is associated with lower risk of developing stroke, the authors found a marginally significant inverse association for yogurt, the RR (95% CI) of stroke was 0.69 (0.34–1.40; p for trend = 0.06) for women who consumed 5 or more servings of yogurt per week in comparison with those who rarely ate yogurt [94]. In a recent metaanalysis on dairy food intakes and incident hypertension, yogurt consumption alone was not associated with hypertension risk, whereas total dairy, low-fat dairy, and milk were associated with a lower risk of developing hypertension [95].

In the Swedish Malmö Diet and Cancer cohort with a mean follow-up time of 12 years (44–74 years; 62% females), overall consumption of dairy products was inversely associated with the risk of CVD. Among the specific dairy products, a statistically significant inverse relationship was observed for fermented milk only (yogurt and cultured sour milk). When comparing to subjects with the lowest intake, the highest consumption of fermented milk was associated with a 15% reduction in incidence of CVD ($p = 0.003$) [96].

Wang and coworkers evaluated the associations of intake of dairy products, calcium, and vitamin D with the incidence of hypertension in a prospective cohort (28,886 US women aged ≥45 years). Consumption of low-fat dairy products, calcium, and vitamin D were inversely associated with the risk of hypertension in middle-aged and older women. A reduction of 10–15% in hypertension risk was observed for all four major low-fat dairy products (comparing the highest to the lowest intake category), however, the reduction was statistically significant only for skim milk and yogurt. Multivariate adjustment substantially attenuated the inverse association for yogurt [97]. Additionally, in the same cohort (Framingham Heart Study Offspring), yogurt consumption was significantly associated with lower levels of triglycerides, healthier levels of circulating glucose, healthy levels of systolic blood pressure within the normal range, and an improved homeostatic model assessment of insulin resistance score [98].

The effects of yogurt or certain probiotics on blood lipids are not consistent. Some studies show that certain probiotics or yogurts might lower total and LDL cholesterol [99] while others do not show any association [100]. As different probiotic strains can have different benefits, these effects are not surprising and probably specific to the probiotic used in each study. A recent metaanalysis of a total of 30 randomized studies reported that the use of probiotics may improve lipid metabolism by decreasing total and LDL cholesterol concentrations although the authors also found heterogeneity in the results [101].

The beneficial effect on CVD observed in some studies following the consumption of yogurt may be attributable to nutrients, such as proteins, vitamin D, calcium, magnesium, and potassium [102], as well as to the presence of different saturated and unsaturated fats which could act in different ways on lipid metabolism [103,104]. In vitro studies have shown that some strains of *Bifidobacterium* and *Lactobacillus* can assimilate cholesterol [105]. It has also been demonstrated that fermentation of milk by *S. thermophilus* and *L. bulgaricus* leads to the reduction of cholesterol content in milk. Additionally, certain bacteria can assimilate cholesterol directly from the gut or impact cholesterol absorption by deconjugating bile salts [106]. Certain bacteria can also ferment indigestible carbohydrates and produce SCFAs in the gastrointestinal tract that can cause a reduction in plasma total cholesterol concentrations either by inhibiting hepatic cholesterol synthesis or by redistributing cholesterol from plasma to the liver [107]. The amount of bioactive peptides increases during fermentation of milk. Peptides with the amino acid sequence isoleucine–proline–proline (IPP) and valine–proline–proline (VPP) are the best characterized peptides found in fermented milk [108]. Many studies evaluating the role of dairy protein/peptides on hypertension focused on the effect of IPP and VPP which have shown angiotensin-converting-enzyme (ACE) inhibitory effect in vitro. ACE is one of the key enzymes in blood pressure regulation because it generates the vasoconstrictor angiotensin-II and inactivates the vasodilator bradykinin. Animal studies suggest that rats fed fermented milk with IPP and VPP showed significant reductions of systolic blood pressure [109]. Moreover, studies in human have shown positive effects following a treatment with fermented food enriched with bioactive peptides on arterial stiffness [110] and blood pressure in hypertensive subjects [111–113]. Finally, some authors have also observed that calcium, potassium, and magnesium, regularly present in dairy products and yogurt, seem to be needed by the body for optimal control of blood pressure [114–116].

YOGURT AND IMMUNE FUNCTION

The immune system protects the body against environmental challenges and is divided into the innate and adaptive immune systems. The main site of the mucosal immune system within the digestive system is referred to as gut-associated lymphoid tissue. Antigens are captured by immune cells in Peyer's patches in the small intestine where several cells are involved in this immune response. Lymphocytes are the main effector cells (T and B cells) of the adaptive immune system. B cells are responsible for antibody production whereas T cells destroy pathogens directly (CD8+T which are cytotoxic) or control the function of other cell types (CD4+T cells named T helper cells). T helper cells can produce different cytokines, such as interferon-gamma (IFN-γ), tumor necrosis factor (TNF-α), interleukins IL-4 and IL-5. The best-defined effector of the mucosal adaptive immune system is secretory immunoglobulin A (sIgA).

A selection of studies performed on the effects of regular yogurt on the immune system is described in the subsequent section. Additional studies done with specific probiotic cultures will be further explored in other chapters of this book. Meyer and coworkers evaluated the effect of a conventional yogurt and a yogurt supplemented with an additional beneficial culture (*L. casei* DN-114 001) in a randomized clinical trial where young healthy women consumed 100 g, then 200 g of either products during 2 weeks. The authors concluded that both conventional and supplemented yogurt enhanced the production of proinflammatory cytokines [117]. Another study performed in healthy subject evaluated the effects of yogurt on the production of cytokines by the circulating blood mononuclear cells. In this case, consumption of yogurt significantly increased 2–5A synthetase activity in mononuclear cells, suggesting increased IFN production [118].

Atopic diseases arise from abnormal immune responses to allergens [119]. The atopic diseases of childhood consist mainly of atopic dermatitis, allergic rhinitis, and asthma. Epidemiological evidence regarding the association between the intake of dairy foods and allergic disorders has been inconsistent [120]. Van de Water and coworkers performed a 1-year study in which they gave young and older adults 200 g/day of yogurt, pasteurized fermented milk or no intervention. The consumption of yogurt was significantly associated with a decrease in allergic symptoms in both age groups ($p < 0.05$). Compared with no yogurt and pasteurized fermented milk, the yogurt consumption was associated with reduced nasal allergies, especially in the young adult population [121]. An epidemiological study in Japanese students showed that subjects who regularly eat yogurt and/or fermented milk experienced a significant reduction in allergy development in comparison with those who do not frequently eat fermented foods [122]. A large study in five European countries showed that the introduction of yogurt and complementary food in the first year of life reduced significantly the risk for atopic dermatitis [123].

Few studies have been performed to evaluate the effects of yogurt in asthma. Some studies show a protective effect while others do not show an association between yogurt consumption and asthma [124,125]. The duration of the study, the dose of yogurt used, and variations in genetic and environmental factors that may differ between populations could explain the discrepancy between the results.

Diarrhea is usually a symptom of an infection in the intestinal tract, which can be caused by a variety of bacterial, viral, and parasitic organisms [126]. Subjects with diarrhea have an imbalance water movement across the gut which can lead, among other consequences, to abdominal pain, dehydration, and weight loss. Antibiotic treatment is also often accompanied by diarrhea and can occur in children and adults given antibiotics [127–129]. *Clostridium difficile*–associated diarrhea also accounts for 10–20% of antibiotic-associated diarrhea [130]. The World Gastroenterology Organization concluded that several clinical studies and metaanalysis support the use of specific beneficial LAB strains in the treatment and prevention of rotavirus diarrhea in infants. However, effects need to be verified for each specific strain in humans [131]. The majority of studies on the use of LAB in the prevention and treatment of diarrhea have been performed in children and with specific singular beneficial strains. Few studies have been performed in healthy adults and with conventional yogurt.

A randomized clinical trial was performed in 78 children aged 3–36 months with confirmed persistent diarrhea to evaluate the effect of substituting yogurt for milk in their diet. Children consumed either milk (infant formula) or yogurt (infant formula fermented with *L. bulgaricus* and *S. thermophilus*). Persistent diarrhea was significantly decreased in children consuming yogurt versus milk [132]. Boudraa and coworkers performed a controlled clinical trial in 112 well-nourished young children aged 3–24 months with acute diarrhea. Children consumed either yogurt or an infant formula. The authors reported that the termination of diarrhea and proper weight gain after 7 days was similar in both groups. However, the children who consumed yogurt had a significant decrease in stool frequency[133]. A metaanalysis of clinical studies on acute infectious diarrhea reported a reduction in diarrhea duration of 0.7 days with intake of yogurt including *Lactobacillus* bacteria in children. Moreover, there was a reduction in the frequency of diarrhea of 1.6 stools on the second day of treatment in the yogurt groups [134]. Beniwal and coworkers conducted a clinical trial to test the effect of yogurt consumption in preventing antibiotic-associated diarrhea. Approximately 230 g of yogurt were given to patients (average 70 years old) receiving oral or intravenous antibiotics for 8 days. Patients receiving yogurt reported less frequent diarrhea (12% vs. 24%; $p = 0.04$), and significantly less total diarrhea. The authors concluded that dietary supplementation with yogurt is a simple, effective, and safe treatment that decreases the incidence and duration of antibiotic-associated diarrhea [130]. de Mattos et al. observed that children (154 male infants aged between 1 and 30 months with persistent diarrhea) fed a yogurt-based diet or an amino acid–based diet had a significant reduction in stool output and in the duration of diarrhea compared to children fed a hydrolyzed protein-based formula [135]. A recent review on persistent diarrhea in children stated that yogurt-based or amino acid–based diets may accelerate their recovery and that zinc supplementation reduces the severity and duration of diarrhea [136].

He and coworkers performed an interventional study in Chinese children aged 3–5 years whose height for age and/or weight for age were lower than the reference level. Children receiving yogurt for a period of 10 months (125 g/day during 5 days/week) showed a decrease in the incidence and duration of diarrhea and upper-respiratory infections [137]. Similarly, in an elderly population, Makino et al. showed that yogurt consumption for 8–12 weeks resulted in a reduced risk of developing colds, together with an enhancement in NK cell cytotoxicity, particularly in subjects with low NK-T cell activity at baseline [138].

The immune modulating and antimicrobial activities of LAB depend on several factors, such as pH, organic acids, carbon dioxide, hydrogen peroxide, bacteriocins, ethanol, diacetyl, depletion of nutrients, and competition for available living space. It has been suggested that the mechanisms by which LAB bacteria modulate gut immune function are linked with effects on the mucosal epithelial lining as well as with the lymphoid cells residing within the gut-associated lymphoid tissue [139]. LAB can modulate specific or nonspecific immune responses [140]. Several studies have shown that consumption of yogurt can increase the production of IFN-γ, which is commonly regarded as the major cytokine of T helper type 1 (Th1)-lymphocytes that have antiinflammatory properties [141]. LAB, have also been described to reinforce phagocytosis, cytokine production, and secretory IgA or IgM in response to infectious antigens. Moreover, some peptides (e.g., opioid-like peptides) present in yogurt may maintain or restore intestinal homeostasis and could play an important role in protecting against damaging agents in the intestinal lumen [142]. The passage of yogurt bacteria in the gut could contribute to the elimination of pathogenic enteric bacteria by the production of SCFAs, hydrogen peroxide, or antimicrobial substances [143,144]. They could also reinforce the intestinal barrier capacity by competing with pathogenic bacteria for adhesion to the enterocytes and competition for nutrients [145,146]. The effects of certain LAB could also be explained by a change in the composition of mucus layer, modulation of the permeability of the gut barrier, and stimulation of the production of defensins [145,147].

YOGURT AND DIGESTIVE FUNCTION

Yogurt and certain fermented milk products may influence the outcome of gastrointestinal health by affecting different intestinal functions, such as motility, secretion of mucus, enzyme activity, antibody production, water and electrolyte balance, nutrient absorption, transit time, and management of different digestive issues, such as bloating, rumbling, and general abdominal discomfort. Yogurt with LAB has also been reported to influence activities of intestinal microbes and potentially shape the intestinal microbiota [41].

Some studies have also investigated the effects of yogurt on different gastrointestinal issues, such as management of intestinal symptoms, improvement in irritable bowel syndrome (IBS), increasing the eradication rate of *Helicobacter pylori* infection, and treatment of constipation, among others. IBS is a functional syndrome in which recurrent abdominal pain or discomfort is associated with a change in bowel habits. IBS is a very common syndrome affecting millions of people worldwide and the prevalence of IBS varies among countries due in part to the use of different criteria to define it. Currently, there is no agreed biological marker of IBS and diagnosis is mainly symptomatic without organic or structural abnormalities. There is increasing evidence supporting an important role of the gut microbiota in IBS. Clinical evidence showed an association between IBS pathology and alterations in the gut microbiota in a distinct subset of patients [148]. The role of the microbiota is also supported by the ability of some beneficial bacteria [149] and antibiotics [150] to alleviate IBS symptoms. However, the extent of the benefit that can be obtained from these bacteria and the most effective species and strains remain uncertain. A recent review investigated the effect of different probiotic species, including *Bifidobacterium* and *Lactobacillus* on IBS management [151], although the authors did not describe studies that specifically investigated the effect of conventional yogurt on IBS, which indicates this can be a future area of research [152].

A case–control study has shown that yogurt consumption may have a protective effect against *H. pylori* seropositivity [153] and that consumption of yogurt containing *Bifidobacterium* species and *L. acidophilus* before and during antibiotic regimens and triple therapy can increase the eradication rate and lower the incidence of adverse effects, although not in all studies [154,155].

Functional constipation is a prevalent problem within the Western population. There is evidence supporting the fact that the inclusion of certain prebiotics and probiotics in the diet can favorably modify intestinal function related to constipation. Several studies have demonstrated that daily consumption of fermented milk containing *Bifidobacterium lactis* DN-173 010 in association with *L. bulgaricus* and *S. thermophilus* improves gastrointestinal transit and also symptoms of minor digestive issues, such as bloating, rumbling, discomfort, and gas [156].

Cancer is one of the leading causes of death in the United States. Risk factors for cancer include both genetic and environmental factors, and the role of diet in the etiology of different forms of cancer has been given greater attention in recent years [41]. Epidemiologic evidence suggests a negative correlation between the incidence of certain cancers, including colon cancer, and the intake of fermented dairy products [157]. A large population-based case–control study performed in California, including patients diagnosed with an invasive adenocarcinoma, found a protective effect of yogurt consumption against colorectal cancer. Interestingly, total calories were associated with excess risk throughout the colon while calcium intake was associated with significantly decreased risk [158]. This observation was corroborated in a recent prospective study among 45,241 volunteers who participated in the EPIC-Italy cohort [157]. In this study, when comparing participants who were in the highest versus the lowest tertile of yogurt consumption at baseline, the RR of developing colorectal cancer was 0.65 (95% CI, 0.48, 0.89), with a significant linear trend (p for trend 0.002) after adjustment of multiple lifestyle and dietary factors. Other studies performed in Japan have not found such correlation, although investigators believe the cause is the relatively low yogurt consumption in Japan [159,160]. Studies performed in the Western world examined the associations between intakes of calcium, vitamin D, and dairy foods and the risk of colon cancer in 47,935 US male, aged 40–75 years and free of cancer at the start of the study. Consumption of fermented dairy products (including yogurt, sour cream, cottage cheese, cream cheese, and hard cheese) and milk was not significantly associated with the risk of colon cancer. Calcium and vitamin D were inversely associated with colon cancer risk, but after adjusting for confounding variables they found that the trend was no longer statistically significant.

More recently, the associations between intakes of specific dairy products and dietary calcium with colorectal cancer risk was investigated in the European Prospective Investigation into Cancer and Nutrition (477,122 men and women; dietary questionnaires administered at baseline; 11 years of follow up, 4,513 incident cases of colorectal cancer). They observed an inverse association between colorectal risk and dietary calcium and total milk consumption (which did not differ by the fat content of the milk). Moreover, inverse associations were observed for yogurt and cheese in the categorical models. However, these associations were nonsignificant in the linear models [161].

Some observational studies have also been conducted to investigate the associations between yogurt consumption and the risk of bladder cancer. In two early retrospective case–control studies, yogurt consumption was consistently associated with a lower risk of developing bladder cancer [162,163]. In the largest investigation to date, Larsson et al. followed 82,002 Swedish women and men for 9 years and identified 485 incident bladder cancer cases [164]. In this analysis, comparing participants who consumed sour milk and yogurt ≥ 2 servings/day to nonconsumers, the RR (95% CI) was 0.62 (0.46, 0.85; p for trend 0.006). Reports for other types of cancer, such as lymphoma, prostate cancer, ovarian cancer, and upper digestive tract cancer, are sporadic, without extensive replication data from other independent investigations [165,166].

The postulated mechanism by which LAB may affect carcinogenesis may involve an effect on the host's immune response, suppression of harmful intestinal bacteria, sequestration of potential mutagens or carcinogens, production of antimutagenic compounds, reduction of pH concentrations in the colon, and modification of cell metabolism or alteration of other physiologic conditions [167]. Moreover, the calcium content of dairy products has been hypothesized to protect against colorectal cancer risk. Calcium may bind proinflammatory secondary bile acids and ionize fatty acids that could reduce cell proliferation and promote cell differentiation [168].

ACKNOWLEDGMENTS

The author would like to thank Amanda Blechman, Nancy Dowling, Kristie Leigh, David Graham, and Randal Shaheen for the fruitful discussions and for helping to review this manuscript.

DISCLAIMER

The views expressed in this chapter are those of the author only and do not necessarily represent those of The Dannon Company Inc.

REFERENCES

[1] United Nations—Food and Agriculture Organization of the United Nations/World. No Title. Guidelines for the Evaluation of Probiotics in Food; 2002.

[2] Greene JD, Klaenhammer TR. Factors involved in adherence of lactobacilli to human Caco-2 cells. Appl Environ Microbiol 1994;60(12):4487–94.

[3] Conway PL, Gorbach SL, Goldin BR. Survival of lactic acid bacteria in the human stomach and adhesion to intestinal cells. J Dairy Sci 1987;70(1):1–12.

[4] Bezkorovainy A. Probiotics: determinants of survival and growth in the gut. Am J Clin Nutr 2001;73(2 Suppl.):399S–405S.

[5] del Campo R, Bravo D, Cantón R, et al. Scarce evidence of yogurt lactic acid bacteria in human feces after daily yogurt consumption by healthy volunteers. Appl Environ Microbiol 2005;71(1):547–9.

[6] Pedrosa MC, Golner BB, Goldin BR, Barakat S, Dallal GE, Russell RM. Survival of yogurt-containing organisms and *Lactobacillus gasseri* (ADH) and their effect on bacterial enzyme activity in the gastrointestinal tract of healthy and hypochlorhydric elderly subjects. Am J Clin Nutr 1995;61(2):353–9.

[7] Mater DDG, Bretigny L, Firmesse O, et al. *Streptococcus thermophilus* and *Lactobacillus delbrueckii* subsp. *bulgaricus* survive gastrointestinal transit of healthy volunteers consuming yogurt. FEMS Microbiol Lett 2005;250(2):185–7.

[8] Zhong Y, Huang C-Y, He T, Harmsen HMJ. Effect of probiotics and yogurt on colonic microflora in subjects with lactose intolerance. Wei Sheng Yan Jiu 2006;35(5):587–91.

[9] García-Albiach R, José M, De Felipe P, et al. Molecular analysis of yogurt containing *Lactobacillus delbrueckii* subsp. *bulgaricus* and *Streptococcus thermophilus* in human intestinal microbiota. Am J Clin Nutr 2008;87(1):91–6.

[10] Webb D, Donovan SM, Meydani SN. The role of yogurt in improving the quality of the American diet and meeting dietary guidelines. Nutr Rev 2014;72(3):180–9.

[11] Wang H, Troy LM, Rogers GT, et al. Longitudinal association between dairy consumption and changes of body weight and waist circumference: the Framingham Heart Study. Int J Obes 2014;38(2):299–305.

[12] US Department of Agriculture, Agricultural Research Service. USDA National Nutrient Database for Standard Reference, Release 26, 2013. Available from: http://ndb.nal.usda.gov/. Updated October 2013.

[13] US Department of Agriculture. Center for Nutrition Policy and Promotion. Report card on the quality of Americans' diets. Nutr Insights. 2002;28.

[14] Dietary Guidelines Advisory Committee. Report of the Dietary Guidelines Advisory Committee on the Dietary Guidelines for Americans, 2010, to the Secretary of Agriculture and the Secretary of Health and Human Services. Washington, DC: US Department of Agriculture, Agricultural Research Service; 2010.

[15] Krebs-Smith SM, Guenther PM, Subar AF, Kirkpatrick SI, Dodd KW. Americans do not meet federal dietary recommendations. J Nutr 2010;140(10):1832–8.

[16] Institute of Medicine. Food and Nutrition Board. Dietary reference intakes for water, potassium, sodium, chloride, and sulfate. The National Academies Press; 2005.

[17] Weaver CM. Yogurt, diet quality, and bone health. Funct Food Rev 2013;5(2):1–10.

[18] Heaney RP, Layman DK. Amount and type of protein influences bone health. Am J Clin Nutr 2008;87(5):1567–70.

[19] Institute of Medicine Dietary Reference Intakes. Dietary reference intakes for calcium and vitamin D. Pediatrics 2012;130(November): e1424.

[20] Bailey RL, Dodd KW, Goldman JA, et al. Estimation of total usual calcium and vitamin D intakes in the United States. J Nutr 2010;140(4):817–22.

[21] Fulgoni VL, Huth PJ, DiRienzo DB, Miller GD. Determination of the optimal number of dairy servings to ensure a low prevalence of inadequate calcium intake in Americans. J Am Coll Nutr 2004;23(6):651–9.

[22] Sahni S, Tucker KL, Kiel DP, Quach L, Casey VA, Hannan MT. Erratum to: milk and yogurt consumption are linked with higher bone mineral density but not with hip fracture: the Framingham Offspring Study. Arch Osteoporos 2013;8(1–2):119.

[23] Ganpule A, Yajnik CS, Fall CH, et al. Bone mass in Indian children—relationships to maternal nutritional status and diet during pregnancy: The Pune Maternal Nutrition study. J Clin Endocrinol Metab 2006;91(8):2994–3001.

[24] Napoli N, Thompson J, Civitelli R, Armamento-Villareal RC. Effects of dietary calcium compared with calcium supplements on estrogen metabolism and bone mineral density. Am J Clin Nutr 2007;85(5):1428–33.

[25] Miller GD, Jarvis JK, McBean LD. The importance of meeting calcium needs with foods. J Am Coll Nutr 2001;20(2 Suppl.): 168S–85S.

[26] Food and Agriculture Organization of the United Nations. Report of an FAO Expert Consultation. Dietary protein quality evaluation

in human nutrition. FAO Food and Nutrition Paper 92. Rome. Available from: http://www.fao.org/ag/humannutrition/35978-02317b979a686a57aa4593304ffc17f06.pdf; 2013.

[27] Institute of Medicine. Food and Nutrition Board. Dietary reference intakes: energy, carbohydrates, fiber, fat, fatty acids, cholesterol, protein, and amino acids; 2002.

[28] Westerterp-Plantenga MS, Lemmens SG, Westerterp KR. Dietary protein—its role in satiety, energetics, weight loss and health. Br J Nutr 2012;108(S2):S105–12.

[29] Abou-Samra R, Keersmaekers L, Brienza D, Mukherjee R, Macé K. Effect of different protein sources on satiation and short-term satiety when consumed as a starter. Nutr J 2011;10(1):139.

[30] Veldhorst MA, Nieuwenhuizen AG, Hochstenbach-Waelen A, et al. Dose-dependent satiating effect of whey relative to casein or soy. Physiol Behav 2009;96(4–5):675–82.

[31] Pal S, Ellis V. The acute effects of four protein meals on insulin, glucose, appetite and energy intake in lean men. Br J Nutr 2010;104:1241–8.

[32] Belza A, Ritz C, Sorensen MQ, Holst JJ, Rehfeld JF, Astrup A. Contribution of gastroenteropancreatic appetite hormones to protein-induced satiety. Am J Clin Nutr. 2013;97(5):980–9.

[33] Jones KW, Eller LK, Parnell JA, Doyle-Baker PK, Edwards AL, Reimer RA. Effect of a dairy- and calcium-rich diet on weight loss and appetite during energy restriction in overweight and obese adults: a randomized trial. Eur J Clin Nutr 2013;67(4):371–6.

[34] Basiotis P, Carlson A, Gerrior A, et al. The Healthy Eating Index, 1999–2000: charting dietary patterns of Americans. Fam Econ Nutr Rev 2004;16(1):39–48.

[35] Keast D, Gallant K, Albertson A, Gugger C, Holschuh N. Associations between yogurt, dairy, calcium, and vitamin D intake and obesity among U.S. children aged 8–18 years: NHANES, 2005–2008. Nutrients 2015;7(3):1577–93.

[36] Wang H, Livingston KA, Fox CS, Meigs JB, Jacques PF. Yogurt consumption is associated with better diet quality and metabolic profile in American men and women. Nutr Res 2013;33:18–26.

[37] Mullie P, Godderis L, Clarys P. Determinants and nutritional implications associated with low-fat food consumption. Appetite 2012;58(1):34–8.

[38] US Department of Agriculture, US Department of Health and Human Services. Dietary guidelines for Americans. 7th ed. Washington, DC: US Government Printing Office; 2010.

[39] Hess J, Slavin J. Snacking for a cause: nutritional insufficiencies and excesses of U.S. children, a critical review of food consumption patterns and macronutrient and micronutrient intake of U.S. children. Nutrients 2014;6(11):4750–9.

[40] Zhu Y, Wang H, Hollis JH, Jacques PF. The associations between yogurt consumption, diet quality, and metabolic profiles in children in the USA. Eur J Nutr 2015;54:543–50.

[41] Hartman C, Shamir R. A general overview of the impact of yogurt on health. Funct Food Rev 2013;5(2):1–13.

[42] Jackson KA, Savaiano DA. Lactose maldigestion, calcium intake and osteoporosis in African-, Asian-, and Hispanic-Americans. J Am Coll Nutr 2001;20(2 Suppl.):198S–207S.

[43] Eaton TK, Zhang C, Savaiano D. Lactose intolerance and yogurt. Funct Food Rev 2013;5(2):62–7.

[44] Mattar R, Mazo DFDC, Carrilho FJ. Lactose intolerance: diagnosis, genetic, and clinical factors. Clin Exp Gastroenterol 2012;5(1):113–21.

[45] Piaia M, Antoine J, Mateos-guardia J, Leplingard A, Lenoir-wijnkoop I. Assessment of the benefits of live yogurt: methods and markers for in vivo studies of the physiological effects of yogurt cultures. Microb Ecol Health Dis 2003;15(2–3):79–87.

[46] Kolars JC, Levitt MD, Aouji M, Savaiano DA. Yogurt—an autodigesting source of lactose. N Engl J Med 1984;310(1):1–3.

[47] Sanders ME, Walker DC, Walker KM, Aoyama K, Klaenhammer TR. Performance of commercial cultures in fluid milk applications. J Dairy Sci 1996;79(6):943–55.

[48] Lomer MCE, Parkes GC, Sanderson JD. Review article: lactose intolerance in clinical practice—myths and realities. Aliment Pharmacol Ther 2008;27(2):93–103.

[49] Labayen I, Forga L, González A, Lenoir-Wijnkoop I, Nutr R, Martínez JA. Relationship between lactose digestion, gastrointestinal transit time and symptoms in lactose malabsorbers after dairy consumption. Aliment Pharmacol Ther 2001;15(4):543–9.

[50] US Department of Health and Human Services, National Institutes of Health, National Institute of Diabetes and Digestive and Kidney Diseases. Overweight and obesity statistics; 2010.

[51] Ogden CL, Carroll MD, Kit BK, Flegal KM. Prevalence of obesity among adults: United States, 2011–2012. NCHS Data Brief. (1941-4927 (Electronic)). Hyattsville, MD: National Center for Health Statistics; 2013. p. 1–8.

[52] Zemel MB. Role of calcium and dairy products in energy partitioning and weight management. Am J Clin Nutr 2004;79(5):907S–12S.

[53] Pereira MA, Jacobs DR, Van Horn L, Slattery ML, Kartashov AI, Ludwig DS. Dairy consumption, obesity, and the insulin resistance syndrome in young adults: the CARDIA Study. JAMA 2002;287(16):2081–9.

[54] Mozaffarian D, Hao T, Rimm EB, Willett WC, Hu FB. Changes in diet and lifestyle and long-term weight gain in women and men. N Engl J Med 2011;364(25):2392–404.

[55] Tsai F, Coyle WJ. The microbiome and obesity: is obesity linked to our gut flora? Curr Gastroenterol Rep 2009;11(4):307–13.

[56] Arora T, Sharma R. Fermentation potential of the gut microbiome: implications for energy homeostasis and weight management. Nutr Rev 2011;69(2):99–106.

[57] Diamant M, Blaak EE, de Vos WM. Do nutrient-gut-microbiota interactions play a role in human obesity, insulin resistance and type 2 diabetes? Obes Rev 2011;12(4):272–81.

[58] Lee H-Y, Park J-H, Seok S-H, et al. Human originated bacteria, *Lactobacillus rhamnosus* PL60, produce conjugated linoleic acid and show anti-obesity effects in diet-induced obese mice. Biochim Biophys Acta 2006;1761(7):736–44.

[59] Parnell JA, Reimer RA. Weight loss during oligofructose supplementation is associated with decreased ghrelin and increased peptide YY in overweight and obese adults. Am J Clin Nutr 2009;89(6):1751–9.

[60] McNulty NP, Yatsunenko T, Hsiao A, et al. The impact of a consortium of fermented milk strains on the gut microbiome of gnotobiotic mice and monozygotic twins. Sci Transl Med 2011;3:106ra106.

[61] Zemel MB, Shi H, Greer B, Dirienzo D, Zemel PC. Regulation of adiposity by dietary calcium. FASEB J 2000;14(9):1132–8.

[62] Shi H, Dirienzo D, Zemel MB. Effects of dietary calcium on adipocyte lipid metabolism and body weight regulation in energy-restricted aP2-agouti transgenic mice. FASEB J 2001;15(2):291–3.

[63] Shi H, Norman AW, Okamura WH, Sen A, Zemel MB. 1Alpha, 25-dihydroxyvitamin D3 modulates human adipocyte metabolism via nongenomic action. FASEB J 2001;15(14):2751–3.

[64] Weaver CM. Calcium. 8th ed. Present knowledge in nutrition. Washington, DC: ILSI Press; 2001. p. 273–80.

[65] Montecucco F, Steffens S, Mach F. Insulin resistance: a proinflammatory state mediated by lipid-induced signaling dysfunction and involved in atherosclerotic plaque instability. Mediators Inflamm 2008;2008:767623.

[66] World Health Organization. Diabetes. Fact sheet N 312. August 2011, vol. 2009. World Health Organization; 2011.

[67] Association AD. Available from: http://www.diabetes.org/diabetes-basics/statistics/cdc-infographic.html

[68] Liu E, McKeown NM, Newby PK, et al. Cross-sectional association of dietary patterns with insulin-resistant phenotypes among adults without diabetes in the Framingham Offspring Study. Br J Nutr 2009;102(4):576–83.

[69] Tong X, Dong J-Y, Wu Z-W, Li W, Qin L-Q. Dairy consumption and risk of type 2 diabetes mellitus: a meta-analysis of cohort studies. Eur J Clin Nutr 2011;65(9):1027–31.

[70] Choi HK, Willett WC, Stampfer MJ, Rimm E, Hu FB. Dairy consumption and risk of type 2 diabetes mellitus in men: a prospective study. Arch Intern Med 2005;165:997–1003.

[71] Liu S, Choi HK, Ford E, et al. A prospective study of dairy intake and the risk of type 2 diabetes in women. Diabetes Care 2006;29(7):1579–84.

[72] Margolis KL, Wei F, de Boer IH, et al. A diet high in low-fat dairy products lowers diabetes risk in postmenopausal women. J Nutr 2011;141(11):1969–74.

[73] Kirii K, Mizoue T, Iso H, et al. Calcium, vitamin D and dairy intake in relation to type 2 diabetes risk in a Japanese cohort. Diabetologia 2009;52(12):2542–50.

[74] Sluijs I, Forouhi NG, Beulens JWJ, et al. The amount and type of dairy product intake and incident type 2 diabetes: results from the EPIC-InterAct Study. Am J Clin Nutr 2012;96(2):382–90.

[75] Aune D, Norat T, Romundstad P, Vatten LJ. Dairy products and the risk of type 2 diabetes: a systematic review and dose-response meta-analysis of cohort studies. Am J Clin Nutr 2013;98(4):1066–83.

[76] Chen M, Sun Q, Giovannucci E, et al. Dairy consumption and risk of type 2 diabetes: 3 cohorts of US adults and an updated meta-analysis. Am J Clin Nutr 2014;94(4):1088–96.

[77] Diaz-Lopez A, Bullo M, Martinez-Gonzalez MA, et al. Dairy product consumption and risk of type 2 diabetes in an elderly Spanish Mediterranean population at high cardiovascular risk. Eur J Nutr 2016;55:349–60.

[78] Snijder MB, van der Heijden AA, van Dam RM, et al. Is higher dairy consumption associated with lower body weight and fewer metabolic disturbances? The Hoorn Study. Am J Clin Nutr 2007;85(4):989–95.

[79] Soedamah-Muthu SS, Masset G, Verberne L, Geleijnse JM, Brunner EJ. Consumption of dairy products and associations with incident diabetes, CHD and mortality in the Whitehall II study. Br J Nutr 2013;109:718–26.

[80] Grantham NM, Magliano DJ, Hodge A, Jowett J, Meikle P, Shaw JE. The association between dairy food intake and the incidence of diabetes in Australia: the Australian Diabetes Obesity and Lifestyle Study (AusDiab). Public Health Nutr 2013;16(2):339–45.

[81] Kim J. Dairy food consumption is inversely associated with the risk of the metabolic syndrome in Korean adults. J Hum Nutr Diet 2013;26(Suppl. 1):171–9.

[82] Abreu S, Moreira P, Moreira C, et al. Intake of milk, but not total dairy, yogurt, or cheese, is negatively associated with the clus-

tering of cardiometabolic risk factors in adolescents. Nutr Res 2014;34(1):48–57.

[83] Rizkalla SW, Luo J, Kabir M, Chevalier A, Pacher N, Slama G. Chronic consumption of fresh but not heated yogurt improves breath-hydrogen status and short-chain fatty acid profiles: a controlled study in healthy men with or without lactose maldigestion. Am J Clin Nutr 2000;72(6):1474–9.

[84] Nazare J-A, de la Perrière AB, Bonnet F, et al. Daily intake of conjugated linoleic acid-enriched yoghurts: effects on energy metabolism and adipose tissue gene expression in healthy subjects. Br J Nutr 2007;97:273–80.

[85] Chang BJ, Park SU, Jang YS, et al. Effect of functional yogurt NY-YP901 in improving the trait of metabolic syndrome. Eur J Clin Nutr 2011;65(11):1250–5.

[86] Asemi Z, Samimi M, Tabassi Z, et al. Effect of daily consumption of probiotic yoghurt on insulin resistance in pregnant women: a randomized controlled trial. Eur J Clin Nutr 2013;67(1):71–4.

[87] Schaafsma G, Meuling WJ, van Dokkum W, Bouley C. Effects of a milk product, fermented by *Lactobacillus acidophilus* and with fructo-oligosaccharides added, on blood lipids in male volunteers. Eur J Clin Nutr 1998;52(6):436–40.

[88] World Health Organization. Cardiovascular disease. Fact sheet; 2011.

[89] EFSA Panel on Dietetic Products Nutrition and Allergies (NDA). Scientific opinion on dietary reference values for fats, including saturated fatty acids, polyunsaturated fatty acids, monounsaturated fatty acids, trans fatty acids, and cholesterol. EFSA J. 2010;8(3):1461.

[90] Appel LJ, Brands MW, Daniels SR, Karanja N, Elmer PJ, Sacks FM. Dietary approaches to prevent and treat hypertension: a scientific statement from the American Heart Association. Hypertension 2006;47(2):296–308.

[91] Mancia G, Laurent S, Agabiti-Rosei E, et al. Reappraisal of European guidelines on hypertension management: a European Society of Hypertension Task Force document. J Hypertens 2009;27(11):2121–58.

[92] Appel LJ, Moore TJ, Obarzanek E, et al. A clinical trial of the effects of dietary patterns on blood pressure. DASH Collaborative Research Group. N Engl J Med 1997;336(16):1117–24.

[93] Tavani A, Gallus S, Negri E, La Vecchia C. Milk, dairy products, and coronary heart disease. J Epidemiol Community Health 2002;56(6):471–2.

[94] Iso H, Stampfer MJ, Manson JE, et al. Prospective study of calcium, potassium, and magnesium intake and risk of stroke in women. Stroke 1999;30(9):1772–9.

[95] Soedamah-Muthu SS, Verberne LDM, Ding EL, Engberink MF, Geleijnse JM. Dairy consumption and incidence of hypertension: a dose-response meta-analysis of prospective cohort studies. Hypertension 2012;60(5):1131–7.

[96] Sonestedt E, Wirfält E, Wallström P, Gullberg B, Orho-Melander M, Hedblad B. Dairy products and its association with incidence of cardiovascular disease: The Malmö diet and cancer cohort. Eur J Epidemiol 2011;26(8):609–18.

[97] Wang L, Manson JE, Buring JE, Lee I-M, Sesso HD. Dietary intake of dairy products, calcium, and vitamin D and the risk of hypertension in middle-aged and older women. Hypertension 2008;51(4):1073–9.

[98] Wang H, Livingston KA, Fox CS, Meigs JB, Jacques PF. Yogurt consumption is associated with better diet quality and metabolic profile in American men and women. Nutr Res 2013;33(1):18–26.

[99] Agerholm-Larsen L, Bell ML, Grunwald GK, Astrup A. The effect of a probiotic milk product on plasma cholesterol: a meta-analysis of short-term intervention studies. Eur J Clin Nutr 2000;54(11): 856–60.

[100] Ivey KL, Lewis JR, Hodgson JM, et al. Association between yogurt, milk, and cheese consumption and common carotid artery intima-media thickness and cardiovascular disease risk factors in elderly women. Am J Clin Nutr 2011;94(1):234–9.

[101] Cho YA, Kim J. Effect of probiotics on blood lipid concentrations: a meta-analysis of randomized controlled trials. Medicine 2015;94(43):e1714.

[102] Cam A, de Mejia EG. Role of dietary proteins and peptides in cardiovascular disease. Mol Nutr Food Res 2012;56(1):53–66.

[103] Huth PJ, Park KM. Influence of dairy product and milk fat consumption on cardiovascular disease risk: a review of the evidence. Adv Nutr 2012;3(3):266–85.

[104] de Oliveira Otto MC, Nettleton JA, Lemaitre RN, et al. Biomarkers of dairy fatty acids and risk of cardiovascular disease in the multi-ethnic study of atherosclerosis. J Am Heart Assoc 2013;2(4):e000092.

[105] Klaver FA, van der Meer R. The assumed assimilation of cholesterol by lactobacilli and *Bifidobacterium bifidum* is due to their bile salt-deconjugating activity. Appl Environ Microbiol 1993;59(4):1120–4.

[106] St-Onge MP, Farnworth ER, Jones PJH. Consumption of fermented and nonfermented dairy products: effects on cholesterol concentrations and metabolism. Am J Clin Nutr 2000;71(3):674–81.

[107] Pereira DI, Gibson GR. Effects of consumption of probiotics and prebiotics on serum lipid levels in humans. Crit Rev Biochem Mol Biol 2002;37(4):259–81.

[108] Boelsma E, Kloek J. Lactotripeptides and antihypertensive effects: a critical review. Br J Nutr 2009;101(6):776–86.

[109] Sipola M, Finckenberg P, Santisteban J, Korpela R, Vapaatalo H, Nurminen ML. Long-term intake of milk peptides attenuates development of hypertension in spontaneously hypertensive rats. J Physiol Pharmacol 2001;52(4 Pt. 2):745–54.

[110] Jauhiainen T, Vapaatalo H, Poussa T, Kyrönpalo S, Rasmussen M, Korpela R. *Lactobacillus helveticus* fermented milk lowers blood pressure in hypertensive subjects in 24-h ambulatory blood pressure measurement. Am J Hypertens 2005;18(12):1600–5.

[111] Mizushima S, Ohshige K, Watanabe J, et al. Randomized controlled trial of sour milk on blood pressure in borderline hypertensive men. Am J Hypertens 2004;17(8):701–6.

[112] Seppo L, Jauhiainen T, Poussa T, Korpela R. A fermented milk high in bioactive peptides has a blood pressure-lowering effect in hypertensive subjects. Am J Clin Nutr 2003;77(2):326–30.

[113] Tuomilehto J, Lindström J, Hyyrynen J, et al. Effect of ingesting sour milk fermented using *Lactobacillus helveticus* bacteria producing tripeptides on blood pressure in subjects with mild hypertension. J Hum Hypertens 2004;18(11):795–802.

[114] Azadbakht L, Mirmiran P, Esmaillzadeh A, Azizi F. Dairy consumption is inversely associated with the prevalence of the metabolic syndrome in Tehranian adults. Am J Clin Nutr 2005;82(3): 523–30.

[115] Zemel MB. Calcium modulation of hypertension and obesity: mechanisms and implications. J Am Coll Nutr 2001;20(5 Suppl.): 428S–35S. discussion 440S–442S.

[116] Lawes CM, Vander Hoorn S, Rodgers A. Global burden of blood-pressure-related disease, 2001. Lancet 2008;371(9623):1513–8.

[117] Meyer AL, Elmadfa I, Herbacek I, Micksche M. Probiotic, as well as conventional yogurt, can enhance the stimulated production of proinflammatory cytokines. J Hum Nutr Diet 2007;20(6):590–8.

[118] Aattouri N, Lemonnier D. Production of interferon induced by *Streptococcus thermophilus*: role of CD4+ and CD8+ lymphocytes. J Nutr Biochem 1997;8(1):25–31.

[119] Stone KD. Atopic diseases of childhood. Curr Opin Pediatr 2003;15(5):495–511.

[120] Miyake Y, Tanaka K, Okubo H, Sasaki S, Arakawa M. Dairy food, calcium and vitamin D intake and prevalence of allergic disorders in pregnant Japanese women. Int J Tuberc Lung Dis 2012;16(2): 255–61.

[121] Van de Water J, Keen CL, Gershwin ME. The influence of chronic yogurt consumption on immunity. J Nutr 1999;129 (7 Suppl.):1492S–5S.

[122] Enomoto T, Shimizu K, Shimazu SI. Suppression of allergy development by habitual intake of fermented milk foods, evidence from an epidemiological study. Jpn J Allergol 2006;55(11):1394–9.

[123] Roduit C, Frei R, Loss G, et al. Development of atopic dermatitis according to age of onset and association with early-life exposures. J Allergy Clin Immunol 2012;130(1):130–6.

[124] Maslova E, Halldorsson TI, Strøm M, Olsen SF. Low-fat yoghurt intake in pregnancy associated with increased child asthma and allergic rhinitis risk: a prospective cohort study. J Nutr Sci 2012;1:e5.

[125] Chatzi L, Mendez M, Garcia R, et al. Mediterranean diet adherence during pregnancy, fetal growth: INMA (Spain) and RHEA (Greece) mother-child cohort studies. Br J Nutr 2012;107(1):135–45.

[126] World Health Organization. WHO diarrhoeal disease. WHO website. Available from: http://www.who.int/mediacentre/factsheets/fs330/en/index.html; 2012.

[127] Turck D, Bernet J-P, Marx J, et al. Incidence and risk factors of oral antibiotic-associated diarrhea in an outpatient pediatric population. J Pediatr Gastroenterol Nutr 2003;37(1):22–6.

[128] Vanderhoof JA, Whitney DB, Antonson DL, Hanner TL, Lupo JV, Young RJ. Lactobacillus GG in the prevention of antibiotic-associated diarrhea in children. J Pediatr 1999;135(5):564–8.

[129] Arvola T, Laiho K, Torkkeli S, et al. Prophylactic Lactobacillus GG reduces antibiotic-associated diarrhea in children with respiratory infections: a randomized study. Pediatrics 1999;104(5):e64.

[130] Beniwal RS, Arena VC, Thomas L, et al. A randomized trial of yogurt for prevention of antibiotic-associated diarrhea. Dig Dis Sci 2003;48(10):2077–82.

[131] WGO. World Gastroenterology Organisation practice guideline: probiotics and prebiotics. Arab J Gastroenterol. 2009;10(1):33–42.

[132] Boudraa G, Touhami M, Pochart P, Soltana R, Mary JY, Desjeux JF. Effect of feeding yogurt versus milk in children with persistent diarrhea. J Pediatr Gastroenterol Nutr 1990;11(4):509–12.

[133] Boudraa G, Benbouabdellah M, Hachelaf W, Boisset M, Desjeux JF, Touhami M. Effect of feeding yogurt versus milk in children with acute diarrhea and carbohydrate malabsorption. J Pediatr Gastroenterol Nutr 2001;33(3):307–13.

[134] Van Niel CW, Feudtner C, Garrison MM, Christakis DA. Lactobacillus therapy for acute infectious diarrhea in children: a meta-analysis. Pediatrics 2002;109(4):678–84.

[135] de Mattos AP, Ribeiro TCM, Mendes PSA, Valois SS, Mendes CMC, Ribeiro HC. Comparison of yogurt, soybean, casein, and amino acid-based diets in children with persistent diarrhea. Nutr Res 2009;29(7):462–9.

[136] Moore SR. Update on prolonged and persistent diarrhea in children. Curr Opin Gastroenterol 2011;27(1):19–23.

[137] He M, Yang Y-X, Han H, Men J-H, Bian L-H, Wang G-D. Effects of yogurt supplementation on the growth of preschool children in Beijing suburbs. Biomed Environ Sci 2005;18(3):192–7.

[138] Makino S, Ikegami S, Kume A, Horiuchi H, Sasaki H, Orii N. Reducing the risk of infection in the elderly by dietary intake of yoghurt fermented with *Lactobacillus delbrueckii* ssp. *bulgaricus* OLL1073R-1. Br J Nutr 2010;104(7):998–1006.

[139] Adolfsson O, Meydani SN, Russell RM. Yogurt and gut function. Am J Clin Nutr. 2004;80(2):245–56.

[140] Perdigón G, Vintiñi E, Alvarez S, Medina M, Medici M. Study of the possible mechanisms involved in the mucosal immune system activation by lactic acid bacteria. J Dairy Sci 1999;82(6):1108–14.

[141] Meydani SN, Ha WK. Immunologic effects of yogurt. Am J Clin Nutr 2000;71(4):861–72.

[142] Plaisancié P, Claustre J, Estienne M, et al. A novel bioactive peptide from yoghurts modulates expression of the gel-forming MUC2 mucin as well as population of goblet cells and Paneth cells along the small intestine. J Nutr Biochem 2013;24(1):213–21.

[143] Heyman M. Effect of lactic acid bacteria on diarrheal diseases. J Am Coll Nutr 2000;19(2 Suppl.):137S–46S.

[144] Rohde CL, Bartolini V, Jones N. The use of probiotics in the prevention and treatment of antibiotic-associated diarrhea with special interest in *Clostridium difficile*-associated diarrhea. Nutr Clin Pract 2009;24(1):33–40.

[145] Freitas M, Tavan E, Cayuela C, Diop L, Catherine S, Trugnan G. Host-pathogens cross-talk. Indigenous bacteria and probiotics also play the game. Biol Cell 2003;95(8):503–6.

[146] de Vrese M, Offick B. Probiotics and prebiotics: effects on diarrhea. Bioactive foods in promoting health. London: Academic Press; 2010. p. 205–227.

[147] Antoine JM. Probiotics: beneficial factors of the defence system. Proc Nutr Soc 2010;69(3):429–33.

[148] Jeffery IB, O'Toole PW, Ohman L, et al. An irritable bowel syndrome subtype defined by species-specific alterations in faecal microbiota. Gut 2012;61(7):997–1006.

[149] Moayyedi P, Ford AC, Talley NJ, et al. The efficacy of probiotics in the treatment of irritable bowel syndrome: a systematic review. Gut 2010;59(3):325–32.

[150] Pimentel M, Lembo A, Chey WD, et al. Rifaximin therapy for patients with irritable bowel syndrome without constipation. N Engl J Med 2011;364(1):22–32.

[151] Clarke G, Cryan JF, Dinan TG, Quigley EM. Review article: probiotics for the treatment of irritable bowel syndrome—focus on lactic acid bacteria. Aliment Pharmacol Ther 2012;35(4):403–13.

[152] Guyonnet D, Chassany O, Ducrotte P, et al. Effect of a fermented milk containing *Bifidobacterium animalis* DN-173 010 on the health-related quality of life and symptoms in irritable bowel syndrome in adults in primary care: a multicentre, randomized, double-blind, controlled trial. Aliment Pharmacol Ther 2007;26(3):475–86.

[153] Ornelas IJ, Galvan-Potrillo M, López-Carrillo L. Protective effect of yoghurt consumption on *Helicobacter pylori* seropositivity in a Mexican population. Public Health Nutr 2007;10(11):1283–7.

[154] Yoon H, Kim N, Kim JY, et al. Effects of multistrain probiotic-containing yogurt on second-line triple therapy for *Helicobacter pylori* infection. J Gastroenterol Hepatol 2011;26(1):44–8.

[155] Deguchi R, Nakaminami H, Rimbara E, et al. Effect of pretreatment with *Lactobacillus gasseri* OLL2716 on first-line *Helicobacter pylori* eradication therapy. J Gastroenterol Hepatol 2012;27(5):888–92.

[156] Marteau P, Guyonnet D, Lafaye de Micheaux P, Gelu S. A randomized, double-blind, controlled study and pooled analysis of two identical trials of fermented milk containing probiotic *Bifidobacterium lactis* CNCM I-2494 in healthy women reporting minor digestive symptoms. Neurogastroenterol Motil 2013;25(4):331-e252.

[157] Pala V, Sieri S, Berrino F, et al. Yogurt consumption and risk of colorectal cancer in the Italian European prospective investigation into cancer and nutrition cohort. Int J Cancer 2011;129(11):2712–9.

[158] Peters RK, Pike MC, Garabrant D, Mack TM. Diet and colon cancer in Los Angeles County, California. Cancer Causes Control 1992;3(5):457–73.

[159] Kojima M, Wakai K, Tamakoshi K, et al. Diet and colorectal cancer mortality: results from the Japan Collaborative Cohort Study. Nutr Cancer 2004;50(1):23–32.

[160] Matsumoto M, Ishikawa S, Nakamura Y, Kayaba K, Kajii E. Consumption of dairy products and cancer risks. J Epidemiol 2007;17(2):38–44.

[161] Murphy N, Norat T, Ferrari P, et al. Consumption of dairy products and colorectal cancer in the European Prospective Investigation into Cancer and Nutrition (EPIC). PLoS One 2013;8(9):e72715.

[162] Ohashi Y, Nakai S, Tsukamoto T, et al. Habitual intake of lactic acid bacteria and risk reduction of bladder cancer. Urol Int 2002;68(4):273–80.

[163] Radosavljevic V, Jankovic S, Marinkovic J, Djokic M. Fluid intake and bladder cancer. A case control study. Neoplasma 2003;50(3):234–8.

[164] Larsson SC, Andersson S-O, Johansson J-E, Wolk A. Cultured milk, yogurt, and dairy intake in relation to bladder cancer risk in a prospective study of Swedish women and men. Am J Clin Nutr 2008;88(4):1083–7.

[165] Rohrmann S, Linseisen J, Jakobsen MU, et al. Consumption of meat and dairy and lymphoma risk in the European Prospective Investigation into Cancer and Nutrition. Int J Cancer 2011;128(3):623–34.

[166] Genkinger JM, Hunter DJ, Spiegelman D, et al. Dairy products and ovarian cancer: a pooled analysis of 12 cohort studies. Cancer Epidemiol Biomarkers Prev 2006;15(2):364–72.

[167] Saikali J, Picard C, Freitas M, Holt P. Fermented milks, probiotic cultures, and colon cancer. Nutr Cancer. 2004;49(1):14–24.

[168] Aune D, Lau R, Chan DSM, et al. Dairy products and colorectal cancer risk: a systematic review and meta-analysis of cohort studies. Ann Oncol 2012;23(1):37–45.

Part D

Basic Physiologic Effects of Microbiota

Chapter 25

Dysbiosis

W.A. Walker*,**

*Harvard Medical School, Boston, MA, United States; **Mucosal Immunology and Biology Research Center, Pediatric Gastroenterology and Nutrition, Massachusetts General Hospital for Children, Boston, MA, United States

INTRODUCTION

It is increasingly clear from epidemiologic, clinical, and basic studies that an appropriate colonization of the gastrointestinal tract is necessary for short- and long-term immunologic and metabolic health [1]. The gut microbiome functions as an ancillary organ to the body providing more cells and genes that exist in the human body and functions as an active metabolic structure contributing to nutrient digestion and absorption, immune function, and metabolic activity necessary for health. Colonizing bacteria communicate with gut structures through direct interaction with enterocytes and immunocytes, or through secretion of metabolically active substances [2,3]. Intestinal microbes also benefit from their interaction with the host. This mutually beneficial relationship is known as symbiosis and its normal function results in absence of inflammation and disease [4].

Microbial-intestinal interaction begins in utero with microorganisms from a variety of sources (maternal gut, mouth, and skin) identified in the placenta, amniotic fluid, and meconium [5]. This interaction continues at birth and during the perinatal period. Initial colonization with appropriate symbiotic commensal bacteria is particularly important after birth when the newborn is adjusting to its extrauterine environment [6]. Since digestive and immunologic function is immature at birth and colonizing bacteria play an important role in these gut functions, a proper initial colonization is necessary to ensure immune and metabolic homeostasis [7].

Normal initial colonization occurs when a full term infant is born by vaginal delivery and exclusively breast-fed for the first 4–6 months of life. Under these conditions, the gut is colonized with microorganisms, which specifically stimulate normal immunologic and metabolic developmental gut function [8]. Normal colonization proceeds over the first 3 years until a permanent gut microbiota is established, which persists throughout life. The significance of normal colonization is as follows. The partially colonized newborn passes through the birth canal where it ingests a healthy bolus of vaginal and colonic maternal microbiota, the first and most important stage of colonization. With the introduction of oral feedings the intestinal microbiota is further stimulated (stage two). The striking difference in stimulus occurs depending on the nature of oral feedings, for example, breast milk versus formula feeding [9]. This will be discussed later when we consider diet and microbial colonization. After 6 months to 1 year with the weaning to solid foods, the microbiota further change (phase three) and by 3 years of life a permanent microbiota is established which is unique to that individual [10]. The establishment of "normal" gut microbiota has a profound effect on short- and long-term health. A "healthy microbiota" is a nebulous term and refers to the gut microbiota in healthy individuals [11]. Modern lifestyle in developed countries with improved sanitation, immunizations, a westernized diet, and excessive use of antibiotics has resulted in a difference in bacterial colonization of the gut, and is implicated in the incidence of allergic and other immune-mediated diseases (autoimmune disease) over the last several decades (*The Hygiene Hypothesis*) [12,13].

DYSBIOSIS

A disruption in "normal" intestinal colonization, for example, dysbiosis is hypothesized to contribute to these diseases. Again this term, dysbiosis, is nebulous. Briefly, dysbiosis is defined as "any change to the components of resident commensal communities relative to the community found in health individuals" [14]. A major unanswered question regarding dysbiosis is microbial disruption, a cause or an effect of the associated disease. This question is still controversial. Dysbiosis has been categorized into various types or combination of types (Table 25.1).

For example, one category could be the loss of known beneficial microorganisms. As previously mentioned, specific organisms present in the initial "normal" colonization of the gut have been shown to influence development of immune homeostasis. For example, *Bacteroides fragilis* or a

TABLE 25.1 Categories of Dysbiosis

- Loss of beneficial microbial organisms
- Expression of pathobionts or potentially beneficial microorganisms
- Loss of overall microbial diversity

Source: Adapted from Cell Microbiol 2014;16:1024–1033.

mixture of *Clostridium* strains has been shown to influence the development and maintenance of T-regulatory cells (Tregs) [14]. Tregs are necessary to prevent inappropriate inflammatory reactions to commensal bacteria and self-antigens, and are thought to be necessary in the prevention of expression of autoimmune diseases. In like manner, other organisms, such as *Lactobacillus acidophilus* or *Bifidobacterium breve* [15] have direct antiinflammatory activity by neutralizing inflammatory cytokines. An absence of these beneficial bacteria can upset the balance between protection against pathogens and tolerance to commensal organisms, and self-antigens.

A second category of dysbiosis can be caused by an expansion of pathobionts. This occurs with an increase in the bacteria phylum, Proteobacterium, and one of its families, Enterobacteriaceae with organisms, such as *Escherichia. coli*, *Shigella,* and *Klebsiella*. These pathobionts can increase with genetic defects in immune function leading to various forms of colitis. This is true with genetic polymorphism defects in nucleotide binding oligomerization domain protein like receptors (NRLs) or toll-like receptors, such as TLR-5 [16]. Both genetic defects lead to increase in proteobacterial organisms and predisposition to colitis.

Finally a third category is the loss of diversity in colonizing bacteria. Very often, analysis of microbiota and specific diseases is associated with less diversity. Bacterial diversity provides overlapping protection by multiple organisms in the prevention of noncommunicable diseases (NCDs). Examples of this category is the observation that single species of specific organisms is not as effective in stimulating protection against disease in germfree animals colonized with the organism from a diseased animal as several members of the same genera. For example, thirty species of *Clostridium* organisms are more effective in stimulating Tregs than a single species [17]. In addition, the elevated IgE levels in germfree animals can more effectively be reduced when colonized with multiple organisms of bacteria rather than one or a few [18]. Diversity, therefore has many features and can be represented by a single category or a mixture of categories.

What is the nature of colonizing microbiota under conditions of dysbiosis? This area has been addressed extensively over the last several years with the advent of metagenomic analysis of gut microbiota. As previously observed in other chapters of this book, most of the bacteria colonizing the gut cannot be identified by culture techniques alone [19]. Accordingly with establishment of molecular techniques for identification, large groups of bacteria (phyla) have been identified and these groups have been broken down into taxa, genera, and specific species of bacteria. Under normal colonizing conditions, a diverse number of species have been identified (greater than 1000) in stool analysis by a variety of bioinformatic techniques [20].

Five to seven bacteria phyla reside in the mammalian gut. However, in the healthy individual 90% of bacteria exists two phyla—Firmicutes and *Bacteroides* with others Proteobacteria, Actinobacteria, and Verrucomicrobia present in much smaller amounts [21]. In fact, Shin et al, have suggested that dysbiosis and inflammatory NCDs are associated with an increase in Proteobacteria phyla [22]. Thus this phyla and its subclasses, such as Enterobacteriaceae have become a marker for dysbiosis and are present in inflammatory diseases, such as inflammatory bowel disease [23] and metabolic diseases, such as type 2 diabetes [24]. As metagenomic bioinformatic programs have improved, specific genera or species of bacteria have been associated with specific diseases [25]. For example, *Enterobacter cloacae* B-29 has been associated with obesity and the organism transplanted into germfree mice has resulted in increased weight gain in these otherwise normal animals. In like manner, *Bilophila wadswothia* has been associated with experimental colitis [26]. As these studies progress, we should have microbial biomarkers for NCDs and could identify specific microorganisms which can be used to prevent or treat these diseases.

CAUSES OF DYSBIOSIS (TABLE 25.2)

As stated previously, the individual is more vulnerable to dysbiosis in infancy. Intestinal metabolic and immunologic development occurs in young infants in the initial stages of adjustment to the extrauterine environment. Therefore, any disruption in the sequence of normal colonization can have profound effects on short- and long-term health. Several environmental factors occurring in the perinatal period can affect initial colonization and cause aberrations in development of metabolic and immunologic function resulting in disease expression throughout the lifespan [27].

TABLE 25.2 Causes of Dysbiosis

- Perinatal disruption of colonization
- Genetic
- Dietary
- Disease
- Stress

PERINATAL CAUSES OF DYSBIOSIS

Disruption of microbial exposure in utero can affect the colonization process of the neonate. For example, excessive weight gain during pregnancy leading to maternal obesity can affect the mother's microbiota and in turn the infant's microbiota as a result of altered placental and birth exposure to mother's dysbiosis [28]. These infants are more like to develop excessive weight gain in infancy leading to obesity later in life [29]. In addition, premature delivery of infants as a result of intrauterine infection, excessive weight, etc. results in a dysbiotic colonization of the premature's intestine and increased susceptibility to excessive intestinal inflammation, such as necrotizing enterocolitis [30]. The dysbiosis is also contributed to by an immature intestinal barrier with birth commonly occurring by Cesarean section delivery.

Infants born by Cesarean section, a common practice in developed countries, do not obtain the initial bacterial bolus from mother's birth canal [10,31] and requires a much longer time to fully colonize the intestine [31]. These children grow up to have much higher incidence of asthma, obesity, and celiac disease [32] suggesting that early appropriate colonization is essential for lifelong health.

In developed countries, there is an excessive, and in many cases, inappropriate use of antibiotics, particularly during the first year of life. Several clinical studies have shown a direct correlation between the number of episodes of use of antibiotics and the nature of antibiotics used (broad spectrum vs. narrow spectrum), and the likelihood of infants developing asthma as adolescents [33]. An interesting research study recently reported shows that low doses of penicillin given to newborn animal models, which transiently disrupt colonizing microbiota resulted in excessive weight gain leading to obesity later in life [34]. These studies suggest that perinatal use of antibiotics can disrupt the sequence of normal colonization of the gut resulting in a loss of immunologic/metabolic developmental homeostasis and expression of disease later in life.

Finally, failure to feed newborn infants exclusive breast milk during the first few months of life alters the nutritional environmental impact of initial colonization and results in an increase in NCDs later in life including allergy, autoimmune disease, and excessive weight gain [35].

Genetic Causes

Disruption in immune or metabolic intestinal function due to a genetic defect, such as polymorphisms in TLR-4 or IL-10 results in a disruption in normal intestinal colonization leading to diseases, such as inflammatory bowel disease [16]. Carvalho et al have published that in a TLR-5 knockout animal, a TLR-5 is a toll receptor that interacts on enterocytes with bacterial flagella, resulting in disrupted intestinal microbiota leading to colitis and metabolic

syndrome [36]. When the microbiota from the TLR-5 knockout animals were transplanted into a germfree mouse, the mouse developed both colitis and metabolic syndrome suggesting the role of dysbiotic microbiota in phenotypic expression of disease. In an animal model for immune deficiency, dysbiotic colonization occurred resulting in a defect in fat absorption [37]. Fat malabsorption is common in immunodeficient patients.

Dietary Causes

A major environmental factor influencing the composition of microbiota is the person's diet. This has been illustrated in the discussion of breast-feeding versus formula feeding in infancy. Other studies have illustrated the observation. A study in by DeFilippo [38] compared the composition of gut microbiota in two adolescent populations (from Florence, Italy where children ate a typical western diet high in animal fat and protein compared to a rural African village where children ate a high fiber, complex carbohydrate diet, with little animal fat and protein). The microbiota analysis was strikingly different suggesting that diet had a major impact on gut microbiota. Although in that study, it is interesting to speculate that diet and intestinal microbiota may account for the striking difference in NCDs between the two populations. A recent study in large populations of healthy adults showed that the composition of diet (high simple or complex carbohydrates, high fat vs. high protein) over a prolonged period affected bacterial enterotypes (enterotypes are groups of bacteria with similar functional activity) [39]. The Western Diet has recently been implemented in altered gut microbiota and disease expression. Diets, such as the Mediterranean Diet cause longer life expectancy and less heart and metabolic disease [40]. Recent publications have implicated emulsifying agents and sugar substitutes in altered gut microbiota, and increased expression of disease [41,42]. These observations suggest that diet at any age may impact gut microbiota and contribute to the increased incidence of NCDs.

Disease Causes

Disease per se can alter intestinal microbiota and may contribute to a dysbiosis leading to clinical problems. It is not uncommon for patients developing inflammatory bowel disease, particularly Crohn disease, to have a history of intestinal infection prior to the onset of the disease. A *Pediatric* review suggests that the composition of intestinal microbiome in patients with disease differs strikingly from age-matched controls (Fig. 25.1) [27]. This observation underscores the dilemma between the microbiota and its effect in contributing to disease. My personal bias is that disease per se disrupts the intestinal microbiota leading to dysbiosis. Dysbiotic bacteria, in turn, because of their enormous

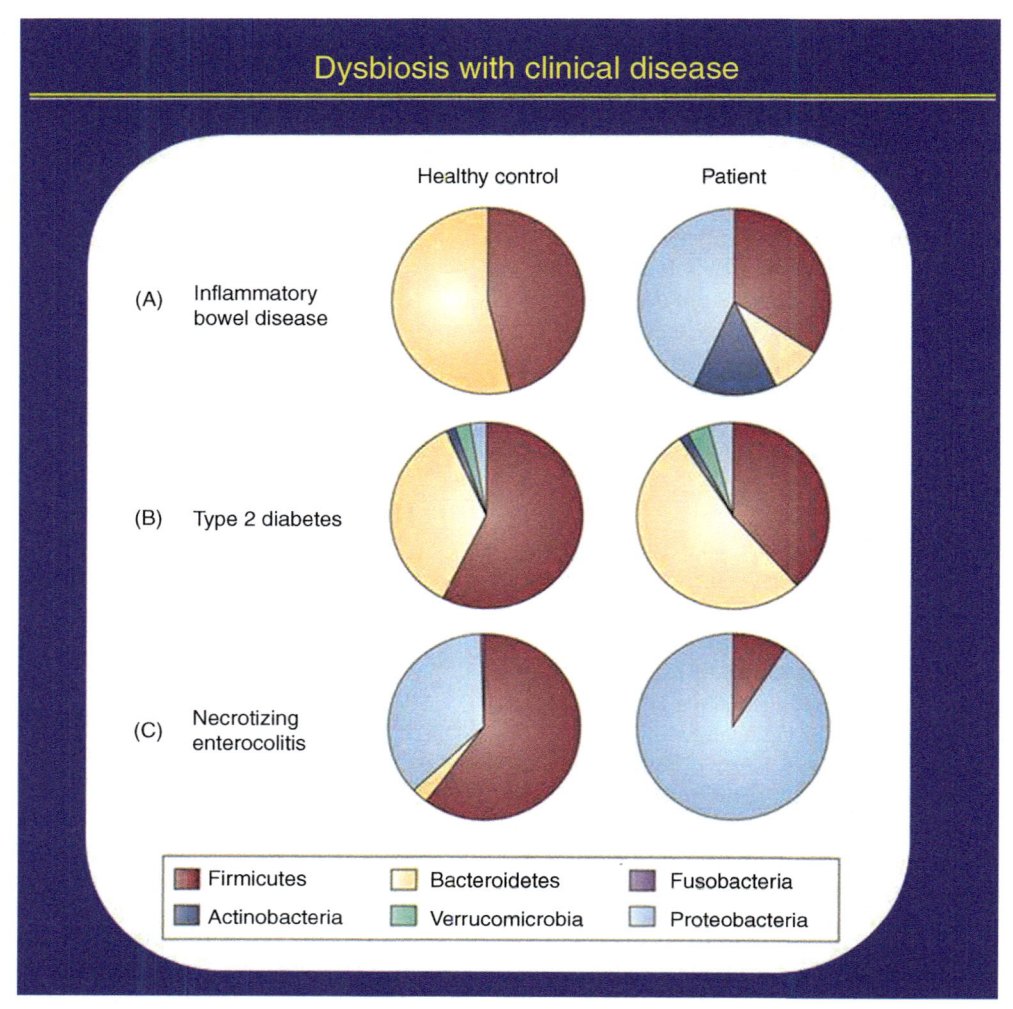

FIGURE 25.1 **Comparisons of bacterial phyla as determined by metaanalysis of patients with disease versus age-matched controls.** (A) Inflammatory bowel disease, (B) type 2 diabetes, (C) necrotizing enterocolitis. *Reproduced with permission from Nat Rev Microbiol 2011;9(4):279–290.*

metabolic activity propagate the phenotypic expression of disease and when the dysbiotic microbiota is transplanted into a germfree mouse, the animal in turn expresses the disease. This observation also helps to explain the combined genetic and metabolic basis for disease. An underlying subtle genetic defect, for example, polymorphisms in a key molecule, can result in dysbiosis and expression of disease. The dysbiotic intestinal microbiota acts to propagate the disease. An example of that is a genetically predisposed microbiota, which metabolizes animal protein in the diet to produce metabolites which enhance the likelihood of cardiovascular disease [43].

Stress Causes

A new area of consideration as a cause of dysbiosis is stress in ones' life. There is both basic and clinical evidence to suggest that stress can influence gut microbiota and in turn expression of neurologic disease [44]. This area is very

new and there are many questions still unanswered. For example, clinical studies with healthy individuals exposed to stressful situations can alter both gut microbiota and brain function [45]. Interestingly, the neurologic changes with stress can be relieved with probiotic use [46]. Disruption of the microenvironment during pregnancy can affect the maternal gut microbiota and possible predispose to dysbiotic neonatal colonization, and ultimately expression of neurologic diseases, such as autism and schizophrenia. These observations are currently associations and require additional basic and clinical studies to show actual cause and effect. However, they are fascinating and should be pursued.

HOW DO WE APPROACH DYSBIOSIS IN THE CONTEXT OF DISEASE?

As we become more efficient in determining the composition of intestinal microbiota in health and disease, we may be able to define phyla, families, genera, and species of bacteria associated with specific disease states. The microbiota signature of a disease can be used both as biomarkers or potential as means of both preventing and treating disease. Clinical studies have suggested that the absence or reduction in specific health promoting bacteria, for example, *Lactobacillus* or *Bifidobacteria* in patients who develop allergic disease [47,48] may be a risk factor. In some studies, use of these microbes as probiotics in infancy has been shown to prevent expression of allergic disease [49]. A specific microbiome has been identified with obesity, including specific organisms. Use of these absent or reduced organisms to treat individuals at risk for disease has been shown to be in part successful [50,51]. When groups of organisms are required for disrupting the clinical impact of dysbiosis, for example, recurrent *Clostridium difficile* diarrhea, approaches, such as fecal transplant may be required. Regardless, more clinical and basic studies about the basis for dysbiosis and disease are needed for further approaches to dealing with NCDs.

SUMMARY AND CONCLUSIONS

In this chapter, we have attempted to comprehensively define the role of dysbiosis and expression of disease. Dysbiosis has been defined and categories of dysbiosis considered. The importance of normal intestinal colonization of the gut during the neonatal period has been emphasized for proper development of immunologic and metabolic intestinal function, and lifelong health. The factors that influence gut colonization and the consequences of improper factors leading to dysbiosis and disease have been considered. Finally, in the context of our understanding of dysbiosis future studies to define dysbiotic signatures of disease and to suggest possible approaches to affecting dysbiosis are necessary.

REFERENCES

[1] Pearce N, Ebrahim S, McKee M, et al. Global prevention and control of NCD: limitations of the standard approach. J Public Health Policy 2015;36:408–25.

[2] Hooper LV, Littman DR, Macpherson AJ. Interactions between the microbiota and the immune system. Science 2012;336:1268–73.

[3] Arpaia N, Campbell C, Fan X, et al. Metabolites produced by commensal bacteria promote peripheral regulatory T-cell generation. Nature 2013;504:451–5.

[4] Rodriquez JM, Murphy K, Stanton C, et al. The composition of the gut microbiota throughout life, with an emphasis on early life. Microb Ecol Health Dis 2015;26:26050.

[5] Aagaard K, Ma J, Antony KM, Ganu R, Peterosino J, Versalovic J. The placenta harbors a unique microbiome. Sci Transl Med 2014;8:e66986.

[6] Mazmanian SK, Round JK, Kasper DL. A microbial synthesis factor prevents intestinal inflammatory disease. Nature 2008;453:620–5.

[7] Houghteling P, Walker WA. Why is initial bacterial colonization of the intestine important to the infant's and child's health. J Pediatr Gastro Nutr 2015;60:294–307.

[8] Round JL, Mazmanian SK. The gut microbiota shapes intestinal immune responses during health and disease. Nat Rev Immunol 2009;9:313–23.

[9] Walker WA, Iyengar RS. Breast milk, microbiota and intestinal immune homeostasis. Pediatr Res 2015;77:220–8.

[10] Avershina E, Rudi K. Confusion about the species richness of human gut microbiota. Benefic Microbes 2015;6:657–9.

[11] Huttenhower C, Gevers D, Knight R, et al. Structure function and diversity of the healthy human microbiome. Nature 2012;486:207–14.

[12] Palmer C, Bik EM, DiGiulio DB, Relman DA, Brown PO. Development of the human infant intestinal microbiota. PLoS Biol 2007;5:e177.

[13] Adlerberth I, Wold AE. Establishment of the gut microbiota in western infants. Acta Paediat 2009;98:229–38.

[14] Maharshak N, Packey CD, Ellermann M, et al. Altered enteric microbiota ecology in interleukin 10-deficient mice during development and progression of intestinal inflammation. Gut Microbes 2013;4:316–24.

[15] Bäckhed F, Ley RE, Sonnenburg JL, Peterson DA, Gordon JI. Host-bacterial mutualism in the human intestine. Science 2005;307:1915–20.

[16] Vijay-Fumar M, Atiken JD, Carvalho FA, et al. Metabolic syndrome and altered gut microbiota in mice lacking toll-like receptor 5. Science 2010;9:228–31.

[17] Cahenzil J, Koller Y, Wyss M, Geuking MB, McCoy KD. Intestinal microbial diversity during early-life colonization shapes long-term IgE levels. Cell host Microbe 2013;14:559–70.

[18] Atarash K, Tanoue T, Oshima K, et al. Treg induction by a rationally selected mixture of *Clostridia* strains from the human microbiota. Nature 2013;500:232–6.

[19] Lauber CL, Hamady M, Knight R, Fierer N. Pyrosequencing-based assessment of soil pH as a predictor of soil bacterial community structure at the continental scale. Appl Environ Microbiol 2009;75:5111–20.

[20] Pascault N, Roux S, Artigas J, et al. A high-throughput sequencing ecotoxicology study of freshwater bacterial communities and their responses to tebuconazole. FEMS Microbiol Ecol 2014;90(3):563–74.

[21] Rakoff-Nahoum S, Medzhitov R. Role of the innate immune system and host-commensal mutualism. Curr Top Microbiol Immunol 2006;308:1–18.

[22] Shin N-R, Whon TW, Bae J-W. Proteobacteria: microbial signature of dysbiosis in gut microbiota. Trends in Biothechnol 2015;33:496–503.

[23] Ogura Y, Bonen DK, Inohara N, et al. A frameshift mutation in NOD2 associated with susceptibility to Crohn's disease. Nature 2001;411603–6.

[24] Karlsson FH, Tremaroli V, Nookaew I, et al. Gut metagenome in European women with normal, impaired and diabetic glucose control. Nature 2013;498:99–103.

[25] Cho I, Blaser MJ. The human microbiome: at the interface of health and disease. Nat Rev Genet 2012;13:260–70.

[26] Garrett W, Lord GM, Punit S, et al. Communicable ulcerative colitis induced by T-bet deficiency in the innate immune system. Cell 2007;131:33–45.

[27] Johnson CK, Versalovic J. The human microbiome and its potential importance to pediatrics. Pediatr 2012;129:950–60.

[28] Collado MC, Isolauri E, Laitinen K, Slaminen S. Effect of mother's weight on infant's microbiota acquisition, composition, and activity during early infancy: a prospective follow-up study initiated in early pregnancy. Am J Clin Nutr 2010;92:1023–30.

[29] Baker JL, Michaelsen KF, Rasmussen KM, Sorensen TI. Maternal prepregnant body mass index, duration of breast-feeding, and timing of complementary food introduction are associated with infant weight gain. Am J Clin Nutr 2004;80:1579–88.

[30] Mshvidadze M, Neu J, Shuster J, Theriaque D, Li N, Mai V. Intestinal microbial ecology in premature infants assessed with non-culture-based techniques. J Pediatr 2010;156:20–5.

[31] Jakobsson HE, Abrahamsson TR, Jenmalum MC, et al. Decreased gut microbiota diversity, delayed Bacteroidetes colonization and reduced Th1 responses in infants delivered by cesarean section. Gut 2014;63:559–66.

[32] Vassallo MF, Walker WA. Neonatal microbial flora and disease outcome. Nestle Nutr Workshop Ser Pediatr Program 2008;61:211–24.

[33] Jakobsson HE, Jernberg C, Andersson AF, Sjolund-Karlsson M, Jansson JK, Engstrand K. Short-term antibiotic treatment has differing long-term impact on the human throat and gut microbiome. PLoS One 2010;5:e9836.

[34] Blaser M. Antibiotic overuse: stop the killing of beneficial bacteria. Nature 2011;476:393–4.

[35] Donnet-Hughes A, Schriffin E, Walker WA. Protective properties of human milk and bacterial colonization of the neonatal gut. In: Duggan C, Koletzko B, Watkins J, Walker WA, editors. Nutrition in pediatrics—basic science clinical aspects. 5th ed. New Haven, CT: Chinese Publications Inc.; 2015. p. 165–250. Chapter 30.

[36] Carvalho FA, Koren O, Goodrich JK, et al. Transient inability to manage Proteobacteria promotes chronic gut inflammation in TRL5-deficient mice. Cell Host Microbe 2012;12:139–52.

[37] Shulzhenko N, Morgun A, Hsiao W, et al. Crosstalk between B lymphocytes, microbiota and the intestinal epithelium governs immunity versus metabolism in the gut. Nat Med 2011;17:1585–93.

[38] DeFilippo C, Cavalieri D, Di Paola M, et al. Impact of diet in shaping gut microbiota revealed by a comparative study in children from Europe and rural Africa. Proc Natl Acad Sci USA 2010;107:14691–6.

[39] Wu GD, Chen J, Hoffmann C, et al. Linking long-term dietary patterns with gut microbial enterotypes. Science 2011;334:105–8.

[40] Azzini E, Polita A, Fumagalli A, et al. Mediterranean diet effects: an Italian picture. Nutr J 2011;10:125.

[41] Chassaing B, Koren O, Goodrich JK, et al. Dietary emulsifiers impact the mouse gut microbiota promoting colitis and metabolic syndrome. Nature 2015;519:92–6.

[42] Suez J, Korem T, Zeevi D, et al. Artificial sweeteners induce glucose intolerance by altering the gut microbiota. Nature 2014;514:181–6.

[43] David LA, Maurice CF, Carmody RN, et al. Diet rapidly alters the human gut microbiome. Gut 2014;505:559–63.

[44] Moya A, Ferrer M. Functional redundancy-induced stability of gut microbiota subjected to disturbance. Trends Microbiol 2016;24:402–13.

[45] Faith JJ, Guruge JL, Charbonneau M, et al. The long-term stability of the human gut microbiota. Science 2013;341:1237439.

[46] Costello EK, Stagaman K, Dethlefsen L, Mohannan BJM, Relman DA. The application of ecological theory towards an understanding of the human microbiome. Science 2012;336:1255–62.

[47] Riedler J, Eder W, Oberfeld G, Schreuer M. Austrian children living on a farm have less hay fever, asthma and allergic sensitization. Clin Exp Allergy 2000;30:194–200.

[48] Wlasiuk G, Vercelli D. The farm effect, or when what and how a farming environment protects from asthma and allergic disease. Curr Opin Allergy Clin Immunol 2012;12:461–6.

[49] Gibson MK, Pesesky MW, Danta G. The yin and yang of bacterial resilience in the human gut microbiota. J Mol Biol 2014;426:3866–76.

[50] Luota R, Collado MC, Salmnien S, Isolauri E. Reshaping the gut microbiota at an early age: functional impact on obesity risk? Ann Nutr Metab 2013;13:17–26.

[51] Kadooka Y, Sato M, Imaisum K, et al. Regulation of abdominal adiposity by robotics (*Lactobacillus gasseri* SBT2055) in adults with obese tendencies in a randomized controlled trial. Eur J Clin Nutr 2010;64:636–43.

Chapter 26

Immunologic Response in the Host

K. Madsen* and H. Park**

*Division of Gastroenterology, The Center of Excellence for Gastrointestinal Inflammation and Immunity Research, University of Alberta, Edmonton, AB, Canada; **Faculty of Medicine and Dentistry, University of Alberta, Edmonton, AB, Canada

INTESTINAL EPITHELIAL CELLS

Intestinal epithelial cells (IECs) are continually exposed to microbes and foreign antigens, and act as immunological sensors to integrate microbial-derived signals into appropriate antimicrobial and immunoregulatory responses [1]. Innate sensing of microbes by IECs involves cross-talk between TLRs (Toll-like receptors), NLRs [nucleotide-binding oligomerization domain (NOD)-like receptor—and leucine-rich repeat (LRR)-containing proteins], and RLRs (RIG-I-like receptors) [2–5]. Stimulation of these receptors by microbial-associated molecular patterns (MAMPs) promotes intestinal homeostasis through induction of intracellular signaling pathways, and cytokine and chemokine release. Enhanced tissue repair is also facilitated through MAMP stimulation and increased expression of molecules, such as cytoprotective heat-shock proteins, trefoil factor 3 (TFF3), and epidermal growth receptor ligands [2,6,7].

TOLL-LIKE RECEPTORS AND INTESTINAL EPITHELIAL CELLS

IECs are polarized cells that have an apical surface facing the intestinal lumen and a basolateral surface facing the lamina propria. TLRs are type 1 transmembrane proteins that recognize conserved microbial or pathogen-associated molecular patterns (PAMPs) [8]. There have been 13 different TLRs identified to date, each receptor binding distinct ligands (Table 26.1) [9]. TLR2 senses bacterial lipopeptides and can heterodimerize either with TLR1 to recognize triacylated lipopeptides or with TLR6 to recognize diacylated lipopeptides. TLR4 recognizes lipopolysaccharide (LPS) found in the cell wall of Gram-negative bacteria and interacts with three different extracellular proteins: LPS-binding protein (LBP), CD14, and myeloid differentiation protein 2 (MD-2) to induce a signaling cascade that results in the activation of NF-κB and release of proinflammatory cytokines [10]. TLR9 recognizes unmethylated double-stranded

CpG motifs, which are found in microbial but not eukaryote DNA [11]. TLR5 binds to flagellin, which is the major protein component of Gran-negative flagella [12]. Intracellular TLRs include TLR3, which recognizes double-stranded RNA, TLR7, and TLR8, which recognize single-stranded RNA, TLR11 and 12 which recognize profilin, a molecular signature of protozoan parasites, and TLR13 which recognizes bacterial 23S rRNA [9,13,14]. TLRs can also be activated by endogenous molecules, such as heat-shock proteins, high-mobility group box 1 (HMGB1), surfactant protein A, extracellular matrix components, and uric acid [15]. Stimulation of TLRs results in the recruitment of adaptor proteins (MyD88, Mal/TIRAP, TRAM, and TRIF) and activation of NF-κB, or interferon regulatory factor signaling pathways [9,16,17]. This can result in the release of chemotactic factors for both underlying myeloid and lymphoid cells.

TLR signaling in IECs is highly regulated in order to maintain epithelial tolerance and hyporesponsiveness to commensal organisms, while allowing IECs to mount an effective immune response to pathogens. This is achieved through subcellular segregation and polarized distribution of TLRs in addition to the expression of negative regulators of microbial-induced proinflammatory signaling pathways [17,20,21]. TLR2 and TLR4 are expressed in very low concentrations on the apical surfaces [22,23], but can increase in expression under inflammatory or infectious conditions, and promote inflammation [24–27]. TLR2, TLR3, TLR4, and TLR5 are expressed on the basolateral cell membrane in colonic cells [4,12,28]. Thus, stimulation of these TLRs will occur in the presence of invasive bacteria, and not by commensal bacteria. This prevents a sustained TLR-mediated immune response by IECs which are in contact with luminal microbes on a continual basis. Stimulation of TLR5 on the basolateral surface triggers an inflammatory response through the production of cytokines and chemokines, such as IL-8, IL-17C, and CCL20 (CC-chemokine ligand) [29]. In IECs, TLR9 recognizes unmethylated CpG motifs in double-stranded DNA and is expressed on both

TABLE 26.1 Toll-Like Receptors and Their Ligands [9,18,19]

TLR	Cell Type	Ligands
TLR1	Monocytes/macrophages DCs B lymphocytes	Triacyl lipopeptides from bacteria
TLR2	Monocytes/macrophages Neutrophils Myeloid DCs Mast cells Epithelial cells Platelets	Peptidoglycan, lipoproteins, LTA from bacteria Zymosan from yeast cell wall Lipoarabinomannan from mycobacteria Hemagglutinin protein from measles virus HSP60, HSP70, Gp96 hyaluronic acid, biglycan
TLR3	DCs B lymphocytes Platelets	ssRNA virus, dsRNA virus, respiratory syncytial virus, murine cytomegalovirus
TLR4	Monocytes/macrophages Neutrophils Myeloid DCs Mast cells B lymphocytes Epithelial cells Platelets	LPS from bacteria RSV fusion protein from viruses Mannan from fungi Glycoinositolphospholipids from protozoa HSP22, HSP 60, HSP70, HSP72, Gp96, HMGB1, oxidized phospholipids heparin sulfate, fibronectin, tenascin-C, β-defensin 2, versican, hyaluronic acid, nickel
TLR5	Monocytes/macrophages DCs Epithelial cells	Flagellin from bacteria
TLR6	Monocytes/macrophages Mast cells B lymphocytes Platelets	Diacyl lipopeptides
TLR7	Monocytes/macrophages Plasmacytoid DCs B lymphocytes Platelets	ssRNA viruses, purine analog compounds (imidazoquinolines), RNA from bacteria from group B streptococcus
TLR8	Monocytes, macrophages DCs Mast cells	ssRNA viruses, purine analog compounds (imidazoquinolines)
TLR9	Monocytes/macrophages Plasmacytoid DCs B lymphocytes Epithelial cells	CpG motifs from bacteria, viruses and protozoa, dsDNA viruses herpes simplex virus, and murine cytomegalovirus
TLR10 (humans)	Monocytes/macrophages Plasmacytoid DCs B lymphocytes Eosinophils, neutrophils Epithelial cells	Listeria monocytogenes
TLR11 (mice)	Macrophages DCs Epithelial cells	Uropathogenic bacteria, profilin-like molecule from *Toxoplasma gondii*
TLR12 (mice)	Macrophages DCs	Profilin-like molecule from *T. gondii*
TLR13 (mice)	Macrophages DCs	23S rRNA from bacteria

DC, Dendritic cell; HMGB, high mobility group box 1; HSP, heat-shock protein; LPS, lipopolysaccharide; LTA, lipoteichoic acid; RSV, respiratory syncytial virus.

apical and basolateral surfaces where stimulation elicits both immune-suppressive and immune-stimulatory functions, depending on location and ligand [30,31]. Activation of TLR9 on the apical surface of IECs induces IκB ubiquitination and accumulation, preventing NFκB activation, whereas activation of basolateral TLR9 induces IκB degradation and NFκB activation [11,31]. Furthermore, apical activation of TLR9 inhibits inflammatory responses to basolateral stimulation, suggesting that cross-tolerance can occur. In addition, TLR9 can respond differentially to different bacterial species, with pathogenic strains evoking a phosphorylation of the ERK pathway and activation of AP-1, while nonpathogenic strains modulate cellular responses through the NF-kB pathway [31,32]. These divergent responses to different bacterial strains and to localization of stimulation reflect the ability of IECs to use TLR sensing to respond in an appropriate noninflammatory manner to commensal organisms but yet maintain the ability to induce an inflammatory response to a break in the functional barrier.

There are several molecules present in IECs that inhibit TLR signaling. These include toll-interacting protein (TOLLIP), which is an intracellular protein that inhibits both TLR2 and TLR4 signaling through effects on IL-1R-associated kinases (IRAKs); single immunoglobulin IL-1R-related molecule (SIGIRR) which is a negative regulator of TLR4 and TLR9 signaling; peroxisome proliferator activated receptor-γ (PPARγ), which is induced by TLR4 signaling and acts as a negative regulator; and A20, which is an ubiquitin modifying enzyme that inhibits NFκB activation by downregulating polyubiquitination-dependent inflammatory mediators [33–36].

TABLE 26.2 Nucleotide-Binding and Oligomerization Domain-Like Receptors [40–42,49]

Family	Receptor	Activators
NLRA	CIITA	IFNγ
NLRB	NAIPs	Flagellin, rod proteins
NLRC	NOD-1	iE-DAP
	NOD-2	MDP
	NLRC3/C5/XI	Unknown
	NLRC4	Flagellin, bacterial type 3 secretion system
NLRP	NLRP1	MDP, bacterial toxins
	NLRP3	LPS, MDP, LTA, bacterial and viral RNA, oxMito-DNA, ceramide, cardiolipin, K$^+$ efflux, mitochondrial lysosomal disruption, ROS, crystals/aggregates, Ca$^+$ signaling, external ATP, silica crystals, and aluminum salts
	NLRP6	Bacterial products
	NLRP7	Bacterial acylated lipopeptides
	NLRP10	Unknown
	NLRP11	Unknown
	NLRP12	Acylated lipid A
NLRX	NLRX1	Poly I:C

ATP, Adenosine triphosphate; CIITA, class II major histocompatibility complex, transactivator; iE-DAP, D-glutamyl-meso-diaminopimelic acid; MDP, muramyl dipeptide; NAIPs, neuronal apoptosis inhibitory protein; NLR, nucleotide-binding oligomerization domain-like receptor; NOD, nucleotide-binding oligomerization domain; ROS, reactive oxygen species.

NOD RECEPTORS AND INTESTINAL EPITHELIAL CELLS

Another key group of intracellular receptors is the nucleotide-binding oligomerization domain-like receptors (NLRs) that are located within the cytoplasm of IECs (Table 26.2). NOD1 and NOD2 sense microbial molecular patterns that are components of bacterial peptidoglycan [37–41]. NOD1-dependent signaling requires muramyl tripeptides that are unique to Gram-negative bacteria while NOD2 responds to muramyl dipeptides found in both Gram-positive and Gram-negative organisms [38,39,42]. Inflammasomes also have a role in IEC immunomodulation and tissue homeostasis through the detection of pathogens and tissue damage [43]. Inflammasomes are multiprotein cytoplasmic complexes that include a sensor protein (a NOD-like receptor, such as NLRP1, NLRP2, NLRP3, NLRP6, NLRP7, NLRC4, and NLRP12), an adaptor protein (ASC: apoptosis-associated speck-like protein containing a CARD domain), and caspase-1 [44,45]. Activation of caspase-1 results in the increased processing and secretion of IL-1β and IL-18,

and the induction of inflammatory cell death (pyroptosis) [43]. Inflammasome sensor proteins are activated by direct interaction with specific PAMPs which are specific for each sensor; for example, NLRP1 is activated by muramyl dipeptide and NLRC4 is activated by flagellin, while NLRP3 is activated by numerous compounds including bacterial and viral nucleic acids, LPS, and damage-associated molecular patterns, such as ATP, uric acid, and amyloid β peptides [45]. Secretion of IL-1β and IL-18 occurs following stimulation with two separate signals—the first signal requires a PAMP recognized by pattern recognition receptors (PRRs) and enhanced expression of pro-IL-1β and pro-IL-18 [46]. The second signal arising from tissue damage is recognized by PRRs or purinergic receptors, and stimulates inflammasome complex formation and activation of caspase-1 which cleaves pro-IL-1β and pro-IL-18. Reactive oxygen species, lysosomal damage, potassium efflux, and oxidized mitochondrial DNA have all been shown to activate the NLRP3 inflammasome [47,48]. This dual signaling allows for the sensing of both microbial pathogenicity and factors released by the host following bacterial invasion.

IMMUNOREGULATORY ROLE OF INTESTINAL EPITHELIAL CELLS

IECs regulate and coordinate innate and adaptive immune cell responses to microbial stimuli. IECs modulate mucosal immune cells through the release of cytokines, such as thymic stromal lymphopoietin (TSLP), transforming growth factor-β (TGFβ), and IL-25, as well as the B-cell stimulating/activating factors APRIL (a proliferation-inducing ligand) and BAFF (B-cell activating factor) [50–54]. The release of TSLP, TGFβ, and retinoic acid from IECs results in the tolerization of underlying dendritic cells (DCs) and macrophages to an antiinflammatory phenotype with enhanced secretion of IL-10 and retinoic acid [50,51]. APRIL is released from IECs in a MyD88-dependent manner and drives T-cell independent IGA$_2$ class switching while BAFF (B-cell activating factor of the TNF family) regulates B-cell maturation, survival, and function [53,54]. Depending upon the stimulus, IECs also release chemokines that act as chemoattractants toward neutrophils (CXCL8, CXCL1, CXCL3, and CXCL5), macrophages and DCs (CCL2), DCs and memory T cells (CCL20), DCs and Th2 cells (CCL22), Th1 cells (CXCL9, CXCL10, CXCL11), plasma cells (CCL28), $\alpha 4\beta 7$ T cells (CCL25), and cytokines (TNFα, GM-CSF) [55].

TRANSCYTOSIS OF IMMUNOGLOBULIN A BY INTESTINAL EPITHELIAL CELLS

B cells are found within the lamina propria in the small and large intestine [56]. B cells secrete dimeric IgA that binds to the polymeric immunoglobulin receptor (pIgR) on the basolateral surface of IECs prior to being transcytosed across the epithelial cell into the lumen. The expression of pIgR on IECs is regulated by TLR signaling and cytokines, such as IL-17 [57,58]. Maturation of naïve B cells into mature IgA-secreting plasma cells requires priming by mucosal DCs carrying antigen and live bacteria [59,60]. DCs are conditioned by IEC-derived signals including nitric oxide, IL-10, retinoic acid, and TGFβ to promote IgA class switching [60,61]. Although antigen-specific IgA increases during intestinal infections, a large amount of basal IgA secretion occurs to help prevent entry of commensal microbes across the epithelium [62,63].

ANTIGEN PRESENTATION IN THE GUT

Antigen can traverse the gut epithelium through different pathways including across M cells in Peyer's patches, through goblet cells (GAPs: goblet cell-associated antigen passages), across IECs, and via paracellular leaks between epithelial cells [64]. M cells are differentiated epithelial cells that are found in the follicle-associated epithelium overlying Peyer's patches and isolated lymphoid follicles

[65]. M cells continually sample luminal contents and transcytose small molecules, as well as intact bacteria for delivery to underlying DCs in the subepithelial dome of the lymphoid tissue where the primary response is a secretion of IgA [66]. Antigen that is delivered through M cells can also be found in mesenteric lymph nodes [60]. Another mechanism of antigen delivery in the intestine is through goblet cells (GAPs) [67]. Studies have shown that small soluble antigens can pass through goblet cells for delivery to DCs present in the lamina propria [67]. GAPs were shown to preferentially delivery antigen to CD103+ DCs, which can migrate to mesenteric lymph nodes to initiate immune responses [68]. Further, CD103+ DCs generate retinoic acid which is required for expression of the gut homing receptors CCR9 and $\alpha 4\beta 7$ on lymphocytes as well as the generation of Foxp3+ Treg cells and IgA production [69,70].

IECs express several antigen-presenting molecules, including MHC class I and MHC class II, and costimulatory molecules, and play a role in the activation and expansion of both CD4+ and CD8+ regulatory T-cell subsets [71,72]. IECs can also present lipid-derived antigen to $\gamma\delta$ T cells and NKT cells through the expression of CD1d [73]. IECs also express transcytotic receptors, such as the neonatal Fc receptor (FcRn) that contributes to a controlled passage of antigen [74]. FcRn transports IgG across the intestinal epithelium into the lumen where it can bind antigen. FcRn then recycles the IgG/antigen complex back across the epithelial cells into the lamina propria where the complex can be processed by DCs and presented to CD4+ T cells [74].

ROLE OF DENDRITIC CELLS

DCs specific to the gut integrate signals derived from IECs and from direct sampling of luminal contents to elicit both regulatory and inflammatory responses. DCs are an important link between innate and adaptive immunity, and are critical to the maintenance of gut homeostasis [75,76]. DCs are found throughout the lamina propria, as well as within gut-associated lymphoid tissues and mesenteric lymph nodes. DCs act as antigen presenting cells and also express PRRs, which modulate their maturation toward different subsets that selectively promote polarized lymphocyte responses. Once activated, DCs will mature into effector or regulatory cells, and subsequently activate naïve T cells toward Th1, Th2, Th9, Th17, or Treg functional phenotypes [77,78].

DCs in the lamina propria are defined by the expression of the surface markers CD103, CD11c, and CD11b [79]. Several subsets of DCs have been identified within the intestine that are characterized by differential surface marker expression and function [79–81]. Subepithelial CD103+ DCs are able to sample luminal contents by passing dendrites between IECs [32,82,83]. CD103+ DCs also traffic to mesenteric lymph nodes and Peyer's patches bringing antigens and live bacteria for presentation to adaptive immune cells where they induce immune tolerance through the differentiation of FOXP3+ (forkhead box P3) regulatory T cells by a TGFβ and retinoic acid-dependent mechanism [60,68,84–86]. This interaction also causes differentiation to FoxP3+ Treg cells that express CCR9, a gut specific homing molecule. In contrast, CD103-DCs promote an increased expression of inflammatory mediators, such as TNFα and IL-6 following stimulation with TLR ligands [87]. Another DC subset characterized by CX$_3$CR1$^+$ do not traffic to lymph nodes but remain within the lamina propria where they function as phagocytes to clear any translocated microbes [68,88].

INNATE LYMPHOID CELLS

Innate lymphoid cells (ILCs) have recently been identified to have a key role in the maintenance of gut homeostasis [89]. ILCs are found within the lamina propria and can be characterized by their differential cytokine expression into three groups. Group 1 ILCs (ILC1) include natural killer cells and are characterized by the production of IFNγ and TNFα in response to IL-12 and IL-15 [90]. Group 2 ILCs (ILC2) produce Th2-type cytokines, such as IL-5 and IL-13, and induce goblet cell hyperplasia and mucus secretion [91,92]. Activation of ILC2s occurs in response to the release of IL-25, IL-33, and TSLP from IECs [93]. Group 3 ILCs (ILC3) respond to stimulation with IL-23 by producing IL-17A and IL-22, which are Th17 and Th22 cell-type cytokines [90]. IL-17 has a primarily proinflammatory effect while IL-22 has tissue protective effects in enhancing repair following injury or infection, but also has been shown to be involved in the initiation of inflammation-induced cancer [94,95]. Microbes, such as segmented filamentous bacteria (SFB) induce ILC3s to secrete IL-22. This increase in IL-22 production induces IEC production of serum amyloid alpha, which causes a Th17 cell differentiation and IgA production [96,97].

T CELLS

Effector and regulatory T cells that exert inflammatory and tolerogenic effects are found throughout the lamina propria while αβ+ and γδ+ T cells (IELs) are located within the epithelium [56,98,99]. IELs adhere to IECs by interactions between CD103 expressed on IELs and E-cadherin expressed on IECs [100]. Following activation, IELs express several effector cytokines and can have both protective and potentially pathogenic effects through the release of both proinflammatory cytokines, such as IFNγ and epithelial-protective signals, such as keratinocyte growth factor [101]. Within the lamina propria in the small intestine are found primarily αβ T-cell receptor-positive T

cells [56]. Activated T-cell subsets include CD4+ CD25+ regulatory T cells which can inhibit T-cell proliferation and cytokine secretion along with CD4+ T cells that can secrete both IL-17 and IL-22 [102–105]. T cells are able to respond to signals received from innate immune cells and IECs, and initiate both pro- and antiinflammatory responses.

MICROBIAL MODULATION OF IMMUNE FUNCTION

Underlying immune cells can be modulated by signals released from IECs as well as by direct contact with microbial structural components and soluble metabolites. These microbial compounds interact with TLRs and c-type lectin receptors expressed intracellular or on the cell surface of immune cells, and modulate cellular phenotypes and functional profiles. Several polysaccharides from bacterial cell walls, such as capsular polysaccharide A (PSA) from *Bacteroides fragilis,* can modulate DC function [106]. PSA from *B. fragilis* interacts with DCs through TLR2 and also with DC-specific intercellular adhesion molecule-3-grabbing nonintegrin (DC-SIGN) receptor expressed on monocyte-derived DCs. These DCs subsequently induce the conversion of CD4+ T cells into Foxp3+ Treg cells that secrete IL-10 and suppress IL-17 production [107]. *B. fragilis* has also been shown to produce α-galactosylceramide, a glycosphingolipid which binds CD1d and activates NKT cells [108]. MHC-II-dependent antigen presentation of segmented filamentous bacterial antigens by CD11c+ DCs results in the local induction of Th17 cells [109]. Several studies have demonstrated that certain species of Clostridia from clusters IV and XIV are able to induce the development of Foxp3+ Treg cells in the large intestine [110,111]. Short-chain fatty acids, such as butyrate, acetate, and propionate are derived from bacterial fermentation [112]. Butyrate in particular has several antiinflammatory and immunoregulatory effects. Butyrate modulates histone deacetylase (HDAC) activity, which acts to limit DC production of proinflammatory cytokines, such as IL-12 and IL-6, and promotes Treg cells. Butyrate has also been shown to induce Treg cells and IL-10 producing T cells through a G-protein coupled receptor (GPR109a) mediated mechanism [113,114]. Butyrate can act directly on immune cells to inhibit IL-12 production and block NF-κβ translocation. Histamine secreted by host cells and by luminal bacteria can interact with histamine 2 receptor (H$_2$R) on DCs to decrease proinflammatory cytokine secretion induced by TLR stimulation and increase IL-10 secretion [115]. Tryptophan metabolites have an important role in maintaining gut homeostasis through aryl hydrocarbon receptor signaling [116]. Stimulation of the aryl hydrocarbon receptor induces Th17 cell differentiation and enhances function of Group 3 ILCs [117].

CONCLUSIONS

Interactions between the gut microbiome and the host act to maintain a balance between tolerance and immune reactivity. Both innate and adaptive immune responses can be modulated by luminal microbes in a strain- and dose-dependent fashion through interactions with mucosal DCs and IECs.

REFERENCES

[1] Peterson LW, Artis D. Intestinal epithelial cells: regulators of barrier function and immune homeostasis. Nat Rev Immunol 2014;14(3):141–53.

[2] Rakoff-Nahoum S, Paglino J, Eslami-Varzaneh F, Edberg S, Medzhitov R. Recognition of commensal microflora by toll-like receptors is required for intestinal homeostasis. Cell 2004;118(2):229–41.

[3] Broquet AH, Hirata Y, McAllister CS, Kagnoff MF. RIG-I/MDA5/MAVS are required to signal a protective IFN response in rotavirus-infected intestinal epithelium. J Immunol 2011;186(3):1618–26.

[4] Abreu MT. Toll-like receptor signalling in the intestinal epithelium: how bacterial recognition shapes intestinal function. Nat Rev Immunol 2010;10(2):131–44.

[5] Elinav E, Henao-Mejia J, Flavell RA. Integrative inflammasome activity in the regulation of intestinal mucosal immune responses. Mucosal Immunol 2013;6(1):4–13.

[6] Podolsky DK, Gerken G, Eyking A, Cario E. Colitis-associated variant of TLR2 causes impaired mucosal repair because of TFF3 deficiency. Gastroenterology 2009;137(1):209–20.

[7] Cario E, Gerken G, Podolsky DK. Toll-like receptor 2 enhances ZO-1-associated intestinal epithelial barrier integrity via protein kinase C. Gastroenterology 2004;127(1):224–38.

[8] Akira S, Hemmi H. Recognition of pathogen-associated molecular patterns by TLR family. Immunol Lett 2003;85(2):85–95.

[9] Kawai T, Akira S. Toll-like receptors and their crosstalk with other innate receptors in infection and immunity. Immunity 2011;34(5):637–50.

[10] Kang JY, Nan X, Jin MS, et al. Recognition of lipopeptide patterns by Toll-like receptor 2-Toll-like receptor 6 heterodimer. Immunity 2009;31(6):873–84.

[11] Lee J, Mo JH, Katakura K, et al. Maintenance of colonic homeostasis by distinctive apical TLR9 signalling in intestinal epithelial cells. Nat Cell Biol 2006;8(12):1327–36.

[12] Gewirtz AT, Navas TA, Lyons S, Godowski PJ, Madara JL. Cutting edge: bacterial flagellin activates basolaterally expressed TLR5 to induce epithelial proinflammatory gene expression. J Immunol 2001;167(4):1882–5.

[13] Koblansky AA, Jankovic D, Oh H, et al. Recognition of profilin by Toll-like receptor 12 is critical for host resistance to *Toxoplasma gondii*. Immunity 2013;38(1):119–30.

[14] Oldenburg M, Kruger A, Ferstl R, et al. TLR13 recognizes bacterial 23S rRNA devoid of erythromycin resistance-forming modification. Science 2012;337(6098):1111–5.

[15] Rifkin IR, Leadbetter EA, Busconi L, Viglianti G, Marshak-Rothstein A. Toll-like receptors, endogenous ligands, and systemic autoimmune disease. Immunol Rev 2005;204:27–42.

[16] Bryant CE, Symmons M, Gay NJ. Toll-like receptor signalling through macromolecular protein complexes. Mol Immunol 2015;63(2):162–5.

[17] Gay NJ, Symmons MF, Gangloff M, Bryant CE. Assembly and localization of Toll-like receptor signalling complexes. Nat Rev Immunol 2014;14(8):546–58.

[18] Asea A. Heat shock proteins and toll-like receptors. Handb Exp Pharmacol 2008;(183):111–27.

[19] Yu L, Wang L, Chen S. Endogenous toll-like receptor ligands and their biological significance. J Cell Mol Med 2010;14(11):2592–603.

[20] Otte JM, Cario E, Podolsky DK. Mechanisms of cross hyporesponsiveness to Toll-like receptor bacterial ligands in intestinal epithelial cells. Gastroenterology 2004;126(4):1054–70.

[21] Cario E, Podolsky DK. Toll-like receptor signaling and its relevance to intestinal inflammation. Ann N Y Acad Sci 2006;1072:332–8.

[22] Melmed G, Thomas LS, Lee N, et al. Human intestinal epithelial cells are broadly unresponsive to Toll-like receptor 2-dependent bacterial ligands: implications for host-microbial interactions in the gut. J Immunol 2003;170(3):1406–15.

[23] Abreu MT, Vora P, Faure E, Thomas LS, Arnold ET, Arditi M. Decreased expression of Toll-like receptor-4 and MD-2 correlates with intestinal epithelial cell protection against dysregulated proinflammatory gene expression in response to bacterial lipopolysaccharide. J Immunol 2001;167(3):1609–16.

[24] Dheer R, Santaolalla R, Davies JM, et al. Intestinal epithelial Toll-like receptor 4 signaling affects epithelial function and colonic microbiota and promotes a risk for transmissible colitis. Infect Immun 2016;84(3):798–810.

[25] Cario E, Podolsky DK. Differential alteration in intestinal epithelial cell expression of toll-like receptor 3 (TLR3) and TLR4 in inflammatory bowel disease. Infect Immun 2000;68(12):7010–7.

[26] Szebeni B, Veres G, Dezsofi A, et al. Increased expression of Toll-like receptor (TLR) 2 and TLR4 in the colonic mucosa of children with inflammatory bowel disease. Clin Exp Immunol 2008;151(1):34–41.

[27] Hausmann M, Kiessling S, Mestermann S, et al. Toll-like receptors 2 and 4 are up-regulated during intestinal inflammation. Gastroenterology 2002;122(7):1987–2000.

[28] Rhee SH, Im E, Riegler M, Kokkotou E, O'Brien M, Pothoulakis C. Pathophysiological role of Toll-like receptor 5 engagement by bacterial flagellin in colonic inflammation. Proc Natl Acad Sci USA 2005;102(38):13610–5.

[29] Im E, Jung J, Rhee SH. Toll-like receptor 5 engagement induces interleukin-17C expression in intestinal epithelial cells. J Interferon Cytokine Res 2012;32(12):583–91.

[30] Lee J, Mo JH, Katakura K, et al. Maintenance of colonic homeostasis by distinctive apical TLR9 signalling in intestinal epithelial cells. Nat Cell Biol 2006;8(12):1327–36.

[31] Jijon H, Backer J, Diaz H, et al. DNA from probiotic bacteria modulates murine and human epithelial and immune function. Gastroenterology 2004;126(5):1358–73.

[32] Akhtar M, Watson JL, Nazli A, McKay DM. Bacterial DNA evokes epithelial IL-8 production by a MAPK-dependent, NF-kappaB-independent pathway. FASEB J 2003;17(10):1319–21.

[33] Garlanda C, Anders HJ, Mantovani A. TIR8/SIGIRR: an IL-1R/TLR family member with regulatory functions in inflammation and T cell polarization. Trends ImmunolV 30 2009;(9):439–46.

[34] Maillard MH, Bega H, Uhlig HH, et al. Toll-interacting protein modulates colitis susceptibility in mice. Inflamm Bowel Dis 2014;20(4):660–70.

[35] Shibolet O, Podolsky DK. TLRs in the gut. IV. Negative regulation of Toll-like receptors and intestinal homeostasis: addition by subtraction. Am J Physiol Gastrointest Liver Physiol 2007;292(6):G1469–73.

[36] Vereecke L, Vieira-Silva S, Billiet T, et al. A20 controls intestinal homeostasis through cell-specific activities. Nat Commun V 5 2014;5103.

[37] Girardin SE, Travassos LH, Herve M, et al. Peptidoglycan molecular requirements allowing detection by Nod1 and Nod2. J Biol Chem 2003;278(43):41702–8.

[38] Girardin SE, Boneca IG, Carneiro LA, et al. Nod1 detects a unique muropeptide from gram-negative bacterial peptidoglycan. Science 2003;300(5625):1584–7.

[39] Inohara N, Ogura Y, Fontalba A, et al. Host recognition of bacterial muramyl dipeptide mediated through NOD2. Implications for Crohn's disease. J Biol Chem 2003;278(8):5509–12.

[40] Chavarria-Smith J, Vance RE. The NLRP1 inflammasomes. Immunol Rev 2015;265(1):22–34.

[41] Vance RE. The NAIP/NLRC4 inflammasomes. Curr Opin Immunol 2015;32:84–9.

[42] Motta V, Soares F, Sun T, Philpott DJ. NOD-like receptors: versatile cytosolic sentinels. Physiol Rev 2015;95(1):149–78.

[43] Zaki MH, Lamkanfi M, Kanneganti TD. The Nlrp3 inflammasome: contributions to intestinal homeostasis. Trends Immunol 2011;32(4):171–9.

[44] Schroder K, Tschopp J. The inflammasomes. Cell 2010;140(6):821–32.

[45] Martinon F, Mayor A, Tschopp J. The inflammasomes: guardians of the body. Annu Rev Immunol 2009;27:229–65.

[46] Latz E, Xiao TS, Stutz A. Activation and regulation of the inflammasomes. Nat Rev Immunol 2013;13(6):397–411.

[47] Shimada K, Crother TR, Karlin J, et al. Oxidized mitochondrial DNA activates the NLRP3 inflammasome during apoptosis. Immunity 2012;36(3):401–14.

[48] Strowig T, Henao-Mejia J, Elinav E, Flavell R. Inflammasomes in health and disease. Nature 2012;481(7381):278–86.

[49] Sellin ME, Maslowski KM, Maloy KJ, Hardt WD. Inflammasomes of the intestinal epithelium. Trends Immunol 2015;36(8):442–50.

[50] Rimoldi M, Chieppa M, Salucci V, et al. Intestinal immune homeostasis is regulated by the crosstalk between epithelial cells and dendritic cells. Nat Immunol 2005;6(5):507–14.

[51] Zeuthen LH, Fink LN, Frokiaer H. Epithelial cells prime the immune response to an array of gut-derived commensals towards a tolerogenic phenotype through distinct actions of thymic stromal lymphopoietin and transforming growth factor-beta. Immunology 2008;123(2):197–208.

[52] Zaph C, Du Y, Saenz SA, et al. Commensal-dependent expression of IL-25 regulates the IL-23-IL-17 axis in the intestine. J Exp Med 2008;205(10):2191–8.

[53] He B, Xu W, Santini PA, et al. Intestinal bacteria trigger T cell-independent immunoglobulin A(2) class switching by inducing epithelial-cell secretion of the cytokine APRIL. Immunity 2007;26(6):812–26.

[54] Xu W, He B, Chiu A, et al. Epithelial cells trigger frontline immunoglobulin class switching through a pathway regulated by the inhibitor SLPI. Nat Immunol 2007;8(3):294–303.

[55] Kagnoff MF. Microbial-epithelial cell crosstalk during inflammation: the host response. Ann N Y Acad Sci 2006;1072:313–20.

[56] Resendiz-Albor AA, Esquivel R, Lopez-Revilla R, Verdin L, Moreno-Fierros L. Striking phenotypic and functional differences in lamina propria lymphocytes from the large and small intestine of mice. Life Sci 2005;76(24):2783–803.

[57] Bruno ME, Frantz AL, Rogier EW, Johansen FE, Kaetzel CS. Regulation of the polymeric immunoglobulin receptor by the classical and alternative NF-kappaB pathways in intestinal epithelial cells. Mucosal Immunol 2011;4(4):468–78.

[58] Cao AT, Yao S, Gong B, Elson CO, Cong Y. Th17 cells upregulate polymeric Ig receptor and intestinal IgA and contribute to intestinal homeostasis. J Immunol 2012;189(9):4666–73.

[59] Cerutti A. The regulation of IgA class switching. Nat Rev Immunol 2008;8(6):421–34.

[60] Macpherson AJ, Uhr T. Induction of protective IgA by intestinal dendritic cells carrying commensal bacteria. Science 2004;303(5664):1662–5.

[61] Mora JR, von Andrian UH. Differentiation and homing of IgA-secreting cells. Mucosal Immunol 2008;1(2):96–109.

[62] Frankel G, Phillips AD, Novakova M, et al. Intimin from enteropathogenic Escherichia coli restores murine virulence to a Citrobacter rodentium eaeA mutant: induction of an immunoglobulin A response to intimin and EspB. Infect Immun 1996;64(12):5315–25.

[63] Suzuki K, Meek B, Doi Y, et al. Aberrant expansion of segmented filamentous bacteria in IgA-deficient gut. Proc Natl Acad Sci USA 2004;101(7):1981–6.

[64] Knoop KA, Miller MJ, Newberry RD. Transepithelial antigen delivery in the small intestine: different paths, different outcomes. Curr Opin Gastroenterol 2013;29(2):112–8.

[65] Jang MH, Kweon MN, Iwatani K, et al. Intestinal villous M cells: an antigen entry site in the mucosal epithelium. Proc Natl Acad Sci USA 2004;101(16):6110–5.

[66] Neutra MR, Frey A, Kraehenbuhl JP. Epithelial M cells: gateways for mucosal infection and immunization. Cell 1996;86(3):345–8.

[67] McDole JR, Wheeler LW, McDonald KG, et al. Goblet cells deliver luminal antigen to CD103+ dendritic cells in the small intestine. Nature 2012;483(7389):345–9.

[68] Schulz O, Jaensson E, Persson EK, et al. Intestinal CD103+, but not CX3CR1+, antigen sampling cells migrate in lymph and serve classical dendritic cell functions. J Exp Med 2009;206(13):3101–14.

[69] Nakamura Y, Watanabe M, Matsuzuka F, Maruoka H, Miyauchi A, Iwatani Y. Intrathyroidal CD4+ T lymphocytes express high levels of Fas and CD4+ CD8+ macrophages/dendritic cells express Fas ligand in autoimmune thyroid disease. Thyroid 2004;14(10):819–24.

[70] Jaensson E, Uronen-Hansson H, Pabst O, et al. Small intestinal CD103+ dendritic cells display unique functional properties that are conserved between mice and humans. J Exp Med 2008;205(9):2139–49.

[71] Lin XP, Almqvist N, Telemo E. Human small intestinal epithelial cells constitutively express the key elements for antigen processing and the production of exosomes. Blood Cells Mol Dis 2005;35(2):122–8.

[72] Bland PW, Warren LG. Antigen presentation by epithelial cells of the rat small intestine. II. Selective induction of suppressor T cells. Immunology 1986;58(1):9–14.

[73] Luoma AM, Castro CD, Adams EJ. gammadelta T cell surveillance via CD1 molecules. Trends Immunol 2014;35(12):613–21.

[74] Yoshida M, Claypool SM, Wagner JS, et al. Human neonatal Fc receptor mediates transport of IgG into luminal secretions for delivery of antigens to mucosal dendritic cells. Immunity 2004;20(6):769–83.

[75] Bar-On L, Zigmond E, Jung S. Management of gut inflammation through the manipulation of intestinal dendritic cells and macrophages? Semin Immunol 2011;23(1):58–64.

[76] MacDonald TT, Monteleone I, Fantini MC, Monteleone G. Regulation of homeostasis and inflammation in the intestine. Gastroenterology 2011;140(6):1768–75.

[77] Chang SY, Ko HJ, Kweon MN. Mucosal dendritic cells shape mucosal immunity. Exp Mol Med 2014;46:e84.

[78] Kaplan MH. Th9 cells: differentiation and disease. Immunol Rev 2013;252(1):104–15.

[79] Merad M, Sathe P, Helft J, Miller J, Mortha A. The dendritic cell lineage: ontogeny and function of dendritic cells and their subsets in the steady state and the inflamed setting. Annu Rev Immunol 2013;31:563–604.

[80] Geissmann F, Manz MG, Jung S, Sieweke MH, Merad M, Ley K. Development of monocytes, macrophages, and dendritic cells. Science 2010;327(5966):656–61.

[81] Satpathy AT, Wu X, Albring JC, Murphy KM. Re(de)fining the dendritic cell lineage. Nat Immunol 2012;13(12):1145–54.

[82] Rescigno M, Urbano M, Valzasina B, et al. Dendritic cells express tight junction proteins and penetrate gut epithelial monolayers to sample bacteria. Nat Immunol 2001;2(4):361–7.

[83] Chieppa M, Rescigno M, Huang AY, Germain RN. Dynamic imaging of dendritic cell extension into the small bowel lumen in response to epithelial cell TLR engagement. J Exp Med 2006;203(13):2841–52.

[84] MacPherson G, Milling S, Yrlid U, Cousins L, Turnbull E, Huang FP. Uptake of antigens from the intestine by dendritic cells. Ann N Y Acad Sci 2004;1029:75–82.

[85] Coombes JL, Siddiqui KR, Arancibia-Carcamo CV, et al. A functionally specialized population of mucosal CD103+ DCs induces Foxp3+ regulatory T cells via a TGF-beta and retinoic acid-dependent mechanism. J Exp Med 2007;204(8):1757–64.

[86] Sun CM, Hall JA, Blank RB, et al. Small intestine lamina propria dendritic cells promote de novo generation of Foxp3 T reg cells via retinoic acid. J Exp Med 2007;204(8):1775–85.

[87] del Rio ML, Rodriguez-Barbosa JI, Bolter J, et al. CX3CR1+ c-kit+ bone marrow cells give rise to CD103+ and CD103- dendritic cells with distinct functional properties. J Immunol 2008;181(9):6178–88.

[88] Niess JH, Brand S, Gu X, et al. CX3CR1-mediated dendritic cell access to the intestinal lumen and bacterial clearance. Science 2005;307(5707):254–8.

[89] Spits H, Cupedo T. Innate lymphoid cells: emerging insights in development, lineage relationships, and function. Annu Rev Immunol 2012;30:647–75.

[90] Spits H, Artis D, Colonna M, et al. Innate lymphoid cells—a proposal for uniform nomenclature. Nat Rev Immunol 2013;13(2):145–9.

[91] Moro K, Yamada T, Tanabe M, et al. Innate production of T(H)2 cytokines by adipose tissue-associated c-Kit(+)Sca-1(+) lymphoid cells. Nature 2010;463(7280):540–4.

[92] Neill DR, Wong SH, Bellosi A, et al. Nuocytes represent a new innate effector leukocyte that mediates type-2 immunity. Nature 2010;464(7293):1367–70.

[93] Mjosberg J, Bernink J, Golebski K, et al. The transcription factor GATA3 is essential for the function of human type 2 innate lymphoid cells. Immunity 2012;37(4):649–59.

[94] Kirchberger S, Royston DJ, Boulard O, et al. Innate lymphoid cells sustain colon cancer through production of interleukin-22 in a mouse model. J Exp Med 2013;210(5):917–31.

[95] Huber S, Gagliani N, Zenewicz LA, et al. IL-22BP is regulated by the inflammasome and modulates tumorigenesis in the intestine. Nature 2012;491(7423):259–63.

[96] Sano T, Huang W, Hall JA, et al. An IL-23R/IL-22 circuit regulates epithelial serum amyloid A to promote local effector Th17 responses. Cell 2015;163(2):381–93.

[97] Atarashi K, Tanoue T, Ando M, et al. Th17 cell induction by adhesion of microbes to intestinal epithelial cells. Cell 2015;163(2):367–80.

[98] Cheroutre H, Lambolez F, Mucida D. The light and dark sides of intestinal intraepithelial lymphocytes. Nat Rev Immunol 2011;11(7):445–56.

[99] Mucida D, Husain MM, Muroi S, et al. Transcriptional reprogramming of mature CD4(+) helper T cells generates distinct MHC class II-restricted cytotoxic T lymphocytes. Nat Immunol 2013;14(3):281–9.

[100] Qiu Y, Yang Y, Yang H. The unique surface molecules on intestinal intraepithelial lymphocytes: from tethering to recognizing. Dig Dis Sci 2014;59(3):520–9.

[101] Cepek KL, Shaw SK, Parker CM, et al. Adhesion between epithelial cells and T lymphocytes mediated by E-cadherin and the alpha E beta 7 integrin. Nature 1994;372(6502):190–3.

[102] Kleinschek MA, Boniface K, Sadekova S, et al. Circulating and gut-resident human Th17 cells express CD161 and promote intestinal inflammation. J Exp Med 2009;206(3):525–34.

[103] Makita S, Kanai T, Nemoto Y, et al. Intestinal lamina propria retaining CD4+CD25+ regulatory T cells is a suppressive site of intestinal inflammation. J Immunol 2007;178(8):4937–46.

[104] Makita S, Kanai T, Oshima S, et al. CD4+CD25bright T cells in human intestinal lamina propria as regulatory cells. J Immunol 2004;173(5):3119–30.

[105] Munoz M, Heimesaat MM, Danker K, et al. Interleukin (IL)-23 mediates Toxoplasma gondii-induced immunopathology in the gut via matrixmetalloproteinase-2 and IL-22 but independent of IL-17. J Exp Med 2009;206(13):3047–59.

[106] Dasgupta S, Erturk-Hasdemir D, Ochoa-Reparaz J, Reinecker HC, Kasper DL. Plasmacytoid dendritic cells mediate anti-inflammatory responses to a gut commensal molecule via both innate and adaptive mechanisms. Cell Host Microbe 2014;15(4):413–23.

[107] Bloem K, Garcia-Vallejo JJ, Vuist IM, Cobb BA, van Vliet SJ, van Kooyk Y. Interaction of the capsular polysaccharide A from Bacteroides fragilis with DC-SIGN on human dendritic cells is necessary for its processing and presentation to T cells. Front Immunol 2013;4:103.

[108] Wieland Brown LC, Penaranda C, Kashyap PC, et al. Production of alpha-galactosylceramide by a prominent member of the human gut microbiota. PLoS Biol 2013;11(7):e1001610.

[109] Goto Y, Panea C, Nakato G, et al. Segmented filamentous bacteria antigens presented by intestinal dendritic cells drive mucosal Th17 cell differentiation. Immunity 2014;40(4):594–607.

[110] Atarashi K, Tanoue T, Shima T, et al. Induction of colonic regulatory T cells by indigenous Clostridium species. Science 2011;331(6015):337–41.

[111] Atarashi K, Tanoue T, Oshima K, et al. Treg induction by a rationally selected mixture of Clostridia strains from the human microbiota. Nature 2013;500(7461):232–6.

[112] Topping DL, Clifton PM. Short-chain fatty acids and human colonic function: roles of resistant starch and nonstarch polysaccharides. Physiol Rev 2001;81(3):1031–64.

[113] Chang PV, Hao L, Offermanns S, Medzhitov R. The microbial metabolite butyrate regulates intestinal macrophage function via histone deacetylase inhibition. Proc Natl Acad Sci USA 2014;111(6):2247–52.

[114] Singh N, Gurav A, Sivaprakasam S, et al. Activation of Gpr109a, receptor for niacin and the commensal metabolite butyrate, suppresses colonic inflammation and carcinogenesis. Immunity 2014;40(1):128–39.

[115] Frei R, Ferstl R, Konieczna P, et al. Histamine receptor 2 modifies dendritic cell responses to microbial ligands. J Allergy Clin Immunol 2013;132(1):194–204.

[116] Zelante T, Iannitti RG, Cunha C, et al. Tryptophan catabolites from microbiota engage aryl hydrocarbon receptor and balance mucosal reactivity via interleukin-22. Immunity 2013;39(2):372–85.

[117] Qiu J, Guo X, Chen ZM, et al. Group 3 innate lymphoid cells inhibit T-cell-mediated intestinal inflammation through aryl hydrocarbon receptor signaling and regulation of microflora. Immunity 2013;39(2):386–99.

Chapter 27

Gastrointestinal Microbiota and the Neural System

V. Philip and P. Bercik

Department of Medicine, Farncombe Family Digestive Health Research Institute, McMaster University, Hamilton, ON, Canada

INTESTINAL MICROBIOTA

The human body is colonized by many different bacteria collectively referred to as microbiota. There are more bacterial cells than human cells in the human body with unique microbiota profiles in specific niches, such as the skin, mouth, respiratory tracts, and the gastrointestinal (GI) tract [1]. Bacterial colonization in the GI tract occurs during birth and shortly thereafter, and over 500 different species of bacteria can colonize the intestines. The vast majority of gut bacteria fall into the Firmicutes and Bacteroidetes phyla, with a much smaller proportion into the Actinobacteria and Proteobacteria phyla [2]. The microbial diversity in an adult GI tract is resilient to change and maintain beneficial influence on the host. This includes proper maturation of the GI mucosal immune system, enhancing nutrient acquisition, strengthening the integrity of the intestinal barrier, and providing defense against pathogenic microorganisms [3].

The intestinal microbiota is commonly referred to as a metabolically active organ, as there exists a dynamic relationship between the host and its resident microbiota, particularly in relation to the health of the host [4]. The abundance of genes contained within this microbial organ far outweighs the genetic content of the host by an estimated factor of 1–2 and it is conceivable that for these reasons the microbiota is able to maintain a powerful influence on the host, particularly on GI tract and GI-related functions [5]. Alterations in the gut microbiota (dysbiosis) due to antibiotics, infections, medications, and long-term dietary changes can potentially lead to detrimental negative health outcomes including diarrhea, opportunistic infections, and obesity [6,7]. Most chronic gut disorders, such as inflammatory bowel disease, irritable bowel syndrome, and celiac disease are associated with dysbiosis [8]. Future studies are needed in order to shed light on whether a particular microbial composition is associated with specific diseases.

MICROBIOTA AND THE BRAIN

It is well established that the gut bacteria can influence the central nervous system (CNS) [9]. The notion that gut bacteria influence brain function and behavior seems implausible but appears feasible when considering the often dramatic improvement in patients with severe hepatic encephalopathy after oral antibiotic administration [10]. Antibiotics have also been anecdotally reported to induce acute psychosis in patients which resolved after withdrawal of the drug [11]. Dysbiosis has been shown to be present in patients with autism and interestingly, some patients with late-onset autism ameliorate their symptoms after antibiotic administration [12]. Furthermore, *Bacteroides fragilis* has improved anxiety-like behavior, communicative, and repetitive behaviors in an animal model of autism with the underlying mechanisms being related to changes in gut microbiota composition and serum metabolomic profile [13].

The initial experiments which linked bacteria to the function of the CNS and behavior involved bacterial pathogens, and comparisons between conventional and germ-free (GF) mice. Lyte et al. were the first to show that mice infected with *Campylobacter jejuni* display anxiety-like behavior before the onset of any discernible inflammatory response to infection [14]. Subsequent studies from this group suggested that the neural pathways involved in this altered behavior include the vagus nerve [15]. Sudo et al. showed that commensal bacteria can affect the postnatal development of the HPA stress response in mice [16]. In his study, the plasma ACTH following restrain stress was elevated in GF mice compared to mice with specific pathogen free (SPF), suggesting that absence of bacteria renders mice more susceptible to stress. Interestingly, plasma ACTH was reduced in ex-GF mice colonized with bacteria, although this was observed only in very young mice, indicating a critical period in which the neural regulation of the stress response is sensitive to input from the microbiota [16]. Several recent studies reported marked behavioral differences between GF mice and mice colonized with an SPF microbiota. Higher exploratory and lower anxiety-like behavior was found in GF mice compared to SPF mice using standard behavioral tests (elevated plus maze, open field and light/dark preference) [17]. Heijtz et al. showed that compared to GF mice, SPF mice had higher central expression of neurotrophins, including nerve growth factor and BDNF, as well as altered expression of genes involved in the secondary messenger pathways and synaptic long-term potentiation in brain regions involved in motor control and anxiety-like behavior [17]. Neufeld et al. demonstrated that the presence of microbiota also affects expression of N-methyl-D-aspartate receptor subunit NR2B in amygdala and serotonin receptor 1A in hippocampus [18]. The observed difference between GF mice and mice colonized with complex microbiota may relate, at least in part, to the ability of bacteria to affect host immunity and metabolism, as well as gut physiology. It has been shown that even monocolonization with *Bacteroides thetaiotaomicron* alters expression of large array of host genes involved in multiple bodily functions [19].

The effect of bacteria on behavior and brain chemistry was also demonstrated in conventionally raised mice. Our group showed that perturbation of healthy intestinal microbiota using oral antimicrobials (ATM) in mice increased exploratory behavior and altered expression of BDNF in the amygdala and hippocampus [20]. No difference in behavior or BDNF levels were observed in SPF mice treated with ATM intraperitoneally or when ATM was administered orally in GF mice.

Specific probiotic bacteria have been shown to affect the CNS. *Bifidobacterium longum* NCC3001 normalized anxiety-like behavior and BDNF expression in the hippocampus of mice with mild to moderate colitis that was induced by a chronic parasitic infection [21]. Similarly, in a chemical model of low-grade inflammation, *B. longum* NCC3001 normalized anxiety-like behavior, the beneficial effect being dependent on the integrity of vagus nerve and independent of immunomodulatory activity [21,22]. *Lactobacillus rhamnosus (JB-1)* reduced stress-induced corticosterone and anxiety- and depression-related behavior in mice [23]. These mice displayed altered GABA receptor expressions in the amygdala and hippocampus regions compared to control-fed mice. The neurochemical and behavioral effects of the probiotic were not found in vagotomized mice, identifying the vagus nerve responsible for these observed changes [23].

Until 2013, there was limited evidence in humans that probiotics have the same neurochemical and behavioral effects as observed in animal models. However, two clinical studies have confirmed beneficial effects of specific bacteria on the CNS activity. Tillish et al. has shown that

consumption of a fermented milk product with several probiotics (*Bifidobacterium animalis* subsp. *lactis, Streptococcus thermophiles, Lactobacillus bulgaricus, and Lactococcus lactis* subsp. *lactis*) by healthy women for 4 weeks affected activity of brain regions that control central processing of emotion and sensation [24]. In a pilot study, we have demonstrated that administration of *B. longum* NCC3001 to patients with IBS and comorbid depression, and/or anxiety improves depression scores and alters activity patterns in multiple areas of the brain involved in mood control, including amygdala [25].

ENTERIC NERVOUS SYSTEM

The enteric nervous system (ENS) is a self-reinforcing neural network that has the ability to function autonomously and govern most of the GI function bidirectionally with the CNS. It encompasses the intrinsic neuroglial network of the gut and is organized into three major plexuses, the myenteric, submucosal, and mucosal plexuses, which controls virtually all aspects of GI physiology [26]. The autonomic nervous system (ANS), with the sympathetic and parasympathetic limbs, drives both afferent signals, arising from the lumen and transmitted through enteric, spinal and vagal pathways to CNS, and efferent signals from CNS to the intestinal wall. With respect to the GI tract, sympathetic, parasympathetic, and spinal afferent nerve fibers are considered to be the extrinsic innervations with the ENS and afford the necessary conduit to maintain bi-directional communication with the CNS through intimate connections with the spinal cord. The vast innervations of the GI tract and the connections between intrinsic and extrinsic fibers allow the CNS to monitor a number of gut parameters, from chemical sensing in the lumen, to sensing mechanical stress along the gut wall [27]. The sympathetic neurons (effector branch of the ANS) have axons that extend along the mesenteric nerves deep into the gut wall to the myenteric, submucosal, and mucosal plexuses of the ENS [28]. The terminals of these axons are composed of numerous neurotransmitters and their associated enzymes, mainly norepinephrine (NE) and tyrosine hydroxylase (TH). New enteric neurons and glia continue to integrate themselves to preexisting neuronal circuits several weeks after birth [29]. This suggests that changes in the environment associated with diet, luminal microflora, and maturation of the mucosal immune system may likely affect postnatal development of the ENS.

Inflammatory signals have a broad effect on the CNS and peripheral nervous system. For example, proinflammatory cytokines IL-1β and TNF-α promoted neurogenesis in a coculture model of enteric neurons and smooth muscle cells through NF-κB–dependent manner to upregulate expression of GDNF [30]. The potential effects of cytokines or other immune products during normal ENS development are unknown, but the ability of enteric neurons to respond to inflammatory cytokines and leukotrienes [31] raises the possibility that members of such families of signaling molecules could play an important role in the development and maturation of the mammalian ENS. Such an effect could be more pronounced in the postnatal gut during the establishment of the microbiota and the maturation of the innate and adaptive immune systems.

MICROBIOTA AND THE ENS

Studies in GF animals have shed light into the importance of bacterial colonization on proper development of the ENS [32]. During embryogenesis the ENS develops within a largely sterile environment, but the postnatal stages of ENS development and maturation take place under strikingly different conditions due to the ingestion of food and the establishment of microbiota, which result in multitude of changes to the maturation of the immune system [33]. The absence of microbiota is associated with alterations of the gut sensory–motor functions, resulting in delayed gastric emptying and intestinal transit, reduced migrating motor complex cyclic recurrence and distal propagation, and enlarged cecal size [34]. GF mice showed altered spontaneous muscle contractions, decreased nerve density in small intestine [35], and reduced sensory neuron excitability, which was restored following bacterial colonization [36].

There is ample evidence that bacteria can affect the function of the ENS also in conventional mice. Chronic infection with Helicobacter *pylori* increased the density of substance P, calcitonin gene-related peptide, and vasoactive intestinal polypeptide immunoreactivity nerves in the gastric myenteric plexus and spinal cord [37], and altered response to gastric distension which only partially improved after bacterial eradication [38].

As mentioned earlier, the anxiolytic effect of *B. longum* requires vagal integrity but does not involve gut immunomodulation [22]. *B. longum* has been shown to alter excitability of the enteric sensory neurons and thus it may signal to the brain by interacting through vagal pathways at the level of the ENS [22,39]. Another well-characterized probiotic, *Lactobacillus reuteri* effects gut motility and pain perception in rats by increasing excitability of the enteric neurons through the inhibition of the calcium-dependent potassium channel opening [40]. In an ex vivo perfusion model, *L. reuteri* was shown to moderate mouse jejunal motor patterns within minutes, but this effect was absent when administering heat-killed *L. reuteri* or another live commensal (*Lactobacillus salivarius*) [41]. Verdu et al. showed that antibiotic-induced perturbation in the gut microbiota produced changes in inflammatory cell infiltrate and substance P levels in the gut resulting in an increase in visceral sensitivity to colorectal distension [42]. These changes were prevented by administration of *Lactobacillus paracasei* suspended in spent culture medium [42]. *L. paracasei*

also prevented partial restraint stress-mediated visceral hyperalgesia in maternally deprived rats [43]. *Lactobacillus farciminis* treatment prevented hyperalgesia to colorectal distension induced by acute stress [44] via a decrease in the stress-induced sensitization of sensory neurons at the spinal and supraspinal level [45].

A number of molecules have been identified through which the microbiota can communicate with the ENS. Metabolites produced by gut microbes include short-chain fatty acids (SCFAs), metabolites of bile acids, and neuroactive substances, such as GABA, tryptophan precursors, serotonin, biologically active forms of catecholamines, [46] and cytokines released during the immune response to microbes [47] can all signal to the host via receptors on nerve ending within the gut. These factors can also signal via neurocrine (through afferent vagal and possibly spinal) pathways to targets well beyond the GI tract, including vagal afferents in the portal vein and receptors in the brain. *Lactobacilli* can produce hydrogen sulfide that modulates gut motility by interacting with the vanilloid receptor on capsaicin-sensitive nerve fibers [48]. Most metabolites identified in the circulation of the host are of gut microbial origin, [49] providing the theoretical basis for the microbiota–gut–brain axis signaling system. SCFAs, such as butyric acid, propionic acid, and acetic acid, are able to stimulate sympathetic nervous system [50] and mucosal serotonin release [51]. Signaling through GPCRs and SCFAs transporter by SLC5A8 and the resulting physiological effects can occur due to the dietary intake of fermentable fiber [52]. Different types of SCFA receptors have been identified on enteroendocrine cells as well as on neurons of the submucosal and myenteric ganglia [53].

What are the potential mechanisms by which the microbiota influences the development and organization of the ENS? The host senses gut bacteria via the patter-recognition receptors, a large family of proteins that have the ability to recognize unique microbial components and play a crucial role in innate immunity, and protection against pathogenic microorganisms. Since intestinal epithelial cells and the innate immune response are the first-line of defense against commensal and pathogens, their products are likely to influence the development and homeostasis of the ENS [54]. Gene expression analysis show that TLR3 and TLR7 (receptors for viral RNA), and TLR4 (LPS receptor) are expressed by enteric neurons and glia, which suggest that ENS lineages can directly sense microbial microbiota [55]. In vitro studies with isolated rat myenteric plexus demonstrated a direct effect of LPS on enteric glial cells, evidenced by the increased production of the proinflammatory factors IL-1 and prostaglandin E2. Anitha et al. showed that GF and antibiotic-treated mice exhibited reduced motility and fewer nNOS$^+$ neurons [56]. This effect was mediated, at least partly, via TLR4, as Tlr4$^{-/-}$ mice exhibited similar deficits in intestinal motility and reduced number of nitrenergic neurons as GF mice. This phenotype was reproduced in mice with ENS specific MyD88 knockout suggesting that TLR4 signaling is critical for the nitrergic neurons within ENS lineages. The same study demonstrated that LPS promoted the survival of cultured enteric neurons in an NF-κB-dependent manner [56]. Enteric neurons and glia also express TLR2 and the myenteric ganglia of Tlr2$^{-/-}$ mice contained fewer neurons compared with their wild-type mice, with reduction in inhibitory nNOS$^+$ neurons being the most notable phenotype [57]. The reduction in nNOS$^+$ neurons is accompanied by intestinal dysmotility and impaired chloride secretion in ileum. Administration of GDNF can correct many of the ENS deficiencies in Tlr2$^{-/-}$ mice and in antibiotic-treated animals, suggesting that one of the roles of the microbiota-TLR2 axis is to promote the expression of neurotrophic factors that are required to maintain the functional organization of the mammalian ENS [57]. Even though there is much to discover regarding the mechanism by which microbiota can influence ENS development and function, the significance of these observations have significant implications. For example, overstimulation of the TLRs by recurrent infection or through early life dysbiosis caused by antibiotic use could hamper the proper development and maturation of the ENS [58]. Data suggest that the developing ENS is capable of responding to environmental cues and adjust to the volatile milieu of the GI lumen. Reverse regulation, in which the ENS contributes to the shaping of the microbiota is also possible, as suggested by a study in which alterations in the composition of colonic and fecal microbiota were observed in a mouse model of congenital aganglionosis [59].

In summary, there are multiple mechanisms by which the microbiota can influence interactions between the gut and the nervous system. Regardless of the sequence of events leading to a state of dysbiosis in a particular disorder, alterations in the microbial community are likely to affect the interactions between the gut, ENS, and CNS. Such influences may occur early in life and thus affect the differentiation and maturation of ENS lineages, and neuronal circuits, which can lead to impaired development of the neural system. In adults, the gut microbial community is considered more resilient to change. Consequently, the structural and/or functional integrity of the ENS is less likely to be impaired unless other systems, such as the immune or neuroendocrine system, are affected which then can contribute to the development of chronic gut disorders.

REFERENCES

[1] Turnbaugh PJ, Ley RE, Hamady M, et al. The human microbiome project. Nature 2007;449(7164):804–10.

[2] Sekirov I, Russell SL, Antunes LC, et al. Gut microbiota in health and disease. Physiol Rev 2010;90(3):859–904.

[3] Hooper LV, Gordon JI. Commensal host-bacterial relationships in the gut. Science 2001;292(5519):115–8.

[4] Lyte M. The microbial organ in the gut as a driver of homeostasis and disease. Med Hypotheses 2010;74(4):634–8.

[5] Preidis GA, Versalovic J. Targeting the human microbiome with antibiotics, probiotics, and prebiotics: gastroenterology enters the metagenomics era. Gastroenterology 2009;136(6):2015–31.

[6] Turnbaugh PJ, Ley RE, Mahowald MA, et al. An obesity-associated gut microbiome with increased capacity for energy harvest. Nature 2006;444(7122):1027–31.

[7] Chang JY, Antonopoulos DA, Kalra A, et al. Decreased diversity of the fecal microbiome in recurrent *Clostridium difficile*-associated diarrhea. J Infect Dis 2008;197(3):435–8.

[8] Nadal I, Donat E, Ribes-Koninckx C, et al. Imbalance in the composition of the duodenal microbiota of children with coeliac disease. J Med Microbiol 2007;56(Pt 12):1669–74.

[9] Collins SM, Surette M, Bercik P. The interplay between the intestinal microbiota and the brain. Nat Rev Microbiol 2012;10(11):735–42.

[10] Victor DW III, Quigley EM. Hepatic encephalopathy involves interactions among the microbiota, gut, brain. Clin Gastroenterol Hepatol 2014;12(6):1009–11.

[11] Mehdi S. Antibiotic-induced psychosis: a link to D-alanine? Med Hypotheses 2010;75(6):676–7.

[12] Sandler RH, Finegold SM, Bolte ER, et al. Short-term benefit from oral vancomycin treatment of regressive-onset autism. J Child Neurol 2000;15(7):429–35.

[13] Hsiao EY, McBride SW, Hsien S, et al. Microbiota modulate behavioral and physiological abnormalities associated with neurodevelopmental disorders. Cell 2013;155(7):1451–63.

[14] Lyte M, Varcoe JJ, Bailey MT. Anxiogenic effect of subclinical bacterial infection in mice in the absence of overt immune activation. Physiol Behav 1998;65(1):63–8.

[15] Goehler LE, Gaykema RP, Opitz N, et al. Activation in vagal afferents and central autonomic pathways: early responses to intestinal infection with *Campylobacter jejuni*. Brain Behav Immun 2005;19(4):334–44.

[16] Sudo N, Chida Y, Aiba Y, et al. Postnatal microbial colonization programs the hypothalamic-pituitary-adrenal system for stress response in mice. J Physiol 2004;558(Pt 1):263–75.

[17] Diaz HR, Wang S, Anuar F, et al. Normal gut microbiota modulates brain development, behavior. Proc Natl Acad Sci USA 2011;108(7):3047–52.

[18] Neufeld KM, Kang N, Bienenstock J, et al. Reduced anxiety-like behavior and central neurochemical change in germ-free mice. Neurogastroenterol Motil 2011;23(3):255–64. e119.

[19] Hooper LV, Wong MH, Thelin A, et al. Molecular analysis of commensal host-microbial relationships in the intestine. Science 2001;291(5505):881–4.

[20] Bercik P, Denou E, Collins J, et al. The intestinal microbiota affect central levels of brain-derived neurotropic factor and behavior in mice. Gastroenterology 2011;141(2):599–609. 609.e1–609.e3.

[21] Bercik P, Verdu EF, Foster JA, et al. Chronic gastrointestinal inflammation induces anxiety-like behavior and alters central nervous system biochemistry in mice. Gastroenterology 2010;139(6):2102–12.

[22] Bercik P, Park AJ, Sinclair D, et al. The anxiolytic effect of *Bifidobacterium longum* NCC3001 involves vagal pathways for gut-brain communication. Neurogastroenterol Motil 2011;23(12):1132–9.

[23] Bravo JA, Forsythe P, Chew MV, et al. Ingestion of *Lactobacillus* strain regulates emotional behavior and central GABA receptor expression in a mouse via the vagus nerve. Proc Natl Acad Sci USA 2011;108(38):16050–5.

[24] Tillisch K, Labus J, Kilpatrick L, et al. Consumption of fermented milk product with probiotic modulates brain activity. Gastroenterology 2013;144(7):1394–401. 1401.e1–1401.e4.

[25] Pinto-Sanchez MI, Hall G, Gajar K, et al. *Bifidobacterium longum* NCC3001 improves depression scores and alters brain activity in patients with irritable bowel syndrome: a randomized, double blind, placebo-controlled trial. UEG Proc J 2015;3(S1.):2050.

[26] Furness JB. Novel gut afferents: intrinsic afferent neurons and intestinofugal neurons. Auton Neurosci 2006;125(1–2):81–5.

[27] Furness JB. Types of neurons in the enteric nervous system. J Auton Nerv Syst 2000;81(1–3):87–96.

[28] Lomax AE, Sharkey KA, Furness JB. The participation of the sympathetic innervation of the gastrointestinal tract in disease states. Neurogastroenterol Motil 2010;22(1):7–18.

[29] Pham TD, Gershon MD, Rothman TP. Time of origin of neurons in the murine enteric nervous system: sequence in relation to phenotype. J Comp Neurol 1991;314(4):789–98.

[30] Gougeon PY, Lourenssen S, Han TY, et al. The pro-inflammatory cytokines IL-1beta and TNFalpha are neurotrophic for enteric neurons. J Neurosci 2013;33(8):3339–51.

[31] Liu S, Hu HZ, Gao C, et al. Actions of cysteinyl leukotrienes in the enteric nervous system of guinea-pig stomach and small intestine. Eur J Pharmacol 2003;459(1):27–39.

[32] Barbara G, Stanghellini V, Brandi G, et al. Interactions between commensal bacteria and gut sensorimotor function in health and disease. Am J Gastroenterol 2005;100(11):2560–8.

[33] Hooper LV, Littman DR, Macpherson AJ. Interactions between the microbiota and the immune system. Science 2012;336(6086):1268–73.

[34] Husebye E, Hellstrom PM, Sundler F, et al. Influence of microbial species on small intestinal myoelectric activity and transit in germ-free rats. Am J Physiol Gastrointest Liver Physiol 2001;280(3):G368–80.

[35] Collins J, Borojevic R, Verdu EF, et al. Intestinal microbiota influence the early postnatal development of the enteric nervous system. Neurogastroenterol Motil 2014;26(1):98–107.

[36] McVey Neufeld KA, Mao YK, Bienenstock J, et al. The microbiome is essential for normal gut intrinsic primary afferent neuron excitability in the mouse. Neurogastroenterol Motil 2013;25(2):183–8.

[37] Bercik P, De GR, Blennerhassett P, et al. Immune-mediated neural dysfunction in a murine model of chronic *Helicobacter pylori* infection. Gastroenterology 2002;123(4):1205–15.

[38] Bercik P, Verdu EF, Foster JA, et al. Role of gut–brain axis in persistent abnormal feeding behavior in mice following eradication of *Helicobacter pylori* infection. Am J Physiol Regul Integr Comp Physiol 2009;296(3):R587–94.

[39] Khoshdel A, Verdu EF, Kunze W, et al. *Bifidobacterium longum* NCC3001 inhibits AH neuron excitability. Neurogastroenterol Motil 2013;25(7):e478–84.

[40] Kunze WA, Mao YK, Wang B, et al. *Lactobacillus reuteri* enhances excitability of colonic AH neurons by inhibiting calcium-dependent potassium channel opening. J Cell Mol Med 2009;13(8B):2261–70.

[41] Wang B, Mao YK, Diorio C, et al. Luminal administration ex vivo of a live *Lactobacillus* species moderates mouse jejunal motility within minutes. FASEB J 2010;24(10):4078–88.

[42] Verdu EF, Bercik P, Verma-Gandhu M, et al. Specific probiotic therapy attenuates antibiotic induced visceral hypersensitivity in mice. Gut 2006;55(2):182–90.

[43] Eutamene H, Lamine F, Chabo C, et al. Synergy between *Lactobacillus paracasei* and its bacterial products to counteract stress-induced gut permeability and sensitivity increase in rats. J Nutr 2007;137(8):1901–7.

[44] Ait-Belgnaoui A, Han W, Lamine F, et al. *Lactobacillus farciminis* treatment suppresses stress induced visceral hypersensitivity: a possible action through interaction with epithelial cell cytoskeleton contraction. Gut 2006;55(8):1090–4.

[45] Ait-Belgnaoui A, Eutamene H, Houdeau E, et al. *Lactobacillus farciminis* treatment attenuates stress-induced overexpression of Fos protein in spinal and supraspinal sites after colorectal distension in rats. Neurogastroenterol Motil 2009;21(5):567–9.

[46] Chey WY, Jin HO, Lee MH, et al. Colonic motility abnormality in patients with irritable bowel syndrome exhibiting abdominal pain and diarrhea. Am J Gastroenterol 2001;96(5):1499–506.

[47] Bailey MT, Dowd SE, Galley JD, et al. Exposure to a social stressor alters the structure of the intestinal microbiota: implications for stressor-induced immunomodulation. Brain Behav Immun 2011;25(3):397–407.

[48] Schicho R, Krueger D, Zeller F, et al. Hydrogen sulfide is a novel prosecretory neuromodulator in the Guinea-pig and human colon. Gastroenterology 2006;131(5):1542–52.

[49] Wikoff WR, Anfora AT, Liu J, et al. Metabolomics analysis reveals large effects of gut microflora on mammalian blood metabolites. Proc Natl Acad Sci USA 2009;106(10):3698–703.

[50] Kimura I, Inoue D, Maeda T, et al. Short-chain fatty acids and ketones directly regulate sympathetic nervous system via G protein-coupled receptor 41 (GPR41). Proc Natl Acad Sci USA 2011;108(19):8030–5.

[51] Grider JR, Piland BE. The peristaltic reflex induced by short-chain fatty acids is mediated by sequential release of 5-HT and neuronal CGRP but not BDNF. Am J Physiol Gastrointest Liver Physiol 2007;292(1):G429–37.

[52] Ganapathy V, Thangaraju M, Prasad PD, et al. Transporters and receptors for short-chain fatty acids as the molecular link between colonic bacteria and the host. Curr Opin Pharmacol 2013;13(6):869–74.

[53] Nohr MK, Pedersen MH, Gille A, et al. GPR41/FFAR3 and GPR43/FFAR2 as cosensors for short-chain fatty acids in enteroendocrine cells vs. FFAR3 in enteric neurons and FFAR2 in enteric leukocytes. Endocrinology 2013;154(10):3552–64.

[54] Prescott D, Lee J, Philpott DJ. An epithelial armamentarium to sense the microbiota. Semin Immunol 2013;25(5):323–33.

[55] Barajon I, Serrao G, Arnaboldi F, et al. Toll-like receptors 3, 4, and 7 are expressed in the enteric nervous system and dorsal root ganglia. J Histochem Cytochem 2009;57(11):1013–23.

[56] Anitha M, Vijay-Kumar M, Sitaraman SV, et al. Gut microbial products regulate murine gastrointestinal motility via Toll-like receptor 4 signaling. Gastroenterology 2012;143(4):1006–16.

[57] Brun P, Giron MC, Qesari M, et al. Toll-like receptor 2 regulates intestinal inflammation by controlling integrity of the enteric nervous system. Gastroenterology 2013;145(6):1323–33.

[58] Nylund L, Satokari R, Salminen S, et al. Intestinal microbiota during early life-impact on health and disease. Proc Nutr Soc 2014;73(4):457–69.

[59] Ward NL, Pieretti A, Dowd SE, et al. Intestinal aganglionosis is associated with early and sustained disruption of the colonic microbiome. Neurogastroenterol Motil 2012;24(9):874-e400.

Chapter 28

Effect on the Host Metabolism

M.H. Sarafian*, N.S. Ding**, E. Holmes* and A. Hart**

*Department of Surgery and Cancer, Division of Computational Systems Medicine, Imperial College London, London, United Kingdom;
**Inflammatory Bowel Disease Unit, St Mark's Hospital, Imperial College London, London, United Kingdom

INTRODUCTION

The gut microbiota is a complex component of the gastrointestinal tract and is comprised of over 100 trillion microorganisms. In health, interaction between the host and gut microbiota is largely symbiotic. The gut microbiota is essential in maintaining host internal functions, such as metabolite digestion and protection of the gastrointestinal barrier. Disruption to gut microbiota integrity (i.e., dysbiosis) has been associated with metabolic disorders and pathological processes, such as inflammatory bowel disease (IBD), obesity, metabolic syndrome, autism, and certain cancers [1–4].

The gut microbiota has a wide range of microbial communities, but identification of its full composition remains incomplete, with many anaerobic organisms being difficult to culture and study in a laboratory environment [5]. Recently, new microorganisms have been discovered by metagenomic which add to the increased microbial diversity and offer new perspectives into understanding the gut microbiota–host axis [6]. It has been estimated that approximately 160 bacteria species colonize the gastrointestinal tract with low diversity at the division level [7]. The evolution of gut microbiota in the gastrointestinal tract occurs gradually in the first few years following birth by adaptation to internal and external factors with strong host selection [8]. Previous studies have emphasized the functional structure of gut microbiota by investigating the effects of factors, such as an individual's diet [9], genetics, age [10], population geography [10,11], and drugs on the microbiome. However, the contribution of these factors to the gut microbial distribution is contentious due to poor characterization of the microbiota and high interindividual variations [12].

Currently, there is limited information available on the gut microbiota. However, key noninvasive methods within global system biology are being use to characterize individual microbial species/taxa and their associated metabolic phenotype [7,13]. For instance, genomic and transcriptomic analysis of the gut microbiota, are captured by metagenomic and 16s RNA sequencing respectively, in fecal samples or intestinal content. Furthermore, proteomic and metabonomic information can be gathered with the use of analytical techniques, such as nuclear magnetic resonance spectroscopy (NMR) and mass spectrometry (MS). Metabonomics assesses the metabolic status of a biological system by profiling or quantifying metabolites, which are considered "low molecular weight molecules," and are typically less than 2 kDa in mass. The scope of metabonomics to detect and quantify subtle metabolic changes is largely due to advances accomplished in modern analytical techniques. For instance, ambient MS characterizes tissue metabolites, such as desorption electrospray ionization MS (DESI-MS) in real time [14]. Furthermore, rapid evaporative ionization MS (REIMS) is a unique technique used in surgery and based on direct in vivo and in situ straightforward analysis compared to DESI-MS which requires sample pretreatment [14–16].

Altogether, applications of these robust analytical techniques can be used to characterize and understand the modulation of metabolic processes by the gut microbiota. Metabonomics has been essential in identifying metabolic signatures associated with metabolic disorders and gut microbiota dysbiosis. This can be seen in hyperlipidemia, hypertension [17], hyperglycaemia, type 1 diabetes [18], insulin resistance, and inflammation, where physiological and pathological events can be triggered by the gut microbiota dysbiosis. Gut microbiota dysbiosis may affect the host by alteration of metabolite levels which have been implicated in numerous diseases. The role of the gut microbiome in disease pathogenesis has been defined in diseases, such as obesity [19], inflammatory bowel diseases (IBD) [20], liver diseases [21], atherosclerosis [22], cardiometabolic diseases [23], and extensively metabolic syndrome [24].

Extensive work has been implemented to understand the interaction that exists between host and gut microbiota. Recent metabonomic research focuses on identifying potential biomarkers that can predict disease progression allowing for personalized diagnosis and treatment of patients, via the regulation of gut microbiota taking into account the subsequent impact on host signaling [25]. This chapter describes and discusses host's metabolic processes and metabolites influenced by the gut microbiota, especially lipid metabolism.

GUT MICROBIOTA FUNCTIONS IN HOST INTEGRITY

Microorganisms are necessary to supply the energy required to maintain vital functions. This energy source is provided by aerobic or anaerobic microorganisms that promote fermentation of dietary fibers, which include resistant starch, inulin, and oligosaccharides. Humans and other vertebrates are normally deficient in enzymatic function, β-1,4-glycosidic bond hydrolysis which is crucial in processing dietary fibers, however, this process is achieved by gastrointestinal microorganisms [26]. Some of the by-products of the fermentation process include the production of short chain fatty acids (SCFAs): acetate (C2:0), butyrate (C3:0), and propionate (C4:0) [27].

These SCFAs are absorbed through the gastrointestinal membrane and transported to various tissues. On a cellular level, SCFAs can enter the tricarboxylic acid cycle (TCA) to produce adenosine triphosphate (ATP) which contains the phosphate group essential to activate physiological processes that require energy. SCFAs are stored as triglycerides (TGs) prior to their hydrolysis and release from adipose tissue as adipocytes, where they are then taken up and oxidised by tissues to prevent energy loss and to maintain vital functions.

Digested SCFAs have a significant impact on a microorganism's proportion and distribution along the gastrointestinal tract. Thus other factors need to be considered, such as gastrointestinal tract pH, which can be affected by a deficiency of SCFAs supplied from diet fiber. Gastrointestinal tract pH is dependent on SCFA levels, which subsequently influences growth of pH-sensitive microorganisms and composition of the gut microbiota [28]. Low pH has been shown to minimize the growth of microorganisms and affect the action of enzymes particularly involved in the fermentation process of dietary fibers [29]. Fluctuation of gastrointestinal pH is autoregulated through SCFAs and bile acid (BA) levels with microorganisms' growth.

Furthermore, SCFAs are known to act as signaling molecules, able to modulate host metabolic pathways to maintain healthy conditions in the intestine. SCFAs have been shown to confer specific metabolic outcomes. For instance, butyrate can downregulate inflammation and colorectal cancer [30,31], whereas propionate downregulates lipolysis, inflammation, carcinogen process, decreases cholesterol

levels, and increases satiety [32]. Additionally, SCFAs activate G-protein coupled receptors (GPR41, GPR43, and GPR109a). However, the specific mechanisms of action are not yet fully understood [33]. Consequently, changes to SCFAs levels can lead to undesirable physiological effects with the main trigger appearing to be gut dysbiosis.

SCFAs, especially acetate, are substrates for long-chain FAs and also cholesterol metabolism by formation of acetyl CoA. Cholesterol is the central compound of vitamin D, steroid hormone, oxysterols, and BA metabolism. Oxysterols are precursors of primary bile acids (i.e., cholic acid and chenodeoxycholic acid) synthesized in the liver. BAs are secreted into the gastrointestinal tract to achieve emulsification and transport of dietary fat, and complete their enterohepatic circulation to the liver. Assembly of these fatty compounds (TGs, cholesterol ester, phospholipids, and apolipoproteins) enables them to access the systemic circulation and enter the lipoprotein metabolism as chylomicrons. During the enterohepatic circulation, around 95% of these compounds are reabsorbed in the terminal ileum trough the apical sodium-dependent bile salt transporter (ASBT; SLC10A2) to portal vein and the liver (≈20 g/day), and around 5% escaped absorption (≈0.2–0.5 g/day).

BAs adjust their own systemic and cellular level mainly through CYP7A1 (cholesterol-7-α-hydroxylase), a microsomal rate-limiting enzyme that converts cholesterol to initiate oxysterol and primary BA synthesis in the liver. Oxysterols mediate the activation of CYP7A1 via Liver X receptor (LXR) and BAs mediate the inhibition of CYP7A1 via the farnesoid X receptor (FXR or Nuclear Receptor Subfamily 1 Group H Member 4, NR1H4) [34]. The actions of BAs also serve to initiate antiinflammatory response, protect against insulin resistance and prevent obesity via intestinal receptors FXR and TGR5 (or G protein-coupled bile acid receptor 1, GPBAR-1) [35,36].

BAs are important intermediates of the host–gut microbiota axis and both entities have enzymes able to modulate BA structure, which may compromise dietary intake and BA pool diversity. Secondary BAs (e.g., deoxycholic acid and lithocholic acid) refers to BAs that are modulated by the gut microbiota on the first enterohepatic circulation [37]. Tertiary BAs refers to BAs that are modified either by the host or gut microbiota on multiple enterohepatic circulation. BA elimination can be mediated by the gut microbiota desulfation activity. In case of BA excess, BAs are sulfated in the liver by sulfotransferase. Sulfated BAs show higher solubility than their unconjugated counterpart and increase bile acid clearance. However, the gut microbiome has desulfatase enzymes that can interfere with the elimination pathway and induce reabsorption of the BAs in the intestine [38]. Cholesterol and BA transport to bypass absorption in the intestine and stimulate fecal or urinary excretion are also regulated via FXR and LXR activation.

GUT MICROBIOTA DYSBIOSIS IMPACT ON HOST METABOLISM

The effects of the gut microbiota have been extensively studied by controlling the colonization of microbes in germ-free mice models. Over the last few decades, improvements around study methodology and review of specific analytical techniques have played a key role in elucidating the mechanisms behind host–gut microbiota interactions. A combination of genomic and metabonomic studies have highlighted microbial species involved in modulation of multiple metabolic disorders.

SCFAs produced by the gut microbiota influence metabolic pathways, such as lipid, glucose, and cholesterol metabolism. As mentioned previously, SCFAs regulate synthesis of many metabolites and associated pathways. In obese mice, an increase of SCFAs in the intestine has been associated with an overgrowth of specific microorganism species in the gastrointestinal tract. This finding is unclear as microorganisms and SCFA levels in obese humans vary between studies [33]. Data from studies in germ-free mice suggested that reduced gut microbiota diversity in obesity, influences microbial enzymatic efficiency by reducing the fermentation process [33]. Also, it has been established that modification of the gut microbiota in obese, compared to lean subjects, has consequences on lipid clearance [39,40]. Trials are ongoing with regard to whether these have implications on therapy. The metabolic signature of human gastrointestinal tumors analyzed by REIMS was characterized by an increase in phosphoethanolamines [15]. Clearly, this finding recognizes REIMS as a powerful tool for diagnosis and more precise resection of the tumour site than endoscopy. However, mechanisms related to the regulation of gut microbiota and their ability to breakdown dietary fibers remains unknown.

Studies using germ-free mice presenting with less body fat than conventional mice has shown that the gut microbiota modulates lipid metabolism [41]. FAs and TGs were shown to play an important role in the host–gut interaction [42]. FA excess has a major impact on cellular pathways, via exacerbated activation of nuclear receptor (i.e., nuclear receptor peroxisome proliferator activated, PPAR), production of reactive oxygen species (ROS) and proinflammatory cytokines (i.e., tumor necrosis factor-α, TNF α). These reactions culminate in chronic inflammation and insulin resistance, which are the main contributing factors of obesity (i.e., adipocytes) and can lead to metabolic syndrome [43]. Excess FAs also increase synthesis of glycerols, diacylglycerols (DGs), and TGs. DGs play a key role in the response of proinflammatory regulators and the inhibition of insulin sensitivity [43]. Insulin homeostasis has been shown to be enhanced in obese subjects with gut microbiota dysbiosis and risk factors of metabolic syndrome [44]. Insulin resistance is an immunological component that contributes to the maintenance of low-grade inflammation in adipose tissue of obese individuals.

It has been recognized that gut microbiota can trigger antiinflammatory response via GPR43 [45]. In particular, the phylum Bacteroidetes has demonstrated benefits in regulating systemic immune responses in germ-free mice. For example, polysaccharide A, similar to peptidoglycan, is a glycan mediator produced from dietary fibers that demonstrate strong antiinflammatory responses [45]. Germ-free mice are markedly depleted of an adaptive immune system and are only able to compensate with exacerbated innate immune system due to the absence of gut microbiota. Other studies have highlighted the advantage of dietary fiber intake and ω-3 fatty acids in IBD patients to overcome any deficiency in the immune system through activation of GPCR43 and GPCR120, respectively [45,46].

The gut microbiota is known to modulate pool size and polarity properties of circulating BAs, which impact on assimilation of dietary lipids and host tissue function [47]. Altered BA enterohepatic circulation is complex and can be explained by both compromised host liver functions [37,48] and gut microbiota dysbiosis [49,50]. Murine models emphasize the role of BAs that are structurally modified by the gut, namely secondary and tertiary BAs. Analytical methods recently developed in metabonomics give more access to BA deriving from gut microbiota activity [51]. Excess of primary BAs were shown to cause a consequent increase in the levels of secondary and tertiary BAs. For example, deoxycholic acid is a product of CA 7α-dehydroxylation by gut microbiota enzymes. Deoxycholic acid has been identified as marker of obesity, type 2-diabetes, and hepatocellular carcinoma and can lead to a dramatic shift toward Firmicutes microorganisms, especially *Clostridium* XIVa [52–55]. Conversely, deoxycholic acid is decreased significantly in IBD patients [56]. This is consistent with the reduced ability of the gut microbiota in IBD to process primary BAs. The tertiary BA, ursodeoxycholic acid is the 7β-epimer of chenodeoxycholic acid, derived from lithocholic acid, and has been shown to have beneficial effect in IBD patients and is used in the treatment of cholestasis [57,58].

Other compounds identified as being modulated by the gut microbiota are also engaged in pathways related to metabolic disorders. For example, choline is degraded by the gut microbiota to form methylamines (e.g., trimethylamine, TMA). These metabolites are further modified by hepatic flavin-containing monooxygenase enzyme 3 (FMO3) that synthesize trimethylamine-N-oxide (TMAO) and dimethylamine (DMA). Increase of TMAO has been associated with impaired FMO3 levels, progression to cardiovascular diseases and to fatty liver disease [59,60]. A further example is ethanol which has been shown to be related to gastrointestinal permeability and fatty liver disease [59]. Studies have associated behavioral health to gut microbiota metabolites that interact with the brain to regulate appetite [61–63]. Neurohormonal gut peptides can trigger both satiety (e.g., cholecystokinin, glucagon-like peptide-l GLP-1, and peptide YY PYY) and appetite stimulation (e.g., ghrelin). Recent evidence suggests that bariatric surgery may modify appetite response and energy homeostasis via changes in the levels of neurohormonal gut peptides signals [64].

CONCLUSIONS

The gut microbiome has been widely investigated and has been shown to be involved in various disease mechanisms. New technologies in the genomic and metabonomic fields, as well as system models, such as germ-free mice, have improved over the past decade and increased our knowledge and understanding regarding the complex interaction of the gut microbiota with host metabolism. However, the ability to translate findings from rodent studies and apply them to human function remains problematic, as distinct microbial and metabolite species sharing similar metabolic functionalities are found in both organisms. Analytical methods are challenged by the high diversity of metabolites. Metabolite levels affected by gut microbiota dysbiosis, such as SCFAs and BAs play a key role in host metabolism. Newly developed techniques are expected to deliver new insights on gut microbiota function by monitoring metabolites influenced by gut function. Clearly, the complementarity of various -omic analyses is an asset in maximizing the detection range of metabolites for the understanding of the host–gut microbiota axis and involvement in gastrointestinal pathophysiology (Table 28.1).

TABLE 28.1 Summary of SCFAs and BAs Metabolism Linked to gut–Host Axis

	Gastrointestinal Metabolites Function	Impact of Gut Microbiota Dysbiosis
SCFAs	• Gastrointestinal tract pH • Impact on ATP, TGs, cholesterol, and BA synthesis • Activate pathways via GPCR (GPR41, GPR43, and GPR109a)	• Activation of PPAR, ROS, and TNFα pathways • Synthesis of DGs and TGs modulated
BAs	• Emulsification and absorption of dietary fat • Gastrointestinal tract pH • Regulate their own synthesis via CYP7A1 with FXR, LXR • Protect against insulin resistance via TGR5	• Variation in secondary and tertiary BA levels

REFERENCES

[1] Nicholson JK, Holmes E, Kinross J, et al. Host-gut microbiota metabolic interactions. Science 2012;336(6086):1262–7.

[2] Mehal WZ. The gordian knot of dysbiosis, obesity, and NAFLD. Nat Rev Gastroenterol Hepatol 2013;10(11):637–44.

[3] Wallace KL, Zheng LB, Kanazawa Y, Shih DQ. Immunopathology of inflammatory bowel disease. World J Gastroenterol 2014;20(1):6–21.

[4] Ley RE, Turnbaugh PJ, Klein S, Gordon JI. Microbial ecology: human gut microbes associated with obesity. Nature 2006;444(7122):1022–3.

[5] Shendure J, Ji H. Next-generation DNA sequencing. Nat Biotechnol 2008;26(10):1135–345.

[6] Luef B, Frischkorn KR, Wrighton KC, et al. Diverse uncultivated ultra-small bacterial cells in groundwater. Nat Commun 2015;6:6372.

[7] Qin J, Li R, Raes J, et al. A human gut microbial gene catalogue established by metagenomic sequencing. Nature 2010;464(7285):59–65.

[8] Backhed F, Roswall J, Peng Y, et al. Dynamics and stabilization of the human gut microbiome during the first year of life. Cell Host Microbe 2015;17(6):852.

[9] David LA, Maurice CF, Carmody RN, et al. Diet rapidly and reproducibly alters the human gut microbiome. Nature 2014;505(7484):559–63.

[10] Yatsunenko T, Rey FE, Manary MJ, et al. Human gut microbiome viewed across age and geography. Nature 2012;486(7402):222–7.

[11] Dehingia M, Thangjam Devi K, Talukdar NC, et al. Gut bacterial diversity of the tribes of India and comparison with the worldwide data. Sci Rep 2015;5:18563.

[12] Ley RE, Lozupone CA, Hamady M, Knight R, Gordon JI. Worlds within worlds: evolution of the vertebrate gut microbiota. Nat Rev Microbiol 2008;6(10):776–88.

[13] Human Microbiome Project Consortium. A framework for human microbiome research. Nature 2012;486(7402):215–21.

[14] Abbassi-Ghadi N, Jones EA, Gomez-Romero M, et al. A comparison of DESI-MS and LC-MS for the lipidomic profiling of human cancer tissue. J Am Soc Mass Spectrom 2015;27:255–64.

[15] Balog J, Kumar S, Alexander J, et al. In vivo endoscopic tissue identification by rapid evaporative ionization mass spectrometry (REIMS). Angew Chem Int Ed Engl 2015;54(38):11059–62.

[16] Golf O, Strittmatter N, Karancsi T, et al. Rapid evaporative ionization mass spectrometry imaging platform for direct mapping from bulk tissue and bacterial growth media. Anal Chem 2015;87(5):2527–34.

[17] DeMarco VG, Aroor AR, Sowers JR. The pathophysiology of hypertension in patients with obesity. Nat Rev Endocrinol 2014;10(6):364–76.

[18] Knip M, Siljander H. The role of the intestinal microbiota in type 1 diabetes mellitus. Nat Rev Endocrinol 2016;12(3):154–67.

[19] Turnbaugh PJ, Hamady M, Yatsunenko T, et al. A core gut microbiome in obese and lean twins. Nature 2009;457(7228):480–4.

[20] Manichanh C, Rigottier-Gois L, Bonnaud E, et al. educed diversity of faecal microbiota in Crohn's disease revealed by a metagenomic approach. Gut 2006;55(2):205–11.

[21] Jiang W, Wu N, Wang X, et al. Dysbiosis gut microbiota associated with inflammation and impaired mucosal immune function in intestine of humans with non-alcoholic fatty liver disease. Sci Rep 2015;5:8096.

[22] Koeth RA, Wang Z, Levison BS, et al. Intestinal microbiota metabolism of L-carnitine, a nutrient in red meat, promotes atherosclerosis. Nat Med 2013;19(5):576–85.

[23] Aron-Wisnewsky J, Clement K. The gut microbiome, diet, and links to cardiometabolic and chronic disorders. Nat Rev Nephrol 2015;12(3):169–81.

[24] Parekh PJ, Balart LA, Johnson DA. The influence of the gut microbiome on obesity, Metabolic syndrome, and gastrointestinal disease. Clin Transl Gastroenterol 2015;6:e91.

[25] Nicholson JK, Holmes E, Wilson ID. Gut microorganisms, mammalian metabolism and personalized health care. Nat Rev Microbiol 2005;3(5):431–8.

[26] Lombard V, Bernard T, Rancurel C, Brumer H, Coutinho PM, Henrissat B. A hierarchical classification of polysaccharide lyases for glycogenomics. Biochem J 2010;432(3):437–44.

[27] Cummings JH, Macfarlane GT. The control and consequences of bacterial fermentation in the human colon. J Appl Bacteriol 1991;70(6):443–59.

[28] Duncan SH, Louis P, Thomson JM, Flint HJ. The role of pH in determining the species composition of the human colonic microbiota. Environ Microbiol 2009;11(8):2112–22.

[29] Palframan RJ, Gibson GR, Rastall RA. Effect of pH and dose on the growth of gut bacteria on prebiotic carbohydrates in vitro. Anaerobe 2002;8(5):287–92.

[30] Berni Canani R, Di Costanzo M, Leone L. The epigenetic effects of butyrate: potential therapeutic implications for clinical practice. Clin Epigenet 2012;4(1):4.

[31] Hamer HM, Jonkers D, Venema K, Vanhoutvin S, Troost FJ, Brummer RJ. Review article: the role of butyrate on colonic function. Aliment Pharmacol Ther 2008;27(2):104–19.

[32] Reichardt N, Duncan SH, Young P, et al. Phylogenetic distribution of three pathways for propionate production within the human gut microbiota. ISME J 2014;8(6):1323–35.

[33] Canfora EE, Jocken JW, Blaak EE. Short-chain fatty acids in control of body weight and insulin sensitivity. Nat Rev Endocrinol 2015;11(10):577–91.

[34] Calkin AC, Tontonoz P. Transcriptional integration of metabolism by the nuclear sterol-activated receptors LXR and FXR. Nat Rev Mol Cell Biol 2012;13(4):213–24.

[35] Fang S, Suh JM, Reilly SM, et al. Intestinal FXR agonism promotes adipose tissue browning and reduces obesity and insulin resistance. Nat Med 2015;21(2):159–65.

[36] Duboc H, TacheY, Hofmann AF. The bile acid TGR5 membrane receptor: from basic research to clinical application. Dig Liver Dis 2014;46(4):302–12.

[37] Russell DW. The enzymes, regulation, and genetics of bile acid synthesis. Annu Rev Biochem 2003;72:137–74.

[38] Alnouti Y. Bile Acid sulfation: a pathway of bile acid elimination and detoxification. Toxicol Sci 2009;108(2):225–46.

[39] Turnbaugh PJ, Gordon JI. The core gut microbiome, energy balance, and obesity. J Physiol 2009;587(Pt 17):4153–8.

[40] Backhed F. Programming of host metabolism by the gut microbiota. Ann Nutr Metab 2011;58(Suppl. 2):44–52.

[41] Backhed F, Ding H, Wang T, et al. The gut microbiota as an environmental factor that regulates fat storage. Proc Natl Acad Sci USA 2004;101(44):15718–23.

[42] Tremaroli V, Backhed F. Functional interactions between the gut microbiota and host metabolism. Nature 2012;489(7415):242–9.

[43] Wymann MP, Schneiter R. Lipid signalling in disease. Nat Rev Mol Cell Biol 2008;9(2):162–76.

[44] Cani PD. Crosstalk between the gut microbiota and the endocannabinoid system: impact on the gut barrier function and the adipose tissue. Clin Microbiol Infect 2012;18(Suppl. 4):50–3.

[45] Maslowski KM, Mackay CR. Diet, gut microbiota and immune responses. Nat Immunol 2011;12(1):5–9.

[46] Oh DY, Talukdar S, Bae EJ, et al. GPR120 is an omega-3 fatty acid receptor mediating potent anti-inflammatory and insulin-sensitizing effects. Cell 2010;142(5):687–98.

[47] Swann JR, Want EJ, Geier FM, et al. Systemic gut microbial modulation of bile acid metabolism in host tissue compartments. Proc Natl Acad Sci USA 2011;108(Suppl. 1):4523–30.

[48] Lake AD, Novak P, Shipkova P, et al. Decreased hepatotoxic bile acid composition and altered synthesis in progressive human nonalcoholic fatty liver disease. Toxicol Appl Pharmacol 2013;268(2):132–40.

[49] Mouzaki M, Comelli EM, Arendt BM, et al. Intestinal microbiota in patients with nonalcoholic fatty liver disease. Hepatology 2013;58(1):120–7.

[50] Schnabl B, Brenner DA. Interactions between the intestinal microbiome and liver diseases. Gastroenterology 2014;146(6):1513–24.

[51] Sarafian MH, Lewis MR, Pechlivanis A, et al. Bile acid profiling and quantification in biofluids using ultra-performance liquid chromatography tandem mass spectrometry. Anal Chem 2015;87(19):9662–70.

[52] Bernstein H, Bernstein C, Payne CM, Dvorakova K, Garewal H. Bile acids as carcinogens in human gastrointestinal cancers. Mutat Res 2005;589(1):47–65.

[53] Yoshimoto S, Loo TM, Atarashi K, et al. Obesity-induced gut microbial metabolite promotes liver cancer through senescence secretome. Nature 2013;499(7456):97–101.

[54] Plotnikoff GA. Three measurable and modifiable enteric microbial biotransformations relevant to cancer prevention and treatment. Glob Adv Health Med 2014;3(3):33–43.

[55] Kuipers F, Bloks VW, Groen AK. Beyond intestinal soap—bile acids in metabolic control. Nat Rev Endocrinol 2014;10(8):488–98.

[56] Duboc H, Rajca S, Rainteau D, et al. Connecting dysbiosis, bile-acid dysmetabolism and gut inflammation in inflammatory bowel diseases. Gut 2013;62(4):531–9.

[57] Chapman CG, Rubin DT. The potential for medical therapy to reduce the risk of colorectal cancer and optimize surveillance in inflammatory bowel disease. Gastrointest Endosc Clin N Am 2014;24(3): 353–65.

[58] Jones H, Alpini G, Francis H. Bile acid signaling and biliary functions. Acta Pharm Sin B 2015;5(2):123–8.

[59] Joyce SA, Gahan CG. The gut microbiota and the metabolic health of the host. Curr Opin Gastroenterol 2014;30(2):120–7.

[60] Dumas ME, Barton RH, Toye A, et al. Metabolic profiling reveals a contribution of gut microbiota to fatty liver phenotype in insulin-resistant mice. Proc Natl Acad Sci USA 2006;103(33):12511–6.

[61] Li JV, Ashrafian H, Bueter M, et al. Metabolic surgery profoundly influences gut microbial-host metabolic cross-talk. Gut 2011;60(9):1214–23.

[62] Tremaroli V, Karlsson F, Werling M, et al. Roux-en-Y gastric bypass and vertical banded gastroplasty induce long-term changes on the human gut microbiome contributing to fat mass regulation. Cell Metab 2015;22(2):228–38.

[63] Neufeld KM, Kang N, Bienenstock J, Foster JA. Reduced anxiety-like behavior and central neurochemical change in germ-free mice. Neurogastroenterol Motil 2011;23(3):255–64. e119.

[64] Ochner CN, Gibson C, Shanik M, Goel V, Geliebter A. Changes in neurohormonal gut peptides following bariatric surgery. Int J Obes 2011;35(2):153–66.

Chapter 29

Relationship Between Gut Microbiota, Energy Metabolism, and Obesity

G.J. Bakker* and M. Nieuwdorp**,†,‡

*Department of Vascular Medicine, Academic Medical Center, Amsterdam, The Netherlands; **Department of Vascular Medicine, Academic Medical Center, Amsterdam, The Netherlands; †Department of Internal Medicine, VUmc Diabetes Center, Free University Medical Center, Amsterdam, The Netherlands; ‡Wallenberg Laboratory, University of Gothenberg, Gothenberg, Sweden

INTRODUCTION

The prevalence of overweight [body mass index (BMI) ≥ 25 kg/m^2] and obesity (BMI ≥ 30 kg/m^2) has been steadily increasing over the past decades. In 2014, more than 1.9 billion adults worldwide were overweight, of whom over 600 million were obese. Although overweight and obesity are most prevalent in high-income countries, the rate of increase has been highest in low- and middle-income countries, causing experts to term obesity a global epidemic.

As obesity is a major risk factor for a plethora of non-communicable diseases, including coronary heart disease, hypertension, stroke, diabetes, and certain types of cancer, the World Health Organization (WHO) has aimed to halt the rise of obesity, focusing especially on childhood obesity.

Underlying the development of obesity are genetic factors and environmental factors, including an increased intake of high-fat diets and a sedentary lifestyle. Recently, there has been increased interest for the role of the gut microbiota in the development of obesity and its associated complications.

The intestinal microbiota is a complex organ, consisting of 10^{14} bacteria distributed across the entire gut, with the density lowest in the acidic environment of the proximal intestine [approximately 10^4 colony forming units (CFU)/mL] and highest in the distal intestine (approximately 10^{12} CFU/mL). It has several important functions, including digestion of otherwise indigestible food substrates, production of hormones, protection against invasive bacterial strains, and training of the immune system, and these functions differ based on intestinal segment. In this chapter we will focus on the mechanisms by which the gut microbiota may influence energy metabolism. Further research into this area may contribute to the development of novel therapeutic targets aimed at preventing obesity and its consequences.

THE GUT MICROBIOTA EXTRACTS ENERGY FROM THE DIET

Humans and their gut microbes have evolved to live in a symbiotic relationship. While our intestines provide a warm, anaerobic environment with a steady flow of nutrients, the trillions of microbes living in our gut play a significant role in digestion and absorption of nutrients. These bacteria help us to metabolize otherwise indigestible food components including fibers into energy-rich substrates, such as short-chain fatty acids (SCFAs) of which butyrate, propionate, and acetate are the most abundant, and that can be utilized by the host. Moreover, the gut microbiota promotes absorption of monosaccharides (Fig. 29.1).

The role of extraction of energy from the food was shown by early experiments using germ-free mice. These animals are born and raised in a sterile environment, and thus have no gut microbiota. Transplanting the gut microbiota from conventionally raised mice into germ-free recipients resulted in an increase in body fat and insulin resistance without an increase in food consumption [1].

Thus, it was concluded that the gut microbiota helps to extract energy from food and may be involved in the development of obesity. These first experiments led to the gut microbiota as a whole new field of research. Metagenomic and biochemical analyses in large cohorts have now established a correlation between the obese phenotype and the composition of the gut microbiota in humans. Specifically, decreases in the Bacteroidetes phylum and increases in the Firmicutes and Proteobacteria phyla are associated with obesity [2]. Moreover, bacterial composition contributes to variations in lipid metabolism [3].

Further research focuses on interventions aimed at manipulating the gut microbiota. For example, both mice fed a high-fed diet and genetically obese ob/ob mice that were

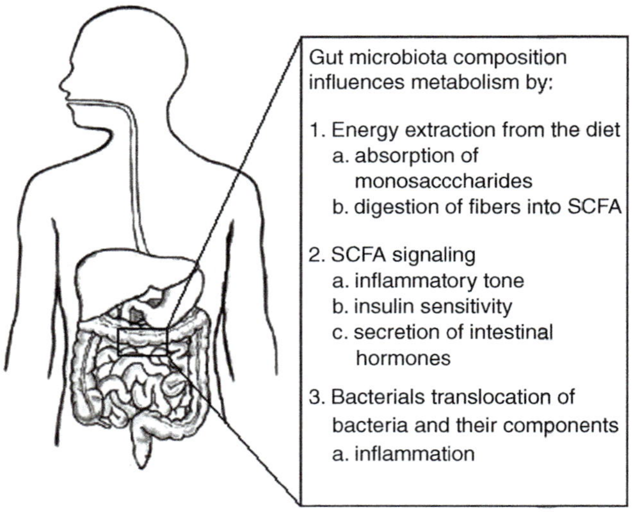

FIGURE 29.1 **The main ways in which the gut microbiota influences metabolism.**

treated with antibiotics for 4 weeks were partially protected against the development of metabolic diseases [4].

Fecal microbiota transplant (FMT) in mice showed that obesity can be transmitted, indicating a causal relationship between the gut microbiota and development of obesity. FMT studies may also be used as a way of finding new correlations. For example, in a randomized controlled trial, allogenic FMT from lean donors increased insulin sensitivity in metabolic syndrome patients in comparison to autologous FMT [5]. The allogenic group had several species-level differences in gut microbiota composition, providing novel correlations between microbiota composition and insulin sensitivity. If specific strains can be correlated to clinical parameters, this could result in novel therapeutic targets. Thus, beneficial bacteria may be given to patients as a dietary supplement (i.e., probiotics). On the other hand, targeting specific harmful bacteria, for example, by using vaccination could be an exciting new strategy to treat or even prevent metabolic disease.

REGULATION OF FAT STORAGE THROUGH PRODUCTION OF SHORT-CHAIN FATTY ACIDS

Apart from direct energy harvesting from the diet, the gut microbiota is involved in energy expenditure via a number of different mechanisms, most importantly via the production of SCFAs. Complex carbohydrates or fibers, such as cellulose, resistant starch, and inulin, cannot be degraded by human enzymes. Instead, they are metabolized by intestinal bacteria into absorbable SCFAs, mainly butyrate ($\pm 15\%$), propionate ($\pm 25\%$), and acetate ($\pm 60\%$) in a process called fermentation. These SCFA not only serve as energy substrates for microbes and host tissues, but also serve as signaling molecules to have a profound effect on host metabolism.

Butyrate is the primary energy substrate for colonic epithelium. Moreover, it has a trophic effect on the mucosa and stimulates intestinal mucus production, leading to improved gut barrier function. Apart from these direct effects on the colon, it has several beneficial metabolic effects, including increased insulin sensitivity and reduced adipose tissue inflammation. Differences in butyrate production are thought to play a role in metabolic regulation. For example, lower fecal concentrations of the butyrate-producing *Roseburia* and *Faecalibacterium prausnitzii* was highly discriminant of prevalent type 2 diabetes [6,7]. Moreover, the improvement in insulin sensitivity seen in metabolic syndrome patients after lean donor FMT was coupled with increases in butyrate-producing *Roseburia* and *Eubacterium hallii* [5].

The beneficial properties of butyrate have led to several studies involving oral administration of either sodiumbutyrate or probiotics containing bacteria that are known to produce butyrate, with promising results in animal studies [8]. Trials in humans are still ongoing at this moment.

The other SCFAs also have several effects on metabolism. For example, propionate and acetate are important substrates for gluconeogenesis and lipogenesis, respectively. Propionate improves insulin sensitivity and inhibits hepatic gluconeogenesis from lactate [9].

The SCFAs also control colonic gene expression by inhibiting histone deacetylases (HDAC). Moreover, SCFAs bind to the G-protein coupled receptors GPR41 and GPR43 (also known as FFAR3 and FFAR 2, respectively), which results in increased expression of peptide YY (PYY), an enteroendocrine hormone that inhibits gut motility, increases intestinal transit time, and decreases appetite [10]. Binding to these G-protein coupled receptors also induces GLP-1 secretion, another enteroendocrine hormone that improves insulin sensitivity and decreases appetite. Interestingly, supplementation with prebiotics blunted the overexpression of GPR43 in high-fat fed mice, resulting in lower fat mass [11]. Thus, the production of SCFAs by the gut microbiota is an important source of energy from the diet as well as an important mediator in gut motility, appetite, and the regulation of fat storage.

BACTERIAL TRANSLOCATION CONTRIBUTES TO OBESITY AND ASSOCIATED DISORDERS THROUGH LOW-GRADE INFLAMMATION

Obesity and type 2 diabetes mellitus (T2DM), a condition strongly associated with obesity, are characterized by chronic low-grade systemic inflammation. Especially visceral adipose tissue inflammation is strongly associated with insulin resistance. Although the triggering factor for the inflammatory response is still unknown, it is now generally accepted that translocation of intestinal bacteria into blood and visceral adipose tissue is an important underlying mechanism in the development of obesity and T2DM.

Translocation of intestinal bacteria encompasses a broad spectrum of mechanisms. Thus, whole bacteria or their metabolites may cross the intestinal barrier and enter the blood stream. For example, bacterial lipopolysaccharide (LPS), a component of the cell wall of Gram-negative bacteria, is a strong inducer of the innate immune system cascade that drives an inflammatory response; LPS binding protein can be readily measured in plasma as marker of translocation and has been associated with development of metabolic syndrome and type 2 diabetes [12].

Another way of measuring translocation of inflammation-inducing bacteria is direct measurement of LPS. Indeed, several studies showed plasma LPS, also called endotoxin, to be associated with insulin resistance. Moreover, chronic infusion with LPS causes inflammation and insulin resistance in mice [13]. Interestingly, both LPS and bacterial DNA have been shown in visceral adipose tissue of obese mice [14], suggesting that bacterial translocation might occur in specific organs where (dead) bacteria drive a local inflammatory response.

The ability of the gut wall to prevent bacterial translocation is termed gut barrier function. Increased intestinal permeability resulting from a decrease in gut barrier function can be measured in a variety of ways. For example, pathologic examination of intestinal mucosa may reveal a decrease in tight junction proteins. Moreover, an Ussing chamber may be used to quantify the passage of ions, nutrients, or drugs across intestinal epithelium. However, routinely sampling multiple mucosal biopsies may be too invasive to be used on large scale in mechanistic clinical trials. Another functional way of testing intestinal permeability in vivo is oral ingestion of oligosaccharides of different sized, such as mannitol and saccharose followed by determination of urinary excretion. The ratio of urinary excretion of these saccharides reflects permeability of the intestinal wall. Unfortunately, the clinical significance and reproducibility of measuring intestinal permeability with these small molecules is currently unknown. Although mono- or disaccharides may readily pass the mucosa under some circumstances, it is unknown to what extend this also applies to bacteria or LPS. Alternatively, direct measurement of bacterial DNA may be used to measure bacterial translocation. Importantly, bacterial DNA isolated from blood or visceral adipose tissue can be sequenced to determine which specific bacterial strains are involved in translocation. Although not yet widely adopted, this approach may lead to novel therapeutic targets for obesity and T2DM including vaccination and phage therapy.

Although bacterial translocation is a well established and often described concept, interestingly it is largely unknown where translocated bacteria cross the intestinal wall. One possibility is the colon, where the concentration of bacteria is up to 100 times higher than in the small intestine. However, the colon is relatively metabolically inactive.

Conversely, in the duodenum a plethora of metabolic activities are regulated. Moreover, the absorption of food particles in the small intestine may present an opportunity for pathogens to cross the intestinal wall.

Dietary intake is an important determinant of gut microbiota composition. After consuming a diet high in fat and low in fibers, large shifts in gut microbiota composition can be seen within 24 h [15]. While excessive fat intake is a clear risk factor for a variety of inflammatory diseases, including obesity and T2DM, the exact mechanisms linking diet-induced changes in gut microbiota composition to increased bacterial translocation and inflammation are largely unknown. However, ingestion of a high-fat meal leads to a systemic inflammatory response. Considering that LPS has a high affinity for chylomicrons, derived from dietary fatty acids and transported through the intestinal wall, it is hypothesized that bacterial translocation can be induced by uptake of dietary lipids. This hypothesis is supported by in vivo and in vitro studies that showed that chylomicron formation promotes intestinal LPS absorption and that plasma LPS levels correlate to postprandial triglycerides changes.

In conclusion, bacterial translocation could represent a very interesting direction for future therapies, especially if specific bacterial strains are found to be more prone to translocation. For example, targeted vaccination against those strains, replenishing gut microbiota balance by pro- or prebiotics, or targeting molecular pathways may be developed in the coming years. However, at this point the causality of specific intestinal microbial strains in obesity and T2DM is not yet proven.

THE ANGPTL4 AND AMPK PATHWAYS ARE INVOLVED IN FAT STORAGE IN GERM-FREE MICE

Angiopoietin-like factor IV (ANGPTL4), previously known as fasting-induced adipose factor, is a circulating enzyme produced by the intestine, liver, and adipose tissue that inhibits lipoprotein lipase (LPL) activity. Colonizing germ-free mice with a gut microbiota of conventionally raised mice (also known as "conventionalization") does not only lead to an increase in fat mass, but also results in a decrease in intestinal expression of ANGPTL4, resulting in increased LPL activity in adipose tissue. Thus, inhibition of ANGPTL4 has been proposed as another mechanism by which the gut microbiota may stimulate accumulation of adipose tissue.

The ANGPTL4 hypothesis has been supported by several experiments. For example, *Lactobacillus paracasei* induced an upregulation of ANGPTL4 in colonic epithelial cells, while oral inoculation of germ-free mice with this bacterium resulted in increased circulating levels of ANGPTL4. Moreover, germ-free ANGPTL4-knockout was able to contain the same amount of total body fat as conventionalized mice [1].

Another pathway that has been proposed to play a role in fat storage involves phosphorylated AMP-activated protein kinase (AMPK), an enzyme that stimulates fatty acid oxidation. Germ-free mice that are resistant to diet-induced obesity have increased skeletal muscle and hepatic AMPK. Moreover, they have increased expression of the downstream targets of AMPK that are involved in fatty acid oxidation.

Despite these interesting early findings, similar experiments investigating these pathways have not reproduced consistent results [16]. Moreover, most of these experiments were done in germ-free mice. Thus, while the ANG-PTL4 and AMPK theories are supported by some evidence, it is unclear to what extend the findings apply to physiological circumstances in humans.

THE GUT MICROBIOTA PRESENTS AN IMPORTANT HOPE FOR FUTURE TREATMENT OPTIONS IN OBESITY

In summary, the gut microbiota plays an important role in energy homeostasis through several mechanisms, most importantly energy extraction from the diet and regulation of fat storage through SCFAs. Moreover, accumulating evidence strongly suggests bacterial translocation and systemic inflammation to be associated with obesity and insulin resistance.

Nevertheless, it should be emphasized that there are still many unknown gaps. Importantly, while many differences in gut microbiota composition between obese and lean humans have been characterized, it is unclear to what extend these differences have a causal origin. Moreover, the impact of different interventions on the gut microbiota is unclear in many cases. Much research has been performed in animals, and the relevance to human metabolism has to be further explored in human trials.

Although the gut microbiota is certainly not the only factor in weight gain and the development of metabolic disease, it represents an important novel therapeutic target. Manipulation of the gut microbiota through various interventions, such as probiotics, prebiotics, or fecal transplantation may yield novel insights into the link between the gut microbiota and disease (Fig. 29.2).

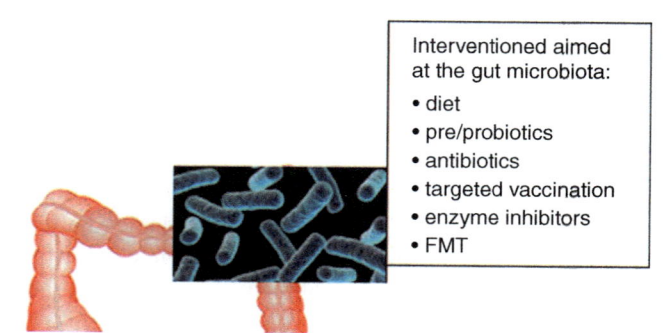

Interventioned aimed at the gut microbiota:
- diet
- pre/probiotics
- antibiotics
- targeted vaccination
- enzyme inhibitors
- FMT

FIGURE 29.2 **Interventions aimed at the gut microbiota that may be investigated in future research.**

Moreover, with more and more research investigating the role of intestinal microbes in the onset of disease, future treatments could be directed at specific bacterial strains. Thus, further exploring the exciting avenues that were investigated in the past years could help fight the growing population of obese patients.

REFERENCES

[1] Backhed F, Ding H, Wang T, et al. The gut microbiota as an environmental factor that regulates fat storage. Proc Natl Acad Sci USA 2004;101(44):15718–23.

[2] Verdam FJ, Fuentes S, de Jonge C, et al. Human intestinal microbiota composition is associated with local and systemic inflammation in obesity. Obesity 2013;21(12):E607–E6015.

[3] Fu J, Bonder MJ, Cenit MC, et al. The gut microbiome contributes to a substantial proportion of the variation in blood lipids. Circ Res 2015;117(9):817–24.

[4] Cani PD, Neyrinck AM, Fava F, et al. Selective increases of bifidobacteria in gut microflora improve high-fat-diet-induced diabetes in mice through a mechanism associated with endotoxaemia. Diabetologia 2007;50(11):2374–83.

[5] Vrieze A, Van Nood E, Holleman F, et al. Transfer of intestinal microbiota from lean donors increases insulin sensitivity in individuals with metabolic syndrome. Gastroenterology 2012;143(4):913–6. e7.

[6] Karlsson FH, Tremaroli V, Nookaew I, et al. Gut metagenome in European women with normal, impaired and diabetic glucose control. Nature 2013;498(7452):99–103.

[7] Qin J, Li Y, Cai Z, et al. A metagenome-wide association study of gut microbiota in type 2 diabetes. Nature 2012;490(7418):55–60.

[8] Gao Z, Yin J, Zhang J, et al. Butyrate improves insulin sensitivity and increases energy expenditure in mice. Diabetes 2009;58(7): 1509–17.

[9] Schwiertz A, Taras D, Schafer K, et al. Microbiota and SCFA in lean and overweight healthy subjects. Obesity 2010;18(1):190–5.

[10] Samuel BS, Shaito A, Motoike T, et al. Effects of the gut microbiota on host adiposity are modulated by the short-chain fatty-acid binding G protein-coupled receptor, Gpr41. Proc Natl Acad Sci USA 2008;105(43):16767–72.

[11] Dewulf EM, Cani PD, Neyrinck AM, et al. Inulin-type fructans with prebiotic properties counteract GPR43 overexpression and PPAR-gamma-related adipogenesis in the white adipose tissue of high-fat diet-fed mice. J Nutr Biochem 2011;22(8):712–22.

[12] Sun L, Yu Z, Ye X, et al. A marker of endotoxemia is associated with obesity and related metabolic disorders in apparently healthy Chinese. Diabetes Care 2010;33(9):1925–32.

[13] Cani PD, Amar J, Iglesias MA, et al. Metabolic endotoxemia initiates obesity and insulin resistance. Diabetes 2007;56(7):1761–72.

[14] Amar J, Chabo C, Waget A, et al. Intestinal mucosal adherence and translocation of commensal bacteria at the early onset of type 2 diabetes: molecular mechanisms and probiotic treatment. EMBO Mol-Med 2011;3(9):559–72.

[15] David LA, Maurice CF, Carmody RN, et al. Diet rapidly and reproducibly alters the human gut microbiome. Nature 2014;505(7484): 559–63.

[16] Fleissner CK, Huebel N, Abd El-Bary MM, et al. Absence of intestinal microbiota does not protect mice from diet-induced obesity. Br J Nutr 2010;104(6):919–29.

Chapter 30

Taxonomic and Metagenomic Alterations of Microbiota in Bariatric Surgery

N. Floch*,**,†,‡

*Department of Surgery, Norwalk Hospital, Norwalk, CT, United States; **Fairfield County Bariatrics and Surgical Specialists, P.C, Norwalk, CT, United States; †Associate Clinical Professor of Surgery, University of Vermont School of Medicine, Burlington, VT, United States; ‡Associate Clinical Professor of Surgery, The Netter School of Medicine, Quinnipiac University, Hamden, CT, United States

INTRODUCTION

Obesity has increased globally over the past 40 years. The obesity epidemic cannot be explained by an increase in the quantity and type of food or a decrease in human activity alone. Evidence has begun to elucidate the multiple factors that have led to the obesity epidemic [1]. Although some genetic, epigenetic, and environmental factors have been identified, the causes of obesity cannot fully be explained. The contribution of each factor to the causation of obesity is a subject of ongoing research. Pertinent to this topic is the discovery of our microbiome as a contributor to obesity and metabolic disease. Over 90% of the gut microbiome is composed of Firmicutes (Gram-positive), Bacteroidetes (Gram-negative), and Actinobacteria (Gram-positive) [2].

Humans experience a symbiotic relationship with their intestinal microbiome. The microbiota react with their hosts to: influence the inflammatory response from the intestine, synthesize proteins and molecules, and ultimately alter the energy absorbed by the gut. Diet dictates the quantity and type of energy available to the gut [3].

TECHNOLOGY

Advances in technology have led to the ability to identify both the taxonomy of our microbiota and their specific metabolic pathways. The standard DNA-based method to identify a human microbiome was using a microarray of microbial 16S ribosomal RNA (rRNA), which is a unique RNA region that is specific for each microbe. This technique was rapid but missed the identification of previously unknown microbiota [4,5]. Whole-genome shotgun sequencing can identify all microbiota, their DNA, the species, and genes present in a sample [6–8].

Metagenomic sequencing can determine a complete qualitative and quantitative analysis of the intestinal microbiome. Species of low abundance can be identified. A microbial gene catalogue of the human gut has now been established, identifying 3.3 million unique microbial genes as taken from the feces of European individuals [7]. Metabolomics may be the most useful technology for analyzing complicated disease processes, such as obesity and type 2 diabetes (T2D) which involve microbiota, metagenomics, and host cross talk [9–11].

OBESITY AND MICROBIOTA

There is a relationship between obesity and the composition of the gut microbiome [2,12,13]. The gut microbial communities are altered in those who are obese compared to lean individuals [14]. Obese people have an increased ratio of Firmicutes to *Bacteroides*, which has been found to lead to lower levels of serum bile acids, elevated inflammation, and a change in levels of gut hormones, such as ghrelin and leptin [15,16]. When weight is lost in both mice and humans, the ratio of Firmicutes to Bacteroidetes decreases or reverses [17]. Continued weight loss in humans will result in increasing quantities of *Bacteroides* [2,13,14,17,18]. Only two studies have refuted the change in Firmicutes to *Bacteroides* ratio [19]. These findings have been repeated numerous times and now establish the most basic relationship between the gut microbiome and obesity [14,17,19,20–32].

An analysis of the microbiome of twins reveals a lower quantity of Bacteroidetes and a larger amount of Actinobacteria in the obese twin compared with the thinner twin. The study revealed no difference in Firmicutes [18]. Thin individuals have markedly different microbiota abundances than those who are obese [18]. Obese individuals have also been found to have less diversity of bacteria and richness of genes [14]. Healthy individuals and those with T2D have different abundances of microbial communities [33]. A less diverse microbiome is more likely to be associated with obesity, insulin resistance, and dyslipidemia [18,33]. Whether the taxonomy of bacteria found in obese mice and humans, as well as their characteristic metagenomic fingerprints are a result of obesity or are a part of the cause is still unclear [2].

Manipulation of the gut microbiome can result in weight gain and obesity. Fortunately, changing diet can lead to an alteration in the abundances of certain microbiota. Bariatric surgery may create dramatic changes in not only the abundances of microbiota, as well as a reversal of the *Bacteroides* to Firmicutes ratio but may increase serum bile acid levels, alter gut hormones and reduce intestinal inflammation. Many of the benefits of bariatric surgery, such as weight loss, the remission of T2D, and improvement of metabolic disorders may be attributed to a change in the gut microbiota communities that result in metagenomic changes which are believed to cause improved host responses.

STUDIES ON BARIATRIC SURGERY AND THE MICROBIOME

Microbiota and Bariatric Surgery

Multiple studies have now been performed that reveal the taxonomic and metagenomic changes that occur in the gut microbiota of humans before and after bariatric surgery. These studies build upon a framework of studies performed on animal models. Although human trials usually mimic animal studies, not all data is consistent. The discovery of changes in the microbiota of postbariatric surgery patients that may induce long-standing weight loss and metabolic improvement are limited on the topic.

The evidence concerning the gut microbial changes after bariatric surgery include mostly studies on roux-en-y gastric bypass (RYGB) [29–32,34–37], but also include data on vertical banded gastroplasty (VBG) [40] and vertical sleeve gastrectomy (VSG) [37,39,40]. There are no studies found on the microbiome and adjustable gastric banding or duodenal switch surgery. The data supports alterations in weight, metabolic disease, and inflammation [29].

Animal Studies on Bariatric Surgery

1. The first animal study involving bariatric surgery was performed on nonobese Wistar rats without any metabolic dysfunction [31]. The post-RYGB rats were compared to a sham group that underwent a gastrostomy and jejunotomy. The experiment revealed that post-RYGB rats lost weight as a result of the surgery and voluntarily decreased caloric intake. The study investigated a population without metabolic dysfunction, looking for metagenomic and metabolic influencing factors. The study was consistent with other human studies and resulted in increased abundance of Gammaproteobacteria in the RYGB rats. Other studies showed similar results in humans [29,30,35,36]. The procedure resulted in increased production of amines and p-cresol, as well as oligosaccharide fermentation.

2. In a similar study performed on the same animals with overlapping data [32], the microbiota that colonized the post-RYGB intestines produced urinary phenylacetylglycine, indoxyl sulfate, fecal gamma-aminobutyric acid, putrescine, tyramine, and uracil. These substances were detected in the animals' urine and stool. They were found to be toxic to the animals and resulted in decreased survival. The study raised the question of the long-term cancer risk of bariatric surgery [32].

3. Mouse stool was analyzed by 16S ribosomal RNA gene sequencing in three groups: post-RYGB surgery, sham surgery, and a sham surgery combined with caloric restriction [34]. Changes in intestinal microbiota post-RYGB appeared to be conserved among humans and animals [34]. Relative abundance of taxonomic groups within the phylum Firmicutes were similar in weight loss controls and post-RYGB. The first week postoperative and sustained was an increase in the relative abundance of Gammaproteobacteria (*Escherichia*) and Verrucomicrobia (genus: *Akkermansia*) [34]. This was also found in other studies [29,30]. Changes in abundance were independent of weight change and caloric restriction. The effect of RYGB on the gut microbiota was the same whether the post-RYGB mice were fed a high fat diet (HFD) or normal chow (NC) indicating that post-RYGB abundance changes dominate in impact on the gut microbiome composition over diet alterations. Microbial abundance alterations were present throughout the gastrointestinal tract but were most pronounced in the distal gut, beyond the furthest surgical anastomosis. Although the mouse control group that lost weight had improvement of fasting glucose and insulin levels, there was a significantly greater level of improvement in insulin sensitivity and glucose tolerance in the post-RYGB mice. The data supports mechanisms beyond weight loss that improve metabolic disease [34].

When the intestinal microbiota from post-RYGB mice was transplanted into nonoperated, germ-free mice the result was weight loss and adipose mass reduction [34]. It was believed to be secondary to an alteration of the microbial production of short-chain fatty acids (SCFAs). The findings were the first evidence that support that alterations of intestinal microbiota are at least partially responsible for the reduction of weight and adiposity after RYGB surgery in the animal host and are independent of diet change or transection of the intestine.

Verrucomicrobium *Akkermansia* were significantly increased in abundance in post-RYGB mice, as well as after transplant to germ-free mice. *Akkermansia* have been found to use their mucous as a source of carbon and nitrogen when calories are very limited [41]. *Akkermansia* was not only increased in abundance in the mucus layer of weight loss control mice, but throughout the entire intestine in the post-RYGB mice [34]. It is possible that *Akkermansia* are responsible for altered weight loss, adipose tissue, inflammation, and insulin sensitivity after RYGB [34,42,43].

4. Farnesoid X receptor (FXR) was investigated as a molecular target to explain the effectiveness of the VSG [39], but the major finding of this mouse study was an association of the receptor with alterations in the gut microbiota. The bile receptor, FXR was found to be essential in causing the reduced body weight and glucose tolerance associated with the VSG. Evidence supports changes in microbial communities to contribute to the benefits of bariatric surgery [29,30,34,36,44].

Bariatric surgery alters the quantity of varied microbial communities so it is unclear which specific species are most influential in the changes that occur in adiposity, weight loss, and metabolic function [39]. The study compared the effects of wild-type (WT) versus FXR knockout (KO) mice to see which microbiota may be responsible for the beneficial effects of FXR [39]. It is known that bile acids have the ability to change the abundance and types of microbiota in the gut either directly or by activating the FXR receptor [44–46]. The gut microbiota can modulate bile acid composition, specifically with the assistance of enzymes produced by the microbes in the intestinal lumen. Through this activity, the microbiota have the ability to alter bile acid homeostasis and FXR receptor signaling [47,48].

The genomic DNA from the cecal contents of high-fat diet (HFD) fed, wild type, KO-sham, and FXR KO mice were compared 14 weeks after VSG surgery. VSG was found to alter the relative abundance of certain microbiota that have been identified as modifiers of systemic metabolism [39]. The abundance of *Bacteroides* was decreased in WT mice compared to sham-operated mice. This reduction was associated with reduced weight, reduced fat, and improved glucose tolerance indicating that the change is associated with the FXR receptor. There was a genus of Porphyromonadaceae family, which was increased in abundance; it is usually increased in the presence of a HFD and type 1 diabetes [49].

Roseburia was 12 times higher in the WT-VSG compared to sham-VSG mice [39]. Metagenomic analyses of both European and Chinese populations revealed decreased amounts of *Roseburia* in humans with T2D [50,51]. Increased *Roseburia* has also been found in probiotic treated patients with weight loss and improved glucose tolerance [52]. The abundance of *Roseburia* was found to be associated with an improved glucose tolerance test in WT mice undergoing VSG, a finding associated with FXR-dependent actions [39]. There were also increased abundance of genera *Lactobacillus* and *Lactococcus* of Enterobacteriaceae. An increase in *E. coli* has been found in most studies on post bariatric subjects [29,30,34]. It is unclear whether the increase in Enterobacteriaceae results is a benefit to the VSG procedure in obese patients [39].

Gut microbiota can help to stimulate the metabolism of the host by the fermentation of fatty acids in the colon. SCFAs have been demonstrated to alter the host metabolism [50,51,53,54]. There was a decrease in the acetate to butyrate ratio, indicating that acetate is more efficiently converted to butyrate after VSG [39]. In the study, VSG KO and WT mice had equivalent size stomachs, yet only the WT mice had demonstrated weight loss and metabolic improvements. The findings contradict traditional thought that the small stomach or restriction is responsible for weight loss and metabolic changes that occurs after VSG [39]. Evidence supports altered FXR signaling for bile acids after VSG to be responsible for weight loss and metabolic changes [39].

Human Studies

1. The initial study of analyzing bariatric surgery and the microbiome in humans compared three groups of three patients: one normal weight, one morbidly obese, and the third group had undergone RYGB [29]. The RYGB patients were compared to the obese patients as their control group, not to themselves prior to surgery. The fecal samples of all groups were analyzed using the 16S rRNA gene technique. The study focused on the taxonomy and diversity of each subject's gut microbiome but not metagenomics. The patients lost weight after RYGB as expected. A significant alteration in the relative abundances of microbiota was found in the post-RYGB patients [29].

The post-RYGB patients had more relative abundant levels of the Enterobacteriaceae, Fusobacteriaceae and *Akkermansia* families. The data identified an increased quantity of Gammaproteobacteria and decreased amount of Clostridia in the post-RYGB patients. The obese controls had higher amounts of microbiota in the domain of Archaea as compared to the nonobese and post-RYGB surgical patients [29]. Verrucomicrobia were found in large quantities in both normal and morbidly obese individuals but were rarely found in the post-RYGB patients [29].

This study established that the microbiota in fecal samples of humans undergoing RYGB were distinctly different from those of obese and nonobese individuals. What was not clear was whether the change in taxonomy and diversity of gut microbiota was the cause or result of the weight loss from bariatric surgery or from alteration in caloric intake or diet. Although this study was helpful in identifying differences in taxonomy, it was limited by a small sampling size of patients. Using the fecal samples of the same group of patients preoperatively and comparing them to the samples collected postoperatively, would have created a much stronger analysis [29].

2. In 2010, Furet et al. compared the microbiota samples of 13 thin and 30 obese patients. Seven of the 30 obese individuals were diagnosed with T2D. The study was the first analysis to demonstrate the change in the type of microbiota that occurred after bariatric surgery using the preoperative patients as a control group [30].

Targeted polymerase reaction was used to determine the presence of 7 different groups of microbiota in the specimens of a possible 1000. The study showed that there was an increase in the ratio of *Bacteroides* to Firmicutes before surgery, which decreased at 3 and 6 months postsurgery, was believed to be directly related to obesity and was depended on the amount of calories consumed [30]. The change in the ratio of *Bacteroides/* Firmicutes was more specifically a change in the ratio of *Bacteroides* to *Prevotella* and was consistent with the findings in previous studies [17,18,30]. The study also shed light on hormone alteration. When the ratio of *Bacteroides/Prevotella* increased, the bmi and leptin levels decreased after 3 months post-RYBG. The ratio of *Bacteroides/Prevotella* stabilized at 6 months while BMI, body fat mass, and the size of adipocytes continued to decrease [30].

E. coli levels also increased in relative quantity consistent with other studies [29,30]. *Bifidobacterium* and *Lactobacillus*, *Leuconostoc*, and *Pediococcus* relative levels decreased [30,55]. The quantity of total microbiota, were not found to be decreased, as the patients lost weight after surgery [30]. Other studies have demonstrated a loss in total bacteria quantity after weight loss [22,23]. Specific microbiota groups were seen to be associated with changes in energy, body weight, and metabolism.

The microbe, *Faecalibacterium prausnitzii* was found to be less abundant in patients with T2D and obesity. When the quantity of this bacteria increased, the amount of inflammatory markers, such as hs-CRP, IL-6, and orsomucoid serum decreased. The trend continued at 3 and 6 months after surgery and it was independent of food or calorie intake. The study was performed in the morbidly obese population so findings in an obese group could be different [30].

3. A study investigated the ability of RYGB to modify the gut microbiota and the gene expression in white adipose tissue (WAT) [35]. Fecal samples were analyzed by pyro sequencing to investigate the genera of the microbiota. Morbidly obese patients were examined at baseline, 3 and 6 months after RYGB. WAT gene expression and clinical markers were studied at baseline and 3 months. A richness of gut microbiota was found after RYGB, with a higher diversity of 58 more bacteria genera [35]. In the postbariatric surgery group compared to the obese group, 37% of the microbiota were in the phylum Proteobacteria. Multiple microbial groups increased: *Bacteroides*, *Escherichia*, and *Alistipes* while four groups decreased: *Bifidobacterium*, *Blautia*, *Dorea*, and *Lactobacillus*. *Escherichia* genus was found to be abundant in the patients after surgery and was independent of changes in calorie intake [35].

The connection between bacterial types and WAT gene expression increased after surgery. In the preoperative control groups there were 8 WAT genes and 28 bacterial groups, which increased significantly to 562 WAT genes and 102 bacterial groups post-RYGB. Also, post-RYGB, 7 dominant and 7 nondominant bacterial genera changed, as well as 202 WAT genes [35]. The changes in microbial genera occurred in direct relationship with the clinical phenotype and expression of the WAT genes. Many of the genes that were increased post-RYGB were those that code for metabolic and inflammatory processes related to transforming growth factor-β signaling pathway. The downregulated genes, post-RYBG were involved in the metabolic pathways, such as 24-dehydrocholesterol reductase [35]. The most abundant microbiota in the postsurgical patients exhibited WAT genes relating to cell transport and binding, signaling, enzymatic activity and the structure of the cells. The study analyzed WAT gene expression and bariatric surgery, successfully elucidating the relationship between the two [35].

4. In 2013, Graessler demonstrated the changes of gut microbial gene content from fecal samples at baseline and 3 months after RYGB in six morbidly obese patients by metagenomic sequencing. Of the six patients, five had T2D and were taking medications. All six patients were hypertensive, four had hypertriglyceridemia, and five had elevated total cholesterol. The preoperative RYGB patients had evidence of inflammation with high–normal CRP values ranging from 2.1 to 5.6 mg/L, high uric acid levels in all patients and increased liver enzymes in three patients [36].

Three months post-RYGB, there was a 15–32% decrease in BMI, a 10–55 mmHg decrease in sbp, decrease in total cholesterol and LDL, and a reduction of triglycerides in five of six cases. There was a postoperative reduction of HbA_{1c}. There was also a decrease from preoperative levels in all patients for γ-glutamyltransferase and uric acid n 5 of 6 patients for CRP, ALAT, and total serum protein. The post-RYGB patients had significant metabolic improvements of glucose and lipid metabolism. There was an overall reduction in the inflammatory state [36].

Taxonomic analysis revealed 1061 species, 729 genera, 44 phyla, and 5127 KO of the KEGG Orthology. Bacteria were found to be diverse and could be grouped into seven bacterial divisions. Twenty-two species and 11 genera were changed in the post-RYGB group. Nine of the species were either increased or decreased in abundance and the change was independent of BMI change [36]. The post-RYGB weight loss occurred concomitant with a varying degree in the reduction of both Bacteroidetes and Firmicutes. The relative amount of Proteobacteria increased while the relative amount of Firmicutes decreased. Therefore, the Proteobacteria/Firmicutes ratio increased significantly as seen in multiple other studies [2,29]. Using component analysis, groups of multiple species of Firmicutes were highly correlated with BMI and C-reactive protein. This group was found to have an increased quantity of Proteobacterium *Enterobacter cancerogenus* and decreased amounts of Firmicutes *F. prausnitzii* and *Coprococcus comes* [36].

When carbohydrate metabolism was functionally analyzed by KO, the study revealed significant effects in 13 KOs assigned to the phosphotransferase system. Blood levels of total cholesterol and LDL were associated with 10 species and 5 species were associated with triglycerides. The presence of *F. prausnitzii* was directly related to the fasting blood glucose level [36].

Metagenomic sequencing of the human gut microbiome before and 3 months after bariatric surgery in obese patients with T2D, correlated with inflammatory and metabolic parameters [36]. Three months post-RYGB, there was a significant decrease in BMI and improvement of inflammatory markers in all six patients, as well as resolution of T2D and evidence of inflammatory activity after RYGB [36]. The findings support that metabolic alterations may not be secondary to reduced calorie intake, secondary to the RYGB but may be as a result of RYGB induced microbiota abundance changes [36]. This was the first study to compare the change in the intestinal microbiota and the metagenomic sequencing in individual patients preoperatively and postoperative after RYGB. There was a direct relationship between microbiota composition changes and gene function with an improvement in metabolic and inflammatory parameters [36].

5. The long-term effects of two types of bariatric surgery, randomized to RYGB or VBG were analyzed for their effects on the microbiome. The patients had matched control groups for weight and fat mass loss. The two surgical procedures were different in relation to changes in the bile and nutrient flow but induced similar and long-lasting alterations of the intestinal microbiome that were independent of BMI. The study resulted in a change in the levels of fecal and circulating metabolites when compared to the obese controls group [38]. The

stools from the postbariatric surgery patients were then used to colonize germ-free mice. The microbiota transplanted from the postbariatric patients into the germ-free mice created an environment where fat deposition was reduced. These mice demonstrated a decreased utilization of carbohydrates as fuel, indicated by a lower respiratory quotient. The results yield strong evidence after RYGB and VBG, that the gut microbiota may play a direct role in the reduction of adipose tissue [38]. Taxonomy changes of RYGB and VBG patients are long lasting, at least over 9 years [38].

6. Metabolomics analyses of metabolic alterations associated with obesity and changes that occur after bariatric surgery demonstrate unique patterns of metabolites that are characteristic of each individual type of bariatric surgery [37]. A longitudinal observational study used H-nuclear magnetic resonance-based global, untargeted metabolomics strategy on preoperative and postoperative human blood samples less than a year after a VSG, proximal RYGB, and distal RYGB. A nonsurgical group of similar BMI was used as a matched control group [37]. Metabolites that were found in severe obesity were elevated levels of aromatic and branched-chain amino acids. Metabolites related to energy metabolism were elevated levels of pyruvate and citrate. The postbariatric surgery gut microbiota were associated with elevated levels of formate, methanol, and isopropanol [37]. RYGB, long-limb gastric bypass and sleeve gastrectomy surgery reverse metabolic alterations associated with obesity and demonstrate significant changes in gut microbiome–host interactions after bariatric surgery [37].

7. Five patients on a very low calorie (VLCD) diet of approximately 800 kcal/d were compared to five patients post-VSG, all nondiabetic. Fecal analysis by metabolomics was performed preoperatively, at 3 and 6 months postoperative. Ten additional subjects in each group were analyzed for SCFA. Three patients in each group were analyzed for gut microbiota by shotgun sequencing. Fecal fatty acids and bile content were measured by mass spectrometry and SCFA by gas chromatography. The taxonomy levels of microbiota were recorded and fecal metabolomics were measured. The two groups of patients lost similar weight: 23.9% relative weight loss (RWL) for VSG group and 24.6% RWL for the VLCD group at 6 months [40]. The energy-harvesting ability of the intestinal microbiota decreased post-VSG. The Bacteroidetes/Firmicutes ratio increased after VSG but decreased after VLCD. The ability of the microbiota to ferment butyrate decreased after VSG and increased after VLCD. VSG resulted in fecal loss of nonesterified fatty acids and bile acids. VSG altered the gut microbiota to abundances to reflect a lean microbiome. There was increased malabsorption due to loss in energy-rich fecal substrates and impairment of bile acid circulation in the VSG patients [40].

Technological advances, such as shotgun sequencing have allowed researchers to compare both the taxonomy and metagenomic characteristics of the microbiome of animals and humans. There exist distinct differences in the relative abundances of microbiota in obese and lean animals and humans. Diet can rapidly change these abundances but bariatric surgery can alter microbial abundances with changes that last beyond 9 years in humans [38]. Further investigations in both animals and humans shall yield a greater understanding of the significance of taxonomy and metagenomic changes as they relate to individual species after bariatric surgery.

REFERENCES

[1] Frank J. Origins of the obesity pandemic can be analysed. Nature 2016;532(7598):149.

[2] DiBaise JK, Zhang H, Crowell MD, Krajmalnik-Brown R, Decker GA, Rittmann BE. Gut microbiota and its possible relationship with obesity. Mayo Clin Proc 2008;83:460–9.

[3] Sanmiguel C, Arapana G, Mayer EA. Gut microbiome and obesity: a plausible explanation for obesity. Curr Obes Rep 2015;4:250–61.

[4] Sweeney TE, Morton JM. The human gut microbiome: a review of the effect of obesity and surgically induced weight loss. JAMA Surg 2013;148(6):563–9.

[5] Marguiles M, Egholm M, Altman WE, et al. Genome sequencing in microfabricated high-density picolitre reactors. Nature 2005;437(7057):376–80. [Published correction appears in Nature 2006;441(7089):120].

[6] Gill SR, Pop M, Deboy RT, et al. Metagenomic analysis of the human distal gut microbiome. Science 2006;312(5778):1355–9.

[7] Qin J, Li R, Raes J, et al. A human gut microbial gene catalogue established by metagenomic sequencing. Nature 2010;464:59–65.

[8] Human Microbiome Project Consortium. Structure, function and diversity of the healthy human microbiome. Nature 2012;486(7402):207–14.

[9] Faber JH, Malmodin D, Toft H, et al. Metabonomics in diabetes research. J. Diabetes Sci Technol 2007;1:549–57.

[10] Llorach R, Garcia-Aloy M, Tulipani S, Vazquez-fresno R, Andreslacueva C. Nutrimetabolomic strategies to develop new biomarkers of intake and health effects. J Agric Food Chem 2012;60:8797–808.

[11] Du F, Virtue A, Wang H, Yang X-F. Metabolomic analyses for atherosclerosis, diabetes, and obesity. Biomark Res 2013;1:17.; Kurland IJ, Accili D, Burant C, et al. Application of combined omics platforms to accelerate biomedical discovery in diabesity. Ann NY Acad Sci 2013;1287:1–16.

[12] Flint HJ. Obesity and the gut microbiota. J Clin Gastroenterol 2011;45(Suppl.):S128–32.

[13] Angelakis E, Armougom F, Million M, Raoult D. The relationship between gut microbiota and weight gain in humans. Future Microbiol 2012;7:91–109.

[14] Ley RE, Backhed F, Turnbaugh P, Lozupone CA, Knight RD, Gordon JI. Obesity alters gut microbial ecology. Proc Natl Acad Sci USA 2005;102:11070–5.

[15] Musso G, Gambino R, Cassader M. Obesity, diabetes, and gut microbiota: the hygiene hypothesis expanded? Diab Care 2010;33:2277–84.

[16] Steinert RE, Peterli R, Keller S, et al. Bile acids and gut peptide secretion after bariatric surgery: a 1-year prospective randomized pilot trial. Obesity 2013;21:E660–668.

[17] Ley RE, Turnbaugh PJ, Klein S, Gordon JI. Microbial ecology: human gut microbes associated with obesity. Nature 2006;444(7122):1022–3.

[18] Turnbaugh PJ, Hamady M, Yatsunenko T, et al. A core gut microbiome in obese and lean twins. Nature 2009;457:480–4.

[19] Schwiertz A, Taras D, Schäfer K, et al. Microbiota and SCFA in lean and overweight healthy subjects. Obesity 2010;18(1):190–5.

[20] Armougom F, Henry M, Vialettes B, Raccah D, Raoult D. Monitoring bacterial community of human gut microbiota reveals an increase in Lactobacillus in obese patients and methanogens in anorexic patients. PLoS One 2009;4(9):e7125.

[21] Balamurugan R, George G, Kabeerdoss J, Hepsiba J, Chandragunasekaran AM, Ramakrishna BS. Quantitative differences in intestinal *Faecalibacterium prausnitzii* in obese Indian children. Br J Nutr 2010;103(3):335–8.

[22] Duncan SH, Lobley GE, Holtrop G, et al. Human colonic microbiota associated with diet, obesity and weight loss. Int J Obes 2008;32(11):1720–4.

[23] Nadal I, Santacruz A, Marcos A, et al. Shifts in Clostridia, *Bacteroides* and immunoglobulin-coating fecal bacteria associated with weight loss in obese adolescents. Int J Obes 2009;33(7):758–67. [Published correction appears in Int J Obes 2012;36(10):1370].

[24] Santacruz A, Marcos A, Wärnberg J, et al. Interplay between weight loss and gut microbiota composition in overweight adolescents. Obesity 2009;17(10):1906–15.

[25] Turnbaugh PJ, Ley RE, Mahowald MA, Magrini V, Mardis ER, Gordon JI. An obesity-associated gut microbiome with increased capacity for energy harvest. Nature 2006;444(7122):1027–31.

[26] Zhang C, Zhang M, Wang S, et al. Interactions between gut microbiota, host genetics and diet relevant to development of metabolic syndromes in mice. ISME J 2010;4(2):232–41.

[27] Turnbaugh PJ, Ridaura VK, Faith JJ, Rey FE, Knight R, Gordon JI. The effect of diet on the human gut microbiome: a metagenomic analysis in humanized gnotobiotic mice. Sci Transl Med 2009;1(6):6–14.

[28] Woodard GA, Encarnacion B, Downey JR, et al. Probiotics improve outcomes after Roux-en-Y gastric bypass surgery: a prospective randomized trial. J Gastrointest Surg 2009;13(7):1198–204.

[29] Zhang H, DiBaise JK, Zuccolo A, et al. Human gut microbiota in obesity and after gastric bypass. Proc Natl Acad Sci USA 2009;106(7):2365–70.

[30] Furet JP, Kong LC, Tap J, et al. Differential adaptation of human gut microbiota to bariatric surgery-induced weight loss: links with metabolic and low-grade inflammation markers. Diabetes 2010;59(12):3049–57.

[31] Li JV, Ashrafian H, Bueter M, et al. Metabolic surgery profoundly influences gut microbial-host metabolic cross-talk. Gut 2011;60(9):1214–23.

[32] Li JV, Reshat R, Wu Q, et al. Experimental bariatric surgery in rats generates a cytotoxic chemical environment in the gut contents. Front Microbiol 2011;2:183.

[33] Le Chatelier E, Nielsen T, Qin J, et al. Richness of human gut microbiome correlates with metabolic markers. Nature 2013;500:541–6.

[34] Liou AP, Paziuk M, Luevano JM Jr, Machineni S, Turnbaugh PJ, Kaplan LM. Conserved shifts in the gut microbiota due to gastric bypass reduce host weight and adiposity. Sci Transl Med 2013;5. 178ra41.

[35] Kong LC, Tap J, Aron-Wisnewsky J, et al. Gut microbiota after gastric bypass in human obesity: increased richness and associations of bacterial genera with adipose tissue genes. Am J Clin Nutr 2013;98:16–24.

[36] Graessler J, Qin Y, Zhong H, et al. Metagenomic sequencing of the human gut microbiome before and after bariatric surgery in obese patients with type 2 diabetes: correlation with inflammatory and metabolic parameters. Pharmacogenom J 2013;13(6):514–22.

[37] Gralka E, Luchinat C, Tenori L, Ernst B, Thurnheer M, Schultes B. Metabolomic fingerprint of severe obesity is dynamically affected by bariatric surgery in a procedure-dependent manner. Am J Clin Nutr 2015;102(6):1313–22.

[38] Tremaroli V, Karlsson F, Werling M, et al. Roux-en-Y gastric bypass and vertical banded gastroplasty induce long-term changes on the human gut microbiome contributing to fat mass regulation. Cell Metab 2015;22(2):228–38.

[39] Ryan KK, Tremaroli V, Clemmensen C, et al. FXR is a molecular target for the effects of vertical sleeve gastrectomy. Nature 2014;509:183–8.

[40] Damms-Machado A, Mitra S, Schollenberger AE, et al. Effects of surgical and dietary weight loss therapy for obesity on gut microbiota composition and nutrient absorption. Biomed Res Int 2015;2015:806248.

[41] Belzer C, de Vos WM. Microbes inside—From diversity to function: he case of *Akkermansia*. ISME J 2012;6:1449–58.

[42] Derrien M, Van Baarlen P, Hooiveld G, Norin E, Müller M, de Vos WM. Modulation of mucosal immune response, tolerance, and proliferation in mice colonized by the mucindegrader *Akkermansia muciniphila*. Front Microbiol 2011;2:166.

[43] Hansen CH, Krych L, Nielsen DS, et al. Early life treatment with vancomycin propagates *Akkermansia muciniphila* and reduces diabetes incidence in the NOD mouse. Diabetologia 2012;55:2285–94.

[44] Parks BW, Nam E, Org E, et al. Genetic control of obesity and gut microbiota composition in response to high-fat, high-sucrose diet in mice. Cell Metab 2013;17:141–52.

[45] Merritt ME, Donaldson JR. Effect of bile salts on the DNA and membrane integrity of enteric bacteria. J Med Microbiol 2009;58(Pt 12):1533–41.

[46] Islam KB, Fukiya S, Hagio M, et al. Bile acid is a host factor that regulates the composition of the cecal microbiota in rats. Gastroenterology 2011;141:1773–81.

[47] Sayin SI, Wahlström A, Felin J, et al. Gut microbiota regulates bile acid metabolism by reducing the levels of tauro-beta-muricholic acid, a naturally occurring FXR antagonist. Cell Metab 2013;17(2):225–35.

[48] Swann JR, Want EJ, Geier FM, et al. Systemic gut microbial modulation of bile acid metabolism in host tissue compartments. Proc Natl Acad Sci USA 2011;108:4523–30.

[49] Wen L1, Ley RE, Volchkov PY, et al. Innate immunity and intestinal microbiota in the development of Type 1 diabetes. Nature 2008;455(7216):1109–13.

[50] Qin J, Li Y, Cai Z, et al. A metagenome-wide association study of gut microbiota in type 2 diabetes. Nature 2012;490:55–60.

[51] Karlsson FH, Tremaroli V, Nookaew I, et al. Gut metagenome in European women with normal, impaired and diabetic glucose control. Nature 2013;498:99–103.

[52] Neyrinck AM, Possemiers S, Verstraete W, De Backer F, Cani PD, Delzenne NM. Dietary modulation of clostridial cluster XIVa gut bacteria (*Roseburia* spp.) by chitin-glucan fiber improves host metabolic alterations induced by high-fat diet in mice. J Nutr Biochem 2012;23(1):51–9.

[53] Ridaura VK, Faith JJ, Rey FE, et al. Gut microbiota from twins discordant for obesity modulate metabolism in mice. Science 2013;341:1241214.

[54] Vrieze A, Van Nood E, Holleman F, et al. Transfer of intestinal microbiota from lean donors increases insulin sensitivity in individuals with metabolic syndrome. Gastroenterology 2012;143. 913–6.e7.

[55] Korner J, Inabnet W, Febres G, et al. Prospective study of gut hormone and metabolic changes after adjustable gastric banding and Roux-en-Y gastric bypass. Int J Obes 2009;33:786–95.

Chapter 31

The Influence of Microbiota on Mechanisms of Bariatric Surgery

N. Floch*,**,†,‡

*Department of Surgery, Norwalk Hospital, Norwalk, CT, United States; **Fairfield County Bariatrics and Surgical Specialists, P.C., Norwalk, CT, United States; †Associate Clinical Professor of Surgery, University of Vermont School of Medicine, Burlington, VT, United States; ‡Associate Clinical Professor of Surgery, The Netter School of Medicine, Quinnipiac University, Hamden, CT, United States

THE ANATOMIC AND PHYSIOLOGIC CHANGES THAT OCCUR AFTER BARIATRIC SURGERY AND HOW THEY ALTER THE GUT MICROBIOME

Bariatric surgery entails altering the stomach and variably altering the small bowel flow, depending upon the procedure. After bariatric surgery the amount of food consumed is limited by the creation of a smaller stomach and the alteration is termed, restriction. In several bariatric procedures, the pathway of the small bowel flow is altered so that nutrients consumed are exposed to bile and pancreatic enzymes along a significantly shorter length of the small bowel. This alteration is referred to as malabsorption. These physical changes have subsequent physiologic ramifications, which have traditionally believed to cause less caloric intake and less nutrients absorption.

The classic mechanisms of bariatric surgery function have been challenged and new mechanisms have been elucidated. After bariatric surgery, the relative abundances of micriobota are changed. The quantities of bacteria carry with them alternate levels of metagenomic pathways that establish a "new normal" level of metabolism. The results of these changes may account in part for the beneficial changes that occur after bariatric surgery, such as weight loss and improvement and many times, resolution of metabolic diseases.

There are now multiple surgeries that are available in which to study the physiologic changes and how they influence host microbiota. The specific surgeries have been evaluated in humans, such as roux-en-y gastric bypass (RYGB) [1–4], vertical-banded gastroplasty (VBG) [4], and vertical sleeve gastrectomy (VSG) [5] in mice. The adjustable gastric band (AGB) and VBG are restrictive procedures that limit the quantity of food that is delivered to the distal

bowel. They may also limit humans from eating large quantities of meat and complex carbohydrate vegetable substances. The AGB may also compress the vagus nerves below the esophagogastric junction. The AGB has not been studied for changes in microbiota. VBG may be most helpful as its effects are the closest to isolated restriction of food intake but the effects of stapling the fundus of the stomach are unclear. The VBG, which is a non-AGB, has been studied in humans, the results of which are compared to post-RYGB patients [4]. The VSG is a combination of both restriction and an alteration of gastric anatomy that results in increase gastric emptying, decreased ghrelin production, altered bile production as influenced by the farnesoid X receptor (FXR) [5].

Post-RYGB patients are the most complex as multiple mechanisms are working concomitantly. The RYGB procedure involves variable ligation of the vagus nerves, creation of a limited stomach with decreased acid exposure to orally induced bacteria, and one that limits the quantity of food introduced into the bowel, increase of the length of small intestine that is exposed to a less acidic environment. After the procedure, the quantity of both bile and gut hormones are also changed. Other bariatric procedure, such as the duodenal switch (DS), long-limb bypass (LLRYGB) have yet to be studied for their influence on the gut microbiome.

In the past, a pneumonic termed, *B.R.A.V.E* had been developed to explain the mechanisms that exert weight loss after RYGB. The mechanisms may be applied in all or part to the multiple different bariatric procedures currently available, changing *B*ile flow, *R*educing the stomach size or creating a pouch, rearranging the *A*natomy of the gut and altering the flow of nutrients, *V*agotomy, and alteration *E*nteric gut hormone production [6–9]. The anatomical changes that result from bariatric surgery create physiologic changes that are directly responsible for the taxonomic, metagenomic, and metabolomic changes in the gut microbiome. Although multiple changes are observed after bariatric surgery, the variables that have been studied include: restriction resulting in reduced caloric intake, alteration in diet, and the relation to colonic fermentation, a lower intestinal pH environment emanating from decreased acidity in a small stomach, altered bile metabolism and quantity, and gut hormonal changes. These changes may contribute to improvements in the host, which relate to both weight loss and metabolic improvements.

Unfortunately, it is difficult to control for all these changes to determine the contribution of each component. Specifically in the RYGB, there are caloric restrictive, diet changes, increased metabolism, increased nutrient transit time, pH alterations, bile metabolism, and gut hormone changes that are all occurring concomitantly [1–4]. The VBG [4] and VSG [5,10] share most of these mechanisms, yet there are fewer investigations available to analyze. Comparison of the data available on bariatric procedures has elucidated some of the effects that each of these procedures incurs on the microbiome [1–4].

Taxonomic And Metagenomic Differences Between "Restriction" And "Malabsorption"

A direct comparison between a "restrictive" operation, VBG and a "malabsorptive" operation, RYGB has elucidated the effects that these techniques incur upon microbial and metagenomic alterations. There are more dramatic long-term changes to the relative abundances of microbiota that occur after RYGB than VBG when compared to obese controls.

Post-RYGB humans demonstrate a long-term difference in multiple species of Gammaproteobacteria [4]. Also in abundance are facultative anaerobes in the Proteobacteria genus, such as *Escherichia, Klebsiella,* and *Pseudomonas.* The post-RYGB humans also demonstrate a lower level of three species of Firmicutes phylum: *Clostridium difficile, Clostridium hiranonis,* and *Gemella sanguinis* [4]. Long-term bariatric surgery evidence reveals that post-RYGB women have more Proteobacteria than obese controls [4]. Although RYGB and VBG are different bariatric procedures with different potential mechanisms, the micriobiota profiles were not significantly different [4]. There was a significant increase in the abundance of three *Escherichia coli* species in women who had VBG surgery [4].

Bariatric patients had larger quantity of Kyoto encyclopedia of genes and genomes (KEGG) pathways or KEGG orthology (KOs) compared to obese control patients. Both RYGB and VBG patients had large difference in KOs. Compared to obese controls, RYGB patients had 928 enriched and 60 depleted KOs, and VBG patients had 682 enriched and 33 depleted KOs. The two surgical groups were similar with only 17 KOs that had a difference in quantity

[4]. The post-RYGB patients exhibited a greater number of KOs for the phosphotransferase system and fluorobenzoate degradation. All postbariatric patients exhibited increased levels of nitrogen metabolism, fatty acid metabolism, and two-component system. The post-VBG patients had higher levels of flagellar assembly, sulfur relay system, glutathione metabolism, glyoxylate and dicarboxylate metabolism, ABC transporters, and phenylalanine metabolism [4].

A two-component regulatory system is a stimulus-response coupling mechanism that allows bacteria to sense and respond to changes in different environmental conditions [11]. Two-component systems were found to be enriched in post-RYGB patients compared to obese controls were: nitrogen availability (*nifA*), phosphoglycerate transport (*pgtCAP*), and short-chain fatty acid (SCFA) metabolism (acetoacetate, *atoCE*). When post-VBG patients were compared to obese controls; salt stress response and twitching motility were enriched. In both postbariatric surgery groups: RYGB and VBG, genes for response to antimicrobial peptides, outer membrane stress, and trimethylamine *N*-oxide (TMAO) were enriched when compared to obese control humans. Also enriched is the gene coding for TMAO reductase (*torA*, K07811), involved in the respiration of TMAO to TMA [4]. The elevated presence of tor genes indicates that there is a larger amount of *E. coli* in the post-RYGB and post-VBG patients. The *E. coli* microbes use TMAO as a terminal electron receptor [12]. There are increased levels of TMAO in only post-RYGB patients. There was no difference in choline and betaine dietary precursors [4].

The increase in TMAO is believed to be a consequence of a shortened small bowel after RYGB, which is conducive to less anaerobic metabolism, and therefore an environment that is more habitable for facultative anaerobes [4]. The finding of increased facultative anaerobes was also found by Furet et al. [1]. It had been believed that TMAO was produced in the intestine by TMA monooxygenase, which is produced by *Pseudomonas* [12]. High TMAO levels are believed to be associated with risk of cardiac disease [13,14], but ironically gastric bypass results in lower risk of cardiac disease [15].

The changes in the composition of the microbiota as well as the bacterial genes of those patients undergoing bariatric surgery are not independently a result of a decreased body mass index (BMI), but are caused by the bariatric surgery itself [4]. BMI matched obese presurgery control group and the matched for decrease weight postsurgery control group had similar microbiota, both of which were markedly different from the postbariatric surgery groups [4]. Multiple previous studies that have concentrated on dietary weight loss without surgery have not found an increase in the amounts of Proteobacteria. Three studies showed an increase in the taxa; *Bacteroides* [16], *Eubacterium/Roseburia* [17], and *Eubacterium rectale* [18].

Reduced Gastric Volume and Caloric Restriction

The reduction of the gastric pouch size as a result of restrictive operations can directly decrease the amount of food that enters into the stomach and ultimately the intestine. Most bariatric surgeries have a restrictive component; AGB, VSG, VBG, RYGB, as well as DS and biliopancreatic diversion (BPD). The RYGB, DS, BPD operations also reduce nutrient quantity intake as well as decrease the vitamin absorption that normally occurs in the proximal duodenum.

Independent of any bariatric surgery, obese patients on low-calorie diets have low *Bacteroides*/Firmicutes ratio that reversed as patients lost significant weight [16]. Despite studies that refute this evidence, it is generally accepted [17]. There are multiple issues of concern involving the creation of a smaller stomach. They include: the quantity and type of food consumed [19,20]. A lower quantity of food may trigger a different mechanism than the alteration in the type of food consumed. Although other mechanisms are involved, the restrictive component of bariatric surgery can reduce the quantity of nutrients eaten and may account for adaptive metabolic responses, which occur as a result of a smaller stomach.

The brainstem anorexia pathway may account for the early effects of RYGB on meal size and food intake. This may lead to adaptive neural and behavioral changes that control nutrient intake and BMI [21]. When there is significant calorie reduction, such as seen in patients with anorexia nervosa, the type of bacteria can change significantly. Stool from an anorexic patient has revealed the discovery of 11 new types of microbiota, 7 species from the phylum Firmicutes, 2 from *Bacteroides*, and 2 from Actinobacteria [22]. A study of nine patients with anorexia nervosa revealed high levels of the Archean, *Methanobrevibacter smithii*. This microbiota is influential in creating more efficient bacterial fermentation processes, a larger energy harvest and therefore it is a response to the severely calorie depleted individual [23].

Diet alteration, Choice, and Fermentation

A smaller gastric pouch limits the quantity of food but may also change the types of foods consumed by the host. A decreased desire to eat, alterations in eating behavior and choice may explain the early weight loss and diabetes remission after bariatric surgery [21]. It is now agreed that diet controls the shift in micriobota in the gut and is ultimately responsible for obesity [24]. Conversely, changes in diet can cause weight loss in obese individuals and increase the gene richness of microbiota resulting in similar microbiomes to that of thin subjects [16,18]. Studies that confirm the ratio between Firmicutes and Bacteroides have also included transplanting mice with human fecal bacteria from those who eat a western diet. The overweight or obese characteristics transfer from the obese human to the thin, germ-free mice, and result in the mice becoming obese [25,26].

Interactions between gut microbiota, host genetics, and diet are relevant to development of metabolic syndromes in mice [25]. Dieting has been demonstrated to reverse many of the metabolic diseases associated with obesity [18–66]. Multiple studies support that types of nutrients are capable of changing the relative abundance of the intestinal microbiota. Most influential are consistent, long-term dietary pattern changes that influence the composition and richness of the gut microbiota [27–30].

Italian and African children have comparatively different relative abundances of microbiota, which is a reflection of the different diets that people eat in separate parts of the world [27]. These two distinct diets result in the growth of greater quantities of either genera *Proventella* or *Bacteroides*. The African diet is composed of more simple and complex carbohydrates while the Italian diet is considered "western" and based on fats and animal protein [27]. Proventella dominated in the African diet while *Bacteroides* is more abundant in the Italian diet. When diets are changed, the microbiome can change within days [31].

Further evidence demonstrates that alterations in carbohydrate, fat, and protein as well as the change from a plant-based to animal fat and protein diet causes rapid changes in the host microbiome [32]. Those who eat a western diet are dependent on a microbiome, that is, capable of fermenting polysaccharides to SCFAs in order to release energy sources to the host [24,33]. Nutrition influences the composition and richness of the gut microbiome, which occurs rapidly after diet change and remains long-term.

The microbiota changes that occur after bariatric surgery, especially RYGB may account either directly or indirectly for the quantity of weight lost. Most postbariatric surgery patients are encouraged to eat high-protein foods in the quantity of 60 daily grams per day and stress less simple carbohydrates and sugars. Unfortunately, complex carbohydrates, such as vegetables are harder to chew into smaller boluses that may pass successfully through the new pouch outlets. In animals given a choice of food options, a study revealed that rats consumed less protein and fat, and slightly more carbohydrates after undergoing VSG [34].

Diet preference is different in post-RYGB patients. Post-RYGB patients choose to eat food that is less energy dense and nutritionally improved at up to 2 years after surgery [35]. When germ-free mice were transplanted with microbiota from post-RYGB and post-VGB human, oxygen consumption and carbon dioxide production were the same but there was a profoundly lower respiratory quotient in post-RYGB patients, indicating that post-RYGB mice consumed more lipids as fuel and less carbohydrate [4]. Food choice appears to improve after bariatric surgery either intentionally or possibly biochemically. The exact mechanisms are unclear but a brainstem anorexia pathway has been suggested [21,36]. The alteration in diet may not be by conscious choice. After bariatric surgery, diet changes irrespective of the mechanism and the gut microbiota are altered. These alterations may directly or indirectly effect colonic fermentation, which results in substrate changes and nutrient utilization as fermentation. The fermentation by the gut bacteria provides nutrition for the host they serve [37].

Alterations in the use of varying degrees of protein, lipid, and carbohydrate occur in the host microbiome after bariatric surgery [4]. There is an enrichment of KOs in the phosphoglycerate, acetoacetate, SCFA, and phosphotransferase systems after RYGB in women, indicating a favoring of sugar metabolism and glycolysis [4]. Another study revealed a significant increase in 13 KOs in the phosphotransferase system 3 months after RYGB [38]. The phosphotransferase system is used by bacteria to uptake sugar where the source of energy is from phosphoenolpyruvate. A possible explanation may be that decreased caloric intake results in the improved ability to process substrate as a compensatory mechanism to decreased nutritional consumption of energy [38]. In contrast, there is a difference in post-VBG patients who exhibit an enrichment of pathways involved in glyoxylate metabolism and amino acid uptake and metabolism [4]. The post-VBG microbiome is conducive to amino acids and acetate as a major energy source [4].

The measurement of fecal SCFA and branched-chain fatty acids (BCFA) was analyzed for both carbohydrate and protein metabolism. The concentrations of acetate, butyrate, and propionate were decreased in post-RYGB and post-VBG patients while BCFAs, isobutyrate, and isovalerate increased mildly [4]. There was a decrease in the SCFA/BCFA ratio for the bariatric surgery patients, most significant with the post-RYGB patients [4]. From a dietary standpoint, the patients had no difference in fiber intake but protein intake was less for the bariatric patients [4]. The lower SCFA/BCFA ratio was not a result of a change in dietary intake. Bariatric surgery patients including RYGB and VBG were both found to shift their microbiome characteristics away from SCFA toward increased amino acid fermentation and appears to be a diet independent phenomenon [4]. The finding of reduced amounts of SCFA in both post-RYGB and post-VBG microbiota transplanted to mice is supportive of the belief that there is a reduction of energy harvest from the diet in post-bariatric surgery patients [4].

Gastric and Colonic pH

The reduction in stomach size results in less parietal cells and less acid production. The microbiota are then exposed to a higher pH environment within the stomach. This was demonstrated from gastric pH samples taken from post-RYGB mice [19,39]. The environment is less likely to destroy and more likely to recolonize, orally ingested bacteria

as seen after RYGB, as a higher pH environment is less toxic to microbiota. It remains unclear if post-VSG animals or humans have a higher gastric pH but this procedure also reduces the amount of parietal cells. Shared mechanisms between RYGB and VBG may include an increase in the intestinal pH from decreased exposure to gastric acid. An elevated pH promotes growth of *E. coli* in postbariatric surgery patients [40].

Ironically, as RYGB results in a higher pH stomach in humans, it has been found to create a lower pH environment in the distal small bowel and colon in rats. Fecal samples from post-RYGB mice had a lower pH than weight-controlled and sham surgical mice [19]. Human studies reveal the contrary as the colonic pH has been found to be higher, which is attributed to less acid production in the smaller stomach [38].

Extrapolating from nonsurgical studies in humans, Gram-negative Bacteroidetes are more sensitive to mildly acidic pH, while Gram-positive Firmicutes and high GC content Actinomycetes are significantly more tolerant to a mildly acidic pH environment [40]. The increased human gut pH is responsible for the alteration in the relative abundance of phyla as well as promoting the growth of *E. coli*. [40]. Increase Proteobacteria found after RYGB in humans have been proposed as a result of increased pH. A suggested mechanism is the increased oxireduction potential in the gut as a result of increased pH, which influences the growth of aerobe or facultative anaerobe phyla, such as Proteobacteria [3,41].

It remains unclear how specifically the gastric bypass procedure creates a change in the microbiota composition and quantity. The small stomach pouch with its decrease in acidity along with an alteration of the flow of bile more rapidly into the distal jejunum then ileum, results in a change in the colonic microbiota populations. Which one of these mechanisms is most dominant is a topic for further investigation.

Gut Hormones

Multiple gut hormones are associated with the intestinal microbiota and are altered post-RYGB surgery in both humans and animals [20]. Ghrelin and leptin have been studied most frequently in association with bariatric surgery [42–45]. Prebiotics can indirectly modulate the intestinal microbiota, which results in a decreased amount of circulating ghrelin [42]. Consumption of the prebiotic inulin has also demonstrated a significant reduction of ghrelin human subjects [46].

Leptin has also been found to cause weight reduction in association to gut microbiota in multiple mice studies [44,45]. Calorie restricted mice have been has 80% reduced leptin levels [45]. Leptin levels studied in obese and weight loss induced, mice revealed that serum leptin levels were associated with the abundance of Mucispirillum,

Lactococcus, and Lachnospiraceae. Mucispirillum has been found to react with intestinal mucin, creating a positive environment for intestinal bacterial growth [45]. Mucispirillum was found in higher abundance in obese mice. Leptin is now seen as a potential regulator of microbiota levels in the intestine by its control over mucin production by Mucispirillum. This is an effect seen in obese mice [44]. Leptin levels are negatively correlated with quantities of *Bacteroides*, *Clostridium*, and *Proventella* species and positively correlated with *Bifidobacterium* and *Lactobacillus*, and populations. Decreases in body weight and leptin levels have been suggested to explain changes that occur to the microbiota when calories are restricted in animals [20].

Analysis of humans after bariatric surgery reveals an association of Leptin reduction and microbial taxonomy changes. When post-RYGB patients' leptin levels decreased, the species of *E. coli*, *Bacteroides*, and *Prevotella* increased in abundance [1]. This finding is associated with a higher *Bacteroides/Prevotella* ratio [2]. These findings are not consistent in all studies [2]. During the 3 months after RYGB, when starch-based foods are often the principal food and solid foods are slowly added with reduced calories, changes in body composition occur and leptin levels drop. Changes in leptin correlate negatively with the ratio of *Bacteroides/Prevotella* and abundance of *E. coli*, while positively with *Bifidobacterium* and *Lactobacillus/Leuconostoc/Pediococcus* [2].

Leptin levels decrease with the start of energy deprivation 3 months after RYGB surgery, less than 50% of baseline, and have found to stabilize 6 months after RYGB, while BMI, adipose mass and size continue to decrease, as found in other studies [2,47]. Changes in leptin blood levels signal the adaptation to starvation [48,49]. The ability of continued fasting to separate the influence of circulating leptin levels upon adipose tissue mass could be a permissive effect of insulin on leptin secretion. This is not a normal occurrence as usually leptin stimulates energy expenditure and inhibits appetite [48].

Bile Acids

Bile acids (BAs) are sterol compounds produced in the liver from cholesterol. The primary conjugated BAs are secreted from the liver into the duodenum, usually after a high lipid meal [50]. Liver excretion is decreased after a meal in obese patients when compared to lean [51,52]. Serum BAs levels are higher in obese individuals, but decrease when they lose weight by change in diet [53–55]. BAs have been found to be responsible for many of the changes that occur in the intestinal microbiota in rats after consuming a high-fat diet [56]. The increased BAs after weight loss from dieting may encourage increased abundances of more bile-tolerant microbiota in the phylum Proteobacter, such as microbe *Bilophila wadsworthia* [32]. BAs also have direct and indirect antimicrobial effects [57].

The BAs are converted to secondary BAs in the distal ileum by 7-alpha-dehydroxylation [58,59]. The microbiota in the distal gut contain bile salt hydrolases that are capable of deconjugating the taurine- and glycine-conjugated BAs to unconjugated free BAs [60]. Humans that enhance their diet with cholic acid have been found to harbor elevated amounts of 7α-dehydroxylating bacteria. As expected, cirrhotic patients, who have lower amounts of BAs carry lower levels of 7α-dehydroxylating bacteria [61]. Deconjugation is catalyzed by the anaerobic bacteria of the genera *Bacteroides*, *Clostridium*, *Eubacterium*, *Lactobacillus*, and *Escherichia*, and the species *Clostridium* and *Ruminococcus* [58,59,62,63]. These BAs can be secreted or reabsorbed into the blood. Some BAs are utilized for their carbon content as energy by aerobic bacteria, such as Actinobacteria and Betaproteobacteria. Enterobacteria have the ability to transform CA and TDCA, which decrease the function of the ileal apical sodium-dependent BA transporter, SLC10A2 [60].

The enterohepatic circulation accounts for most of the bile to be reabsorbed in the distal ileum that occurs with the assistance of the apical sodium-dependent bile acid transporter (ASBT) and returns to the liver [64]. In the enterohepatic circulation 5–10% of BAs are altered or destroyed by the gut bacteria [59]. When Enterobacteria are decreased, more BAs are absorbed by the host. Treatment with antibiotics or studies in germ-free mice have demonstrated that the gene expression in the liver result in alterations of the hepatic synthesis of BAs, steroids, and cholesterol as well as BA conjugation in other parts of the hosts' body [65].

Post-RYGB and post-VBG humans do not exhibit different blood BA levels in the fasting state compared to obese controls [4]. In the postprandial state, the BA response is blunted in obesity and is generally enhanced after bariatric surgery [52]. Previous studies on restrictive bariatric surgery other than RYGB, such as AGB and SG have variable results [50], revealing elevated serum BA [4,66–68], decreased [69,70], and unchanged [71,72]. One study on VBG revealed modestly elevated BA levels [4–38]. The limited studies on VSG [10] in humans and mice [67,68] suggest an elevation in BA levels [10]. Most studies of post-RYGB humans reveal an increase in a type of serum BAs [70,72,73]. Only one study revealed a decrease followed by a late increase [69,74–79].

Postoperative RYGB patients have increased levels of both primary and secondary BAs [51,70,74,78]. It was believed that excreted BAs had less time to interact with nutrients through a functionally shorter small bowel, leaving more BAs available to be reabsorbed in the distal ileum [50]. The finding of increased BA production has been suggested as one possible mechanism to explain the satiety, decreased nutrient intake, and weight loss associated with bariatric surgery [52].

BAs are able to regulate energy metabolism once activated through their receptors: TGR5 and FXR [80,81]. The FXR receptor is responsible for the signaling that enhances the increase in enterohepatic circulation of bile [5].

BAs bind to the FXR receptor in the small intestine. FXR activates fibroblast growth factor-19 (FGF-19) that then travels to the liver where FGF-19 activates CYP7A1. The end result is the production BA from cholesterol. FGF-19 is decreased in diabetic patients. In post-RYGB with diabetes, FGF-19 expression and BA production both increase to a greater degree compared to obese patients [52,76]. Other studies do not show an increase in FGF-19 post-RYGB [78].

Bariatric surgery alters the gut microbiome by changing FXR signaling and the enterohepatic circulation. These mechanisms may in turn influence one another [5]; as BAs have the ability to change the abundance and types of microbiota in the gut either directly or by activating the FXR receptor [56,82]. The converse is also true as the microbiota can regulate BA composition with enzymes they produce and have the ability to alter both BA levels and FXR receptor signaling [65,83].

Well known is the alteration to a higher abundance of the phylum Firmicutes and a lower relative abundance of *Bacteroidetes* after RYGB in mice [84]. Human studies have confirmed the initial animal studies [16]. The changes in the composition of gut microbiota and BAs that occur together in humans and animals have been believed to be secondary to diet change and weight loss. It was thought that the change in diet that is forced to occur by a smaller stomach pouch after bariatric surgery was responsible for the alteration of the relative abundances of microbiota [84].

Instead, the gut microbiota have been demonstrated to change the obesity phenotype of the host as demonstrated in a transplantation study in twins. Germ-free mice transplanted with fecal bacteria from identical twins, one obese and one lean, result in only the germ-free recipient from the obese mouse becoming obese. Furthermore, the differences in body composition correlated to differences in BA metabolism and FXR signaling [85]. Thin mice that have an elevated abundance of the phylum Bacteroidetes, which have the capability of processing BAs. The presence of *Bacteroidetes* correlates with altered FXR signaling and is believed to be a mechanism that reduces adiposity [85].

Following RYGB, the gut microbiota change in both rats and humans. The most significant changes are increased overall richness and increased abundance of Proteobacteria and decreased Firmicutes [1–3,86]. Human studies in post-RYGB patients have shown an increase in *Bacteroidetes*, which conflict with a decrease in rats [2,86,87].

Alterations in the gut microbiome occur not only after RYGB, but also after VSG in humans [88]. FXR signaling controls the effects of VSG through modulation of the gut microbiota [5]. In mice, *Bacteroides* are significantly reduced after VSG surgery while Porphyromonadaceae and *Roseburia* are greatly increased. The increased abundance of *Roseburia* was associated with less glucose intolerance. There is decreased body fat and improvement of glucose intolerance in wild type compared with sham-operated mice. FXR signaling is believed to be responsible for the weight loss associated with VSG as well as decreased eating, and improved insulin sensitivity [5]. The metabolic benefits of VSG are a result of having the FXR gene as demonstrated by FXR KO mice that do not exhibit sustained weight and body fat loss after 5 weeks [5]. Despite this finding, the benefits of the gut microbial alteration of VSG surgery have a greater beneficial influence than not having the FXR genotype [5].

Intestinal microbiota may alter BA metabolism in the intestine [32]. Gut microbiota can directly regulate the production of the BA pool by conjugating them from primary to secondary [76]. The secondary BAs produced by the microbiota may in turn regulate the FXR receptor, which activates FGF-19 [57]. FGF-19 is an intestinal factor that regulates BA, carbohydrate, lipid, and energy metabolism through the BA activation of the FXR gene [89].

After bariatric surgeries, such as RYGB and VBG, the increase in abundance of microbiota may influence reactions that increase the amount of bile or specifically secondary BAs in the blood. As mentioned, the activation of FGF-19 by FXR results in increased bile production from cholesterol in the liver. Consequently, there is an increased FGF-19 activation in post-RYGB patients compared to obese female controls, supporting evidence that there is increased FXR signaling in humans [5]. The exact mechanisms of how the FXR receptor is influenced by VSG surgery is not yet fully understood. However, BAs activate FXR, the G-coupled protein receptor and G-protein-coupled bile acid receptor 1 (TGR5) [90,91]. Post-VSG, the small intestine increases the villi length and total surface area, as well as the active surface of the ileal ASBT protein, which may explain the finding of increased blood levels of BAs as found in many postbariatric and nonsurgical studies [67,92–96].

Post-RYGB, human studies reveal an increase in Gammaproteobacteria, of which Enterobacteria is a species [1,2,86]. The intestinal microbiota regulate BA metabolism. Enterobacteria are capable of deconjugating taurocholic acid, which results in heightened FXR signaling in the ileum. The unconjugated cholic acid that is produced activates the ileal FXR genes at an increased potency compared to tauro-beta-muricholic acid, which is a naturally occurring FXR antagonist. The effect does not occur in FXR KO mice [83]. Compared to obese control patients, post-RYGB patients have increased amount of microbiota genes for 7-alpha-of primary BAs. These genes were baiB, baiCD, baiE, baiF, and baiG [4].

There is an increase postprandial BA production in post-RYGB patients compared to obese controls. The increase occurs in total conjugated, as well as taurine and glycine-conjugated BAs. Unconjugated BAs are not increased [97]. The post-RYGB patients exhibited an increase in both primary and secondary conjugated BAs [4]. There is an increased abundance of post-RYGB microbiota containing genes involved in dehydroxylation of primary to secondary BAs, explaining an increase in secondary BAs in post-RYGB patients. These changes are maintained up to an *n* average of 9.3 years in humans.

Bariatric Surgery Improves Metabolic Disease and Inflammation by Altering Microbiota Abundances

Adipose tissue lies at the epicenter of the obesity disease. When activated, it secretes substances that alter human metabolism [98–100]. The intestinal microbiome has the ability to regulate the balance of adipose tissue and is, therefore a factor in the control of obesity [101–104]. Evidence that microbiota manipulates the host metabolism was originally exhibited in mice studies [105]. Intestinal dysbiosis can cause not only obesity but also insulin resistance and type 2 diabetes (T2D), according to two large prevalence studies [101,102].

Germ-free mice colonized with microbiota causing them to become insulin resistant and gain significant weight gain despite decreasing their calorie intake by 30%. The mice were found to have a decrease in the *Bacteroides*/Firmicutes ratio [84]. The Bacteroidetes to Firmicutes ratio significantly correlates with the level of plasma glucose concentration but not with the BMI [105]. There is a decreased quantity of both Firmicutes and Clostridia in diabetic patients compared to those that are nondiabetic [105]. Patients with T2D have less relative abundance of *Roseburia* species and *faecalibacterium prausnitzii* [101,102]. At the level of phylum, there is a higher relative abundance of Proteobacteria inT2D [105]. A Chinese study showed a higher level of *E. coli* in type two diabetics [101]. Insulin resistance may also be improved in humans after the transplantation of lean gut microbiota [106].

Gut microbiota is implicated in the development of obesity and T2D as found in the evidence of metabolomic investigations that reveal alterations in symbiotic, host and microbial metabolism of BAs, branched fatty acids, choline, niacin, purines, and phenolic compounds that are associated with the state of obesity and T2D [107]. The risk of developing T2D is 40 times higher at a BMI higher than 35 kg/m^2 [108]. There are multiple mechanisms that may be involved in the role of microbiota and the development of obesity and T2D. Gut microbiota exert their effects on glucose and lipid metabolism through many complementary mechanisms but a common feature to all of them is the altered intestinal integrity of the gut epithelium wall [109].

1. There is a control of energy homeostasis by influencing the amount of energy that is harvested from diet, adipose tissue storage and production, as well as the metabolism of host energy through fatty acid oxidation [109,110]. Gut microbiota cross talk with the host may induce a low-grade inflammatory state that alters energy homeostasis and eventually glucose metabolism [109].

2. There is a control of the intestinal epithelium barrier, gut motility, and the binding of SCFAs to G-protein-coupled receptors (GPR) to feed colonocytes [111]. SCFA: butyrate, acetate, and propionate are fermented from dietary fiber. Butyrate has the capability of improving insulin sensitivity [112]. Butyrate binds to G-protein-coupled receptors GPR41 and GPR43. When bound to immune cells, butyrate modulates decreased inflammation in the enteroendocrine L-cells which causes increased glucagon like peptide-1 (GLP-1) and peptide YY (PYY), leading to improved insulin sensitivity [112–115].

3. There is modulation of gut peptide hormones by decreasing lipoprotein lipase inhibitor, modulating adipose mass, and control of the release from muscle of fatty acids from circulating triglycerides and lipoproteins in muscle and adipose tissue [116].

4. There is control over inflammation that can occur in the host with increasing quantities of lipopolysaccharide (LPS) that lead to insulin insensitivity by metabolic endotoxemia [115–120]. The inflammatory state occurs as bacterial LPS from the outer membrane of Gram-negative bacteria create a metabolic endotoxemia as they induce proinflammatory cytokines [121,122]. It is a G-protein dependent mechanism of decreased permeability and reduced LPS levels [43]. The microbial molecules activate the immune system by binding to toll-like receptor, toll-like receptor 4 (TLR4) resulting in low-grade inflammation and decreased insulin sensitivity [43].

5. Secondary BA bind to G-protein-coupled receptor, TGR5 which results in increased energy expended in muscles and increased GLP-1 secretion in the enteroendocrine L-cells, both of these lead to improved insulin sensitivity [123,124]. Bariatric surgery is most successful to reduce and maintain weight loss. The evidence that is less appreciated is that RYGB has a dramatic effect on the inflammatory processes of T2D patients and a capability of regaining glycemic control that results in placing the majority of patients into remission [125,126]. The success of bariatric surgery, and especially RYGB in treating T2D has prompted extensive research into the mechanisms that result in the improvement of insulin sensitivity and resolution of T2D postoperatively [127]. The factors that are simultaneously altered after RYGB that improve T2D include caloric restriction, decreased absorption of vitamins and nutrients, decreased adipose mass, altered intestinal hormone levels, as well

as glucose metabolism [128,129]. After RYGB, T2D is improved by increased gut hormone levels, such as ghrelin, GLP-1, and PYY that improve insulin sensitivity [130,131] altered gut gluconeogenesis, decreased muscular lipid levels [132], and increasing serum BA levels which indirectly elevate gut hormones [5].

After RYGB in type 2 diabetics, the relative abundances of microbiota in the gut are altered. Most significant are dramatic increases in Proteobacteria and decreased quantities of Firmicutes and Bacteroidetes [38]. Post-RYGB, normal weight rodents had a decrease in the Firmicutes to *Bacteroides* ratio but were found to have a 52 time increase in the amount of Proteobacteria [1]. Similar results have been established in multiple previous studies [1,2,16,86,133–135]. Metagenomic sequencing of fecal microbiota in morbidly obese, T2D before and 3 months after RYGB demonstrated relative increased abundances of Proteobacteria and decreased levels of Firmicutes and Bacteroidetes [38].

In the Proteobacteria phylum most species are facultative anaerobes, in contrast Firmicutes are most likely anaerobic bacteria. After RYGB there is an increase in dissolved oxygen, which may favor the growth of facultative anaerobe microbiota and a decreased in anaerobes, favoring Proteobacteria. Another explanation may be the increased availability of substrates [136]. Analysis of species-specific changes has demonstrated to be more beneficial than broad phylum changes in the understanding of potential causes for improvement in metabolic and inflammatory parameters. Alterations in the composition of the gut microbiota correlate with gene function and result in an improvement in metabolism and inflammation [38].

Akkermansia

The abundances of Verrucomicrobium *Akkermansia* have been found to increase in post-RYGB patients in multiple studies including mice and humans [1,2,19]. When the gut microbiota from post-RYGB mice was transplanted into nonoperated, germ-free mice, the result was weight loss and adipose mass reduction. It was believed to be secondary to an alteration of the microbial production of SCFAs. It has been suggested [19] that *Akkermansia* could regulate weight loss and adipose tissue but this is not proven [19]. The study supports mechanisms beyond weight loss that improve metabolic and inflammatory diseases [19].

Akkermansia has not been found to significantly correlate with C-reactive protein (CRP) changes [38]. *Akkermansia* may be responsible for altered inflammation and insulin sensitivity after RYGB but its repeat increase abundance in multiple studies has only been found to be associated with an improvement in metabolic disease and inflammation, and has not withstood the rigors of multivariate analysis [19,38,137]. Metformin treatment has been associated with increased levels of *Akkermansia* species in high-fat diet fed

mice leading to improved glucose homeostasis [138]. Direct supplementation with *Akkermansia* as a probiotic also results in improvement of glucose intolerance, metabolic endotoxemia, and tissue inflammation [138,139]. Despite clear evidence in the post-RYGB population, treatment of young mice with vancomycin has revealed an increase population of *Akkermansia muciniphila* with a resultant decrease incidence of T2D, further supporting the antiinflammatory benefits of this microbiota [140].

Faecalibacterium prausnitzii

F. prausnitzii is as a dominant species in the microbiome of healthy hosts. It is a butyrate-producing species that is preventative in acute inflammatory disease [141]. Reduced abundance of *F. prausnitzii* has been found in inflammatory bowel disease as well as in infectious colitis [142]. *F. prausnitzii* is involved in low-grade inflammatory disease, such as obesity and diabetes [2,143–145]. The relative abundances of *F. prausnitzii* are significantly lower in type 2 diabetic patients that develop more severe inflammation and worse insulin resistance [146]. Changes in intestinal hormones may be related to changes in the abundance of *F. prausnitzii* [2]. Modulation of urinary metabolites by *F. prausnitzii* has provided evidence that the species is active in many metabolic pathways in the host microbiome [147].

The microbe, *F. prausnitzii*, is capable of blocking nuclear factor-κB activation, which results in decreased secretion of proinflammatory mediators and an overall, antiinflammatory effect [148,149]. Consumption of *F. prausnitzii* or the substance produced by *F. prausnitzii* cultures has been found to increase interleukin-10 (IL-10) by blood mononuclear cells, which is antiinflammatory and decrease IL-12, which is proinflammatory in the colon [150,151].

Post-RYGB Faecalibacterium prausnitzii

At 3 months after RYGB surgery, patients reduce both BMI and improve inflammatory parameters [38]. Metagenomic alterations correlate with improvements in blood serum lipids and carbohydrate metabolism [38]. One study identified 22 microbial species significantly affected that are associated with metabolic changes [38]. An analysis by Spearmen's rank correlation reveals an association of 10 species with plasma total cholesterol and low-density lipoprotein cholesterol. Five species changes were associated with decreased triglycerides and two with HbA$_{1c}$ level [38].

The alteration of the gut microbiota can account for many of the changes that occur in the host withT2D [152]. Different microbiota have been found to have BMI dependent and independent changes [38]. The BMI independent changes may account for the immediate metabolic affects that occur after RYGB and other bariatric surgeries [38]. Some studies reported normalization of glucose homeostasis in patients with T2D long before a noticeable weight reduction was achieved [153].

Univariate analysis has revealed a decrease in the abundance of *F. prausnitzii*, which is associated with diabetes and was found to be associated with improved glucose levels after RYGB [2,38]. There is a negative correlation between the relative abundance of *F. prausnitzii* and homeostatic model assessment (HOMA-IR), which is consistent with the finding of improved glucose metabolism in post-RYGB diabetic patients.[2]. Adiponectin, another marker of insulin sensitivity, has not been found to correlate, as the discrepancy between HOMA-IR and adiponectin with weight loss has been documented in the past [2,153]. Though *F. prausnitzii*, RYGB has the similar ability to statins and salicylates, to modulate nuclear factor-κB and potentially improve insulin sensitivity in T2D patients [150,151].

F. prausnitzii was both negatively associated with inflammatory markers before RYGB and throughout the postoperative period, indicating its beneficial contribution to a healthy gut. The post-RYGB benefits of *F. prausnitzii* of improving T2D and metabolic disease were originally believed to be independent of BMI [2]. A study was valid for severely obese humans but may not be valid in those who are moderately obese [2]. Other studies suggest that the inflammatory benefits, such as T2D remission are BMI dependent [38]. Host-microbial cross talk may explain the weight-dependent changes in metabolic disease after RYGB [38].

F. prausnitzii and CRP are found to directly correlate [38]. The nature of the correlation conflicts with another study that established a negative association between the two [2]. When *F. prausnitzii* increases in abundance, CRP, IL-6, and orosomucoid serum decreases up to 6 months after RYGB. The change was independent of food or calorie intake [2]. The presence of *F. prausnitzii* directly or indirectly correlates with 10 other species associated with BMI, but not all with CRP. Of these species, six significantly correlated with BMI as well as CRP, indicating that these variables may be dependent upon each other [38]. The cross talk that occurs between species of microbiota confuses the pathophysiologic contribution that each microbe contributes.

A multivariate analysis of the association of microbiota abundance changes in post-RYGB patients with significantly decreased CRP levels and BMI has been performed. Findings indicate that most post-RYGB changes in the abundance of gut microbiota that affect inflammation, represented by CRP, are strongly related to a decreasing BMI. The microbiota that decrease CRP and are dependent upon BMI include: *F. prausnitzii, C. comes, Lactobacillus acidophilus, L. succinogenes,* and *Treponema pallidum* which all decreased in abundance, and *Enterobacter cancerogenus, Veillonella dispar, Trichothecium roseum* which increased in abundance after RYGB [38]. Only *Staphylococcusepidermidis*, which was associated with CRP and was not associated with BMI, decreased in abundance [38].

F. prausnitzii, C. comes, and *Anaerostipes caccae* are all butyrate-producing species that decreased in abundance [38], indicating that substrate may be associated with these improvements.

Other human studies support the findings of BMI independent changes that occur with the microbiota resulting in inflammatory and metabolic improvement. Further supporting the weight-independent changes that occur in type 2 diabetic patients that undergo RYGB and are put into remission of their disease [38]. In mice with weight-induced T2D, increased levels of *E. coli* and *Bifidobacterium* improve metabolic disease parameters and may be associated with the BMI-independent improvements that occur after RYGB [122,154].

VSG and Inflammation

In post-VSG mice, the abundance of *Bacteroides* was reduced as weight was lost and glucose tolerance improved, indicating that metabolic benefits of VSG are the responsibility of the FXR receptor [5]. After VSG in humans, *E. rectale* was decreased in relative abundance. *E. rectale* has previously been shown to positively correlate with obesity-related comorbidities and metabolic status [10,155]. *Roseburia* was 12 times higher in post-VSG mice and the abundance was associated with an improved glucose tolerance test [5]. Metagenomic diet studies in humans reveal decreased amounts of *Roseburia* in with T2D [101,102].

White Adipose Fat

Post-RYGB patients increase the gene expression in white adipose tissue (WAT) for at least 6 months after surgery. The gene increase was from 8 WAT genes before to 562 WAT genes post-RYGB; it was in direct relationship to weight loss and change in metabolism but independent of caloric intake [3]. The most significant changes in genes were those coding for metabolic and inflammatory processes, such as the transforming growth factor-β signaling pathway. The down-regulated genes were the metabolic pathways, such as 24-dehydrocholesterol reductase [3]. The most abundant microbiota in the post-RYGB patients exhibited WAT genes relating to cell transport and binding, signaling, enzymatic activity, and the structure of the cells [3]. Gut microbiota and WAT genes or clinical variables were dependent on the variation in calorie intake in nearly 50% of the associations after RYGB. This finding lends further support of BMI-independent alterations of bariatric surgery [38].

ENERGY HARVEST AND FXR SIGNALING HYPOTHESES

In the past, it was believed that alteration in gut microbiota may directly or indirectly lead to decreased energy removed from a stable amount of nutrient that is consumed.

As bariatric surgery alters the microbiota in the gut, it is also believed that surgery leads to decreased energy harvest, which explains the weight loss that occurs in these patients.

There is evidence that energy harvested from food that traverses the gut may be modulated by the metagenomic characteristics of different types of gut microbiota [33]. There remains a stable quantity of microbial genes that control major functions including metabolism and the growth of cells in both obese and lean, mice, and humans. Variations in the quantity of genes that control harvesting of energy and the metabolism of polysaccharides may explain differences in energy extraction from nutrients that are consumed [33,135,156,157]. Microbiota from the obese are believed to be capable of harvesting a higher quantity of energy from the diet consumed [33]: they are especially more efficient at extracting energy from fat and carbohydrates [84,158].

It is believed that changes that occur in the types of microbiota present in patients after post-RYGB are a result of the patient adapting to maximize the amount of energy harvested from the gut in a situation that is similar to starvation [159]. The microbiota in post-RYGB patients remains stable at both 3 and 6 months despite improvements in weight and metabolic parameters supporting a permanent change in their microbiota abundances [2]. Another study demonstrates exchanges in fecal microbiota 1 week after RYGB [3].

The type of species that develop may be a response to calorie restriction. Patients with anorexia nervosa have relatively high levels of *M smithii*. This microbiota has been associated with a more efficient bacterial fermentation processes, leading to a larger energy harvest [23]. Alterations in individual microbiota may have an increase or decrease effect on the amount of energy presented to the host. Mice that were colonized with *M. smithii* increased their energy harvest by the ability to metabolize fructans to SCFAs. The end result is that these mice gained weight [160,161]. Microbiota may also work more efficiently together. There is higher polysaccharide metabolism efficiency when two bacteria species: eubacteria and *Bacteroides* are both present in mice compared to one alone [162]. It is possible that the complete opposite mechanisms occur after bariatric surgery.

Reduced caloric intake post-RYGB may force the host to use an alternate quantity of substrate in an attempt to compensate for decreased energy available as demonstrated by functional genomic analysis revealing 13 orthologs for the phosphotransferase system [2,38]. The finding of reduced amounts of SCFA in both post-RYGB and post-VBG microbiota transplanted to mice indicates that a reduced energy harvest from the ingested diet may be the mechanism to explain a reduction in fat mass gain [4]. Bariatric surgery patients including RYGB and VBG were both found to shift their microbiome characteristics away from SCFA toward increased amino acid fermentation and appears to

be a diet independent phenomenon [4]. Decreased SCFA fermentation is supportive of the belief that there is a reduction of energy harvest from the diet in postbariatric surgery patients [4].

The differences in the quantity of fatty acids and metabolites in WT and $FXR^{-/-}$ mice post-VSG do not fully support the energy harvest hypothesis [5]. The mechanism of more efficient extraction of energy from macronutrients by microbiota has been contested as too simple. Studies in mice support a connection of FXR signaling and the alterations of gut microbiota with the metabolic consequences of bariatric surgery [5]. There is some evidence of the effects of VSG on humans [10].

After VSG, the ability of the gut microbiota to harvest energy decreased. The microbiota were less able to ferment butyrate and, therefore there was a fecal loss of both fatty acids and BAs [10]. VSG altered the gut microbiota to abundances that reflect a lean microbiome. As the fecal substrates were lost in the stool of post-VBG patients, malabsorption occurred along with impairment of BA circulation [10]. Whether the direct mechanism of weight loss in the bariatric surgery patient is a decrease in energy harvested or an alteration of the FXR signaling system, or a combination of both mechanisms, has not been clearly defined. Further research shall elucidate the workings that create an environment of weight loss, improved metabolism, and decreased inflammation.

LESSONS LEARNED FROM BARIATRIC SURGERY TO CREATE NONSURGICAL WEIGHT LOSS AND METABOLIC TREATMENTS

Efforts to resolve obesity with alterations in the microbiome may focus on determining the host phenotype, genetics, microbial taxonomy, and metagenomics as well as metabolomics that are associated with obesity and metabolic disorders. It is evident that bariatric surgery, such as RYGB, VBG, and VSG can alter the microbiota to cause weight loss and decreased metabolic disease. Other clinically available bariatric surgeries have yet to be evaluated from a microbial standpoint. Future applications of the findings from bariatric surgery may apply to better usage of more specific prebiotics and antibiotics for the treatment of obesity without surgery. One controlled study in humans who have undergone bariatric surgery has revealed increase weight loss when treated with probiotics after gastric bypass [163].

In the experimental setting, studies have revealed that the transplantation of microbiota from the gut of mice to gnotobiotic mice results in weight gain [26,33]. Transplantation of microbiota has now become a treatment for severe *C. difficile* colitis. In an unexpected consequence of treatment, weight gain has been demonstrated in a lean human treated for *C. difficile* infection by an obese donor [164].

The corollary has been observed in the experimental setting. The transplantation of microbiota from the gut of post-RYGB mice is associated with reduced weight when compared to sham-operated mice [19]. The effect of weight loss is more dramatic when the microbiota from post-RYGB mice were transplanted into obese mice causing a 26% weight loss than when post-RYGB, mouse microbiota are transplanted in lean mice resulting in only 5% weight loss. Irrespective of the starting weight of the animal, the transplanted post-RYGB microbiota cause weight loss supporting that gut microbiota may modulate weight and adiposity [19].

The microbiota in the feces of post-RYGB, post-VBG, and obese female humans were transplanted into germ-free mice [5]. The mice which had been transplanted by post-RYGB patients, accumulated 43% less body fat and while the post-VBG patients accumulated 26% less body fat than mice that were colonized by the control group of obese human microbiota [5]. When the post-RYGB and post-VGB groups were compared, the post-RYGB mice had a higher average increase in lean mass but body weight and food intake was not different in the two groups [5].

Much can be learned from analysis of postbariatric patients that may be applied to nonsurgical patients. The most significant ability demonstrated may be the potential to transmit the human adiposity phenotype by transplanting the gut microbiota from one human to another [5]. This finding may someday lead to the ability to transplant anti-obesity phenotypes through microbiota from one human to another in the clinical setting. Determining the proper mix of species to dwell in the human gut, and how to colonize and sustain them without having to perform bariatric surgery on the host remains the ultimate challenge of bariatric surgery research. These treatments would not have been imaginable without bariatric surgery.

REFERENCES

[1] Zhang H, DiBaise JK, Zuccolo A, et al. Human gut microbiota in obesity and after gastric bypass. Proc Natl Acad Sci USA 2009;106(7):2365–70.

[2] Furet JP, Kong LC, Tap J, et al. Differential adaptation of human gut microbiota to bariatric surgery-induced weight loss: links with metabolic and low-grade inflammation markers. Diabetes 2010;59(12):3049–57.

[3] Kong LC, Tap J, Aron-Wisnewsky J, et al. Gut microbiota after gastric bypass in human obesity: increased richness and associations of bacterial genera with adipose tissue genes. Am J Clin Nutr 2013;98:16–24.

[4] Tremaroli V, Karlsson F, Werling M, et al. Roux-en-Y gastric bypass and vertical banded gastroplasty induce long-term changes on the human gut microbiome contributing to fat mass regulation. Cell Metab 2015;22(2):228–38.

[5] Ryan KK, Tremaroli V, Clemmensen C, et al. FXR is a molecular target for the effects of vertical sleeve gastrectomy. Nature 2014;509:183–8.

[6] Ashrafian H, Ahmed K, Rowland SP, et al. Metabolic surgery and cancer: protective effects of bariatric procedures. Cancer 2011;117(9):1788–99.

[7] Ashrafian H, Athanasiou T, Li JV, et al. Diabetes resolution and hyperinsulinaemia after metabolic Roux-en-Y gastric bypass. Obes Rev 2011;12(5):e257–72.

[8] Ashrafian H, Bueter M, Ahmed K, et al. Metabolic surgery: an evolution through bariatric animal models. Obes Rev 2010;11(12):907–20.

[9] Ashrafian H, Darzi A, Athanasiou T. Autobionics: a new paradigm in regenerative medicine and surgery. Regen Med 2010;5:279–88.

[10] Damms-Machado A, Mitra S, Schollenberger AE, et al. Effects of surgical and dietary weight loss therapy for obesity on gut microbiota composition and nutrient absorption. Biomed Res Int 2015;2015:806248.

[11] Mascher T, Helmann JD, Unden G. Stimulus perception in bacterial signal-transducing histidine kinases. Microbiol Mol Biol Rev 2006;70:910–38.

[12] Barrett EL, Kwan HS. Bacterial reduction of trimethylamine oxide. Annu Rev Microbiol 1985;39:131–49.

[13] Tang WH, Wang Z, Levison BS, et al. Intestinal microbial metabolism of phosphatidylcholine and cardiovascular risk. N Engl J Med 2013;368:1575–84.

[14] Wang Z, Klipfell E, Bennett BJ, et al. Gut flora metabolism of phosphatidylcholine promotes cardiovascular disease. Nature 2011;472:57–63.

[15] Sjöström L, Peltonen M, Jacobson P, et al. Bariatric surgery and long-term cardiovascular events. JAMA 2012;307:56–65.

[16] Ley RE, Turnbaugh PJ, Klein S, Gordon JI. Microbial ecology: human gut microbes associated with obesity. Nature 2006;444(7122):1022–3.

[17] Duncan SH, Lobley GE, Holtrop G, et al. Human colonic microbiota associated with diet, obesity and weight loss. Int J Obes (Lond) 2008;32(11):1720–4.

[18] Cotillard A, Kennedy SP, Kong LC, et al. ANR MicroObes consortium dietary intervention impact on gut microbial gene richness. Nature 2013;500:585–8.

[19] Liou AP, Paziuk M, Luevano JM Jr, Machineni S, Turnbaugh PJ, Kaplan LM. Conserved shifts in the gut microbiota due to gastric bypass reduce host weight and adiposity. Sci Transl Med 2013;5:178ra41.

[20] Peat CM, Kleiman SC, Bulik CM, Carroll IM. The intestinal microbiome in bariatric surgery patients. Eur Eat Disord Rev 2015;23:496–503.

[21] Mumphrey MB, Hao Z, Townsend RL, et al. Eating in mice with gastric bypass surgery causes exaggerated activation of brainstem anorexia circuit. Int J Obes 2016;40(6):921–8.

[22] Pfleiderer A, Lagier JC, Armougom F, Robert C, Vialettes B, Raoult D. Culturomics identified 11 new bacterial species from a single anorexia nervosa stool sample. Eur J Clin Microbiol Infect Dis 2013;32(11):1471–81.

[23] Armougom F, Henry M, Vialettes B, Raccah D, Raoult D. Monitoring bacterial community of human gut microbiota reveals an increase in *Lactobacillus* in obese patients and Methanogens in anorexic patients. PLoS One 2009;4(9):e7125.

[24] Schwiertz A, Taras D, Schäfer K, et al. Microbiota and SCFA in lean and overweight healthy subjects. Obesity (Silver Spring) 2010;18(1):190–5.

[25] Zhang C, Zhang M, Wang S, et al. Interactions between gut microbiota, host genetics and diet relevant to development of metabolic syndromes in mice. ISME J 2010;4(2):232–41.

[26] Turnbaugh PJ, Ridaura VK, Faith JJ, Rey FE, Knight R, Gordon JI. The effect of diet on the human gut microbiome: a metagenomic analysis in humanized gnotobiotic mice. Sci Transl Med 2009;1(6):6ra14.

[27] De Filippo C, Cavalieri D, Di Paola M, et al. Impact of diet in shaping gut microbiota revealed by a comparative study in children from Europe and rural Africa. Proc Natl Acad Sci USA 2010;107(33):14691–6.

[28] Muegge BD, Kuczynski J, Knights D, et al. Diet drives convergence in gut microbiome functions across mammalian phylogeny and within humans. Science 2011;332(6032):970–4.

[29] Walker AW, Ince J, Duncan SH, et al. Dominant and diet-responsive groups of bacteria within the human colonic microbiota. ISME J 2011;5(2):220–30.

[30] Wu GD, Chen J, Hoffmann C, et al. Linking long-term dietary patterns with gut microbial enterotypes. Science 2011;334(6052):105–8.

[31] Fava F, Gitau R, Griffin BA, Gibson GR, Tuohy KM, Lovegrove JA. The type and quantity of dietary fat and carbohydrate alter faecal microbiome and short-chain fatty acid excretion in a metabolic syndrome 'at-risk' population. Int J Obes 2013;37(2):216–23.

[32] David LA, Maurice CF, Carmody RN, et al. Diet rapidly and reproducibly alters the human gut microbiome. Nature 2014;505(7484):559–63.

[33] Turnbaugh PJ, Ley RE, Mahowald MA, Magrini V, Mardis ER, Gordon JI. An obesity-associated gut microbiome with increased capacity for energy harvest. Nature 2006;444(7122):1027–31.

[34] Wilson-Pérez HE1, Chambers AP, Sandoval DA, et al. The effect of vertical sleeve gastrectomy on food choice in rats. Int J Obes (Lond) 2013;37(2):288–95.

[35] Laurenius A, Larsson I, Melanson KJ, et al. Decreased energy density and changes in food selection following Roux-en-Y gastric bypass. Eur J Clin Nutr 2013;67(2):168–73.

[36] Münzberg H, Laque A, Yu S, Rezai-Zadeh K, Berthoud HR. Appetite and body weight regulation after bariatric surgery. Obes Rev 2015;16(Suppl. 1):77–90.

[37] Flint HJ, Scott KP, Louis P, Duncan SH. The role of the gut microbiota in nutrition and health. Nat Rev Gastroenterol Hepatol 2012;9:577–89.

[38] Graessler J, Qin Y, Zhong H, et al. Metagenomic sequencing of the human gut microbiome before and after bariatric surgery in obese patients with type 2 diabetes: correlation with inflammatory and metabolic parameters. Pharmacogenomics J 2013;13(6):514–22.

[39] Melissas J, Kampitakis E, Schoretsanitis G, et al. Does reduction in gastric acid secretion in bariatric surgery increase diet-induced thermogenesis? Obes Surg 2002;12:236–40.

[40] Duncan SH, Louis P, Thomson JM, Flint HJ. The role of pH in determining the species composition of the human colonic microbiota. Environ Microbiol 2009;11:2112–22.

[41] Sandoval D. Bariatric surgeries: beyond restriction and malabsorption. Int J Obes (Lond) 2011;35(Suppl. 3):S45–9.

[42] Cani PD, Delzenne NM. The role of the gut microbiota in energy metabolism and metabolic disease. Curr Pharm Des 2009;15(13):1546–58.

[43] Cani PD, Possemiers, Van de Wiele T, et al. Changes in gut microbiota control inflammation in obese mice through a mechanism involving GLP-2-driven improvement of gut permeability. Gut 2009;58(8):1091–103.

[44] Queipo-Ortuno MI, Seoane LM, Murri M, et al. Gut microbiota composition in male rat models under different nutritional status and physical activity and its association with serum leptin and ghrelin levels. PLoS One 2013;8(5):e65465.

[45] Ravussin Y, Koren O, Spor A, et al. Responses of gut microbiota to diet composition and weight loss in lean and obese mice. Obesity (Silver Spring) 2012;20(4):738–47.

[46] Tarini J, Wolever TM. The fermentable fibre inulin increases postprandial serum short-chain fatty acids and reduces free-fatty acids and ghrelin in healthy subjects. Appl Physiol Nutr Metab 2010;35(1):9–16.

[47] Korner J, Inabnet W, Febres G, et al. Prospective study of gut hormone and metabolic changes after adjustable gastric banding and Roux-en-Y gastric bypass. Int J Obes (Lond) 2009;33:786–95.

[48] Stylopoulos N, Hoppin AG, Kaplan LM. Roux-en-Y gastric bypass enhances energy expenditure and extends lifespan in diet-induced obese rats. Obesity (Silver Spring) 2009;17:1839–47.

[49] Ahima RS, Prabakaran D, Mantzoros C, et al. Role of leptin in the neuroendocrine response to fasting. Nature 1996;382:250–2.

[50] Sweeney TE, Morton JM. Metabolic surgery: action via hormonal milieu changes, changes in bile acids or gut microbiota? A summary of the literature. Best Pract Res Clin Gastroenterol 2014;28(4):727–40.

[51] Glicksman C, Pournaras DJ, Wright M, et al. Postprandial plasma bile acid responses in normal weight and obese subjects. Ann Clin Biochem 2010;47(Pt 5):482–4.

[52] Pournaras DJ, Glicksman C, Vincent RP, et al. The role of bile after Roux-en-Y gastric bypass in promoting weight loss and improving glycaemic control. Endocrinology 2012;153:3613–9.

[53] Halmy L, Fehér T, Steczek K, Farkas A. High serum bile acid level in obesity: its decrease during and after total fasting. Acta Med Hung 1986;43(1):55–8.

[54] Mok HY, von Bergmann K, Crouse JR, Grundy SM. Bilary lipid metabolism in obesity. Effects of bile acid feeding before and during weight reduction. Gastroenterology 1979;76(3):556–67.

[55] Trouillot TE, Pace DG, McKinley C, et al. Orlistat maintains biliary lipid composition and hepatobiliary function in obese subjects undergoing moderate weight loss. Am J Gastroenterol 2001;96(6):1888–94.

[56] Islam KB, Fukiya S, Hagio M, et al. Bile acid is a host factor that regulates the composition of the cecal microbiota in rats. Gastroenterology 2011;141:1773–81.

[57] Nie YF1, Hu J, Yan XH. Cross-talk between bile acids and intestinal microbiota in host metabolism and health. J Zhejiang Univ Sci B 2015;16(6):436–46.

[58] Akao T, Akao T, Hattori M, Namba T, Kobashi K. Enzymes involved in the formation of $3\beta,7\beta$-dihydroxy-12-oxo-5β-cholanic acid from dehydrocholic acid by *Ruminococcus* sp. obtained from human intestine. Biochim Biophys Acta 1987;921(2):275–80.

[59] Jones BV, Begley M, Hill C, Gahan CG, Marchesi JR. Functional and comparative metagenomic analysis of bile salt hydrolase activity in the human gut microbiome. Proc Natl Acad Sci USA 2008;105:13580–5.

[60] Miyata M, Yamakawa H, Hamatsu M, Kuribayashi H, Takamatsu Y, Yamazoe Y. Enterobacteria modulate intestinal bile acid transport and homeostasis through apical sodium-dependent bile acid transporter (SLC10A2) expression. J Pharmacol Exp Ther 2011;336:188–96.

[61] Kakiyama G, Pandak WM, Gillevet PM, et al. Modulation of the fecal bile acid profile by gut microbiota in cirrhosis. J Hepatol 2013;58:949–55.

[62] Edenharder R, Pfutzner M. Partial purification and char-acterization of an NAD-dependent 3 beta-hydroxysteroid dehydrogenase from *Clostridium innocuum*. Appl Environ Microbiol 1989;55(6):1656–9.

[63] Edenharder R, Pfutzner M, Hammann R. NADP-dependent 3β-, 7α- and 7β-hydroxysteroid dehydrogenase activities from a lecithinase-lipase-negative *Clostridium* species 25.11.c. Biochim Biophys Acta 1989;1002(1):37–44.

[64] Motino AD, Hoffman T, Dawson PA, et al. Increased expression of ileal apical sodium-dependent bile acid transporter in postpartum rats. Am J Physiol Gastrointest Liver Physiol 2002;282:G41–50.

[65] Swann JR, Want EJ, Geier FM, et al. Systemic gut microbial modulation of bile acid metabolism in host tissue compartments. Proc Natl Acad Sci USA 2011;108:4523–30.

[66] Nakatani H, Kasama K, Oshiro T, Watanabe M, Hirose H, Itoh H. Serum bile acid along with plasma incretins and serum high-molecular weight adiponectin levels are increased after bariatric surgery. Metabolism 2009;58(10):1400–7.

[67] Myronovych A, Kirby M, Ryan KK, et al. Vertical sleeve gastrectomy reduces hepatic steatosis while increasing serum bile acids in a weight-loss-independent manner. Obesity (Silver Spring) 2014;22:390–400.

[68] Kholi R1, Myronovych A, Tan BK, et al. Bile acid signaling: mechanism for bariatric surgery, Cure for NASH? Dig Dis 2015;33(3):440–6.

[69] Steinert RE, Peterli R, Keller S, et al. Bile acids and gut peptide secretion after bariatric surgery: a 1-year prospective randomized pilot trial. Obesity (Silver Spring) 2013;21(12):E660–8.

[70] Kohli R, Bradley D, Setchell KD, Eagon JC, Abumrad N, Klein S. Weight loss induced by Roux-en-Y gastric bypass but not laparoscopic adjustable gastric banding increases circulating bile acids. J Clin Endocrinol Metab 2013;98(4):E708–12.

[71] Haluzíková D, Lacinová Z, Kaválková P, et al. Laparoscopic sleeve gastrectomy differentially affects serum concentrations of FGF-19 and FGF-21 in morbidly obese subjects. Obesity (Silver Spring) 2013;21(7):1335–42.

[72] Scholtz S, Miras AD, Chhina N, et al. Obese patients after gastric bypass surgery have lower brain-hedonic responses to food than after gastric banding. Gut 2014;63(6):891–902.

[73] Werling M, Vincent RP, Cross GF, et al. Enhanced fasting and post-prandial plasma bile acid responses after Roux-en-Y gastric bypass surgery. Scand J Gastroenterol 2013;48(11):1257–64.

[74] Ahmad NN, Pfalzer A, Kaplan LM. Roux-en-Y gastric bypass normalizes the blunted postprandial bile acid excursion associated with obesity. Int J Obes (Lond) 2013;37(12):1553–9.

[75] Ashrafian H, Li JV, Spagou K, et al. Bariatric surgery modulates circulating and cardiac metabolites. J Proteome Res 2014;13(2):570–80.

[76] Gerhard GS, Styer AM, Wood GC, et al. A role for fibroblast growth factor 19 and bile acids in diabetes remission after Roux-en-Y gastric bypass. Diabetes Care 2013;36(7):1859–64.

[77] Jansen PL, van Werven J, Aarts E, et al. Alterations of hormonally active fibroblast growth factors after Roux-en-Y gastric bypass surgery. Dig Dis 2011;29(1):48–51.

[78] Patti ME, Houten SM, Bianco AC, et al. Serum bile acids are higher in humans with prior gastric bypass: potential contribution to improved glucose and lipid metabolism. Obesity (Silver Spring) 2009;17(9):1671–7.

[79] Simonen M, Dali-Youcef N, Kaminska D, et al. Conjugated bile acids associate with altered rates of glucose and lipid oxidation after Roux-en-Y gastric bypass. Obes Surg 2012;22(9):1473–80.

[80] Hylemon PB, Zhou H, Pandak WM, Ren S, Gil G, Dent P. Bile acids as regulatory molecules. J Lipid Res 2009;50:1509–20.

[81] Lefebvre P, Cariou B, Lien F, Kuipers F, Staels B. Role of bile acids and bile acid receptors in metabolic regulation. Physiol Rev 2009;89:147–91.

[82] Merritt ME, Donaldson JR. Effect of bile salts on the DNA and membrane integrity of enteric bacteria. J Med Microbiol 2009;58(Pt 12):1533–41.

[83] Sayin SI, Wahlström A, Felin J, et al. Gut microbiota regulates bile acid metabolism by reducing the levels of tauro-beta-muricholic acid, a naturally occurring FXR antagonist. Cell Metab 2013;17(2):225–35.

[84] Ley RE, Backhed F, Turnbaugh P, Lozupone CA, Knight RD, Gordon JI. Obesity alters gut microbial ecology. Proc Natl Acad Sci USA 2005;102:11070–5.

[85] Ridaura VK, Faith JJ, Rey FE, et al. Gut microbiota from twins discordant for obesity modulate metabolism in mice. Science 2013;341:1241214.

[86] Li JV, Ashrafian H, Bueter M, et al. Metabolic surgery profoundly influences gut microbial-host metabolic cross-talk. Gut 2011;60(9):1214–23.

[87] Larsson E, Tremaroli V, Lee YS, et al. Analysis of gut microbial regulation of host gene expression along the length of the gut and regulation of gut microbial ecology through MyD88. Gut 2012;61(8):1124–31.

[88] MacHado AD, Forster-Fromme K, Mitra S, et al. Competence network obesity: an integrated multi -omics approach to study the guest-host metabolic interaction during restrictive obesity intervention. Obes Facts 2012;5:4–5. (abstract FV 2.5).

[89] Beenken A, Mohammadi M. The FGF family: biology, pathophysiology and therapy. Nat Rev Drug Discov 2009;8:235–53.

[90] Thomas C, Pellicciari R, Pruzanski M, Auwerx J, Schoonjans K. Targeting bile-acid signalling for metabolic diseases. Nat Rev Drug Discov 2008;7:678–9.

[91] Rajendra R. Ménage-à-trois of bariatric surgery, bile acids and the gut microbiome. World J Diabetes 2015;6(3):367–70.

[92] Kohli R, Kirby M, Setchell KD, et al. Intestinal adaptation after ileal interposition surgery increases bile acid recycling and protects against obesity-related comorbidities. Am J Physiol Gastrointest Liver Physiol 2010;299:G652–60.

[93] Kohli R, Setchell KD, Kirby M, et al. A surgical model in male obese rats uncovers protective effects of bile acids post-bariatric surgery. Endocrinology 2013;154:2341–51.

[94] Habegger KM, Al-Massadi O, Heppner KM, et al. Duodenal nutrient exclusion improves metabolic syndrome and stimulates villus hyperplasia. Gut 2014;63(8):1238–46.

[95] Perkins WJ, Hale ER, Fernandez AZ, Dawson P, Weinberg RB. Mo2046 differential effects of laparoscopic sleeve gastrectomy and Roux-en-Y gastric bypass on dietary fatty acid absorption, bile acid absorption, and post-prandial gut hormone secretion. Gastroenterology 2014;146:S-726–7.

[96] Kang K, Schmahl J, Lee JM, et al. Mouse ghrelin-O-acyltransferase (GOAT) plays a critical role in bile acid reabsorption. FASEB 2012;26:259–71.

[97] Kuribayashi H, Miyata M, Yamakawa H, Yoshinari K, Yamazoe Y. Enterobacteria-mediated deconjugation of taurocholic acid

enhances ileal farnesoid X receptor signaling. Eur J Pharmacol 2012;697:132–8.

[98] Clement K, Langin D. Regulation of inflammation-related genes in human adipose tissue. J Intern Med 2007;262:422–30.

[99] Pradhan A. Obesity, metabolic syndrome, and type 2 diabetes: inflammatory basis of glucose metabolic disorders. Nutr Rev 2007;65:S152–6.

[100] Bäckhed F, Ding H, Wang T, et al. The gut microbiota as an environmental factor that regulates fat storage. Proc Natl Acad Sci USA 2004;101:15718–23.

[101] Qin J, Li Y, Cai Z, et al. A metagenome-wide association study of gut microbiota in type 2 diabetes. Nature 2012;490:55–60.

[102] Karlsson FH, Tremaroli V, Nookaew I, et al. Gut metagenome in European women with normal, impaired and diabetic glucose control. Nature 2013;498:99–103.

[103] Smith MI, Yatsunenko T, Manary MJ, et al. Gut microbiomes of Malawian twin pairs discordant for kwashiorkor. Science 2013;339:548–54.

[104] Tremaroli V, Bäckhed F. Functional interactions between the gut microbiota and host metabolism. Nature 2012;489:242–9.

[105] Larsen N, Vogensen FK, van den Berg FW, Nielsen DS, Andreasen AS, Pedersen BK, et al. Gut microbiota in human adults with type 2 diabetes differs from non-diabetic adults. PLoS One 2010;5:e9085.

[106] Vrieze A, Van Nood E, Holleman F, et al. Transfer of intestinal microbiota from lean donors increases insulin sensitivity in individuals with metabolic syndrome. Gastroenterology 2012;143:913–916e7.

[107] Palau-Rodriguez M, Tulipani S, Isabel Queipo-Ortuño M, Urpi-Sarda M, Tinahones FJ, Andres-Lacueva C. Metabolomic insights into the intricate gut microbial–host interaction in the development of obesity and type 2 diabetes. Front Microbiol 2015;6:1151.

[108] Mokdad AH, For ES, Bowman BA, et al. Prevalence of obesity, diabetes, and obesity-related health risk factors, 2001. JAMA 2003;289:76–9.

[109] Tilg H, Moschen AR, Kaser A. Obesity and the microbiota. Gastroenterology 2009;136:1476–83.

[110] Musso G, Gambino R, Cassader M. Obesity, diabetes, and gut microbiota: the hygiene hypothesis expanded? Diabetes Care 2010;33:2277–84.

[111] Samuel BS, Shaito A, Motoik T, et al. Effects of the gut microbiota on host adiposity are modulated by the short-chain fatty-acid binding G protein-coupled receptor, Gpr41. Proc Natl Acad Sci USA 2008;105:16767–72.

[112] Gao Z, Yin J, Zhang J, et al. Butyrate improves insulin sensitivity and increases energy expenditure in mice. Diabetes 2009;58:1509–17.

[113] Lin HV, Frassetto A, Kowalik EJ Jr, et al. Butyrate and propionate protect against diet-induced obesity and regulate gut hormones via free fatty acid receptor 3-independent mechanisms. PLoS One 2012;7:e35240.

[114] Cani PD, Bibiloni R, Knauf C, et al. Changes in gut microbiota control metabolic endotoxemia-induced inflammation in high-fat diet-induced obesity and diabetes in mice. Diabetes 2008;57:1470–81.

[115] Cani PD, Amar J, Iglesias MA, et al. Metabolic endotoxemia initiates obesity and insulin resistance. Diabetes 2007;56:1761–72.

[116] Bäckhed F, Manchester JK, Semenkovich CF, Gordon JI. Mechanisms underlying the resistance to diet-induced obesity in germ-free mice. Proc Natl Acad Sci USA 2007;104:979–84.

[117] Cani PD, Osto M, Geurts L, Everard A. Involvement of gut microbiota in the development of low-grade inflammation and type 2 diabetes associated with obesity. Gut Microbes 2012;3:279–88.

[118] Sun L, Yu Z, Ye X, et al. A marker of endotoxemia is associated with obesity and related metabolic disorders in apparently healthy Chinese. Diabetes Care 2010;33:1925–32.

[119] Vrieze A, Out C, Fuentes S, et al. Impact of oral vancomycin on gut microbiota, bile acid metabolism, and insulin sensitivity. J Hepatol 2014;60:824–31.

[120] Shen J, Obin MS, Zhao L. The gut microbiota, obesity and insulin resistance. Mol Aspects Med 2013;34:39–58.

[121] Cani PD, Delzenne NM. Gut microflora as a target for energy and metabolic homeostasis. Curr Opin Clin Nutr Metab Care 2007;10:729–34.

[122] Cani PD, Neyrinck AM, Fava F, et al. Selective increases of bifidobacteria in gut microflora improve high-fat-diet-induced diabetes in mice through a mechanism associated with endotoxaemia. Diabetologia 2007;50:2374–83.

[123] Le Chatelier E, Nielsen T, Qin J, et al. Richness of human gut microbiome correlates with metabolic markers. Nature 2013;500:541–6.

[124] Thomas C, Gioiello A, Noriega L, et al. TGR5-mediated bile acid sensing controls glucose homeostasis. Cell Metab 2009;10:167–77.

[125] Mingrone G, Panunzi S, De Gaetano A, et al. Bariatric surgery versus conventional medical therapy for type 2 diabetes. N Engl J Med 2012;366:1577–85.

[126] Schauer PR, Kashyap SR, Wolski K, et al. Bariatric surgery versus intensive medical therapy in obese patients with diabetes. N Engl J Med 2012;366:1567–76.

[127] Couzin J. Medicine: bypassing medicine to treat diabetes. Science 2008;320:438–40.

[128] Buchwald H, Avidor Y, Braunwald E, et al. Bariatric surgery: a systematic review and meta-analysis. JAMA 2004;292:1724–37.

[129] Buchwald H, Estok R, Fahrbach K, et al. Weight and type 2 diabetes after bariatric surgery: systematic review and meta-analysis. Am J Med 2009;122:248–256.e5.

[130] Korner J, Bessler M, Cirilo LJ, et al. Effects of Roux-en-Y gastric bypass surgery on fasting and postprandial concentrations of plasma ghrelin, peptide YY, and insulin. J Clin Endocrinol Metab 2005;90:359–65.

[131] Morínigo R, Moizé V, Musri M, et al. Glucagon-like peptide-1, peptide YY, hunger, and satiety after gastric bypass surgery in morbidly obese subjects. J Clin Endocrinol Metab 2006;91:1735–40.

[132] Houmard JA, Tanner CJ, Yu C, et al. Effect of weight loss on insulin sensitivity and intramuscular long-chain fatty acyl-CoAs in morbidly obese subjects. Diabetes 2002;51:2959–63.

[133] DiBaise JK, Zhang H, Crowell MD, Krajmalnik-Brown R, Decker GA, Rittmann BE. Gut microbiota and its possible relationship with obesity. Mayo Clin Proc 2008;83:460–9.

[134] Angelakis E, Armougom F, Million M, Raoult D. The relationship between gut microbiota and weight gain in humans. Future Microbiol 2012;7:91–109.

[135] Turnbaugh PJ, Hamady M, Yatsunenko T, Cantarel BL, Duncan A, Ley RE, et al. A core gut microbiome in obese and lean twins. Nature 2009;457:480–4.

[136] Hartman AL, Lough DM, Barupal DK, et al. Human gut microbiome adopts an alternative state following small bowel transplantation. Proc Natl Acad Sci USA 2009;106:17187–92.

[137] Derrien M, Van Baarlen P, Hooiveld G, Norin E, Müller M, de Vos WM. Modulation of mucosal immune response, tolerance, and proliferation in mice colonized by the mucindegrader *Akkermansia muciniphila*. Front Microbiol 2011;2:166.

[138] Shin NR, Lee JC, Lee HY, et al. An increase in the *Akkermansia* spp. population induced by metformin treatment improves glucose homeostasis in diet-induced obese mice. Gut 2014;63:727–35.

[139] Everard A, Belzer C, Geurts L, et al. Cross-talk between *Akkermansia muciniphila* and intestinal epithelium controls diet-induced obesity. PNAS 2013;110:9066–71.

[140] Hansen CH, Krych L, Nielsen DS, et al. Early life treatment with vancomycin propagates *Akkermansia muciniphila* and reduces diabetes incidence in the NOD mouse. Diabetologia 2012;55:2285–94.

[141] Tap J, Mondot S, Levenez F, et al. Towards the human intestinal microbiota phylogenetic core. Environ Microbiol 2009;11:2574–84.

[142] Sokol H, Seksik P, Furet JP, et al. Low counts of Faecalibacterium prausnitzii in colitis microbiota. Inflamm Bowel Dis 2009;15:1183–9.

[143] Hotamisligil GS, Shargill NS, Spiegelman BM. Adipose expression of tumor necrosis factor-alpha: direct role in obesity-linked insulin resistance. Science 1993;259:87–91.

[144] Sartipy P, Loskutoff DJ. Monocyte chemoattractant protein 1 in obesity and insulin resistance. Proc Natl Acad Sci USA 2003;100:7265–70.

[145] Maachi M, Piéroni L, Bruckert E, et al. Systemic low-grade inflammation is related to both circulating and adipose tissue TNFalpha, leptin and IL-6 levels in obese women. Int J Obes Relat Metab Disord 2004;28:993–7.

[146] Akbay E, Yetkin I, Ersoy R, Kulaksizog˘lu S, Törüner F, Arslan M. The relationship between levels of alpha1-acid glycoprotein and metabolic parameters of diabetes mellitus. Diabetes Nutr Metab 2004;17:331–5.

[147] Li M, Wang B, Zhang M, et al. Symbiotic gut microbes modulate human metabolic phenotypes. Proc Natl Acad Sci USA 2008;105:2117–22.

[148] Sokol H1, Pigneur B, Watterlot L, et al. Faecalibacterium prausnitzii is an anti-inflammatory commensal bacterium identified by gut microbiota analysis of Crohn disease patients. Proc Natl Acad Sci USA 2008;105(43):16731–6.

[149] Seksik P, Langella P. *Faecalibacterium prausnitzii* is an anti-inflammatory commensal bacterium identified by gut microbiota analysis of Crohn disease patients. Proc Natl Acad Sci USA 2008;105:16731–6.

[150] Weitz-Schmidt G. Statins as anti-inflammatory agents. Trends Pharmacol Sci 2002;23:482–6.

[151] Fleischman A, Shoelson SE, Bernier R, Goldfine AB. Salsalate improves glycemia and inflammatory parameters in obese young adults. Diabetes Care 2008;31:289–94.

[152] Reed MA, Pories WJ, Chapman W, et al. Roux-en-Y gastric bypass corrects hyperinsulinemia implications for the remission of type 2 diabetes. J Clin Endocrinol Metab 2011;96:2525–31.

[153] Keogh JB, Brinkworth GD, Noakes M, Belobrajdic DP, Buckley JD, Clifton PM. Effects of weight loss from a very-low-carbohydrate diet on endothelial function and markers of cardiovascular disease risk in subjects with abdominal obesity. Am J Clin Nutr 2008;87:567–76.

[154] Santacruz A, Marcos A, Wärnberg J, et al. Interplay between weight loss and gut microbiota composition in overweight adolescents. Obesity (Silver Spring) 2009;17(10):1906–15.

[155] Munukka E, Wiklund P, Pekkala S, et al. Women with and without metabolic disorder differ in their gut microbiota composition. Obesity 2012;20(5):1082–7.

[156] Arumugam M, Raes J, Pelletier E, et al. Enterotypes of the human gut microbiome. Nature 2011;473(7346):174–80.

[157] Greenblum S, Turnbaugh PJ, Borenstein E. Metagenomic systems biology of the human gut microbiome reveals topological shifts associated with obesity and inflammatory bowel disease. Proc Natl Acad Sci USA 2012;109(2):594–9.

[158] Parks BW, Nam E, Org E, et al. Genetic control of obesity and gut microbiota composition in response to high-fat, high-sucrose diet in mice. Cell Metab 2013;17:141–52.

[159] Bajzer M, Seeley RJ. Physiology: obesity and gut flora. Nature 2006;444:1009–10.

[160] Samuel BS, Gordon JI. A humanized gnotobiotic mouse model of host-archaeal-bacterial mutualism. Proc Natl Acad Sci USA 2006;103(26):10011–6.

[161] Samuel BS, Hansen EE, Manchester JK, et al. Genomic and metabolic adaptations of *Methanobrevibacter smithii* to the human gut. Proc Natl Acad Sci USA 2007;104(25):10643–8.

[162] Mahowald MA, Rey FE, Seedorf H, et al. Characterizing a model human gut microbiota composed of members of its two dominant bacterial phyla. Proc Natl Acad Sci USA 2009;106(14):5859–64.

[163] Woodard GA, Encarnacion B, Downey JR, et al. Probiotics improve outcomes after Roux-en-Y gastric bypass surgery: a prospective randomized trial. J Gastrointest Surg 2009;13(7):1198–204.

[164] Alang N, Kelly CR. Weight gain after fecal microbiota transplantation. Open Forum Infect Dis 2015;2(1):ofv004.

Part E

Management of Disease and Disorders by Prebiotics and Probiotic Therapy

Chapter 32

Allergic and Immunologic Disorders

H. Szajewska* and A. Nowak-Wegrzyn**

*Department of Paediatrics, The Medical University of Warsaw, Warsaw, Poland; **Pediatrics, Allergy and Immunology, Icahn School of Medicine at Mount Sinai, Jaffe Food Allergy Institute, New York, NY, United States*

INTRODUCTION

Allergic diseases, such as asthma, allergic rhinitis, food allergy, and atopic dermatitis (eczema), defined as adverse, immune-mediated reactions that are reproducible on a subsequent exposure [1], are common in all age groups [2]. The pathophysiology is multifactorial. The disease is triggered by environmental factors in individuals with genetic susceptibility. Globally, the prevalence of asthma and allergies has increased over the last few decades. However, the International Study of Asthma and Allergies in Childhood (ISAAC), the largest, relevant epidemiological study, found variations in the prevalence of asthma, allergic rhinoconjunctivitis, and eczema between different centers throughout the world [3]. The highest prevalence not only of asthma, but also that for many other allergic diseases, remains in certain industrialized countries, such as Australia, Ireland, New Zealand, the United Kingdom, and the United States [3]. For example, 1997–2013 US data from the National Health Interview Survey, which included more than 207,000 children and adolescents, showed prevalence rates of 17.1% for hay fever, 12.8% for asthma, 9.5% for eczema, and 4.2% for food allergies [4]. However, the ISAAC suggests that the prevalence of allergic diseases is also increasing in less affluent countries [3]. The rising number of children and adults with allergic disorders is a major public health concern. The origins of this increase are still not well understood. Factors, such as lifestyle, dietary habits, socioeconomic differences, and environmental factors may contribute to the development of allergic diseases [3]. Recent evidence demonstrated that, among other factors, disturbances in gut microbiota (dysbiosis) may be relevant. Likewise, some data suggest that gut microbiota alterations may be implicated in other immune-related disorders, such as type 1 diabetes and celiac disease.

This chapter provides an update on the current knowledge in the area of gut microbiota and allergy and other immune-mediated disorders. The potential for manipulating the gut microbiota of patients with these disorders via the provision of probiotics and prebiotics is assessed. To identify current data, an electronic database search of MEDLINE (via PubMed) was performed with specific key words in January 2016. Additionally, the same database and the Cochrane Library were searched to locate randomized controlled trials (RCTs) or their metaanalyses, with a special focus on those published in the last 5 years. The MEDLINE database was also searched for evidence-based clinical practice guidelines developed by scientific societies.

GUT MICROBIOTA

The human microbiota is defined as organisms (bacteria, viruses, or eukaryotes) that are present in an environmental habitat [5], but mainly in the gut. Table 32.1 presents a summary of the predominant bacterial phyla in the human body based on Ref. [6]. The intestinal microbiota has a profound impact on immunologic, nutritional, physiological, and protective processes [7] and thus, is considered by some to represent a new organ in the human body [5].

Gut Microbiota and Allergy

Early gut microbial colonization plays an important role in the development of the innate and the adaptive immune systems. Among other processes [8], contact with microbes is crucial for the development of oral tolerance [9], which is the specific suppression of cellular and/or humoral immune responses to an antigen. There are several proposed theories and mechanisms to explain how alterations in the gut microbiota could lead to the development of allergy, as briefly summarized later.

TABLE 32.1 Bacterial Phyla in the Human Body

Phylum	Class	Examples
Firmicutes	Bacilli; Clostridia	*Lactobacillus, Ruminococcus, Clostridium, Staphylococcus, Enterococcus, Faecalibacterium*
Bacteroidetes	Bacteroidetes	*Bacteroidetes, Prevotella*
Proteobacteria	Gammaproteobacteria; Betaproteobacteria	*Escherichia, Pseudomonas*
Actinobacteria	Actinobacteria	*Bifidobacterium, Streptomyces, Nocardia*

Source: Adapted from Johnson CL, Versalovic J. The human microbiome and its potential importance to pediatrics. Pediatrics 2012;129(5):950–60 [6].

Hygiene Hypothesis

It has been suggested that improved hygiene and the reduced exposure of the immune system to the microbial stimulus during infancy and early childhood predisposes to impaired immunoregulation in later life, leading to either Th2 diseases (such as allergy) or Th1 diseases (such as type 1 diabetes) [10–12]. However, over the years, this so called "hygiene hypothesis" has been challenged [13–15]. While exposure to some pathogens protects against atopy, other exposures promote allergic diseases. Factors, such as the timing of exposure and the properties of the infectious agent, host genetic susceptibility, and other environmental factors also may be important [16,17].

Differences in Gut Colonization

There are differences in the neonatal gut microbiota that may precede or coincide with the early development of atopy. Atopic subjects have more clostridia and tend to have fewer bifidobacteria than nonatopic subjects [18]. Reduced diversity of gut microbiota is associated with an increased risk of atopic eczema [19–22]. At least two studies showed that reduced total gut microbiota diversity during the first 3 months of life was also linked to asthma at 7 years of age [23,24].

Mode of Delivery

Over many years, there have been multiple observations that the mode of delivery contributes to the development of allergic diseases [25]. Recent data using culture-independent methods confirmed that colonization patterns differ between infants born by cesarean section compared with infants born vaginally. Infants delivered by cesarean section had lower total microbiota diversity, as well as lower abundance and diversity of Bacteroidetes phylum during the first 2 years of life. Moreover, reduced levels of Th1-associated chemokines, with a shift of the Th1/Th2 balance toward a more allergic Th2 response, were documented in infants delivered by cesarean section compared with those born vaginally [26].

Use of Antibiotics

A further well-known factor responsible for the shifts in gut microbiota is use of antibiotics during the early neonatal period [27]. The latest data show that not only antibiotic use by the infant [28], but also maternal antibiotic intake during birth, alters the microbiota of newborns [29]. A 2015 study documented that the use of antibiotics during delivery, regardless of whether it was a cesarean or vaginal delivery, was associated with dysbiosis at 3 months of age, and that breastfeeding modified some of these effects when infants were 12 months old [30].

Mode of Feeding

A number of studies have underscored the importance of the mode of feeding [28,31,32]. Compared with breastfed infants, exclusively formula-fed infants were more often colonized with *Escherichia coli*, *Clostridium difficile*, Bacteroides, and lactobacilli. Term infants who were born vaginally at home and were breastfed exclusively seemed to have the most "beneficial" gut microbiota (highest numbers of bifidobacteria and lowest numbers of *C. difficile* and *E. coli*) [28].

Other Factors

Other important determinants of the gut microbiota composition in infants include country of origin, gestational age, infant hospitalization, and time of weaning [28,33,34].

Gut Microbiota and Other Immune-Related Disorders

In addition to allergic diseases, the gut microbiota disturbances are implicated in other immune-related disorders, such as celiac disease (discussed also in Chapter 39) and type 1 diabetes [35,36].

Celiac disease, which affects approximately 1–3% of the general population in most parts of the world, is an immune-mediated enteropathy triggered by ingestion of gluten in genetically predisposed subjects (HLA-DQ2 and/or DQ8 positive) [37]. The factors leading to the breakdown of tolerance to gluten are not known; however, environmental factors, such as mode of delivery, perinatal infections, or gastrointestinal infections may also play a role [38]. Whether and how microbiota changes these environmental factors is a matter of debate. Since 2004, at least 20 studies have been published that assessed microbial alterations using conventional culturing techniques, as well as culture-independent molecular biological techniques, in subjects with celiac disease, as summarized in detail elsewhere [39]. In brief, subjects with celiac disease exhibit a decreased Bacteroidetes/Firmicutes ratio, increased *Bacteroides fragilis* and *Staphylococcus* spp., and the reduced presence of *Bifidobacterium longum* and *Bifidobacterium* spp. However, no typical "celiac microbiota signature" has been demonstrated.

Type 1 diabetes is caused by the immune-mediated, pancreatic beta-cell destruction in genetically susceptible individuals [40]. Although geographical variations exist, globally, the incidence of type 1 diabetes is increasing [41]. The role of environmental factors, such as mode of birth, early feeding, hygiene, or antibiotics use, in the pathogenesis of type 1 diabetes has been suggested. Compared to healthy controls, subjects with type 1 diabetes exhibit a less diverse and less stable gut microbiota [42–46]. Changes in gut microbiota are reflected by an abnormal ratio between Firmicutes and Bacteroidetes [42,43]. Additionally, *Faecalibacterium prausnitzii*, which has antiinflammatory effects, is reduced in children with type 1 diabetes–associated autoimmunity. For a detailed review of studies evaluating the gut microbiota in these patients, see the review by Gulden et al. [47].

Taken together, these studies add to a growing body of evidence that suggests that the gut microbiota, especially its diversity, plays a role in the establishment and development of allergic and other immune-related disorders, although the exact mechanisms remain unclear. However, for none of these diseases a specific "microbiota signature" has been identified. It also remains to be determined whether the gut microbiota alterations are a cause or a consequence of these disorders.

MICROBIOTA MODULATION STRATEGIES

With the growing recognition of the role of gut microbiota in health and disease, it is clear that gut microbiota may be a target for improving outcomes in subjects affected or at risk for certain diseases. To date, modification of gut microbiota via the provision of probiotics and/or prebiotics is the most extensively studied strategy.

Probiotics

A 2014 definition by the International Scientific Association for Probiotics and Prebiotics (ISAPP) defines probiotics as "live microorganisms that, when administered in adequate amounts, confer a health benefit on the host" [48]. In humans, by far, the most commonly used probiotics are bacteria from the genus *Lactobacillus* or *Bifidobacterium* and a yeast, *Saccharomyces boulardii*. However, novel probiotics are an area of current investigation [48].

Prebiotics

Prebiotics are "nondigestible food components that beneficially affect the host by selectively stimulating the growth and/or activity of one or a limited number of bacteria in the colon and thereby improving host health" [49]. In humans, nondigestible carbohydrates, oligosaccharides or short polysaccharides, such as inulin, oligofructose, fructooligosaccharides, galactofructose, galactooligosaccharides, and xylooligosaccharides, are the most intensively studied and commonly used prebiotics, as they increase the fecal counts of bacteria thought to be beneficial, such as bifidobacteria or certain butyrate producers.

Synbiotics

The term "synbiotic" is used for products containing both probiotics and prebiotics [50].

The mechanism(s) of action of pro/prebiotics remain(s) unclear. Nevertheless, recently, the ISAPP experts proposed three possible main mechanisms of probiotic action, discussed in detail elsewhere [48] and summarized in Table 32.2. Protolerogenic action of pro/prebiotics in the gastrointestinal tract is summarized in Fig. 32.1.

Other Therapies

Among other potential therapies aimed at modulation of the gut microbiota is fecal microbiota transplantation, defined as the administration of fecal material containing distal gut microbiota from a healthy donor to a patient with a disease/condition related to dysbiosis [5]. The goal is to treat disease by restoring the phylogenetic diversity and microbiota to that more typical of a healthy person [5]. Fecal microbiota transplantation has been consistently shown to be effective in the treatment of recurrent *C. difficile* infection [51]. While there is significant interest in its use for the treatment of other diseases, including allergic diseases, further research is needed to support its use in patients with allergy or other immune-related disorders.

TABLE 32.2 Mechanisms of Action of Probiotics

Mechanism[a]	Examples
Widespread (thought to be common among probiotic genera)	• colonization resistance, • acid and short-chain fatty acid production, • regulation of intestinal transit, • normalization of perturbed microbiota, • increased turnover of enterocytes, • competitive exclusion of pathogens.
Frequent (common among probiotic strains)	• vitamin synthesis, • direct antagonism, • gut barrier reinforcement, • bile salt metabolism, • enzyme activity, • neutralization of carcinogens.
Rare (strain-specific)	• immunological effects, • production of specific bioactives, • endocrinological effects, • neurological effects.

[a]It is likely that several mechanisms operate simultaneously; however, it is considered unlikely that a given individual probiotic might exert all three mechanisms.
Source: Based on Hill C, Guarner F, Reid G, et al. Expert consensus document. The International Scientific Association for Probiotics and Prebiotics consensus statement on the scope and appropriate use of the term probiotic. Nat Rev Gastroenterol Hepatol 2014;11:506–14 [48].

CLINICAL EFFICACY OF PROBIOTICS/PREBIOTICS

Methodological Issues

A number of relevant RCTs and metaanalyses that have assessed the effects of administering probiotics/prebiotics for allergy prevention and/or treatment are available, which are summarized later. In the hierarchy of research designs, the results of a metaanalysis of RCTs are considered to be the evidence of the highest grade. It is known that various probiotic strains (similarly, various prebiotics) differ in their effects, thus, pooling data on different probiotics has been repeatedly questioned [52]. Issues likely to contribute to problems with the metaanalytical approach include differences in the study populations (e.g., high-risk vs. unselected population); optimal strain selection (all probiotics/prebiotics are not equal); differences in definitions of outcomes (ideally, the diagnosis should be based on widely agreed-upon criteria); differences in the timing and duration of the intervention (e.g., prenatal and/or postnatal supplementation); and a lack of repeat studies. A strain-specific metaanalysis seems to be more appropriate; however, there is a paucity of such analyses. Only recently, criteria that may be acceptable for combining different probiotic strains into the same "class of intervention" for a specific outcome have been proposed [53].

In summary, caution should be exercised so one does not overinterpret the results of metaanalyses when all probiotics (or prebiotics) have been evaluated together.

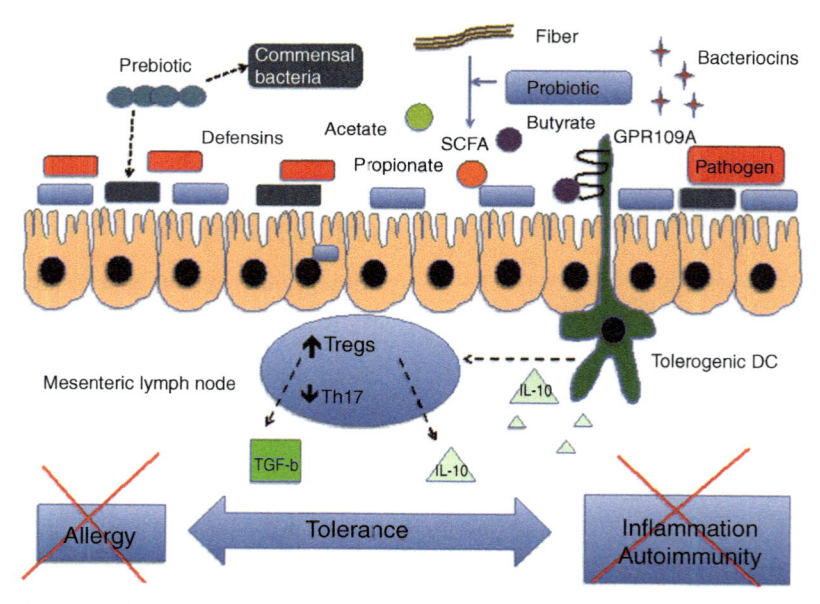

FIGURE 32.1 **Protolerogenic action of pro/prebiotics in the gastrointestinal tract.** Probiotics compete with pathogenic organisms for nutrients and binding sites on the intestinal epithelium. Prebiotics support the endogenous colonic commensal bacteria. Probiotics secrete bacteriocins and induce intestinal epithelium to secrete defensins, natural antimicrobial peptides. Probiotics ferment fiber to short-chain fatty acid *(SCFA)*: butyrate, acetate, and propionate. SCFA activates G protein-coupled receptors (GPCRs) that stimulate colonic dendritic cells and macrophages to secrete IL-10 and promote development of regulatory T lymphocytes *(Tregs)* in the mesenteric lymph nodes. Tregs are a source of tolerogenic cytokines: IL-10 and TGF-beta that inhibit allergic and inflammatory responses.

TABLE 32.3 Summary of the Metaanalyses on the Effects of Probiotics/Prebiotics in the Prevention of Allergic Diseases

References	Trials (Participants)	Outcome	Effect Measure/Size (95% Confidence Interval)	Authors' Conclusions
Probiotics				
Doege et al. [54]	7 RCTs (n = 2843)	Atopic eczema [mothers taking *Lactobacillus* during pregnancy (4 RCTs)]	RR 0.82 (0.71–0.95), I^2 0%	Probiotics, especially lactobacilli, reduce the risk of atopic eczema.
Pelucchi et al. [55]	18 publications based on 14 RCTs	Atopic dermatitis	RR 0.79 (0.71–0.88), I^2 24%	Moderate role of probiotics in the prevention of atopic dermatitis and IgE-associated atopic dermatitis in infants.
		IgE-associated atopic dermatitis	RR 0.80 (0.66–0.96)	
Elazab et al. [56]	25 RCTs (n = 4031)	Atopic sensitization (prenatal administration)	RR 0.88 (0.78–0.99)	Prenatal and/or early-life probiotic administration reduces the risk of atopic sensitization and decreases the total IgE level in children but may not reduce the risk of asthma/wheeze.
		Atopic sensitization (postnatal administration)	RR 0.86 (0.75–0.98)	
		Asthma/wheeze	RR 0.96 (0.85–1.07	
Azad et al. [57]	20 RCTs (n = 4866)	Asthma (9 RCTs, n = 3257)	RR 0.99 (0.81–1.21), I^2 0%	No protective association between perinatal use of probiotics and doctor-diagnosed asthma or childhood wheeze.
		Wheeze (9 RCTs, n = 1949)	RR 0.97 (0.87–1.09), I^2 0%	
Cuello-Garcia et al. [58]	29 RCTs	Eczema (administration during pregnancy)	RR 0.71 (0.60–0.84)	Probiotics used by pregnant women or breastfeeding mothers and/or given to infants reduced the risk of eczema in infants.
		Eczema (probiotics used by breastfeeding mothers)	RR 0.61 (0.50–0.74)	
		Eczema (probiotics given to infants)	RR 0.81 (0.70–0.94)	
Prebiotics				
Osborn and Sinn (Cochrane review) [59]	4 RCTs (n = 1428)	Asthma (2 RCTs, n = 226)	NS	Some evidence that a prebiotic supplement added to infant feeds may prevent eczema.
		Eczema (4 RCTs, n = 1218)	RR 0.68 (0.48–0.97)	

NS, Not significant; RCTs, randomized controlled trials; RR, relative risk.

PREVENTION OF ALLERGIC DISEASES

Table 32.3 presents a summary of metaanalyses on the effects of probiotics/prebiotics in the prevention of allergic diseases [54–59].

Probiotics

A 2013 systematic review [56] (search date: 2001–12) investigated the effects of probiotic administration on asthma and atopic sensitization, and it included 25 RCTs (n = 4031). Early probiotic administration reduced the risk of atopic sensitization, which was usually defined in the included studies as a positive skin prick test and/or IgE level >0.35 kU/L to any food or inhalant allergen [prenatal administration: relative risk (RR), 0.88, 95% confidence interval (CI), 0.78–0.99; postnatal administration: RR

0.86, 95% CI 0.75–0.98]. Administration of *Lactobacillus acidophilus*, compared with other strains, was associated with an increased risk of atopic sensitization (P = 0.002). Probiotics did not significantly reduce the risk of asthma/wheeze (RR 0.96, 95% CI 0.85–1.07). Thus, this review suggested that probiotic administration reduces atopic sensitization, but not the risk of diseases, such as asthma. Also, another 2013 systematic review found no consistent evidence that the perinatal administration of probiotics reduces the risk of childhood asthma or wheeze [57].

A 2015 systematic review [58] (search date: December 2014) involved a search of the Cochrane Central Register of Controlled Trials, MEDLINE, and EMBASE, and it included 29 publications in which 12 various probiotics, single or in combinations, were used; however, except for *Lactobacillus* GG, none were studied in more than one trial. The reviewers concluded that there are significant benefits of probiotic supplements in reducing the risk of eczema when administered to women during the last trimester of pregnancy (14 RCTs, n = 3109, RR 0.71, 95% CI 0.60–0.84) or during breastfeeding (10 RCTs, n = 1595, RR, 0.61, 95% CI 0.50–0.74); however, no such effect was observed when probiotics were used exclusively during breastfeeding (1 RCT, n = 88, RR 0.57, 95% CI 0.29–1.11). Probiotics given to infants also reduced the risk of eczema (15 RCTs, n = 3447, RR 0.81, 95% CI 0.7–0.94). In contrast to the effect on eczema, probiotics compared with no probiotics had no effect on the risk of other allergies, such as asthma/wheezing, food allergy, and allergic rhinitis as well as no effect on the nutritional status or incidence of adverse effects. Overall, the quality of evidence was low or very low due to the risk of bias, inconsistency and imprecision of the results, and the indirectness of available research. On the basis of their findings, the authors concluded that probiotics used by pregnant women or breastfeeding mothers and/or given to infants reduced the risk of eczema in infants; however, the certainty of the evidence was low.

Prebiotics

A 2013 Cochrane review [59] (search date: August 2012) included 4 RCTs and enrolled 1428 infants. The authors of this Cochrane review concluded that there is some evidence that prebiotic supplementation may reduce the risk of eczema. However, it is unclear whether the use of prebiotics should be restricted to infants at high risk of allergy or whether prebiotics may have an effect in low-risk populations. It is also unclear whether prebiotic supplementation may have an effect on other allergic diseases including asthma. As with probiotics, the effects of different prebiotics are not equivalent. Thus, the most relevant data are those related to specific prebiotics.

Synbiotics

No systematic review focusing specifically on synbiotics was found. In one recent, double-blind RCT conducted in Finland, 1223 pregnant women carrying high-risk infants were randomly assigned to receive *Lactobacillus rhamnosus*, *Bifidobacterium breve*, and *Propionibacterium freudenreichii* or placebo for 2–4 weeks before delivery. Their infants received the same preparation plus galactooligosaccharides (i.e., a synbiotic) or placebo for 6 months after delivery. In the group of infants fed synbiotics, a significantly higher proportion of infants had a reduction in eczema (RR 0.81, 95% CI 0.66, 0.99) and in atopic eczema (RR 0.70, 95% CI 0.51, 0.96). However, there was no significant difference between groups in the cumulative incidence of all allergic diseases (RR 0.90, 95% CI 0.75–1.08). It conferred protection only to cesarean-delivered children [60,61].

Current Guidelines

Recently, two independent guidelines were published yielding contradictory recommendations. In 2014, the European Academy of Allergy and Clinical Immunology (EAACI), based on the results of a systematic review of RCTs (search date: September 2012) [62], concluded that there is no evidence to support the use of probiotics (also prebiotics) for food allergy prevention [63].

In 2015, the World Allergy Organization (WAO) guidelines were published [64]. These guidelines are based on the findings from the systematic review (search date: December 2014) by Cuello-Garcia et al. [58] discussed earlier. In line with the EAACI, the WAO experts agreed that probiotic supplementation cannot be recommended for reducing the risk of allergy in children. However, the WAO considered that there is likely a net benefit from using probiotics for preventing eczema. Specifically, the WAO suggests: "a) using probiotics in pregnant women at high risk for having an allergic child; b) using probiotics in women who breastfeed infants at high risk of developing allergy; and c) using probiotics in infants at high risk of developing allergy." All recommendations were conditional and supported by a very low quality of evidence.

One important limitation of the WAO guidelines is the lack of answers to the most important practical questions. Which probiotic(s) should be used to reduce the risk of eczema? When should one start the administration of probiotics with proven efficacy? When should one stop? What is the dose of an effective probiotic? In summary, probiotics as a group reduce the risk of eczema. However, it would be premature to support the routine use of probiotics for preventing eczema. Data regarding which probiotic products should be administered, at what dosages, and the most effective dosing schedule are needed. With one exception (*Lactobacillus* GG), there is no single probiotic that has been studied in more than one RCT [65–69]. Even with regard to *Lactobacillus* GG, while the pooled results of four RCTs [66–69] indicate a trend toward a reduction in the risk of eczema at 12–24 months, the difference between the LGG and control groups was not statistically significant (Fig. 32.2) [70].

TREATMENT OF ALLERGIC DISEASES

Table 32.4 presents a summary of the metaanalyses on the effects of probiotics for the treatment of allergic diseases [71–80].

Eczema

This condition is also referred to as atopic dermatitis or atopic eczema. No cure has been found for eczema, hence, there is interest in novel therapeutic options. A number of systematic reviews [72–74], including a Cochrane review

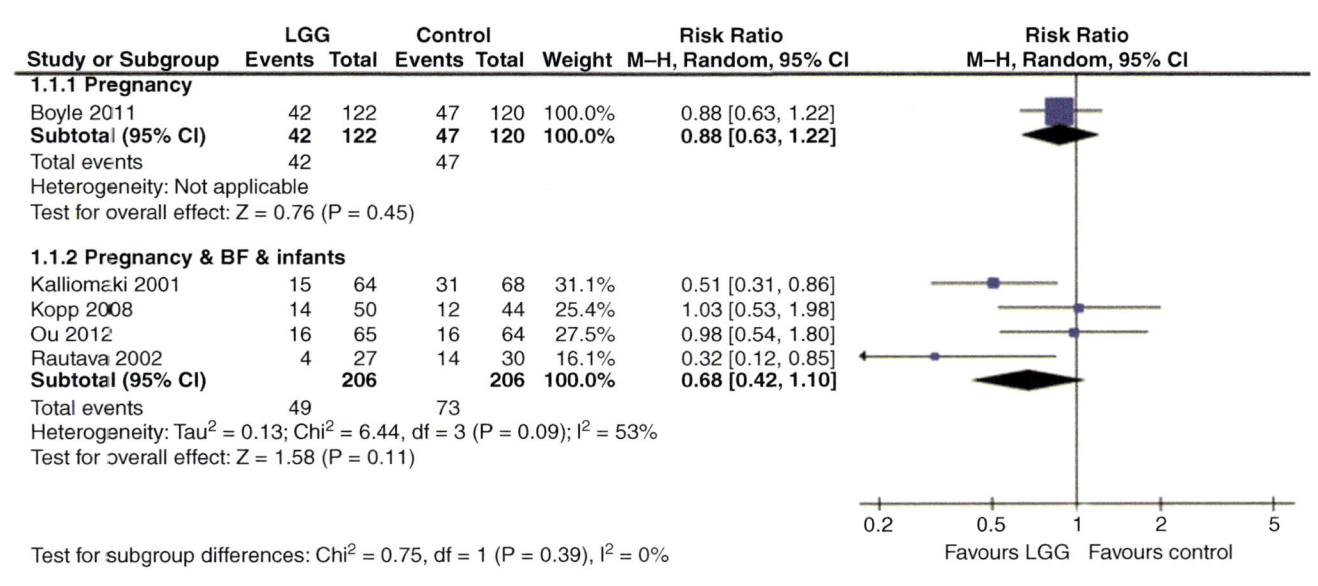

Study or Subgroup	LGG Events	Total	Control Events	Total	Weight	Risk Ratio M–H, Random, 95% CI	Risk Ratio M–H, Random, 95% CI
1.1.1 Pregnancy							
Boyle 2011	42	122	47	120	100.0%	0.88 [0.63, 1.22]	
Subtotal (95% CI)	**42**	**122**	**47**	**120**	**100.0%**	**0.88 [0.63, 1.22]**	
Total events	42		47				
Heterogeneity: Not applicable							
Test for overall effect: Z = 0.76 (P = 0.45)							
1.1.2 Pregnancy & BF & infants							
Kalliomaki 2001	15	64	31	68	31.1%	0.51 [0.31, 0.86]	
Kopp 2008	14	50	12	44	25.4%	1.03 [0.53, 1.98]	
Ou 2012	16	65	16	64	27.5%	0.98 [0.54, 1.80]	
Rautava 2002	4	27	14	30	16.1%	0.32 [0.12, 0.85]	
Subtotal (95% CI)		**206**		**206**	**100.0%**	**0.68 [0.42, 1.10]**	
Total events	49		73				
Heterogeneity: Tau^2 = 0.13; Chi^2 = 6.44, df = 3 (P = 0.09); I^2 = 53%							
Test for overall effect: Z = 1.58 (P = 0.11)							

Test for subgroup differences: Chi^2 = 0.75, df = 1 (P = 0.39), I^2 = 0%

FIGURE 32.2 The effect of *Lactobacillus* GG supplementation versus placebo on eczema at 12–24 months. *(From Szajewska H, Shamir R, Turck D, et al. Recommendations on probiotics in allergy prevention should not be based on pooling data from different strains. J Allergy Clin Immunol 2015;136(5):1422 [70])*

[75], have investigated the efficacy of probiotics in the treatment of pediatric atopic dermatitis/eczema.

A 2014 systematic review [71] (search date: December 2013) with a metaanalysis identified 25 RCTs involving 1599 participants. Compared with placebo, the use of probiotics (in some trials together with prebiotics) significantly reduced scoring of atopic dermatitis (SCORAD) values overall (weighted mean difference (WMD) −4.51, 95% CI −6.78 to −2.24), in adults (WMD −8.26, 95% CI −13.28 to −3.25, I^2 57%), in children 1–18 years of age (WMD −5.74, 95% CI −7.27 to −4.20, I^2 28%), but not in infants younger than 1 year (WMD 0.52, 95% CI −1.59 to 2.63, I^2 48%). Subgroup analysis showed that compared with placebo, administration of the mixture of different probiotics was effective (7 RCTs, n = 436, WMD −6.6, 95% CI −10.42 to −2.79, I^2 80%), followed by treatment with the *Lactobacillus* species (17 RCTs, n = 1066, WMD −3.81, 95% CI −6.42 to −1.21, I^2 71%); however, the administration of *Bifidobacterium* species was not effective (4 RCTs, n = 234, WMD 1.75, 95% CI 1.10–2.40, I^2 0%). Another subgroup analysis by treatment duration showed that while treatment for <8 weeks was not effective (4 RCTs, n = 358, WMD −2.2, 95% CI −6.75 to 2.36, I^2 75%), treatment for ≥8 weeks was effective (20 RCTs, n = 1219, WMD −4.98, 95% CI −7.69 to −2.77, I^2 88%). Finally, the treatment was effective in patients with moderate/severe atopic dermatitis (19 RCTs, n = 1293, WMD −4.66, 95% CI −7.33 to −1.99, I^2 89%), but not in those with mild atopic dermatitis (5 RCTs, n = 239, WMD −0.81, 95% CI −4.21 to 2.6, I^2 26%).

In summary, current evidence suggests a modest role for probiotics in the treatment of atopic dermatitis in children and adults but not in infants. This effect was seen in patients with moderate/severe disease rather than in those with mild disease, as well as in those with prolonged treatment. Findings should be interpreted with caution due to methodological limitations of some of the included trials and to clinical and statistical heterogeneity. It remains unclear which probiotic strains (single or combinations) are the most effective. Moreover, in some trials, probiotics were administered together with prebiotics, thus, precluding one from concluding which substance(s) was/were effective.

Allergic Rhinitis

Commonly known as "hay fever," allergic rhinitis (perennial when allergic symptoms occur throughout the year or seasonal when allergic symptoms occur during a particular season) is an inflammation of the lining of the nose caused by inhaling an allergen.

A 2008 systematic review identified 11 RCTs and concluded that administration of probiotics may have a beneficial effect on allergic rhinitis by reducing symptom severity and medication use. In most of the included trials, compared with placebo or no intervention, the administration of probiotics resulted in clinical improvement, including quality of life. As different measurements of effectiveness were used, the authors abstained from pooling data [79].

More recently, two systematic reviews evaluating the effects of probiotics (single or in combinations) were published. One of them (search date: March 2014) identified 11 RCTs (n = 306). While the authors concluded that there are significant benefits of probiotics in improving the quality of life scores and nasal symptom scores when used by patients

TABLE 32.4 Summary of the Metaanalyses on the Effects of Probiotics in the Treatment of Allergic Diseases

References	Trials (Participants)	Outcome	Effect Measure/Size (95% Confidence Interval)		Authors' Conclusions
Eczema					
Kim et al. [71]	25 RCTs (n = 1599)	SCORAD overall (18 RCTs)	MD	−4.51 (−6.78 to −2.24)	Probiotics could be an option for treatment of moderate to severe AD in children and adults, but not in infants.
		SCORAD children 1–18 years (8 RCTs)	MD	−5.74 (−7.27 to −4.2)	
		SCORAD adults (4 RCTs)	MD	−8.26 (−13.28 to −3.25)	
		SCORAD <1 year (6 RCTs)	MD	NS (no data)	
Betsi et al. [72]	10 (n = 1115)	Narrative review. Three studies found a statistically significant reduction in SCORAD values after treatment with *Lactobacillus* GG or *B. lactis* Bb12 for 1 or 2 months compared with placebo in infants with AD with or without a cow's milk allergy. Another study reported a significant reduction in SCORAD values after 12 weeks of treatment with the same probiotics compared with placebo in food-sensitized children. Other studies reported significant reductions in the probiotic groups, but no significant differences between those treated with probiotics and placebo. Probiotics had little effect on most of the inflammatory markers measured.			
Michail et al. [73]	10 RCTs (n = 678)	SCORAD	WMD	−3.05 (−5.36 to −0.66) (doubtful clinical significance)	A modest role for probiotics in pediatric atopic dermatitis (in moderately severe rather than mild disease).
Lee et al. [74]	4 RCTs (n = 299)	SCORAD	WMD	No statistically significant changes in intergroup SCORAD scores. Significant therapeutic effect in sensitized (IgE status) patients.	Current evidence is more convincing for probiotics' efficacy in prevention rather than treatment of pediatric atopic dermatitis.
Boyle et al. [75]	5 RCTs (n = 313)	Participant or parent-rated symptoms of eczema (SCORAD part C) at the end of treatment	MD	0.90 (−2.84, 1.04)	Probiotics are not an effective treatment for eczema, and probiotic treatment carries a small risk of adverse events.
	3 RCTs (n = 150)	Participant or parent-rated overall eczema severity	OR	0.40 (0.14, 1.15)	
	7 RCTs (n = 588)	Investigator rated eczema severity	MD	2.46 (−2.53 to 7.45)	
Allergic Rhinits					
Zajac et al. [76]	21 RCTs + 2 crossover (n = 1919)	RQLQ Global Score (4 RCTs, n = 622)	SMD	2.23 (−4.13 to −2.65)	Probiotics may be beneficial in improving symptoms and quality of life in patients with allergic rhinitis. Evidence remains limited due to study, heterogeneity, and variable outcome measures.
		RQLQ Nose Score (4 RCTs, n = 622)	SMD	−1.21 (−1.42 to −0.99)	
		RQLQ Eye Score (4 RCTs, n = 622)	SMD	−1.45 (−3.04 to 0.15)	
		RTSS-Global (4 RCTs, n = 533)	SMD	−0.36 (−0.83 to 0.10)	
		RTSS-Nose (3 RCTs, n = 513)	SMD	−0.82 (−2.41 to 0.78)	
		RTSS-Eye (3 RCTs, n = 513)	SMD	−0.10 (−0.26 to 0.07)	
		Total IgE (8 RCTs, n = 446)	SMD	0.01 (−0.18 to 0.19)	
		Antigen-specific IgE (7 RCTs, n = 359)	SMD	0.20 (−0.01 to 0.41)	

Peng et al. [77]	11 RCTs, n = 306	Nose symptom score in frequency (2 RCTs, n = 140)	MD	−0.96 (−3.78 to 1.86)	Probiotics may improve the overall quality of life and nasal symptom scores. As the available data were generated from only a few trials with a high degree of heterogeneity, routine use of probiotics for prevention and treatment in patients with AR cannot be recommended.
		Nose symptom score in severity (2 RCTs, n = 140)	MD	−1.11 (−3.38 to 1.17)	
		QoL score in frequency (2 RCT, n = 140)	MD	−5.60 (−16.92 to 5.72)	
		QoL score in severity	MD	−4.40 (−9.84 to 1.04)	
Das et al. [78]	7 RCTs, n = 610 (children and adults)	Change in frequency (2 RCTs, n = 140)	SMD	−0.90 (−1.34 to −0.45)	Probiotic therapy might be useful in allergic rhinitis, but data do not allow any treatment recommendations.
		Change in level of bother (2 RCTs, n = 140)	SMD	−1.40 (−1.82 to −0.98)	
		Time free from the episodes (months)		Exp. 4.1 (3.1–5); contr. 3.3 (2.4–4.3); P = 0.9	
		Episodes per year	OR	0.39 (0.19–0.82)	
Vliagoftis et al. [79]	12 RCTs (children and adults)	Narrative review. Nine of the 12 RCTs showed an improvement due to the use of probiotics. All the RCTs that studied perennial AR showed lower symptom scoring and medication use with the use of probiotics compared with placebo. Five of the 8 RCTs that referred to seasonal AR suggested an improvement in clinical outcomes.			
Asthma					
He et al. [80]	5 RCTs	Free episodes of asthma (2 RCTs, n = 260)	SMD	1.48 (1.20–1.76)	Probiotics may prolong free episodes of asthma and seem to have an effect on improving lung function, but fail to reduce the acute onset of asthma and offer no advantage in improving immune function.
Vliagoftis et al. [79]	4 RCTs	Narrative review. No consistent effect of probiotics on asthma treatment.			

AD, Atopic dermatitis; AR, allergic rhinitis; MD, mean difference; NS, not significant; OR, odds ratio; RCT, randomized controlled trial; RQLQ, Rhinitis Quality of Life Questionnaire; RTSS, rhinitis total symptom score; SCORAD, Scoring of atopic dermatitis; SMD, standardized mean difference; WMD, weighted mean difference.

with allergic rhinitis, data from various studies were not appropriately pooled. In contrast, there were no differences when these scores were evaluated separately [77].

A 2015 review (search date: 2000–14) identified 23 studies, including 21 RCTs and 2 crossover studies involving 1919 participants. Compared with placebo, the use of probiotics showed a benefit in at least 1 outcome measure in 17 trials, whereas 6 trials showed no benefit of probiotic use. The pooled results showed that compared with placebo, the use of probiotics significantly improved the global Rhinitis Quality of Life Questionnaire (RQLQ) score (4 RCTs, $n = 622$, SMD -2.23, 95% CI -4.13 to -2.65); however, there was no difference between groups in the global Rhinitis Total Symptom Score (RTSS) (4 RCTs, $n = 533$, SMD -0.36, 95% CI -0.83 to 0.1) [76].

In summary, while probiotics may be beneficial in improving symptoms and quality of life in patients with allergic rhinitis, caution is needed when interpreting current evidence due to methodological issues, such as significant heterogeneity between studies as well as wide variations in the probiotic strains used, study populations, and outcomes assessed. None of the probiotics have been assessed in more than one study.

Asthma

A 2008 systematic review [79] identified four RCTs ($n = 257$) carried out in adults and in children [81–84] and found no effect of the probiotics studied on asthma.

A 2012 systematic review [80] (search date: August 2011) identified five RCTs ($n = 430$) that evaluated the effects of probiotics in the treatment of asthma [84–88]. Compared with placebo, the administration of probiotics prolonged free episodes of asthma (2 RCTs, $n = 260$, SMD 1.48, 95% CI 1.20–1.76). Differences in the probiotics studied and outcome measures assessed do not allow one to accurately define the role of probiotics in the treatment of asthma.

In summary, there is no proven effect of probiotics for the treatment of asthma.

Cow's Milk Allergy

Cow's milk allergy (CMA) is a common diagnosis in infants and children. The majority of affected children have one or more symptoms involving one or more organ systems, mainly the gastrointestinal tract and/or skin. Strict avoidance of the offending allergen is the only therapeutic option. A number of studies have evaluated whether supplementation of therapeutic formulas with specific probiotics for the management of CMA is beneficial.

A 2008, double-blind, placebo-controlled RCT performed in 119 infants with CMA found that *Lactobacillus casei* CRL431 and *Bifidobacterium lactis* Bb12 added to

extensively hydrolyzed formula did not significantly affect clinical tolerance at 6 and 12 months of treatment. At 12 months, the cumulative tolerance was 81% in the placebo group and 77% in the probiotics group (OR 1.1; 95% CI 0.6–1.9) [89].

Results from a more recent study suggest that formula selection and the choice of probiotics influence the rate of acquisition of tolerance in children with CMA. Berni-Canani et al. [90] randomly allocated infants with CMA (while still receiving intact protein formula) to either a group that received extensively hydrolyzed casein formula or a group that received the same extensively hydrolyzed formula containing *L. rhamnosus* GG. After 6 months of an exclusion diet, a double-blind, placebo-controlled, milk challenge was performed in 55 patients, and evidence of tolerance was seen in 21.4 and 59.3% of infants, respectively. The difference in acquisition of immunotolerance was significant only for those children with non-IgE-mediated CMA ($P = 0.017$).

Another open-label, non-RCT evaluated the acquisition of tolerance in a total of 260 infants aged 1–12 months with confirmed CMA fed five different formulas: extensively hydrolyzed casein formula, extensively hydrolyzed casein formula with *Lactobacillus* GG, hydrolyzed rice formula, soy formula, or amino acid–based formula. Development of tolerance was assessed after a 12-month period with repeat skin prick test, atopy patch test, and a double-blind, placebo-controlled food challenge. The rate of oral tolerance after 1 year of treatment, as determined by the food challenge, was significantly higher in the groups that received extensively hydrolyzed casein formula, whether it was with *Lactobacillus* GG (78.9%) or without (43.6%), compared with the other groups that received hydrolyzed rice formula (32.6%), soy formula (23.6%), and amino acid–based formula (18.2%) [91].

In summary, while these data are promising, larger RCTs are needed to confirm these findings and to define the mechanisms of action and to evaluate the potential factors influencing the response in subjects with CMA.

SYNBIOTICS

Eczema

A 2014 systematic review (search date: December 2013) identified five RCTs that investigated the efficacy of synbiotics in the treatment of eczema in children (3 RCTs), infants (1 RCT), and in an uncategorized group of participants (1 RCT). Due to significant heterogeneity (I^2 90%), only one subgroup analysis was performed (for the children population). Compared with control, the use of synbiotics significantly reduced the SCORAD (a clinical tool used to assess the extent and severity of eczema) value in children with atopic dermatitis (3 RCTs, $n = 183$, WMD -7.02, 95% CI -10.33 to -3.71, I^2 32%) [71].

In summary, as none of the synbiotics has been assessed in more than one study, the role of synbiotics for the treatment of eczema remains unclear.

TYPE 1 DIABETES

No RCTs have investigated the effects of probiotics in patients with type 1 diabetes. However, one, recent, prospective, cohort study, which was carried out as part of the TEDDY (The Environmental Determinants of Diabetes in the Young) study, aimed to examine the association between early probiotic exposure and islet autoimmunity (positive antibodies to insulin, glutamic acid decarboxylase, or insulinoma antigen 2 on at least two consecutive visits) in children genetically at increased risk for type 1 diabetes. This study found that early (i.e., during the first 27 days of life) administration of probiotics (mainly *Lactobacillus* and *Bifidobacterium* given either as a supplement or in infant formula supplemented with probiotics) may be associated with a reduced risk of islet autoimmunity [hazard ratio (HR) 0.66, 95% CI 0.45–0.96], especially in children with the highest-risk HLA genotype of *DR3/4* (HR 0.4, 95% CI 0.21–0.74). Of note, no reduction was seen in children with moderately higher-risk genotypes (HR 0.97, 95% CI 0.62–1.54) [92].

In summary, given the limitations of the study design (observational study), these findings need confirmation in well-designed and executed RCTs. It is worth noting that, currently, scientific societies do not recommend the routine use of probiotic-supplemented formulas [93]. Among others, one issue to be resolved is which probiotics should be used.

CELIAC DISEASE

The mainstay of treatment for celiac disease is a life-long, gluten-free diet. Due to the constraints of a gluten-free diet, the role of other therapies, either as an adjunct or as an alternative to a gluten-free diet, is an area of research interest (see also Chapter 39).

Several studies have assessed the ability of probiotics to digest or alter gluten [94,95]. However, only a limited number of relevant clinical trials have been performed. The first of them aimed to evaluate the effects of probiotic administration in patients with celiac disease. This was a double-blind, placebo-controlled RCT carried out in 22 adult patients, with newly diagnosed celiac disease, who were still consuming a gluten-containing diet. Compared with placebo, the addition of *Bifidobacterium infantis* natren life start strain to their gluten-containing diet had no effect on intestinal permeability (using the lactulose/mannitol fractional excretion ratio) at the end of 3 weeks of treatment (the primary outcome measure). While the authors reported some significant differences within the probiotic group, there were no differences between the study groups in any of the immunological parameters. There were also no differences between the study groups for five syndromes of the Gastrointestinal Symptom Rating Scale. The authors' conclusion that *B. infantis* may alleviate symptoms in untreated CD should be interpreted with extreme caution due to limitations of the study [96].

A 2014 double-blind, placebo-controlled RCT performed in 36 children aged 2–17 years with newly diagnosed celiac disease found that compared with placebo, administration of *B. longum* CECT 7347 daily for 3 months together with a gluten-free diet resulted in greater height percentile increases ($P = 0.048$) but similar weight percentile increases ($P = 0.234$) [88]. There was no difference in gastrointestinal symptoms (diarrhea, constipation, abdominal pain, and vomiting) during the 3 months of the intervention between the study groups. The investigators also found that compared with placebo, the administration of *B. longum* significantly decreased peripheral CD3$^+$ lymphocytes ($P = 0.004$), had no effect on the TNF-α concentration ($P = 0.067$), reduced the numbers of *B. fragilis* group ($P = 0.02$), and reduced the content of sIgA in stools ($P = 0.011$). On the basis of their findings, the authors concluded that *B. longum* CECT 7347 as an adjunct to a gluten-free diet could help improve the health status of patients with newly diagnosed celiac disease [97]. However, their conclusions have been challenged by others [98].

In summary, while the authors of both studies consider probiotics to be promising for the management of patients with celiac disease, caution needs to be expressed, as these conclusions are based on very limited data. Further mechanistic studies are needed to define the optimal probiotic strain and dose.

CONCLUSIONS

There is a growing interest in the relationship between gut microbiota and human health and disease. However, despite extensive research, the role of gut microbiota in allergy and immune-related disorders has not been fully clarified. Moreover, it remains to be determined if the gut microbiota alterations are a cause or a consequence. Modifications of gut microbiota employing probiotics, prebiotics, and synbiotics are used in attempts to prevent and treat these diseases. There is growing evidence that specific probiotics may be beneficial for preventing eczema. However, further research is needed to clarify which probiotic strain(s) and dosages should be used. More data are needed regarding when the administration of probiotics with proven efficacy should start and when it should stop. Currently, there is no evidence to support the administration of probiotics to reduce the risk of other allergic diseases. For treating allergy, there is no clear evidence that any of the specific probiotics or prebiotics is effective. A lack of evidence regarding the efficacy of a certain probiotics/prebiotics does not mean that future studies will not establish health benefit(s). Further research is warranted to accurately define the role of prebiotics/probiotics in the prevention and treatment of allergy and other immune-related disorders. Ideally, the investigators should agree on methodological standards and a core outcomes set. Differences in outcome definitions and measurements make it difficult to synthesize the results of trials and apply them in a meaningful way.

REFERENCES

[1] Boyce JA, Assa'ad A, Burks AW, et al. Guidelines for the diagnosis and management of food allergy in the United States: summary of the NIAID-Sponsored Expert Panel Report. J Allergy Clin Immunol 2010;126(6):1105–18.

[2] Muraro A, Werfel T, Hoffmann-Sommergruber K, et al. EAACI food allergy and anaphylaxis guidelines: diagnosis and management of food allergy. Allergy 2014;69(8):1008–25.

[3] Asher MI, Montefort S, Björkstén B, et al. Worldwide time trends in the prevalence of symptoms of asthma, allergic rhinoconjunctivitis, and eczema in childhood: ISAAC Phases One and Three repeat multicountry cross-sectional surveys. Lancet 2006;368(9537):733–43. Erratum in: Lancet 2007;370(9593):1128.

[4] Drury KE, Schaeffer M, Silverberg JI. Association between atopic disease and anemia in US Children. JAMA Pediatr 2015;30:1–6.

[5] Marchesi JR, Adams DH, Fava F, et al. The gut microbiota and host health: a new clinical frontier. Gut 2016;65:330–9.

[6] Johnson CL, Versalovic J. The human microbiome and its potential importance to pediatrics. Pediatrics 2012;129(5):950–60.

[7] Sommer F, Bäckhed F. The gut microbiota—masters of host development and physiology. Nat Rev Microbiol 2013;211(4):227–38.

[8] Arrieta MC, Stiemsma LT, Amenyogbe N, Brown EM, Finlay B. The intestinal microbiome in early life: health and disease. Front Immunol 2014;5:427.

[9] Sansonetti PJ, Medzhitov R. Learning tolerance while fighting ignorance. Cell 2009;138(3):416–20.

[10] Strachan DP. Hay fever, hygiene, and household size. BMJ 1989;299(6710):1259–60.

[11] Prescot SL. Allergy: the price we pay for cleaner living? Ann Allergy Asthma Immunol 2003;90:64–70.

[12] Bach JF, Chatenoud L. The hygiene hypothesis: an explanation for the increased frequency of insulin-dependent diabetes. Cold Spring Harb Perspect Med 2012;2(2):a007799.

[13] Renz H, von Mutius E, Brandtzaeg P, Cookson WO, Autenrieth IB, Haller D. Gene-environment interactions in chronic inflammatory disease. Nat Immunol 2011;12(4):273–7.

[14] Hanski I, von Hertzen L, Fyhrquist N, et al. Environmental biodiversity, human microbiota, and allergy are interrelated. Proc Natl Acad Sci USA 2012;109(21):8334–9.

[15] Haahtela T, Laatikainen T, Alenius H, et al. Hunt for the origin of allergy—comparing the Finnish and Russian Karelia. Clin Exp Allergy 2015;45(5):891–901.

[16] Fishbein AB, Fuleihan RL. The hygiene hypothesis revisited: does exposure to infectious agents protect us from allergy? Curr Opin Pediatr 2012;24(1):98–102.

[17] Guarner F, Bourdet-Sicard R, Brandtzaeg P, et al. Mechanisms of disease: the hygiene hypothesis revisited. Nat Clin Pract Gastroenterol Hepatol 2006;3(5):275–84.

[18] Kalliomaki M, Kirjavainen P, Eerola E, et al. Distinct patterns of neonatal gut microflora in infants in whom atopy was and was not developing. J Allergy Clin Immunol 2001;107:12934.

[19] Abrahamsson TR, Jakobsson HE, Andersson AF, Björkstén B, Engstrand L, Jenmalm MC. Low diversity of the gut microbiota in infants with atopic eczema. J Allergy Clin Immunol 2012;129:434–40.

[20] Wang M, Karlsson C, Olsson C, et al. Reduced diversity in the early fecal microbiota of infants with atopic eczema. J Allergy Clin Immunol 2008;121:129–34.

[21] Forno E, Onderdonk AB, McCracken J, et al. Diversity of the gut microbiota and eczema in early life. Clin Mol Allergy 2008;6:11.

[22] West CE, Rydén P, Lundin D, Engstrand L, Tulic MK, Prescott SL. Gut microbiome and innate immune response patterns in IgE-associated eczema. Clin Exp Allergy 2015;45(9):1419–29.

[23] Abrahamsson TR, Jakobsson HE, Andersson AF, Björkstén B, Engstrand L, Jenmalm MC. Low gut microbiota diversity in early infancy precedes asthma at school age. Clin Exp Allergy 2014;44(6):842–50.

[24] Arrieta MC, Stiemsma LT, Dimitriu PA, et al. Early infancy microbial and metabolic alterations affect risk of childhood asthma. Sci Transl Med 2015;7(307):307ra152.

[25] Dominguez-Bello MG, Costello EK, Contreras M, et al. Delivery mode shapes the acquisition and structure of the initial microbiota across multiple body habitats in newborns. Proc Natl Acad Sci USA 2010;107:11971–5.

[26] Jakobsson HE, Abrahamsson TR, Jenmalm MC, et al. Decreased gut microbiota diversity, delayed Bacteroidetes colonisation and reduced Th1 responses in infants delivered by caesarean section. Gut 2014;63(4):559–66.

[27] Fouhy F, Guinane CM, Hussey S, et al. High-throughput sequencing reveals the incomplete, short-term recovery of infant gut microbiota following parenteral antibiotic treatment with ampicillin and gentamicin. Antimicrob Agents Chemother 2012;56(11):5811–20.

[28] Penders J, Thijs C, Vink C, et al. Factors influencing the composition of the intestinal microbiota in early infancy. Pediatrics 2006;118(2):511–21.

[29] Arboleya S, Sánchez B, Milani C, et al. Intestinal microbiota development in preterm neonates and effect of perinatal antibiotics. J Pediatr 2015;166(3):538–44.

[30] Azad MB, Konya T, Persaud RR, et al. Impact of maternal intrapartum antibiotics, method of birth and breastfeeding on gut microbiota during the first year of life: a prospective cohort study. BJOG 2016;123:983–93.

[31] Bezirtzoglou E, Tsiotsias A, Welling GW. Microbiota profile in feces of breast- and formula-fed newborns by using fluorescence in situ hybridization (FISH). Anaerobe 2011;17:478–82.

[32] Le Huërou-Luron I, Blat S, Boudry G. Breast- v. formula-feeding: impacts on the digestive tract and immediate and long-term health effects. Nutr Res Rev 2010;23:23–36.

[33] Yatsunenko T, Rey FE, Manary MJ, et al. Human gut microbiome viewed across age and geography. Nature 2012;486(7402):222–7.

[34] Bäckhed F, Roswall J, Peng Y, et al. Dynamics and stabilization of the human gut microbiome during the first year of life. Cell Host Microbe 2015;17:690–703.

[35] Prescot SL. Early-life environmental determinants of allergic diseases and the wider pandemic of inflammatory noncommunicable diseases. J Allergy Clin Immunol 2013;131:23–30.

[36] West CE, Renz H, Jenmalm MC, et al. The gut microbiota and inflammatory noncommunicable diseases: associations and potentials for gut microbiota therapies. J Allergy Clin Immunol 2015;135(1):3–13.

[37] Husby S, Koletzko S, Korponay-Szabó IR, et al. European Society for Pediatric Gastroenterology, Hepatology, and Nutrition guidelines for the diagnosis of coeliac disease. J Pediatr Gastroenterol Nutr 2012;54(1):136–60. Erratum in: J Pediatr Gastroenterol Nutr 54(4):572..

[38] Mårild K, Ludvigsson JF, Størdal K. Current evidence on whether perinatal risk factors influence coeliac disease is circumstantial. Acta Paediatr 2016;105:366–75.

[39] Verdu EF, Galipeau HJ, Jabri B. Novel players in coeliac disease pathogenesis: role of the gut microbiota. Nat Rev Gastroenterol Hepatol 2015;12(9):497–506.

[40] Atkinson MA, Eisenbarth GS, Michels AW. Type 1 diabetes. Lancet 2014;383(9911):69–82.

[41] Dabelea D. The accelerating epidemic of childhood diabetes. Lancet 2009;373(9680):1999–2000.

[42] Giongo A, Gano KA, Crabb DB, et al. Toward defining the autoimmune microbiome for type 1 diabetes. ISME J 2011;5(1):82–91.

[43] Brown CT, Davis-Richardson AG, Giongo A, et al. Gut microbiome metagenomics analysis suggests a functional model for the development of autoimmunity for type 1 diabetes. PLoS One 2011;6(10):e25792.

[44] Murri M, Leiva I, Gomez-Zumaquero JM, et al. Gut microbiota in children with type 1 diabetes differs from that in healthy children: a case–control study. BMC Med 2013;11:46.

[45] de Goffau MC, Luopajärvi K, Knip M, et al. Fecal microbiota composition differs between children with β-cell autoimmunity and those without. Diabetes 2013;62(4):1238–44.

[46] Endesfelder D, zu Castell W, Ardissone A, et al. Compromised gut microbiota networks in children with anti-islet cell autoimmunity. Diabetes 2014;63:2006–14.

[47] Gülden E, Wong FS, Wen L. The gut microbiota and type 1 diabetes. Clin Immunol 2015;159(2):143–53.

[48] Hill C, Guarner F, Reid G, et al. Expert consensus document. The International Scientific Association for Probiotics and Prebiotics consensus statement on the scope and appropriate use of the term probiotic. Nat Rev Gastroenterol Hepatol 2014;11:506–14.

[49] Gibson GR, Roberfroid MB. Dietary modulation of the human colonic microbiota. Introducing the concept of prebiotics. J Nutr. 1995;125:1401–12.

[50] Schrezenmeir J, de Vrese M. Probiotics, prebiotics, and synbiotics—approaching a definition. Am J Clin Nutr 2001;73(2 Suppl.):361S–4S.

[51] Bowman KA, Broussard EK, Surawicz CM. Fecal microbiota transplantation: current clinical efficacy and future prospects. Clin Exp Gastroenterol 2015;8:285–91.

[52] Szajewska H, Guarino A, Hojsak I, et al. Use of probiotics for management of acute gastroenteritis: a position paper by the ESPGHAN Working Group for Probiotics and Prebiotics. J Pediatr Gastroenterol Nutr 2014;58(4):531–9.

[53] Glanville J, King S, Guarner F, Hill C, Sanders ME. A review of the systematic review process and its applicability for use in evaluating evidence for health claims on probiotic foods in the European Union. Nutr J 2015;14:16.

[54] Doege K, Grajecki D, Zyriax BC, et al. Impact of maternal supplementation with probiotics during pregnancy on atopic eczema in childhood—a meta-analysis. Br J Nutr 2012;107(1):1–6.

[55] Pelucchi C, Chatenoud L, Turati F, et al. Probiotics supplementation during pregnancy or infancy for the prevention of atopic dermatitis: a meta-analysis. Epidemiology 2012;23(3):402–14.

[56] Elazab N, Mendy A, Gasana J, Vieira ER, Quizon A, Forno E. Probiotic administration in early life, atopy, and asthma: a meta-analysis of clinical trials. Pediatrics 2013;132(3):e666–76.

[57] Azad MB, Coneys JG, Kozyrskyj AL, et al. Probiotic supplementation during pregnancy or infancy for the prevention of asthma and wheeze: systematic review and meta-analysis. BMJ 2013;347:f6471.

[58] Cuello-Garcia CA, Brożek JL, Fiocchi A, et al. Probiotics for the prevention of allergy: a systematic review and meta-analysis of randomized controlled trials. J Allergy Clin Immunol 2015;136(4):952–61.

[59] Osborn DA, Sinn JK. Prebiotics in infants for prevention of allergy. Cochrane Database Syst Rev 2013;3:CD006474.

[60] Kukkonen K, Savilahti E, Haahtela T, et al. Probiotics and prebiotic galacto-oligosaccharides in the prevention of allergic diseases: a randomized, double-blind, placebo-controlled trial. J Allergy Clin Immunol 2007;119(1):192–8.

[61] Kuitunen M, Kukkonen K, Juntunen-Backman K, et al. Probiotics prevent IgE-associated allergy until age 5 years in cesarean-delivered children but not in the total cohort. J Allergy Clin Immunol 2009;123(2):335–41.

[62] de Silva D, Geromi M, Halken S, et al. Primary prevention of food allergy in children and adults: systematic review. Allergy 2014;69:581–9.

[63] Muraro A, Halken S, Arshad SH, et al. EAACI food allergy and anaphylaxis guidelines. Primary prevention of food allergy. Allergy 2014;69:590–601.

[64] Fiocchi A, Pawankar R, Cuello-Garcia C, et al. World Allergy Organization-McMaster University Guidelines for Allergic Disease Prevention (GLAD-P): probiotics. World Allergy Organ J 2015;8:4.

[65] Boyle RJ, Ismail IH, Kivivuori S, et al. *Lactobacillus* GG treatment during pregnancy for the prevention of eczema: a randomized controlled trial. Allergy 2011;66:509–16.

[66] Kalliomäki M, Salminen S, Arvilommi H, Kero P, Koskinen P, Isolauri E. Probiotics in primary prevention of atopic disease: a randomized placebo-controlled trial. Lancet 2001;357:1076–9.

[67] Kopp MV, Hennemuth I, Heinzmann A, Urbanek R. Randomized, double-blind, placebo-controlled trial of probiotics for primary prevention: no clinical effects of Lactobacillus GG supplementation. Pediatrics 2008;121:e850–6.

[68] Ou CY, Kuo HC, Wang L, et al. Prenatal and postnatal probiotics reduces maternal but not childhood allergic diseases: a randomized, double-blind, placebo-controlled trial. Clin Exp Allergy 2012;42:1386–96.

[69] Rautava S, Kalliomäki M, Isolauri E. Probiotics during pregnancy and breast-feeding might confer immunomodulatory protection against atopic disease in the infant. J Allergy Clin Immunol 2002;109:119–21.

[70] Szajewska H, Shamir R, Turck D, et al. Recommendations on probiotics in allergy prevention should not be based on pooling data from different strains. J Allergy Clin Immunol 2015;136(5):1422.

[71] Kim SO, Ah YM, Yu YM, Choi KH, Shin WG, Lee JY. Effects of probiotics for the treatment of atopic dermatitis: a meta-analysis of randomized controlled trials. Ann Allergy Asthma Immunol 2014;113(2):217–26.

[72] Betsi GI, Papadavid E, Falagas ME. Probiotics for the treatment or prevention of atopic dermatitis: a review of the evidence from randomized controlled trials. Am J Clin Dermatol 2008;9(2):93–103.

[73] Michail SK, Stolfi A, Johnson T, Onady GM. Efficacy of probiotics in the treatment of pediatric atopic dermatitis: a meta-analysis of randomized controlled trials. Ann Allergy Asthma Immunol 2008;101(5):508–16.

[74] Lee J, Seto D, Bielory L. Meta-analysis of clinical trials of probiotics for prevention and treatment of pediatric atopic dermatitis. J Allergy Clin Immunol 2008;121(1):116–21. e11.

[75] Boyle RJ, Bath-Hextall FJ, Leonardi-Bee J, Murrell DF, Tang ML. Probiotics for treating eczema. Cochrane Database Syst Rev 2008;8(4):CD006135.

[76] Zajac AE, Adams AS, Turner JH. A systematic review and meta-analysis of probiotics for the treatment of allergic rhinitis. Int Forum Allergy Rhinol 2015;5(6):524–32.

[77] Peng Y, Li A, Yu L, Qin G. The role of probiotics in prevention and treatment for patients with allergic rhinitis: a systematic review. Am J Rhinol Allergy 2015;29(4):292–8.

[78] Das RR, Singh M, Shafiq N. Probiotics in treatment of allergic rhinitis. World Allergy Organ J 2010;3(9):239–44.

[79] Vliagoftis H, Kouranos VD, Betsi GI, Falagas ME. Probiotics for the treatment of allergic rhinitis and asthma: systematic review of randomized controlled trials. Ann Allergy Asthma Immunol 2008;101(6):570–9.

[80] He M, Wang T, Zhang HP, Jia CE, Xiong XY, Wang G, Ji YL. Probiotics for prevention and treatment of bronchial asthma: a systematic review. Chin J Evid Based Med 2012;12(4):460–9.

[81] Helin T, Haahtela S, Haahtela T. No effect of oral treatment with an intestinal bacterial strain, Lactobacillus rhamnosus (ATCC 53103), on birch-pollen allergy: a placebo-controlled double-blind study. Allergy 2002;57:243–6.

[82] Stockert K, Schneider B, Porenta G, Rath R, Nissel H, Eichler I. Laser acupuncture and probiotics in school age children with asthma: a randomized, placebo-controlled pilot study of therapy guided by principles of Traditional Chinese Medicine. Pediatr Allergy Immunol 2007;18:160–6.

[83] Wheeler JG, Shema SJ, Bogle ML, et al. Immune and clinical impact of Lactobacillus acidophilus on asthma. Ann Allergy Asthma Immunol 1997;79:229–33.

[84] Giovannini M, Agostoni C, Riva E, et al. A randomized prospective double blind controlled trial on effects of long-term consumption of fermented milk containing Lactobacillus casei in pre-school children with allergic asthma and/or rhinitis. Pediatr Res 2007;62:215–20.

[85] Rose MA, Stieglitz F, Köksal A, et al. Efficacy of probiotic Lactobacillus GG on allergic sensitization and asthma in infants at risk. Clin Exp Allergy 2010;67(40):1398–405.

[86] Gutkowski P, Madalinski K, Grek M, et al. Effect of orally administered probiotic strains Lactobacillus and Bifidobacterium in children with atopic asthma. Centr Eur J Immunol 2010;35(4): 233–8.

[87] Chen YS, Jan RL, Lin LY, et al. Randomized placebo-controlled trial of lactobacillus on asthmatic children with allergic rhinitis. Pediatr Pulmonol 2010;54(45):1111–20.

[88] van der Aa LB, van Aalderen WM, Heymans HS, et al. Synbiotics prevent asthma-like symptoms in infants with atopic dermatitis. Allergy 2011;66(2):170–7.

[89] Hol J, van Leer EH, Elink Schuurman BE, et al. The acquisition of tolerance toward cow's milk through probiotic supplementation: a randomized, controlled trial. J Allergy Clin Immunol 2008;121: 1448–54.

[90] Berni Canani R, Nocerino R, Terrin G, et al. Effect of Lactobacillus GG on tolerance acquisition in infants with cow's milk allergy: a randomized trial. J Allergy Clin Immunol 2012;129:580–2.

[91] Berni CR, Nocerino R, Terrin G, et al. Formula selection for management of children with cow's milk allergy influences the rate of acquisition of tolerance: a prospective multicenter study. J Pediatr 2013;163:771–7.

[92] Uusitalo U, Liu X, Yang J, et al. Association of Early Exposure of Probiotics and Islet Autoimmunity in the TEDDY Study. JAMA Pediatr 2015;9:1–9.

[93] Braegger C, Chmielewska A, Decsi T, et al. Supplementation of infant formula with probiotics and/or prebiotics: a systematic review and comment by the ESPGHAN Committee on Nutrition. J Pediatr Gastroenterol Nutr 2011;52(2):238–50.

[94] De Angelis M, Rizzello CG, Fasano A, et al. VSL#3 probiotic preparation has the capacity to hydrolyze gliadin polypeptides responsible for Celiac Sprue. Biochim Biophys Acta 2006;1762(1):80–93.

[95] Di Cagno R, Barbato M, Di Camillo C, et al. Gluten-free sourdough wheat baked goods appear safe for young celiac patients: a pilot study. J Pediatr Gastroenterol Nutr 2010;51(6):777–83.

[96] Smecuol E, Hwang HJ, Sugai E, et al. Exploratory, randomized, double-blind, placebo-controlled study on the effects of Bifidobacterium infantis natren life start strain super strain in active celiac disease. J Clin Gastroenterol 2013;47(2):139–47.

[97] Olivares M, Castillejo G, Varea V, Sanz Y. Double-blind, randomised, placebo-controlled intervention trial to evaluate the effects of Bifidobacterium longum CECT 7347 in children with newly diagnosed coeliac disease. Br J Nutr 2014;112(1):30–40.

[98] Smecuol E, Pinto-Sánchez MI, Bai JC. Understanding the role of probiotics in coeliac disease. Br J Nutr 2015;113(10):1664–5.

Chapter 33

Probiotics Use in Infectious Disease (Respiratory, Diarrhea, and Antibiotic-Associated Diarrhea)

S. Guandalini and N. Sansotta

Section of Gastroenterology, Hepatology and Nutrition, University of Chicago Medicine—Comer Children's Hospital, Chicago, IL, United States

INTRODUCTION

Infectious disorders represent an intuitively logical field of application of probiotics, given their effects on promoting a healthy immune system, as well as an area where sufficient data for solid recommendations can be found. In this chapter we will focus on the most common of them, namely respiratory infections, acute infectious diarrhea (AGE), and antibiotic-associated diarrhea (AAD). In addition, we will briefly discuss a rapidly emerging application for the use of probiotics: the prevention of necrotizing enterocolitis (NEC), a severe and potentially fatal newborn-specific disease.

In these areas, a number of data have been accumulated and we applied the pyramid of evidence in order to obtain indications on the use of probiotics for those conditions. When available, we focused on guidelines and metaanalyses or systematic reviews (including Cochrane reviews). Original data from single randomized controlled trials (RCT) are utilized when metaanalyses may not be available. However, one should also remember that the application of recommendations suffers of major limits, including the often found heterogeneity of data and suboptimal level of quality of evidence.

ACUTE RESPIRATORY INFECTIONS

Acute upper respiratory infections (URTI) represent the most frequent reason for parental concern and medical visits in preschool and elementary school children. In this age group, URTI occur at an annual rate of 5–6 infections [1], account for 22–26.7% of all hospitalizations [2,3] and 33.5–59% of all general practitioner consultations [4].

Furthermore, respiratory (as well as gastrointestinal) infections are a predominant cause of morbidity and mortality in the elderly whose ageing immune system contributes significantly to poor outcomes [5]. Ageing is, in fact, associated with a decline of innate and adaptive immune responses. Moreover, it has been also shown that the age-dependent modifications of the composition of the gut microbiota also contribute to the defective local and systemic immune defenses in the elderly population [6,7]. Some studies have shown that in the elderly probiotics may help maintain immune function either by direct interaction with the host immune system or indirectly by reequilibrating the gut microbiota [8–10].

Thus, there has been an increased interest in recent years in testing the potential benefit of probiotics in the prevention of respiratory infections, even though these studies must be still considered at an early stage and results so far obtained have been somewhat inconsistent [11].

MECHANISMS OF ACTION

Probiotics have been suggested to protect against infectious diseases by several strain-dependent mechanisms [12] including secretion of antipathogen substances, competitive exclusion of pathogens, maintenance of mucosal integrity, and stimulation of systemic or mucosal immune responses [12–15].

As in other conditions, for respiratory infections too, *Lactobacillus* and *Bifidobacterium* are the most commonly used bacterial probiotics. Previous controlled clinical trials have shown that the intake of such probiotics stimulates mucosal immune systems by enhancing secretory IgA [16,17].

In support of human clinical studies, animal studies involving the use of *Lactobacillus* in the prevention and treatment of respiratory viral infections have also yielded potentially promising results. In influenza-infected mice, oral intake of *Lactobacillus plantarum* enhanced type I interferon production [18]. As a result, mice that had received probiotics had lower pulmonary viral titers [18].

In a similar animal model, *Lactobacillus* strains administered orally reduced influenza virus titers in the lungs with the fast concomitant rise of mRNA expression of interferon-γ, tumor necrosis factor, interleukin (IL)-12, and IL-2,39 upregulation of interferon-γ, IL-10 and IL-6 levels, and the increase in number of T helper cells in the pulmonary parenchyma [19]. Some strains of *Lactobacillus rhamnosus* and *Bifidobacterium lactis* increased interferon-γ, IL-4, IL-6, and IL-10 levels in the bronchoalveolar lavage, number and activity of phagocytic cells, and cytotoxic potential of natural killer cells [20,21].

Phagocytes and natural killer cells are the most powerful and important parts of the innate host defense that can operate without delay against the invading viruses [22]. Their mediating role could explain the shortening of the duration of URTI; however, their exact role has not been studied yet [23].

Bacillus species have been shown to produce antimicrobial substances [24], to enhance epithelial gut barrier functions [25,26] and stimulate cytokine [26–28].

Supplementation with *Bacillus subtilis* CU1 stimulated systemic immune response in seniors by significantly increasing serum IFN gamma in the probiotic group following first *B. subtilis* supplementation [29].

Huang et al. [30] also found that *Bacillus* strains could stimulate systemic and intestinal IFN-gamma production in mice. It is a Pro-Th1 cytokine and it plays a role in the host defense against several infectious diseases, including viral infection and has a variety of immune functions, such as stimulation of macrophages and natural killer cells [31,32]. Different studies have emphasized the importance of IFN-gamma production for the protective effect of probiotics against influenza infection [33,34].

Further investigations in *B. subtilis* CU1's capability in increasing serum IFN-gamma levels and the strengthening of the systemic antiviral and antibacterial immune defenses in the elderly population would be interesting.

CLINICAL IMPACT

There are contradictory data on probiotic use in the prevention of URTI. The variability in the outcomes between clinical trials studying probiotics' role in URTI may be explained by variability in probiotic strains, bacterial doses, and matrices. In addition, it should be noted that the effects of probiotics are highly strain-specific and the adequate amount of bacteria transferred into the effector sites in the gut may be crucial [35].

In Table 33.1, we report the effects of probiotics on respiratory tract infections in various clinical setting.

Prevention

A variety of probiotic strains have been used in the clinical trials, most of them belonging to the genus *Lactobacillus*, with many studies using specifically *L. rhamnosus* GG (LGG) [36–43,77].

Investigating the role of LGG in the prevention of respiratory infections in children, several authors showed that it has the potential to reduce the risk of upper RTI, episodes that lasted >3 days [37], days of absence from day care because of illness [36], and incidence of acute otitis media [43]. However, there were no significant differences between the LGG and the control groups in the hospitalization duration [37], in the number of antibiotic treatments or respiratory symptom episodes [39], and in the incidence of lower RTIs [77]. Of note, in preterm infants, LGG reduced the incidence of RTI and specifically the incidence of rhinovirus-induced episodes, which comprised 80% of all RTI episodes, was found to be significantly lower in the probiotic group compared with the placebo group [40].

Several studies were conducted with *Lactobacillus* strains other than LGG: *L casei* CRL431 [44], *L. casei* DN 1140001 [45–47], *L reuteri* DSM17938 [44], *L. casei rhamnosus* [48], *L. paracasei* CBA L74 [49], *L. fermentum* CECT5716 [50].

No effects on the reduction of the incidence of URTIs were shown by using *L. casei* CRL431, *L. reuteri* DSM17938 in healthy children aged 1–6 years in an Indonesian setting [44] or in the administration of *L. reuteri* 55730 in healthy term-infants aged 4–10 months at child care in Israel [51].

L. casei rhamnosus, *L. paracasei* CBA L74, and *L. fermentum* CECT5716 determined a reduction of URTI [48,50], and prevention of common cold in children [49].

Contradictory results have been documented about the efficacy of *L. casei* DN114001 on the reduction of the incidence of RTI [45–47]. In children, this probiotic decreased the duration and the incidence of only lower RTI (bronchitis and pneumonia), but not upper RTI [47]; while in elderly it decreased the duration of RTI [45,46] but had no effect on their incidence [45].

Bifidobacteria have also been studied for a possible use in the prevention of common respiratory infections.

B. animalis subsp. *lactis* (BB-12) was unable to reduce duration of respiratory tract infections, absence from day care center due to infections, use of antibiotics in healthy children who attend day care centers [51,52].

On the other hand, the use of this probiotic in healthy newborns was able to reduce the number of RTIs, but it was ineffective in reducing the occurrence or symptoms of acute otitis media [53].

TABLE 33.1 Effects of Probiotics on RTIs in Various Clinical Settings

Author (References)	Subjects	Probiotic Used	Outcomes
Hatakka and Savilahti [36]	Healthy children	*L. rhamnosus* GG	Days with respiratory symptoms = No. of children with RTI↓
Hojsak et al. [37]	Healthy children	*L. rhamnosus* GG	No. of children with RTI↓ No. of URTI episodes↓ No. of RTI lasting >3 days↓
Hojsak et al. [38]	Hospitalized children	*L. rhamnosus* GG	Risk for RTI↓ Risk for duration of RTI lasting >3 days ↓ Duration of hospitalization =
Kumpu et al. [39]	Healthy children	*L. rhamnosus* GG	Days with respiratory symptoms = Antibiotic treatment =
Luoto et al. [40]	Preterm infants	*L. rhamnosus* GG	Incidence of RTI↓ Duration/severity of RTI =
Kumpu et al. [41]	Healthy children	*L. rhamnosus* GG	Days with respiratory symptoms ↓ Occurrence of virus in nasopharynx ↓
Kukkonen et al. [42]	Pregnant mothers	*L. rhamnosus* GG + LC705, *Bifidobacterium breve* Bb99, + *Propionibacterium freudenreichii* ssp. *shermanii*	Incidence of RTI in newborns ↓ Antibiotic treatment ↓
Rautava et al. [43]	Healthy newborns	*L. rhamnosus* GG + *Bifidobacterium animalis* ssp. *lactis* Bb12	No. of RTI episodes = Antibiotic treatment ↓ Incidence of AOM ↓
Agustina et al. [44]	Healthy children	*Lactobacillus casei* CRL43 or *Lactobacillus reuteri* DSM 17938	No. of RTI episodes = Duration of RTI =
Turchet et al. [45]	Healthy elderly	*L. casei* DN 114001	Duration of RTI ↓ Incidence of RTI =
Guillemard et al. [46]	Healthy elderly	*L. casei* DN 114001	Duration per episode of RTI ↓
Cobo Sanz et al. [47]	Healthy school children	*L. casei* DN 114001	Incidence/duration (days) of RTI = Duration of lower RTI ↓ Incidence of lower RTI ↓
Lin et al. [48]	Healthy children	*L. casei rhamnosus*	Incidence of bacterial infections ↓
Nocerino et al. [49]	Healthy children	*Lactobacillus paracasei* CBA L74	Incidence of URTI ↓
Maldonado et al. [50]	Healthy infants	*Lactobacillus fermentum* CECT5716	Incidence of upper and lower RTI ↓
Weizman et al. [51]	Healthy infants	*L. reuteri* SD 112 or *B. animalis* ssp. *lactis* Bb12	Incidence/duration of RTI =
Hojsak et al. [52]	Healthy children	*B. animalis* subsp. *lactis*	Duration of symptoms of RTI = Incidence of RTI = Absence from day care =
Taipale et al. [53]	Healthy newborns	*B. animalis* ssp. *lactis* Bb12	No of RTI episodes ↓ Occurrence of AOM =
Garaiova et al. [54]	Healthy children	*L. acidophilus* CUL21 (NCIMB 30156), *L. acidophilus* CUL60 (NCIMB 30157), *Bifidobacterium bifidum* CUL20 (NCIMB 30153) + *B. animalis* subsp. *lactis* CUL34 (NCIMB 30172)	Incidence of URTI ↓ No. of days with URTI symptoms ↓ Absence from school ↓
Winkler et al. [55]	Healthy adults	*L. gasseri* PA16/8+ *Bifidobacterium longum* SP07/3+ *B. bifidum* MF 20/5	Incidence and duration of RTI = No. of days with fever ↓
De Vrese et al. [56]	Healthy adults	*Lactobacillus gasseri* PA16/8+ *B. longum* SP07/3+ *B. bifidum* MF 20/5	Duration of RTI ↓ Severity of RTI = No. of days with fever ↓

(Continued)

TABLE 33.1 Effects of Probiotics on RTIs in Various Clinical Settings (*cont.*)

Author (References)	Subjects	Probiotic Used	Outcomes
De Vrese et al. [57]	Healthy adults	*L. gasseri* PA16/8+ *B. longum* SP07/3+ *B. bifidum* MF 20/5	No of RTI = Duration of RTI ↓ Severity of RTI =
Hatakka et al. [58]	Otitis prone children	*L. rhamnosus* GG, *L. rhamnosus* LC705, *B. breve* 99, *P. freudenreichii* JS	Occurrence of AOM/RTI =
Boge et al. [59]	Healthy elderly	*L. casei* DN-114 001	Influenza vaccination immune response ↑
Bosch et al. [60]	Healthy elderly	*L. plantarum* CECT7315 and CECT7316	Influenza vaccination immune response ↑
Davidson et al. [61]	Healthy adults	*Lactobacillus* GG	Influenza vaccination immune response ↑
Van Puyenbroeck et al. [62]	Healthy elderly nursing home residents	*L. casei* Shirota	Duration of RTI = No. of people with RTI = Influenza vaccination immune response =
Berggren et al. [63]	Healthy adults	*L. plantarum* HEAL 9 (DSM 15312) + *L. paracasei* 8700:2 (DSM 13434)	Incidence of URTI ↓ No. of days with symptoms of URTI ↓
Fujita et al. [64]	Elderly	*L. casei* Shirota	No. of acute URTI events = Severity of acute URTI events = No. of people with acute URTI = Mean duration of URTI =
Makino et al. [65]	Elderly	*Lactobacillus delbrueckii* ssp. *bulgaricus* OLL1073R-1	Incidence of URTI ↓
Merestein et al. [66]	Healthy children	*L. casei* DN 114001	Incidence of URTI ↓ Missed day care/school =
Rerksuppaphol et al. [67]	Healthy children	*L. acidophilus* + *B. bifidum*	Duration of URTI ↓ Risk of URTI symptoms(e.g., fever, cough) ↓
Rio et al. [68]	Healthy children	*L. acidophilus* + *L. casei*	Incidence of RTI ↓ Severity of RTI ↓
Smith et al. [69]	Healthy college students	*L. rhamnosus* GG + *B. animalis* ssp. *lactis* Bb12	Mean duration of URTI ↓ Severity of URTI ↓ Missed school days ↓
Hojsak et al. [70]	Hospitalized children	*B. animalis* subsp. *lactis*	Incidence of RTI = Duration of RTI = Duration of hospitalization=
Lehtoranta et al. [71]	Otitis prone children	*L. rhamnosus* GG, *L. rhamnosus* LC705, *B. breve* 99, *P. freudenreichii* JS	Human bocavirus load in the nasopharynx ↓
Mane et al. [72]	Institutionalized elderly	*L. plantarum* CECT 7315 and CECT 7316	Incidence of RTI ↓ Risk of mortality for RTI↓
Gilbey et al. [73]	Hospitalized adults with severe pharyngotonsillitis	*Streptococcus salivarius* K12	Duration of hospitalization = Symptoms severity =
Di Pierro et al. [74]	Children with a history of recurrent oral streptococcal pathology	*S. salivarius* K12	Incidence of RTI/AOM ↓
Di Pierro et al. [75]	Adults with a history of recurrent oral streptococcal pathology	*S. salivarius* K12	Incidence of RTI ↓
Di Pierro et al. [76]	Children with a history of recurrent oral streptococcal pathology	*S. salivarius* K12	Incidence of RTI ↓ Missed school days ↓ Antibiotic treatment ↓

AOM, Acute otitis media; URTI, upper respiratory tract infections.

B. subtilis CU1 did not significantly decrease the mean number of days of reported common infection disease symptoms in elderly subjects (aged 60–74) [29].

The effectiveness of several combinations of probiotics on the prevention of RTIs has been investigated as well.

The combination of *Lactobacillus acidophilus* CUL21 (NCIMB 30156), *L. acidophilus* CUL60 (NCIMB 30157), *B. bifidum* CUL20 (NCIMB 30153), and *B. animalis* subsp. *lactis* CUL34 (NCIMB 30172) significantly reduced the incidence rate of URTI, the number of days with URTI symptoms, and the incidence of absence from preschool in healthy children aged 3–6 years attending preschool facilities [54].

Similarly, the combined use of *L. acidophilus* DDS-1 and *B. lactis* UABLA-12 decreased the incidence of URTIs versus placebo in children at an increased risk of infection due to extended indoor exposure to a sick household [23].

Furthermore, the combination of LGG and *B. lactis* BB-12 administered to infants reduced the risk of acute otitis media when compared with placebo ($p = 0.014$) [43].

In addition, a combination of *L. gasseri* PA16/8, *Bifidobacterium longum* SP07/3, and *B. bifidum* MF20/5 reduced the duration of RTI symptoms [55], as well as the duration of RTI episodes [56,57].

Conversely, the combination of LGG, *L. rhamnosus* LC 705, *B. breve* 99, and *P. freudenreichii* spp. *shermanii* JS failed to decrease occurrence, recurrence, or duration of acute otitis media episodes in children aged between 10 months and 6 years [58].

Finally, some trials showed that *Lactobacillus* and *Bifidobacterium* probiotics could increase influenza vaccination immune responses in the elderly [59–62].

Most of these and other studies [36,43,47,55,58,63–69] were included in two recent reviews and metaanalyses [78,79] aiming to assess the effectiveness and safety of probiotics compared to placebo, in the prevention of URTI in people of various age groups.

Cochrane reviewers extracted and pooled data from 12 trials with a total of 3720 participants [78]. Six studies were in children, three in adults, and two in elderly people, involving different types of probiotics [36,43,47,55,58,63–69]. These authors found probiotics to be better than placebo in reducing the odds of experiencing an episode of URTI, the mean duration of an URTI episode, antibiotic use associated with URTI episodes, and cold-related school absence. In fact, the number of people who develop an URTI was significantly reduced when taking probiotics [odds ratio (OR) 0.57, 95% confidence interval (CI) 0.37–0.76]. In absolute terms, 11 fewer people out of 100 who may develop a URTI when taking probiotics for 3–8 months. In addition, the duration of an URTI was reduced by approximately 2 days (mean difference: 1.89 fewer days from 2.03 to 1.75 fewer days) [78,80,81].

Of note, *Lactobacillus* spp. and *Bifidobacterium* spp. strains brought by food products or supplements significantly lowered the number of days of acute respiratory infections in a healthy population of children and adults, and shortened acute respiratory infective periods [79].

Unfortunately, the quality of evidence on individual outcomes was generally judged as "low" or "very low." Some studies had methodological flaws, and probiotic manufacturers funded others. The totality of the evidence on individual questions was often sparse (leading to imprecision of effect estimates), reliant on small studies (leading to questions about publication bias), or heterogeneous (reflecting inconsistency). The heterogeneity of the strains and doses of probiotics contributes to the difficulties in fully evaluating safety profiles of probiotics [81].

We can conclude that despite the low quality of evidence combined with the discussed limitations previously, probiotics should still be considered for URTI prevention in clinical practice.

Treatment

In hospitalized children, LGG reduced the risk of RTIs and duration of RTI episodes [38] while *B. animalis* subsp. *lactis* instead was ineffective to reduce the duration, the incidence, the severity of common nosocomial (respiratory and gastrointestinal) infections [70].

Of interest, in otitis-prone children, a combination of LGG, *L. rhamnosus* Lc705, *B. breve* Bb99, and *P. freudenreichii* ssp. *shermanii* JS reduced human bocavirus load in the nasopharynx, a factor thought to plays an important role in favoring the onset of acute otitis media [71].

Mañé et al. showed significant trends in reducing infection incidence and mortality due to pneumonia in institutionalized elderly subjects treated with two *L. plantarum* strains [72].

To our knowledge, there is only one prospective, randomized, placebo-controlled, double-blinded study compared treatment with *S. salivarius* to placebo in addition to antibiotics in adults who were hospitalized with severe pharyngotonsillitis. The probiotic *S. salivarius* was ineffective in the treatment of acute infection [73]. Conversely, the same probiotic was found effective in the prophylaxis for pharyngotonsillitis [74–76].

ACUTE INFECTIOUS DIARRHEA

In the vast majority of cases and especially in children, the acute onset of diarrhea is due to an episode of infectious gastroenteritis and is generally defined as a decrease in the consistency of stools (loose or liquid) and/or an increase in the frequency of evacuations (typically ≥ 3 in 24 h), with or without fever, or vomiting and usually lasting no more than 7 days [82].

AGE is probably the main, certainly the original field of application for probiotics. A large number of data have been obtained since the paper that in 1991 firstly provided relevant evidence for the efficacy of LGG in the treatment of AGE [83].

In the last years, an increasing number of RCTs have been published on this issue in various settings and with different outcomes. Overall, the data available have progressively provided compelling evidence that probiotic administration is effective against AGE.

It should be remembered that, as in other clinical settings, there is a wide heterogeneity in definitions, settings, and outcome parameters. In trials on AGE, Johnston et al. [84] found 64 different definitions of acute diarrhea in children and 69 definitions of healing. This may significantly affect the reproducibility of the efficacy of any intervention aimed at reducing diarrheal outcomes.

To overcome this situation, a group of experts proposed definitions as well as the parameters to be used in future trials on nutritional interventions for acute diarrhea [85].

Although the indications for those outcome parameters are not directly related with probiotic use, they provide valuable indications for future studies as well as for the interpretation of homogeneous available data.

According to these indications, the most important outcome is the duration of diarrhea for therapeutic studies and the incidence of episodes of diarrhea for prevention studies. Irrespective of the definition of diarrhea used, probiotics have evidence to reduce both clinical outcomes.

A large number of clinical trials, and subsequently metaanalyses and guidelines, have been published on the use of probiotics in AGE, in community acquired versus nosocomial diarrhea, developing and developed countries, viral or bacterial etiology, in children exposed or not to risk conditions (e.g., those admitted to day care centers).

Prevention

Healthy Infants and Children in Day Care Centers

Due to the excellent safety profile of probiotics, the administration of functional foods added to infant formulas and other foods, and given over a long period of time for the prevention of acute intestinal infections in healthy and at-risk children has been proposed and studied.

The vast majority of prevention studies have been performed with probiotic strains added to a milk-based feeding. The first such trials documented a reduction in the incidence or severity of acute diarrheal disease [86,87]. However, subsequent studies provided evidence of a very modest effect of some probiotic strains (LGG, *B. lactis*, alone or in combination with *Streptococcus thermophilus*, and *L. reuteri, L. rhamnosus*—not GG—and *L. acidophilus*) on the prevention of community-acquired diarrhea,

and some authors—although the results were statistically significant—questioned their clinical relevance [88,89]. Gutierrez-Castrellon et al. demonstrated that the daily administration of the probiotic strain *L. reuteri DSM 17938* to healthy children may reduce episodes of acute diarrhea, resulting in addition in a cost-saving benefit to families and to the healthcare system [90].

These somewhat conflicting results may in part be explained by variability in the dose used and in the viability of the probiotics administered in feeding. In fact, it is likely that many bacteria get killed during the preparation of infant foods, whenever the use of boiled water, as recommended by the WHO [91] is implemented for hygienic reasons.

As for the use of probiotics added in formulas for healthy children, both European [92] and North American Pediatric Societies [93] agree on the paucity of evidence in support of such use. The American Academy of Pediatrics, however, considered as beneficial the use of probiotics in special circumstances, such as children in long-term healthcare facilities or care centers [93].

Hospital-Acquired Diarrhea

Hospital-acquired infections, and especially diarrhea, are quite common, with rotavirus being, especially in children, a leading cause of nosocomial gastroenteritis. Hand washing and isolation of infected children represent the first and inexpensive procedures to prevent the nosocomial spreading of diarrhea, but the effectiveness of these measures is still unsatisfactory [94]. Despite some positive evidence, it is unclear whether LGG is also effective in preventing nosocomial diarrhea. According to the European Food and Safety Authority (EFSA), only one of the three available RCTs showed LGG to be efficacious in decreasing the incidence or duration of diarrhea in hospitalized children [95].

In a recent metaanalysis, the administration of LGG compared with placebo to hospitalized children was shown to have the potential to reduce the overall incidence of healthcare-associated diarrhea, including rotavirus gastroenteritis [two RCTs, $n = 823$, Risk Ratio (RR) 0.37, 95% CI 0.23–0.59] [14].

Supplementation with bifidobacteria to hospitalized infants significantly prevented the incidence of diarrhea and the onset of hospital-acquired diseases. As an example, a beneficial effect of *B. breve* strain *Yakult* has been reported in immunocompromised children on chemotherapy. These young patients suffered from infectious complications; following probiotic administration, the use of antibiotics to cure infections was lower and the gut habitation of anaerobes was enhanced [96].

Treatment

Therapy of AGE is arguably the main indication for probiotics use in childhood and has been discussed in several

TABLE 33.2 Guidelines Reporting Recommendations of Probiotic Use in the Treatment of Acute Infectious Diarrhea

Author, Year (References)	Place	Conclusion	Evidence	Strain	Dose/Time
CDC, 2003 [98]	USA	Not recommended	+	—	—
Bhatnagar et al. (IAP 2006) [97]	India	Insufficient evidence to recommend probiotics	+	—	—
NICE, 2009 [102]	UK	Not recommended	+	—	—
Cincinnati Children's Hospital, Medical Center, 2011 [101]	USA	Recommended	++	LGG	10^{10} CFU/day (typically 5–7 days)
Wittenberg, 2012 [100]	South Africa	Can be considered in specific situations	+	—	—
ESPGHAN, 2014 [103]	Europe	Strongly recommended	++	LGG	10^{10} CFU/day (typically 5–7 days)
				S. boulardii	250–750 mg/day (typically 5–7 days)
		Weakly recommended	+	L. reuteri	
				Lactobacillus LB	
Salazar-Lindo et al., 2014 [99]	Peru	Strongly recommended	++	LGG S. boulardii L. reuteri	

CDC, Center for Diseases Control and Prevention; ESPGHAN, European Society for Pediatric Gastroenterology, Hepatology, and Nutrition; IAP, Indian Association of Pediatrics; LGG, *Lactobacillus rhamnosus* GG strain; NICE, National Institute for Clinical Health and Clinical Excellence.

documents based on a large volume of data. Of note, the evidence obtained is consistent along the years and led to similar conclusions that, therefore appear to be applicable to a general scale. It is, in other words, fair to state that we currently have conclusive evidence that selected probiotics do reduce the severity and duration of acute diarrhea, albeit by a modest but still clinical significant degree.

Overall, LGG and *Saccharomyces. boulardii* are the two most studied strains and have obtained consistent evidence of efficacy.

From a review of guidelines available worldwide [82,97–102], several—produced either in developed or developing countries—do consider the use of probiotic as a therapeutic option in addition to rehydration [82,99–101]. Different guidelines vary according to the settings and the availability of products on the market: see also Table 33.2.

Two authoritative documents have been developed in 2014 by the ESPGHAN: one is a position paper that specifically addresses the use of probiotics in children with AGE [103] and the other is a more complete document on the overall management of AGE in children [82]. Those documents provide evidence-based recommendations for the use of probiotics in the treatment of acute infectious diarrhea. Of interest, the assessment was made specifically for individual strains or preparation of multiple strains. Recommendations were provided only if at least two distinct RCTs were available. Briefly, the bulk of evidence from available data shows that selected probiotics reduce the duration of symptoms by approximately 24 h (without substantial

differences in efficacy among effective strains), while also reducing the risk of complications. A total of four strains were recommended for active treatment of gastroenteritis, in adjunct to oral rehydration therapy: LGG, *S. boulardii*, *L. reuteri*, and *Lactobacillus* LB. The first two (LGG and *S. boulardii*) received a "strong recommendation," supported, however, by a "low quality of evidence" because of some methodological weaknesses that partially limited the consistency. *L. reuteri* received a "weak recommendation" with a "very low" quality of evidence; this is mainly due to the lack of data supporting the use of the new strain DSM 17938 developed after the elimination of the plasmid responsible of transporting of antibiotic resistance identified in the original strain *L. reuteri* ATCC 55730 [104]. In a recent metaanalysis on the efficacy of *L. reuteri*, Szajewska et al. [105] concluded that in hospitalized children, use of both strains of *L. reuteri* reduced the duration of diarrhea, and more children were cured within 3 days, although the authors did also point out that data from outpatients as well as country-specific cost-effectiveness analyses were needed.

As for *Lactobacillus* LB, it should be noted that this preparation actually does not meet the standard definition of probiotic, being available as a heat-inactivated preparation.

Of note, a strong recommendation was made against the administration of *Enterococcus faecium* strain SF68 for the risk of spreading plasmids carrying vancomycin resistance.

Guidelines produced in the United States recognize the efficacy of some probiotic strains (including LGG) in reducing the severity and duration of diarrhea, and recommend to share the decision of probiotic prescription with child's family [101].

Is there a "right" dose of probiotics to be used in these settings? There is indeed evidence of a dose-response curve for Lactobacilli: a metaanalysis published already in 2002 [106] on data available then on efficacy of various Lactobacilli on reducing the duration of acute diarrhea first showed that unless a dose of at least 10^9 CFUs/day is reached, there is no significant effect. With the subsequent availability of a much larger pool it could be established that, in the case of LGG, high doses of this probiotic ($\geq 10^{10}$ CFUs/day) are definitely more effective than low doses in reducing the duration of diarrhea [107].

Since the demonstration of a dose-dependent effect by *L. rhamnosus* in the reduction of rotavirus shedding in stools in children [108], other documents reported that probiotics overall seems to have a stronger effect in rotavirus-positive diarrhea rather than in other etiology. It should be considered that the spreading of rotavirus immunization might change in part the current scenario.

ANTIBIOTIC-ASSOCIATED AND *Clostridium difficile*-ASSOCIATED DIARRHEA

A substantial percentage of individuals undergoing antibiotic treatment do develop diarrhea during the course of the treatment or shortly after its discontinuation. AAD is observed in about 10% of hospitalized patients, especially elderly patients [109], a population at a disproportionate risk of developing symptomatic disease and associated complications when the cause of the AAD is *C. difficile* (CD) [110]. It is thought that the diarrhea is not the expression of an acute infection, but rather the direct consequence of a disruption of gut microbiota (dysbiosis) with a reduction in bacterial richness and a prevalence/overgrowth of pathogen species (e.g., Clostridia).

CD is an agent typically linked to AAD in adults as well as in children. However, it may also be responsible for secretory diarrhea in subjects who did not receive antibiotics or cause severe conditions, such as pseudomembranous colitis, toxic megacolon, or intestinal perforation in at risk populations (e.g., patients with inflammatory bowel diseases, transplantation, or neoplasia) [111,112].

Probiotics are able to recolonize the mucosa, correct the dysbiosis, and enhance mucosal and systemic immunity. Thus, they have been proposed for the prevention and treatment of both AAD and *C. difficile*-associated diarrhea (CDAD).

Since quite a number of RCTs have been performed, there have been several metaanalyses conducted on the efficacy of probiotics in preventing AAD and CDAD both in children and/or in adults [113–119]. Table 33.3 reports conclusions from the main metaanalyses.

A 2013 systematic review with a metaanalysis assessed the efficacy and safety of probiotics for preventing CDAD in adults and children [114]. The analysis showed that, compared with placebo or no treatment, administration of probiotics reduced the risk of CDAD by 64% (23 RCTs, $n = 4213$, RR 0.36, 95% CI 0.26–0.51) in adults and children.

When restricting the analysis to the especially vulnerable population of hospitalized patents, another metaanalysis of 2013 showed a significant reduction in the risks of AAD (RR 0.61, 95% CI 0.47–0.79) and CD infection rates (RR 0.37, 95% CI 0.22–0.61) among patients randomly assigned to coadministration of probiotics.

TABLE 33.3 Metaanalyses on the Efficacy of Probiotics in Preventing the Onset of Antibiotic-Associated Diarrhea (AAD)

Author, Year (References)	Strain(s)	Outcomes	Conclusion
Johnston et al., 2011 [116]	Lactobacilli spp. *Saccharomyces* spp. *Bifidobacterium* spp *Leuconostoc cremoris* *Bacillus* spp	Reduced incidence of AAD (RR 0.52, 95% CI 0.38–0.72); subgroup analysis for LGG (RR 0.35, 95% CI 0.22–0.56)	Overall evidence suggests a protective effect of probiotics in preventing AAD, in particular with LGG strains
Videlock and Cremonini, 2012 [119]	Lactobacilli	Reduced incidence of AAD (RR 0.48, 95% CI 0.35–0.65) (10/24 RCTs in pediatric population)	This study supports the preventive effect of probiotic administration in AAD, independently from the indication and type of antibiotic
	Bifidobacteria		
	Enterococci		
	Streptococci		
	S. boulardii		

TABLE 33.3 Metaanalyses on the Efficacy of Probiotics in Preventing the Onset of Antibiotic-Associated Diarrhea (AAD) (*cont.*)

Author, Year (References)	Strain(s)	Outcomes	Conclusion
Hempel et al., 2012 [115]	Lactobacilli spp *Bifidobacterium* *Saccharomyces* *Streptococcus* *Enterococcus* *Bacillus*	Reduced incidence of AAD (RR 0.55, 95% CI 0.38–0.80) (16/82 RCTs in pediatric population)	Probiotics are associated with a reduced risk of AAD. Evidence is insufficient to determine the effectiveness among specific populations, antibiotics, and probiotic strains used
Goldenberg et al., 2013 [114]	LGG *L. acidophilus* and *L. casei* *S. boulardii* *L. casei* immunitas, *Lactobacillus bulgaricus*, and *S. thermophilus* *Clostridium butyricum* *L. plantarum* *Bifidobacterium* and *L. acidophilus* *L. acidophilus*, *L. bulgaricus*, *B. bifidum*, and *S. thermophilus* *L. acidophilus* VSL#3 *L. paracasei* spp. LGG, *L. acidophilus*, and *Bifidobacterium*	Reduced incidence of CDAD (RR 0.36, 95% CI 0.26, 0.51) No effect on incidence of *C. difficile* infection (RR 0.89, 95% CI 0.64, 1.24) Protective effect of probiotics in preventing AAD (−84 AAD episodes per 1000 patients treated)	Moderate quality evidence supports the preventive role of probiotics in CDAD, but not in *C. difficile* infection Low-quality evidence supports the preventive role of probiotics in AAD
Pattani et al., 2013 [113]	LGG *L. acidophilus* and *L. casei* *L. reuteri* *Enterococcus* *S. boulardii*	Significant reductions in the risks of AAD (RR 0.61, 95% CI 0.47–0.79) and CDI (RR 0.37, 95% CI 0.22–0.61) among patients randomly assigned to coadministration of probiotics. The number needed to treat for benefit was 11 (95% CI 8–20) for AAD and 14 (95% CI 9–50) for CD	Probiotics used concurrently with antibiotics reduce the risk of AAD and CDI
Szajewska and Kolodziej, 2015 [117]	*Lactobacillus GG*	Treatment with LGG compared with placebo reduced the risk of AAD in patients treated with antibiotics from 22.4% to 12.3% (RR 0.49, 95% CI 0.29–0.83, low QoE). However, the difference was significant in children only (five RCTs, n = 445, RR 0.48, 95% CI 0.26–0.89; moderate QoE). In adults, the difference was not significant.	*L. rhamnosus* GG is effective in preventing AAD in children and adults treated with antibiotics for any reason
Szajewska and Kolodziej, 2015 [118]	*S. boulardii*	Administration of *S. boulardii* compared with placebo reduced the risk of AAD in patients treated with antibiotics from 18.7% to 8.5% (RR 0.47, 95% CI 0.38–0.57). In children, *S. boulardii* reduced the risk from 20.9% to 8.8%; in adults, from 17.4% to 8.2%. Moreover, *S. boulardii* reduced the risk of CDAD; however, this reduction was significant only in children	*S. boulardii* is effective in reducing the risk of AAD in children and adults

CD, *Clostridium difficile*; CDAD, *Clostridium difficile*-associated diarrhea; CI, confidence interval; LGG, *Lactobacillus rhamnosus* GG strain; RCTs, randomized controlled trials.

More recently, a comprehensive metaanalysis by a working group of ESPGHAN (thus, only restricting analysis and recommendations to children) was published [120]. The pooled results of 21 RCTs showed that compared with placebo or no intervention, probiotics as a class reduced the risk of AAD by 52% (21.2% vs. 9.1%, respectively). Compared with placebo, the administration of probiotics also reduced the risk of CDAD (4 RCTs, $n = 938$, RR 0.34, 95% CI 0.15–0.76). The authors were thus able to conclude that for children: "If the use of probiotics for preventing AAD is considered because of the existence of risk factors, such as class of antibiotic(s), duration of antibiotic treatment, age, need for hospitalization, comorbidities, or previous episodes of AAD diarrhea, the working group recommends using LGG (moderate quality of evidence, strong recommendation) or *S. boulardii* (moderate quality of evidence, strong recommendation). If the use of probiotics for preventing CDAD is considered, the working group suggests using *S. boulardii* (low quality of evidence, conditional recommendation). Other strains or combinations of strains have been tested, but sufficient evidence is still lacking."

Of note, a paucity of data is available on the potential benefit of probiotics in the treatment of AAD or CDAD. Several experts gave different opinions in favor or against the use of probiotic as treatment in addition to standard antibiotic treatment for CD infections. In essence, there is a lack of adequate evidence to support any recommendation for the use of probiotics and in line with this, the American Academy of Pediatrics does not include this therapeutic option in its recent pediatric guidelines for the management of CD infections [111].

It is, however, worth mentioning here that a new therapeutic option, based on the similar general principle of restoring a proper microbial balance, is rapidly gaining ground as a valuable treatment option for recurrent and severe CD infections both in adults and more recently also in children: that of fecal microbiota transplantation [121].

NECROTIZING ENTEROCOLITIS

NEC is a severe and potentially fatal inflammatory disease that affects preterm neonates. Disease onset is usually characterized by abdominal distention, loose and/or bloody stools, and poor feeding, but rapidly progresses to systemic involvement (mainly in very low birth weight neonates) with hypotension, low tissue perfusion, disseminated intravascular coagulation, and evidence of septic shock or gastrointestinal hemorrhage; in many cases these events lead to bowel necrosis.

The role of gut microbiota in the pathogenesis of NEC is not well defined and the data currently available are conflicting [122]. In any case, NEC has become another potential field of application of probiotics. In fact, early administration of probiotics to premature newborns has been found to possibly have a preventative effect on NEC and also reduce the mortality due to other conditions in this at-risk population. Several clinical trials and observational studies have been conducted, mostly either in Asia or Europe. A very recent review and metaanalysis [123] included 12 studies with 10,800 premature neonates (5,144 receiving prophylactic probiotics and 5,656 controls). The metaanalysis showed a significantly decreased incidence of NEC [risk ratio (RR) 0.55, 95% CI 0.39–0.78; $p = 0.0006$] and mortality (RR 0.72, 95% CI 0.61–0.85; $p < 0.0001$). Sepsis, however, did not differ significantly between the two groups (RR 0.86, 95% CI 0.74–1.00; $p = 0.05$). The authors concluded that probiotic supplementation reduces the risk of NEC and mortality in preterm infants, with the effect sizes similar to findings in metaanalyses of RCTs. However, they also could not avoid noticing that the optimal strain, dose and timing need further investigation.

In this regard, a recent European study on almost 600 premature newborns [124] focused on the combination of 2 strains (*L. acidophilus* plus *Bifidobacterium infantis*) and found that such supplementation reduced NEC rates from 5.2% to 1.4%, while mortality was reduced from 5.2% to 3.5%. An open question thus is: are we at the point of identifying which probiotics are most effective? A metaanalysis by Aceti et al. [125] included 26 studies and in their strain-specific submetaanalyses the author noticed a significant effect for bifidobacteria [RR 0.24 (95% CI 0.10–0.54); $p = 0.0006$] and even more significant for probiotic mixtures [RR 0.39 (95% CI 0.27–0.56); $p < 0.00001$]. However, they also noticed that "there are still insufficient data on the specific probiotic strain to be used and on the effect of probiotics in high-risk populations, such as extremely-low-birth-weight infants, before a widespread use of these products can be recommended." Of interest, even though almost all of the studies have been conducted in Asia or Europe, one such study in North America (Canada) [126] confirmed both safety and efficacy of this intervention in more than 600 premature newborns.

CONCLUSIONS

All in all, it seems that some conclusions can be tentatively drawn at this point in this rapidly evolving field.

Respiratory infections have an interesting conceptual support by basic pathophysiology studies; their clinical applications are already promising, with recent metaanalyses showing significant, albeit modest, efficacy in children and adults with URTI. Clearly, more studies are needed regarding doses and specific indications in various populations, but it can already be seen that the future looks bright for this field.

Acute diarrheas represent arguably the principal filed of application of probiotics. Here, besides strong pathophysiological bases, hundreds of randomized clinical trials conducted over the past quarter of a century allow us to reach some firm conclusions.

Two strains dominate the arena: *Lactobacillus* GG and *S. boulardii*. When given in proper doses, they both can be expected to reduce the duration of gastroenteritis by approximately 24 h. Their widespread availability and strong safety profile should allow their use for acute gastroenteritis in the otherwise healthy child without hesitation. Although much less data are available in adults, where the clinical relevance of acute gastroenteritis is obviously less evident, the general safety profile of probiotics and the clear-cut results from pediatric studies would seem to be a reasonable basis for their use in this age too.

For AAD, in spite of several clinical trials showing significant benefit and encouraging metaanalyses, we still lack official recommendations for their use. In our opinion, again the optimal safety profile and the conclusions of recent metaanalyses, however, should allow the practitioner to recommend the use of either *Lactobacillus* GG or *S. boulardii* for their patients about to undergo antibiotic treatment, especially when dealing with young children and elderly persons, where the threat of devastating CD infections is highest.

Lastly, the challenge of NEC, where we have the paradox of a life-threatening condition characterized by a high morbidity that appears to be largely preventable by a relatively inexpensive and safe intervention, supported by dozens of studies and several metaanalyses; and for which no authority has so far explicitly recommended their use. The impasse appears to be largely due to concerns about "providing a product that is not contaminated, adulterated, or inconsistent with what we intend to prescribe" [127] given the regulatory structure for probiotics in the USA: one can only hope that urgent attention is paid to this issue by US regulatory agencies, as procrastinating the fruition of this intervention puts a considerable number of premature babies at a greater, unnecessary risk.

ABBREVIATIONS

AAD   Antibiotic-associated diarrhea
AGE   Acute gastroenteritis
AOM   Acute otitis media
CDAD   *Clostridium difficile*-associated diarrhea
LGG   *Lactobacillus rhamnosus* GG strain
NEC   Necrotizing enterocolitis
RCT   Randomized controlled trials
URTI   Upper respiratory tract infections

REFERENCES

[1] Chonmaitree T, Revai K, Grady JJ, et al. Viral upper respiratory tract infection and otitis media complication in young children. Clin Infect Dis 2008;46:815–23.

[2] Nicholson KG, McNally T, Silverman M, Simons P, Stockton JD, Zambon MC. Rates of hospitalisation for influenza, respiratory syncytial virus and human metapneumovirus among infants and young children. Vaccine 2006;24:102–8.

[3] Massin MM, Montesanti J, Gerard P, Lepage P. Spectrum and frequency of illness presenting to a pediatric emergency department. Acta Clinica Belg 2006;61:161–5.

[4] Sauro A, Barone F, Blasio G, Russo L, Santillo L. Do influenza and acute respiratory infective diseases weigh heavily on general practitioners' daily practice? Eur J Gen Pract 2006;12:34–6.

[5] Gavazzi G, Krause KH. Ageing and infection. Lancet Infect Dis 2002;2:659–66.

[6] Claesson MJ, Cusack S, O'Sullivan O, et al. Composition, variability, and temporal stability of the intestinal microbiota of the elderly. Proc Natl Acad Sci USA 2011;108(Suppl. 1):4586–91.

[7] Duncan SH, Flint HJ. Probiotics and prebiotics and health in ageing populations. Maturitas 2013;75:44–50.

[8] Gill HS, Rutherfurd KJ, Cross ML, Gopal PK. Enhancement of immunity in the elderly by dietary supplementation with the probiotic *Bifidobacterium lactis* HN019. Am J Clin Nutr 2001;74:833–9.

[9] Kotani Y, Shinkai S, Okamatsu H, et al. Oral intake of *Lactobacillus pentosus* strain b240 accelerates salivary immunoglobulin A secretion in the elderly: a randomized, placebo-controlled, double-blind trial. Immun Ageing 2010;7:11.

[10] Lahtinen SJ, Forssten S, Aakko J, et al. Probiotic cheese containing *Lactobacillus rhamnosus* HN001 and *Lactobacillus acidophilus* NCFM(R) modifies subpopulations of fecal lactobacilli and *Clostridium difficile* in the elderly. Age 2012;34:133–43.

[11] Caffarelli C, Cardinale F, Povesi-Dascola C, Dodi I, Mastrorilli V, Ricci G. Use of probiotics in pediatric infectious diseases. Expert Rev Anti Infect Ther 2015;13:1517–35.

[12] Foligne B, Deutsch SM, Breton J, et al. Promising immunomodulatory effects of selected strains of dairy propionibacteria as evidenced in vitro and in vivo. Appl Environ Microbiol 2010;76:8259–64.

[13] Howarth GS, Wang H. Role of endogenous microbiota, probiotics and their biological products in human health. Nutrients 2013;5:58–81.

[14] Rijkers GT, Bengmark S, Enck P, et al. Guidance for substantiating the evidence for beneficial effects of probiotics: current status and recommendations for future research. J Nutr 2010;140:671S–6S.

[15] Sanchez B, Arias S, Chaignepain S, et al. Identification of surface proteins involved in the adhesion of a probiotic *Bacillus cereus* strain to mucin and fibronectin. Microbiology 2009;155:1708–16.

[16] Fukushima Y, Kawata Y, Hara H, Terada A, Mitsuoka T. Effect of a probiotic formula on intestinal immunoglobulin A production in healthy children. Int J Food Microbiol 1998;42:39–44.

[17] Kabeerdoss J, Devi RS, Mary RR, et al. Effect of yogurt containing *Bifidobacterium lactis* Bb12(R) on faecal excretion of secretory immunoglobulin A and human beta-defensin 2 in healthy adult volunteers. Nutr J 2011;10:138.

[18] Pang IK, Iwasaki A. Control of antiviral immunity by pattern recognition and the microbiome. Immunol Rev 2012;245:209–26.

[19] Villena J, Chiba E, Tomosada Y, et al. Orally administered *Lactobacillus rhamnosus* modulates the respiratory immune response triggered by the viral pathogen-associated molecular pattern poly(I:C). BMC Immunol 2012;13:53.

[20] Gill HS, Rutherfurd KJ, Cross ML. Dietary probiotic supplementation enhances natural killer cell activity in the elderly: an investigation of age-related immunological changes. J Clinical Immunol 2001;21:264–71.

[21] Sheih YH, Chiang BL, Wang LH, Liao CK, Gill HS. Systemic immunity-enhancing effects in healthy subjects following dietary consumption of the lactic acid bacterium *Lactobacillus rhamnosus* HN001. J Am Coll Nutr 2001;20:149–56.

[22] Pulendran B, Maddur MS. Innate immune sensing and response to influenza. Curr Top Microbiol Immunol 2015;386:23–71.

[23] Gerasimov SV, Ivantsiv VA, Bobryk LM, Tsitsura OO, Dedyshin LP, Guta NV, Yandyo BV. Role of short-term use of *L. acidophilus* DDS-1 and *B. lactis* UABLA-12 in acute respiratory infections in children: a randomized controlled trial. Eur J Clin Nutr 2015;70(4):463–9.

[24] Pinchuk IV, Bressollier P, Verneuil B, et al. In vitro anti-*Helicobacter pylori* activity of the probiotic strain *Bacillus subtilis* 3 is due to secretion of antibiotics. Antimicrob Agents Chemother 2001;45:3156–61.

[25] Fujiya M, Musch MW, Nakagawa Y, et al. The *Bacillus subtilis* quorum-sensing molecule CSF contributes to intestinal homeostasis via OCTN2, a host cell membrane transporter. Cell Host Microbe 2007;1:299–308.

[26] Hosoi T, Hirose R, Saegusa S, Ametani A, Kiuchi K, Kaminogawa S. Cytokine responses of human intestinal epithelial-like Caco-2 cells to the nonpathogenic bacterium *Bacillus subtilis* (natto). Int J Food Microbiol 2003;82:255–64.

[27] Ciprandi G, Tosca MA, Milanese M, Caligo G, Ricca V. Cytokines evaluation in nasal lavage of allergic children after *Bacillus clausii* administration: a pilot study. Pediatr Allergy Immunol 2004;15:148–51.

[28] Urdaci MC, Bressollier P, Pinchuk I. *Bacillus clausii* probiotic strains: antimicrobial and immunomodulatory activities. J Clin Gastroenterol 2004;38:S86–90.

[29] Lefevre M, Racedo SM, Ripert G, et al. Probiotic strain *Bacillus subtilis* CU1 stimulates immune system of elderly during common infectious disease period: a randomized, double-blind placebo-controlled study. Immun Ageing 2015;12:24.

[30] Huang JM, La Ragione RM, Nunez A, Cutting SM. Immunostimulatory activity of *Bacillus* spores. FEMS Immunol Med Microbiol 2008;53:195–203.

[31] Mehrad B, Standiford TJ. Role of cytokines in pulmonary antimicrobial host defense. Immunol Res 1999;20:15–27.

[32] Sadler AJ, Williams BR. Interferon-inducible antiviral effectors. Nat Rev Immunol 2008;8:559–68.

[33] Hori T, Kiyoshima J, Shida K, Yasui H. Augmentation of cellular immunity and reduction of influenza virus titer in aged mice fed *Lactobacillus casei* strain Shirota. Clin Diagn Lab Immunol 2002;9:105–8.

[34] Park MK, Ngo V, Kwon YM, et al. *Lactobacillus plantarum* DK119 as a probiotic confers protection against influenza virus by modulating innate immunity. PLoS One 2013;8:e75368.

[35] Lehtoranta L, Pitkaranta A, Korpela R. Probiotics in respiratory virus infections. Eur J Clin Microbiol Infect Dis Microbiol 2014;33:1289–302.

[36] Hatakka K, Savilahti E, Ponka A, et al. Effect of long term consumption of probiotic milk on infections in children attending day care centres: double blind, randomised trial. BMJ 2001;322:1327.

[37] Hojsak I, Snovak N, Abdovic S, Szajewska H, Misak Z, Kolacek S. *Lactobacillus* GG in the prevention of gastrointestinal and respiratory tract infections in children who attend day care centers: a randomized, double-blind, placebo-controlled trial. Clin Nutr 2010;29:312–6.

[38] Hojsak I, Abdovic S, Szajewska H, Milosevic M, Krznaric Z, Kolacek S. *Lactobacillus* GG in the prevention of nosocomial gastrointestinal and respiratory tract infections. Pediatrics 2010;125:e1171–7.

[39] Kumpu M, Kekkonen RA, Kautiainen H, et al. Milk containing probiotic *Lactobacillus rhamnosus* GG and respiratory illness in children: a randomized, double-blind, placebo-controlled trial. Eur J Clin Nutr 2012;66:1020–3.

[40] Luoto R, Ruuskanen O, Waris M, Kalliomaki M, Salminen S, Isolauri E. Prebiotic and probiotic supplementation prevents rhinovirus infections in preterm infants: a randomized, placebo-controlled trial. J Allergy Clin Immunol 2014;133:405–13.

[41] Kumpu M, Lehtoranta L, Roivainen M, et al. The use of the probiotic *Lactobacillus rhamnosus* GG and viral findings in the nasopharynx of children attending day care. J Med Virol 2013;85:1632–8.

[42] Kukkonen K, Savilahti E, Haahtela T, et al. Long-term safety and impact on infection rates of postnatal probiotic and prebiotic (synbiotic) treatment: randomized, double-blind, placebo-controlled trial. Pediatrics 2008;122:8–12.

[43] Rautava S, Salminen S, Isolauri E. Specific probiotics in reducing the risk of acute infections in infancy—a randomised, double-blind, placebo-controlled study. Br J Nutr 2009;101:1722–6.

[44] Agustina R, Kok FJ, van de Rest O, et al. Randomized trial of probiotics and calcium on diarrhea and respiratory tract infections in Indonesian children. Pediatrics 2012;129:e1155–64.

[45] Turchet P, Laurenzano M, Auboiron S, Antoine JM. Effect of fermented milk containing the probiotic *Lactobacillus casei* DN-114001 on winter infections in free-living elderly subjects: a randomised, controlled pilot study. J Nutr Aging 2003;7:75–7.

[46] Guillemard E, Tondu F, Lacoin F, Schrezenmeir J. Consumption of a fermented dairy product containing the probiotic *Lactobacillus casei* DN-114001 reduces the duration of respiratory infections in the elderly in a randomised controlled trial. Br J N 2010;103:58–68.

[47] Cobo Sanz JM, Mateos JA, Munoz Conejo A. Effect of *Lactobacillus casei* on the incidence of infectious conditions in children. Nutr Hosp 2006;21:547–51.

[48] Lin JS, Chiu YH, Lin NT, et al. Different effects of probiotic species/strains on infections in preschool children: a double-blind, randomized, controlled study. Vaccine 2009;27:1073–9.

[49] Nocerino R, Paparo L, Terrin G, et al. Cow's milk and rice fermented with *Lactobacillus paracasei* CBA L74 prevent infectious diseases in children: a randomized controlled trial. Clin Nutr 2015.

[50] Maldonado J, Canabate F, Sempere L, et al. Human milk probiotic *Lactobacillus fermentum* CECT5716 reduces the incidence of gastrointestinal and upper respiratory tract infections in infants. J Pediatr Gastroenterol Nutr 2012;54:55–61.

[51] Weizman Z, Asli G, Alsheikh A. Effect of a probiotic infant formula on infections in child care centers: comparison of two probiotic agents. Pediatrics 2005;115:5–9.

[52] Hojsak I, Mocic Pavic A, Kos T, Dumancic J, Kolacek S. *Bifidobacterium animalis* subsp. *lactis* in prevention of common infections in healthy children attending day care centers—randomized, double blind, placebo-controlled study. Clin Nutr 2015;35:587–91.

[53] Taipale TJ, Pienihakkinen K, Isolauri E, Jokela JT, Soderling EM. *Bifidobacterium animalis* subsp. *lactis* BB-12 in reducing the risk of infections in early childhood. Pediatr Res 2016;79:65–9.

[54] Garaiova I, Muchova J, Nagyova Z, et al. Probiotics and vitamin C for the prevention of respiratory tract infections in children

attending preschool: a randomised controlled pilot study. Eur J Clin Nutr 2015;69:373–9.

[55] Winkler P, de Vrese M, Laue C, Schrezenmeir J. Effect of a dietary supplement containing probiotic bacteria plus vitamins and minerals on common cold infections and cellular immune parameters. Int J Clin Pharmacol Ther 2005;43:318–26.

[56] de Vrese M, Winkler P, Rautenberg P, et al. Effect of *Lactobacillus gasseri* PA 16/8, *Bifidobacterium longum* SP 07/3, *B. bifidum* MF 20/5 on common cold episodes: a double blind, randomized, controlled trial. Clin Nutr 2005;24:481–91.

[57] de Vrese M, Winkler P, Rautenberg P, et al. Probiotic bacteria reduced duration and severity but not the incidence of common cold episodes in a double blind, randomized, controlled trial. Vaccine 2006;24:6670–4.

[58] Hatakka K, Blomgren K, Pohjavuori S, et al. Treatment of acute otitis media with probiotics in otitis-prone children-a double-blind, placebo-controlled randomised study. Clin Nutr 2007;26:314–21.

[59] Boge T, Remigy M, Vaudaine S, Tanguy J, Bourdet-Sicard R, van der Werf S. A probiotic fermented dairy drink improves antibody response to influenza vaccination in the elderly in two randomised controlled trials. Vaccine 2009;27:5677–84.

[60] Bosch M, Mendez M, Perez M, Farran A, Fuentes MC, Cune J. *Lactobacillus plantarum* CECT7315 and CECT7316 stimulate immunoglobulin production after influenza vaccination in elderly. Nutr Hosp 2012;27:504–9.

[61] Davidson LE, Fiorino AM, Snydman DR, Hibberd PL. *Lactobacillus* GG as an immune adjuvant for live-attenuated influenza vaccine in healthy adults: a randomized double-blind placebo-controlled trial. Eur J Clin Nutr 2011;65:501–7.

[62] Van Puyenbroeck K, Hens N, Coenen S, et al. Efficacy of daily intake of *Lactobacillus casei* Shirota on respiratory symptoms and influenza vaccination immune response: a randomized, double-blind, placebo-controlled trial in healthy elderly nursing home residents. Am J Clin Nutr 2012;95:1165–71.

[63] Berggren A, Lazou Ahren I, Larsson N, Onning G. Randomised, double-blind and placebo-controlled study using new probiotic lactobacilli for strengthening the body immune defence against viral infections. Eur J Nutr 2011;50:203–10.

[64] Fujita R, Iimuro S, Shinozaki T, et al. Decreased duration of acute upper respiratory tract infections with daily intake of fermented milk: a multicenter, double-blinded, randomized comparative study in users of day care facilities for the elderly population. Am J Infect Control 2013;41:1231–5.

[65] Makino S, Ikegami S, Kume A, Horiuchi H, Sasaki H, Orii N. Reducing the risk of infection in the elderly by dietary intake of yoghurt fermented with *Lactobacillus delbrueckii* ssp. *bulgaricus* OLL1073R-1. Br J Nutr 2010;104:998–1006.

[66] Merenstein D, Murphy M, Fokar A, et al. Use of a fermented dairy probiotic drink containing *Lactobacillus casei* (DN-114 001) to decrease the rate of illness in kids: the DRINK study. A patient-oriented, double-blind, cluster-randomized, placebo-controlled, clinical trial. Eur J Clin Nutr 2010;64:669–77.

[67] Rerksuppaphol S, Rerksuppaphol L. Randomized controlled trial of probiotics to reduce common cold in schoolchildren. Pediatr Int 2012;54:682–7.

[68] Rio ME, Zago Beatriz L, Garcia H, Winter L. The nutritional status change the effectiveness of a dietary supplement of lactic bacteria on the emerging of respiratory tract diseases in children. Arch Latinoam Nutr 2002;52:29–34.

[69] Smith TJ, Rigassio-Radler D, Denmark R, Haley T, Touger-Decker R. Effect of *Lactobacillus rhamnosus* LGG(R) and *Bifidobacterium animalis* ssp. *lactis* BB-12(R) on health-related quality of life in college students affected by upper respiratory infections. Br J Nutr 2013;109:1999–2007.

[70] Hojsak I, Tokic Pivac V, Mocic Pavic A, Pasini AM, Kolacek S. *Bifidobacterium animalis* subsp. *lactis* fails to prevent common infections in hospitalized children: a randomized, double-blind, placebo-controlled study. Am J Clin Nutr 2015;101:680–4.

[71] Lehtoranta L, Soderlund-Venermo M, Nokso-Koivisto J, et al. Human bocavirus in the nasopharynx of otitis-prone children. Int J Pediatr Otorhinolaryngol 2012;76:206–11.

[72] Mane J, Pedrosa E, Loren V, et al. A mixture of *Lactobacillus plantarum* CECT 7315 and CECT 7316 enhances systemic immunity in elderly subjects. A dose-response, double-blind, placebo-controlled, randomized pilot trial. Nutr Hosp 2011;26:228–35.

[73] Gilbey P, Livshits L, Sharabi-Nov A, Avraham Y, Miron D. Probiotics in addition to antibiotics for the treatment of acute tonsillitis: a randomized, placebo-controlled study. Eur J Clin Microbiol Infect Dis 2015;34:1011–5.

[74] Di Pierro F, Donato G, Fomia F, et al. Preliminary pediatric clinical evaluation of the oral probiotic *Streptococcus salivarius* K12 in preventing recurrent pharyngitis and/or tonsillitis caused by *Streptococcus pyogenes* and recurrent acute otitis media. Int J Gen Med 2012;5:991–7.

[75] Di Pierro F, Adami T, Rapacioli G, Giardini N, Streitberger C. Clinical evaluation of the oral probiotic *Streptococcus salivarius* K12 in the prevention of recurrent pharyngitis and/or tonsillitis caused by *Streptococcus pyogenes* in adults. Expert Opin Biol Ther 2013;13:339–43.

[76] Di Pierro F, Colombo M, Zanvit A, Risso P, Rottoli AS. Use of *Streptococcus salivarius* K12 in the prevention of streptococcal and viral pharyngotonsillitis in children. Drug Healthc Patient Saf 2014;6:15–20.

[77] Liu S, Hu P, Du X, Zhou T, Pei X. *Lactobacillus rhamnosus* GG supplementation for preventing respiratory infections in children: a meta-analysis of randomized, placebo-controlled trials. Indian Pediatr 2013;50:377–81.

[78] Hao Q, Dong BR, Wu T. Probiotics for preventing acute upper respiratory tract infections. Cochrane Database Syst Rev 2015;2:CD006895.

[79] King S, Glanville J, Sanders ME, Fitzgerald A, Varley D. Effectiveness of probiotics on the duration of illness in healthy children and adults who develop common acute respiratory infectious conditions: a systematic review and meta-analysis. Br J Nutr 2014;112:41–54.

[80] Santesso N. A Summary of a Cochrane Review. Probiotics to prevent acute upper respiratory tract infections. Glob Adv Health Med 2015;4:18–9.

[81] Quick M. Cochrane commentary: probiotics for prevention of acute upper respiratory infection. Explore 2015;11:418–20.

[82] Guarino A, Ashkenazi S, Gendrel D, et al. European Society for Pediatric Gastroenterology, Hepatology, and Nutrition/European Society for Pediatric Infectious Diseases evidence-based guidelines for the management of acute gastroenteritis in children in Europe: update. J Pediatr Gastroenterol Nutr 2014;59:132–52.

[83] Isolauri E, Juntunen M, Rautanen T, Sillanaukee P, Koivula T. A human *Lactobacillus* strain (*Lactobacillus casei* sp strain GG) promotes recovery from acute diarrhea in children. Pediatrics 1991;88:90–7.

[84] Johnston BC, Shamseer L, da Costa BR, Tsuyuki RT, Vohra S. Measurement issues in trials of pediatric acute diarrheal diseases: a systematic review. Pediatrics 2010;126:e222–31.

[85] Karas J, Ashkenazi S, Guarino A, et al. A core outcome set for clinical trials in acute diarrhoea. Arch Dis Child 2015;100:359–63.

[86] Saavedra JM, Tschernia A. Human studies with probiotics and prebiotics: clinical implications. Br J Nutr 2002;87(Suppl. 2):S241–6.

[87] Saran S, Gopalan S, Krishna TP. Use of fermented foods to combat stunting and failure to thrive. Nutrition 2002;18:393–6.

[88] Guandalini S. Probiotics for prevention and treatment of diarrhea. J Clin Gastroenterol 2011;45(Suppl.):S149–53.

[89] Szajewska H, Setty M, Mrukowicz J, Guandalini S. Probiotics in gastrointestinal diseases in children: hard and not-so-hard evidence of efficacy. J Pediatr Gastroenterol Nutr 2006;42:454–75.

[90] Gutierrez-Castrellon P, Lopez-Velazquez G, Diaz-Garcia L, et al. Diarrhea in preschool children and *Lactobacillus reuteri*: a randomized controlled trial. Pediatrics 2014;133:e904–9.

[91] Vandenplas Y, De Greef E, Hauser B, Devreker T, Veereman-Wauters G. Probiotics and prebiotics in pediatric diarrheal disorders. Expert Opin Pharmacother 2013;14:397–409.

[92] Braegger C, Chmielewska A, Decsi T, et al. Supplementation of infant formula with probiotics and/or prebiotics: a systematic review and comment by the ESPGHAN committee on nutrition. J Pediatr Gastroenterol Nutr 2011;52:238–50.

[93] Thomas DW, Greer FR. American Academy of Pediatrics Committee on Nutrition, American Academy of Pediatrics Section on Gastroenterology, Hepatology, and Nutrition. Probiotics and prebiotics in pediatrics. Pediatrics 2010;126:1217–31.

[94] Posfay-Barbe KM, Zerr DM, Pittet D. Infection control in paediatrics. Lancet Infect Dis 2008;8:19–31.

[95] EFSA Panel on Dietetic Products NaAN. Scientific Opinion on the substantiation of health claims related to non-characterised microorganisms pursuant to Article 13(1) of Regulation (EC) No 1924/20061. EFSA J 2009;7:1247.

[96] Wada M, Nagata S, Saito M, et al. Effects of the enteral administration of *Bifidobacterium breve* on patients undergoing chemotherapy for pediatric malignancies. Support Care Cancer 2010;18:751–9.

[97] Bhatnagar S, Lodha R, Choudhury P, et al. IAP Guidelines 2006 on management of acute diarrhea. Indian Pediatr 2007;44:380–9.

[98] King CK, Glass R, Bresee JS, Duggan C. Centers for Disease C, Prevention. Managing acute gastroenteritis among children: oral rehydration, maintenance, and nutritional therapy. MMWR Recomm Rep 2003;52:1–16.

[99] Salazar-Lindo E, Polanco Allue I, Gutierrez-Castrellon P. Grupo Ibero-Latinoamericano sobre el Manejo de la Diarrea A. Ibero-Latin American guide clinical practice on the management of acute gastroenteritis in children under 5 years: pharmacological treatment. An Pediatr (Barc) 2014;80(Suppl. 1):15–22.

[100] Wittenberg DF. Management guidelines for acute infective diarrhoea/gastroenteritis in infants. S Afr Med J 2012;102:104–7.

[101] Acute Gastroenteritis Guideline Team CCsHMC, Dec 21, 2011. Evidence-based care guideline for prevention and management of acute gastroenteritis in children age 2 months to 18 years. http://wwwcincinnatichildrensorg/service/j/anderson-center/evidence-based-care/gastroenteritis/Guideline 5 2011:1–20.

[102] Khanna R, Lakhanpaul M, Burman-Roy S, Murphy MS. Guideline Development Group and the technical team. Diarrhoea and vomiting caused by gastroenteritis in children under 5 years: summary of NICE guidance. BMJ 2009;338:b1350.

[103] Szajewska H, Guarino A, Hojsak I, et al. Use of probiotics for management of acute gastroenteritis: a position paper by the ESPGHAN Working Group for Probiotics and Prebiotics. J Pediatr Gastroenterol Nutr 2014;58:531–9.

[104] Rosander A, Connolly E, Roos S. Removal of antibiotic resistance gene-carrying plasmids from *Lactobacillus reuteri* ATCC 55730 and characterization of the resulting daughter strain, *L. reuteri* DSM 17938. Appl Environ Microbiol 2008;74:6032–40.

[105] Szajewska H, Urbanska M, Chmielewska A, Weizman Z, Shamir R. Meta-analysis: *Lactobacillus reuteri* strain DSM 17938 (and the original strain ATCC 55730) for treating acute gastroenteritis in children. Benef Microbes 2014;5:285–93.

[106] Van Niel CW, Feudtner C, Garrison MM, Christakis DA. *Lactobacillus* therapy for acute infectious diarrhea in children: a meta-analysis. Pediatrics 2002;109:678–84.

[107] Szajewska H, Skorka A, Ruszczynski M, Gieruszczak-Bialek D. Meta-analysis: *Lactobacillus* GG for treating acute gastroenteritis in children—updated analysis of randomised controlled trials. Aliment Pharmacol Ther 2013;38:467–76.

[108] Fang SB, Lee HC, Hu JJ, Hou SY, Liu HL, Fang HW. Dose-dependent effect of *Lactobacillus rhamnosus* on quantitative reduction of faecal rotavirus shedding in children. J Trop Pediatr 2009;55:297–301.

[109] Elseviers MM, Van Camp Y, Nayaert S, et al. Prevalence and management of antibiotic associated diarrhea in general hospitals. BMC Infect Dis 2015;15:129.

[110] Keller JM, Surawicz CM. *Clostridium difficile* infection in the elderly. Clin Geriatr Med 2014;30:79–93.

[111] Schutze GE, Willoughby RE. Committee on Infectious D., American Academy of P. *Clostridium difficile* infection in infants and children. Pediatrics 2013;131:196–200.

[112] Lo Vecchio A, Zacur GM. *Clostridium difficile* infection: an update on epidemiology, risk factors, and therapeutic options. Curr Opin Gastroenterol 2012;28:1–9.

[113] Pattani R, Palda VA, Hwang SW, Shah PS. Probiotics for the prevention of antibiotic-associated diarrhea and *Clostridium difficile* infection among hospitalized patients: systematic review and meta-analysis. Open Med 2013;7:e56–67.

[114] Goldenberg JZ, Ma SS, Saxton JD, et al. Probiotics for the prevention of *Clostridium difficile*-associated diarrhea in adults and children. Cochrane Database Syst Rev 2013;5. CD006095.

[115] Hempel S, Newberry SJ, Maher AR, et al. Probiotics for the prevention and treatment of antibiotic-associated diarrhea: a systematic review and meta-analysis. JAMA 2012;307:1959–69.

[116] Johnston BC, Ma SS, Goldenberg JZ, et al. Probiotics for the prevention of *Clostridium difficile*-associated diarrhea: a systematic review and meta-analysis. Ann Intern Med 2012;157:878–88.

[117] Szajewska H, Kolodziej M. Systematic review with meta-analysis: *Lactobacillus rhamnosus* GG in the prevention of antibiotic-associated diarrhoea in children and adults. Aliment Pharmacol Ther 2015;42:1149–57.

[118] Szajewska H, Kolodziej M. Systematic review with meta-analysis: *Saccharomyces boulardii* in the prevention of antibiotic-associated diarrhoea. Aliment Pharmacol Ther 2015;42:793–801.

[119] Videlock EJ, Cremonini F. Meta-analysis: probiotics in antibiotic-associated diarrhoea. Aliment Pharmacol Ther 2012;35:1355–69.

[120] Szajewska H, Canani RB, Guarino A, et al. Probiotics for the prevention of antibiotic-associated diarrhea in children. J Pediatr Gastroenterol Nutr 2015;62(3):495–506.

[121] Rao K, Safdar N. Fecal microbiota transplantation for the treatment of *Clostridium difficile* infection. J Hosp Med 2016;11:56–61.

[122] Elgin TG, Kern SL, McElroy SJ. Development of the neonatal intestinal microbiome and its association with necrotizing enterocolitis. Clin Ther 2016;38(4):706–15.

[123] Olsen R, Greisen G, Schroder M, Brok J. Prophylactic probiotics for preterm infants: a systematic review and meta-analysis of observational studies. Neonatology 2016;109:105–12.

[124] Guthmann F, Arlettaz Mieth RP, Bucher HU, Buhrer C. Short courses of dual-strain probiotics appear to be effective in reducing necrotising enterocolitis. Acta Paediatr 2016;105:255–9.

[125] Aceti A, Gori D, Barone G, et al. Probiotics for prevention of necrotizing enterocolitis in preterm infants: systematic review and meta-analysis. Ital J Pediatr 2015;41:89.

[126] Janvier A, Malo J, Barrington KJ. Cohort study of probiotics in a North American neonatal intensive care unit. J Pediatr 2014;164:980–5.

[127] Chan LN, Soltani H, Hazlet TK. Probiotics for neonates: safety for prime time questioned without regulatory changes. J Pediatr 2015;166:502.

Chapter 34

FMT in *Clostridium difficile* and Other Potential Uses

S. Fine* and C.R. Kelly**

*Department of Gastroenterology, Warren Alpert School of Medicine, Brown University, Providence, RI, United States; **Warren Alpert Medical School of Brown University, Miriam Hospital, and Lifespan Hospital System, Providence, RI, United States*

INTRODUCTION

The relationship between bacteria and people was once thought only to be one of "opportunistic" pathogens, rather than a mutually beneficial relationship. The rise in antimicrobial resistance and decrease in the number of available antibiotics [1] has led to efforts to improve knowledge around the bacteria–host interplay. We are beginning to understand that the vast number of microorganisms within the human gastrointestinal (GI) tract plays a critical role in maintaining gut homeostasis, a healthy immune function, and metabolism [2]. However, the idea of bacteria being beneficial to our well-being or the usefulness of restoring disrupted intestinal microbiota at times of illness is not new. In the 4th century, during the Dong-jin dynast in China, a traditional medicine doctor Ge Hong successfully used human fecal suspensions for patients who were stricken with food poisoning or severe diarrhea [3]. Many centuries later, an Italian anatomist by the name of Fabricius Aquapendente used fecal transplant for ruminal disorders in veterinary medicine [4]. However, the idea of using fecal bacteria to cure a patient's illness was one that did not gain significant medical attention for quite some time. With the advent and use of the first commercially available antibiotic in the United States in 1942 [5], Penicillin provided hope for previously fatal ailments. The following decade, an American surgeon described the use of fecal enemas in the treatment of pseudomembranous enterocolitis, a consequence of antibiotics [6]. The first case of fecal microbiota transplantation (FMT) for the treatment of recurrent *Clostridium difficile* infection (CDI) was reported in 1983 [7]. This "new look at an old idea" was recently brought to the forefront, when a randomized control trial demonstrated efficacy of donor stool administered via nasoduodenal infusion in patients with recurrent CDI [8]. Mainstream media headlines have promoted this "natural" remedy to eager readers, one recent article proclaiming, *"One man's poop is another's medicine* [9]." Many patients suffering from recurrent or refractory cases of CDI undergo FMT as an alternative treatment to prolonged antibiotic tapers with a mean cure rate between 87% and 90% [8,10,11]. There is a hope that FMT will prove beneficial in other conditions associated with dysbiosis, such as inflammatory bowel diseases (IBD) and the metabolic syndrome.

Clostridium difficile

C. difficile is a gram positive, spore-forming bacteria which was originally described in 1935 as a normal component of the intestinal microbiota in newborn infants [12]. Interestingly, early studies showed that despite having high rates of *C. difficile* colonization and positive stool toxin, there was no evidence of clinical symptoms in neonates [13]. The reason why infants appear to be unaffected by the toxin is unclear, but may, in part, be related to the lack of receptors for *C. difficile* toxins on the infant colonocyte [14]. The effects of CDI range from mild diarrhea to pseudomembranous colitis (PMC), toxic megacolon, and death. With the implementation of antibiotic use in the 1950s, PMC was a feared complication in postoperative patients, but the etiology of the syndrome was unclear [15]. This disease process was initially felt related to the pathogen *Staphylococcus aureus* and was treated with oral vancomycin [16]. The correlation between antibiotics and PMC came into the spotlight when the first study using endoscopy to aid in the diagnosis showed a strong association with the use of clindamycin [17]. These reports spurred interest into further unraveling the cause of the "antibiotic-associated" disease, and strategies to avoid and better manage patients who were afflicted. In 1974, a group of scientists who were studying penicillin-induced death in guinea pigs found that stool specimens contained cytopathic changes that, in retrospect, were a consequence of CDI [18]. Further scientific work in hamsters detected *C. difficile* and established it as the pathogen responsible for the disease [19]. The production of enterotoxin (Toxin A) and cytotoxin (Toxin B) by the organism leads to inflammation and clinical manifestations [20]. The transmission of *C. difficile* is highest in healthcare facilities due to contamination of environmental surfaces and hand carriage by workers and patients. Antibiotic use, increasing age, and exposure to the healthcare environment are the biggest risk factors for disease acquisition [14]. Infection is typically confirmed by stool PCR, which tests for toxin genes, or the presence of pseudomembranes on lower endoscopy. Treatment regimens are based on the severity of the disease and initial agents include metronidazole or oral vancomycin [21,22]. Treatment failures with metronidazole have risen from 2.5% to 18% over the past decade [23]. Furthermore, after two or more episodes of *C. difficile,* the risk for future recurrence after another course of antibiotic therapy exceeds 60% [24]. This high recurrence rate led to the exploration of alternative therapies that have recently been focused on using healthy donor stool to restore bacterial diversity and beneficial anaerobes in an infected individual.

FECAL MICROBIOTA TRANSPLANTATION

FMT is considered a drug and a biologic and is not currently FDA approved, though the agency has chosen to exercise enforcement discretion, permitting providers to use FMT in cases of CDI not responding to standard treatment [25]. The majority of experience with FMT has been published as case reports and uncontrolled case series. There have been several small, open-label, randomized controlled trials demonstrating excellent outcomes. The first divided 43 patients with recurrent *C. difficile* into 3 groups: donor stool via nasoduodenal tube after bowel lavage, standard course of oral vancomycin, or vancomycin plus bowel lavage, with resolution rates of 81, 31, and 21%, respectively [8]. The second compared two different methods for delivery of frozen donor stool, nasogastric tube (NGT), and colonoscopy, in a total of 20 patients. Resolution rates were high in both groups, and no difference in efficacy was found between the two delivery methods [26]. A third study randomized patients with recurrent *C. difficile* to receive a short course of oral vancomycin followed by one or more fecal infusions via colonoscopy or a prolonged course of oral vancomycin and demonstrated a 90% resolution rate versus 26%, respectively [27]. A double-blind, randomized trial in patients with recurrent *C. difficile* demonstrated that frozen-and-thawed FMT was noninferior to fresh FMT in regards to clinical efficacy [28]. Finally, Kelly et al. [29] conducted the first placebo-controlled trial in FMT for recurrent CDI, randomizing patients who had received at least a 10-day course of oral vancomycin to donor FMT or autologous "placebo" FMT (subject's own stool) infused by

colonoscopy, and found the rate of clinical cure at 8 weeks was 91% after FMT versus 63% with placebo.

A working group [30] comprised of members representing several professional societies met in 2010 and suggested appropriate indications for FMT including:

1. Recurrent or relapsing CDI:
   a. Three or more episodes of mild-to-moderate CDI and failure to respond to a 6- to 8-week taper with vancomycin with or without an alternative antibiotic (e.g., rifaximin, nitazoxanide, or fidaxomicin).
   b. At least two episodes of CDI resulting in hospitalization and associated with significant morbidity.
2. Moderate CDI with no response to standard therapy (vancomycin or fidaxomicin) for at least 1 week and, possibly.
3. Severe (even fulminant) CDI with no response to standard therapy for 48 h.

Patients hospitalized with severe or fulminant CDI, who are not responding to standard antibiotic treatment, pose difficult management questions. Surgical intervention with colectomy is an option, and timely intervention has been shown to improve outcomes in otherwise hopeless cases [31–33]. However, the surgical risk of performing a colectomy in these situations carriers a high mortality rate. More experience with FMT in these patients has demonstrated excellent efficacy and success rates, and appears safe [34–36]. A study performed by Fischer et al. [37] in a cohort of patients with severe or severe-complicated CDI at high risk for colectomy demonstrated that continued oral vancomycin use in conjunction with sequential FMT lead to 93% treatment response in severe cases and 89% in severe complicated cases. Fidaxomicin, a narrow-spectrum antibiotic, was used successfully in a patient with refractory CDI at high risk for colectomy in a case report by Pecere et al. [38]. Randomized, controlled clinical trials confirming these findings in this complicated group of patients with severe or complicated CDI are warranted.

DONOR SELECTION AND TESTING

Stool donors may be a family member, friend, or a healthy volunteer. However, since the initial use of FMT, recruitment criteria have become more stringent, with the focus of optimizing safety. A recent study on recruitment of potential stool donors, reported that only 10% were eligible, to participate [39]. Half of the screened potential donors were unwilling or unable to meet the commitment requirements of long-term donation due to the attendant inconvenience. A substantial portion of the remaining potential donors were excluded due to underlying infections upon workup, high body mass index, and illicit drug use.

There are potential advantages and disadvantages to consider when choosing donors for FMT. Intimate household contacts share similar environmental risk factors that may decrease the risk of transmission of infectious agents. Maternal-line first-degree relatives may offer the potential benefit of sharing the greatest number of microbial species in the intestinal microbiota with the recipient [40]. FMT using a donor who hypothetically shares a similar GI flora profile, may allow for better tolerance of the transplanted bacteria by the intestinal immune system. This tolerance is derived from a mechanism acquired early on in life by commensal microbes inducing "primitive" T-cell independent IgA sufficient for the management and homeostasis of commensal bacteria [41]. Age- and sex-matched donors may also pose a theoretical advantage, but there is no data to support this hypothesis. There may also be an important advantage to using carefully screened, healthy donors who are genetically unrelated to recipients when treating conditions, such as IBD, where genetics are felt to play a role in disease pathology. Interestingly, a higher rate of CDI resolution by FMT was reported in unrelated donors (93%) than in related donors (84%) [42], though this observation has not been confirmed in randomized controlled trials. Furthermore, unrelated donors may be preferable to family members, who may feel coerced to donate and deny underlying infectious risk factors. With an increasing number of stool transplants being performed, stool banks have emerged to source fecal material from healthy donors. The first and largest of these, OpenBiome, a nonprofit organization in Medford, Massachusetts, is currently working to make safe donor stool available for providers, though given current regulatory limitations, the future of stool banks is uncertain.

Stringent questioning of potential donors is important to ensure that the donor is in good health, the donation process is safe for the donor, and that any known risk factors for infections possibly transmissible by stool are identified. The questionnaire also provides a way to identify important risk factors for infections that tests are not sensitive for, are unable to detect, or are in an early stage or window period at the time of donation. The current recommendations by the FMT Workgroup [30] suggest using the same screening that are employed to screen potential blood donors, including questions about high-risk sexual behaviors, drug use, travel and other factors which may place a donor at high risk of communicable disease: http://www.fda.gov/BiologicsBlood-Vaccines/BloodBloodProducts/ApprovedProducts/Licensed-ProductsBLAs/BloodDonorScreening/ucm164185.htm.

Suggested donor exclusion criteria for FMT include [30]:

- Known human immunodeficiency virus (HIV), hepatitis B or C infections
- Known exposure to HIV or viral hepatitis (within the previous 12 months)
- High-risk sexual behaviors (e.g., sexual contact with anyone with HIV/acquired immune deficiency syndrome or hepatitis, men who have sex with men, sex for drugs or money)

- Use of illicit drugs
- Tattoo or body piercing within 6 months
- Incarceration or history of incarceration
- Known current communicable disease
- Risk factors for variant Creutzfeldt–Jakob disease
- Travel (within the last 6 months) to areas of the world where diarrheal illnesses are endemic or risk of traveler's diarrhea is high
- History of IBD
- History of IBS, idiopathic chronic constipation, or chronic diarrhea
- History of GI malignancy or known polyposis
- Antibiotics within the preceding 3 months
- Major immunosuppressive medications
- Systemic antineoplastic agents

There are further some relative exclusion criteria that have been proposed that include: history of weight loss surgeries, metabolic syndrome, atopic disease, autoimmune diseases, and a history of chronic pain disorders.

Testing donors for relevant communicable diseases is important. It is recommended that donor stool be tested for *C. difficile*, ova and parasites, cryptosporidium, *Cyclospora*, giardia, *Isospora*, and bacterial cultures for enteric pathogens. Serum studies on donors should include HIV1 and 2, hepatitis A IgM, hepatitis B surface antigen, hepatitis core antibody, hepatitis B surface antibody, hepatitis C antibody, and rapid plasma regain, and fluorescent treponemal antibody absorption test.

Although complete testing to exclude unfit donors is important, there are situations where extensive pretreatment testing may hinder prompt treatment. Theoretically, partners who are sexually intimate would have previously shared bodily fluids and exposure to relevant communicable diseases. In cases where FMT must be performed expeditiously, such as severe/fulminant CDI, the treating physician might weigh the risks and benefits of performing abbreviated donor screening and testing protocol prior to FMT.

PREPARATION OF FECAL MATERIAL AND DELIVERY

There is not yet standardized methodology that exists for stool preparation. The feces are typically dissolved in normal saline or water, homogenized, and filtered (e.g., gauze, coffee filter, strainer) to make a liquid slurry [43]. The dilutant, such as tap water, milk, or nonbacteriostatic normal saline has not been shown to make a difference in regards to success rates. Once the specimen is processed, the material is either infused for treatment or frozen and stored for future use. Studies have demonstrated that frozen fecal preparations are as effective as fresh for the treatment of recurrent CDI [44,45].

Limited evidence is available to support the most efficacious route for administration of fecal microbiota to recipients. Current options for infusing donor material include the upper GI tract via NGT, nasointestinal tube, endoscopy, or ingestion of capsules [8,45–49]; the proximal colon via colonoscopy [50–57]; or the distal colon by enema, rectal tube, or sigmoidoscopy [58–61]; or a combination of upper and lower GI infusions [62]. A systematic review suggested that use of the lower GI route led to the achievement of higher eradication rates than the upper GI route (84–93% vs. 81–86%, respectively) [10]. Administration via nasoenteric tubes may not be as acceptable or tolerable as a treatment option for patients [63] in that the placement of the tube may be uncomfortable and there is a small but significant risk of aspiration [64] and vomiting. The presence of ileus precludes upper GI administration [65]. The volume of stool infused into the upper GI tract is generally less than 100 cm^3, whereas higher volumes of 100–500 cm^3 are administered via colonoscopy, NGT infusion of the fecal suspension may have the added benefit of providing more exposure of the recipient's GI tract to the donor's fecal bacterial flora [66]. Despite the noted association of proton pump inhibitors (PPI) with CDI [67], some practitioners temporarily treat with a PPI prior to upper GI FMT to decrease the vulnerability of infused donor fecal bacterial flora to acidic conditions in the recipient's stomach. For hypervirulent *C. difficile*, BI/NAP1/027, the use of combined early FMT via NGT and antibiotics (oral metronidazole, vancomycin, or fidaxomicin) was shown to significantly reduce mortality in one small study [68]. NGT administration also eliminates the need for sedation and the associated anesthetic risk, which may be greater in older patients [45]. Retention enemas of donor fecal material are inexpensive with minimal procedural risk, but it may be difficult for patients to retain and donor material does not reach the most proximal colon. The colonoscopic approach may offer better therapeutic potential than any other modalities by allowing for examination of the colonic mucosa and exclusion of other pathology, such as IBD. Furthermore, the capacity to deliver fecal material directly to the colon with adjunct lavage treatment may be more effective in the treatment of CDI, since bowel irrigation may help in clearing active *C. difficile* organisms and spores [69]. Endoscopic delivery of FMT does carry potential procedural risk as well as increasing healthcare utilization and cost. However, a cost-effectiveness study showed that FMT, even when administered endoscopically, was less costly and more effective when compared with vancomycin for initial CDI [70].

EFFICACY OF FMT

FMT has been examined in pediatric, and older patients with minimal comorbidities, as well as more complicated patients who are immunocompromised [71], and appears safe, well tolerated, and effective. The experience with FMT continues to grow, with well over 500 cases reported to date [10]. A recent metaanalysis showed an efficacy between 83–90% of FMT for recurrent CDI [11,42,46,48,52,55,57,72,73]. Furthermore, an open-label study demonstrated good efficacy for encapsulated fecal material administered orally [45]. Fecal material from healthy donors was encapsulated and frozen, and then administered as 15 frozen pills on 2 consecutive days to patients with relapsing CDI. The overall response rate was 90% demonstrating oral administration is a viable alternative. Not all CDI patients respond favorably to FMT. Fisher et al. recently reported a retrospective study investigating response rates in patients after a single FMT [74]. Overall success rates for FMT were 76% and three risk factors were associated with failure in multivariable analysis: FMT performed in an inpatient setting, greater number of CDI-related hospitalizations prior to FMT, and immunocompromise [74]. Given the growing experience with FMT and high efficacy, professional society guidelines recommend FMT for the use in recurrent CDI [22,75].

MECHANISM OF FMT

The GI commensal microbiota that is acquired from birth is estimated to be approximately 100 trillion organisms, most of which are bacteria, with the greatest concentration in the distal ileum and colon [76]. Certain dietary changes, invasive pathogens, and antibiotic use lead to alterations in the composition of the microbiota favoring the growth of potentially pathogenic constituents [77]. *C. difficile* is an opportunistic organism that generally arises as a result of antibiotic use and decreased diversity of the intestinal microflora. It has been demonstrated that patients with recurrent CDI are deficient in phyla of bacteria normally dominant in the colon [78]. This may be due to changes that are induced by the use of antibiotics. One study demonstrated a significant decrease in butyrate-producing bacteria, the major source of energy for colonocytes, in a patient who had been taking amoxicillin–clavulanic acid [79]. An additional study using 16 S rRNA sequencing to characterize bacterial populations in the distal gut of patients recently exposed to ciprofloxacin, showed reduced microbial richness, evenness, and diversity [80]. An earlier study, evaluating microbial changes in relationship to antibiotic use, described decreased fecal tryptic and urobilinogen activity as well as decreased conversion of cholesterol to coprostanol, with restoration of normal function after FMT from healthy donors [58]. Interestingly, a study in which the third generation cephalosporin, cefoperazone, was administered to mice demonstrated alterations to the metabolome of the microbiota [81]. The mice exhibited increased levels of primary bile acids, increased levels of sugar alcohols, decreased levels of free short-, medium-, and long-chain fatty acids, increased levels of amino acids, and decreased levels of branched-chain fatty acids; all of which favor *C. difficile* growth [81]. A higher proportion of primary bile acids to secondary bile acids in the colon have been shown to favor germination of *C. difficile* spores, suggesting that bile acids stimulate colony formation from *C. difficile* spores and commensal GI flora likely influence levels of these bile acids [82]. FMT for recurrent CDI may work via this mechanism as demonstrated by Weingarden et al., who reported an increase in secondary bile acids after FMT compared with pretransplant stool samples [34]. A more recent study by Buffie et al. showed that the intestinal bacterium *Clostridium scindens*, which harbors the gene 7-hydroxysteroid dehydrogenase, required for secondary bile acid synthesis, is associated with resistance to CDI, and administration enhances resistance to infection in a secondary bile acid dependent fashion [83].

An important factor determining the success of FMT is the restoration of microbial diversity and, with the advent of new technology, these changes in microbiota post-FMT are evident [84]. FMT appears to result in colonization of the recipient with new species of bacteria from the donor as well as augmentation of species that were present at minute levels in the infected individual prior to FMT [50]. Further evidence suggest that the absence of specific species of bacteria, such as Lachnospiraceae, predisposes to more severe cases of *C. difficile* [85] and that replenishing this community of bacteria cures murine models with *C. difficile* [86]. Lastly, there are likely several species of bacteria that secrete bacteriocins, which inhibit growth of similar or closely related bacterial strains; *Bacillus thuringiensis* secretes thuricin CD, with narrow-spectrum activity against gram-positive bacteria including *C. difficile* [87]. In summary, recolonization of the colon with healthy donor microbiota likely eliminates *C. difficile* via several mechanisms including niche exclusion, production of antimicrobial peptides that inhibit growth of *C. difficile*, and an increase in secondary bile acid production.

SAFETY OF FMT

FMT appears to be a safe treatment method, with few published reports of complications to date. However, due to the early implementation of the procedure without data from large trials and careful follow-up for adverse events, it is difficult to be certain that there are not untoward long-term effects. Mild symptoms that have been reported post-FMT include diarrhea, constipation, flatulence, abdominal pain, vomiting, borborygmus, and fever [8,88]. Some potential complications are related to the endoscopic procedure itself including aspiration [71] secondary to sedation, and bleeding or perforation due to instrumentation. There have been surprisingly few reported cases of suspected pathogen transmission after FMT. Norovirus infection was documented in two patients after FMT, though the donor tested

negative, and transmission was felt more likely related to contamination from unit staff or interval exposure postprocedure [89]. Another patient with small bowel Crohn's disease (CD), diverticulitis, and recurrent episodes of *C. difficile* was reported to have *Escherichia coli* bacteremia after FMT, though he had suffered six prior episodes of *E. coli* bacteremia prior to FMT [90]. Importantly, in a large retrospective study of FMT for treatment of CDI in patients who were immunocompromised, and theoretically at highest risk, there were no infectious complications related to FMT [71].

Of greatest long-term concern is the possibility of altering the microbial flora in a way that would lead to the development of a chronic illness. A number of diseases have been associated with alterations in the microbial flora including obesity, nonalcoholic steatohepatitis (NASH), multiple sclerosis (MS), IBD, diabetes, irritable bowel syndrome (IBS), and atherosclerosis [40]. The theoretical risk of inducing disease development in an individual as a consequence of FMT is a powerful notion. Long-term clinical follow-up in all patients who have received FMT is important to detect a higher rate of diseases that may be related to alterations in the microbial flora. This may be best accomplished via a large registry of FMT-treated patients.

FMT FOR OTHER CONDITIONS

Inflammatory Bowel Disease

IBD encompasses two major phenotypes, CD and ulcerative colitis (UC). The etiology of the disease processes remain poorly understood but are thought to arise from a combination of genetics, environmental factors, barrier dysfunction, and a dysregulated immune response toward luminal bacteria [91]. It has been postulated that alterations in the gut microbiota in genetically susceptible patients may evoke an aberrant immune reaction in the gut, resulting in IBD [92]. Advanced molecular techniques have identified that microbiota in IBD patients are different than that of healthy individuals [93–95]. A decrease in the bacterial diversity in IBD patients is seen, with increases in Proteobacteria and Actinobacteria and decreases in the Bacteroidetes phylum and the Lachnospiraceae group within the Firmicutes phylum [96]. A recent large study in pediatric patients with CD demonstrated an imbalance in the microflora with an increased abundance of bacteria that included *Enterobacteriaceae*, *Pasteurellacaea*, *Veillonellaceae*, and *Fusobacteriaceae*; and a decreased abundance in Erysipelotrichales, Bacteroidales, and Clostridiales; correlating strongly with disease status [97]. This study alluded to the concept of detectable microbial signatures early on in the diagnosis of CD. Currently, medical treatments are aimed at controlling the dysregulated inflammatory response. However, these medications do carry risks, such as infection, allergic reaction, and malignancy. The idea of having a "natural" treatment option that could revert the altered microflora back to a homeostatic state without toxic effects has spurred further investigation of FMT in patients with IBD.

The first reported use of FMT for IBD was in a case reported by Bennet and Brinkman [98]. This experiment came about when Bennet, who was suffering from UC refractory to standard treatment, self-administered large-volume retention enemas of healthy donor stool. Follow-up colonic biopsies showed resolution of inflammation and symptoms, and he remained off of IBD-medications for years. Additional FMTs to patients with IBD ensued, with durable responses reported in a patient with UC and another with ileal CD [99]. Small uncontrolled studies continued to suggest benefit of FMT for treatment of IBD. Six patients with confirmed UC, who were failing to improve with 5-aminosalicylic acid, steroids, and azathioprine were treated with FMT, and all reportedly achieved a durable response [100]. One study used FMT as rescue treatment in a patient with refractory CD complicated by a fistula and large intraperitoneal inflammatory mass, with sustained response for greater than 9 months [101]. A similar case report detailed a patient with CD who had failed medical therapy and achieved clinical, endoscopic, and histologic remission after a single fecal infusion [102]. This report analyzed pre- and post-FMT distal gut microbiota and showed that alterations induced by FMT were associated with improved clinical outcomes, however, these changes did not persist, suggesting that patients with IBD may require continual or repeated FMT as maintenance therapy. Further experience with FMT in children and adults with UC, CD, and pouchitis have demonstrated mixed results [103–108]. Limitations in these case reports/series include absence of control groups, variation in disease severity and phenotype, and lack of uniformity in treatment protocols. Many did not employ standardized measures for outcomes.

A systematic review and metaanalysis of 18 studies (9 cohort, 8 case studies, 1 RCT) including 122 patients with IBD who underwent FMT found a clinical remission rate of 45% during follow-up [109]. Among analysis of the cohort studies, patients achieving clinical remission fell to 36%. Subgroup pooled analyses of response to FMT in UC and CD showed clinical remission of 22% and 60%, respectively. There were no reported serious adverse events during short-term follow-up and the authors concluded that FMT appeared safe but varied in efficacy for treatment of IBD.

There have been two recent randomized placebo-controlled trials of FMT in UC. In the first study by Rossen et al. [88], a total of 50 patients with mild-to-moderate UC were randomized to receive either healthy donor or autologous FMT by nasoduodenal tube at the start of the study and 3 weeks later. Unfortunately, only 37 out of the 50 patients initially enrolled completed the primary endpoint assessment, and there was no difference in clinical or endoscopic remission between the two groups. Moayyedi et al. [110] recruited 75 patients with active UC who were treated with weekly FMT or water enema in a blinded fashion for 6 weeks. The safety committee stopped the trial early due to futility, however, they were permitted to treat subjects who were already enrolled, resulting in a statistically significant difference in remission between the groups, with 24% of patients in the FMT group and only 5% of the placebo-treated patients in remission at the 7 week endpoint. Interestingly, seven of the nine patients in the FMT group achieving remission had received fecal material from a single donor, suggesting a "donor effect." Furthermore, 75% of patients with UC less than 1 year in duration treated with FMT entered remission compared with only 17% of patients who had suffered greater than 1 year of disease. There was no difference in reported adverse events between the groups.

Given the mixed results and further ongoing investigation, FMT is not currently recommended for induction therapy or maintenance of remission in IBD. There have been safety concerns raised in this population. One case reported a flare in a patient with UC who had been in deep remission and was treated with FMT for recurrent *C. difficile* [111]. Another reported a new diagnosis of CD several weeks after transplant [112]. Furthermore, a study investigating FMT for cases of recurrent, refractory, or severe cases of *C. difficile* in immunocompromised patients found that 14% of IBD patients experienced a posttransplant IBD flare. The disease process in IBD is more complex than in CDI, and patients may respond differently to FMT. A future treatment option for IBD may be the infusion of specific bacterial populations based on a patient's unique microbial flora profile to correct imbalances and restore homeostasis.

Irritable Bowel Syndrome

IBS is among the most commonly diagnosed GI conditions, affecting anywhere from 7% to 21% of the general population [113]. The disease appears to be more prevalent among women in the United States, Canada, and Israel while it is equally distributed among genders in Asia [114]. In the majority of patients, IBS is a chronic relapsing disease with varying symptoms over time and accounts for 3 million ambulatory visits and close to 6 million annual prescriptions [115]. The pathogenesis of the disease remains complex and likely involves a number of different mechanisms including altered motility, psychosocial distress, and increased visceral sensitivity. More recently, the gut microbiome and a dysregulated gut immune response have been postulated as potential mechanisms [116]. Studies have identified differences in the microbial flora composition between IBS patients and healthy controls [117–122]. Similar to alterations in microflora seen in *C. difficile* and IBD, IBS patients lack microbial diversity. A study by Parkes et al. performed rectal biopsies from IBS and healthy controls and found greater numbers of mucosal-associated microbiota using fluorescent in situ hybridization, but microbial diversity was less than when compared to controls [123]. Another study looked at fecal samples from 62 patients with IBS and 46 controls and found a significant difference in the composition of the microbial flora, with samples from IBS patients having decreased Bacteroidetes, Actinobacteria, and Bifidobacteria and increased Firmicutes species [124].

RCTs evaluating FMT in patients with IBS are lacking. A case report by Zoller et al. [125] reported symptom improvement with a single FMT administered by colonoscopy and demonstrated sustained changes in the microbiome of the recipient that came to resemble that of the donor. A recent systematic review [126] included two small case series. In one, three patients with chronic constipation were treated with FMT and had resolution of their symptoms (regular bowel movements without the use of laxatives). The second reported 13 patients with IBS who demonstrated an improvement in symptoms of abdominal pain, bloating, bowel habit, and flatus after FMT. IBS subtypes may require specific bacterial communities for treatment since, as was demonstrated in the small study by Lyra et al. [127] different subtypes of IBS may have differences in intestinal microbial compositions. Furthermore, a recent study investigating 16s rRNA of healthy controls and patients with IBS found significant differences in the microbial taxa between the groups [128]. The association between alterations/dysbiosis in the microbial flora and IBS cannot be denied. Further prospective controlled trials will be the only way to confirm whether FMT is a treatment option for IBS.

Metabolic Disease

Metabolic syndrome and obesity are a rising epidemic. Dysbiosis has been linked to metabolic syndrome and obesity and evidence has alluded to the potential pivotal role the microbiota may play in shaping the diseases [129,130]. One study showed that germ-free mice fed a high fat, high carbohydrate diet did not gain weight, suggesting that diet is not the only factor required to induce obesity [131]. However, when germ-free mice were colonized with microbiota from obese mouse donors, the obese phenotype was transmissible and there was a significant gain in body fat [130]. Translocation of lipopolysaccharide (LPS), a component of gram-negative bacterial cell membranes, from the intestinal to the portal circulation in mice has been shown to decrease hepatic insulin sensitivity and trigger weight gain [132]. Furthermore, obese individuals have been shown to have increased plasma levels of bacteria and their metabolic products, which may be related to increased intestinal permeability [133]. Vrieze et al. [134] performed a double-blinded placebo-controlled trial in 18 men with the metabolic syndrome. Nine of the subjects were administered FMT from healthy lean donors via nasoduodenal tube and the other nine were infused with their own feces. Six weeks postinfusion, recipients of lean donor FMT were found to have decreased serum triglyceride levels and increased peripheral and hepatic insulin sensitivity, suggesting an important role of the intestinal microbiota in the pathogenesis of insulin sensitivity and obesity.

Nonalcoholic fatty liver disease (NAFLD) is frequently associated with underlying obesity or metabolic syndrome. NAFLD is currently emerging worldwide as one of the leading causes of end-stage liver disease and indications for liver transplant [135]. NAFLD encompasses two different disease entities: steatosis that usually does not clinically progress and nonalcoholic steatosis (NASH) that can lead to cirrhosis or hepatocellular carcinoma [136]. The exact mechanisms underlying the disease pathophysiology are incompletely understood. The physiology of the hepatic portal circulation may lead to hepatocyte exposure to bacteria or bacterial products via increased intestinal permeability secondary to dysbiosis [137,138]. Mouzaki et al. [139] found an inverse correlation between NASH and the presence of Bacteroidetes in the stool, suggesting a potential role of intestinal microbiota in disease development. This imbalance in microbial flora has been similarly demonstrated in murine models [140]. NAFLD has been associated with small intestinal bacterial overgrowth (SIBO) and inflammation may occur secondary to increased gut permeability and hepatic expression of TLR4 resulting in elevated TNF-α levels [141]. Pediatric cases of NAFLD/NASH when compared to healthy controls were found to have elevated levels of endogenous ethanol, a by-product of intestinal flora [142]. The hypotheses that microbial flora play a significant role in NAFLD led investigators to attempt to modulate the disease via administration of probiotics. A recent metaanalysis including 4 randomized trials of 134 patients with NAFLD/NASH found probiotics resulted in decreased aminotransferase levels and TNF-α production [143]. Coffee consumption has been shown to reverse fat deposition in the liver via a mechanism that may result in increased levels of the beneficial bacteria, *Bifidobacterium* [144]. Vitamin E is currently the only medicine used to treat NASH that is refractory to lifestyle changes. Cheng et al. [145] showed that NAFLD/NASH patients who responded to Vitamin E treatment had detectable microbial derived metabolites indole-propionic acid and phenyl-propionic acid. These metabolites are free radical scavengers with mitochondrial-protective properties and these microbial products may explain the benefits of Vitamin E in this population. It appears that the intestinal microflora play an important role in the development of NAFLD/NASH, and better understanding of the mechanisms involved may lead to the beneficial manipulation of microflora as a treatment option.

Potential Future Indications of FMT for Chronic Illnesses

FMT may prove to be beneficial in the treatment of other diseases. In murine models for MS, there has been some evidence that specific bacterial species can protect against the development of the disease [146]. Interestingly, symptoms associated with MS in three patients completely resolved for a range of 2–15 years after FMT for the treatment of constipation [147]. A similar observation was made in a patient with Parkinson's disease who underwent FMT for constipation and had a dramatic improvement in neurologic symptoms [148].

Application of FMT has also been postulated in the treatment of autism. A small case series of 13 children with autism found a greater number of clostridial species when compared with controls [149], and there is some data to suggest that FMT in autistic children improved symptoms [150].

Other chronic disease, such as rheumatoid arthritis, Sjogren's syndrome, Hashimoto's thyroiditis, allergic atopy, atherosclerotic disease, hepatic encephalopathy, and even depression have been associated with dysregulated immune responses against the intestinal microflora and future treatments may specifically target these interactions. The experience with FMT in these illnesses remains limited to case series, and randomized controlled trials with larger numbers of patients are required in order to confirm efficacy and safety.

CONCLUSIONS

The complex role the intestinal microflora plays in maintaining a healthy and homeostatic environment is undeniable. Our understanding of this important mutually beneficial relationship is continuing to grow with ongoing research and clinical trials. A 4th century homeopathic use of "fecal suspensions" for intestinal illnesses in traditional Chinese medicine has now emerged as the most-effective treatment of recurrent *C. difficile*. These findings have stirred desires to investigate FMT for treatment of other disease processes associated with dysbiosis. Although FMT efficacy is high in the treatment of *C. difficile*, further research is needed in these more complex diseases. DNA sequencing technologies will likely identify bacterial communities that are deficient or are in excess and future treatment will be directed at reestablishing the correct balance. In the future, specific bacterial isolates, rather than whole stool transplants, will likely be used to reestablish gut homeostasis based on individualized intestinal microflora characteristics. Long-term follow-up of patients treated with FMT is critical to identify unforeseen safety concerns.

REFERENCES

[1] Doron S, Davidson LE. Antimicrobial stewardship. Mayo Clinic Proc 2011;86(11):1113–23.

[2] Backhed F, Ley RE, Sonnenburg JL, et al. Host-bacterial mutualism in the human intestine. Science 2005;307(5717):1915–20.

[3] Zhang F. Luo W, Shi Y, et al. Should we standardize the 1,700-year-old fecal microbiota transplantation? Am J Gastroenterol 2012;107(11):1755. author reply p. 6.

[4] Borody TJ, Warren EF, Leis SM, et al. Bacteriotherapy using fecal flora: toying with human motions. J Clin Gastroenterol 2004;38(6):475–83.

[5] Grossman CM. The first use of penicillin in the United States. Ann Inter Med 2008;149(2):135–6.

[6] Eiseman B, Silen W, Bascom GS, et al. Fecal enema as an adjunct in the treatment of pseudomembranous enterocolitis. Surgery 1958;44(5):854–9.

[7] Schwan A, Sjolin S, Trottestam U, et al. Relapsing *Clostridium difficile* enterocolitis cured by rectal infusion of homologous faeces. Lancet 1983;2(8354):845.

[8] van Nood E, Vrieze A, Nieuwdorp M, et al. Duodenal infusion of donor feces for recurrent *Clostridium difficile*. New Engl J Med 2013;368(5):407–15.

[9] Bonifield J. One man's poop is another's medicine; 2015. Available from: http://www.cnn.com

[10] Cammarota G, Ianiro G, Gasbarrini A. Fecal microbiota transplantation for the treatment of *Clostridium difficile* infection: a systematic review. J Clin Gastroenterol 2014;48(8):693–702.

[11] Kassam Z, Lee CH, Yuan Y, et al. Fecal microbiota transplantation for *Clostridium difficile* infection: systematic review and meta-analysis. Am J Gastroenterol 2013;108(4):500–8.

[12] Hall I, O'Toole E. Intestinal flora in newborn infants with a description of a new pathogenic anaerobe, *Bacillus difficilis*. Am J Dis Child 1935;49:390–402.

[13] Viscidi R, Willey S, Bartlett JG. Isolation rates and toxigenic potential of *Clostridium difficile* isolates from various patient populations. Gastroenterology 1981;81(1):5–9.

[14] Bartlett JG. Historical perspectives on studies of *Clostridium difficile* and *C. difficile* infection. Clin Infect Dis 2008;46(Suppl. 1): S4–S11.

[15] Hummel RP, Altemeier WA, Hill EO. Iatrogenic staphylococcal enterocolitis. Ann Surg 1964;160:551–60.

[16] Khan MY, Hall WH. Staphylococcal enterocolitis—treatment with oral vancomycin. Ann Intern Med 1966;65(1):1–8.

[17] Tedesco FJ, Barton RW, Alpers DH. Clindamycin-associated colitis. A prospective study. Ann Intern Med 1974;81(4):429–33.

[18] Green RH. The association of viral activation with penicillin toxicity in guinea pigs and hamsters. Yale J Biol Med 1974;47(3): 166–81.

[19] Bartlett JG, Chang TW, Gurwith M, et al. Antibiotic-associated pseudomembranous colitis due to toxin-producing clostridia. New Engl J Med 1978;298(10):531–4.

[20] Taylor NS, Thorne GM, Bartlett JG. Comparison of two toxins produced by *Clostridium difficile*. Infect Immun 1981;34(3):1036–43.

[21] Cohen SH, Gerding DN, Johnson S, et al. Clinical practice guidelines for *Clostridium difficile* infection in adults: 2010 update by the society for healthcare epidemiology of America (SHEA) and the infectious diseases society of America (IDSA). Infect Control Hosp Epidemiol 2010;31(5):431–55.

[22] Surawicz CM, Brandt LJ, Binion DG, et al. Guidelines for diagnosis, treatment, and prevention of *Clostridium difficile* infections. Am J Gastroenterol 2013;108(4):478–98. quiz 499.

[23] Kelly CP, LaMont J. *Clostridium difficile*-more difficult than ever. N Engl J Med 2008;359:1932–40.

[24] Petrella LA, Sambol SP, Cheknis A, et al. Decreased cure and increased recurrence rates for *Clostridium difficile* infection caused by the epidemic *C. difficile* BI strain. Clin Infect Dis 2012;55(3):351–7.

[25] Enforcement policy regarding investigational new drug requirements for use of fecal microbiota for transplantation to treat *Clostridium difficile* infection not responsive to standard therapies: U.S. Department of Health and Human Services Food and Drug Administration Center for Biologics Evaluation and Research; 2013. Available from: http://www.fda.gov/downloads/BiologicsBlood-Vaccines/GuidanceComplianceRegulatoryInformation/Guidances/Vaccines/UCM361393.pdf

[26] Youngster I, Sauk J, Pindar C, et al. Fecal microbiota transplant for relapsing *Clostridium difficile* infection using a frozen inoculum from unrelated donors: a randomized, open-label, controlled pilot study. Clin. Infect Dis 2014;58(11):1515–22.

[27] Cammarota G, Masucci L, Ianiro G, et al. Randomised clinical trial: faecal microbiota transplantation by colonoscopy vs. vancomycin for the treatment of recurrent *Clostridium difficile* infection. Aliment Pharmacol Therapeut 2015;41(9):835–43.

[28] Lee CH, Steiner T, Petrof EO, et al. Frozen vs fresh fecal microbiota transplantation and clinical resolution of diarrhea in patients with recurrent *Clostridium difficile* infection: a randomized clinical trial. JAMA 2016;315(2):142–9.

[29] Kelly CBL, Abd M, Alani M, Bakow B, Curran P, Tisch A, Reinert S. A multicenter, randomized, placebo-controlled, double-blind study to evaluate the efficacy and safety of fecal microbiota transplantation in patients with recurrent *Clostridium difficile* infection. ACG 2015 Annual Scientific Meeting Abstracts. 2015. Program 44.

[30] Bakken JS, Borody T, Brandt LJ, et al. Treating *Clostridium difficile* infection with fecal microbiota transplantation. Clin Gastroenterol Hepatol 2011;9(12):1044–9.

[31] Osman KA, Ahmed MH, Hamad MA, et al. Emergency colectomy for fulminant *Clostridium difficile* colitis: Striking the right balance. Scand J Gastroenterol 2011;46(10):1222–7.

[32] Byrn JC, Maun DC, Gingold DS, et al. Predictors of mortality after colectomy for fulminant *Clostridium difficile* colitis. Arch Surg 2008;143(2):150–4. discussion 155.

[33] Longo WE, Mazuski JE, Virgo KS, et al. Outcome after colectomy for *Clostridium difficile* colitis. Dis Colon Rectum 2004;47(10):1620–6.

[34] Weingarden AR, Hamilton MJ, Sadowsky MJ, et al. Resolution of severe *Clostridium difficile* infection following sequential fecal microbiota transplantation. J Clin Gastroenterol 2013;47(8):735–7.

[35] You DM, Franzos MA, Holman RP. Successful treatment of fulminant *Clostridium difficile* infection with fecal bacteriotherapy. Ann Inter Med 2008;148(8):632–3.

[36] Gweon TG, Lee KJ, Kang DH, et al. A case of toxic megacolon caused by *Clostridium difficile* infection and treated with fecal microbiota transplantation. Gut Liver 2015;9(2):247–50.

[37] Fischer M, Sipe BW, Rogers NA, et al. Faecal microbiota transplantation plus selected use of vancomycin for severe-complicated *Clostridium difficile* infection: description of a protocol with high success rate. Aliment Pharmacol Therapeut 2015;42(4):470–6.

[38] Pecere S, Sabatelli M, Fantoni M, et al. Letter: faecal microbiota transplantation in combination with fidaxomicin to treat severe complicated recurrent *Clostridium difficile* infection. Aliment Pharmacol Therapeut 2015;42(8):1030.

[39] Paramsothy S, Borody TJ, Lin E, et al. Donor recruitment for fecal microbiota transplantation. Inflammat Bowel Dis 2015;21(7):1600–6.

[40] Kelly CR, Kahn S, Kashyap P, et al. Update on fecal microbiota transplantation 2015: indications, methodologies, mechanisms, and outlook. Gastroenterology 2015;149(1):223–37.

[41] Slack E, Balmer ML, Fritz JH, et al. Functional flexibility of intestinal IgA—broadening the fine line. Front Immunol 2012;3:100.

[42] Gough E, Shaikh H, Manges AR. Systematic review of intestinal microbiota transplantation (fecal bacteriotherapy) for recurrent *Clostridium difficile* infection. Clin Infect Dis 2011;53(10):994–1002.

[43] Smits LP, Bouter KE, de Vos WM, et al. Therapeutic potential of fecal microbiota transplantation. Gastroenterology 2013;145(5):946–53.

[44] Hamilton MJ, Weingarden AR, Sadowsky MJ, et al. Standardized frozen preparation for transplantation of fecal microbiota for recurrent *Clostridium difficile* infection. Am J Gastroenterol 2012;107(5):761–7.

[45] Youngster I, Russell GH, Pindar C, et al. Oral, capsulized, frozen fecal microbiota transplantation for relapsing *Clostridium difficile* infection. JAMA 2014;312(17):1772–8.

[46] Aas J, Gessert CE, Bakken JS. Recurrent *Clostridium difficile* colitis: case series involving 18 patients treated with donor stool administered via a nasogastric tube. Clin Infect Dis 2003;36(5):580–5.

[47] Kronman MP, Nielson HJ, Adler AL, et al. Fecal microbiota transplantation via nasogastric tube for recurrent *Clostridium difficile* infection in pediatric patients. J Pediatr Gastroenterol Nutr 2015;60(1):23–6.

[48] Rubin TA, Gessert CE, Aas J, et al. Fecal microbiome transplantation for recurrent *Clostridium difficile* infection: report on a case series. Anaerobe 2013;19:22–6.

[49] Russell G, Kaplan J, Ferraro M, et al. Fecal bacteriotherapy for relapsing *Clostridium difficile* infection in a child: a proposed treatment protocol. Pediatrics 2010;126(1):e239–42.

[50] Khoruts A, Dicksved J, Jansson JK, et al. Changes in the composition of the human fecal microbiome after bacteriotherapy for recurrent *Clostridium difficile*-associated diarrhea. J Clin Gastroenterol 2010;44(5):354–60.

[51] Hamilton MJ, Weingarden AR, Unno T, et al. High-throughput DNA sequence analysis reveals stable engraftment of gut microbiota following transplantation of previously frozen fecal bacteria. Gut Microbes 2013;4(2):125–35.

[52] Persky SE, Brandt LJ. Treatment of recurrent *Clostridium difficile*-associated diarrhea by administration of donated stool directly through a colonoscope. Am J Gastroenterol 2000;95(11):3283–5.

[53] Yoon SS, Brandt LJ. Treatment of refractory/recurrent *C. difficile*-associated disease by donated stool transplanted via colonoscopy: a case series of 12 patients. J Clin Gastroenterol 2010;44(8):562–6.

[54] Brandt LJ, Aroniadis OC, Mellow M, et al. Long-term follow-up of colonoscopic fecal microbiota transplant for recurrent *Clostridium difficile* infection. Am J Gastroenterol 2012;107(7):1079–187.

[55] Kelly CR, de Leon L, Jasutkar N. Fecal microbiota transplantation for relapsing *Clostridium difficile* infection in 26 patients: methodology and results. J Clin Gastroenterol 2012;46(2):145–9.

[56] Mattila E, Uusitalo-Seppala R, Wuorela M, et al. Fecal transplantation, through colonoscopy, is effective therapy for recurrent *Clostridium difficile* infection. Gastroenterology 2012;142(3):490–6.

[57] Rohlke F, Surawicz CM, Stollman N. Fecal flora reconstitution for recurrent *Clostridium difficile* infection: results and methodology. J Clin Gastroenterol 2010;44(8):567–70.

[58] Gustafsson A, Berstad A, Lund-Tonnesen S, et al. The effect of faecal enema on five microflora-associated characteristics in patients with antibiotic-associated diarrhoea. Scand J Gastroenterol 1999;34(6):580–6.

[59] Kassam Z, Hundal R, Marshall JK, et al. Fecal transplant via retention enema for refractory or recurrent *Clostridium difficile* infection. Arch Inter Med 2012;172(2):191–3.

[60] Lee CH, Belanger JE, Kassam Z, et al. The outcome and long-term follow-up of 94 patients with recurrent and refractory *Clostridium difficile* infection using single to multiple fecal microbiota transplantation via retention enema. Eur J Clin Microbiol Infect Dis 2014;33(8):1425–8.

[61] Silverman MS, Davis I, Pillai DR. Success of self-administered home fecal transplantation for chronic *Clostridium difficile* infection. Clin Gastroenterol Hepatol 2010;8(5):471–3.

[62] Dutta SK, Girotra M, Garg S, et al. Efficacy of combined jejunal and colonic fecal microbiota transplantation for recurrent *Clostridium difficile* Infection. Clin Gastroenterol Hepatol 2014;12(9):1572–6.

[63] Zipursky JS, Sidorsky TI, Freedman CA, et al. Patient attitudes toward the use of fecal microbiota transplantation in the treatment of recurrent *Clostridium difficile* infection. Clin Infect Dis 2012;55(12):1652–8.

[64] Baxter M, Ahmad T, Colville A, et al. Fatal aspiration pneumonia as a complication of fecal microbiota transplant. Clin Infecti Dis 2015;61(1):136–7.

[65] Brandt LJ, Borody TJ, Campbell J. Endoscopic fecal microbiota transplantation: "first-line" treatment for severe *Clostridium difficile* infection? J Clin Gastroenterol 2011;45(8):655–7.

[66] Brandt LJ, Reddy SS. Fecal microbiota transplantation for recurrent *Clostridium difficile* infection. J Clin Gastroenterol 2011;45(Suppl.):S159–67.

[67] McCarthy DM. Proton pump inhibitor use and *Clostridium difficile* colitis: cause or coincidence? J Clin Gastroenterol 2012;46(5):350–3.

[68] Lagier JC, Delord M, Million M, et al. Dramatic reduction in *Clostridium difficile* ribotype 027-associated mortality with early fecal transplantation by the nasogastric route: a preliminary report. Eur J Clin Microbiol Infect Dis 2015;34(8):1597–601.

[69] Liacouras CA, Piccoli DA. Whole-bowel irrigation as an adjunct to the treatment of chronic, relapsing *Clostridium difficile* colitis. J Clin Gastroenterol 1996;22(3):186–9.

[70] Varier RU, Biltaji E, Smith KJ, et al. Cost-effectiveness analysis of treatment strategies for initial *Clostridium difficile* infection. Clin Microbiol Infect 2014;20(12):1343–51.

[71] Kelly CR, Ihunnah C, Fischer M, et al. Fecal microbiota transplant for treatment of *Clostridium difficile* infection in immunocompromised patients. Am J Gastroenterol 2014;109(7):1065–71.

[72] Guo B, Harstall C, Louie T, et al. Systematic review: faecal transplantation for the treatment of *Clostridium difficile*-associated disease. Aliment Pharmacol Therapeut 2012;35(8):865–75.

[73] MacConnachie AA, Fox R, Kennedy DR, et al. Faecal transplant for recurrent *Clostridium difficile*-associated diarrhoea: a UK case series. Quarter J Med 2009;102(11):781–4.

[74] Fischer M, Kao D, Mehta SR, Martin T, Dimitry J, Keshteli AH, Cook GK, Phelps E, Sipe BW, Xu H, Kelly CR. Predictors of early failure after fecal microbiota transplantation for the therapy of clostridium difficile infection: a multicenter study. Am J Gastroenterol 2016;117(7):1024–31.

[75] Debast SB, Bauer MP, Kuijper EJ, et al. European Society of Clinical Microbiology and Infectious Diseases: update of the treatment guidance document for *Clostridium difficile* infection. Clin Microbiol Infect 2014;20(Suppl. 2):1–26.

[76] Ley RE, Peterson DA, Gordon JI. Ecological and evolutionary forces shaping microbial diversity in the human intestine. Cell 2006;124(4):837–48.

[77] Maynard CL, Elson CO, Hatton RD, et al. Reciprocal interactions of the intestinal microbiota and immune system. Nature 2012;489(7415):231–41.

[78] Chang JY, Antonopoulos DA, Kalra A, et al. Decreased diversity of the fecal microbiome in recurrent *Clostridium difficile*-associated diarrhea. J Infect Dis 2008;197(3):435–8.

[79] Young VB, Schmidt TM. Antibiotic-associated diarrhea accompanied by large-scale alterations in the composition of the fecal microbiota. J Clin Microbiol 2004;42(3):1203–6.

[80] Dethlefsen L, Huse S, Sogin ML, et al. The pervasive effects of an antibiotic on the human gut microbiota, as revealed by deep 16S rRNA sequencing. PLoS Biol 2008;6(11):e280.

[81] Theriot CM, Koenigsknecht MJ, Carlson PE Jr, et al. Antibiotic-induced shifts in the mouse gut microbiome and metabolome increase susceptibility to *Clostridium difficile* infection. Nat Commun 2014;5:3114.

[82] Giel JL, Sorg JA, Sonenshein AL, et al. Metabolism of bile salts in mice influences spore germination in *Clostridium difficile*. PLoS One 2010;5(1):e8740.

[83] Buffie CG, Bucci V, Stein RR, et al. Precision microbiome reconstitution restores bile acid mediated resistance to *Clostridium difficile*. Nature 2015;517(7533):205–8.

[84] Shahinas D, Silverman M, Sittler T, et al. Toward an understanding of changes in diversity associated with fecal microbiome transplantation based on 16S rRNA gene deep sequencing. MBio 2012;3(5). e00338-12.

[85] Reeves AE, Koenigsknecht MJ, Bergin IL, et al. Suppression of *Clostridium difficile* in the gastrointestinal tracts of germfree mice inoculated with a murine isolate from the family Lachnospiraceae. Infect Immun 2012;80(11):3786–94.

[86] Lawley TD, Clare S, Walker AW, et al. Targeted restoration of the intestinal microbiota with a simple, defined bacteriotherapy resolves relapsing *Clostridium difficile* disease in mice. PLoS Pathog 2012;8(10):e1002995.

[87] Rea MC, Sit CS, Clayton E, et al. Thuricin CD, a posttranslationally modified bacteriocin with a narrow spectrum of activity against *Clostridium difficile*. Proc Natl Acad Sci USA 2010;107(20):9352–7.

[88] Rossen NG, Fuentes S, van der Spek MJ, et al. Findings from a randomized controlled trial of fecal transplantation for patients with ulcerative colitis. Gastroenterology 2015;149(1). 110-8 e4.

[89] Schwartz M, Gluck M, Koon S. Norovirus gastroenteritis after fecal microbiota transplantation for treatment of *Clostridium difficile* infection despite asymptomatic donors and lack of sick contacts. Am J Gastroenterol 2013;108(8):1367.

[90] Quera R, Espinoza R, Estay C, et al. Bacteremia as an adverse event of fecal microbiota transplantation in a patient with Crohn's

disease and recurrent *Clostridium difficile* infection. J Crohn Colitis 2014;8(3):252–3.

[91] Kucharzik T, Maaser C, Lugering A, et al. Recent understanding of IBD pathogenesis: implications for future therapies. Inflamm Bowel Dis 2006;12(11):1068–83.

[92] Sartor RB. Microbial influences in inflammatory bowel diseases. Gastroenterology 2008;134(2):577–94.

[93] Nagalingam NA, Lynch SV. Role of the microbiota in inflammatory bowel diseases. Inflamm Bowel Dis 2012;18(5):968–84.

[94] Kostic AD, Xavier RJ, Gevers D. The microbiome in inflammatory bowel disease: current status and the future ahead. Gastroenterology 2014;146(6):1489–99.

[95] Marichanh C, Borruel N, Casellas F, et al. The gut microbiota in IBD. Nat Rev Gastroenterol Hepatol 2012;9(10):599–608.

[96] Frank DN, St Amand AL, Feldman RA, et al. Molecular-phylogenetic characterization of microbial community imbalances in human inflammatory bowel diseases. Proc Natl Acad Sci USA 2007;104(34):13780–5.

[97] Gevers D, Kugathasan S, Denson LA, et al. The treatment-naive microbiome in new-onset Crohn's disease. Cell Host Microbe 2014;15(3):382–92.

[98] Bennet JD, Brinkman M. Treatment of ulcerative colitis by implantation of normal colonic flora. Lancet 1989;1(8630):164.

[99] Borody TJ, George L, Andrews P, et al. Bowel-flora alteration: a potential cure for inflammatory bowel disease and irritable bowel syndrome? Med J Austr 1989;150(10):604.

[100] Borody TJ, Warren EF, Leis S, et al. Treatment of ulcerative colitis using fecal bacteriotherapy. J Clin Gastroenterol 2003;37(1):42–7.

[101] Zhang FM, Wang HG, Wang M, et al. Fecal microbiota transplantation for severe enterocolonic fistulizing Crohn's disease. World J Gastroenterol 2013;19(41):7213–6.

[102] Kao D, Hotte N, Gillevet P, et al. Fecal microbiota transplantation inducing remission in Crohn's colitis and the associated changes in fecal microbial profile. J Clin Gastroenterol 2014;48(7):625–8.

[103] Angelberger S, Reinisch W, Makristathis A, et al. Temporal bacterial community dynamics vary among ulcerative colitis patients after fecal microbiota transplantation. Am J Gastroenterol 2013;108(10):1620–30.

[104] Kump PK, Grochenig HP, Lackner S, et al. Alteration of intestinal dysbiosis by fecal microbiota transplantation does not induce remission in patients with chronic active ulcerative colitis. Inflamm Bowel Di 2013;19(10):2155–65.

[105] Kunde S, Pham A, Bonczyk S, et al. Safety, tolerability, and clinical response after fecal transplantation in children and young adults with ulcerative colitis. J Pediatr Gastroenterol Nutr 2013;56(6):597–601.

[106] Damman CBM, Hayden H, et al. Single colonoscopically administered fecal microbiota transplant for ulcerative colitis—a pilot study to determine therapeutic benefit and graft stability (abstr). Gastroenterology 2014;146:S-460.

[107] Vaughn BPGD, Ting A, et al. Fecal microbiota transplantation induces early improvement in symptoms in patients with active Crohn's disease (abstr). Gastroenterology 2014;146. S-591-S2.

[108] Vermeire SJM, Verbeke K, et al. Pilot study on the safety and efficacy of faecal microbiota transplantation in refractory Crohn's disease (abstr). Gastroenterology 2012;142:S-360.

[109] Colman RJ, Rubin DT. Fecal microbiota transplantation as therapy for inflammatory bowel disease: a systematic review and meta-analysis. J Crohn Colitis 2014;8(12):1569–81.

[110] Moayyedi P, Surette MG, Kim PT, et al. Fecal microbiota transplantation induces remission in patients with active ulcerative colitis in a randomized controlled trial. Gastroenterology 2015;149(1). 102-9 e6.

[111] De Leon LM, Watson JB, Kelly CR. Transient flare of ulcerative colitis after fecal microbiota transplantation for recurrent *Clostridium difficile* infection. Clin Gastroenterol Hepatol 2013;11(8): 1036–8.

[112] Kelly CZH, Kahn S. New diagnosis of Crohn's colitis 6 weeks after fecal microbiota transplantation (abstr). Inflamm Bowel Dis 2014;20:S21.

[113] Lovell RM, Ford AC. Global prevalence of and risk factors for irritable bowel syndrome: a meta-analysis. Clin Gastroenterol Hepatol 2012;10(7). 712-21 e4.

[114] Lovell RM, Ford AC. Effect of gender on prevalence of irritable bowel syndrome in the community: systematic review and meta-analysis. Am J Gastroenterol 2012;107(7):991–1000.

[115] Chey WD, Kurlander J, Eswaran S. Irritable bowel syndrome: a clinical review. JAMA 2015;313(9):949–58.

[116] Simren M, Barbara G, Flint HJ, et al. Intestinal microbiota in functional bowel disorders: a Rome foundation report. Gut 2013;62(1):159–76.

[117] Matto J, Maunuksela L, Kajander K, et al. Composition and temporal stability of gastrointestinal microbiota in irritable bowel syndrome—a longitudinal study in IBS and control subjects. FEMS Immunol Med Microbiol 2005;43(2):213–22.

[118] Jalanka-Tuovinen J, Salojarvi J, Salonen A, et al. Faecal microbiota composition and host-microbe cross-talk following gastroenteritis and in postinfectious irritable bowel syndrome. Gut 2014;63(11):1737–45.

[119] Codling C, O'Mahony L, Shanahan F, et al. A molecular analysis of fecal and mucosal bacterial communities in irritable bowel syndrome. Digest Dis Sci 2010;55(2):392–7.

[120] Carroll IM, Ringel-Kulka T, Siddle JP, et al. Alterations in composition and diversity of the intestinal microbiota in patients with diarrhea-predominant irritable bowel syndrome. Neurogastroenterol Motil 2012;24(6):521–30. e248.

[121] Mayer EA, Savidge T, Shulman RJ. Brain-gut microbiome interactions and functional bowel disorders. Gastroenterology 2014;146(6):1500–12.

[122] Kassinen A, Krogius-Kurikka L, Makivuokko H, et al. The fecal microbiota of irritable bowel syndrome patients differs significantly from that of healthy subjects. Gastroenterology 2007;133(1): 24–33.

[123] Parkes GC, Rayment NB, Hudspith BN, et al. Distinct microbial populations exist in the mucosa-associated microbiota of sub-groups of irritable bowel syndrome. Neurogastroenterol Motil 2012;24(1):31–9.

[124] Rajilic-Stojanovic M, Biagi E, Heilig HG, et al. Global and deep molecular analysis of microbiota signatures in fecal samples from patients with irritable bowel syndrome. Gastroenterology 2011;141(5):1792–801.

[125] Zoller V, Laguna AL, Prazeres Da Costa O, et al. Fecal microbiota transfer (FMT) in a patient with refractory irritable bowel syndrome. Deutsche Med Wochenschrift 2015;140(16):1232–6.

[126] Rossen NG, MacDonald JK, de Vries EM, et al. Fecal microbiota transplantation as novel therapy in gastroenterology: a systematic review. World J Gastroenterol 2015;21(17):5359–71.

[127] Lyra A, Rinttila T, Nikkila J, et al. Diarrhoea-predominant irritable bowel syndrome distinguishable by 16S rRNA gene phylotype quantification. World J Gastroenterol 2009;15(47):5936–45.

[128] Ringel-Kulka T, Benson AK, Carroll IM, et al. Molecular characterization of the intestinal microbiota in patients with and without abdominal bloating. Am J Physiol Gastrointest Liver Physiol 2015;23:00044.

[129] Ley RE, Turnbaugh PJ, Klein S, et al. Microbial ecology: human gut microbes associated with obesity. Nature 2006;444(7122):1022–3.

[130] Turnbaugh PJ, Ley RE, Mahowald MA, et al. An obesity-associated gut microbiome with increased capacity for energy harvest. Nature 2006;444(7122):1027–31.

[131] Backhed F, Manchester JK, Semenkovich CF, et al. Mechanisms underlying the resistance to diet-induced obesity in germ-free mice. Proc Natl Acad Sci USA 2007;104(3):979–84.

[132] Cani PD, Amar J, Iglesias MA, et al. Metabolic endotoxemia initiates obesity and insulin resistance. Diabetes 2007;56(7):1761–72.

[133] Teixeira TF, Collado MC, Ferreira CL, et al. Potential mechanisms for the emerging link between obesity and increased intestinal permeability. Nutr Res 2012;32(9):637–47.

[134] Vrieze A, Van Nood E, Holleman F, et al. Transfer of intestinal microbiota from lean donors increases insulin sensitivity in individuals with metabolic syndrome. Gastroenterology 2012;143(4). 913-6 e7.

[135] Kemmer N, Neff GW, Franco E, et al. Nonalcoholic fatty liver disease epidemic and its implications for liver transplantation. Transplantation 2013;96(10):860–2.

[136] Hashimoto E, Taniai M, Tokushige K. Characteristics and diagnosis of NAFLD/NASH. J Gastroenterol Hepatol 2013;28(Suppl. 4):64–70.

[137] Raman M, Ahmed I, Gillevet PM, et al. Fecal microbiome and volatile organic compound metabolome in obese humans with nonalcoholic fatty liver disease. Clin Gastroenterol Hepatol 2013;11(7):868–75. e1–e3.

[138] Spencer MD, Hamp TJ, Reid RW, et al. Association between composition of the human gastrointestinal microbiome and development of fatty liver with choline deficiency. Gastroenterology 2011;140(3):976–86.

[139] Mouzaki M, Comelli EM, Arendt BM, et al. Intestinal microbiota in patients with nonalcoholic fatty liver disease. Hepatology 2013;58(1):120–7.

[140] Ley RE, Backhed F, Turnbaugh P, et al. Obesity alters gut microbial ecology. Proc Natl Acad Sci USA 2005;102(31):11070–5.

[141] Ferolla SM, Armiliato GN, Couto CA, et al. The role of intestinal bacteria overgrowth in obesity-related nonalcoholic fatty liver disease. Nutrients 2014;6(12):5583–99.

[142] Zhu L, Baker SS, Gill C, et al. Characterization of gut microbiomes in nonalcoholic steatohepatitis (NASH) patients: a connection between endogenous alcohol and NASH. Hepatology 2013;57(2):601–9.

[143] Ma YY, Li L, Yu CH, et al. Effects of probiotics on nonalcoholic fatty liver disease: a meta-analysis. World J Gastroenterol 2013;19(40):6911–8.

[144] Cowan TE, Palmnas MS, Yang J, et al. Chronic coffee consumption in the diet-induced obese rat: impact on gut microbiota and serum metabolomics. J Nutr Biochem 2014;25(4):489–95.

[145] Cheng J, Joyce A, Yates K, et al. Metabolomic profiling to identify predictors of response to vitamin E for non-alcoholic steatohepatitis (NASH). PloS One 2012;7(9):e44106.

[146] Ochoa-Reparaz J, Mielcarz DW, Ditrio LE, et al. Central nervous system demyelinating disease protection by the human commensal Bacteroides fragilis depends on polysaccharide A expression. J Immunol 2010;185(7):4101–8.

[147] Borody TJLS, Campbell J, Torres M, Nowak A. Fecal microbiota transplantation (FMT) in multiple sclerosis (MS) [abstract]. Am J Gastroenterol 2011;106:S352.

[148] Ananthaswamy A. Faecal transplant eases symptoms of Parkinson's disease. New Sci 2011;209:8–9.

[149] Finegold SM, Molitoris D, Song Y, et al. Gastrointestinal microflora studies in late-onset autism. Clin Infect Dis 2002;35(Suppl 1): S6–S16.

[150] Aroniadis OC, Brandt LJ. Fecal microbiota transplantation: past, present and future. Curr Opin Gastroenterol 2013;29(1): 79–84.

Chapter 35

Probiotics in the Treatment of Pouchitis

M. Guslandi

Clinical Hepato-Gastroenterology Unit, Division of Gastroenterology & Digestive Endoscopy, S. Raffaele University Hospital, Milan, Italy

INTRODUCTION

The surgical treatment of choice in patients with ulcerative colitis (UC) either refractory to medical treatment or showing histological signs of dysplasia as well as in subjects with familial adenomatous polyposis is restorative proctocolectomy with ileal pouch–anal anastomosis (IPAA) [1,2]. Inflammation of the ileal reservoir (pouch) is called pouchitis and develops only in patients who undergo the surgical procedure because of UC. The incidence of pouchitis occurs in 24–60% of cases [2–6] with a tendency to recur in about 60% of cases, of whom up to 32% will develop chronic pouchitis, which, in turn, may lead to pouch excision in 10% of cases [2,3,5,7].

Symptoms include rectal bleeding, increased stool frequency often with urgency, tenesmus and/or incontinence, and occasionally general systemic symptoms, such as fever, fatigue, and arthralgia [4,8].

Endoscopic features are represented by mucosal reddening, contact bleeding, petechiae, erosions, or frank ulcerations, while histological examination shows inflammatory infiltrate, crypt hyperplasia and/or abscesses, and villous atrophy.

A pouchitis disease activity index (PDAI) based on both clinical symptoms as well as endoscopic and histological findings has been developed [9].

The etiology of pouchitis remains poorly understood. However, due to its occurrence only after the diverting ileostomy is closed and to the good response to antibiotic treatment, a major role of the gut microbiota can be postulated. The bacterial population, which appears to be normal in the healthy pouch has been shown in various studies to be altered in pouchitis, with a decrease in anaerobes, bifidobacteria and lactobacilli [10,11], and Firmicutes [12], and high levels of *Clostridium perfringens* [10,11] and Proteobacteria [12] as well as a rise in local pH [10] which makes the protective mucus layer of the reservoir more prone to bacterial degradation. Only a few studies failed to observe changes in the bacterial population [13,14] or alterations in bacterial diversity [15,16].

Antibacterial agents, such as metronidazole and/or ciprofloxacin are widely and successfully employed in the treatment of acute pouchitis.

PROBIOTICS IN POUCHITIS

Considering the postulated role of intestinal bacteria in the development of pouchitis, probiotics may provide a safer therapeutic alternative to the use of antibiotics. The reasons for employing probiotics in the management of pouchitis are their various effects: strengthening of barrier function with a decrease in gut permeability, inhibitory activity against pathogenic bacteria, and antiinflammatory effect through modulation of cytokine production [17].

The reduced amount of lactobacilli and the high pH values observed in the pouch [10] suggest a possible benefit of products containing strains of lactobacilli, which are also known to attenuate both spontaneous colitis in interleukin 10 knock-out mice and in experimentally induced colitis [18–21].

Lactobacilli-containing probiotics have been found to reduce the expression of proinflammatory cytokines IL-1beta and IL-8, as well as tissue influx of polymorphonuclear cells in subjects given the products to prevent the development of pouchitis [22].

Similarly probiotic supplementation with a multispecies product containing various strains of lactobacilli and bifidobacteria was able to reduce transmucosal passage of *Escherichia coli* K12 and mucosal permeability as assessed by horseradish peroxidase [23,24].

Primary Prevention of Pouchitis

An uncontrolled, retrospective study compared *Lactobacillus rhamnosus* GG with an untreated control group [25]. The incidence of pouchitis at 3 years was 7% and 29%, respectively ($p = 0.011$).

Two trials [26,27] have been carried out with the probiotic mixture VSL#3, which contains four strains of

The Microbiota in Gastrointestinal Pathophysiology

327

lactobacilli (*Lactobacillus casei, Lactobacillus plantarum, Lactobacillus. acidophilus, and Lactobacillus bulgaricus*), three strains of bifidobacteria (*Bifidobacterium longum, Bifidobacterium breve, and Bifidobacterium infantis*), and *Streptococcus salivarius* subsp. *thermophilus*. In a double-blind, placebo-controlled study versus placebo [26], the product was administered for 12 months to 40 IPPA patients at a dose of 1 sachet/day (i.e., 900 billion bacteria), with development of pouchitis in only 10% of cases compared with 40% in the placebo group ($p < 0.05$).

In a randomized trial versus no treatment enrolling 31 IPPA patients [27], only subjects receiving VSL#3 showed at 3, 6, and 12 months a significant reduction in the values of PDAI.

A randomized, prospective study versus placebo employing a different probiotic mixture (*L. acidophilus, L. bulgaricus* and *Bifidobacterium bifidus*) for 9 months in 43 patients recently submitted to restorative proctocolectomy found a significant reduction in the number of patients with pouchitis, as well as in PDAI scores and levels of fecal calprotectin in the probiotic group [28]. The trial was enrolling both patients without pouchitis (primary prevention) and subjects with active pouchitis.

Recently, a very small, pilot study versus placebo from Japan has shown promising results in the primary prevention of pouchitis with *Clostridium butyricum* MIYARI [29] over a follow-up period of 24 months.

Treatment of Active Pouchitis

L. rhamnosus GG given for 3 months has been compared with placebo in a randomized double-blind trial [23]. The probiotic agent changed the pouch intestinal flora by increasing the ratio of total fecal lactobacilli to total fecal anaerobes, but was found unable to decrease the values of the PDAI index.

Fermented milk containing lactobacilli and bifidobacteria (Cultura) was administered for 4 weeks to IPAA patients with either UC or FAP for 4 weeks. A significant improvement of clinical and endoscopic scores was observed in IPPA/UC patients [30].

A pilot, uncontrolled study has suggested a possible benefit by *E. coli* strain Nissle 1917 in inducing and maintaining remission of pouchitis [31], but definitive data are still lacking.

After exhibiting favorable results in preventing recurrence of pouchitis (see later), the probiotic cocktail VSL#3

has been also tested for its ability to promote remission in mildly active pouchitis [32]. In a pilot trial, the product when administered for 4 weeks in high doses (2 sachets b.i.d., i.e., 3600 billion bacteria/day) induced a significant ($p < 0.01$) reduction in the median PDAI values, and complete remission in 69% of cases. A subsequent prolonged treatment with VSL#3 one sachet b.i.d. for 6 months maintained remission in all healed subjects [32].

Maintenance Treatment

The efficacy of VSL#3 in maintaining antibiotic-induced remission of pouchitis has been repeatedly investigated [33–35].

Patients whose pouchitis in clinical and endoscopic remission after 1 month of ciprofloxacin 1 g/day plus rifaximin 2 g/day were randomized to receive for 9 months either VSL#3 1 sachet b.i.d. (1800 billion bacteria/day) or placebo. All patients in the placebo group relapsed, whereas 85% of the probiotic-treated patients remained in remission [33].

Similar results have been obtained in a placebo-controlled randomized study where patients in remission after a 4-week treatment with metronidazole 500 mg b.i.d and ciprofloxacin 500 mg b.i.d. were shifted to a daily dose of VSL#3 (1800 billion bacteria) [34]. After 12 months, 85% of patients remained in remission, compared with only 6% in the placebo-treated subjects ($p < 0.0001$).

Oddly enough, in a small open-label study employing VSL as a maintenance treatment for 8 months after ciprofloxacin-induced remission of pouchitis, 80.6% of patients dropped out mostly because recurrence of symptoms [35]. Different type and length of the initial antibiotic therapy and concomitant NSAID intake, as well as inadequate compliance due to the cost of the probiotic product, may explain the discordant findings reported in that study [35].

CONCLUSIONS

The number of controlled studies on the use of probiotics in the management of pouchitis is still limited. Most of the available data concern the probiotic mixture VSL#3 which, in placebo-controlled trials, proved to be significantly effective in preventing development of pouchitis in IPAA patients and in maintaining antibiotic-induced remission of pouchitis for up to 1 year (Table 35.1).

TABLE 35.1 Randomized Placebo-Controlled Trials of VSL#3 in Pouchitis

Author/Year (References)	Duration (Months)	Product	Protocol	% Response	*p* Value
Gionchetti et al., 2003 [26]	12	VSL#3	Primary prevent.	90% (PL 60%)	<0.05
Gionchetti et al., 2000 [33]	9	VSL#3	Maintenance	85% (PL 0%)	<0.001
Mimura et al., 2004 [34]	12	VSL#3[a]	Maintenance	85% (PL 6%)	<0.0001

PL, Placebo.
[a]*High dose (1.8 g daily).*

As for the possible efficacy of probiotics in promoting remission of pouchitis, both VSL#3 and Cultura (a fermented milk containing lactobacilli and bifidobacteria) have provided promising results, but further controlled trials are needed to confirm the preliminary data.

A metaanalysis, pooling the results of the clinical studies where probiotics have been compared with placebo in order to either induce and maintain remission of pouchitis or to prevent its development, has observed a significantly superior effect ($p < 0.0001$) of probiotics [36].

A second metaanalysis reached similar conclusions by showing that probiotics are significantly more effective than placebo in reducing the incidence of pouchitis and in maintaining remission [37]. According to a third metaanalysis [38], this statement would appear to be true only for VSL#3 in the prevention of relapses of pouchitis.

All things considered, the use of probiotics in the management of pouchitis seems to be promising and further controlled trials are warranted, so much so because probiotics are much better tolerated than antibiotics, especially in long-term treatment.

In particular, on the basis of the available evidence, some probiotics are useful only in the management of either antibiotic-responsive or antibiotic-dependent pouchitis, whereas in cases refractory to antibiotic therapy, where the role of the gut microbiota is not pivotal, other therapeutic options should be employed.

REFERENCES

[1] Johnson E, Carlsen E, Nazir M, Nygaard K. Morbidity and functional outcome after restorative proctocolectomy for ulcerative colitis. Eur J Surg 2007;167:40–5.

[2] Fazio VW, Ziv Y, Church JM, et al. Ileal pouch-anal anastomoses complications and function in 1005 patients. Ann Surg 1995;222:120–7.

[3] Meagher AP, Farouk R, Dozois M, et al. J-ileal pouch-anal anastomosis for chronic ulcerative colitis: complications and long-term outcome in 1310 patients. Br J Surg 1998;85:800–3.

[4] Sandborn WJ. Pouchitis following ileal pouch-anal anastomosis: definition, pathogenesis and treatment. Gastroenterology 1994;107:1856–60.

[5] Pardi DS, Sandborn WJ. Management of pouchitis. Aliment Pharmacol Ther 2006;23:1087–96.

[6] Simchuck E, Thirlby R. Risk factors and true incidence of pouchitis in patients after ileal pouch-anal anastomosis. World J Surg 2000;24:851–6.

[7] Mowschenson PM, Critchlow JF, Peppercorn MA. Ileo-anal pouch operation: long-term outcome with or without diverting ileostomy. Arch Surg 2000;135:463–6.

[8] Madden MV, Farthing MJ, Nicholls RJ. Inflammation in ileal reservoirs: 'pouchitis'. Gut 1990;31:247–9.

[9] Sandborn WJ, Tremaine WJ, Batts KP, et al. Pouchitis after ileal pouch-anal anastomosis: a Pouchitis Disease Activity Index. Mayo Clin Proc 1994;69:409–15.

[10] Ruseler-van Embden JG, Schouten WR, van Lieshout LM. Pouchitis: result of microbial imbalance? Gut 1994;35:658–64.

[11] Iwaya A, Iiai T, Okamoto H, et al. Change in the bacterial flora of pouchitis. Hepatogastroenterology 2006;53:55–9.

[12] Komanduri S, Gillevet PM, Sikaroodi M, et al. Dysbiosis in pouchitis: evidence of unique microfloral patterns in pouch inflammation. Clin Gastroenterol Hepatol 2007;5:352–60.

[13] Kmiot WA, Youngs D, Tudor R, et al. Mucosal morphology, cell proliferation and faecal bacteriology in acute pouchitis. Br J Surg 1993;80:1445–9.

[14] McLauglin SD, Walker AW, Churcer C, et al. The bacteriology of pouchitis: a molecular phylogenetic analysis using 16S rRNA gene cloning and sequencing. Ann Surg 2010;252:90–8.

[15] Kuchbacher T, Ott SJ, Helwig U, et al. Bacterial and fungal microbiota in relation to probiotic therapy (VSL#3) in pouchitis. Gut 2006;55:833–41.

[16] Tannock GW, Lawley B, Munro K, et al. Comprehensive analysis of the bacterial content of stool from patients with chronic pouchitis, normal pouches or familial adenomatous polyposis pouches. Inflamm Bowel Dis 2011;18:925–34.

[17] Jones JL, Foxx-Orenstein AE. The role of probiotics in inflammatory bowel disease. Dig Dis Sci 2007;52:607–11.

[18] Schultz M, Veltkamp C, Dieleman LA, et al. *Lactobacillus plantarum* 299Vs in the treatment and prevention if spontaneous colitis in interleukin-10-deficient mice. Inflamm Bowel Dis 2002;8:71–80.

[19] Madsen KL, Doyle J, Jewell LD, et al. *Lactobacillus* species prevents colitis in interleukin 10-gene deficient mice. Gastroenterology 1999;116:1107–14.

[20] Fabia R, Ar'Rajab A, Johansson ML. The effect of exogenous administration of *Lactobacillus reuteri* R2LC and oat fiber on acetic acid-induced colitis in the rat. Scand J Gastroenterol 1993;28:155–62.

[21] Mao Y, Nobaek S, Kasravi B, et al. The effects *of Lactobacillus* strains and oat fiber on methotrexate-induced enterocolitis in rats. Gastroenterology 1996;111:334–44.

[22] Lammers KM, Vergopoulos A, Babel N, et al. Probiotic therapy in the prevention of pouchitis onset: decreased interleukin-1Beta, interleukin-8 and interferon-gamma gene expression. Inflamm Bowel Dis 2005;11:447–54.

[23] Kujisma J, Mentula S, Jarvinen H, et al. Effect of *Lactobacillus rhamnosus* GG on ileal pouch inflammation and microbial flora. Aliment Pharmacol Ther 2003;17:509–15.

[24] Persborn M, Gerritsen J, Wallon C, et al. The effects of probiotics on barrier function and mucosal pouch microbiota during maintenance treatment for severe pouchitis in patients with ulcerative colitis. Aliment Pharmacol Ther 2013;38:772–3.

[25] Elahi B, Nifkar S, Derakshani S, et al. On the benefit of probiotics in the management of pouchitis in patients underwent ileal pouch anal anastomosis: a meta-analysis of controlled clinical trials. Dig Dis Sci 2008;53:1278–84.

[26] Gionchetti P, Rizzello F, Helwig U, et al. Prophylaxis of pouchitis onset with prebiotic therapy: a double-blind, placebo-controlled trial. Gastroenterology 2003;124:1202–9.

[27] Pronio A, Montesani C, Butteroni C, et al. Probiotic administration in patients with ileal-pouch-anal anastomosis for ulcerative colitis is associated with expansion of mucosal regulatory cells. Inflamm Bowel Dis 2008;14:662–8.

[28] Tomasz B, Zoran S, Jaroslaw W, et al. Long-term use of probiotics *Lactobacillus* and *Bifidobacterium* has a prophylactic effect on the occurrence and severity of pouchitis; a randomized prospective study. BioMed ResInt 2014;. 208064.

[29] Yasueda A, Mizushima T, Nezu R, et al. The effect of *Clostridium butyricum* MIYAIRI on the prevention of pouchitis and alteration of the microbiota profile in patients with ulcerative colitis. Surg Today 2015;46:939–49.

[30] Laake KO, Bjrneklett A, Aamodt G. Outcome of four weeks' intervention with probiotics on symptoms and endoscopic appearance after surgical reconstruction with a J-configurated ileal-pouch-anal anastomosis in ulcerative colitis. Scand J Gastroenterol 2005;40:43–51.

[31] Kuzela L, Kascak M, Vavrecka A. Induction and maintenance of remission with nonpathogenic *Escherichia coli* in patients with pouchitis. Am J Gastroenterol 2001;96:3218–9.

[32] Gionchetti P, Rizzello F, Morselli C, et al. High-dose probiotics for the treatment of active pouchitis. Dis Colon Rectum 2007;50:2075–84.

[33] Gionchetti P, Rizzello F, Venturi A, et al. Oral bacteriotherapy as maintenance treatment in patients with chronic pouchitis: a double-blind, placebo-controlled trial. Gastroenterology 2000;119:305–9.

[34] Mimura T, Rizzello F, Helwig U, et al. Once daily high dose probiotic therapy (VSL #3) for maintaining remission in recurrent or refractory pouchitis. Gut 2004;53:108–14.

[35] Shen B, Brezinski A, Fazio VW, et al. Maintenance therapy with a probiotic in antibiotic-dependent pouchitis: experience in clinical practice. Aliment Pharmacol Ther 2005;22:721–8.

[36] Elahi B, Nikfar S, Derakhshani S, et al. On the benefit of probiotics in the management of pouchitis in patients underwent ileal pouch anal anastomosis: a meta-analysis of controlled clinical trials. Dig Dis Sci 2008;53:1278–84.

[37] Ashraf I, Sohail U, Arif M, et al. Probiotic use for management of pouchitis: a systematic review and meta-analysis. Am J Gastroenterol 2014;109(Suppl. 2):S677–8.

[38] Shen J, Zuo ZX, Mao AP. Effects of probiotics on inducing remission and maintaining therapy in ulcerative colitis, Crohn's disease and pouchitis: meta-analysis of randomized controlled trials. Inflamm Bowel Dis 2014;20:21–35.

Chapter 36

Probiotic Treatment in Crohn's Disease

K.I. Kroeker and C. Lu

Department of Medicine, Division of Gastroenterology, University of Alberta, Edmonton, AB, Canada

INTRODUCTION

Crohn's disease (CD) is chronic inflammatory disease that affects all areas of the GI tract and is characterized by aphthous ulcers, transmural inflammation, skip lesions, and granuloma formation. Pathogenesis of CD is complex and believed to be due to an interplay between genetic susceptibility, the environment, enteric microflora, and the host immune response. Traditionally, treatment of CD has focused on attempted modification of the host immune response and surgical resection of diseased areas. Smoking cessation is encouraged to modify the environment. Antibiotics, to alter intestinal flora, have been used to treat abscesses, perianal fistula and prevent postoperative recurrence.

Food and naturally occurring probiotics have long been used to promote health and wellness and a broad range of folk and traditional medicinal remedies. However, only recently, probiotics have been considered as potential mainstream therapeutics for inflammatory bowel disease (IBD). This transition is a result of recent, well-designed, clinical trials and research in both experimental systems and animal models. Due to this, anecdotal observations in support of probiotic therapies for IBD are being slowly replaced by experimental insight into the complex interactions of probiotics with the gut microflora, the epithelial lining, and the host's mucosal and systemic immune system. Taken together, this mechanistic understanding is building the foundation to support the randomized controlled clinical trials that are needed for probiotics to enter mainstream medical therapy algorithms in the treatment of IBD.

RATIONALE FOR USING PROBIOTICS IN CROHN'S DISEASE

Numerous studies have identified several mechanisms by which probiotics can have a direct effect on epithelial cell function and intestinal health, including enhancing epithelial barrier function, modulating epithelial cytokine secretion into an antiinflammatory dominant profile, altering mucus production, changing bacterial luminal flora, modifying the innate and systemic immune system, and inducing regulatory T-cell effects (Fig. 36.1).

A series of review articles have been published outlining the efficacy of probiotics in human health [3–14]. The production of antibacterial substances by probiotics inhibits the growth of pathogens. Lactobacilli lower the pH by the production of acetic, lactic, and propionic acid which can inhibit the growth of gram-negative bacteria [15]. Lactic acid produced by *Lactobacillus rhamnosus* GG has been identified as primary antimicrobial compound in the inhibition of growth of *Salmonella typhimurium* [16]. Bacteriocin-like compounds produced by gram-positive bacteria including Bifidobacteria are toxic to gram-positive and gram-negative bacteria [17]. The action of bacteriocins is the formation of pores in the cytoplasmic membrane of certain bacteria [15]. Probiotics also play a role in the inhibition of pathogenic bacteria through competition via decreasing adhesion of bacteria and their toxins. Strains of *Lactobacillus* and *Bifidobacteria* have been shown to bind to intestinal mucus and to inhibit and displace the adhesion of *Bacteroides*, *Clostridium*, *Staphylococcus*, and *Enterobacter* [18]. The demonstration that immune and epithelial cells can discriminate between different microbial species has extended the known mechanism(s) of action of probiotics beyond simple barrier and antimicrobial effects [1].

Aligning the mechanism of action of the individual probiotics with pathogenic mechanisms of the diseases, including CD, will assist clinicians in understanding the action of probiotics, test these hypotheses in randomized controlled clinical trials and eventually, develop a therapeutic algorithm for their effective use in IBD and other disorders.

The Microbial-Related Pathogenesis of Crohn's Disease

Although closer than ever to understanding the pathogenesis of CD, it is acutely apparent that there is a myriad of complex associations between an individual's genetics and their luminal microbial environment. New concepts, developed out of this appreciation, suggest

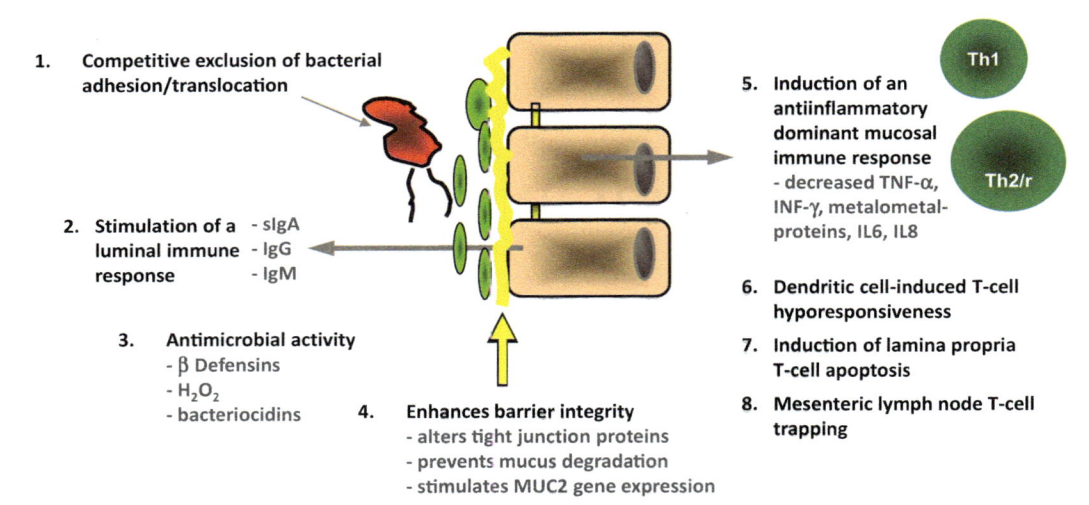

1. **Competitive exclusion of bacterial adhesion/translocation**

2. **Stimulation of a luminal immune response** - sIgA, - IgG, - IgM

3. **Antimicrobial activity**
   - β Defensins
   - H_2O_2
   - bacteriocidins

4. **Enhances barrier integrity**
   - alters tight junction proteins
   - prevents mucus degradation
   - stimulates MUC2 gene expression

5. **Induction of an antiinflammatory dominant mucosal immune response**
   - decreased TNF-α, INF-γ, metalometal-proteins, IL6, IL8

6. **Dendritic cell-induced T-cell hyporesponsiveness**

7. **Induction of lamina propria T-cell apoptosis**

8. **Mesenteric lymph node T-cell trapping**

Th1

Th2/r

FIGURE 36.1 Schematic illustration of the various mechanisms by which probiotics exert therapeutic effects at the cellular level in the intestinal tract [1,2]

the involvement of dysfunctional mucosal and systemic immune cells that are: (1) unable to detect and/or eradicate potentially injurious microbes, (2) detecting normal enteric flora as "foreign" and setting up an inflammatory response that cannot be shut down, and/or (3) altering the enteric microbiota composition thereby disrupting the otherwise platonic relationship with the host's tissues [19]. Appreciation for the transmetabolic functioning of the human intestinal tract is a relatively new concept. Bacterial strain ratios in otherwise healthy humans are being linked to a range of whole-body manifestations, most notably obesity [20]. With respect to IBD, recent studies have identified significant strain variations between healthy and diseased individuals. For instance, Kotlowski et al. determined that patients with either CD or ulcerative colitis (UC) had 3–4 logs greater amounts of Enterobacteriaceae in their tissues compared to healthy individuals ($p = 0.05$) [21]. A recent study has found significant differences regarding both the number and genetic composition of bacteria associated with intestinal biopsies between UC and CD patients [22]. As well, it was found that patients with CD had more bacteria from Bacteroidetes phylum than those with UC. In another study, inflamed and noninflamed ileal mucosa from CD patients were assessed. No significant differences regarding bacteria composition or numbers were found suggesting that the systemic gut's biome is altered as opposed to a localized response [23]. Conversely, a difference in both leukocyte count and the concentration of *Faecalibacterium prausnitzii* have been identified and the stability of variance is such that it could be considered as a noninvasive marker to identify IBD and help discriminate between CD and UC [24]. Furthermore *F. prausnitzii*, has been found to be markedly reduced

in CD patients who have rapid recurrence following surgical resection, again, implying a strong bacterial association between disease and health [25].

Increased intestinal permeability may have an important role in the pathogenesis of CD. Garcia Vilela et al. have demonstrated that in 31 patients with CD in remission have a higher intestinal permeability, measured by lactulose/mannitol ratio, as compared to healthy controls [26]. Administration of *Saccharomyces boulardii* (4×10^8 CFU q8h) resulted in improved intestinal permeability over the 3-month study period [26]. In in vitro experiments of the human intestinal epithelial cell line T84 and enterocyte preparations from the ileal resection specimens of CD patients, *Saccharomyces cerevisiae* CNCM I-3856 inhibited the adhesion of adherent-invasive *Escherichia coli* (AIEC) [27].

Currently, the exact roles of bacteria within the pathogenesis of CD are unknown but emerging evidence confirms that the gut's biome of patients is measurably distinct from that of healthy individuals. Assuming that IBD occurs as a consequence of an altered response to the mucosal microflora, it is conceivable that the introduction of probiotic organisms as therapeutic agents will change the luminal microflora and/or enhance epithelial barrier function, and/or mollify the response of the mucosal and systemic immune systems in such a way as to attenuate the intestinal inflammatory response.

Probiotics and Animal Models of IBD

In order to understand biological processes, it is essential to develop both experimental systems and animal models. Such models have been proven invaluable in the overall advancement of our understanding of IBD; recently, they have been essential to advance the study of probiotics both at the cellular and subcellular levels as well as to identify their localized and systemic effects. As early as 1992, a murine model was used to examine the effect of *Bifidobacterium breve* on antibody production [28]. Since then, other animal models have been used, such as rats, pigs, and cats. The development of animals that are deficient in key proteins that have been associated with IBD have enhanced the research options enabling an increased understanding of the physiology. The first such study explored the ameliorating effect of *Lactobacillus reuteri* enemas administered to interleukin (IL)-10 gene-deficient mice which, when raised under nonsterile conditions, normally develop colitis similar to CD [29]. Since then multiple other gene knockout and transgenic animal models have been used for the study of probiotics.

In studying probiotics for the treatment of IBD, the murine model is the most popular. The majority of these studies assessed the effect of probiotic therapy on various inflammatory markers in animals with colitis induced using either trinitrobenzenesulfonic acid (TNBS) or dextran sulfate sodium (DSS). Mucosal integrity was enhanced by *E. coli* Nissle 1917 via upregulating the junction-associated protein, zonula occludens 1 (ZO-1) in intestinal epithelial cells [30]. In IL-10–deficient mice, secreted bioactive factors from *Bifidobacterium infantis* have been shown to attenuate inflammation, normalize colonic permeability, and decrease IFN-γ secretion [31]. The commercial probiotic mixture, VSL#3, ameliorated colitis through immune modulation involving colonic inhibitory κβ-α, IL-1β, and myeloperoxidase.

Optimum dosing of the probiotic mixture to attenuate colitis was also explored; this was also a concern in previous human trials as dose-finding studies with probiotics is very limited [32]. Kim et al. presented convincing evidence that *Bifidobacterium bifidum* BGN4 suppressed cytokine production in colitis-induced mice concluding that it should help restore dysfunctional immune responses in intestinal tissues [33]. In a comparative study, *Lactobacillus casei*, *Lactobacillus acidophilus*, and *Bifidobacterium lactis* yielded similar therapeutic benefits in colitis-induced rats but each species had its own antiinflammatory profile [34]. With respect to *L. casei*, Chung et al. found that this probiotic modulates the expression of inflammatory cytokines and is dependent upon the signaling of the Toll-like receptor (TLR)-4 complex determined by using TLR-4 mutant mice [35]. The effect of two putative probiotics on ameliorating inflammation in colitis-induced murine models were investigated. *Enterococcus durans*, isolated from healthy human feces, decreased the colonic cDNA concentration of IL-1β and tumor necrosis factor (TNF)-α suggesting that further trials are warranted to explore possible therapeutic benefits for UC patients [36]. A comparison of the antiinflammatory properties of a previously studied probiotic, *L. reuteri*, and *Lactobacillus fermentum* (both of which were isolated from breast milk) concluded that the ability of *L. fermentum*

to significantly lower TNF-α relative to *L. reuteri* suggests that it possesses immunomodulatory properties beneficial to humans and merits additional study [36].

Further evidence of the role of intestinal permeability in the pathogenesis of CD is the recent study of gut colonization of AIEC LF82 in a transgenic mouse model treated with DSS to induce colitis. In this study, *S. cerevisiae* CNCM I-3856 decreased the AIEC colonization and colitis severity. This was via a restoration in barrier function as evidenced by claudin-2 immunofluorescence staining [27].

PROBIOTIC AGENTS IN THE TREATMENT OF CROHN'S DISEASE

An overview of the recommendations for the use of probiotics in the treatment of CD is summarized in Table 36.1.

The individual results of both open-label and randomized controlled clinical trials investigating probiotic use in treating humans with CD are discussed following sections and summarized in Table 36.2.

Induction of Crohn's Disease Remission

There are limited data available on the treatment of active CD with probiotics. To date there have only been a few open-label studies published that include only 50 patients. An open-label trial by Fujimori et al. examined the effect of synbiotic therapy in 10 CD patients with active disease, primarily diarrhea and abdominal pain, refractory to aminosalicylates and prednisolone [47]. After 13 months (±4.5), two of the six responders successfully discontinued steroid therapy while four were able to decrease their dosages. Values for both the Crohn's disease activity index (CDAI) and the International Organization for the Study of Inflammatory Bowel Disease (IOIBD) index were significantly reduced (255–136, $p = 0.009$ and 3.5–2.1, $p = 0.03$, respectively). Concomitant therapy with prednisolone, aminosalicylate, and home enteral nutrition was allowed.

Similarly, in another study with a synbiotic comprised of *B. longum* and Synergy 1 (inulin and oligofructose), the CDAI of 35 patients with active CD also significantly improved pre- and posttreatment ($p = 0.0020$). This was the first randomized, double-blind placebo-controlled trial to show benefit of synbiotic therapy in active CD patients. In addition to CDAI, rectal biopsies from these patients were collected at the beginning of the study, and at 3- and 6-month intervals where significant improvement in histological scores over time were also obtained ($p = 0.018$) [49]. Interestingly, this study also demonstrated that with synbiotic consumption, there was a significant reduction in expression of TNF-α, a proinflammatory cytokine, but no meaningful change in transcription levels of other proinflammatory immune markers (IL-18, IFN-γ, and IL-1β). Likewise, Borruel et al. has shown that probiotics with

TABLE 36.1 Recommendation for Probiotic Use in Crohn's Disease (CD)

	Level of Recommendation	Organism	Dose (Daily)	References
Induction of Remission	C	Synbiotic mixture		[36]
		B. breve	3×10^{10} CFU	
		Bifidobacterium longum	1.5×10^{10} CFU	
		L. casei	3×10^{10} CFU	
	C	Lactobacillus salivarius	1×10^{10} CFU	[37]
	C	Lactobacillus GG	2×10^{10} CFU	[38]
Maintenance of medically induced remission	Not recommended (B)	L. rhamnosus GG	4×10^{10} CFU	[39]
	B	S. boulardii (with mesalamine 2 g/d)	1 g	[40]
	Not recommended (B)	E. coli Nissle 1917	5×10^{10} CFU	[41]
Maintenance of surgically induced remission	A	VSL#3**	1 sachet BID (3×10^{11} CFU)	[42,43]
	Not recommended (B)	Lactobacillus johnsonii LA1, Nestle	2×10^{9}–1×10^{10} CFU	[44,45]
	Not recommended (B)	Synbiotic 2000*	Single dose	[46]

Levels of Recommendation are based upon The UK National Health Service categories.

- Level A: Consistent Randomized Controlled Clinical Trial, cohort study, clinical decision rule validated in different populations.
- Level B: Consistent Retrospective Cohort, Exploratory Cohort, Ecological Study, Outcomes Research, case–control study; or extrapolations from Level A studies.
- Level C: Case-series study or extrapolations from Level B studies.
- Level D: Expert opinion without explicit critical appraisal, or based on physiology, bench research or first principles.

*Synbiotic 2000, commercial product containing 1×10^{10} CFU each of Pediococcus pentosaceus, L. raffinolactis, Lactobacillus paracasei subsp. Paracasei 19, Lactobacillus plantarum 2362 and fermentable fibers (prebiotic component) consisting of 2.5 g of each of β-glucans, inulin, pectin, resistant starch.
**VSL#3, commercial product containing B. longum, B. infantis, B. breve. L. acidophilus, L. casei, Lactobacillus delbrueckii subsp. bulgaricus, L. plantarum, and Streptococcus salivarius subsp. thermophilus.

TABLE 36.2 Summary of Studies Investigating the Effect of Probiotic Treatment on the Induction and Maintenance of Remission in CD

1st Author Date	Design Duration	Group (Dose/Day) Probiotic	Comparator	Concomitant Therapy	Results
Induction of remission					
Fujimori [47] 2007	O 13 ± 4.5 months	B. breve (3×10^{10} CFU), L. casei (3×10^{10} CFU), B. longum (1.5×10^{10} CFU), Psyllium (9.9 g) n = 10	None	Aminosalicylate, prednisolone, home enteral nutrition	Improved CDAI and IOIBD scores compared with baseline (255–136, $p = 0.009$ and 3.5–2.1, $p = 0.03$, respectively). 60% (6/10) achieved remission
Schultz [48] 2004	DB, R, C 6 months	Lactobacillus GG (2×10^{9} CFU) n = 5	Placebo n = 6	Ciprofloxacin, metronidazole, corticosteroids	No difference in remission rates
McCarthy [38] 2001	Open-label 6 weeks	L. salivarius (1×10^{10} CFU) n = 25	None	Not listed	Reduced disease activity compared with baseline
Gupta [41] 2000	Open-label 6 months	Lactobacillus GG (2×10^{10} CFU) n = 4	None	Predisone, immunomodulatory agents, metronidazole	Improvement in CDAI scores compared with baseline ($p < 0.05$)

TABLE 36.2 Summary of Studies Investigating the Effect of Probiotic Treatment on the Induction and Maintenance of Remission in CD (*cont.*)

1st Author Date	Design Duration	Group (Dose/Day) Probiotic	Comparator	Concomitant Therapy	Results
Steed [49] 2010	Single center, DB, R, C Start, 3 and 6 months	*Synbiotic* with *B. longum* and Synergy 1 *N* = 35	Placebo	Stable doses of CD medication at initiation of trial	Significant improvement CDAI ($p = 0.020$) and histological scores ($p = 0.018$) Decreased TNF-α levels from rectal mucosa at 3 months ($p = 0.041$)
Bourreile [37] 2013	DB, R, C 52 weeks	*S. boulardii*, *N* = 165	Placebo	Weaned off corticosteroids or aminosalicylates by 12 weeks	No difference in CDAI, ESR, or CRP
Maintenance of medically induced remission					
Bousvaros [50] 2005	DB, R, C 2 year	*L. rhamnosus* GG (4×10^{10} CFU) *n* = 39	Placebo *n* = 36	Aminosalicylates, 6-MP, azathioprine, corticosteroids	NSD in relapse rate or time to relapse
Guslandi [51] 2000	R, C 6 months	*S. boulardii* (1 g/day) + mesalamine (2 g) *n* = 16	Mesalamine (3 g/day) *n* = 16	Not listed	Significant prolongation of remission ($p < 0.05$)
Malchow [52] 1997	DB, R, C 1 year	*E. coli* Nissle 1917 (5×10^{10} CFU) *n* = 16	Placebo *n* = 12	Prednisolone	No difference in remission rates
Maintenance of surgically induced remission					
Fedorak [42] 2015	DB, R, C 90 days	VSL#3 *n* = 59	Placebo *n* = 60	None	Reduced IL-1β, TNF-α, IFN-γ and increased TGF-β ($p < 0.05$)
Chermesh [45] 2007	DB, R, C 2 years	Synbiotic 2000 (single dose) *n* = 7	Placebo *n* = 2	Not indicated	NSD regarding postoperative recurrence of CD High dropout rate (*n* = 21)
Van Gossum [44] 2007	DB, R, C 12 weeks	*L. johnsonii* LA1, Nestle (1×10^{10} CFU) *n* = 27	Placebo *n* = 22	None	NSD for endoscopic score postileocecal resection High dropout rate/violations (*n* = 21)
Marteau [40] 2006	DB, R, C 6 months	*L. johnsonii* LA1, Nestle (2×10^{9} CFU) *n* = 43	Placebo *n* = 47	Loperamide, cholestyramine, corticosteroids tapered to nil by week 3	NSD for endoscopic scores Dropout rate (*n* = 8)
Prantera [53] 2002	DB, R, C 1 year	*Lactobacillus* GG (1.2×10^{10} CFU) (*n* = 23)	Placebo *n* = 22	Loperamide, cholestyramine	No significant difference in remission
Campieri [54] 2000	R, C 1 year	VSL#3 (3×10^{11} CFU) (*n* = 20)	Mesalamine (4 g/day) (*n* = 20)	None listed	Equivalent to mesalamine in preventing recurrence

Abbreviations: C, controlled trial; CDAI, Crohn's disease activity index; CFU, colony forming units; DB, double-blind; CRP, C-reactive protein IOIBD, International Organization for the Study of Inflammatory Bowel Disease; NSD, no significant difference; O, open-label; R, randomized.

L. casei and *Lactobacillus bulgaricus* can also downregulate TNF-α production in inflamed ileal specimens from CD patients ex vivo, and have no effect on inducing negative changes in noninflamed mucosa [55].

With respect to *Lactobacilli*, there have been several studies investigating this probiotic in induction of CD remission. In a small, double-blinded trial of 11 patients, Schultz et al. treated CD patients with antibiotics and a tapering course of glucocorticoids [48]. When the antibiotics were stopped at 2 weeks, the subjects were randomized to receive either *Lactobacillus* GG (2×10^{9} CFU/day) or placebo. The study found no difference in remission rates between the groups (4/5 *Lactobacillus* GG vs. 5/6 placebo). This continues to be one of the two studies [37] comparing a probiotic to placebo in a randomized controlled trial for induction of remission of active CD.

In an open-label uncontrolled trial of 25 patients with mild/moderately active CD taking 5-ASA, McCarthy et al. showed that oral administration of *L. salivarius* UCC118 (1×10^{10} CFU/day for 6 weeks) caused a significant drop in disease activity (217 vs. 150 at the end of the trial; $p < 0.05$) and kept 70% of the study group off of steroids for 3 months [38].

Gupta et al. conducted a 6-month open-label trial using *Lactobacillus* GG (2×10^{10} CFU/day for 6 months) to treat four children with mild to moderate CD [41]. All patients were also receiving concomitant prednisone and immunomodulatory agent therapy. A significant improvement in pediatric CDAI scores ($p < 0.05$) were noted as early as 1 week after starting therapy. This improvement was sustained throughout the study and for three patients it was possible to taper their dose of steroids.

Overall, in one of the most recent metaanalyses from 2013 which included 12 randomized trials studying remission induction in active IBD, only 3 studies with CD met inclusion criteria [48,49,52]. Subgroup analyses for CD showed no significant benefit with probiotics for inducing remission or response in active disease ($p = 0.35$, RR = 0.89) [56]. Taken as a whole, due to insufficient data, probiotics cannot be recommended for use to induce remission in patients with active CD.

In addition to probiotic therapy for the treatment of CD, prebiotics have also been studied. To our knowledge, only two randomized controlled studies have used prebiotics for therapy in CD. Using fructooligosaccharides, a study by Benjamin et al. assessed the CDAI of 103 patients (49 controls, 54 intervention group) showing no clinical benefit at the end of 4 weeks of therapy [57]. Similarly, in a study using lactulose as a prebiotic, no improvement in CDAI was observed for this group [39]. Interestingly, both prebiotic studies found a better response in the placebo groups.

Only one randomized placebo-controlled study has evaluated the use of synbiotics in the induction of remission in CD [49]. Steed et al.'s study showed a significant improvement in CDAI ($p = 0.02$) and improvement in histological scores ($p = 0.018$) at 6 months in the synbiotic group [49].

Maintenance of Crohn's Disease Remission

The quantity of literature that exists for evaluating probiotics for the maintenance of CD remission is greater than for induction of active CD. Classically, induction of remission is divided into medically and surgically induced remission. We will follow this convention in the discussion regarding maintenance of remission later. Since 1997, there have been seven randomized controlled trials reported examining the efficacy

of single or dual agent probiotic therapy to maintain remission [37,42,48,50–53]. Although there is considerable variation in probiotics and concomitant therapy, the evidence with respect to trial size, duration, and results strongly suggest that probiotics examined had negligible impact on the duration of maintenance for patients with CD. Two metaanalyses by independent groups also arrived at the same conclusion for these single or dual agent probiotic therapies [51,58].

It important to recognize that not all probiotics are the same. In this regard, Fedorak et al. conducted a postoperative prevention trial with VSL#3, a combination of eight probiotics demonstrating prevention of recurrence when the probiotic was administered immediately after surgery, but not when it was administered later [42]. This study implies that probiotics do not have class effect therapeutics and must be individualized to the therapeutic requirement.

Maintenance of Medically Induced Remission

In 2005, Bousvaros et al. completed a 2-year trial in which pediatric CD patients in remission (PCDAI < 10) were randomly assigned to receive either *L. rhamnosus* strain GG (LGG) ($n = 39$) or placebo ($n = 36$) [50]. Concomitant medications (5-ASA, azathioprine, 6-MP, corticosteroids <0.5 mg/kg every other day) were permitted throughout the study. There were no significant differences between the groups with respect to the median time to relapse (9.8 vs. 11.0 months for the LGG and placebo groups, respectively) or the number of patients who relapsed (12/39 vs. 6/36, respectively).

To date, with regard to *Lactobacillus*, there continues to be only one randomized, placebo-controlled trial in the adult population to evaluate if *Lactobacillus* GG is effective in inducing or maintaining medically induced remission [48]. All patients received a 2-week course of ciprofloxacin and metronidazole, along with a 12-week tapering course of corticosteroids starting at 60 mg. Eleven patients with moderate to active CD were initially enrolled to receive probiotic, LGG (2×10^9 CFU/day), or placebo at week 2 of the study for 6 months. The primary endpoint was sustained remission defined as absence of relapse at the 6-month follow-up visit. Relapse was defined as an increase in CDAI of >100 points. This study did not identify a benefit of *Lactobacillus* GG in maintaining remission in CD. However, a limitation of this study was inadequate power as the sample size was only 11 patients with only 5/11 patients completing the study. Of the five patients who remained in the study, two patients in each of the placebo and the probiotic groups had sustained remission. Systematic reviews [58] and metaanalyses [56,59], have also identified no benefits of *Lactobacillus* as a single probiotic agent in maintaining remission or preventing clinical or endoscopic relapses.

Thus far, the only study to demonstrate a statistically significant prolongation of medically induced remission in CD was that of Guslandi et al., who investigated *S. boulardii* and mesalamine treatment (1 g/day + 2 g/day mesalamine, 6 months) versus mesalamine treatment alone (3 g/day, 6 months) [51]. Clinical relapses were observed in 37.5% of patients receiving mesalamine alone and in 6.25% of patients in the group treated with mesalamine plus *S. boulardii* ($p = 0.04$).

To our knowledge, Bourreille et al. has conducted the only randomized controlled trial of a probiotic since the publication of this original chapter in 2009, showing no significant clinical benefits with *S. boulardii* [37]. These investigators found that there were no significant differences in mean CDAI, erythrocyte sedimentation rate, or median C- reactive protein (CRP) levels when compared to the placebo group. This trial (FLORABEST) is one of the largest and involved 165 patients with active CD who after achieving clinical remission after 4 weeks with steroids or aminosalicylates were subsequently randomly assigned to receive the probiotic yeast or placebo for 52 weeks. Of the 80 CD patients who relapsed in the study, 47.5% were in the *S. boulardii* group and 53.2% were in the placebo group ($p > 0.05$) with no difference in the median time to relapse. These patients were not on any maintenance medications as all patients had to be weaned off of corticosteroids and/or salicylates in the 12 weeks after randomization. This study also excluded patients with fistulizing or perianal CD. Limitations of this study included that disease activity was assessed by CDAI and not histological healing. In addition, there were a number of active CD patients who were treated with salicylates which is generally not the recommended medication for first line therapy in maintenance of CD.

With respect to *E. coli* Nissle 1917, in a randomized, double-blind, placebo-controlled study by Malchow, it was demonstrated that *E. coli* Nissle 1917 (5×10^{10} CFU/day for 1 year) reduced disease relapse rates in a group of 28 patients with active colonic CD for whom remission had been induced by corticosteroid treatment (70% still in remission in the probiotic group at 1 year vs. 30% placebo); however, these results did not achieve statistical significance [52].

Overall, in one of the most recent metaanalysis from 2014 by Shen et al., subgroup analyses assessing seven studies recruiting CD patients revealed no significant difference in maintaining clinical remission with probiotics and placebo. The strains assessed included *E. coli* Nissle [52] and *B. longum* [49]. In addition, three of these seven studies did assess endoscopic relapse rates and there remained no benefit of using probiotics ($p = 0.75$, RR = 1.08). Limitations of all studies are the lack of consistency with probiotic agent, dose, concurrent IBD medications (immunomodulators and corticosteroids), and absence of endoscopic assessment of remission. Thus, there remains inconclusive evidence to support the use of probiotics to maintain remission in CD and well-designed studies are required.

Likewise, Ghouri et al. conducted a systematic review of randomized controlled trials of probiotics assessing maintenance of medically induced remission, but also included studies assessing prebiotics and synbiotics in IBD where 14 studies with CD patients met inclusion criteria [58]. Overall, none of the probiotics containing *Lactobacillus* in these randomized controlled trials have shown significant effects on endoscopic remission or on CDAI [40,44,48,50,53]. Furthermore, this systematic review also included four trials using *S. boulardii* and one study with *E. coli* strain Nissle 1917. Unfortunately, there was also no clinical benefit in relapse rates as measured by CDAI for these probiotics [37].

The second study assessing synbiotic therapy was by Chermesh et al. which demonstrated no statistically significant benefit in CDAI, Rutgeerts' scores on endoscopy, or laboratory measurements between the synbiotic and placebo groups at 3- and 24-month follow up [45].

In summary, based on current evidence, probiotics do not have an impact on maintenance or induction of remission. Probiotics do not significantly change the diverse microflora of the colon. Ahmed et al. studied colonic microflora in eight patients with Crohn's colitis and eight patients with UC in a randomized "crossover" study [46]. Patients received either a synbiotic preparation for 1 month and subsequently crossed over to receive a placebo for 1 month, or vice versa. No patients had a significant alteration in their microbial composition as measured by stool samples. Moreover, patients with IBD are known to have a decreased number of bacterial species within the phylum Firmicutes [60]. Targeting specific deficient species with probiotics may provide more long standing positive benefits. Eeckhaut et al. found that the average number of *Butyricicoccus pullicaecorum* in stools from patients with active and inactive UC and CD were significantly less than healthy controls [60]. In a mouse model of colitis, *B. pullicaecorum* culture supernatant provided a significant epithelial protective effect based on macroscopic and histologic criteria. The supernatant of such a bacterial culture prevented an increase in cytokine-induced epithelial permeability and the subsequent loss of transepithelial resistance. Extrapolating further, fecal microbiota transplantation (FMT) may be a feasible alternative to probiotics in the future. FMT is the delivery or infusion of fecal material from a healthy donor into the gastrointestinal tract of a recipient. As probiotics contain only one or a few strains of bacteria, FMT offers multiple and diverse strains of bacteria that could replenish species (Bacteroidetes and Firmicutes) known to be deficient in IBD [58]. FMT holds potential but similar to probiotics, large well-designed RCTs in CD are needed to show benefit.

Maintenance of Surgically Induced Remission

Recently, an encouraging study demonstrated that probiotic therapy with a combination product significantly reduced CD postoperative recurrence when the probiotic was administered immediately after surgery but not when administered some months after surgery. In this study 120 patients were randomly assigned to receive 1 sachet of VSL#3 (900 billion bacteria) or matching placebo twice daily within 30 days postresection. After 90 days of randomized treatment, all patients demonstrating either no or mild endoscopic recurrence could receive VSL#3 until Day 365. Colonoscopy was performed at Days 90 and 365 to evaluate the neoterminal ileum for disease recurrence and to obtain mucosal biopsies for cytokine analysis. The primary outcome was the proportion of patients with severe endoscopic recurrence (Rutgeerts Grades 3 or 4). At Day 90 the proportion of patients with severe lesions was numerically lower for those taking VSL#3 (10.0%) relative to placebo (17.4%) ($p = 0.36$). However, the 90-day recurrence rate was markedly lower than predicted, leading to statistical underpowering. Nevertheless, at Day 365 the proportion of patients with severe endoscopic recurrence was significantly lower in patients who received VSL#3 immediately after surgery (11.5%) compared to those who received placebo immediately after surgery followed by VSL#3 from Days 90 to 365 (36.6%), ($p = 0.01$). Patients receiving VSL#3 had significantly reduced levels of ileal mucosal proinflammatory cytokine levels ($p < 0.05$) [42].

In 2007, Chermesh et al. conducted a small trial to examine the efficacy of Synbiotic 2000 (a commercial mixture containing four probiotics and four prebiotics) to extend the duration of remission following surgery for CD [45]. A total of 30 subjects were randomized 2:1 into either the active treatment group or the placebo. During the 2-year study, a total of 21 subjects dropped out for various reasons, some of which were unrelated to the study (e.g., pregnancy). When considering the data that were collected, there were no significant differences with respect to either endoscopic or clinical relapses. It is noteworthy to mention that the majority of subjects had the fistulizing form of CD and their response to the active treatment may differ substantially than those who have the inflammatory disease form.

In 2006, Marteau et al. conducted a larger trial ($n = 98$) over 6 months to investigate the efficacy of a single probiotic strain (*L. johnsonii*, LA1, Nestle) to prolong the time to relapse in CD patients [40]. Eligibility required that the individual had undergone resection of less than 1 m of the intestinal tract within the previous 21 days in which all macroscopic lesions had been removed and were in a state of disease remission. At the conclusion of the 6 months, a total of eight subjects had dropped out of the study in both the active treatment and placebo groups. Nevertheless, the per protocol analysis confirmed that there was no significant difference between the two cohorts regarding endoscopic recurrence of disease at 6 months or the distribution of the endoscopic score.

Similarly, Van Gossum et al. examined the efficacy of *L. johnsonii*, LA1, Nestle, to prolong the time to relapse following elective ileocecal resection to alleviate CD symptoms [44]. Subjects were randomly assigned to either the active treatment ($n = 34$) or placebo ($n = 36$) cohorts. At the conclusion of the 12-week study, 7 and 14 subjects had dropped out of the respective groups for various reasons rendering the study slightly underpowered. The intention to treat analyses showed that there were no significant differences between the groups with respect to the endoscopic recurrence of CD at 12 weeks following ileocecal resection. Although hampered by high dropout rates, the evidence indicated that *L. johnsonii* is ineffective for preventing postoperative recurrence in CD patients.

In a randomized, double-blind trial by Prantera et al., 40 patients received either *Lactobacillus* GG (dose not listed) or placebo following surgical resection for their CD [53]. After 52 weeks of treatment, there were no significant differences in clinical or endoscopic remission between the two groups.

Campieri et al. reported that a combination of antibiotic and VSL#3 was more effective in preventing postoperative recurrence of CD than was mesalamine [54]. In that study, 40 patients were randomized to receive either rifaximin (1.8 g/day) for 3 months followed by VSL#3 (3×10^{11} CFU/day) for 9 months or mesalamine for 12 months. After 1 year, the antibiotic/VSL#3 group had an endoscopic remission rate of 80% compared with 60% in the mesalamine group.

There is only one study assessing synbiotic therapy to prevent postoperative recurrence of CD [45]. This demonstrated no statistically significant benefit in CDAI, Rutgeerts' scores on endoscopy, or laboratory measurements between the synbiotic and placebo groups at 3- and 24-month follow up.

To date, there have not been any new studies assessing probiotic use in preventing recurrence of CD in the postoperative setting. The last study that was a randomized controlled trial addressing this issue was in 2015 by Fedorak et al. The current evidence shows that the benefit of probiotics in CD is much lower than in pouchitis and UC [43]. It is speculated that this may be due to factors, such as the transmural nature of CD, poor design of studies, or other unknown factors. Overall, probiotics will not prevent both postoperative clinical and endoscopic recurrence in CD [61,62].

CONCLUSIONS

Probiotic research continues to mature. This is evidenced by the increasing number of trials that are supported by bench research investigating the cellular and subcellular mechanisms of how probiotics produce a therapeutic effect. Results from metagenomic research initiatives will further expand the knowledge regarding the gut's biome through the identification of unculturable organisms that are estimated to range from 15,000 to 36,000, depending upon the conservatism of species classification [63–66]. The continued use of animal models and development of new mutant lines will provide an efficient, reliable method to study these new and putative probiotic organisms.

For probiotics to have a therapeutic role in the management of CD, their therapeutic mechanism of action must be aligned with the pathogenic mechanism of action of the disease. In this regard, the role of probiotics for the clinical treatment of CD is emerging as both the mechanisms and pathogenesis are being unraveled. Working in concert, results from the bench and animal models will better inform which probiotics should graduate to clinical trials. As well, the variety of immunomodulatory mechanisms used by probiotic species is being exploited to provide new drug delivery systems for conventional drugs. An example of this is the IL-10-producing *Lactococcus lactis* for CD that has a novel recombinant protein anchored to its bacterial membrane capable of binding cells expressing certain epitopes [67].

It remains clear that probiotics are able to reduce gastrointestinal inflammation by exerting positive effects on epithelial cell and mucosal immune dysfunction in animal models. However, the evidence for use of probiotics in the treatment of CD is limited. The majority of studies show no significant difference in the induction or maintenance of remission of CD. As a result, there is no conclusive evidence for the use of probiotics in CD. Future research will need to strive to align individual probiotics and their unique mechanism(s) of action with specific IBD populations. Not all probiotics will have equal efficacy or similar therapeutic value and, thus, incorporation of probiotics into the physicians' therapeutic armamentarium will need to be tempered by appropriately designed randomized clinical trials. Information that has been gained from probiotic work with dysbiosis and intestinal inflammation has been translated over to the field of FMT. FMT is an upcoming field for CD and we expect there to be a greater shift of focus from probiotics to FMT in the future.

REFERENCES

[1] Boirivant M, Strober W. The mechanism of action of probiotics. Curr Opin Gastroenterol 2007;23(6):679–92.

[2] Llopis M, Antolin M, Guarner F, Ahrne S, Molin GMJ. Downregulation of TNF-alpha gene expression in human intestinal mucosa by non-adherent probiotic bacteria. Gastroenterology 2006;130(4 Suppl. 2):A369–70.

[3] Bengmark S. Use of some pre-, pro- and synbiotics in critically ill patients. Best Pract Res Clin Gastroenterol 2003;17(5):833–48.

[4] Ewaschuk JB, Dieleman LA. Probiotics and prebiotics in chronic inflammatory bowel diseases. World J Gastroenterol 2006;12(37):5941–50.

[5] Gionchetti P, Amadini C, Rizzello F, Venturi A, Poggioli G, Campieri M. Probiotics for the treatment of postoperative complications following intestinal surgery. Best Pract Res Clin Gastroenterol 2003;17(5):821–31.

[6] Kruis W. Review article: antibiotics and probiotics in inflammatory bowel disease. Aliment Pharmacol Ther 2004;20(Suppl. 4):75–8.

[7] Marteau P, Seksik P, Jian R. Probiotics and intestinal health effects: a clinical perspective. Br J Nutr 2002;88(Suppl. 1):S51–7.

[8] O'Hara AM, Shanahan F. Mechanisms of action of probiotics in intestinal diseases. ScientificWorldJournal 2007;7:31–46.

[9] Quigley EMM, Flourie B. Probiotics and irritable bowel syndrome: a rationale for their use and an assessment of the evidence to date. Neurogastroenterol Motil 2007;19(3):166–72.

[10] Sazawal S, Hiremath G, Dhingra U, Malik P, Deb S, Black RE. Efficacy of probiotics in prevention of acute diarrhoea: a meta-analysis of masked, randomised, placebo-controlled trials. Lancet Infect Dis 2006;6(6):374–82.

[11] Shanahan F. Probiotics in inflammatory bowel disease—therapeutic rationale and role. Adv Drug Deliv Rev 2004;56(6):809–18.

[12] Tuohy KM, Rouzaud GCM, Brück WM, Gibson GR. Modulation of the human gut microflora towards improved health using prebiotics—assessment of efficacy. Curr Pharm Des 2005;11(1):75–90.

[13] Zigra PI, Maipa VE, Alamanos YP. Probiotics and remission of ulcerative colitis: a systematic review. Neth J Med 2007;65(11):411–8.

[14] Isaacs K, Herfarth H. Role of probiotic therapy in IBD. Inflamm Bowel Dis 2008;14(11):1597–605.

[15] Vanderpool C, Yan F, Polk DB. Mechanisms of probiotic action: implications for therapeutic applications in inflammatory bowel diseases. Inflamm Bowel Dis 2008;14(11):1585–96.

[16] De Keersmaecker SCJ, Verhoeven TLA, Desair J, Marchal K, Vanderleyden J, Nagy I. Strong antimicrobial activity of *Lactobacillus rhamnosus* GG against *Salmonella typhimurium* is due to accumulation of lactic acid. FEMS Microbiol Lett 2006;259(1):89–96.

[17] Collado MC, Hernández M, Sanz Y. Production of bacteriocin-like inhibitory compounds by human fecal Bifidobacterium strains. J Food Prot 2005;68(5):1034–40.

[18] Collado MC, Meriluoto J, Salminen S. Role of commercial probiotic strains against human pathogen adhesion to intestinal mucus. Lett Appl Microbiol 2007;45(4):454–60.

[19] Sartor RB, Muehlbauer M. Microbial host interactions in IBD: implications for pathogenesis and therapy. Curr Gastroenterol Rep 2007;9(6):497–507.

[20] Cani PD, Bibiloni R, Knauf C, et al. Changes in gut microbiota control metabolic endotoxemia-induced inflammation in high-fat diet-induced obesity and diabetes in mice. Diabetes 2008;57(6):1470–81.

[21] Kotlowski R, Bernstein CN, Sepehri S, Krause DO. High prevalence of *Escherichia coli* belonging to the B2+D phylogenetic group in inflammatory bowel disease. Gut 2007;56(5):669–75.

[22] Bibiloni R, Mangold M, Madsen KL, Fedorak RN, Tannock GW. The bacteriology of biopsies differs between newly diagnosed, untreated, Crohn's disease and ulcerative colitis patients. J Med Microbiol 2006;55(Pt. 8):1141–9.

[23] Vasquez N, Mangin I, Lepage P, et al. Patchy distribution of mucosal lesions in ileal Crohn's disease is not linked to differences in the dominant mucosa-associated bacteria: a study using fluorescence in situ hybridization and temporal temperature gradient gel electrophoresis. Inflamm Bowel Dis 2007;13(6):684–92.

[24] Swidsinski A, Loening-Baucke V, Theissig F, et al. Comparative study of the intestinal mucus barrier in normal and inflamed colon. Gut 2007;56(3):343–50.

[25] Sokol H, Pigneur B, Watterlot L, et al. *Faecalibacterium prausnitzii* is an anti-inflammatory commensal bacterium identified by gut microbiota analysis of Crohn disease patients. Proc Natl Acad Sci USA 2008;105(43):16731–6.

[26] Garcia Vilela E, De Lourdes De Abreu Ferrari M, Oswaldo Da Gama Torres H, et al. Influence of *Saccharomyces boulardii* on the intestinal permeability of patients with Crohn's disease in remission. Scand J Gastroenterol 2008;43(7):842–8.

[27] Sivignon A, de Vallée A, Barnich N, et al. *Saccharomyces cerevisiae* CNCM I-3856 prevents colitis induced by AIEC bacteria in the transgenic mouse model mimicking Crohn's disease. Inflamm Bowel Dis 2015;21(2):276–86.

[28] Kaila M, Isolauri E, Soppi E, Virtanen E, Laine S, Arvilommi H. Enhancement of the circulating antibody secreting cell response in human diarrhea by a human *Lactobacillus* strain. Pediatr Res 1992;32(2):141–4.

[29] Madsen KL, Doyle JS, Jewell LD, Tavernini MM, Fedorak RN. *Lactobacillus* species prevents colitis in interleukin 10 gene-deficient mice. Gastroenterology 1999;116(5):1107–14.

[30] Ukena SN, Singh A, Dringenberg U, et al. Probiotic *Escherichia coli* Nissle 1917 inhibits leaky gut by enhancing mucosal integrity. PLoS One 2007;2(12):e1308.

[31] Ewaschuk JB, Diaz H, Meddings L, et al. Secreted bioactive factors from *Bifidobacterium infantis* enhance epithelial cell barrier function. Am J Physiol 2008;295(5):G1025–34.

[32] Matthes H, Krummenerl T, Giensch M, Wolff C, Schulze J. Clinical trial: probiotic treatment of acute distal ulcerative colitis with rectally administered *Escherichia coli* Nissle 1917 (EcN). BMC Complement Altern Med 2010;10:13.

[33] Kim N, Kunisawa J, Kweon M-N, Eog Ji G, Kiyono H. Oral feeding of *Bifidobacterium bifidum* (BGN4) prevents CD4(+) CD45RB(high) T cell-mediated inflammatory bowel disease by inhibition of disordered T cell activation. Clin Immunol 2007;123(1):30–9.

[34] Peran L, Camuesco D, Comalada M, et al. A comparative study of the preventative effects exerted by three probiotics, *Bifidobacterium lactis, Lactobacillus casei* and *Lactobacillus acidophilus*, in the TNBS model of rat colitis. J Appl Microbiol 2007;103(4):836–44.

[35] Chung YW, Choi JH, Oh T-Y, Eun CS, Han DS. *Lactobacillus casei* prevents the development of dextran sulphate sodium-induced colitis in Toll-like receptor 4 mutant mice. Clin Exp Immunol 2008;151(1):182–9.

[36] Raz I, Gollop N, Polak-Charcon S, Schwartz B. Isolation and characterisation of new putative probiotic bacteria from human colonic flora. Br J Nutr 2007;97(4):725–34.

[37] Bourreille A, Cadiot G, Le Dreau G, et al. *Saccharomyces boulardii* does not prevent relapse of Crohn's disease. Clin Gastroenterol Hepatol 2013;11(8):982–7.

[38] McCarthy J, O'Mahony LDC. An open trial of a novel probiotic as an alternative to steroids in mild/moderately active Crohn's disease. Gut 2001;49(Suppl. III):A2447.

[39] Hafer A, Krämer S, Duncker S, Krüger M, Manns MP, Bischoff SC. Effect of oral lactulose on clinical and immunohistochemical parameters in patients with inflammatory bowel disease: a pilot study. BMC Gastroenterol 2007;7:36.

[40] Marteau P, Lémann M, Seksik P, et al. Ineffectiveness of *Lactobacillus johnsonii* LA1 for prophylaxis of postoperative recurrence in Crohn's disease: a randomised, double blind, placebo controlled GETAID trial. Gut 2006;55(6):842–7.

[41] Gupta P, Andrew H, Kirschner BS, Guandalini S. Is *Lactobacillus* GG helpful in children with Crohn's disease? Results of a preliminary, open-label study. J Pediatr Gastroenterol Nutr 2000;31(4):453–7.

[42] Fedorak RN, Feagan BG, Hotte N, et al. The probiotic VSL#3 has anti-inflammatory effects and could reduce endoscopic recurrence after surgery for Crohn's disease. Clin Gastroenterol Hepatol 2015;13(5):928–35. e2.

[43] Veerappan GR, Betteridge J, Young PE. Probiotics for the treatment of inflammatory bowel disease. Curr Gastroenterol Rep 2012;14(4):324–33.

[44] Van Gossum A, Dewit O, Louis E, et al. Multicenter randomized-controlled clinical trial of probiotics (*Lactobacillus johnsonii*, LA1) on early endoscopic recurrence of Crohn's disease after Ileo-caecal resection. Inflamm Bowel Dis 2007;13(2):135–42.

[45] Chermesh I, Tamir A, Reshef R, et al. Failure of Synbiotic 2000 to prevent postoperative recurrence of Crohn's disease. Dig Dis Sci 2007;52(2):385–9.

[46] Ahmed J, Reddy BS, Mølbak L, Leser TD, MacFie J. Impact of probiotics on colonic microflora in patients with colitis: a prospective double blind randomised crossover study. Int J Surg 2013;11(10):1131–6.

[47] Fujimori S, Tatsuguchi A, Gudis K, et al. High dose probiotic and prebiotic cotherapy for remission induction of active Crohn's disease. J Gastroenterol Hepatol 2007;22(8):1199–204.

[48] Schultz M, Timmer A, Herfarth HH, Sartor RB, Vanderhoof JA, Rath HC. *Lactobacillus* GG in inducing and maintaining remission of Crohn's disease. BMC Gastroenterol 2004;4:5.

[49] Steed H, Macfarlane GT, Blackett KL, et al. Clinical trial: the microbiological and immunological effects of synbiotic consumption—a randomized double-blind placebo-controlled study in active Crohn's disease. Aliment Pharmacol Ther 2010;32(7):872–83.

[50] Bousvaros A, Guandalini S, Baldassano RN, et al. A randomized, double-blind trial of *Lactobacillus* GG versus placebo in addition to standard maintenance therapy for children with Crohn's disease. Inflamm Bowel Dis 2005;11(9):833–9.

[51] Guslandi M, Mezzi G, Sorghi M, Testoni PA. *Saccharomyces boulardii* in maintenance treatment of Crohn's disease. Dig Dis Sci 2000;45(7):1462–4.

[52] Malchow HA. Crohn's disease and *Escherichia coli*. A new approach in therapy to maintain remission of colonic Crohn's disease? J Clin Gastroenterol 1997;25(4):653–8.

[53] Prantera C, Scribano ML, Falasco G, Andreoli A, Luzi C. Ineffectiveness of probiotics in preventing recurrence after curative resection for Crohn's disease: a randomised controlled trial with *Lactobacillus* GG. Gut 2002;51(3):405–9.

[54] Campieri M, Rizzello FVA. Combination of antibiotic and probiotic treatment is efficacious in prophylaxis of post-operative recurrence of Crohn's disease: a randomised controlled trial vs mesalamine. Gastroenterology 2000;118:A781.

[55] Borruel N, Carol M, Casellas F, et al. Increased mucosal tumour necrosis factor alpha production in Crohn's disease can be downregulated ex vivo by probiotic bacteria. Gut 2002;51(5):659–64.

[56] Shen J, Zuo Z-X, Mao A-P. Effect of probiotics on inducing remission and maintaining therapy in ulcerative colitis, Crohn's disease, and pouchitis: meta-analysis of randomized controlled trials. Inflamm Bowel Dis 2014;20(1):21–35.

[57] Benjamin JL, Hedin CRH, Koutsoumpas A, et al. Randomised, double-blind, placebo-controlled trial of fructo-oligosaccharides in active Crohn's disease. Gut 2011;60(7):923–9.

[58] Ghouri YA, Richards DM, Rahimi EF, Krill JT, Jelinek KA, DuPont AW. Systematic review of randomized controlled trials of probiotics, prebiotics, and synbiotics in inflammatory bowel disease. Clin Exp Gastroenterol 2014;7:473–87.

[59] Rahimi R, Nikfar S, Rahimi F, et al. A meta-analysis on the efficacy of probiotics for maintenance of remission and prevention of clinical and endoscopic relapse in Crohn's disease. Dig Dis Sci 2008.53(9):2524–31.

[60] Eeckhaut V, Machiels K, Perrier C, et al. *Butyricicoccus pullicaecorum* in inflammatory bowel disease. Gut 2013;62(12):1745–52.

[61] Papi C, Fasci Spurio F, Margagnoni G, Aratari A. Randomized controlled trials in prevention of postsurgical recurrence in Crohn's disease. Rev Recent Clin Trials 2012;7(4):307–13.

[62] van Loo ES, Dijkstra G, Ploeg RJ, Nieuwenhuijs VB. Prevention of postoperative recurrence of Crohn's disease. J Crohns Colitis 2012;6(6):637–46.

[63] Health NI of. Office of Portfolio 66. Analysis and Strategic Initiatives. Roadmap 1.5 Update. Office of Portfolio.

[64] Human Metagenome Consortium J (HMGJ). Metagenome.jp. Available from: http://mg.bio.titech.ac.jp/mg/; 2007.

[65] INRA. Sequencing the human intestinal flora: launching of the European project MetaHIT coordinated by INRA; 2008.

[66] Frank DN, St Amand AL, Feldman RA, Boedeker EC, Harpaz N, Pace NR. Molecular-phylogenetic characterization of microbial community imbalances in human inflammatory bowel diseases. Proc Natl Acad Sci USA 2007;104(34):13780–5.

[67] Yuvaraj S, Peppelenbosch MP, Bos NA. Transgenic probiotica as drug delivery systems: the golden bullet? Expert Opin Drug Deliv 2007;4(1):1–3.

Chapter 37

Treatment of Inflammatory Bowel Disease in Ulcerative Colitis

R. Chibbar, A. Alahmadi and L.A. Dieleman

Department of Medicine, Division of Gastroenterology, University of Alberta, Edmonton, AB, Canada

INTRODUCTION

Inflammatory bowel disease (IBD) is a chronic, progressive, and relapsing inflammatory disorder of the gastrointestinal tract, primarily categorized as Crohn's disease (CD) and ulcerative colitis (UC). Though both are unique and independent entities, they share similar clinical features, and the same treatment goal: to induce and maintain deep remission in a safe and efficient manner. However, the specific regimen to achieve this target differs. The current available therapeutic options include corticosteroids, 5-aminosalicylic acid (5-ASA), immune-modulators including azathioprine or 6-mercaptopurine (6-MP) and methotrexate, cyclosporine A, and the antitumor necrosis factor antibodies, such as infliximab, adalimumab, and golimumab. Newer agents that are now available include the antiadhesion molecules, such as vedolizumab.

As research rapidly progresses to understand the pathophysiology of IBD and develop new therapies, there is also an interest in disease prevention and medication safety, focusing on the use of nonpharmacological options, such as prebiotics, probiotics, and fecal microbiota transplantation (FMT). Both probiotics and prebiotics have been studied in the induction and maintenance of remission of UC. Though there is a paucity of randomized controlled trials, the available data suggests that specific formulations of both probiotics and prebiotics can be efficacious adjuvant therapeutic agents. Their safety profile is an added advantage in contrast to more conventional agents, such as immunomodulators, like 6-MP or azathioprine. FMT has been mostly studied in case series and recently in small clinical trials, with inconclusive benefit in IBD treatment. The following review will focus on the role of prebiotics, probiotics, FMT, and the gut microbiota in the prevention and management of UC.

The etiology of IBD is not completely understood. The established hypothesis is that it results from a dysregulated mucosal immune response to commensal gut microflora in genetically susceptible individuals. Many of these susceptibility genes are related to impaired innate immunity and primary defense systems to enteric bacteria, supporting the evolving interaction between the gut microbiome and IBD. The human gut microflora or gut microbiota consists of several intestinal microorganisms including bacterial, fungal, and viral species that synergistically interact with the host [1,2]. The exact composition is influenced by the interaction between the host and its environment. The human gut is sterile at birth, and becomes colonized during childbirth and the first feed. Additional factors, such as infant's genotype, gestational age, method of delivery, perinatal and postnatal antibiotic use, age, sex, state of immune maturation, and the environment play a role in achieving a steady state, that is reached at approximately 2 years of age. In the early years of life, the microbiota stimulates both innate and acquired immunity by stimulating the gut-associated lymphoid system and systemic and local immune responses, respectively [2,3]. This first line of communication with the environment results in normal, low-grade physiologic inflammation, but continuous inflammation in the presence of pathogenic organisms [2]. Loss of homeostasis of the gut microbiota, or dysbiosis, plays an important role in the development of adverse changes, such as chronic inflammation [4,5]. The majority of the gut microbiota resides in the colon. Despite variability in composition, the function remains the same, to ensure nutrient and mineral absorption, regulation of the immune system, synthesis of vitamins and amino acids, decomposition of proteins, inhibition of pathogenic bacteria, maintain mucosal homeostasis, protection of the mucosal barrier, and synthesis of short-chain fatty acids (SCFAs).

The microbiota in IBD has reduced stability and bacterial diversity [6–9]. Specifically, there is a reduction in the dominant commensal bacteria with an associated increase in more pathogenic strains. This has been demonstrated in CD with a loss of Firmicutes (especially *Clostridium*) and *Bacteroides*, and an increase of Proteobacteria and Actinobacteria. In UC, there is a decreased population of *Roseburia hominis* and *Faecalibacterium prausnitzii*, both of which play an important role in maintaining gastrointestinal tract homeostasis [10]. Most Clostridia species produce SCFAs, that provide essential nutrients for the colonic epithelial cells and energy for colonocytes, induce regulatory T cells, and also have an antiinflammatory effect by enhancing epithelial barrier integrity [2,4,6]. Butyrate inhibits proinflammatory cytokine release, and increases production of mucin and antimicrobial peptides. Sulfate-reducing bacteria increase hydrogen sulfide and decrease butyrate concentrations, and are found in higher numbers in patients with IBD [2]. Both *R. hominis* and *F. prausnitzii* also produce butyrate; and it has been shown that *F. prausnitzii* numbers were reduced in UC patients with short-term remission and increased during disease reactivation [11,12].

Loss of the resident gastrointestinal bacterial species Firmicutes and *Bacteroides* is associated with dysbiosis of the gut microbiota and concomitant overexpression of proinflammatory Th_1 and Th_{17} cells, which is associated with CD in susceptible hosts. This model is not as well characterized in UC. However, studies in twins with UC did show evidence of dysbiosis, with increased numbers of Proteobacteria and Actinobacteria, in addition to an increase of sulfate-reducing bacteria [10,13,14]. Similarly an exaggerated adaptive immune response to commensal organisms induces disease reactivation from environmental triggers and further worsening of breaks in the intestinal barrier resulting in chronic intestinal inflammation. This concept of dysbiosis has led to numerous studies to examine the role of prebiotics and probiotics in modulation of the gut microbiota, and to determine their efficacy and safety as adjuvants in the induction and maintenance of UC.

PROBIOTICS

Probiotics are live microorganisms, when administered in adequate quantities confer a health benefit to the host [15]. The purpose of their use in UC is to positively influence the gut microflora by altering the type and number of bacteria as well their function in the gastrointestinal tract. They have been increasingly studied in the induction and maintenance of remission for active and inactive IBD; however, there is a lack of systematic randomized controlled trials. The studies in the literature vary in design, including the type and dosage of probiotic used, as well as the presence and choice of concomitant therapies.

ACTIVE ULCERATIVE COLITIS

Escherichia coli Nissle 1917

For UC in particular, nonlactic acid bacteria probiotics, such as *E. coli* Nissle 1917 (EcN) and the combination probiotic

cocktail VSL#3 have been found to be the most beneficial, and received "A" recommendations by the American Recommendations for probiotic use [16]. EcN is a nonpathogenic strain of *E. coli*, studied in both active and quiescent disease, as well as during induction and maintenance of remission of UC. It is associated with reduction of mucosal damage and colonic epithelial permeability, thus improving healing of colonic tissue. This organism affects mucosal immunity in IBD, either primarily or secondarily; it decreases myeloperoxidase activity, TNF-α, IFN-gamma, and IL-10 levels; and it promotes the down-regulation of IL-12 levels, attenuating the Th_1 response, thus lowering production of IFN-gamma. With reduced apoptosis of colonic epithelial cells, it downregulates T cell, B cell, and Toll cell receptor (TLR) signaling, resulting in decreased expression of Th_1 transcription factors (TNF-α and its induced chemokines). These functions attenuate colonic expression of proinflammatory mediators, such as IL-6, TNF-α, IFN-gamma, ultimately diminishing colonic neutrophil infiltration. EcN has also been shown to decrease the number of *Enterococcus* species and stimulates production of streptococci and lactobacilli [16].

EcN has demonstrated beneficial effect in the treatment of active UC. Rembacken et al. [17] conducted a single-center, randomized, double-dummy study in 116 patients, comparing EcN in combination with standard medical therapy, specifically mesalazine (Asacol 800 mg 3 times daily). At study entry, all patients received a 1-week course of oral gentamicin 80 mg 3 times daily to suppress native *E. coli* microflora. The primary endpoint was time to relapse and rate of relapse postremission induction, with a second endpoint of rate and time to remission. 44/59 (75%) in the 5-ASA group and 39/57 (68%) in the EcN group achieved remission. Mean time to remission showed a significant benefit in the EcN group, with 44 days (median 42) in the mesalazine group and 42 days (median 37) in the EcN group ($P = 0.0092$). Once remission was achieved, patients continued either mesalazine or EcN for 1 year. The mean duration of remission was also significantly better in the EcN treatment arm (206 days vs. 221 days in the EcN group, $P = 0.0174$). EcN was noninferior to oral mesalazine (Asacol formulation 2.4 g daily), and could be as effective in maintaining remission of UC [16]. However, the maintenance dose of mesalazine was lower than recommended in current clinical practice. In the EcN group, 11 patients reported nonserious adverse effects, with abdominal bloating being the most common complaint. Whereas, two serious events were reported in the mesalazine cohort: perforated sigmoid diverticulum requiring laparotomy and respiratory failure in a patient with severe emphysema. Eight patients reported nonserious reported adverse effects, with abdominal bloating again the most documented ailment, followed by headache and nausea. Not only was EcN efficacious, it also proved to be a safe agent in the management of UC.

VSL#3

VSL#3 consists of eight bacterial strains—*Lactobacillus acidophilus, Lactobacillus bulgaricus, Lactobacillus casei, Lactobacillus plantarum, Streptococcus thermophiles, Bifidobacterium breve, Bifidobacterium infantis, Bifidobacterium longum*. Studies in active UC have demonstrated that VSL#3 is an effective and safe therapeutic modality. Tursi et al. [18] first compared low-dose balsalazide 2.25 g/day plus VSL#3 900 billion CFU/day to concomitant 4.5 g/day balsalazide and mesalamine 2.4 g/day over 8 weeks in 90 patients with mild to moderate UC. Efficacy was assessed clinically, endoscopically, and histologically. The combination therapy group was found to be statistically significantly superior to both balsalazide and mesalamine groups ($P < 0.02$). It also achieved remission more quickly versus the monotherapy treatment arms, moreover in those with left-sided colitis compared to pancolitis. This study effectively demonstrated that lower doses (2.25 g/day) of balsalazide in combination with a probiotic (VSL#3) provided better results than the higher balsalazide dose required to maintain remission, and also had a better compliance rate. The combination arm had less adverse effects, although these results were not statistically significant. Balsalzide was better tolerated than mesalamine, regardless of concomitant VSL#3 use. The overall conclusion was that balsalazide plus VSL#3 was superior to mesalamine or balsalazide alone in mild-moderate left-sided or distal colitis.

The same authors then compared VSL#3 to placebo in a double blind, randomized placebo-controlled trial with concomitant 5-ASA/immunosuppressive therapies in 144 patients for duration of 8 weeks [19]. There was a trend to benefit in the VSL#3 group, with a decrease in the UC-DAI score of at least 50%, as well as a higher rate of remission ($s= 0.069$). There was also a statistically significant decrease in rectal bleeding ($P = 0.010$), but no change in stool frequency or endoscopic scores. Additionally, five patients in the placebo group worsened clinically. In this trial 3600 billion CFU/day VSL#3 was used (a higher dose than the prior study) to induce remission. The authors concluded that the dose of VSL#3 combined with standard therapy, was not only beneficial, but also delayed initiation of immunomodulatory agents.

Socd et al. [20] confirmed these findings in a randomized placebo-controlled trial with 147 patients on concomitant mesalazine, azathioprine, or 6-MP, comparing VSL#3 3.6×10^{12} CFU administered twice daily versus placebo for 12 weeks. The primary endpoint of a decrease in UCDAI score by 50% was achieved in the treatment arm, with a significant benefit (32.5% vs. 10%, $P = 0.001$). This effect was maintained at 12 weeks, 51.9% in the VSL#3 group and 18.6% in the placebo arm, and the results were statistically significant ($P < 0.001$). Regarding the secondary endpoint of remission induced at 12 weeks, 33 of 77 patients in the VSL#3 group and 11 of 70 patients in the control group achieved remission (42.9% vs. 15.7%, $P < 0.001$). Mucosal healing was also significantly better at 12 weeks in the VSL#3 group ($P < 0.28$). VSL#3 was well tolerated and minimal adverse effects, such as abdominal bloating and unpleasant taste were observed.

Bibiloni et al. [21] used high dose VSL#3 (900 billion bacteria, 4 sachets twice daily) in a small open study with 34 patients with mild to moderate active UC, over 6 weeks. The primary outcome of this intention-to-treat study was induction of remission with secondary outcomes of treatment response. Remission was achieved in 53%, while 24% of patients responded to therapy and treatment ineffectiveness was seen in 9% ($P < 0.001$), and another 9% of patients developed worsening disease. Only 10 patients (29%) reported increased abdominal bloating, but did not require discontinuation of therapy. Though small, this trial further supported the use and efficacy of VSL#3 in active UC, and continued to highlight its desired safety profile.

These findings are consistent with literature in the pediatrics population [22]. A 1-year prospective, placebo-controlled, double-blind trial was performed with 29 children, ranging from 1.7 to 16.1 years old (mean 9.8 years old) with newly diagnosed UC. Participants received weight-based VSL#3 or placebo and concomitant therapy with glucocorticoids and/or mesalamine therapy. Endoscopic and histologic assessment occurred at baseline, 6 and 12 months, or at the time of relapse. All 29 participants responded to induction therapy, with 92.8% in the VSL#3 group and 36.4% in the placebo group achieving remission ($P < 0.001$). Furthermore, 21.4% in the VSL#3 arm relapsed at 1 year versus 73.3% in the placebo group ($P = 0.014$). The VSL#3 group also showed significantly reduced endoscopic and histologic scores at each time point compared to the control group ($P < 0.05$). No adverse effects were noted in the VSL#3 group [16,22]. This small study demonstrated the efficacy of probiotics in the management of induction and maintenance of UC in children. Larger randomized controlled trials are needed to further support these findings.

BIO-THREE

BIO-THREE, a multispecies probiotic preparation of *Streptococcus faecalis, Clostridium butyricum*, and *Bacillus mesentericus,* has also demonstrated effectiveness in mild to moderate distal UC refractory to standard therapy. Tsuda et al. [23] conducted a small, open-label 4-week study of BIO-THREE comparing pre- and posttreatment UCDAI scores, in addition to clinical and endoscopic findings. A total of 20 patients were included in the trial. Forty-five percent of patients achieved remission, whereas 10% showed a decrease in UCDAI scores >3 points. However, 40% of the patients had no response and 5% worsened. Terminal restriction fragment length polymorphism (T-RFLP) showed increased bifidobacteria counts in fecal samples, which was associated with symptom improvement and UCDAI scores [16,23]. BIO-THREE appears to show clinical benefit; however, further studies are required to support its mainstream use.

Saccharomyces boulardii

S. boulardii is nonpathogenic yeast, that showed promise in a small noncontrolled trial [24]. Twenty-five patients with mild to moderate UC received S. boulardii 250 mg 3 times daily 4 weeks in combination with mesalamine. Twenty-four patients completed the study, of which, 17 achieved clinical and endoscopic remission [16]. One patient dropped out due to worsening symptoms. Though the data are promising, the results are difficult to validate given the low participant numbers and lack of randomized controlled trials.

BIFICO (Bifid Triple Viable Capsule)

BIFICO is a bifid triple viable capsule, composed of bacterial strains enterococci, bifidobacteria, and lactobacilli. It has demonstrated effectiveness in small studies to maintain disease remission. Cui et al. [25] gave 30 patients active UC sulfasalazine and glucocorticoids with BIFICO (bifid triple viable capsule 1.26 g/day) versus placebo for 8 weeks. The treatment arm showed a significant benefit clinically, endoscopically, and histologically. Three patients in the BIFICO group relapsed compared to 14 in the placebo group (20% vs. 93.3%, $P < 0.01$). There was also an increased amount of fecal lactobacilli and bifidobacteria in the probiotic group, as well as decreased NFκB DNA binding activity and increased mRNA expression of antiinflammatory cytokines ($P < 0.05$) [16]. Thus, BIFICO therapy showed effectiveness as an adjunct therapy for prevention of UC disease flares, but the study is limited by small patient numbers.

Lactobacillus reuteri

L. reuteri ATCC 55730 is a heterofermentary probiotic, which has shown clinical efficacy and safety in reducing incidence and severity of diarrhea, and functional disorders in children [26]. Oliva et al. [26] piloted a randomized, placebo-controlled study with L. reuteri ATCC 55730-containing enemas in 40 pediatric patients with active distal UC. Patients received 10^{10} CFU or placebo for 8 weeks, with concomitant mesalazine. They were evaluated with pre- and postclinical, endoscopic, histologic scores, and rectal mucosa expression of pro- and antiinflammatory cytokines. The treatment arm showed a significant reduction in total Mayo scores ($P < 0.01$). IL-10 levels also increased significantly in the L. reuteri group, with statistically significant reductions of TNF-α and IL-8 levels ($P < 0.01$) compared to placebo group. These cytokine changes were not observed in the placebo group. A total of nine patients prematurely discontinued the trial, mostly due to lack of compliance; however, two patients were excluded for infectious complications. More importantly, no serious adverse effects

were documented. In this small placebo-controlled study, L. reuteri ATCC 55730 enema significantly improved Mayo scores and mucosal healing in children with active distal UC (Table 37.1).

MAINTENANCE OF ULCERATIVE COLITIS

VSL#3

VSL#3 has also proven efficacy in maintenance of remission in UC. Venturi et al. [27] performed a small open label in 20 patients in remission and intolerant of 5-ASA therapy. Patients received VSL#3 3 g 2 times daily for 12 months. The primary outcome was that, VSL#3 significantly altered the microbiota composition in a positive manner. Increased fecal concentrations of Streptococcus salivarius, lactobacilli, and bifidobacteria were noted at day 20, and continued throughout study duration ($P < 0.05$). However, probiotic numbers returned to baseline 15 days postprobiotic cessation. Patients 15/20 (75%) maintained remission at 12 months, while 4 relapsed (confirmed with endoscopy and histology). No serious adverse effects were reported, but three patients did experience constipation. VSL#3 appears to be a favorable adjunctive therapeutic modality for mild to moderate UC; however, a large placebo-controlled study will be needed to assess its full benefit in maintenance of disease remission.

Escherichia coli (Nissle 1917)

EcN has been compared to both standard medical therapy and placebo for the maintenance of UC. Kruis et al. [28] conducted a double-blind, double-dummy study, in which 120 patients in remission were randomized to mesalazine 500 mg 3 times daily or to EcN for 12 weeks. End points of the trial included the clinical activity index (CAI, with remission defined as <4 points), relapse rates, relapse-free times, and physician global assessment. The start and end CAI scores showed no statistically significant benefit between groups. In addition, relapse rates and relapse-free time were similar (EcN 16% and 115 days vs. mesalazine 11.3% and 110 days). Both were also well tolerated, with five patient in the EcN arm and eight in the mesalazine group reporting adverse events. Two patients in the EcN and one patient on mesalazine withdrew from the study. Both groups noted similar symptoms, with diarrhea being the most common in the EcN group compared to nausea and vomiting in the mesalazine group. Other complaints included flatulence, nausea/vomiting, and headache.

A larger trial by the same group assessed both clinical and endoscopic parameters in another double-blind, double-dummy trial over 12 months, comparing EcN 200 mg daily to mesalazine 500 mg 3 times daily. A total of 327 patients were included, 162 in the probiotic arm and 165 in the mesalazine group. In the EcN group, 36.4% relapsed

TABLE 37.1 Probiotic Intervention Studies in Patients With Active Ulcerative Colitis

Probiotic, Dose	Design, Duration, Number of Patients	Concomitant Therapy	Clinical Outcome	References
E. coli Nissle 1917	RCT, double dummy study, 1 year, 116 patients	Tapering steroids, gentamicin	Remission rate 68% versus 75% (P 0.0508) Time to remission 42 days versus 44 days (P 0.0092) Duration of remission 206 versus 221 days (P 0.0174)	Rembacken et al. [17]
VSL#3 (9×10^{11})	RCT versus standard therapy, 8 weeks, 90 patients	Balsalazide	Remission rates 80% versus 77% ($P < 0.02$)	Tursi et al. [18]
VSL#3 (3.6×10^{12})	RCT versus PL, 8 weeks, 144 patients	5-ASA and/or azathioprine, or 6-MP	Response rates 63.1% versus 40.8% (P 0.01)	Tursi et al. [19]
VSL#3 (3.6×10^{12})	RCT versus PL, 12 weeks, 147 patients	Mesalazine, azathioprine or 6-MP	Remission rates 42.9% versus 15.7% ($P < 0.001$)	Sood et al. [20]
VSL#3 (3.6×10^{12}) twice daily	Open-label, 6 weeks, 34 patients	5-ASA, 6-MP, azathioprine	Remission rates 53%, 9% no response, 9% worsened ($P < 0.001$)	Biblioni et al. [21]
BIO-THREE	Open-label, 4 weeks, 20 patients	None	Remission rates 45%, 40% no response, 5% worsened	Tsuda et al. [23]
S. boulardii (250 mg tid)	Noncontrolled pilot study, 4 weeks, 25 patients	Mesalazine	Clinical and endoscopic remission rate 70.8%	Guslandii et al. [24]
BIFICO (1.26 g/day)	RCT versus PL, 8 weeks, 30 patients	Glucocorticoids, sulfasalazine	Relapse rates 20% versus 93.3% ($P < 0.01$)	Cui et al. [25]
L. reuteri ATCC 55730 enema	Prospective RCT versus PL, 8 weeks, 40 patients	Mesalamine	Remission rates 31% versus 0% ($P < 0.05$) Decreased Mayo score ($P < 0.01$)	Olivia et al. [26]
VSL#3 (pediatrics) weight based	Prospective, RCT versus PL, 12 months, 29 patients	Glucocorticoids, mesalamine	Remission rates 92.8% versus 36.4% ($P < 0.001$) Relapse rates 21.4% versus 73.3% (P 0.014)	Miele et al. [22]

5-ASA, 5-Aminosalicylic acid; 6-MP, 6-mercaptopurine.
$P < 0.05$ significant.

compared to 33.9% on mesalazine ($P = 0.003$). Subgroup analysis showed no differences in relapse rates with regard to duration, localization, or pretrial treatment. Both treatments were well tolerated, with 11 (EcN 6.7% vs. mesalazine 6.9%) patients in each arm withdrawing from the study for symptoms of worsening disease. This study again highlights that EcN is comparable to mesalazine, and provides an alternative therapeutic option to maintain remission. Additionally, EcN is more cost-effective and less harmful versus conventional medications [16,29]. Of note, mesalazine 1.5 g daily is lower than the recommended 5-ASA maintenance dose used in clinical practice.

Lactobacillus GG

Lactobacillus GG (LGG) is an alternative probiotic preparation proven beneficial in maintenance of remission. In clinical trials, LGG induces Th1 chemokine production and downregulates some proinflammatory cytokines. It has also been shown to modulate expression of genes involved in immune responses, cell adhesion, and cell–cell signaling, all of which function to strengthen the epithelial barrier [30]. Zocco et al. [31] performed a single center, prospective, open-label randomized trial in 187 patients with confirmed clinical and endoscopic remission for at least 12 months. Patients were randomized to receive LGG 18×10^9 CFU/day, mesalamine 2400 mg/day, or *Lactobacillus* plus mesalamine. The primary objective was to determine efficacy of LGG by comparing relapse rates between the different treatment arms, and secondarily to assess relapse-free duration. There was no significant difference in UCDAI scores, endoscopic or histologic findings, or maintenance of remission at 6 and 12 months (6 months $P = 0.44$, 12 months $P = 0.77$). However, both the probiotics and combination treatment groups showed a significantly longer relapse-free survival compared to standard therapy with mesalamine ($P < 0.001$ and $P < 0.03$, respectively) [16]. Although there was no significant clinical improvement, LGG demonstrated benefit in delaying time to relapse. Additional research is required to promote and advance its abilities as an adjuvant therapy in maintaining remission.

Dual Probiotic Therapies

Dual probiotic combinations have also been studied in maintenance of UC remission. Bifidobacteria-fermented milk (BFM), consisting of *B. breve, Bifidobacterium bifidum,* and *Lactobacillus*, showed promise as an adjunctive therapy in addition to preventing disease relapse. Ishikawa et al. [32] compared BFM 100 mL/day with standard medical therapy versus a control group. Twenty-one patients were assessed over 1 year with routine bloodwork, colonoscopy, and fecal analysis. The treatment arm showed significant reduction in exacerbation rate compared to the control

group (BFM 27.3% vs. control 90%, $P = 0.0184$). BFM demonstrates usefulness in maintaining disease remission; however, the encouraging potential is limited by a paucity of participants and trials.

Probio-Tec-AB-25, a blend of *L. acidophilus* and *Bifidobacterium animalis*, was examined in 32 patients with left-sided UC in remission, in a randomized, double-blind, placebo-controlled trial for 52 weeks [33]. In this specific trial, no additional standard medications were allowed. The primary outcome was maintenance of remission, with secondary objectives including time to relapse and probiotic safety/tolerability. There was no statistically significant difference in maintenance of remission [25% (5/20) in the probiotics group versus 8% (1/12) in the placebo group, P 0.37] or median time to relapse that was 125 days in the probiotics group compared to 104 days in the placebo group ($P = 0.683$). Probio-Tec-AB-25 was well tolerated; gastrointestinal side effects (e.g., flatulence) were the most commonly reported. No serious adverse effects were noted, and no patients in the probiotics arm withdrew from the study. Overall, the probiotics arm achieved a higher rate or remission and longer time to relapse. However, these findings were not significant, likely limited by the small sample size. Furthermore, a lack of combination therapy with standard medications may have contributed to the higher than expected rate of relapse. However, this study provides promising evidence that probiotics are noninferior to 5-ASA therapies and provide a safe, and well-tolerated alternative option in maintaining remission of UC (Table 37.2).

Prebiotics

Prebiotics are food substances not digested in the small bowel that provide benefit to the host by demonstrating fermentability and by serving as a substrate for several beneficial gut commensal organisms to flourish [4,34]. This ability of selectivity allows for modulation of the microbiota by stimulating the growth and metabolism of protective commensal enteric bacteria, and more importantly, increases production of SCFAs. The literature well documents that both prebiotic and probiotic supplementation increases fecal SCFA levels, with associated reduction in the production, and potentially the release of TNF-α in animal models and attenuation of IL-6 and IL-8 levels in both human and experimental models [35,36]. Diets high in fermentable fiber are associated with increased bacterial diversity and *Bacteroides* counts, and increase butyrate concentrations, allowing these fibers to act as an antiinflammatory substances by downregulating the proinflammatory NFκB pathway [3,16]. The exact mechanism of action is unclear, but has been associated with specific growth of bifidobacteria. Multiple theories have been postulated, including: bifidobacteria use of oligosaccharides and complex carbohydrates as their energy source, and their higher growth rate in the presence of non-

TABLE 37.2 Probiotic Intervention Studies in Inactive Ulcerative Colitis

Probiotic, Dose	Design, Duration, Number of Patients	Concomitant Therapy	Clinical Outcome	References
VSL#3	Noncontrolled trial, 12 months, 20 patients	None	Remission rates 75%	Venturi et al. [27]
E. coli Nissle 1917 (5×10^{10})	RCT versus standard therapy, 12 weeks, 103 patients	None	Similar relapse rates 16.0% versus 11.3%	Kruis et al. [28]
E. coli Nissle 1917 (5×10^{10})	RCT versus mesalazine, 12 months, 327 patients	None	Similar relapse rates 36.4% versus 33.9% (*P* 0.003)	Kruis et al. [29]
E. coli Nissle 1917 (5×10^{10})	RCT versus standard therapy, 12 months, 83 patients	None	Remission rates 68% versus 75% Time to remission 42 days versus 44 days (*P* 0.0092) Relapse rates 67% versus 73%	Rembacken et al. [17]
Lactobacillus rhamnosus (18×109 CFU/day)	Open-label RCT, 12 months, 187 patients	None	Relapse rates at 6 months (*P* 0.44), 12 months (*P* 0.77)	Zocco et al. [31]
BFM 100 mL/day	RCT versus standard, 21 patients, 1 year	Baseline therapy	Reduction in exacerbation rate 27.3% versus 90% (*P* 0.0184)	Ishikawa et al. [32]
Probio-Tec-AB-25	RCT versus PL, 52 weeks	None	Remission at 1 year 25% versus 8% (*P* 0.37) Median time to relapse·125 versus 104 days (*P* 0.683)	Wildt et al. [33]

BFM, Bifidobacteria-fermented milk.

digestible oligosaccharides versus that of disease-inducing organisms [34].

There is a paucity of clinical trials examining prebiotics in UC; however, experiments in animal models show an apparent benefit, with some being dose- and duration-dependent [37]. To date, the most studied prebiotics are nondigestible carbohydrates, such as fructo-oligosaccharides (FOS) inulin, galacto-oligosaccharides (GOS), *trans*-galacto-oligosaccharides (TOS), and lactulose. Other nondigestible carbohydrates undergoing investigation are soya bean oligosaccharides, isomalto-oligosaccharides, xylo-oligosaccharides, polydextrose, glucans, cereal-derived arabinoxylans, and arabinoxylan oligosaccharides.

Germinated barley foodstuff (GBF) is derived from aleurone and scutellum fractions of germinated barley, and consists of insoluble glutamine-rich protein and dietary fiber. In experimental colitis induced with dextran sulfate sodium (DSS), GBF administration increased butyrate production, markedly reduced colonic inflammation, and accelerated colonic epithelial repair [38,39]. GBF also contains glutamine, which has three major roles in treating and preventing UC. Glutamine is also a major nutrient of enterocytes and colonocytes, and promotes mucosal healing in animal models, stimulates proliferation of intestinal epithelial cells, prevents bacterial translocation from the gut lumen into the internal milieu, and it is stable in the presence of gastric acid. Though studies in IBD are inconclusive, surgical patients supplemented with TPN (with glutamine added) showed decreased IL-6 levels, suggesting it

plays a role in proper functioning of cells in the immune system [40,41].

GBF has demonstrated efficacy in two trials, and was well tolerated. Mitsuyama et al. [38] conducted a pilot non-randomized, open-label trial in ten patients nonresponsive to standard therapy with 5-ASA and low-dose prednisolone for 4 weeks. One patient had pan-colitis, six with left-sided disease, and three with proctitis, all characterized as mild to moderate disease activity. All participants were given GFB (30 g) daily, in three doses for 4 weeks. Results were measured using CAI and endoscopic index scores, laboratory values, and stool SCFA measurements. Both CAI and endoscopic index scores showed a significant reduction (*P* < 0.05 and < 0.0001 respectively), and were regardless of disease extent. C-reactive protein (CRP) and erythrocyte sedimentation rate (ESR) levels decreased, but not significantly. Stool SCFA analysis; however, showed a significant increase in butyrate and acetate concentrations (*P* < 0.05 for both). Importantly, no adverse effects on patients were documented. These findings were confirmed in multicenter, randomized, open-label study in18 patients with mild to moderate activity UC. The treatment group received 20–30 g GBF daily for 4 weeks in addition to baseline therapy, while the control group received standard treatment. At 4 weeks, the GBF group had a significant improvement in the CAI score (*P* < 0.05), but no significant change in serum CRP parameters or endoscopic score. No adverse effects were documented in this study [42]. Though low-powered, these trials show promising benefit for GBF as an adjunctive therapy in the treatment and prevention of mild to moderately active UC.

Another more recent randomized controlled trial with GBF examined its role in production of proinflammatory cytokines [35]. This study was slightly larger, with 41 patients enrolled with varying degrees of disease severity, from remission to mild to moderate disease activity. Two-thirds of the participants had left-sided disease, versus one-third with pan-UC. Twenty-one patients in the control group received standard medical therapy, while the GBF group (20 patients) received 30 g of GBF per day, 3 times daily plus standard medical therapy for 2 months. Both pre- and posttreatment TNF-α, IL-6, and IL-8 levels were measured. Pretreatment values showed no statistically significant difference. The control group demonstrated nonstatistically significant increased TNF-α, IL-6, and IL-8 levels, IL-6 (P 0.08, 0.46, and 0.35, respectively). The GBF group, on the other hand, showed decreased TNF-α, IL-6, and IL-8 levels, with reduction in both IL-6 and IL-8 values achieving statistical significance ($P = 0.034$ and 0.013, respectively). No adverse effects were noted in the treatment group. This study established that prebiotics, specifically GBF, attenuate proinflammatory cytokine level production. However, larger randomized controlled trials are needed to completely characterize its efficacy in the treatment and maintenance of UC (Table 37.3).

Synbiotics

Synbiotics are essentially a synergistic combination of probiotics and prebiotics, which serve to improve the therapeutic benefits of probiotics by combining them with prebiotics to enhance their growth in the colon [4]. A pilot randomized controlled trial with a prebiotic (Synergy 1) and a probiotic *B. longum*, which has been shown to decrease IL-1α mRNA and protein levels, in 18 patients demonstrated potential benefit compared to placebo [43]. The treatment group received 2×10^{11} freeze dried viable *B. longum* gelatin capsule and a sachet with 6 g of the prebiotic, FOS/inulin mixture (Synergy 1) twice daily for 4 weeks. Clinical disease activity was monitored using a daily bowel habit diary and CRP level. The synbiotic group showed a decreased CRP level, (mean 1.8, SD 3.9), while no patients in the placebo group had an elevated CRP level. The treatment arm was found to have a significant reduction in TNF-α and IL-1α mRNA levels ($P = 0.0175$ and 0.0379, respectively), and no statistically significant difference in the regulatory cytokine IL-10 mRNA expression between the two groups. Sigmoidoscopy score demonstrated a trend in improvement in the synbiotic group ($P = 0.06$). The bowel habit index score increased by 70.4% in the placebo group, but showed a 20.4% reduction in the treatment group. Furthermore, representative biopsies from the synbiotic group displayed resolution of acute inflammatory changes, with disappearance of crypt abscesses and regeneration of the epithelial barrier, compared to the placebo group, with larger abscesses with crypt rupture and increased number of infiltrating cells identified. Two patients from the placebo group withdrew from the study due to colitis exacerbation. Synergy 1 appeared to be beneficial in this small study; however, larger controlled studies are needed fully optimize its use as an adjunctive therapy.

Earlier studies have shown the benefits of *B. breve* strain Yakult, a probiotic in BFM, in patients with UC. This study by Ishikawa et al. [44] builds on this work with the addition of a prebiotic, GOS. The primary endpoint of his study was to determine if synbiotic use improved endoscopic scores graded by the Matts classification (grade 1, normal; grade 2, mild granularity of the mucosa, with mild contact bleeding; grade 3, marked granularity and edema of the mucosa, contact bleeding and spontaneous bleeding; grade 4, severe ulceration of the mucosa with hemorrhage) in six regions—cecum, ascending colon, transverse colon, descending colon, sigmoid colon, and rectum (the highest score determined endoscopic score). Additionally, fecal microbiological analysis was performed prior to the start of the study and at 1 year. Twenty-one patients were assigned to the synbiotic group and 20 to the control group; however, two patients

TABLE 37.3 Prebiotic Intervention Studies in Inactive Ulcerative Colitis

Prebiotic, Dose	Design, Duration, Number of Patients	Concomitant Therapy	Clinical Outcome	References
GBF (30 g/day)	Open-label,10 patients, 4 weeks	5-ASA, low dose prednisolone	Significant reduction in clinical activity index and endoscopic index scores ($P < 0.05$; <0.0001)	Mitsuyama et al. [38]
GBF (20–30 g/day)	Randomized open-label versus control, 18 patients, 4 weeks	Baseline therapy	Significant improvement only in clinical activity index score ($P < 0.05$)	Kanauchi et al. [42]
GBF (30 g/day)	RCT versus standard, 41 patients, 2 months	Baseline therapy	Significant reduction in IL-6 and IL-8 levels (P 0.034)	Faghforri et al. [35]

GBF, Germinated barley foodstuff.

from the control group withdrew from the study—one underwent total colectomy and one had a disease flare. The synbiotics group received freeze-dried powder of *B. breve* strain Yakult 10^9 CFU/g 3 times daily and 5.5 g GOS once daily. At the end of the study, the treatment arm demonstrated a significantly decreased endoscopic score compared to the control group ($P < 0.05$). Fecal microbiological analysis showed a statistically significant decrease in fecal numbers of Bacteriodaceae ($P < 0.05$) after treatment but not in bifidobacteria numbers. Fecal pH was also significantly decreased ($P < 0.05$) in the synbiotic group, suggesting increased production of luminal SCFAs or lactate. In addition to significant endoscopic improvement, there was a positive modulation of the gut microbiota to promote a more beneficial environment for the host.

Synbiotics appear to also have an added benefit of improved quality of life (QOL). Fujimoro et al. [45] conducted a randomized controlled trial with 120 patients over 4 weeks, with three treatment groups: prebiotics, probiotics, and synbiotics. Participants continued on their standard therapeutic regimens. The probiotics group received *B. longum* 2×10^9 colony forming units/capsule once daily, the prebiotics group 4 g of psyllium dissolved in 100 mL of water twice daily, while the synbiotics group was given both. Efficacy was measured using the inflammatory bowel disease questionnaire at three intervals: onset of the trial, 2-week midpoint, and at the end of the 4-week study. Overall, only the synbiotic group showed a significantly improved score ($P = 0.03$). In the probiotics arm, emotional function increased significantly but there was no improvement in bowel function. Prebiotics, on the other hand, showed a significant improvement bowel function only. Breakdown from the synbiotic group found statistically significant improvements in social, emotional, and functional facets (P 0.02, 0.05, 0.008 respectively); however, these changes did not extend to bowel function (P 0.12). Serum analysis showed significant increase in hemoglobin and hematocrit levels in the probiotic arm (P 0.04, 0.04 respectively), while there was a significant decrease in CRP level in the synbiotic group (0.04). No changes were noted in the prebiotics group. Though QOL measurements improved significantly in the synbiotic group for patients in

disease remission, several limitations were present. These include the short study duration, open-label design versus placebo-controlled trial, lower dose *B. longum*, and lack of standard disease evaluation with endoscopic and histologic assessment. In addition, psyllium is not being regarded as a prebiotic, as this is a nonfermentable fiber (Table 37.4).

Fecal Microbiota Therapy

Fecal microbiota therapy (FMT), or fecal transplant, is a modality in which stool from normal, healthy individuals is implanted into a diseased gastrointestinal tract with therapeutic intent. It has joined the available armamentarium as another novel approach to altering the gut microbiota. It has demonstrated excellent efficacy in refractory *Clostridium difficile* infection; however, the data from case series in IBD has been less conclusive. A pilot double-blind, randomized controlled trial of FMT versus placebo in 75 patients with active UC assessed induction of remission at week 7 [defined as total Mayo score <3 and complete mucosal healing (Mayo score 0)], as well as symptom improvement, Mayo and IBD-Q scores [46]. Thirty-eight patients in the treatment group received 50 mL FMT, while 37 participants were on placebo (water) via retention enema, given once weekly. Stool samples were collected weekly (prior to treatment administration) for fecal microbiota analysis. Patients continued on their concomitant medications. At week 7, the FMT arm showed significant improvement in inducing remission (24% vs. 5%, $P = 0.03$). Neither group achieved an endoscopic Mayo score 0. Of the nine patients in remission in the treatment group, seven had no histologic evidence of active inflammation, while the others displayed mild patchy rectal inflammation and normal mucosa in the sigmoid and descending colon. Patients on immunosuppressive therapy appeared to derive more benefit with FMT, though not statistically significant (46% vs. 11%, $P = 0.09$). However, patients with a more recent diagnosis (<1 year) responded more significantly to FMT (75% vs. 18% chronic, $P = 0.04$). There was no statistically significant difference in symptoms, QOL scores, or serious adverse events between the groups. Five patients suffered a serious adverse event—one patient in the placebo group

TABLE 37.4 Synbiotic Intervention Studies in Inactive Ulcerative Colitis

Synbiotic, Dose	Design, Duration, Number of Patients	Concomitant Therapy	Clinical Outcome	References
B longum + Synergy 1 (2×10^{11}, 6 g)	RCT versus PL, 18 patients, 1 month	Corticosteroids, 5-ASA, Immunomodulators	Sigmoidoscopy score not statistically significant (P 0.06)	Furrie et al. [43]
B. breve + GOS (1 g/day tid, 5.5 g/day)	RCT versus PL, 1 year, 41 patients, 1 year	5-ASA, corticosteroids	Endoscopic improvement ($P < 0.05$)	Ishikawa et al. [44]

GOS, Galacto-oligosaccharides.

developed worsening colitis requiring urgent colectomy and three patients (two FMT, one placebo) developed patchy colonic inflammation and rectal abscess formation (resolved with antibiotic use). One patient in the FMT arm withdrew from the study with abdominal discomfort and a positive *C. difficile* toxin result. Eight of the nine patients in remission in the FMT group maintained remission at week 52, without relapse. Regarding microbiota composition change, the treatment group showed a significant increase in microbial diversity at week 6 ($P = 0.02$). Furthermore, fecal microbial analysis found that the treatment samples resembled more their specific donor rather than a control sample ($P = 0.04$). This study demonstrated that FMT is significantly beneficial in inducing remission, especially earlier in the UC disease course. However, certain factors still remain to be clarified, such as donor-dependent efficacy, route of administration, and pretransplantation antibiotics.

Moayyedi et al. [47] study provided a foundation for FMT as an alternative therapy in UC. A Dutch group performed a single-center, double-blind, placebo-controlled, randomized, proof-of-concept phase 2 trial that examined efficacy of FMT between control and autologous stool samples in 80 patients [47]. Participants continued their baseline treatment regimens. Patients were pretreated with bowel lavage both the evening before and morning of scheduled nasoduodenal administration. A median amount of 120 g was used (range 85–20 g). Participants received a second dose 3 weeks later. Clinical remission (SCCAI score < 2 and >1 point improvement on the combined Mayo endoscopic score) was assessed at 12 weeks after the first treatment. There was no statistically significant difference in clinical remission or endoscopic response at week 6 or 12 between the healthy (FMT-D) or autologous (FMT-A) donor recipient groups [26.1% (FMT-D) vs. 32% (FMT-A), $P = 0.76$, 30.4% vs. 32%, $P = 1$; 21.7% vs. 36%, $P = 0.35$, 34.7% vs. 36% $P = 1$, respectively). At week 6 and 12, the autologous treatment group showed improved clinical response, but it was not significant (52% vs. 43.5%, $P = 0.58$; 52% vs. 47.8%). In the healthy donor transplant group, one patient lost response at week 12, but two nonresponders at week 6 did show response at week 12. The FMT-arm showed loss of response in two patients at week 12, while two participants gained response by week 12. Three patients in each treatment arm required rescue therapy for ongoing disease flare. Most patients reported mild adverse effects, with transient borborygmus was most commonly reported (FMT-D 78.3% vs. FMT-A 64%, $P = 0.28$); and no patient was required to withdraw from the trial. Two patients in the FMT-D arm reported serious adverse events—one patient suffered a small bowel perforation secondary to small bowel CD and was treated conservatively, while one patient in the FMT-A group developed CMV infection 7 weeks post initial treatment infusion. One patient

underwent operation for cervical carcinoma. Overall, none of these occurrences was thought to be secondary to FMT therapy. Fecal microbiota analysis at 12 weeks showed a significant increase in bacterial diversity in study responders (FMT-D $P = 0.06$, FMT-A $P = 0.04$), but no change in the nonresponders. FMT-D responders showed changes more consistent with their donor's microbiota composition. Specifically, the FMT-D group had higher *Clostridium* and lower *Bacteroides* concentrations, while the FMT-A arm displayed increased numbers of bacilli, Proteobacteria, and *Bacteroides*. FMT has been shown to be a beneficial technique for modifying the gut microbiota to produce an environment capable of mucosal healing. However, these trials emphasize that pretreatment factors, and especially the donors, can affect efficacy, and that larger trials are needed to enhance its potential as an alternative therapeutic modality.

CONCLUSIONS

In summary, IBD is a disorder characterized by dysregulation of the mucosal immune response to commensal gut microflora in a genetically susceptible individual. Treatment of IBD first focuses on inducing remission, followed by maintenance of remission to prevent disease progression. Though the available antiinflammatory and immune modulatory therapies are effective, they are not without toxicity. Alternative therapies, such as probiotics, prebiotics, and FMT could help to alter gut dysbiosis and improve the function of the microbiota to offset the development of inflammation, especially when the disease is in remission to prevent another relapse or during mild inflammation. They provide an alternative or adjunct therapeutic option with a more optimal safety profile than current standard medications. Specifically, in larger controlled trials only specific probiotics, such as VSL#3 and EcN at their studied doses were found to show a significant benefit in induction or maintenance of remission of UC, respectively. Regarding prebiotics, there are promising data with GBF, but more work is needed to enhance its full potential in the use of treatment of UC. Two clinical trials with FMT have proven it effective in favorably shifting the gut microbiota composition, and potentially promoting mucosal healing. And importantly, each of these agents is well tolerated with ideal safety profiles.

All of these treatments have also demonstrated capability in functioning as nonpharmacologic adjunct therapies; however, further progress is required before their mainstream use. Their ability to positively modify the gut microbiota allows them to have an integral role in individualized therapy of IBD. Progress in understanding disease phenotype has shown us that each patient is unique, and thus requires personalized care to treat and prevent disease flares. Future directions include the administration or improving the function of more abundant beneficial microbes, and selecting targets based on the individuals' specific endogenous gut microbiota composition, if these correlate with clinical response. With these specific alterations of the gut microbiome, current and new nonpharmacological options may play a major role in tailored adjunct therapies for UC.

REFERENCES

[1] Bellavia M, Tomasello G, Romeo M, et al. Gut microbiota imbalance and chaperoning system malfunction are central to ulcerative colitis pathogenesis and can be counteracted with specifically designed probiotics: a working hypothesis. Med Microl Immunol 2013;202: 393–406.

[2] Scalcaferri F, Gerardi V, Lopetuso LR, et al. Gut microbial flora, prebiotics, and probiotics in IBD: their current usage and utility. Biomed Res Int 2013;10:1–9.

[3] Barnes D, Yeh AM. Bugs and guts: practical applications of probiotics for gastrointestinal disorders in children. Nutr Clin Pract 2015;30:747–59.

[4] Bernstein C. Antibiotics, probiotics, and prebiotics. Nutrition, gut microbiota and immunity. Therapeutic targets for IBD. Nestle Nutr Inst Workshop Ser 2014;79:83–100.

[5] Bellavia M, Tomasello G, Romeo M, et al. Gut microbiota imbalance and chaperoning system malfunction are central to ulcerative colitis pathogenesis and can be counteracted with specifically designed probiotics: a working hypothesis. Med Microbiol Immunol 2013;202:393–406.

[6] Sartor RB. The intestinal microbiota in inflammatory bowel diseases. Nutrition, gut microbiota and immunity. Therapeutic targets for IBD. Nestle Nutr Inst Workshop Ser 2014;79:29–39.

[7] Kostic AD, Xavier RJ, Gevers D. The microbiome in inflammatory bowel disease: current status and the future ahead. Gastroenterology 2014;146:1489–99.

[8] Jonkers D, Penders J, Masclee A, Pierik M. Probiotics in the management of inflammatory bowel disease. A systematic review of intervention studies in the adult patients. Drugs 2012;72(6):803–23.

[9] Lavelle A, Lennon G, O'Sullivan O, et al. Spatial variation of the colonic microbiota in patients with ulcerative colitis and control volunteers. Gut 2015;64(10):1553–61.

[10] Machiels K, Joossens M, Sabino J, et al. A decrease of the butyrate-producing species Roseburia hominis and Faecalibacterium prausnitzii defines dysbiosis in patients with ulcerative colitis. Gut 2014;63:1275–83.

[11] Eeckhaut V, Machiels K, Perrier C, et al. Butyricicoccus pullicaecorum in inflammatory bowel disease. Gut 2013;62:1745–52.

[12] Varela E, Manichanh C, Gallart M, et al. Colonisation by Feacalibacterium prausnitzii and maintenance of clinical remission in patients with ulcerative colitis. Aliment Pharmacol Ther 2013;38:151–61.

[13] Ghouri YA, Richards DM, Rahimi EF, et al. Systematic review of randomized controlled trials of probiotics, prebiotics, and synbiotics in inflammatory bowel disease. Clin Exp Gastroenterol 2014;7: 473–87.

[14] Chen W-X, Ren L-H, Shi R-H. Enteric microbiota leads to new therapeutic strategies for ulcerative colitis. World J Gastroenterol 2014;20(42):15657–63.

[15] Fedorak R, Demeria D. Probiotic bacteria in the prevention and the treatment of inflammatory bowel disease. Gastroenerol Clin N Am 2012;41:821–42.

[16] Orel R, Trop TK. Intestinal microbiota, probiotics, and prebiotics in inflammatory bowel disease. World J Gastroenterol 2014;20(33): 11505–24.

[17] Rembacken BJ, Snelling AM, Hawkey PM, et al. Non-pathogenic Escherichia coli versus mesalazine for the treatment of ulcerative colitis: a randomized trial. Lancet 1999;354:635–9.

[18] Tursi A, Brandimarte G, Giorgetti GM, et al. Low-dose balsalazide plus a high-potency probiotic preparation is more effective than balsalazide alone or mesalamine in the treatment of acute mild-to-moderate ulcerative colitis. Med Sci Momit 2004;10:P1126–31.

[19] Tursi A, Brandimarte, Papa A, et al. Treatment of relapsing mild to moderately active ulcerative colitis with the probiotic VSL#3 as adjunctive to a standard pharmaceutical treatment: a double-blind, randomized, placebo controlled study. Am J Gastroenterol 2010;105:2218–27.

[20] Sood A, Midha V, Makharia GK, et al. The probiotic preparation, VSL#3 induces remission in patients with mild to moderately active ulcerative colitis. Clin Gastroenterol Hepatol 2009;7:1202–9.

[21] Biblioni R, Fedorak RN, Tannock GW, et al. VSL#3 probioitc-mixture induces remission in patients with active ulcerative colitis. Am J Gastroenterol 2005;100:1539–46.

[22] Miele E, Pascarell F, Giannetti E, et al. Effect of a probiotic preparation (VSL#3) on induction and maintenance of remission in children with ulcerative colitis. Am J Gastroenterol 2009;104:437–43.

[23] Tsuda Y, Yoshimatsu Y, Aoki H, et al. Clinical effectiveness of probiotics therapy (BIO-THREE) in patients with ulcerative colitis refractory to conventional therapy. Scand J Gastroenterol 2007;42:1306–11.

[24] Guslandi M, Giollo P, Testoni PA. A pilot trial of Saccharomyces boulardii in ulcerative colitis. Eur J Gastroenterol Hepatol 2003;15:697–8.

[25] Cui HH, Chen CL, Wang JD, et al. Effects on probiotic on intestinal mucosa of patients with ulcerative colitis. World J Gastroenterol 2004;10:1521–5.

[26] Olivia S, Di Nardo G, Ferrari F, et al. Randomised controlled trial: the effectiveness of Lactobacillus reuteri ATCC 55730 rectal enema in children with active distal ulcerative colitis. Aliment Pharamcol Ther 2012;35:327–34.

[27] Venturi A, Gionchetti P, Rizzello F, et al. Impact on the composition of the faecal flora by a new probiotic preparation: preliminary data on maintenance treatment of patients with ulcerative colitis. Aliment Pharmacol Ther 1999;13:1103–8.

[28] Kruis W, Schutz E, Fric P, et al. Double-blind comparison of an oral Escherichia coli preparation and mesalazine in maintaining remission of ulcerative colitis. Aliment Pharmacol Ther 1997;11:853–8.

[29] Kruis W, Fric P, Pokrotneiks J, et al. Maintaining remission of ulcerative colitis with the probiotic Escherichia coli Nissle 1917 is as effective as with standard mesalamine. Gut 2004;53:1617–23.

[30] Di Cafo S, Tao H, Gillo A, et al. Effects of Lactobacillus GG on genes expression patter in small bowl mucosa. Dig Liver Dis 2005;37:320–9.

[31] Zocco MA, dal Verme LZ, Cremonini F, et al. Efficacy of Lactobacillus GG in maintaining remission of ulcerative colitis. Aliment Pharmacol Ther 2006;23:1567–74.

[32] Ishikawa H, Akedo I, Umesaki Y, et al. Randomized controlled trial of the effect of bifidobacteria-fermented milk on ulcerative colitis. J Am Coll Nutr 2003;22:56–63.

[33] Wildt S, Nordgaard I, Hansen U, et al. A double-blind placebo-controlled trial with Lactobacillus acidophilus La-5 and Bifidobacterium animalis subspecies lactis BB-12 for maintenance of remission in ulcerative colitis. J Crohns Colitis 2011;5:115–21.

[34] Martinez RCR, Bedaini R, Saad SMI. Scientific evidence for health effects attributed to the consumption of probiotics and prebiotics: an update for current perspectives and future challenges. Br J Nutr 2015;114:1993–2015.

[35] Faghfoori Z, Navai L, Shakerhosseini R, et al. Effects of oral supplementation of germinated barley foodstuff on serum tumor necrosis factor-a, interleukin-6 and -8 in patients with ulcerative colitis. Ann Clin Biochem 2011;48:233–7.

[36] Hafer A, Kramer S, Duncker S, et al. Effect of oral lactulose on clinical and immunohistochemical parameters in patients with inflammatory bowel disease: a pilot study. BMC Gastroenterol 2007;7:36–47.

[37] Soldavini J, Kaunitz JD. Pathobiology and potential therapeutic value of intestinal short-chain fatty acids in gut inflammation and obesity. Dig Dis Sci 2013;58:2756–66.

[38] Mitsuyama K, Saiki T, Kanauchi O, et al. Treatment of ulcerative colitis with germinated barley foodstuff feeding: a pilot study. Ailment Pharmacol Ther 1998;12:1225–30.

[39] Kanauchi O, Iwanaga T, Andoh A, et al. Dietary fiber fraction of germinated barley foodstuff attenuated mucosal damage and diarrhea, and accelerated the repair of colonic mucosa in an experimental colitis. J Gastroenterol Hepatol 2001;16:160–8.

[40] Triantafillidis JK, Papalois AE. The role of total parental nutrition in inflammatory bowel disease: current aspects. Scand J Gastroenterol 2014;49(1):3–14.

[41] Bamba T, Kanauchi O, Andoh A, et al. A new prebiotic from germinated barley for nutraceutical treatment of ulcerative colitis. J Gastroenterol Hepatol 2002;17:818–24.

[42] Kanauchi O, Suga T, Tochihara M, et al. Treatment of ulcerative colitis by feeding with germinated barley foodstuff: first report of a multicenter open control trial. J Gastroenterol 2002;37(Suppl. 14):67–72.

[43] Furrie E, Macfarlane S, Kennedy A, et al. Synbiotic therapy (Bifidobacerium longum/Synergy 1) initiates resolution of inflammation in patients with active ulcerative colitis: a randomized controlled pilot trial. Gut 2005;54:242–9.

[44] Ishikawa H, Matsumoto S, Ohashi Y, et al. Beneficial effects of probiotic Bifidobacterium and galacto-oligosaccharide in patients with ulcerative colitis: a randomized controlled study. Digestion 2011;84:128–33.

[45] Fujimori S, Gudis K, Mitsui K, et al. A randomized controlled trial on the efficacy of synbiotic versus probiotic or prebiotic treatment to improve the quality of life in patients with ulcerative colitis. Nutrition 2009;25:520–5.

[46] Moayyedi P, Surette M, Kim P, et al. Fecal microbiota transplantation induces remission in patients with active ulcerative colitis in a randomized controlled trial. Gastroenterology 2015;149:102–9.

[47] Rossen NG, Fuentes S, van der Spek MJ, et al. Findings from a randomized controlled trial of fecal transplantation for patients with ulcerative colitis. Gastroenterology 2015;149:110–8.

Chapter 38

Treatment of Functional Bowel Disorders With Prebiotics and Probiotics

K. Hod*,** and Y. Ringel†,‡

*Department of Epidemiology and Preventive Medicine, School of Public Health, Sackler Faculty of Medicine, Tel Aviv University, Tel Aviv, Israel; **Research Division, Epidemiological Service, Assuta Medical Center, Tel Aviv, Israel; †Department of Gastroenterology, Rabin Medical Center, Petach Tikva, Israel; ‡Department of Medicine, University of North Carolina School of Medicine at Chapel Hill, Chapel Hill, NC, United States

FUNCTIONAL BOWEL DISORDERS— DEFINITIONS, EPIDEMIOLOGY, AND CLINICAL CONDITIONS

Functional bowel disorders (FBDs) are a group of functional gastrointestinal (GI) disorders characterized by symptoms attributed to the middle and lower GI tract not explained by unified structural and/or biochemical abnormalities [1,2]. FBDs include irritable bowel syndrome (IBS), functional bloating (FB), chronic idiopathic constipation (CIC), functional diarrhea (FD), and unspecified FBD [1]. In the absence of identifiable unified etiopathophysiology the diagnosis of FBDs relies on clinical presentation, use of symptom-based criteria, and limited investigations to exclude possible other causes [2]. Important factor in the diagnosis of FBDs is the chronicity of the symptoms which helps separate these conditions from acute, transient, self-limited conditions. Accordingly, the commonly used diagnostic criteria for IBS (Rome III criteria) require the presence of the typical symptoms for at least 6 months prior to diagnosis. FBD are highly prevalent in Western countries with IBS being the most prevalent (10–20%) and best studied condition [3]. Although FBDs have not been found to have significant impact on life expectancy, they account for significant morbidity, utilization of healthcare, and socioeconomic burden [4,5].

Traditionally, IBS and other FBDs have been considered as disorders arising from abnormal function along the brain–gut axis associated with GI hypersensitivity (which may lead to discomfort and pain) and GI motor dysfunction (which may lead to diarrhea, constipation, or alternating bowel movements). However, despite intensive research over the years, no single ethological factor with a defined pathogenic mechanism has been identified thus the pathophysiology of these disorders is still not completely understood. Nevertheless, the research in this area has implicated new theories and suggested additional new underlying pathophysiological mechanisms including genetic predeterminants [6], peripheral GI factors [7], and extraintestinal neurohormonal and central factors [8].

The Intestinal Microbiota in Functional Bowel Disorders

The intestinal microbiota, the complex community of microorganisms residing in the GI tract is believed to contain greater than 1000 different bacterial species that can reach viable numbers of 10^{14} bacteria per gram of luminal content. The highest density of the human intestinal microbiota is in the colon and is dominated by two main bacterial genera *Firmicutes* (64%) and *Bacteroidetes* (23%) followed by *Proteobacteria* (8%) and *Actinobacteria* (3%) [9,10]. At the intestinal level the human host can respond to commensal and pathogenic bacteria via multiple mechanisms including epithelial receptor-mediated signaling or direct stimulation of enteric neurons and immune cells. In addition, there is growing evidence for a broader interaction with the host including bidirectional communications between the intestinal microbiota the peripheral (enteric) and central (brain) nervous systems [11]. Indeed, studies in animal models have shown that products of microbiota have the potential to affect the excitability of enteric and vagal afferents neurons [12], as well as brain functions and behavior [13]. Conversely, the enteric microbiota can be influenced though the brain effects on intestinal motility, secretion, and immune function [14] thus creating the microbiota–gut–brain axis [15].

The intestinal microbiota plays an important role in maintaining the normal function of the GI tract. The importance of the intestinal microbiota in the pathogenesis and/or clinical manifestation of FBDs has been demonstrated by epidemiological, physiological, and microbiological studies. Epidemiological observations have demonstrated that IBS symptoms are often developed following acute gastroenteritis (i.e., postinfectious IBS) [16,17] or use of antibiotics; although for the latter the association is less established and based only on a few small retrospective cross-sectional studies [18]. An association between IBS and small intestinal bacterial overgrowth has also been suggested but the data on this association is nonconclusive and the relationship is debatable [19–21].

From the microbiology perspective, accumulating body of evidence suggests that the intestinal microbiota differ between healthy controls and at least some patients with IBS [22–24]. The current data indicate compositional differences including reduced microbial diversity [22,25,26] and differences in specific bacterial taxa between patients with IBS and healthy controls, and between clinically relevant subtypes of IBS [24–27]. In general patients with IBS seem to have increased levels of *Firmicutes* and decreased levels of *Bacteroidetes* [27] however, the differences in the abundance of specific species within these two microbial phyla are not consistent [26–29]. Most of the studies that investigated the intestinal microbiota in FBD were conducted in patients with IBS and assessed the fecal microbiota [30].

A recent study investigating both fecal and colonic mucosal microbiotas in patients with chronic constipation has found that the colonic mucosal microbiota in constipated patients differ than that of controls with higher abundance of genera from Bacteroidetes in the constipation group. However, while the profile of the colonic mucosal microbiota discriminated between patients with constipation and controls with 94% accuracy, this association was independent of the physiological measure of colonic transit. In contrast, the profile of the fecal microbiota was associated with colonic transit and methane production but not constipation [31]. This study demonstrates several important points: it demonstrates that the intestinal microbiota may have role in the pathogenesis of various FBDs beyond IBS, it can affect relevant physiological functions as well as functional GI symptoms, and that both fecal and intestinal mucosal microbiota are relevant and may have different effects on the human host.

Currently most of the data on the intestinal microbiota in FBD is based on association studies and the causality role of intestinal dysbiosis in these disorders is not established. Nevertheless, the growing evidence for compositional and functional changes in the intestinal microbiota in FBD and the increasing recognition of the association between intestinal dysbiosis and altered gut–brain functions have led to increased interest in manipulation of the intestinal microbiota for the treatment of these disorders. Several therapeutic interventions targeting the intestinal microbiota are increasingly used in FBDs including low fermentable oligo-, di-, monosaccharides, and polyols (FODMAP) diet [32]; antibiotics [33]; prebiotics [34]; and probiotics [35].

Prebiotics in Functional Bowel Disorders

A dietary prebiotic is defined as "selectively fermented ingredients that results in specific changes in the composition and/or activity of the gastrointestinal microbiota, thus conferring benefit(s) upon host health" [36]. The compounds identified as having the evidence of prebiotic effects are the inulin-type fructans (fructooligosaccharides, inulin, oligofructose) and galactooligosaccharides, many of which are widely present in normal diet predominantly in grains, vegetables, and legumes [37].

At this time, the data on the use of prebiotics in FBD are sparse with only a few randomized control trials (RCT) investigating the use of prebiotic in patients with IBS (Table 38.1).

The largest and the most recent RCT of prebiotics in IBS included 121 patients who randomly received either partially hydrolyzed guar gum (PHGG) (Sunfiber; Taiyo Kagaku Co., Ltd., Japan) in a dosage of 3 g/d for the first week and then 6 g/d for 11 weeks or placebo [67]. Bloating score significantly decreased in the PHGG group compared to placebo (-4.1 ± 13.4 vs. -1.2 ± 11.9, $P = 0.03$), as well as bloating + gasses score (-4.3 ± 10.4 vs. -1.12 ± 10.5, $P = 0.035$). Interestingly, the beneficial effect lasted for at least 4 weeks after the last PHGG administration. However, there was no effect with PHGG on other IBS-related symptoms (e.g., abdominal pain and change in bowel habits), nor on IBS severity and quality of life (QOL) scores. In another study 44 IBS patients were treated with either 3.5 g/d, 7 g/d of a β-galactooligosaccharide or placebo for 4 weeks. The 3.5 g/d group demonstrated improvement in a number of IBS-related symptoms including stool consistency, flatulence, bloating, and subjective global assessment (SGA), compared to the placebo group, and the higher dose group reported improvement in SGA and anxiety scores. From the microbiota perspective, this study demonstrated a bifidogenic effect with increase of bifidobacteria counts from baseline to the end of 4-week intervention period in patients receiving either 3.5 g/d ($3.25 \pm 0.51\%$ vs. $5.51 \pm 0.43\%$, $P < 0.05$, respectively) or 7 g/d ($3.01 \pm 0.38\%$ vs. $7.48 \pm 0.59\%$, $P < 0.005$, respectively) [69]. In another RCT, 105 patients with new-onset, minor, functional bowel symptoms were randomized to receive either 5 g/d oligofructose or placebo for 6 weeks [70]. Intensity of abdominal pain and frequency of digestive symptoms were reduced compared with placebo. However, the reported studies in this area have considerable limitation. The studies are relatively small, the study populations are heterogeneous, and the clinical effect is often reported only on patients that completed the protocol and not by intention to treat [36,37]. In addition, the reported studies on the effects of prebiotics in IBS are not always consistent. Two studies have found no effect of prebiotic supplementation of 6 g/d oligofructose for 2 weeks [71] or 20 g/d fructooligosaccharides for 12 weeks [72] in IBS compared with placebo. In fact, in the latter study symptoms were worsened compared to placebo at 4 weeks.

The effect of prebiotics on CIC had been investigated in a single RCT. In this trial, 60 patients were randomized to receive 3 weeks of a 15-g/d mixture of inulin and PHGG or placebo. There was no difference in satisfaction in relief of constipation between groups [9 out of 28 (32.1%) patients satisfied in the prebiotic vs. 10 out of 32 (31.3%) patients satisfied in the placebo group]. The mean number of bowel movements per week was also not statistically different between the groups (5.95 ± 2.50 in the prebiotic group vs. 6.70 ± 3.83 in the control group) [68].

Overall the current evidence for effectiveness of prebiotics in the management of FBDs is not sufficient to draw definite conclusions or recommendations. In addition, a withdrawal rate of 25–50% in some of the studies raises a question regarding the tolerability and acceptability of prebiotic intervention in this group of patents.

Probiotics in Functional Bowel Disorders

Probiotics are "Live microorganisms which when administered in adequate amount confer a health benefit on the host" [73]. The use of probiotics in FGID is based on the assumption that restoration of the disrupted balance of the intestinal microbiota with the right types and numbers of live probiotic microorganisms can improve the intestinal function and reduce GI symptoms [74–76].

Clinical Effects of Probiotics in IBS

Most of the studies investigating the effects of probiotics on FGIDs focused on IBS [38–40,52–58] and multiple systematic reviews and metaanalyses on this topic have been published [77–82]. However, it should be noted that comparing and summarizing the individual studies is difficult due to heterogeneous selection of patients, differences in probiotic species, dosing and regimens, and considerable differences in study design and reported clinical end points. Thus, it is not surprising that less than half of the studies that were reviewed for the published metaanalyses satisfied the selection criteria and were eventually included in final analyses. Furthermore, even after careful selection some of the studies that were included suffered from suboptimal study design as recommended by the design guidelines of treatment trials for FGID [83–85]. Several additional points need to be paid attention to when evaluating the published metaanalyses and systematic reviews in this area. First, there seem to be differences in heterogeneity for different clinical outcome. For example, additional analyses of the data of IBS studies have indicated a significant heterogeneity for the effect of probiotics on persistence of symptoms ($I^2 = 72\%$; >25% indicate high levels of heterogeneity, $P < 0.00001$), but low and nonsignificant level of heterogeneity for the effect of probiotics on global symptoms or abdominal pain ($I^2 = 27\%$, $P = 0.11$), bloating ($I^2 = 16\%$, $P = 0.26$), and flatulence ($I^2 = 0\%$, $P = 0.63$) [86]. Second, the reported magnitude of effect of probiotic intervention may depend on the quality of the studies. For example, a metaanalysis of 19 RCTs among 1650 IBS patients found that the higher quality studies reported a more modest beneficial effect (RR = 0.86; 95% CI 0.72–1.03) compared to higher beneficial effect in lower quality studies (RR = 0.52; 95% CI 0.35–0.77), suggesting there may be a publication bias, with an overrepresentation of small positive studies in the published literature [86]. Nevertheless, despite these limitations most of the metaanalysis and systematic reviews have concluded that overall probiotics have positive effects in IBS. However, it was noticeable that the effect is strain specific and that the magnitude of effect is generally modest.

Careful review of individual published studies investigating the effects of probiotics in IBS (Table 38.1) reveals that most of them were relatively small, of short duration, and used various primary end points, often not clinically applicable. In addition, many of the published studies reported within-group improvement in clinical end points comparing baseline to the end of intervention and only a few were able to demonstrate between-group differences with significant improvement over placebo. This may relate to the large range of placebo effect using subjective symptom-based outcome measures in patients with FGID [83].

Five probiotic interventions seem to have better data in IBS. *B. infantis* 35624 has been studied in two well-designed clinical trials [50,51]. The first study compared *B. infantis* 35624 or *L. salivarius* UCC4331 versus placebo in 77 patients with IBS and found significant reduction in pain, bloating, bowel movement difficulty, composite score (study of primary end points), and normalization of interleukin levels in the group receiving *B. infantis* 35624 ($P < 0.05$) compared with placebo [50]. A larger follow up multicenter trial further supported these findings by demonstrating significant improvement in pain (the study of primary end point) at 4 weeks in *B. infantis* 35624 (10^8 CFU) group versus placebo group (1.73 vs. 1.48, respectively, $P < 0.03$). There was also improvement in global relief of IBS symptoms (62.3 vs. 42.0, respectively, $P < 0.02$) and other individual IBS symptoms including composite score and bloating [51]. The multispecies probiotic product containing *L. rhamnosus* GG (LGG), *L. rhamnosus* LC705, *B. breve* Bb99, and *P. freudenreichii* spp. *shermanii* JS; equal amount of each strain with a total amount of bacteria 8–9 × 10^9 CFU/day; (Valio Ltd., Helsinki, Finland) has also been tested in two clinical trials [64,65]. Both studies have shown a significant reduction in IBS symptom (abdominal pain, distension, flatulence, borborygmi) score over placebo (42% vs. 6%, respectively, $P = 0.015$ [64]; and 37% vs. 9%, respectively, $P = 0.012$ [65]). Another multispecies probiotic product containing a mixture of *L. acidophilus, L. plantarum, L. rhamnosus, B. breve, B. lactis,*

TABLE 38.1 Important Randomized Controlled Trials Investigating Probiotics and Prebiotics Versus Placebo in FBD

Probiotic/Prebiotics Intervention	No. of Studies	No. of Patients	Intervention Duration	Study Outcome
Single Probiotics				
Escherichia coli Nissle 1917	Kruis et al. [38]	120 IBS	12 weeks	Higher responder rate; highest response rate was in subgroup with gastroenteritis or antibiotics prior to IBS onset
Saccharomyces cerevisiae	Pinetone de Chambrun et al. [39]	179 IBS	8 weeks	Higher responder rate; no improvement in individual IBS symptoms
Lactobacillus plantarum 299v (DSM 9843)	Ducrotte et al. [40]	214 IBS	4 weeks	Reduction in pain severity and daily frequency, but no improvement in bloating
Lactobacillus casei Shirota (LcS)	Koebnick et al. [41]	70 CIC	4 weeks	Improvement in severity of constipation and stool consistency, but no change in flatulence or bloating
	Mazlyn et al. [42]	90 CIC	4 weeks	No benefit over placebo
	Sakai et al. [43]	40 CIC	4 weeks	Reduction in proportion of hard or lumpy stools and improvement in Bristol score
Bifidobacterium lactis DN-173010 with two yogurt starters (*Streptococcus thermophilus* and *Lactobacillus bulgaricus*)	Yang et al. [44]	135 CIC	2 weeks	Improvement in stool frequency, defecation condition, and stool consistency
	Agrawal et al. [45]	41 IBS-C	4 weeks	Reduction in distension; acceleration of orocecal and colonic transit; improvement in overall symptom severity
	Roberts et al. [46]	179 IBS-C and IBS-M	12 weeks	No benefit over placebo
	Guyonnet et al. [47]	274 IBS-C	6 weeks	Improvement in HRQoL discomfort score and bloating
Lactobacillus acidophilus NCFM	Ringel-Kulka et al. [48]	60 nonconstipation FGD	8 weeks	Improvement in bloating
Bacillus coagulans GBI-30, 6086	Kalman et al. [49]	61 FB	4 weeks	Improvements in total GSRS score, abdominal pain, and improvement trend for distension
Bifidobacterium infantis 35624, or *Lactobacillus salivarius* UCC4331	O'Mahony et al. [50]	77 IBS	8 weeks	Reduction in abdominal pain/discomfort, bloating/distention, bowel movement difficulty, and normalization of the ratio of an antiinflammatory to a proinflammatory cytokine in the *B. infantis* 35624 group
B. infantis 35624	Whorwell et al. [51]	362 IBS	4 weeks	Improvement in abdominal pain, bloating, bowel dysfunction, incomplete evacuation, straining, passage of gas, global symptom and composite score at a dose of 1×10^8 CFU
Lactobacillus rhamnosus GG (LGG)	Gawronska et al. [52]	104 children IBS, FAP, functional dyspepsia	4 weeks	Greater treatment success (no pain). NNT = 7
	Francavilla et al. [53]	141 children IBS, FAP	8 weeks	Reduction in frequency and severity of abdominal pain from baseline, but not over placebo; greater treatment success; improvement of the gut barrier
	Bauserman et al. [54]	50 children IBS	6 weeks	No benefit over placebo except for a lower incidence of perceived abdominal distention
Mixed probiotics				
L. acidophilus, *L. plantarum*, *L. rhamnosus*, *Bifidobacterium breve*, *B. lactis*, *Bifidobacterium longum*, *S. thermophilus*	Ki Cha et al. [55]	50 IBS-D	8 weeks	Higher AR and responders rates, and improvement of stool consistency not in individual symptom scores or QOL

L. rhamnosus NCIMB 30174, *L. plantarum* NCIMB 30173, *L. acidophilus* NCIMB 30175, *E. faecium* NCIMB 30176	Sisson et al. [56]	186 IBS	12 weeks	Reduction in IBS symptom severity score, but no improvement in IBS-QOL
B. longum, B. bifidum, B. lactis, L. acidophilus, L. rhamnosus, S. thermophilus	Yoon JS et al. [57]	49 IBS	4 weeks	Higher proportion of patients whose IBS symptoms were relieved; improvement in abdominal pain/discomfort and bloating
L. acidophilus, L. rhamnosus, B. breve, B. lactis, B. longum, S. thermophilus	Yoon H et al. [58]	81 IBS	4 weeks	Higher AR rate; no differences in total symptom score; increasing fecal concentration of most probiotic strains and improving diarrhea
VSL#3	Kim SE et al. [59]	30 CIC, 30 controls	2 weeks	Improvement in consistency, complete spontaneous bowel movements
	Guandalini et al. [60]	59 children with IBS	12 weeks	Reduction in subject's global assessment of relief, abdominal pain/discomfort, bloating/gassiness, and family assessment of life disruption
	Michail et al. [61]	24 IBS-D	8 weeks	The gut microbiota was not affected and no clinical benefit over placebo
	Kim HJ et al. [62]	25 IBS-D	8 weeks	Reduction in bloating; no effect on other individual symptoms: abdominal pain, gas, and urgency
	Kim HJ et al. [63]	48 IBS with bloating	8 weeks	Reduction in flatulence; no effect on other individual symptoms: proportions of responders for satisfactory relief of bloating, stool-related symptoms, abdominal pain, and bloating scores
L. rhamnosus GG, *L. rhamnosus* Lc705, *B. breve* Bb99, and *Propionibacterium freudenreichii* ssp. *shermanii* JS	Kajander et al. [64]	103 IBS	6 months	Improvement in total symptom score (abdominal pain + distension + flatulence + borborygmi)
L. rhamnosus GG, *L. rhamnosus* Lc705, *P. freudenreichii* ssp. *shermanii* JS, and *B. animalis* ssp. *lactis* Bb12	Kajander et al. [65]	86 IBS	5 months	Improvement in composite IBS symptoms score (abdominal pain + distension + flatulence + rumbling); stabilization of the microbiota; no differences were seen in CRP
S. thermophilus, L. bulgaricus, L. acidophilus, B. longum	Zeng et al. [66]	30 IBS-D	4 weeks	Improvement in mucosal barrier function; reduction in global IBS scores from baseline, but not over placebo
Prebiotics				
PHGG	Niv et al. [67]	121 IBS	12 weeks	Improvement in bloating score and in bloating + gasses score
I-PHGG supplementation	Linetzky et al. [68]	60 CIC	3 weeks	No benefit over placebo except for a decreased amount of *Clostridium* sp.
Trans-galactooligosaccharide (3.5 g/d or 7 g/d)	Silk et al. [69]	44 IBS	12 weeks	Prebiotic at 3.5 g/d changed stool consistency, improved flatulence, bloating, composite score of symptoms, and SGA; the prebiotic at 7 g/d improved SGA and anxiety scores
sc-FOS	Paineau et al. [70]	105 with minor FBD	6 weeks	Intensity of digestive disorders decreased and daily activities were significantly improved
Oligofructose	Hunter et al. [71]	21 IBS	4 weeks	No benefit over placebo
Fructooligosaccharides	Olesen et al. [72]	98 IBS	12 weeks	No benefit over placebo

I-PHGG, Inulin and partially hydrolyzed guar gum; PHGG, partially hydrolyzed guar gum; QOL, quality of life; sc-FOS, short-chain fructooligosaccharides.

B. longum, and *S. thermophilus* 1.0×10^1 CFU (Duolac7, Cell Biotech, Co., Ltd., Seoul, Korea) was found to be effective in providing adequate relief in IBS symptoms over placebo (48% vs. 12%, respectively, $P = 0.01$), and improvement of stool consistency in IBS-D patients compared with placebo throughout the study period [55]. Another probiotic mixture, VSL#3, containing *B. longum*, *B. infantis*, *B. breve*, *L. acidophilus*, *L. casei*, *L. bulgaricus*, *L. plantarum*, and *Streptococcus salivarius* subspecies *thermophilus*, was evaluated in IBS patients in several RCTs [60–63]. For example, a clinical trial using VSL#3 in 59 children (age 4–15 years) with IBS showed a significantly higher magnitude of improvement in relieving IBS symptoms (primary end point) in the VSL#3 group compared with placebo ($P < 0.05$) [60].

Most of the trials investigating the effects of probiotics in IBS patients were conducted on heterogeneous mixed IBS population and only a few studies were done on specific IBS clinical subtypes for example, diarrhea-predominant IBS (IBS-D), constipation-predominant IBS (IBS-C), or mix-IBS (IBS-M). For example, in a recent metaanalysis of 15 RCTs that examined the effectiveness of probiotics in 1793 patients with IBS [81], only two studies were done on patients with IBS-D [55,66] and in one study the majority of patients (63.89% in the probiotic group and 60.3% in the placebo group) were IBS-D patients [40]. Review of the individual studies in the IBS-C suggests that supplementation of probiotics may be of benefit in this subgroup of patients. A study using fermented dairy product containing *B. lactis* DN-173010 in 312 IBS-C patients have demonstrated acceleration of GI transit and improvement in abdominal distention and overall IBS symptoms severity after 4- [45] and 6-week [47] treatment periods. Another multicenter trial using the same fermented milk probiotic product for 12 weeks in 179 IBS-C or IBS-M patients found significant improvements in pre- to postintervention clinical SGA of symptom relief, IBS symptom scores, abdominal pain, abdominal bloating, flatulence, stool frequency, stool consistency, ease of bowel movement, and QOL. However, the reported clinical improvements in this study were limited to within-group changes between baseline and end of intervention but there were no between-groups differences in the clinical effects of probiotic and placebo [46]. In addition, this study suffered from a 55% dropout rate thus the effect of the active product may have been missed due to lack of power.

The few probiotics RCTs on patients with IBS-D have demonstrated that certain commercially available multispecies probiotic products for example, VSL#3, (Sigma-Tau, Gaithersburg, Maryland, USA) and probiotic fermented milk, AB100 Jianneng (Bright Dairy, Shanghai, China) have beneficial effect on relevant physiological factors, such as mucosal barrier function [66], bowel functions including stool consistency [61], and GI symptoms for example, global IBS score, abdominal pain, sensation of flatulence, and abdominal bloating [62,66]. In addition, a RCT using *L. plantarum*

299v (DSM 9843) for 4 weeks in 214 IBS patients of whom 63.89% were IBS-D, found significant reduction in the frequency of abdominal pain over placebo (51.9% vs. 13.6%, respectively, $P < 0.05$) and bloating ($P < 0.05$) [40].

Clinical Effect of Probiotics in Other (Non-IBS) FBDs

The data on the effect of probiotics in patients with non-IBS FGID are very limited with only a few studies that were done on CIC [41–44,59], FD [48], FB [48,49], and functional abdominal pain (FAP) [52,53].

With regard to CIC, we identified five trials evaluating probiotics in 365 CIC patients [41–44,59]. Three of these trials examined the efficacy of *L. casei* Shirota (LcS) and found a significant improvement in severity of constipation and stool consistency from baseline [41,42] and over placebo [43]. Similarly, other probiotics products including VSL#3 and *B. lactis* DN-173010 appeared to have a beneficial effect on stool consistency, complete spontaneous bowel movements (both $P < 0.001$) from baseline [59] and on stool frequency, ease of defecation, and stool consistency over placebo [44].

The data on probiotics in patients with FAP are limited mainly to children. *L. rhamnosus* GG (LGG) was studied in 2 RCTs ($n = 290$) [52,53]. A study using LGG in 104 children age 6–16 years with IBS, FAP, and functional dyspepsia demonstrated a significant treatment benefit of LGG over placebo. Those in the LGG group were more likely to have treatment success (defined as complete resolution of abdominal pain) than those in the placebo group (33% vs. 5%, respectively; 95% CI, 1.2–38; $P = 0.04$) and reduction of frequency of pain from baseline to end of 4-weeks intervention (3.3 ± 1.3 vs. 2.2 ± 1.7 in the LGG group compare with 3.7 ± 1.0 vs. 2.6 ± 1.4 in the placebo group; $P = 0.02$) with a number needed to treat (NNT) of 4 (95% CI, 2–36 [52]. Another study with LGG in 141 children aged 5–14 years with IBS and FAP showed that the LGG group had lower frequency (1.6 vs. 3.2, respectively; $P < 0.05$) and severity (2.5 vs. 3.6, respectively, $P < 0.01$ on a 0–10 scale) of abdominal pain compared with placebo [53]. However, it should be noted that another study in 50 children (mean age of 12 years) with IBS found that supplementation of LGG (10^{10} CFU) for 6 weeks was not superior to placebo in relieving abdominal pain (40.0% response rate in the placebo group vs. 44.0% in the LGG group, $P = 0.774$) although it was beneficial in alleviating other symptoms, such as perceived abdominal distention ($P = 0.02$) [54].

Regarding functional bloating, one clinical trial using *B. coagulans* (GanedenBC *B. coagulans* GBI-30, 6086) in 61 patients with postmeal intestinal gas symptoms found a significant improvement in total Gastrointestinal Symptom Rating Scale (GSRS) ($P = 0.048$) and abdominal pain GSRS subscore ($P = 0.046$) in the intervention group compare to the placebo group [49]. Another clinical trial with probiotic combination of *Lactobacillus*-NCFM and *Bifidobacteria*-LBi07 in 60 nonconstipation IBS patients for 8 weeks has found a significant improvement in abdominal bloating at 4 and 8 weeks in the intervention group compared to the placebo group [48].

CONCLUSIONS AND CLINICAL IMPLICATIONS OF CURRENT DATA

Emerging data from epidemiological, physiological, and microbiome research in patients with FBDs indicate an association between alterations in the intestinal microbiota and both abnormal GI functions and functional GI symptoms. The causality role of these associations is not yet clear however, colonization of germ-free mice with human feces obtained from IBS patients have shown to be able to affect the animal GI function (e.g., transit) suggesting that alterations in intestinal microbiota may indeed have a role in the pathogenesis of FBDs [87].

The rationale of using probiotics in the treatments of patients with FBDs is based on the accumulating evidence, mainly from animal studies, showing that probiotics has the potential to modulate a number of physiological functions that are relevant to FBDs including motility, sensation, and neuronal gut–brain function [30,88].

This together with the growing interest of the general public in healthy food, direct-to-consumer advertisement, and clinical data on beneficial effects of probiotics in various GI conditions (e.g., as infectious diarrhea and some forms of inflammatory bowel disorders) has led to the increased use of probiotics in patients with FBDs.

Evaluating the effects of pre- and probiotics in clinical setting is difficult due to the limited number of large high quality clinical trials and a possibility of publication bias. The data from metaanalyses should be evaluated with great caution considering that the effect of different strains of probiotics may have different effects on the host. Indeed, different probiotic preparations appear to improve different functional GI symptoms with some reducing pain and discomfort, others improving abdominal bloating and distention, and other relieving bowel movements or a range of symptoms.

Despite the evidence that probiotics may be beneficial in the treatment of FBDs, several points need to be remembered when considering this treatment option for patients with FGIDs. First, since the effect of probiotics is strain and possibly product specific the selection of an appropriate probiotic product for a specific condition should be based on clinical evidence for the specific product. However, probiotic products are usually marketed as either dietary supplements or medical food (only one product in the United States; VSL#3). As such, these products are not required to go through rigorous clinical trials to show efficacy prior to

marketing. This together with US regulatory requirements for IND for conduction of clinical trials led to limited high quality clinical data on the use of many of the marketed probiotic products in FBDs. Second, probiotic products were mostly investigated in IBS patients irrespective of the pattern of their symptoms or bowel movements. However, IBS and FBDs are heterogeneous disorders comprising different clinical subtypes and different underlying pathophysiology. Therefore, it is not always possible to extrapolate the evidence from the published literature to an individual patient with a specific condition or symptom (e.g., IBS-C, IBS-D, CIC, pain, or abdominal bloating). However, the limited effective treatments for these patients and the relatively safe profile of probiotics support trying probiotic products in patients with FBDs. The selection of a probiotic product for a specific patient should be based on evidence of efficacy of a specific product for a specific condition. However, in view of the limitations of the current clinical data trying different probiotic products in an individual patient may still be needed in order to find the probiotic strain or combination of strains with the best clinical effect. Additional important questions remain to be answered. Examples are identification of patients most likely to respond to probiotic intervention, the place of probiotics in the treatment algorithm of FBDs, and the effectiveness of combining probiotics with other approaches (e.g., FODMAP diet or antibiotics). Additional high quality clinical trials with available products would help direct physician and patients in a more effective use of probiotics in FBDs.

ABBREVIATIONS

CIC	Chronic idiopathic constipation
FAP	Functional abdominal pain
FB	Functional bloating
FBD	Functional bowel disorders
FD	Functional diarrhea
FODMAP	Fermentable oligo-, di-, monosaccharides, and polyols
GI	Gastrointestinal
IBS	Irritable bowel syndrome
IBS-C	IBS-constipation
IBS-D	IBS-diarrhea
IBS-M	Mixed IBS
NNT	Number needed to treat
QOL	Quality of life
RCT	Randomized control trials
SGA	Subjective global assessment

REFERENCES

[1] Longstreth GF, Thompson WG, Chey WD, Houghton LA, Mearin F, Spiller RC. Functional bowel disorders. Gastroenterology 2006;130(5):1480–91.

[2] Ringel Y, Drossman D. Irritable bowel syndrome. In: Netter's textbook of internal medicine. 2nd ed. Philadelphia, PA: Sauders Elsevier; 2009. p. 419–25.

[3] Saito YA, Schoenfeld P, Locke GR III. The epidemiology of irritable bowel syndrome in North America: a systematic review. Am J Gastroenterol 2002;97(8):1910–5.

[4] Chang JY, Locke GR III, McNally MA, et al. Impact of functional gastrointestinal disorders on survival in the community. Am J Gastroenterol 2010;105(4):822–32.

[5] Peery AF, Crockett SD, Barritt AS, et al. Burden of gastrointestinal, liver, and pancreatic diseases in the United States. Gastroenterology 2015;149(7):1731–41. e3.

[6] Saito YA, Petersen GM, Locke GR III, Talley NJ. The genetics of irritable bowel syndrome. Clin Gastroenterol Hepatol 2005;3(11): 1057–65.

[7] Camilleri M. Peripheral mechanisms in irritable bowel syndrome. N Engl J Med 2012;367(17):1626–35.

[8] Camilleri M. Physiological underpinnings of irritable bowel syndrome: neurohormonal mechanisms. J Physiol 2014;592(Pt. 14):2967–80.

[9] Gill SR, Pop M, Deboy RT, et al. Metagenomic analysis of the human distal gut microbiome. Science 2006;312(5778):1355–9.

[10] Backhed F, Fraser CM, Ringel Y, et al. Defining a healthy human gut microbiome: current concepts, future directions, and clinical applications. Cell Host Microbe 2012;12(5):611–22.

[11] Cryan JF, Dinan TG. Mind-altering microorganisms: the impact of the gut microbiota on brain and behaviour. Nat Rev Neurosci 2012;13(10):701–12.

[12] Bravo JA, Julio-Pieper M, Forsythe P, et al. Communication between gastrointestinal bacteria and the nervous system. Curr Opin Pharmacol 2012;12(6):667–72.

[13] Yarandi SS, Peterson DA, Treisman GJ, Moran TH, Pasricha PJ. Modulatory effects of gut microbiota on the central nervous system: how gut could play a role in neuropsychiatric health and diseases. J Neurogastroenterol Motil 2016;22(2):201–12.

[14] Kelly JR, Kennedy PJ, Cryan JF, Dinan TG, Clarke G, Hyland NP. Breaking down the barriers: the gut microbiome, intestinal permeability and stress-related psychiatric disorders. Front Cell Neurosci 2015;9:392.

[15] Moloney RD, Johnson AC, O'Mahony SM, Dinan TG, Greenwood-Van Meerveld B, Cryan JF. Stress and the microbiota-gut-brain axis in visceral pain: relevance to irritable bowel syndrome. CNS Neurosci Ther 2016;22(2):102–17.

[16] Halvorson HA, Schlett CD, Riddle MS. Postinfectious irritable bowel syndrome—a meta-analysis. Am J Gastroenterol 2006;101(8): 1894–9. quiz 1942.

[17] Schwille-Kiuntke J, Enck P, Zendler C, et al. Postinfectious irritable bowel syndrome: follow-up of a patient cohort of confirmed cases of bacterial infection with Salmonella or Campylobacter. Neurogastroenterol Motil 2011;23(11):e479–88.

[18] Maxwell PR, Rink E, Kumar D, Mendall MA. Antibiotics increase functional abdominal symptoms. Am J Gastroenterol 2002;97(1):104–8.

[19] Ford AC, Spiegel BM, Talley NJ, Moayyedi P. Small intestinal bacterial overgrowth in irritable bowel syndrome: systematic review and meta-analysis. Clin Gastroenterol Hepatol 2009;7(12):1279–86.

[20] Yu D, Cheeseman F, Vanner S. Combined oro-caecal scintigraphy and lactulose hydrogen breath testing demonstrate that breath testing detects oro-caecal transit, not small intestinal bacterial overgrowth in patients with IBS. Gut 2011;60(3):334–40.

[21] Posserud I, Stotzer PO, Bjornsson ES, Abrahamsson H, Simren M. Small intestinal bacterial overgrowth in patients with irritable bowel syndrome. Gut 2007;56(6):802–8.

[22] Rajilic-Stojanovic M, Jonkers DM, Salonen A, et al. Intestinal microbiota and diet in IBS: causes, consequences, or epiphenomena? Am J Gastroenterol 2015;110(2):278–87.

[23] Ringel Y, Carroll IM. Alterations in the intestinal microbiota and functional bowel symptoms. Gastrointest Endosc Clin N Am 2009;19(1):141–50.

[24] Simren M, Barbara G, Flint HJ, et al. Intestinal microbiota in functional bowel disorders: a Rome foundation report. Gut 2013;62(1):159–76.

[25] Carroll IM, Ringel-Kulka T, Keku TO, et al. Molecular analysis of the luminal- and mucosal-associated intestinal microbiota in diarrhea-predominant irritable bowel syndrome. Am J Physiol Gastrointest Liver Physiol 2011;301(5):G799–807.

[26] Carroll IM, Ringel-Kulka T, Siddle JP, Ringel Y. Alterations in composition and diversity of the intestinal microbiota in patients with diarrhea-predominant irritable bowel syndrome. Neurogastroenterol Motil 2012;24(6):521–30. e248.

[27] Jeffery IB, O'Toole PW, Ohman L, et al. An irritable bowel syndrome subtype defined by species-specific alterations in faecal microbiota. Gut 2012;61(7):997–1006.

[28] Jalanka-Tuovinen J, Salojarvi J, Salonen A, et al. Faecal microbiota composition and host-microbe cross-talk following gastroenteritis and in postinfectious irritable bowel syndrome. Gut 2014;63(11):1737–45.

[29] Rajilic-Stojanovic M, Biagi E, Heilig HG, et al. Global and deep molecular analysis of microbiota signatures in fecal samples from patients with irritable bowel syndrome. Gastroenterology 2011;141(5):1792–801.

[30] Ringel Y, Ringel-Kulka T. The intestinal microbiota and irritable bowel syndrome. J Clin Gastroenterol 2015;49(Suppl. 1):S56–9.

[31] Parthasarathy G, Chen J, Chen X, et al. Relationship between microbiota of the colonic mucosa vs feces and symptoms, colonic transit, and methane production in female patients with chronic constipation. Gastroenterology 2016;150(2):367–79. e1.

[32] Staudacher HM, Irving PM, Lomer MC, Whelan K. Mechanisms and efficacy of dietary FODMAP restriction in IBS. Nat Rev Gastroenterol Hepatol 2014;11(4):256–66.

[33] Dupont HL. Review article: evidence for the role of gut microbiota in irritable bowel syndrome and its potential influence on therapeutic targets. Aliment Pharmacol Ther 2014;39(10):1033–42.

[34] Saulnier DM, Ringel Y, Heyman MB, et al. The intestinal microbiome, probiotics and prebiotics in neurogastroenterology. Gut Microbes 2013;4(1):17–27.

[35] Floch MH, Walker WA, Sanders ME, et al. Recommendations for probiotic use-2015 update: proceedings and consensus opinion. J Clin Gastroenterol 2015;49(Suppl. 1):S69–73.

[36] Roberfroid M, Gibson GR, Hoyles L, et al. Prebiotic effects: metabolic and health benefits. Br J Nutr 2010;104(Suppl. 2):S1–S63.

[37] Gibson GR, Roberfroid MB. Dietary modulation of the human colonic microbiota: introducing the concept of prebiotics. J Nutr 1995;125(6):1401–12.

[38] Kruis W, Chrubasik S, Boehm S, Stange C, Schulze J. A double-blind placebo-controlled trial to study therapeutic effects of probiotic Escherichia coli Nissle 1917 in subgroups of patients with irritable bowel syndrome. Int J Colorectal Dis 2012;27(4):467–74.

[39] Pineton de Chambrun G, Neut C, Chau A, et al. A randomized clinical trial of Saccharomyces cerevisiae versus placebo in the irritable bowel syndrome. Dig Liver Dis 2015;47(2):119–24.

[40] Ducrotte P, Sawant P, Jayanthi V. Clinical trial: Lactobacillus plantarum 299v (DSM 9843) improves symptoms of irritable bowel syndrome. World J Gastroenterol 2012;18(30):4012–8.

[41] Koebnick C, Wagner I, Leitzmann P, Stern U, Zunft HJ. Probiotic beverage containing *Lactobacillus casei* Shirota improves gastrointestinal symptoms in patients with chronic constipation. Can J Gastroenterol 2003;17(11):655–9.

[42] Mazlyn MM, Nagarajah LH, Fatimah A, Norimah AK, Goh KL. Effects of a probiotic fermented milk on functional constipation: a randomized, double-blind, placebo-controlled study. J Gastroenterol Hepatol 2013;28(7):1141–7.

[43] Sakai T, Makino H, Ishikawa E, Oishi K, Kushiro A. Fermented milk containing *Lactobacillus casei* strain Shirota reduces incidence of hard or lumpy stools in healthy population. Int J Food Sci Nutr 2011;62(4):423–30.

[44] Yang YX, He M, Hu G, et al. Effect of a fermented milk containing *Bifidobacterium lactis* DN-173010 on Chinese constipated women. World J Gastroenterol 2008;14(40):6237–43.

[45] Agrawal A, Houghton LA, Morris J, et al. Clinical trial: the effects of a fermented milk product containing *Bifidobacterium lactis* DN-173 010 on abdominal distension and gastrointestinal transit in irritable bowel syndrome with constipation. Aliment Pharmacol Ther 2009;29(1):104–14.

[46] Roberts LM, McCahon D, Holder R, Wilson S, Hobbs FD. A randomised controlled trial of a probiotic 'functional food' in the management of irritable bowel syndrome. BMC Gastroenterol 2013;13:45.

[47] Guyonnet D, Chassany O, Ducrotte P, et al. Effect of a fermented milk containing *Bifidobacterium animalis* DN-173 010 on the health-related quality of life and symptoms in irritable bowel syndrome in adults in primary care: a multicentre, randomized, double-blind, controlled trial. Aliment Pharmacol Ther 2007;26(3):475–86.

[48] Ringel-Kulka T, Palsson OS, Maier D, et al. Probiotic bacteria *Lactobacillus acidophilus* NCFM and *Bifidobacterium lactis* Bi-07 versus placebo for the symptoms of bloating in patients with functional bowel disorders: a double-blind study. J Clin Gastroenterol 2011;45(6):518–25.

[49] Kalman DS, Schwartz HI, Alvarez P, Feldman S, Pezzullo JC, Krieger DR. A prospective, randomized, double-blind, placebo-controlled parallel-group dual site trial to evaluate the effects of a *Bacillus coagulans*-based product on functional intestinal gas symptoms. BMC Gastroenterol 2009;9:85.

[50] O'Mahony L, McCarthy J, Kelly P, et al. Lactobacillus and bifidobacterium in irritable bowel syndrome: symptom responses and relationship to cytokine profiles. Gastroenterology 2005;128(3):541–51.

[51] Whorwell PJ, Altringer L, Morel J, et al. Efficacy of an encapsulated probiotic *Bifidobacterium infantis* 35624 in women with irritable bowel syndrome. Am J Gastroenterol 2006;101(7):1581–90.

[52] Gawronska A, Dziechciarz P, Horvath A, Szajewska H. A randomized double-blind placebo-controlled trial of *Lactobacillus* GG for abdominal pain disorders in children. Aliment Pharmacol Ther 2007;25(2):177–84.

[53] Francavilla R, Miniello V, Magista AM, et al. A randomized controlled trial of *Lactobacillus* GG in children with functional abdominal pain. Pediatrics 2010;126(6):e1445–52.

[54] Bauserman M, Michail S. The use of *Lactobacillus* GG in irritable bowel syndrome in children: a double-blind randomized control trial. J Pediatr 2005;147(2):197–201.

[55] Ki Cha B, Mun Jung S, Hwan Choi C, et al. The effect of a multispecies probiotic mixture on the symptoms and fecal microbiota in diarrhea-dominant irritable bowel syndrome: a randomized, double-blind, placebo-controlled trial. J Clin Gastroenterol 2012;46(3):220–7.

[56] Sisson G, Ayis S, Sherwood RA, Bjarnason I. Randomised clinical trial: a liquid multi-strain probiotic vs. placebo in the irritable bowel syndrome—a 12 week double-blind study. Aliment Pharm Ther 2014;40(1):51–62.

[57] Yoon JS, Sohn W, Lee OY, et al. Effect of multispecies probiotics on irritable bowel syndrome: a randomized, double-blind, placebo-controlled trial. J Gastroenterol Hepatol 2014;29(1):52–9.

[58] Yoon H, Park YS, Lee DH, Seo JG, Shin CM, Kim N. Effect of administering a multi-species probiotic mixture on the changes in fecal microbiota and symptoms of irritable bowel syndrome: a randomized, double-blind, placebo-controlled trial. J Clin Biochem Nutr 2015;57(2):129–34.

[59] Kim SE, Choi SC, Park KS, et al. Change of fecal flora and effectiveness of the short-term vsl#3 probiotic treatment in patients with functional constipation. J Neurogastroenterol Motil 2015;21(1):111–20.

[60] Guandalini S, Magazzu G, Chiaro A, et al. VSL#3 improves symptoms in children with irritable bowel syndrome: a multicenter, randomized, placebo-controlled, double-blind, crossover study. J Pediatr Gastroenterol Nutr 2010;51(1):24–30.

[61] Michail S, Kenche H. Gut microbiota is not modified by randomized, double-blind, placebo-controlled trial of VSL#3 in diarrhea-predominant irritable bowel syndrome. Probiotics Antimicrob Proteins 2011;3(1):1–7.

[62] Kim HJ, Camilleri M, McKinzie S, et al. A randomized controlled trial of a probiotic, VSL#3, on gut transit and symptoms in diarrhoea-predominant irritable bowel syndrome. Aliment Pharmacol Ther 2003;17(7):895–904.

[63] Kim HJ, Vazquez Roque MI, Camilleri M, et al. A randomized controlled trial of a probiotic combination VSL# 3 and placebo in irritable bowel syndrome with bloating. Neurogastroenterol Motil 2005;17(5):687–96.

[64] Kajander K, Hatakka K, Poussa T, Farkkila M, Korpela R. A probiotic mixture alleviates symptoms in irritable bowel syndrome patients: a controlled 6-month intervention. Aliment Pharmacol Ther 2005;22(5):387–94.

[65] Kajander K, Myllyluoma E, Rajilic-Stojanovic M, et al. Clinical trial: multispecies probiotic supplementation alleviates the symptoms of irritable bowel syndrome and stabilizes intestinal microbiota. Aliment Pharmacol Ther 2008;27(1):48–57.

[66] Zeng J, Li YQ, Zuo XL, Zhen YB, Yang J, Liu CH. Clinical trial: effect of active lactic acid bacteria on mucosal barrier function in patients with diarrhoea-predominant irritable bowel syndrome. Aliment Pharmacol Ther 2008;28(8):994–1002.

[67] Niv E, Halak A, Tiommny E, et al. Randomized clinical study: partially hydrolyzed guar gum (PHGG) versus placebo in the treatment of patients with irritable bowel syndrome. Nutr Metab 2016;13:10.

[68] Linetzky Waitzberg D, Alves Pereira CC, Logullo L, et al. Microbiota benefits after inulin and partially hydrolized guar gum supplementation: a randomized clinical trial in constipated women. Nut Hosp 2012;27(1):123–9.

[69] Silk DB, Davis A, Vulevic J, Tzortzis G, Gibson GR. Clinical trial: the effects of a trans-galactooligosaccharide prebiotic on faecal microbiota and symptoms in irritable bowel syndrome. Aliment Pharmacol Ther 2009;29(5):508–18.

[70] Paineau D, Payen F, Panserieu S, et al. The effects of regular consumption of short-chain fructo-oligosaccharides on digestive comfort of subjects with minor functional bowel disorders. Br J Nutr 2008;99(2):311–8.

[71] Hunter JO, Tuffnell Q, Lee AJ. Controlled trial of oligofructose in the management of irritable bowel syndrome. J Nutr 1999;129(7 Suppl.):1451S–3S.

[72] Olesen M, Gudmand-Hoyer E. Efficacy, safety, and tolerability of fructooligosaccharides in the treatment of irritable bowel syndrome. Am J Clin Nutr 2000;72(6):1570–5.

[73] Hill C, Guarner F, Reid G, et al. Expert consensus document. The International Scientific Association for Probiotics and Prebiotics consensus statement on the scope and appropriate use of the term probiotic. Nat Rev Gastroenterol Hepatol 2014;11(8):506–14.

[74] Reid G, Younes JA, Van der Mei HC, Gloor GB, Knight R, Busscher HJ. Microbiota restoration: natural and supplemented recovery of human microbial communities. Nat Rev Microbiol 2011;9(1):27–38.

[75] Varankovich NV, Nickerson MT, Korber DR. Probiotic-based strategies for therapeutic and prophylactic use against multiple gastrointestinal diseases. Front Microbiol 2015;6:685.

[76] Pandey V, Berwal V, Solanki N, Malik NS. Probiotics: healthy bugs and nourishing elements of diet. J Int Soc Prev Community Dent 2015;5(2):81–7.

[77] Ortiz-Lucas M, Tobias A, Saz P, Sebastian JJ. Effect of probiotic species on irritable bowel syndrome symptoms: a bring up to date meta-analysis. Rev Esp Enferm Dig 2013;105(1):19–36.

[78] Tiequn B, Guanqun C, Shuo Z. Therapeutic effects of *Lactobacillus* in treating irritable bowel syndrome: a meta-analysis. Intern Med 2015;54(3):243–9.

[79] McFarland LV, Dublin S. Meta-analysis of probiotics for the treatment of irritable bowel syndrome. World J Gastroenterol 2008;14(17):2650–61.

[80] Hoveyda N, Heneghan C, Mahtani KR, Perera R, Roberts N, Glasziou P. A systematic review and meta-analysis: probiotics in the treatment of irritable bowel syndrome. BMC Gastroenterol 2009;9:15.

[81] Didari T, Mozaffari S, Nikfar S, Abdollahi M. Effectiveness of probiotics in irritable bowel syndrome: updated systematic review with meta-analysis. World J Gastroenterol 2015;21(10):3072–84.

[82] Hungin AP, Mulligan C, Pot B, et al. Systematic review: probiotics in the management of lower gastrointestinal symptoms in clinical practice—an evidence-based international guide. Aliment Pharmacol Ther 2013;38(8):864–86.

[83] Irvine EJ, Whitehead WE, Chey WD, et al. Design of treatment trials for functional gastrointestinal disorders. Gastroenterology 2006;130(5):1538–51.

[84] Miller LE. Study design considerations for irritable bowel syndrome clinical trials. Ann Gastroenterol 2014;27(4):338–45.

[85] Corsetti M, Tack J. FDA and EMA end points: which outcome end points should we use in clinical trials in patients with irritable bowel syndrome? Neurogastroenterol Motil 2013;25(6):453–7.

[86] Ford AC, Quigley EM, Lacy BE, et al. Efficacy of prebiotics, probiotics, and synbiotics in irritable bowel syndrome and chronic idiopathic constipation: systematic review and meta-analysis. Am J Gastroenterol 2014;109(10):1547–61. quiz 1546, 1562.

[87] Kashyap PC, Marcobal A, Ursell LK, et al. Complex interactions among diet, gastrointestinal transit, and gut microbiota in humanized mice. Gastroenterology 2013;144(5):967–77.

[88] Santos AR, Whorwell PJ. Irritable bowel syndrome: the problem and the problem of treating it—is there a role for probiotics? Proc Nutr Soc 2014;73(4):470–6.

Chapter 39

Celiac Disease, the Microbiome, and Probiotics

S. Krishnareddy and P.H.R. Green
Celiac Disease Center, Columbia University, NY, United States

INTRODUCTION

Celiac disease (CD) is a common chronic lifelong autoimmune enteropathy triggered by the consumption of specific proteins by genetically predisposed individuals [1]. These proteins, known as gluten proteins, are most commonly present in cereals and receive specific names according to the food source, such as gliadin (present in wheat), hordein (present in barley), and secalin (present in rye) [2]. As is well established, the genetic predisposition is an important aspect of CD and is mostly associated with the human leukocyte antigen (HLA-DQ) system, which participates in the recognition of self and nonself molecules by the immune system. The variants HLA-DQ2 and/or DQ8 are commonly observed in patients with CD and are necessary for the immune cell activation and autoimmunity that are hallmarks in disease pathogenesis [3]. While nearly 40% of the population may carry one of these genes, only 1–2% go on to develop CD [4]. However, essentially all individuals who go on to develop CD carry one or both of these genotypes. This allows for uniquely informative control groups. Patients who develop CD can be compared to those who don't, but carry the same genetic makeup, as well as family members who don't but carry the same genetic makeup and for all intense and purposes have similar environmental exposures. This allows for deeper understanding of the other factors that contribute to disease development.

The coevolution of the microbiota with humans has generated complex networks of bacterial reactivity and tolerance, some for the benefit of the host and some for its detriment. This complexity is illustrated by the presence of tenfold more microbial cells than eukaryotic cells in the human body, and these bacterial cells contain 100 times as many genes as the entire human genome [5]. Certain clostridial species have been associated with increased numbers of T-regulatory cells in the mouse colon [6], while segmented filamentous bacteria have been associated with the development of T-helper 17 (Th17) cell lineage in the murine small intestine [7], and the role of *Bacteroides fragilis* in inducing the differentiation of Treg cells and promoting an antiinflammatory response [8]. The human immune system has developed different mechanisms to tolerate commensal microbes and prevent pathogens invading the host [9]. In this respect, the microbiota increases the epithelial barrier function through the production of different metabolites, such as short-chain fatty acids and mucus. The microbiota also promotes the production of antimicrobial molecules, such as regenerating islet-derived protein III (RegIII-γ) and REGIII-β by epithelial cells in the intestine [9]. This host–microbiota relationship ensures the establishment of immune homeostasis so that the host's immune system does not attack the commensal microbes and cause unchecked inflammation. This derangement in recognition is one possible mechanism for the role of the microbiome in CD pathogenesis.

Increasing evidence strongly indicates that intestinal microbiota plays a pivotal role in maintaining the homeostasis of the human body. Its integrity and its dynamic interaction with the body are essential for maintaining a healthy state while its alterations may contribute to the development of diseases, not only those related to the gastrointestinal tract, but also on a metabolic and systemic level [10]. CD is an ideal model to study autoimmune diseases, as it is the only autoimmune disease for which the trigger (gluten) is known, as well as the genes (HLA) are known. But, despite progress made in understanding the adaptive immunological aspects of CD pathogenesis, the early steps following intestinal mucosal exposure to gluten that lead to the loss of tolerance and development of the autoimmune process are still unknown. While gluten is the trigger in CD, the loss of gluten tolerance does not necessarily occur at the time of its introduction into the diet in individuals genetically at risk but can occur at any time in life as a consequence of other unknown stimuli [11]. The gut microbiome composition and consequent changes in specific metabolomic pathways have been implicated in the switch from tolerance to immune response to gluten. Preliminary data suggest that particular changes in both the microbiome and metabolomics profiles precede the onset of CD in genetically susceptible individuals [12].

MICROBIOTA AND CELIAC DISEASE

The human gastrointestinal tract is a complex and dynamic environment, sheltering a vast number and variety of commensal microorganisms [13]. This balanced microecosystem provides a natural defense against invasion of pathogens. Recently, much research has focused on the role of the human microbiome in health and disease, and the ability to harness the power of the human microbiome for treatment of these diseases. It has been suggested that dysbiosis may affect autoimmunity by altering the balance between tolerogenic and inflammatory members of the microbiota, and therefore the host immune response.

Bacterial products on the intestinal surface are detected by specific receptors for the pathogen-associated molecular pattern (PAMP) called Toll like receptors (TLRs). Each of these is able to recognize a specific bacterial product, which can lead to the activation of various intracellular cascades [14]. Previous work by Szebeni et al. [15] has shown altered expression of TLR4 and TLR2 in patients with CD, suggesting that dysbiosis plays a role in initiation of the immune response to gluten. The intestinal microbiota plays an important role in modulating immunity, inflammation, and allergic responses to foreign antigens including food [16] and this complex relationship between the gut and its microbial content allows the identification of dangerous and harmless bacteria as well as food antigens. In addition, the microbiota contributes to maintaining both the intestinal barrier function, by increasing the proliferation of epithelial cells, and the integrity of the intestinal epithelium through the translocation of proteins forming tight junctions and stimulating the genes involved in the maintenance of desmosomes, as well as the vascular architecture of the villi [17].

CD is a complex multifactorial disorder involving both genetic and environmental factors. For many years, the only securely established genetic factors contributing the CD risk were various genetic variants located within the HLA region (those encoding the HLA-DQ2/DQ8 heterodimers) [1]. With the introduction of GWAS (genome -wide association studies) and the Immunochip study, an additional 39 non-HLA regions of susceptibility have been associated with CD development, some of which share with other autoimmune diseases [18–20]. Interestingly, most of those chromosome regions associated with CD predisposition contain genes with immune-related functions and some CD susceptibility genes, and/or their altered expression play a role in bacterial colonization and sensing. Studies have also shown an altered expression of nonspecific CD risk-genes involved host–microbiota interactions in the intestinal mucosa of CD patients, such as those of TLRs and their regulators [21]. Furthermore, 81% of the CD-associated genetic variants are located in noncoding regions of the genome suggesting that one of the main mechanisms by which genetic variation could have an impact on CD is by affect gene expression levels [22]. Altered expression of CD-risk genes may contribute to disturbing the host–microbiota interaction, and shift immune balance in CD subjects, therefore inducing the inflammatory response by gluten that is pathognomonic to CD.

Studies of the role the microbiome plays in CD are still in its infancy, and as with most studies of the microbiome, most studies have shown descriptive data, but lack cause and effect. Indeed, although CD is prevalent in both adults and children, most of the microbiome data in CD comes from studies done in children [23–26]. Studies characterizing the microbiota of adult CD patients only began in 2012, and a single study of both children and adults reported a slight difference in the percentages of the main phyla between subjects and also a more diverse profile in duodenal biopsy specimens from adults [27]. The Firmicutes are the most abundant bacteria in CD adults, while Proteobacteria are present mainly in CD children. Other phyla shared between CD adults and CD children belong to the Bacteroidetes and Actinobacteria. Regarding bacterial genera, CD adults harbor larger numbers of *Mycobacterium* spp. and *Methylobacterium* spp. While *Neisseria* spp. and *Haemophilus* spp. are more abundant in CD children. While these studies have given us information about the general makeup of the microbiome of patients with CD, they do little to answer the questions if these changes precede disease onset, if they are a consequence of inflammation, or if the changes seen in the microbiome are associated with changes in immune cell phenotype. Future studies need to focus on causality, and possibly a specific bacterial group that could be pathogenic or protective in this group of patients, and that could be targeted for treatment.

Although it is unclear whether the altered microbiome is a cause of or consequence of disease, it is hypothesized Gram-negative bacteria in genetically susceptible individuals may contribute to the loss of gluten tolerance. If modified bacteria are a result of disease, the disrupted mucosa inundated with immature enterocytes could lead to conditions favoring Gram-negative instead of Gram-positive bacterial colonization. While this theory has not been proved, early studies have shown a propensity toward higher Gram-negative colonization in duodenal samples of pediatric patients with CD compared to healthy controls, in which case the dysbiosis seen seems to be of importance [27].

As previously stated, it is not clear whether an altered microbiota in CD patients could be cause or consequence of disease, however, CD offers a unique disease in which to study the microbiome as many other factors can be controlled for, including genetic makeup, environment, and trigger as these are all known, and the effect of the microbiome on disease pathogenesis can be further explored. Also, since the genetic makeup can be determined prior to a subject acquiring CD, it is possible to do longitudinal studies in these patients and observing the change in microbiome to see if the alterations noted are a cause of or consequence of disease [28]. The possibility that unfavorable bacteria may colonize the intestinal mucosal indicates the need to evaluate the microbiota from this site.

GLUTEN-FREE DIET AND MICROBIOME

To date, a gluten-free diet (GFD) is the only therapy for celiac patients, and a GFD reduces symptoms and restores the well-being of the individual and heals the mucosal damage [29]. Several studies have compared the gut microbiota of CD patients following a GFD with CD on gluten containing diet and controls. In CD patients, even after following a GFD (for at least 2 years), the duodenal mucosa microbiota was not completely restored and showed a less abundant bacterial richness compared to healthy and untreated subjects, with a persistent imbalance of the ratio of potentially harmful/beneficial bacteria [28]. Species-specific analysis has shown that while *Escherichia coli* and *Staphylococcus* counts are restored after a GFD, bifidobacteria counts remain lower in the feces of patients on a GFD compared to controls. A targeted study on bifidobacteria composition from CD patients on both a gluten containing and a GFD, and from healthy controls showed a correlation between the levels of *Bifidobacterium* and *Bifidobacterium longum* species in the fecal and tissue samples. Moreover, a generalized reduction in these bacterial populations was found in CD patients as compared to healthy children overall [30].

Additional information comes from a study, which evaluated the effect of a GFD on healthy subjects [23] using fluorescence in situ hybridization (FISH) and quantitative PCR. In this study, it was noted that the GFD led to a decrease in *B. longum*, *Clostridium lituseburense*, *Lactobacillus*, and *Faecalibacterium prausnitzii*, and an increase in Enterobacteriaceae and *E. coli* strains. This was thought to be due to reduced production of proinflammatory and regulatory cytokines due to a generalized reduction in the total luminal bacterial load of the large intestine caused by the GFD. The main finding was that a GFD-influenced gut microbial composition and immune activation (as measured by cytokine production) regardless of the presence of disease, and these effects were directly related to reduction in polysaccharide intake.

Few studies have followed the same patients pre- and post-GFD to test the effect of gluten on the microbiome in the presence of CD. An Italian study showed that the *Lactobacillus* community was lower before than after a GFD and lower in CD patients than in healthy controls. There was also a lower ratio of *Bifidobacterium* to *Bacteroides* and enterobacteria as compared to healthy controls [31].

These studies show that a GFD only partially restores fecal microbiota balances in CD patients. The reason is still unclear, although some suggest that genetic influences in those predisposed to CD affects the colonization of the microbiome which persists despite a GFD; furthermore since gluten has a prebiotic action, its absence in the GFD induces a different gut microbiota even in healthy individuals [32].

CD GENETICS AND MICROBIOME

CD is a disorder with a complex non-Mendelian pattern of inheritance, involving major histocompatibility complex (MHC) and non-MHC genes. The main genetic risk factor for CD falls within the MHC regions, a region located on 6p21 responsible for the strongest association signals observed in most immune-mediated diseases. The alleles encoding human leukocyte antigen (HLA)-DQ2 have been identified as a key modulator in the genetic risk associated with the MHC region in CD and is found in patients with CD much more frequently than the general population [33]. The main function of the MHC II molecules is to present bacterial antigens to T cells and to activate the immune system [34].

Recently, a prospective study in a cohort of 164 infants with a family history of CD and HLA-DQ genotype was studied to assess their microbiome composition. In this study, HLA-DQ2/8 genotype and the type of feeding (maternal or formula) were shown to influence the intestinal microbial composition. Specific decreases in beneficial species, such as *Bifidobacterium* spp. and *B. longum*, and an in *Staphylococcus* spp. were seen regardless of mild-feeding type [35]. A subanalysis of 22 of these infants, all breast-fed and vaginally delivered, confirmed that the HLA-DQ genotype itself influenced the microbial composition. The high-risk infants, those carrying the HLA-DQ2 genotype, were shown to carry an increased proportion of "harmful bacteria" species belonging to the Firmicutes and Proteobacteria phyla [36].

We have limited knowledge as to the mechanisms by which the HLA-DQ genotype could selectively influence colonization and composition of the gut microbiota. The main function of MHC II molecules is to activate MHC restricted T cells. We can speculate that expression of these genes can influence T cell activation, thereby influencing tissue resident T cells and could contribute to regulating the gut microbiota by regulating TLR expression or other commensal specific interactions.

With the advent of GWAS and Immunochip studies we have been better able to identify the genes associated with the development of CD, although the function of these genes is not all known [37]. Currently, it is established that 39 non-MHC loci are associated with the risk of developing CD, and that some of these 39 non-MHC loci harbor genes related to bacterial colonization, and likely contribute to "the celiac microbiome."

PREBIOTICS AND CD

Currently, the only established therapy for CD is a strict, lifelong adherence to the GFD. Adherence to a diet represents a difficult problem for many patients since traces of gluten are found in the majority of processed foods, and a strict GFD inevitably limits the social activities of patients [2]. In the last decade, research has been focused on new therapies to either improve compliance to the GFD or replace the GFD [38]. The theory of dysbiosis leading to immune cell activation and CD pathogenesis has shifted the focus of new therapies for CD to those that modulate the microbiome, specifically pre- and probiotics.

Prebiotics are substances that induce the growth or activity of microorganisms (e.g., bacteria and fungi) that contribute to the well-being of the host. Dietary prebiotics are typically nondigestible, fiber compounds that stimulate the growth of advantageous bacteria, although they do not target a specific bacterial group. Several foods are rich in prebiotics including raw garlic, leeks, chicory root, and whole wheat (although not relevant to CD patients). However, the ideal daily serving is not agreed upon.

Current research is ongoing as to the possibility of altering gluten-free products with prebiotics. Some early evidence has suggested that adding prebiotic inulin type fructans to gluten-free breads can provide benefits for patients with CD, as these are ingredients that can increase calcium absorption as possibly other nutrients as well [39].

PROBIOTICS AND CD

There has been some preliminary evidence as to the efficacy of prebiotics in the treatment of Crohn's disease, ulcerative colitis [40], calcium absorption [41], but to date no studies have been done looking at the use of prebiotics for the treatment of CD.

While prebiotics refer to the nutritional components found in food sources, probiotics are microorganisms that are believed to provide health benefits when consumed [42]. Live probiotic cultures are available in fermented dairy products and probiotic-fortified foods. Tablets, capsules, powders, and sachets also contain the bacteria in freeze-dried formulations. According to the FAO/WHO a probiotic is defined as "live microorganism, which when administered in adequate amounts confers a health benefit on the host." [43].

Recent research has linked the role of the microbiome to loss of gluten tolerance and increased recruitment of T cells [44,45]. This model for pathogenesis suggests that probiotic administration could modulate the composition and functions of the intestinal microbiota, both to defer and to avoid the onset of CD, and could also be used following a GFD when the normal microbial composition has not been restored. Potential benefits of probiotic therapy could come from the regulation of the immune response, degradation of toxin receptors, competition for nutrients, blockage of adhesion sites, and the production of inhibitory substances against pathogens [46]. The molecular mechanism of probiotic action is yet to be characterized. Most data regarding the use of probiotics for CD come from in vitro experimental models, although few studies have been conducted in humans.

Some probiotics have been found to digest or alter gluten polypeptides. A study done analyzing the role of VSL#3 (a cocktail of eight strains belonging to the species *Bifidobacterium breve, B. longum, Bifidobacterium infantis, Lactobacillus plantarum, Lactobacillus acidophilus, Lactobacillus casei, Lactobacillus delbrueckii* and *Streptococcus thermophiles*) in decreasing the toxic properties of wheat flour, showed that it was highly effective in hydrolyzing gliadin peptides compared to commercial probiotic products. However, the capacity of VSL#3 to degrade gliadin was disabled when the probiotic strains were tested individually, suggesting the importance of the dysbiosis in CD and treatment aimed toward that rather than a specific disease causing species [47].

Many studies in other disease conditions have shown the benefits of specific *Lactobacillus* and bifidobacterial strains in improving gut health. A study done in vitro evaluating the immunomodulatory properties of *B. bifidum* strain IATA-ES2 and *B. longum* strain ATCC 15707 versus *B. fragilis* strain DSM2451, *E. coli* strain CBL2, and *Shigella* sp. strain CBD8 on peripheral blood mononuclear cells under the effect of gliadin and IFN-γ highlights the differences in the ability of Gram negative bacteria and bifidobacteria species in inducing an immune response. *B. bifidum* and *B. longum* strains were able to induce lower levels of interleukin (IL)-12 and IFN-γ compared to *E. coli* and *Shigella*. The highest level of IL-10, a potent antiinflammatory cytokine was observed in the presence of *B. longum*. These data suggest that Gram-negative bacteria, such as *E. coli* and *Shigella* trigger higher levels of proinflammatory cytokines which contribute to disease pathogenesis, while species, such as *B. longum* act to maintain the gut integrity and maintain and antiinflammatory environment [45]. These data also highlight the importance of multistrain probiotics for treatment as opposed to single species treatment.

Another study looking at the effect of *Bifidobacterium lactis* and its role in gluten-induced tissue damage showed that *B. lactis* exerted a protective effect on epithelial cells against cellular damage induced by gliadin. Furthermore, dose escalation studies done in this study showed that a minimal concentration is needed to observe the protective effect, outlining the important of dosing in regards to probiotic administration [48].

More recently, a study using a gliadin-induced enteropathy animal model was developed to observe the effect of *B. longum* CECT 7347 in CD pathogenesis and treatment. The administration of *B. longum* enhanced villus width and enterocyte height in animals sensitized with IFN-γ and fed gliadin; and moderately diminished some of the alteration in jejunal structure. In addition, on cytokine analysis, it reduced levels of TNF-α and increased levels of IL-10 demonstrating an overall antiinflammatory response in the gut mucosa. This effect could contribute to improvement in gliadin-induced tissue damage and gut barrier function [49].

Studies evaluating the role of probiotics and CD in humans are scarce. In a randomized, double-blind, placebo-controlled study, *B. infantis* and its effects on gut permeability, occurrence of symptoms, and the presence of inflammatory cytokines in untreated CD patients were evaluated. In this study it was noted that probiotic administration was unable to modify gut barrier function; however, there was a marked improvement in digestion and a reduction in constipation. Abdominal pain and diarrheal symptom scores were also diminished although not significantly. There was no difference in inflammatory markers [50].

Another study evaluating the effect of *B. longum* CECT 7347 for 3 months in addition to a GFD in children newly diagnosed with CD showed a decrease in CD3 T cells, improving symptoms, and greater height percentile in those on a probiotic and GFD compared to those on the diet alone [51].

Another study in children, studying the effect of *B. breve* BR03 and B632 on serum cytokine production showed a decreased production of proinflammatory cytokine production after administration of probiotics compared to diet alone. The effect on proinflammatory cytokine TNF-α was only seen while receiving the probiotic, whereas antiinflammatory cytokine, IL-10 levels were undetectable throughout the study period. Suggesting that continuous probiotic supplementation is necessary and intermittent administration does not affect microbial milieu.

Alternatively, members of the Firmicutes phylum, specifically lactobacilli are thought to play a role in CD pathogenesis as well. A study identified a significant lack of *Lactobacilli* in symptom-free CD children. Thus the authors isolated five different *Lactobacilli* in the stool of healthy children, and proposed that *Lactobacillus rhamnosus* and *Lactobacillus paracasei* as potential targets [52].

Although preliminary research has suggested a possible role for probiotics in the treatment of CD, the relatively poor regulation of these supplements makes this treatment relatively hard to monitor. A study done testing 22 of the top selling probiotics, labeled gluten-free, and using chromatography to check for presence of gluten showed that 12 of the 22 (55%) probiotics contained >20 ppm of gluten, the acceptable cutoff for labeling a food product as gluten free [46].

To date, the evidence regarding the use of probiotics in patients with CD is still insufficient to justify their use in clinical practice, and until the FDA places stricter regulations on these supplements, their use can be considered dangerous for patients with CD. The recent evidence that probiotics do not alter the fecal microbiome of healthy subjects adds to the question of their applicability to widespread use [53].

CONCLUSIONS

In recent years, an increasing amount of attention has been paid to, as evidenced by the growing number of publications, the microbiome in health and disease. Although the majority of publications on the microbiome in CD have been conducted using different models, study populations, small sample sizes, most of the studies have seen differences in the populations of bifidobacteria and *Lactobacilli* in the gut microbial concentrations of patients with CD. In addition, patients with CD seem to have an increased number of Gram-negative bacteria, specifically Proteobacteria. In vitro data have suggested that dysbiosis in CD can lead to modification of the mucosal barrier and persistent immune activation or sensitization to activation by gliadin causing clinical symptoms. Additional studies dissecting out the role of the microbiome in immune cell activation and T cell priming will help further clarify the role of the microbiome in autoimmune disease pathogenesis and possibly the role of microbiome manipulation as treatment for CD.

As far as the GFD diet is concerned, it is currently the only accepted treatment for patients with CD. However, as evidenced by several studies, in regards to the microbiome, complete "normalization" is not achieved with this diet. In this setting is where probiotic therapy might be beneficial. Treatment with *Bifidobacterium* and/or *Lactobacilli* might be helpful in restoring altered gut microbiota and dampening immune activation, although further studies are needed to understand the dosing and proportion in which these bacteria need to be given in order for this to be achieved.

Finally, if considering the microbiome as a possible environmental activator for CD pathogenesis, it is possible to consider probiotics as a modulator of risk in those with high-risk risk factors, such as the DQ2 or DQ8 phenotype. In these subjects, probiotic administration might have a role in primary prevention, however, no study has been conducted using probiotics for this purpose, and so much research needs to be done in this area before any conclusions can be made.

REFERENCES

[1] Green PH, Cellier C. Celiac disease. N Engl J Med 2007;357(17):1731–43.

[2] Roma E, Roubani A, Kolia E, Panayiotou J, Zellos A, Syriopoulou VP. Dietary compliance and life style of children with coeliac disease. J Hum Nutr Diet 2010;23(2):176–82.

[3] Spurkland A, Sollid LM, Ronningen KS, et al. Susceptibility to develop celiac disease is primarily associated with HLA-DQ alleles. Hum Immunol 1990;29(3):157–65.

[4] Mazzilli MC, Ferrante P, Mariani P, et al. A study of Italian pediatric celiac disease patients confirms that the primary HLA association is to the DQ(alpha 1*0501, beta 1*0201) heterodimer. Hum Immunol 1992;33(2):133–9.

[5] Gill SR, Pop M, Deboy RT, et al. Metagenomic analysis of the human distal gut microbiome. Science 2006;312(5778):1355–9.

[6] Atarashi K, Tanoue T, Shima T, et al. Induction of colonic regulatory T cells by indigenous *Clostridium* species. Science 2011;331(6015):337–41.

[7] Ivanov II, Atarashi K, Manel N, et al. Induction of intestinal Th17 cells by segmented filamentous bacteria. Cell 2009;139(3):485–98.

[8] Round JL, Lee SM, Li J, et al. The Toll-like receptor 2 pathway establishes colonization by a commensal of the human microbiota. Science 2011;332(6032):974–7.

[9] Kamada N, Seo SU, Chen GY, Nunez G. Role of the gut microbiota in immunity and inflammatory disease. Nat Rev Immunol 2013;13(5):321–35.

[10] Festi D, Schiumerini R, Eusebi LH, Marasco G, Taddia M, Colecchia A. Gut microbiota and metabolic syndrome. World J Gastroenterol 2014;20(43):16079–94.

[11] Catassi C, Kryszak D, Bhatti B, et al. Natural history of celiac disease autoimmunity in a USA cohort followed since 1974. Ann Med 2010;42(7):530–8.

[12] Sellitto M, Bai G, Serena G, et al. Proof of concept of microbiome-metabolome analysis and delayed gluten exposure on celiac disease autoimmunity in genetically at-risk infants. PLoS ONE 2012;7(3):e33387.

[13] Kau AL, Ahern PP, Griffin NW, Goodman AL, Gordon JI. Human nutrition, the gut microbiome and the immune system. Nature 2011;474(7351):327–36.

[14] Round JL, Mazmanian SK. The gut microbiota shapes intestinal immune responses during health and disease. Nat Rev Immunol 2009;9(5):313–23.

[15] Szebeni B, Veres G, Dezsofi A, et al. Increased mucosal expression of Toll-like receptor (TLR)2 and TLR4 in coeliac disease. J Pediatr Gastroenterol Nutr 2007;45(2):187–93.

[16] Sharma R, Young C, Neu J. Molecular modulation of intestinal epithelial barrier: contribution of microbiota. J Biomed Biotechnol 2010;2010:305879.

[17] Stappenbeck TS, Hooper LV, Gordon JI. Developmental regulation of intestinal angiogenesis by indigenous microbes via Paneth cells. Proc Natl Acad Sci USA 2002;99(24):15451–5.

[18] van Heel DA, Franke L, Hunt KA, et al. A genome-wide association study for celiac disease identifies risk variants in the region harboring IL2 and IL21. Nat Genet 2007;39(7):827–9.

[19] Hunt KA, Zhernakova A, Turner G, et al. Newly identified genetic risk variants for celiac disease related to the immune response. Nat Genet 2008;40(4):395–402.

[20] Dubois PC, Trynka G, Franke L, et al. Multiple common variants for celiac disease influencing immune gene expression. Nat Genet 2010;42(4):295–302.

[21] Kalliomaki M, Satokari R, Lahteenoja H, et al. Expression of microbiota, Toll-like receptors, and their regulators in the small intestinal mucosa in celiac disease. J Pediatr Gastroenterol Nutr 2012;54(6):727–32.

[22] Castellanos-Rubio A, Fernandez-Jimenez N, Kratchmarov R, et al. A long noncoding RNA associated with susceptibility to celiac disease. Science 2016;352(6281):91–5.

[23] De Palma G, Nadal I, Collado MC, Sanz Y. Effects of a gluten-free diet on gut microbiota and immune function in healthy adult human subjects. Br J Nutr 2009;102(8):1154–60.

[24] De Palma G, Nadal I, Medina M, et al. Intestinal dysbiosis and reduced immunoglobulin-coated bacteria associated with coeliac disease in children. BMC Microbiol 2010;10:63.

[25] De Palma G, Capilla A, Nadal I, et al. Interplay between human leukocyte antigen genes and the microbial colonization process of the newborn intestine. Curr Issues Mol Biol 2010;12(1):1–10.

[26] Nadal I, Santacruz A, Marcos A, et al. Shifts in clostridia, bacteroides and immunoglobulin-coating fecal bacteria associated with weight loss in obese adolescents. Int J Obes (Lond) 2009;33(7):758–67.

[27] Nistal E, Caminero A, Herran AR, et al. Differences of small intestinal bacteria populations in adults and children with/without celiac disease: effect of age, gluten diet, and disease. Inflamm Bowel Dis 2012;18(4):649–56.

[28] Collado MC, Donat E, Ribes-Koninckx C, Calabuig M, Sanz Y. Specific duodenal and faecal bacterial groups associated with paediatric coeliac disease. J Clin Pathol 2009;62(3):264–9.

[29] Guandalini S, Assiri A. Celiac disease: a review. JAMA Pediatr 2014;168(3):272–8.

[30] Collado MC, Donat E, Ribes-Koninckx C, Calabuig M, Sanz Y. Imbalances in faecal and duodenal *Bifidobacterium* species composition in active and non-active coeliac disease. BMC Microbiol 2008;8:232.

[31] Di Cagno R, Rizzello CG, Gagliardi F, et al. Different fecal microbiotas and volatile organic compounds in treated and untreated children with celiac disease. Appl Environ Microbiol 2009;75(12):3963–71.

[32] Jackson FW. Effects of a gluten-free diet on gut microbiota and immune function in healthy adult human subjects—comment by Jackson. Br J Nutr 2010;104(5):773.

[33] Spurkland A, Sollid LM, Polanco I, Vartdal F, Thorsby E. HLA-DR and -DQ genotypes of celiac disease patients serologically typed to be non-DR3 or non-DR5/7. Hum Immunol 1992;35(3):188–92.

[34] Cenit MC, Olivares M, Codoner-Franch P, Sanz Y. Intestinal microbiota and celiac disease: cause, consequence or co-evolution? Nutrients 2015;7(8):6900–23.

[35] Palma GD, Capilla A, Nova E, et al. Influence of milk-feeding type and genetic risk of developing coeliac disease on intestinal microbiota of infants: the PROFICEL study. PLoS One 2012;7(2):e30791.

[36] Olivares M, Neef A, Castillejo G, et al. The HLA-DQ2 genotype selects for early intestinal microbiota composition in infants at high risk of developing coeliac disease. Gut 2015;64(3):406–17.

[37] Trynka G, Wijmenga C, van Heel DA. A genetic perspective on coeliac disease. Trends Mol Med 2010;16(11):537–50.

[38] Kaukinen K, Lindfors K, Maki M. Advances in the treatment of coeliac disease: an immunopathogenic perspective. Nat Rev Gastroenterol Hepatol 2014;11(1):36–44.

[39] Capriles VD, Areas JA. Effects of prebiotic inulin-type fructans on structure, quality, sensory acceptance and glycemic response of gluten-free breads. Food Funct 2013;4(1):104–10.

[40] Hedin C, Whelan K, Lindsay JO. Evidence for the use of probiotics and prebiotics in inflammatory bowel disease: a review of clinical trials. Proc Nutr Soc 2007;66(3):307–15.

[41] Scholz-Ahrens KE, Schrezenmeir J. Inulin and oligofructose and mineral metabolism: the evidence from animal trials. J Nutr 2007 137(11 Suppl.):2513S–23S.

[42] Rijkers GT, de Vos WM, Brummer RJ, Morelli L, Corthier G, Marteau P. Health benefits and health claims of probiotics: bridging science and marketing. Br J Nutr 2011;106(9):1291–6.

[43] Food and Agriculture Organization of the United Nations WHO. Guidelines for the evaluation of probiotics in food. Geneva, Switzerland: World Health Organization; 2002.

[44] Sanchez E, Ribes-Koninckx C, Calabuig M, Sanz Y. Intestinal *Staphylococcus* spp. and virulent features associated with coeliac disease. J Clin Pathol 2012;65(9):830–4.

[45] De Palma G, Cinova J, Stepankova R, Tuckova L, Sanz Y. Pivotal advance: bifidobacteria and Gram-negative bacteria differentially influence immune responses in the proinflammatory milieu of celiac disease. J Leukoc Biol 2010;87(5):765–78.

[46] Vanderpool C, Yan F, Polk DB. Mechanisms of probiotic action: implications for therapeutic applications in inflammatory bowel diseases. Inflamm Bowel Dis 2008;14(11):1585–96.

[47] De Angelis M, Rizzello CG, Fasano A, et al. VSL#3 probiotic preparation has the capacity to hydrolyze gliadin polypeptides responsible for Celiac Sprue. Biochim Biophys Acta 2006;1762(1):80–93.

[48] Lindfors K, Blomqvist T, Juuti-Uusitalo K, et al. Live probiotic *Bifidobacterium lactis* bacteria inhibit the toxic effects induced by wheat gliadin in epithelial cell culture. Clin Exp Immunol 2008;152(3):552–8.

[49] Laparra JM, Olivares M, Gallina O, Sanz Y. *Bifidobacterium longum* CECT 7347 modulates immune responses in a gliadin-induced enteropathy animal model. PLoS One 2012;7(2):e30744.

[50] Smecuol E, Hwang HJ, Sugai E, et al. Exploratory, randomized, double-blind, placebo-controlled study on the effects of *Bifidobacterium infantis* natren life start strain super strain in active celiac disease. J Clin Gastroenterol 2013;47(2):139–47.

[51] Olivares M, Castillejo G, Varea V, Sanz Y. Double-blind, randomised, placebo-controlled intervention trial to evaluate the effects of *Bifidobacterium longum* CECT 7347 in children with newly diagnosed coeliac disease. Br J Nutr 2014;112(1):30–40.

[52] Lorenzo Pisarello MJ, Vintini EO, Gonzalez SN, Pagani F, Medina MS. Decrease in lactobacilli in the intestinal microbiota of celiac children with a gluten-free diet, and selection of potentially probiotic strains. Can J Microbiol 2015;61(1):32–7.

[53] Kristensen NB, Bryrup T, Allin KH, Nielsen T, Hansen TH, Pedersen O. Alterations in fecal microbiota composition by probiotic supplementation in healthy adults: a systematic review of randomized controlled trials. Genome Med 2016;8(1):52.

Chapter 40

Probiotics for the Treatment of Liver Disease

C. Punzalan* and A. Qamar*,**

*Lahey Hospital and Medical Center, Burlington, MA, United States; **Tufts University School of Medicine, Boston, MA, United States

DYSBIOSIS AND LIVER DISEASE

The intestinal microbiota (IM) differs between healthy individuals and patients with liver disease [1–6]. The Human Microbiome Project demonstrated that IM in healthy individuals vary among subjects but Bacteroides and Firmicutes were the two dominant species in most individuals [7]. Other studies showed that in healthy individuals, almost all of IM are from the phyla Bacteroidetes, Firmicutes, Proteobacteria, or Actinobacteria [8]. Other studies demonstrated that in normal subjects, the IM primarily consists of *Clostridium coccoides*, *Clostridium leptum* group, *Bacteroides* spp., *Bifidobacterium* spp., and *Enterobacter* spp. [9,10]. In patients with cirrhosis and various liver diseases, the IM is altered, also known as dysbiosis. Several studies show that depending on the cause of the liver disease, certain bacteria are increased and others are decreased (Table 40.1).

Cirrhosis

Studies have demonstrated a difference in the gut microbiota between healthy individuals and patients with compensated and decompensated cirrhosis who were not infected [1–4]. Previous studies have noted that patients with cirrhosis have a decrease in nonpathogenic bacteria including Lachnospiraceae, Ruminococcaceae, and Clostridiales XIV and an increase in pathogenic bacteria including Enterococcus, Enterobacteriaceae, and Bacteroidaceae [1,3]. In one study, the ratio of "good" versus "bad" bacteria was quantified as the cirrhosis dysbiosis ratio (CDR). Overall, the study found a significant correlation of the CDR with MELD score and presence of endotoxin. Endotoxin was negatively associated with Clostridiales XIV, Lachnospiraceae, Ruminococcaceae, but positively associated with MELD score, Enterobacteriaceae, and Bacteroidaceae. Compared to all cirrhotic patients, the CDR for controls was significantly higher [3]. The same study demonstrated a significant change in IM after development of HE. Patients with HE had an increase in Enterobacteriaceae [3]. Compared to cirrhotic patients with cirrhosis without HE, patients with HE were also found to have an increase in gut Veillonellaceae and higher serum endotoxin and inflammatory markers [2]. Interestingly, patients who developed infections had lower CDR and higher endotoxin than matched patients without infections [3]. Another study demonstrated that patients with cirrhosis from either alcohol or hepatitis B compared to healthy subjects have a decrease in Bacteroidetes, and increase in Proteobacteria, Fusobacteria, Enterobacteriaceae, Veillonellaceae, and Streptococcaceae. Lachnospiraceae was also decreased and correlated negatively with patients' CTP scores [1].

Alcohol

In animal and mouse models, patients with chronic alcoholic liver disease have been shown to have dysbiosis and increased bacterial translocation (BT) [10–14]. Animal studies have demonstrated that chronic alcohol exposure decreases bacterial diversity and increases hepatic markers of inflammation. Chronic alcohol decreases the Clostridium and Bacteroidetes and increases Proteobacteria [13]. Patients with alcoholic cirrhosis also have an increase in fecal Enterobacteriaceae, Halomonadaceae, and lower Lachnospiraceae, Ruminococcaceae, and Clostridiales XIV [1,3,10]. Unlike the Chen study, another study demonstrated that despite similar MELD and BMI, compared to patients with cirrhosis from other causes of liver disease, those with alcoholic cirrhosis have an increase in endotoxemia and lower CDR [3,13]. An acute alcoholic binge in healthy individuals leads to an increase in serum endotoxin and 16S rDNA, a marker of bacteria [3].

TABLE 40.1 Liver Disease and Dysbiosis Bacteria

Liver Disease	Decreased Bacteria	Increased Bacteria
Alcoholic liver disease	Clostridium[a] Bacteroidetes[a] Lachnospiraceae[a] Ruminococcaceae[a] Clostridiales XIV[a]	Proteobacteria[a] Enterobacteriaceae[a] Halomonadaceae[a]
Nonalcoholic fatty liver disease	Faecalibacterium[a] Anaerosporobacter[a] Bifidobacterium[a] Veillonellaceae[c] Firmicutes[d]	Parabacteroides[a] Allisonella[a] Porphyromonadaceae[c] Bacteriodaceae[c] Bacteroidetes[a] Proteobacteria[a]
Hepatitis B	Bifidobacterium[a] Lactobacillus rhamnosus[a] Lactobacillus fermentum[a] Firmicutes[a] Bacteroidetes[a]	Enterobacteriaceae[a]
Hepatitis C	Alphaproteobacteria[a]	Bacilli[a] Gammaproteobacteria[a]
Cirrhosis and HE	Lachnospiraceae[a] Ruminococcaceae[a] Clostridiales XIV[a] Bacteroidetes[a]	Enterococcus[a] Enterobacteriaceae[a,b] Bacteroidaceae[a] Veillonellaceae[b] Proteobacteria[a] Fusobacteria[a] Streptococcaceae[a] Escherichia coli[a] Staphylococcus[a]

[a]Compared to normal subjects.
[b]Compared to cirrhotics without HE.
[c]Compared to cirrhotics from non-NASH patients.
[d]Compared to alcoholic cirrhotics.

NASH

Animal and human studies have also demonstrated that IM changes with fat deposition [3,15,16]. One study demonstrated that compared to healthy controls, patients with NASH have decreased Faecalibacterium and Anaerosporobacter but increased Parabacteroides and Allisonella [15]. Another study demonstrated that patients with NASH cirrhosis have higher Porphyromonadaceae, Bacteriodaceae, and lower Veillonellaceae than cirrhosis patients with other liver diseases, including alcohol. CDR and endotoxin levels, however, were similar [2]. Compared to healthy individuals, patients with NASH had decreased Bifiobacterium and increased Bacteroidetes. Similar to the alcohol group, NASH patients also had decreased Firmicutes, especially Lachnospiraceae and Ruminococcaceae [16]. Another study also showed that NASH patients have increased Bacteroidetes and Proteobacteria, and less Firmicutes [16]. Obese patients have a higher prevalence of small intestinal bacterial overgrowth (SIBO) than healthy individuals and the presence of SIBO is correlated with severe hepatic steatosis [17,18].

Hepatitis B

Studies have also shown that patients with chronic liver disease secondary to hepatitis B may have changes in their IM compared to healthy individuals [19–21]. One study found that Bifidobacterium, "healthy" bacteria was lower in number in patients with chronic hepatitis B (CHB) and cirrhosis compared to normal controls. When specific fecal Bifidobacterium species was further evaluated, the study found that the composition was different in patients with CHB with or without cirrhosis compared to healthy individuals [20]. Another study evaluating fecal Lactobacillus found decreased diversity in patients with cirrhosis from CHB, especially a decrease in L. rhamnosus and L. fermentum compared to healthy controls [19]. Another study evaluated the fecal microbiota of healthy controls, asymptomatic HBV carriers, patients with CHB, and patients with decompensated HBV cirrhosis [21]. The quantitative PCR results found a higher count of Enterobacteriaceae and lower Firmicutes and Bacteroidetes in decompensated HBV cirrhotic patients compared to healthy controls and HBV carriers. Patients in both the chronic HBV group and cirrhosis group had less bifidobacteria and lactic acid bacteria including *Lactobacillus* compared to the healthier groups. The Bifidobacteria/Enterobacteriaceae ratio (B/E) was 1.15 ± 0.11 in healthy controls and decreased significantly in the asymptomatic hepatitis B carrier group (0.99 ± 0.09), CHB group (0.76 ± 0.08), and cirrhotic group (0.64 ± 0.09) [21].

Hepatitis C

There is limited data on the IM of hepatitis C positive patients [22]. In a cross-sectional study of 51 patients with chronic HCV, stool microbiota was compared to healthy controls and it was found that patients with chronic hepatitis C with and without cirrhosis had decreased stool diversity. These patients also had increased bacilli and gammaproteobacteria and decreased alphaproteobacteria. Stool was also evaluated posttransplant and the microbiota had returned to be more similar to the healthy individuals [22].

PSC

There is some evidence, primarily in animal models, that changes in IM may play a role in the pathogenesis of PSC. Bacteria products, such as lipopolysaccharide (LPS) (endotoxin) and peptidoglycan have been proposed as possible triggers of inflammation and fibrosis [23]. Mouse models with SIBO or injected with bacterial products were shown to develop hepatobiliary inflammation. Data linking dysbiosis to PSC in human subjects is limited [23].

PROBIOTICS AND LIVER DISEASE

There have been multiple studies evaluating the use of probiotics in the treatment of various liver diseases. Probiotics may protect or treat liver disease by a number of different mechanisms [15]. Probiotics have been shown to restore the gastrointestinal barrier function, modulate the immune system, and decrease the number of harmful bacteria [24–26]. Repopulating the intestine with beneficial bacteria corrects the dysbiosis, especially in patients with SIBO, thereby reducing the inflammatory response [27]. Besides the decrease in inflammatory cytokines, probiotics have also been shown to increase factors that increase mucus layer protection [14]. The change in the IM may decrease the urease-producing bacteria thereby improving HE. The most promise has been in the use of probiotics in the treatment of fatty liver disease and hepatic encephalopathy.

NAFLD

Several studies have evaluated the role of bacterial microbiota in the pathogenesis of NAFLD [18,28,29]. Patients with NAFLD have dysbiosis, increased gut permeability, and SIBO [18]. NASH patients also have increased gut permeability associated with an increased prevalence of moderate or severe steatosis [18]. IM has also been shown to cause acetaldehyde production and choline deficiency leading to NAFLD [28,29]. Intestinal microbiota has been shown to produce ethanol that is converted to acetaldehyde by gut bacteria. The acetaldehyde is absorbed and causes hepatic injury [29]. Animal models have shown that IM plays a role in choline metabolism that can lead to choline deficiency and NAFLD [28].

As there seems to be a role for microbiota in NAFLD, a number of studies investigated the efficacy of various probiotics in NAFLD. Although the duration of treatment, microbial species, and outcomes measured differ between studies, they all demonstrate improved markers of liver disease with probiotic use (Table 40.2).

TABLE 40.2 NAFLD and Probiotics

References	Treatment	No. of Patients	Duration of Treatment (Months)	Outcome
Loguercio	VSL#3	22	3	Improved liver enzymes Decreased malondialdehyde and 4-hydroxynonenal
Aller	*Lactobacillus bulgaricus* and *Streptococcus thermophilus*	30	3	Improved liver enzymes
Wong	*Lactobacillus plantarum, Lactobacillus delbrueckii, Lactobacillus acidophilus, L. rhamnosus*, and *Bifidobacterium bifidum*	20	6	Decreased liver fat Decreased AST
Malaguarnera	*Bifidobacterium longum*, FOS, and B2	66	6	Decreased TNF-a, CRP, AST, bilirubin, LDL HOMA-IR, and serum endotoxins Decreased NASH activity index Decreased steatosis on liver biopsy
Vajro	*L. rhamnosus* strain GG	20	2	Decreased ALT

A pilot study examining the effect of VSL#3, which contains 450 billion bacteria (*S. thermophilus, Bifidobacterium breve, B. longum, Bifidobacterium infantis, L. acidophilus, L. plantarum, Lactobacillus casei*, and *L. bulgaricus*), in patients with NAFLD ($n = 22$), alcoholic cirrhosis ($n = 20$), and chronic HCV ($n = 36$) found that liver enzymes improved in all patients who received the probiotic for 3 months [27]. Both the NAFLD and AC group also had a significant decrease in markers of lipid peroxidation, malondialdehyde and 4-hydroxynonenal, as well as S-nitrosothiols (S-NO, a reactive oxygen intermediate).

A double-blind randomized controlled study of 30 patients with biopsy-proven NAFLD studied the effect of *L. bulgaricus* and *S. thermophilus* for 3 months. Aminotransferases improved in the patients who were treated with probiotics compared to the placebo group [30].

Another study randomized 20 patients with biopsy-proven NASH to receive standard of care or a probiotic containing *L. plantarum, L. delbrueckii, L. acidophilus. L. rhamnosus*, and *B. bifidum*. Both groups were instructed to lose weight and participated in a diet and exercise regimen. After 6 months of treatment, the probiotic group had a statistically significant decrease in their intrahepatic triglyceride content as measured by spectroscopy compared to the standard care group who did not have a change in their liver fat. Aspartate transaminase (AST) also decreased in the probiotic group [31].

A randomized, double-blind, placebo-controlled study compared lifestyle modifications and a synbiotic formula of *B. longum*, fructooligosaccharides (FOS), a prebiotic that increases the growth of beneficial bacteria, and vitamin B2, 6, 11, and 12 to lifestyle modifications alone in 66 patients with biopsy-proven NASH. After 24 weeks, the patients who were treated with the synbiotic formula had significantly decreased TNF-a, CRP, AST, bilirubin, low-density lipoprotein (LDL) homeostasis model assessment of insulin resistance (HOMA-IR), and serum endotoxins. Compared to the placebo group, the synbiotic group also had significantly decreased NASH activity index and steatosis on repeat liver biopsy [32].

Finally, a study of 20 obese children were diagnosed with NAFLD by abnormal liver chemistries and imaging with hepatic steatosis randomized for treatment with placebo or *L. rhamnosus* strain GG for 8 weeks. The children who received the probiotic had a significant decrease in ALT compared to placebo [33].

A metaanalysis published in 2013 included the latter four randomized trials that assessed the effects of probiotics on NAFLD [34]. All four RCTs demonstrated an improvement in ALT [-23.71 weighted mean difference (WMD)] in patients who were treated with probiotics compared to placebo. Three of the RCTs also demonstrated a significant improvement in AST (WMD -19.77) and total cholesterol (WMD -0.28). Only one RCT, however evaluated

posttreatment histology. Body mass index, glucose, and LDL was not associated with change [34].

Alcoholic Liver Disease

BT has been shown to play a role in the pathogenesis of alcoholic liver disease [12]. Patients with alcoholic liver disease have been shown to have increased gut permeability and increased LPS or endotoxin levels in the portal and systemic circulation [11]. LPS stimulates Kupffer cell production of inflammatory cytokines, such as TNF-a [14]. Another theory involves chronic ethanol's reduction in saturated long chain fatty acids which leads to further dysbiosis [35].

Animal models have demonstrated that probiotics improved alcohol-induced hepatic steatosis and hepatic TNF-a production [14]. In a mouse study, supplementation with *L. rhamnosus* GG decreased ethanol-mediated endotoxemia, hepatic inflammation, and steatosis [36]. The study of VSL#3 mentioned earlier found improvement in aminotransferases, markers of lipid peroxidation, and S-NO in patients with alcoholic cirrhosis but also found improvement in albumin and bilirubin, as well as the cytokines TNF-a, IL-6, and IL-10 only in the group with alcoholic cirrhosis [27].

Probiotics have also been studied in the treatment of alcoholic hepatitis. Two studies have demonstrated improvement in markers of liver disease but clinical outcomes were not measured. A randomized, prospective pilot clinical trial demonstrated that after 7 days of *B. bifidum* and *L. plantarum* 8PA3, patients with mild alcoholic hepatitis had a significant reduction in bilirubin, ALT, and AST. When compared with a group which received only standard therapy (diazepam for alcohol withdrawal and vitamins), however, the only statistically significant difference was the reduction in the ALT in the probiotic group [37]. A multicenter, randomized, controlled clinical trial which evaluated probiotics with *Lactobacillus subtilis/Streptococcus faecium* versus placebo for 7 days in patient with alcoholic hepatitis found that liver enzymes, bilirubin, and prothrombin time improved in both groups. The improvement, however, was likely due to abstinence from alcohol [38]. In the probiotic group, the number of colony-forming units of *E. coli* was significantly reduced while in the placebo group, the stool flora was unchanged. Neither study evaluated outcomes including mortality [37,38].

CIRRHOSIS

Bacterial Translocation

Dysbiosis as well as small bowel bacterial overgrowth (SIBO) and increased intestinal permeability allow BT which appears to play a significant role in the complications of cirrhosis including spontaneous bacterial peritonitis

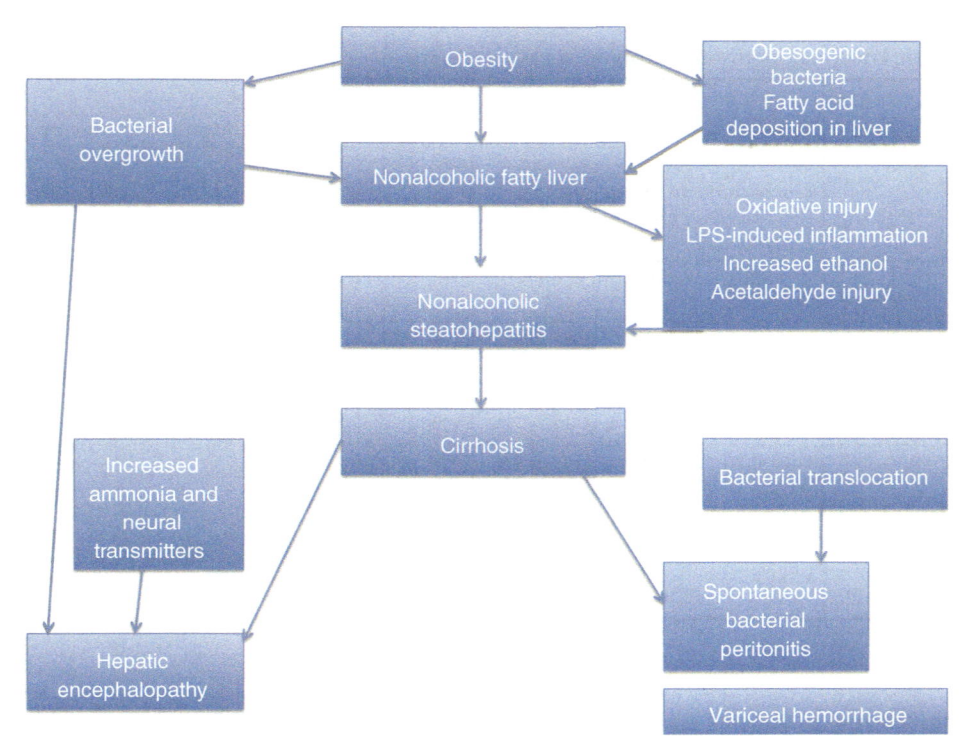

FIGURE 40.1 Potential mechanisms by which probiotics may affect liver disease.

(SBP), portal hypertension, and hepatic encephalopathy [5,12,18,39–42] (Fig. 40.1). Several studies have demonstrated SIBO in cirrhotic animals and humans [5,43,44]. Rat models have shown that SIBO likely plays a significant role in BT [45]. Prevalence of SIBO increases with severity of cirrhosis, including level of bilirubin and presence of ascites [44]. SBO and small bowel dysmotility have also been shown to be increased in cirrhotic patients with a history of SBP versus cirrhotic patients without a history of SBP [46]. Animal models provide further evidence for the role of BT in the development of SBO. Mouse models have demonstrated that BT is associated with positive ascitic fluid culture [47]. The same bacteria have been found in the gut, mesenteric lymph nodes, and ascitic fluid [47,48]. Several epithelial changes have been described in cirrhosis which allows BT [5]. These changes include widening of intercellular spaces, dilated extracellular spaces, vascular congestion, and intestinal mucosal oxidative damage [5]. Furthermore, the innate immune response and possibly acquired immune response have been shown to be decreased in cirrhotics which may further enable BT. Endotoxins lead to the production of proinflammatory cytokines, such as TNF-a and IL-6 [27].

Several studies have evaluated the effect of probiotics on these complications of cirrhosis. Promising results came from a double-blind trial of 130 patients with decompensated cirrhosis randomized patients to receive VSL#3 or placebo for 6 months. After 24 weeks of treatment, the probiotic group had decreased hospitalizations for complications of cirrhosis (24%) compared to the placebo group (45%). Furthermore, in the treatment group, the MELD scores decreased from a baseline of 17–13 while the MELD scores of the placebo group remained static [49]. A smaller study assessing outcomes with *E. coli* Nissle versus placebo found an improvement in intestinal colonization but there was no statistical significance in the level of endotoxin and Child-Pugh Score [50].

A few studies have evaluated the effect of VSL#3 on portal hypertension with less favorable results [51–53]. One study compared the use of propranolol plus placebo, norfloxacin, or VSL#3 in patients with large nonbleeding esophageal varices. Fifty eight percent of patients in the probiotic arm versus 31% in the propranolol only arm achieved a response defined as a hepatic venous pressure gradient (HVPG) decrease of 20% from baseline or to <12 mmHg, however, the differences in the mean decrease in HVPG across the groups were not statistically significant [52]. A small, randomized placebo-controlled trial on the use of VSL3# versus placebo found that while HVPG decreased 11.6% in the treatment group, the results were not statistically significant [51]. Another small study found that 67% of 12 cirrhotic patients who received VSL#3 for 6 weeks had at least a 10% decrease in HVPG however the HVPG decreased only from a mean of 21.8–19.6 mmHg [53].

Although BT increases the risk of SBP, probiotics have not been successful in prevention of SBP. A double-blind,

TABLE 40.3 Hepatic Encephalopathy and Probiotics

References	HE Type	Type of Probiotic	No. of Patients	Duration of Treatment (Month)	Outcome
Liu	MHE	*Pediococcus pentoseceus* *Leuconostoc mesenteroides* *Lactobacillus paracasei* *L. plantarum*	58	1	Increase in *Lactobacillus* Decrease in *E. coli* and *Staphylococcus* MHE reversal in 50%
Mittal	MHE	110 billion colony forming units bd	160	3	MHE recovered in 35% Decrease in SIP
Sharma	MHE	*Streptococcus faecalis, Clostridium butycricum, Bacillus mesentericus,* lactic acid bacillus	190	1	MHE reversed in 50%
Malaguarnera	HE	Bifidobacterium and FOS	125	2	Improved psychometric tests
Agarwal	HE	Four strains of *Lactobacillus,* three strains of *Bifidobacterium,* and one strain of *Streptococcus salivarius* subsp. *thermophilus*	235	12	Decrease in HE recurrence
Lunia	MHE	VSL#3	33	3	Decrease in development of overt HE

randomized controlled trial in patients who had recovered from an episode of SBP or were at high risk based on low ascitic fluid protein or serum bilirubin >2.5 mg/dL found that adding probiotics to norfloxacin did not improve its efficacy in prevention of SBP [54].

Hepatic Encephalopathy

The data on probiotics and hepatic encephalopathy appear more promising than results with other complications of cirrhosis [55–58]. In a randomized, placebo-controlled trial evaluating 97 cirrhotics, 58 of whom had minimal hepatic encephalopathy (MHE) by psychometric and neurophysiological modalities, treatment with a fermentable fiber alone was of benefit in a substantial proportion of patients [55]. Cirrhotic patients with MHE were found to have significant fecal overgrowth with *E. coli* and *Staphylococcus* species. After 30 days of supplementation with synbiotics (fermentable fiber and probiotics—four nonurease-producing bacteria), there was a significant reduction of the overgrowth species and significant increase in nonurease-producing *Lactobacillus* spp. The change in bacteria was evident 14 days after the cessation of supplementation. Fifty percent of patients in the groups that received synbiotics and fermentable fibers also had a reversal of MHE compared to only 13% of the placebo group [55]. Also, 47% of patients in the synbiotic/fermentable fiber group improved their Child-Turcotte-Pugh score compared to only 8% in the placebo group [55] (Table 40.3).

Another prospective trial evaluated outcomes in patients who received lactulose, probiotics, or L-ornithine L-aspartate (LOLA, a salt of the two amino acids) compared to placebo [56]. The prospective, randomized trial found that

MHE resolved in 47.5% of patients on lactulose and 35% of patients who received probiotics (or LOLA) compared to 10% of patients on placebo ($P = 0.006$). The groups who were treated had a statistically significant decrease in their Sickness Impact Profile (SIP) questionnaire compared to placebo although there was no statistically significant difference in the score improvement between the three groups [56].

An open-label randomized study compared 1 month of treatment with lactulose, probiotics (*S. faecalis, C. butycricum, B. mesentericus,* lactic acid bacillus), or lactulose with probiotics in the treatment of MHE which was diagnosed by psychometric tests. Treatment was equally effective and about 50% of patients in all three arms had normalization of abnormal parameters by psychometric testing [57].

In another study, 71% of nonalcoholic cirrhotic patients who ate 12 oz of yogurt daily for 12 days had resolution of their MHE compared to no patients in the no-treatment arm. The yogurt contained *L. bulgaricus* and *S thermophilus.* Furthermore, 25% of patients who were not treated developed overt hepatic encephalopathy versus 0% of the yogurt patients [59].

A metaanalysis published in 2011 evaluated the use of probiotics on minimal or subclinical hepatic encephalopathy [60]. Four randomized controlled trials on MHE were included. Two studies used synbiotic preparations (pre- and probiotic) and two studies only used probiotics. Five additional studies which used lactulose (a prebiotic) were also included in the metaanalysis. Compared with placebo, pre-, pro-, and synbiotics significantly reduced the risk of no improvement of MHE (RR 0.4, 95% CI 0.32–0.5; $P < 0.001$). While lactulose use had a 67% decrease in the risk of no improvement of MHE compared to 50 and 60% decrease with synbiotics and probiotics, respectively, synbiotics and probiotics were better tolerated [60].

Probiotics has also been shown to improve mild, overt hepatic encephalopathy [61]. In a prospective, randomized study of 125 patients with grade 1-2 hepatic encephalopathies (West Haven criteria), treatment with Bifidobacterium and FOS was compared to lactulose alone. At 30 and 60 days, the group treated with Bifidobacterium and FOS had significantly improved neuropsychiatric test scores compared to the lactulose group. The probiotic group also tolerated treatment with no adverse effects observed. Abdominal pain, cramping, and diarrhea were observed in the lactulose group [61].

Probiotics have also been shown to be as effective as primary and secondary prophylaxis for hepatic encephalopathy. In an open-label randomized controlled trial, patients who recovered from an episode of overt hepatic encephalopathy received lactulose or probiotics (one capsule 3 times a day, each containing four strains of *Lactobacillus,* three strains of *Bifidobacterium,* and one strain of *S. salivarius* subsp. *thermophilus*). Patients who received either treatment had a significant decrease in HE recurrence compared to no treatment. Furthermore, there was no difference in recurrence between the patients who received lactulose and those who received probiotics [58]. An open-label randomized controlled trial studied the use of VSL#3 for 3 months versus no treatment in primary prophylaxis against overt HE. Patients with MHE based on psychometric testing were included. On intention to treat analysis, a significantly smaller number of patients in the group who received VSL#3 developed overt HE compared to the untreated group (15% vs. 25.7% $P = 0.04$) [62]. A detailed review of the use of probiotics in the treatment of HE included seven trials and found that each study differed in type of probiotic use and duration of treatment. Their analysis failed to find an advantage of probiotics compared to no treatment or lactulose on lack of recovery, mortality, quality, of life, and adverse events. They also found that the trials included had a high risk of systemic and random errors. Therefore, the authors of the review were unable to recommend probiotics as a therapy for hepatic encephalopathy [63].

Postliver Transplant

Dysbiosis has also been demonstrated in liver transplant patients [19,64,65]. Studies have evaluated whether probiotics decrease infection rates in this patient population. A prospective, randomized placebo-controlled trial of 95 liver transplant patients found that patients who received fiber plus living *L. plantarum* had fewer bacterial infections than patients who had selective bowel decontamination (antibiotics) (13% vs. 48% $P < 0.017$) [66]. The group of patients who received fiber and the probiotic also decreased ICU and hospital length of stay but the difference did not receive statistical significance [66]. The same group then evaluated outcomes using four instead of one lactic acid bacteria. In a prospective randomized double-blind trial in 66 liver transplant patients, postoperative bacterial infections were significantly reduced to 3% in the group who received four lactic acid bacteria and four fibers compared to 48% of patients who received fibers alone. Length of hospital stay did not differ between the two groups but the length of antibiotic use was shorter in patients who had received the probiotics [67]. A small study examined found that probiotics decreased infection in living donor liver transplant recipients (4% vs. 24% $P < 0.033$) [68]. An Australian study compared fiber and probiotic use (*L. acidophilus, L. plantarum, B. lactis, L. casei,* and *L. rhamnosus*) to fiber only after liver transplantation. The study found that only 30.3% of probiotic patients versus 8.8% of the fiber only patients developed a postoperative infection [69]. There was no statistically significant difference in length of hospital stay or mortality [69]. A recent metaanalysis on the four mentioned studies found improved outcomes with probiotics use. Compared to the control group, patients who received perioperative probiotics had decreased infection rate (35% vs. 7%), decreased length of stay (mean difference 1.41 d), shortened ICU stay (MD 1.41 d), and shortened duration of antibiotic use (MD 3.89). There was no significant difference in acute graft rejection, mortality, or side effects [70].

Other

A small study of 14 patients with primary sclerosing cholangitis had no changes in pruritus, fatigue, or liver enzymes or function tests after treatment with 3 months of four *Lactobacillus* and two *Bifidobacillus* strains [71].

CONCLUSION AND PROBIOTICS IN CLINICAL PRACTICE

Intestinal microbiota and BT play a vital role in the development of NAFLD and the complications of cirrhosis; there may be an increasing role for use of probiotics in the management of liver disease particularly NASH and hepatic encephalopathy. Fig. 40.1 summarizes the possible mechanisms by which probiotics may play a role in management of liver disease. The most robust results for probiotic use are in the treatment of NAFLD and hepatic encephalopathy.

REFERENCES

[1] Chen Y, Yang F, Lu H, et al. Characterization of fecal microbial communities in patients with liver cirrhosis. Hepatology 2011;54: 562–72.

[2] Bajaj JS, Ridlon JM, Hylemon PB, et al. Linkage of gut microbiome with cognition in hepatic encephalopathy. Am J Physiol 2012;302:G168–75.

[3] Bajaj J, Hylemon P, Ridlon J, et al. Colonic mucosal microbiome differs from stool microbiome in cirrhosis and hepatic encephalopathy and is linked to cognition and inflammation. Am J Physiol 2012;303:G675–85.

[4] Liu J, Wu D, Ahmed A, et al. Comparison of the gut microbe profiles and numbers between patients with liver cirrhosis and healthy individuals. Curr Microbiol 2012;65:7–13.

[5] Wiest R, Garcia-Tsao G. Bacterial translocation (BT) in cirrhosis. Hepatology 2005;41:422–33.

[6] Kim A. Dysbiosis: a review highlighting obesity and inflammatory bowel disease. J Clin Gastroenterol 2015;49:S20–4.

[7] Human Microbiome Project Consortium. Structure, function and diversity of the healthy human microbiome. Nature 2012;486:207–14.

[8] Frank DN, Amand ALS, Feldman RA, et al. Molecular-phylogenetic characterization of microbial community imbalances in human inflammatory bowel diseases. Proc Natl Acad Sci 2007;104:13780–5.

[9] Eckburg PB, Bik EM, Bernstein CN, et al. Diversity of the human intestinal microbial flora. Science 2005;308:1635–8.

[10] Tuomisto S, Pessi T, Collin P, Vuento R, Aittoniemi J, Karhunen PJ. Changes in gut bacterial populations and their translocation into liver and ascites in alcoholic liver cirrhotics. BMC Gastroenterol 2014;14:40.

[11] Bala S, Marcos M, Gattu A, Catalano D, Szabo G. Acute binge drinking increases serum endotoxin and bacterial DNA levels in healthy individuals. PLoS One 2014;9:e96864.

[12] Szabo G, Bala S. Alcoholic liver disease and the gut-liver axis. World J Gastroenterol 2010;16:1321–9.

[13] Mutlu EA, Gillevet PM, Rangwala H, et al. Colonic microbiome is altered in alcoholism. Am J Physiol 2012;302:G966–78.

[14] Wang Y, Liu Y, Zirpich I, et al. Lactobacillus rhamnosus GG reduces hepatic TNFa production and inflammation in chronic alcohol-induced liver injury. J Nutr Biochem 2013;24:1609–15.

[15] Wong VW-S, Tse C-H, Lam TT-Y, et al. Molecular characterization of the fecal microbiota in patients with nonalcoholic steatohepatitis—a longitudinal study. PLoS One 2013;8:e62885.

[16] Zhu L, Baker SS, Gill C, et al. Characterization of gut microbiomes in nonalcoholic steatohepatitis (NASH) patients: a connection between endogenous alcohol and NASH. Hepatology 2013;57:601–9.

[17] Sabaté JM, Jouët P, Harnois F, et al. High prevalence of small intestinal bacterial overgrowth in patients with morbid obesity: a contributor to severe hepatic steatosis. Obesity Surg 2008;18:371–7.

[18] Miel L, Valenza V, La Torre G, et al. Increased intestinal permeability and tight junction alterations in nonalcoholic fatty liver disease. Hepatology 2009;49:1877–87.

[19] Wu ZW, Lu HF, Wu J, et al. Assessment of the fecal lactobacilli population in patients with hepatitis B virus-related decompensated cirrhosis and hepatitis B cirrhosis treated with liver transplant. Microb Ecol 2012;63:929–37.

[20] Xu M, Wang B, Fu Y, et al. Changes of fecal Bifidobacterium species in adult patients with hepatitis B virus-induced chronic liver disease. Microb Ecol 2012;63:304–13.

[21] Lu H, Wu Z, Xu W, Yang J, Chen Y, Li L. Intestinal microbiota was assessed in cirrhotic patients with hepatitis B virus infection. Microb Ecol 2011;61:693–703.

[22] Heidrich B, Plumeier I, Kahl S, et al. Intestinal microbiota in chronic hepatitis C: differences in microbial communities in patients with and without liver cirrhosis. J Viral Hepat 2015;22:71–3.

[23] Tabibian JH, Talwalkar JA, Lindor KD. Role of the microbiota and antibiotics in primary sclerosing cholangitis. Biomed Res Int 2013;2013:389537.

[24] Zareie M, Johnson-Henry K, Jury J, et al. Probiotics prevent bacterial translocation and improve intestinal barrier function in rats following psychological stress. Gut 2006;55:1553–60.

[25] Madsen K, Cornish A, Soper P, et al. Probiotic bacteria enhance murine and human intestinal epithelial barrier function. Gastroenterology 2001;121:580–91.

[26] Resta-Lenart S, Barrett K. Live probiotics protect intestinal epithelial cells from the effects of infection with enteroinvasive Escheichia coli (EIEC). Gut 2003;52:988–97.

[27] Loguercio C, Federico A, Tuccillo C, et al. Beneficial effects of a probiotic VSL#3 on parameters of liver dysfunction in chronic liver disease. J Clin Gastroenterol 2005;29:540–3.

[28] Spencer M, Hamp T, Reid R, et al. Association between composition of the human gastroenterology microbiome and development of fatty liver with choline deficiency. Gastroenterology 2011;140:976–86.

[29] Cope K, Risby T, Diehl A. Increased gastrointestinal ethanol production and obese mice: implications for fatty liver disease pathogenesis. Gastroenterology 2000;199:1340–7.

[30] Aller R, De Luis D, Izaola O, et al. Effect of a probiotic on liver aminotransferases in nonalcoholic fatty liver disease patient, a double blind randomized clinical trial. Eur Rev Med Pharmacol Sci 2011;15:1090–5.

[31] Wong VW, Won GL, Chim AM, et al. Treatment of nonalcoholic steatohepatitis with probiotics. A proof-of-concept study. Ann Hepatol 2013;12:256–62.

[32] Malaguarnera M, Vacante M, Antic T, et al. Bifidobacterium longum with fructo-oligosaccharides in patients with non alcoholic steatohepatitis. Dig Dis Sci 2012;57:545–53.

[33] Vajro P, Mandato C, Licenziati MR, et al. Effects of Lactobacillus rhamnosus strain GG in pediatric obesity-related liver disease. J Pediatr Gastroenterol Nutr 2011;52:740–3.

[34] Ma YY, Lin L, Yu CH, Shen Z, Chen LH, Li YM. Effects of probiotics on nonalcoholic fattiy liver disease: a meta-analysis. World J Gastroenterol 2013;40:6911–8.

[35] Brenner D, Paik YH, Schnabl B. Role of gut microbiotia in liver disease. J Clin Gastroenterol 2015;49:S25–7.

[36] Bull-Otterson L, Feng W, Kirpich I, et al. Metagenomic analyses of alcohol induced pathogenic alterations in the intestinal microbiome and the effect of Lactobacillus rhamnosus GG treatment. PLoS One 2013;8:e53028.

[37] Kirpich I, Solovieva N, Leikhter S, et al. Probiotics restore bowel flora and improve liver enzymes in human alcohol-induced liver injury: a pilot study. Alcohol 2008;42:675–82.

[38] Han S, Suk K, Kim D, et al. Effects of probiotics (cultured Lactobacillus subtilis/Streptococcus faecium) in the treatment of alcoholic hepatitis: randomized-controlled multicenter study. Eur J Gastroenterol Hepatol 2015;27:1300–6.

[39] Campillo B, Pernet P, Bories P, et al. Intestinal permeability and liver cirrhosis: relationships with severe septic complications. Eur J Gastroenterol Hepatol 1999;11:755–9.

[40] Thalheimer U, Tirantos C, Samonakis D, et al. Infection, coagulation, and variceal bleeding cirrhosis. Gut 2005;54:556–63.

[41] Cirera I, Bauer TM, Navasa M, et al. Bacterial translocation of enteric organisms in patients with cirrhosis. J Hepatol 2001;34:32–7.

[42] García-Tsao G, Albillos A, Barden GE, West AB. Bacterial translocation in acute and chronic portal hypertension. Hepatology 1993;17:1081–5.

[43] Bauer TM, Schwacha H, Steinbrückner B, et al. Small intestinal bacterial overgrowth in human cirrhosis is associated with systemic endotoxemia. Am J Gastroenterol 2002;97:2364–70.

[44] Pande C, Kumar A, Sarin SK. Small-intestinal bacterial overgrowth in cirrhosis is related to the severity of liver disease. Aliment Pharmacol Ther 2009;29:1273–81.

[45] Guarner C, Runyon BA, Young S, Heck M, Sheikh MY. Intestinal bacterial overgrowth and bacterial translocation in cirrhotic rats with ascites. J Hepatol 1997;26:1372–8.

[46] Chang CS, Chen GH, Lien HC, Yeh HZ. Small intestine dysmotility and bacterial overgrowth in cirrhotic patients with spontaneous bacterial peritonitis. Hepatology 1998;28:1187–90.

[47] Llovet JM, Bartoli R, Planas RE, et al. Bacterial translocation in cirrhotic rats. Its role in the development of spontaneous bacterial peritonitis. Gut 1994;35:1648–52.

[48] Runyon BA, Squier S, Borzio M. Translocation of gut bacteria in rats with cirrhosis to mesenteric lymph nodes partially explains the pathogenesis of spontaneous bacterial peritonitis. J Hepatol 1994;21:792–6.

[49] Dhiman RK, Rana B, Agrawal S, et al. Probiotic VSL# 3 reduces liver disease severity and hospitalization in patients with cirrhosis: a randomized, controlled trial. Gastroenterology 2014;147:1327–37.

[50] Lata J, Novotny I, Pribramska V, Juránková J, et al. The effect of probiotics on gut flora, level of endotoxin and Child-Pugh score in cirrhotic patients: results of a double-blind randomized study. Eur J Gastroenterol Hepatol 2007;19:111–1113.

[51] Jayakumar S, Carbonneau M, Hotte N, et al. VSL# 3® probiotic therapy does not reduce portal pressures in patients with decompensated cirrhosis. Liver Int 2013;33:1470–7.

[52] Gupta N, Kumar A, Sharma P, Garg V, Sharma BC, Sarin SK. Effects of the adjunctive probiotic VSL# 3 on portal haemodynamics in patients with cirrhosis and large varices: a randomized trial. Liver Int 2013;33:1148–57.

[53] Rincón D, Vaquero J, Hernando A, et al. Oral probiotic VSL# 3 attenuates the circulatory disturbances of patients with cirrhosis and ascites. Liver Int 2014;34:1504–12.

[54] Pande C, Kumar A, Sarin S. Addition of probiotics to norfloxacin does not improve efficacy in the prevention of spontaneous bacterial peritonitis: a double-blind placebo-controlled randomized-controlled trial. Eur J Gastroenterol Hepatol 2012;24:831–9.

[55] Liu Q, Duan ZP, Ha DK, Bengmark S, Kurtovic J, Riordan SM. Synbiotic modulation of gut flora: effect on minimal hepatic encephalopathy in patients with cirrhosis. Hepatology 2004;39:1441–9.

[56] Mittal VV, Sharma BC, Sharma P, Sarin SK. A randomized controlled trial comparing lactulose, probiotics, and L-orinthine L-aspartate in treatment of minimal hepatic encephalopathy. Eur J Gastroenterol Hepatol 2011;23:725–32.

[57] Sharma P, Sharma B, Puri V, Sarin S. An open-label randomized controlled trial of lactulose and probiotics in the treatment of minimal hepatic encephalopathy. Eur J Gastroenterol Hepatol 2007;20:506–11.

[58] Agrawal A, Sharma BC, Sharma P, Sarin SK. Secondary prophylaxis of hepatic encephalopathy in cirrhosis: an open-label, randomized controlled trial of lactulose, probiotics, and no therapy. Am J Gastroenterol 2012;107:1043–50.

[59] Bajaj J, Saeian K, Christensen K, et al. Probiotic yogurt for the treatment of minimal hepatic encepahlopathy. Am J Gastroneterol 2008;103:1707–15.

[60] Shukla S, Shukla A, Mehboob S, Guha S. Meta-analysis: the effects of gut flora modulation using prebiotics, probiotics and synbiotics on minimal hepatic encephalopathy. Alimet Pharacol Ther 2011;33:662–71.

[61] Malaguarnera M, Gargante M, Malaguarner G, et al. Bifidobacterium combined with fructo-oligosaccharide versus lactulose in the treatment of patients with hepatic encephalopathy. Eur J Gastroenterol Hepatol 2009;22:199–206.

[62] Lunia MK, Sharma BC, Sharma P, Sachdeva S, Srivastava S. Probiotics prevent hepatic encephalopathy in patients with cirrhosis: a randomized controlled trial. Clin Gastroenterol Hepatol 2014;12:1003–8.

[63] McGee RG, Bakens A, Wiley K, Riordan SM, Webster AC. Probiotics for patients with hepatic encephalopathy. Cochrane Database Syst Rev 2011;11:CD008716.

[64] Wu ZW, Ling ZX, Lu HF, et al. Changes of gut bacteria and immune parameters in liver transplant recipients. Hepatobiliary Pancreat Dis Int 2012;11:40–50.

[65] Lu H, He J, Wu Z, et al. Assessment of microbiome variation during the perioperative period in liver transplant patients: a retrospective analysis. Microb Ecol 2013;65:781–91.

[66] Rayes N, Seehofer D, Hansen S, et al. Early enteral supply of lactobacillus and fiber versus selective bowel contamination: a controlled trial in liver transplant recipients. Transplantation 2002;74:123–8.

[67] Rayes N, Seehofer D, Theruvath T, et al. Supply of pre- and probiotics reduces bacterial infection rates after liver transplantation—a randomized, double-blind trial. Am J Transpl 2004;5:125–30.

[68] Eguchi S, Takatsuki M, Hidaka M, Soyama A, Ichikawa T, Kanematsu T. Perioperative synbiotic treatment to prevent infectious complications in patients after elective living donor liver transplantation. A prospective randomized study. Am J Surg 2011;4:498–502.

[69] Zhang Y, Chen J, Wu J, Chalson H, Merigan L, Mitchell A. Probiotic use in preventing postoperative infection in liver transplant patients. Hepatobiliary Surg Nutr 2013;2:142.

[70] Sawas T, Al Halabi S, Hernaez R, Carey WD, Cho WK. Patients receiving prebiotics and probiotics before liver transplantation develop fewer infections than controls: a systematic review and meta-analysis. Clin Gastroenterol Hepatol 2015;13:1567–74.

[71] Vleggaar F, Monkelbaan J, van Erpecum K. Probiotics in primary sclerosing cholangitis: a randomized placebo-controlled crossover pilot study. Eur J Gastoenterol Hepatol 2007;20:688–92.

Chapter 41

The Prevention and Treatment of Radiation and Chemotherapy-Induced Intestinal Mucositis

M.A. Ciorba

Division of Gastroenterology, Washington University in Saint Louis, Saint Louis, MO, United States

MUCOSITIS PATHOPHYSIOLOGY AND THE CLINICAL PROBLEM

Gastrointestinal (GI) mucosal injury from radiation therapy and chemotherapy is due to direct toxicity on epithelial stem cells, as well as indirect effects from apoptosis of microvascular endothelial cells [1]. Cytokines and transcription factors also play a role in mediating cytotoxic radiation and chemotherapy-induced injury [2,3]. Damage to the intestinal villi leads reduces the ability to absorb nutrients and water, resulting in diarrhea and potentially malnutrition. Novel oral agents that act directly in the gut to modulate the epithelial response to ionizing radiation and chemotherapy have the potential to reduce GI mucositis.

Mucositis complicates therapy in more than half of patients treated for an abdominal or pelvic malignancy [4]. Symptoms of mucositis include GI bleeding, abdominal pain, nausea, diarrhea, and malnutrition. Secondary infections can also result from intestinal mucositis [5]. These acute toxicities prevent optimal cancer treatment and can lead to chronic complications in survivors [4]. Delays in therapy lead to worse clinical outcomes [6,7]. For example, in anal cancer, where GI side effects are common, local progression is worse in patients who have significant delays in chemoradiation (89% vs. 42%, $p < 0.01$) [8]. In turn, reduced local tumor response correlates with lower overall survival rates (52% vs. 72%) [9]. An effective prophylactic agent that prevents or reduces treatment-related GI toxicity could potentially extend the length or quality of life for tens of thousands of people each year in the United States. Thus, there remains an important unmet need for prophylactic agents, which can prevent this collateral damage without impacting tumor cytotoxicity and cancer outcomes.

RATIONALE FOR USING PROBIOTICS

There is a complex interplay between cytotoxic cancer therapy and gut microbiota. The response of the GI tissues to injury, including cytotoxic cancer therapy, is influenced by resident intestinal bacteria [10]. Historical data indicate that germ-free mice are more resistant to lethal radiation enteritis than wild type mice, in part due to a lack of pathogenic bacteria to breach the compromised epithelial barrier and cause fatal sepsis [11,12]. Recently, however, it has been demonstrated that certain bacteria and bacterial products [toll-like receptor (TLR 2, 4, and 5 agonists)] are capable of mitigating epithelial cell apoptosis induced by irradiation [13–15].

Cytotoxic therapies also lead to a change in gut microbial communities. Montassier and coworkers recently illustrated that chemotherapy-induced GI mucositis is associated with a severe compositional and functional imbalance in the gut microbial community [16]. Additionally, in a small study analyzing fecal microbial DNA from patients undergoing pelvic radiation therapy found that the commensal microbial population preradiation might be able to predict susceptibility to developing acute mucositis symptoms [17].

Probiotics are microorganisms, typically bacteria, which have been demonstrated to confer health benefit when consumed in sufficient quantity. Probiotic therapies have been clinically evaluated and found safe as treatment for several GI conditions [18–22]. Probiotics can enhance epithelial barrier function, modulate local immune response, and can directly decrease epithelial cell apoptosis [23–26] Based on these properties, we and others have demonstrated that probiotics and other methods of manipulating the microbiome hold great promise as therapeutics to address the unmet need for therapies to prevent or treat radiation and/or chemotherapy-induced mucositis. These data have been reviewed in detail recently [10,27–29]. Key points are summarized later.

Studies in humans to date have used live probiotics; however, microbial derived products and particularly components of the bacterial cell wall may mediate the protective benefits of probiotics in the area of intestinal mucositis. Cario and coworkers found that microbial signaling through TLR can modulate the side effects of anticancer therapy in the small intestine [30]. In the study, the investigators nicely identify a central role for TLR2 as a regulator of xenobiotic defense via multidrug transporters. We previously demonstrated that the TLR universal adapter protein myeloid differentiation primary response 88 (MyD88) and TLR2 signaling was important to *Lactobacillus* spp. mediated small intestinal radioprotection. Others have shown that TLR4 and TLR5 signaling can mediate intestinal radioprotective benefits, although these were not specifically attributed to probiotic bacteria [13,14,31].

PROBIOTICS AND INTESTINAL CYTOPROTECTION IN HUMANS

Several clinical studies have evaluated probiotics in patients undergoing cytotoxic cancer therapy (reviewed in [27,28]). Most studies to date have focused on preventing or treating mucositis resulting from radiation alone or radiation with adjuvant chemotherapy rather than on chemotherapy-induced mucositis alone. In these studies *Lactobacillus* spp. containing probiotics in particular show a reduction (or a trend toward reduction) in radiation and/or chemotherapy-induced diarrhea or use of antidiarrheal agents among those taking the probiotic. However, all of these studies were conducted outside of the United States, most were small, and in some the primary endpoints were not clearly defined at study outset [32–36]. Still, the encouraging results of these preliminary studies led the Multinational Association for Supportive Care in Cancer to recommend probiotics for prevention of radiotherapy-induced intestinal adverse events [5,37].

A recently reported randomized, double-blind, placebo-controlled trial of 229 patients receiving pelvic radiation used a commercially available combination probiotic product containing (*Lactobacillus acidophilus*-361 and *Bifidobacterium longum*) [38]. The comparison groups included placebo, standard-dose (1.3 billion CFU), and high dose probiotic (10 billion CFU). The primary endpoint, which was not met, assessed the ability of the probiotic to prevent or delay moderate to severe diarrhea. However, some significant differences were noted on subgroup and secondary analysis including a reduction in moderate to severe diarrhea at day 60 in the standard dose, but not high dose group compared to placebo.

Synbiotics, probiotic bacteria plus a prebiotic substance, which fosters the growth of the probiotics, may be another rational approach to reduce mucositis. As an example, a synbiotic containing *Lactobacillus* reuteri and soluble fiber was reported to reduce proctitis symptoms and improve the quality of life in a small cohort of prostate cancer patients [39]. Of note, therapy for prostate cancer typically does not typically lead to small bowel mucositis, and thus this may be a different mechanism.

LIMITATIONS OF AVAILABLE CLINICAL DATA AND PROPOSALS FOR FUTURE STUDIES

Limitations in trial design are present in all studies published to date evaluating probiotics for prevention of cytotoxic therapy-induced mucositis. The frequently encountered major limitations include lack of standard primary endpoint, small sample size, and the inclusion of highly heterogeneous patient populations. With regard to the latter, studies frequently enrolled patients with a variety of different cancers who in turn received numerous different cytotoxic therapy regimens. This heterogeneity of cancers and cytotoxic regimens severely limits the ability to evenly distribute patients between probiotic and placebo groups based on anticipated side effect profile of the cancer therapy.

We recently reviewed these limitations in greater detail and proposed a strategy for future trials [29]. A summary of our recommendations included the following.

- *Probiotic selection*: Probiotic species vary greatly in their biologic effect. Early studies selected probiotic strains without preclinical evidence of efficacy or based on commercial interests. Future studies should examine probiotic species with a proven safety record in humans and demonstrated efficacy in preclinical models. Preclinical studies from our group (unpublished observation) indicate that single strain probiotics may be superior to multistrain products and published work indicates that commercially available *Lactobacillus* strains are effective while a bifidobacteria-based probiotic was not [15].
- *Probiotic dosage and delivery*: A pill form appears to be the preferred method of probiotic delivery rather than as part of a milk-based formula which could cause flatulence in some patients with lactose intolerance [32]. Combining a probiotic with a prebiotic may have advantages [39]. A higher bacterial count is not necessarily better and may be worse [38].
- *Study design considerations*: If probiotics are to gain traction as a viable therapy for clinical practice, their efficacy must be demonstrated in an appropriately powered, double-blind placebo-controlled trial. We propose that the goal should be prevention of radiation and/or chemotherapy-induced mucositis, rather than treatment based on clinical studies to date reporting

positive results only when the probiotic was started before cytotoxic therapy [27,28]. Preclinical data also supports this approach [15]. Cancer type and therapeutic approaches should be as homogeneous as possible. Primary endpoints should include standard symptom-based assessments as well as quantifiable biomarkers. Potential options include serum citrulline and fecal calprotectin [40]. Finally, long-term follow-up to examine the prevention of delayed GI consequences should be incorporated.

- *Regulatory approval*: The Food and Drug Administration (FDA) and local institutional review boards must approve of and oversee any trial conducted in the United States. The FDA in particular is interested in having a probiotic product, which is independently tested for strain identity, product purity, and potency.

Based on these recommendations and supporting preclinical data [15,41], we initiated a clinical trial examining the probiotic *Lactobacillus rhamnosus* GG in the prevention of GI toxicity for patients with GI malignancy receiving 5-fluoropyrimidine (5FU)-based chemotherapy and radiation [42]. This is currently being completed as an FDA mandated Phase I safety trial in this population. Nearing completion, no serious adverse events have occurred in this intervention study and the probiotic intervention has been well tolerated by patients.

PREBIOTICS AND THE PREVENTION OF GI TOXICITY IN RADIOTHERAPY

Prebiotics are dietary components (such as inulin or lactulose) that selectively enhance the growth, composition, or function of beneficial microbiota to promote health. Supplementation with prebiotics alone is a less well-studied strategy for preventing or treating cytotoxic cancer therapy-associated GI side effects attributed to mucositis. However, some of the studies previously described used prebiotics with probiotics (synbiotics).

A recent systematic review of nutritional interventions to counteract acute GI toxicity during therapeutic pelvic radiotherapy provides some insight to this area. The authors found that supplementation with dietary fiber (increase in food-based fiber or supplementation with psyllium) alone was not supported by evidence, and may even have a negative impact [43]. Lactose restriction also provided no clear benefit to during pelvic radiotherapy.

Most recently, an open label study of 40 patients examined a nutritional formula containing prebiotics (galacto-oligosaccharides), probiotics [tyndalized (heat-treated) *L. acidophilus* and *Lactobacillus casei*] and vitamins (B_1, B_2, B_6, and nicotinamide). The treatment was deemed beneficial as 17 patients experienced only Grade 1 or 2 treatment-related GI toxicity and no patients experienced Grade 3 or greater toxicity [44]. A randomized controlled trial is needed to confirm the benefit of this product and approach.

CHEMOTHERAPY AND GUT MICROBIOTA

Intestinal microbiota is also thought to play a role in the development and severity of chemotherapy-induced intestinal mucositis [45]. One study found that development of chemotherapy-associated diarrhea with mucositis correlated with decreases in stool quantities of *Lactobacillus* spp., *Bifidobacterium* spp., *Bacteroides* spp., and *Enterococcus* spp., and concurrent increases in *Escherichia coli* and *Staphylococcus* spp. [46] The authors here suggested that changes in microbiota may be biomarkers of chemotherapy-induced diarrhea.

On the prevention side, a recent study found that a probiotic suspension of *L. casei* variety *rhamnosus* or *L. acidophilus,* and *Bifidobacterium bifidum* ameliorated 5-fluorouricil chemotherapy-induced intestinal mucositis in a mouse model [47]. No human study to date has specifically focused on the effect of probiotics and prebiotics on chemotherapy-associated GI side effects.

INFLUENCE ON PROBIOTICS ON CANCER THERAPEUTICS

Among the most intriguing recent advances in the area of probiotics and cancer therapy include observations that specific microbial populations may enhance the efficacy of anticancer therapies. Viaud et al. reported that intestinal microbiota could modulate the anticancer immune effects of cyclophosphamide, a commonly used chemotherapeutic [48]. More recently, investigators found that the efficacy of antitumor immune checkpoint inhibitors (CTLA-4 and PD-L1) was dependent on gut microbiota [49,50]. These studies examined mouse models of melanoma, a cancer target that has shown the most success for immune checkpoint inhibitor drugs in humans to date. Interestingly, supplementation with common bifidobacteria probiotic species was sufficient to enhance the antitumor effects of the PD-L1 inhibitor. Both of these studies were in mouse models where the gut microbiota is assuredly more uniform. However, these studies open up important new avenues for clinical investigations in humans.

SUMMARY

The intimate relationship between gut microbiota and cancer therapy is an exciting and rapidly developing field. Recent studies clearly demonstrate that each have the power to influence the other's dynamics. Probiotic also bacteria influence the response of the normal intestine to radiation and chemotherapy in mice. There is now an opportunity for well-designed studies to clarify this effect in humans and to strengthen the argument for their use in patients receiving cytotoxic cancer therapy. Future studies in oncology will need to consider not only the impact of the microbiota on cancer therapy-related GI side effects, but also on long-term patient outcomes and therapeutic effectiveness. While more studies are needed to clarify safety and strain-specific efficacy, the best evidence to date supports probiotic agents containing *Lactobacillus* species for the prevention of chemotherapy and radiation-induced diarrhea in patients with pelvic malignancy [5]. With regard to prebiotics, there currently is not sufficient evidence to support their use as a clinical strategy to prevent or treat cytotoxic therapy-associated intestinal mucositis.

REFERENCES

[1] Paris F, Fuks Z, Kang A, et al. Endothelial apoptosis as the primary lesion initiating intestinal radiation damage in mice. Science 2001;293:293–7.

[2] Epstein J, Silverman S, Paggiarino D, et al. Benzydamine HCl for prophylaxis of radiation-induced oral mucositis. Cancer 2001;92:875–85.

[3] Sonis ST, Peterson RL, Edwards LJ, et al. Defining mechanisms of action of interleukin-11 on the progression of radiation-induced oral mucositis in hamsters. Oral Oncol 2000;36:373–81.

[4] Hauer-Jensen M, Denham JW, Andreyev HJ. Radiation entercpathy—pathogenesis, treatment and prevention. Nat Rev Gastroenterol Hepatol 2014;11:470–9.

[5] Lalla RV, Bowen J, Barasch A, et al. MASCC/ISOO clinical practice guidelines for the management of mucositis secondary to cancer therapy. Cancer 2014;120:1453–61.

[6] Elting LS, Cooksley C, Chambers M, et al. The burdens of cancer therapy. Clinical and economic outcomes of chemotherapy-induced mucositis. Cancer 2003;98:1531–9.

[7] Trotti A, Bellm LA, Epstein JB, et al. Mucositis incidence, severity and associated outcomes in patients with head and neck cancer receiving radiotherapy with or without chemotherapy: a systematic literature review. Radiother Oncol 2003;66:253–62.

[8] Huang K, Haas-Kogan D, Weinberg V, et al. Higher radiation dose with a shorter treatment duration improves outcome for locally advanced carcinoma of anal canal. World J Gastroenterol 2007;13:895–900.

[9] Chapet O, Gerard JP, Riche B, et al. Prognostic value of tumor regression evaluated after first course of radiotherapy for anal canal cancer. Int J Radiat Oncol Biol Phys 2005;63:1316–24.

[10] Packey CD, Ciorba MA. Microbial influences on the small intestinal response to radiation injury. Curr Opin Gastroenterol 2010;26:88–94.

[11] Crawford PA, Gordon JI. Microbial regulation of intestinal radiosensitivity. Proc Natl Acad Sci USA 2005;102:13254–9.

[12] McLaughlin MM, Dacquisto MP, Jacobus DP, et al. Effects of the germfree state on responses of mice to whole-body irradiation. Radiat Res 1964;23:333–49.

[13] Burdelya LG, Krivokrysenko VI, Tallant TC, et al. An agonist of toll-like receptor 5 has radioprotective activity in mouse and primate models. Science 2008;320:226–30.

[14] Riehl T, Cohn S, Tessner T, et al. Lipopolysaccharide is radioprotective in the mouse intestine through a prostaglandin-mediated mechanism. Gastroenterology 2000;118:1106–16.

[15] Ciorba MA, Riehl TE, Rao MS, et al. *Lactobacillus* probiotic protects intestinal epithelium from radiation injury in a TLR-2/cyclooxygenase-2-dependent manner. Gut 2012;61:829–38.

[16] Montassier E, Gastinne T, Vangay P, et al. Chemotherapy-driven dysbiosis in the intestinal microbiome. Aliment Pharmacol Ther 2015;42:515–28.

[17] Manichanh C, Varela E, Martinez C, et al. The gut microbiota predispose to the pathophysiology of acute postradiotherapy diarrhea. Am J Gastroenterol 2008;103:1754–61.

[18] Williams MD, Ha CY, Ciorba MA. Probiotics as therapy in gastroenterology: a study of physician opinions and recommendations. J Clin Gastroenterol 2010;44:631–6.

[19] Sartor RB. Microbial influences in inflammatory bowel diseases. Gastroenterology 2008;134:577–94.

[20] Snydman DR. The safety of probiotics. Clin Infect Dis 2008;46(Suppl. 2):S104–11. discussion S144–S151.

[21] Ciorba MA. A gastroenterologist's guide to probiotics. Clin Gastroenterol Hepatol 2012;19:960–8.

[22] Floch MH, Walker WA, Sanders ME, et al. Recommendations for probiotic use—2015 update: proceedings and consensus opinion. J Clin Gastroenterol 2015;49(Suppl. 1):S69–73.

[23] Mennigen R, Nolte K, Rijcken E, et al. Probiotic mixture VSL#3 protects the epithelial barrier by maintaining tight junction protein expression and preventing apoptosis in a murine model of colitis. Am J Physiol Gastrointest Liver Physiol 2009;296:G1140–9.

[24] Madsen K, Cornish A, Soper P, et al. Probiotic bacteria enhance murine and human intestinal epithelial barrier function. Gastroenterology 2001;121:580–91.

[25] Seth A, Yan F, Polk DB, et al. Probiotics ameliorate the hydrogen peroxide-induced epithelial barrier disruption by a PKC- and MAP kinase-dependent mechanism. Am J Physiol Gastrointest Liver Physiol 2008;294:G1060–9.

[26] Yan F, Cao H, Cover TL, et al. Soluble proteins produced by probiotic bacteria regulate intestinal epithelial cell survival and growth. Gastroenterology 2007;132:562–75.

[27] Touchefeu Y, Montassier E, Nieman K, et al. Systematic review: the role of the gut microbiota in chemotherapy- or radiation-induced gastrointestinal mucositis—current evidence and potential clinical applications. Aliment Pharmacol Ther 2014;40:409–21.

[28] Ciorba MA, Stenson WF. Probiotic therapy in radiation-induced intestinal injury and repair. Ann NY Acad Sci 2009;1165:190–4.

[29] Ciorba MA, Hallemeier CL, Stenson WF, et al. Probiotics to prevent gastrointestinal toxicity from cancer therapy: an interpretive review and call to action. Curr Opin Support Palliat Care 2015;9:157–62.

[30] Frank M, Hennenberg EM, Eyking A, et al. TLR signaling modulates side effects of anticancer therapy in the small intestine. J Immunol 2015;194:1983–95.

[31] Jones RM, Sloane VM, Wu H, et al. Flagellin administration protects gut mucosal tissue from irradiation-induced apoptosis via MKP-7 activity. Gut 2011;60:648–57.

[32] Salminen E, Elomaa I, Minkkinen J, et al. Preservation of intestinal integrity during radiotherapy using live *Lactobacillus acidophilus* cultures. Clin Radiol 1988;39:435–7.

[33] Delia P, Sansotta G, Donato V, et al. Use of probiotics for prevention of radiation-induced diarrhea. World J Gastroenterol 2007;13:912–5.

[34] Delia P, Sansotta G, Donato V, et al. Prevention of radiation-induced diarrhea with the use of VSL#3, a new high-potency probiotic preparation. Am J Gastroenterol 2002;97:2150–2.

[35] Giralt J, Regadera JP, Verges R, et al. Effects of probiotic *Lactobacillus casei* DN-114 001 in prevention of radiation-induced diarrhea: results from multicenter, randomized, placebo-controlled nutritional trial. Int J Radiat Oncol Biol Phys 2008;71:1213–9.

[36] Chitapanarux I, Chitapanarux T, Traisathit P, et al. Randomized controlled trial of live *Lactobacillus acidophilus* plus *Bifidobacterium bifidum* in prophylaxis of diarrhea during radiotherapy in cervical cancer patients. Radiat Oncol 2010;5:31.

[37] Gibson RJ, Keefe DM, Lalla RV, et al. Systematic review of agents for the management of gastrointestinal mucositis in cancer patients. Support Care Cancer 2013;21:313–26.

[38] Demers M, Dagnault A, Desjardins J. A randomized double-blind controlled trial: impact of probiotics on diarrhea in patients treated with pelvic radiation. Clin Nutr 2014;33:761–7.

[39] Nascimento M, Aguilar-Nascimento JE, Caporossi C, et al. Efficacy of synbiotics to reduce acute radiation proctitis symptoms and improve quality of life: a randomized, double-blind, placebo-controlled pilot trial. Int J Radiat Oncol Biol Phys 2014;90: 289–95.

[40] Wedlake L, McGough C, Hackett C, et al. Can biological markers act as non-invasive, sensitive indicators of radiation-induced effects in the gastrointestinal mucosa? Aliment Pharmacol Ther 2008;27: 980–7.

[41] Jones RM, Desai C, Darby TM, et al. Lactobacilli modulate epithelial cytoprotection through the Nrf2 pathway. Cell Rep 2015;12: 1217–25.

[42] Ciorba MA. Probiotic LGG for prevention of side effects in patients undergoing chemoradiation for gastrointestinal cancer, https://clinicaltrials.gov/ct2/show/NCT01790035; 2016.

[43] Wedlake LJ, Shaw C, Whelan K, et al. Systematic review: the efficacy of nutritional interventions to counteract acute gastrointestinal toxicity during therapeutic pelvic radiotherapy. Aliment Pharmacol Ther 2013;37:1046–56.

[44] Scartoni D, Desideri I, Giacomelli I, et al. Nutritional supplement based on zinc, prebiotics, probiotics and vitamins to prevent radiation-related gastrointestinal disorders. Anticancer Res 2015;35:5687–92.

[45] van Vliet MJ, Harmsen HJ, de Bont ES, et al. The role of intestinal microbiota in the development and severity of chemotherapy-induced mucositis. PLoS Pathog 2010;6:e1000879.

[46] Stringer AM, Al-Dasooqi N, Bowen JM, et al. Biomarkers of chemotherapy-induced diarrhoea: a clinical study of intestinal microbiome alterations, inflammation and circulating matrix metalloproteinases. Support Care Cancer 2013;21:1843–52.

[47] Yeung CY, Chan WT, Jiang CB, et al. Amelioration of chemotherapy-induced intestinal mucositis by orally administered probiotics in a mouse model. PLoS One 2015;10:e0138746.

[48] Viaud S, Saccheri F, Mignot G, et al. The intestinal microbiota modulates the anticancer immune effects of cyclophosphamide. Science 2013;342:971–6.

[49] Vetizou M, Pitt JM, Daillere R, et al. Anticancer immunotherapy by CTLA-4 blockade relies on the gut microbiota. Science 2015;350:1079–84.

[50] Sivan A, Corrales L, Hubert N, et al. Commensal *Bifidobacterium* promotes antitumor immunity and facilitates anti-PD-L1 efficacy. Science 2015;350:1084–9.

Chapter 42

The Role of the Brain–Gut–Microbiome in Mental Health and Mental Disorders

G.J. Treisman

Departments of Psychiatry and Behavioral Sciences and Internal Medicine, Johns Hopkins University School of Medicine, Baltimore, MD, United States

The evolutionary history of organized life is directed at controlling the internal and external environment to extend and maintain integrity of the organism. The first real step in evolution was the development of the cell, a tiny unit of organized chemical reactions that became capable of reproducing identical units. Over time, cells began to cooperate to control larger environmental areas. Competition for environments resulted in better and more comprehensive organizations, initially of identical cells, but over time with the development of related cells that developed specialized functions, and also with the development of different types of cells that came to cooperate.

In all known life, the regulation of internal and external environment involves a mechanism for sensing the environment, a mechanism for integrating sensory information, and a mechanism for activating a change. All behaviors of a single cell, such as division, movement, spore formation, spore emergence, phagocytosis, and exchange of genetic material must sense the internal or external environment, process the information, and activate a process. Single cells sense food and toxins, friendly and hostile environments and reproductive substrates. They process this information by activating cascades of enzymes, changes in protein conformations, binding to RNA or DNA, or deflection of physical structures. The same loop of sensory input, integration, and effector output occur at every level of organized life.

The first complex life-forms almost certainly consisted of groups of related cooperative cells that were more able to survive because of improved manipulation of their environment. The most phylogenetically ancient organisms allow nutrients or foods into a chamber and then process them. They developed sensory communication webs that allowed organism-wide concerted efforts. The primitive gut was born. The gut needed a sensory web, an integration center, and a way to expel unwanted contents. As the gut started to move around to look for food, it also had to find a way to avoid becoming food. The gut developed ways to sense the internal and external environment.

Primitive animal organisms were essentially motile guts, with increasing specialization of cells directed at regulating internal events (eventually evolving into the parasympathetic nervous system), and the external environment (eventually evolving into the sympathetic nervous system that could take over in danger). The next billion years of evolution led to the development of the sensory and integrative parts of the gut and complex structures to improve sensory input and integration and more effective control over getting around. The "gut brain" became more and more elaborate and finally developed a control center with increasing specialization for swimming and for vision and vibratory sense and finally memory and cognition. Complex brains developed the ability to have internal representations of the world and finally to be able to manipulate them. In our neuroscience, we consider the brain at the top of the control system, but we are increasingly finding that the gut may have a role in programming the brain. The sympathetic and parasympathetic nervous system may be the drivers of the goals for the brain and hierarchical decision-making.

During this process of evolution, a second parallel set of evolutionary events was taking place. The cooperation and competition across unrelated cells was developing complex organizational relationships that might be called "superorganisms." The vast rich diversity of microbes that populate our gut have been with us since those first primitive guts evolved. They rapidly colonized eukaryotic environments, even colonizing the inside of eukaryotic cells as mitochondrial ancestors. Due to their rapid division and rapid evolution they could rapidly adapt compared to the ponderous slow evolution of eukaryotes. They learned to signal the gut when they needed something, to send signals to the remote parts of the gut using the gut's own transmission machinery, and to control the organism in a variety of ways. The degree of complexity and control of these interactions have been opaque to us until recently due to technical barriers to the research needed to illuminate them. The diversity of the gut biome is just becoming apparent to us with powerful amplification and sequencing techniques. The actions of the nervous system and the circuitry and transmission within it are similarly reluctantly giving up some of their mystery to new technical methods. Although we have known about the interaction between GI function and brain function for thousands of years, it is only recently that we are beginning to unlock these secrets. The early studies of gut activity observed by William Beaumont, a surgeon caring for a soldier with a GI fistula that allowed direct observation of gut activity, revealed a relationship between gut activity and mood. Unfortunately, most of the work connecting the gut to the brain is early and has not yet had time to develop a rigorous set of investigative data. This chapter will review a broad but very superficial view of what we know and try to describe some of the directions for current and future work.

COMPONENTS OF THE BRAIN–GUT AXIS

The human brain is the main sensory-integration-effector system of the human cell lineage of the human superorganism. It controls voluntary muscle movement and has sympathetic and parasympathetic effectors that regulate endocrine, smooth muscle, and other autonomic functions. The inputs into the brain include the usually described sensory inputs of smell, taste, hearing, vision, and touch, as well as proprioception, autonomic afferent input (parasympathetic from the vagus and sympathetic from the splanchnic sympathetic nerves), and receptors for soluble humoral transmitters produced by the body, the gut microbiome, and the immune system.

The longstanding awareness of the role of the brain in regulation of the metabolic activity of the body stems in part form experiments by Claude Bernard in the mid-1800s in which he was able to create diabetes in a rabbit by puncturing the floor of the 4th cerebral ventricle of the brain. This "brain-centered glucose regulating system" appears to have profound effects on glucose regulation and is an example of the complexity of brain control mechanisms that impact gut activity [1]. The processor in the brain is so complex as to be only vaguely understood at our current level of being able to model it. The brain is the key center to understanding the complex interactions of the microbiome with behavior. There has been an explosion of new findings looking at the effects of the microbiome, the enteric nervous system, and the immune system on the brain, and the mechanisms by which the brain influences all of these systems.

While the human brain contains approximately 100 billion neurons (86 billion as per Azevedo et al. [2]), the enteric nervous system has on the order of 200 million neurons and has complex integrative capacity as well. It receives input from the brain, from a variety of visceral sensors, from the direct humoral input of the microbiome, the immune system, and from food as well. The enteric nervous system is imbedded in the gut, but other components of the gut are also involved in the integrative function of the system. The normal endothelium of the gut does not allow the entry of bacteria into the gut wall, but under certain circumstances the wall of the gut becomes leaky, with provocation of both chronic inflammatory and immune activation, as well as changes in the nerve sensitivity locally. Referred to as the "leaky gut" syndrome, there is increasing evidence that the immune system, the microbiota, the enteric nervous system, and even the brain may play a role in the regulation of the integrity of the endothelial barrier.

The gut microbiome is made up of a huge number of bacteria, as well as fungus, archaea, protozoans, helminthes, and the viruses that infect all of these species. This ecosystem has been hard to assess until newer methods of detection and amplification have made the complexity of the gut microbiome more possible. The human genome has approximately 20,000 coding genes, while the genetic complexity of the gut microbiome with the vast diversity of bacteria (and other life-forms) and the remarkable amount of bacteriophage make the genetic information present in the gut nearly unfathomable. Bacteria that digest cellulose have been identified in populations dependent on Sorghum [3] and bacteria that digest seaweed in populations with a seaweed-intensive diet [4]. The bacteria in a human gut have been evolving for thousands of generations within a single human gut during childhood and adolescence (assuming a generation time of 12–24 h in stable competitive environments, a 10-year-old gut has had 3000–8000 generations of bacteria). Assuming five generations of humans per century, the bacteria in the gut of 10-year-old have been evolving for the same number of generations as humans have existed as a species. This allows the gut microbiome to adapt to the human host and perfect communication and control of the host in a profound (and as yet poorly characterized) way.

Finally, the human immune system can be thought of as a control arm of the superorganism as well. The immune system senses the environment, processes information, and reacts in much of the same ways at the brain, the gut, and the microbiome. It receives humoral and direct cellular contact–based information, processes it through complex cell–cell interactions, and directs effector actions of immune cells. The immune system is highly focused on the gut, and both regulates and is regulated by the gut microbiome and enteric integration centers. Each of these four systems is highly connected to the other three, sending and receiving complex signals and integrating the information in a way that expresses the goals of that system (Fig. 42.1).

Shown in the Fig. 42.1 are the four control arms of the microbiome–gut–brain axis; the brain, the microbiota, the enteric nervous system and gut, and the immune system.

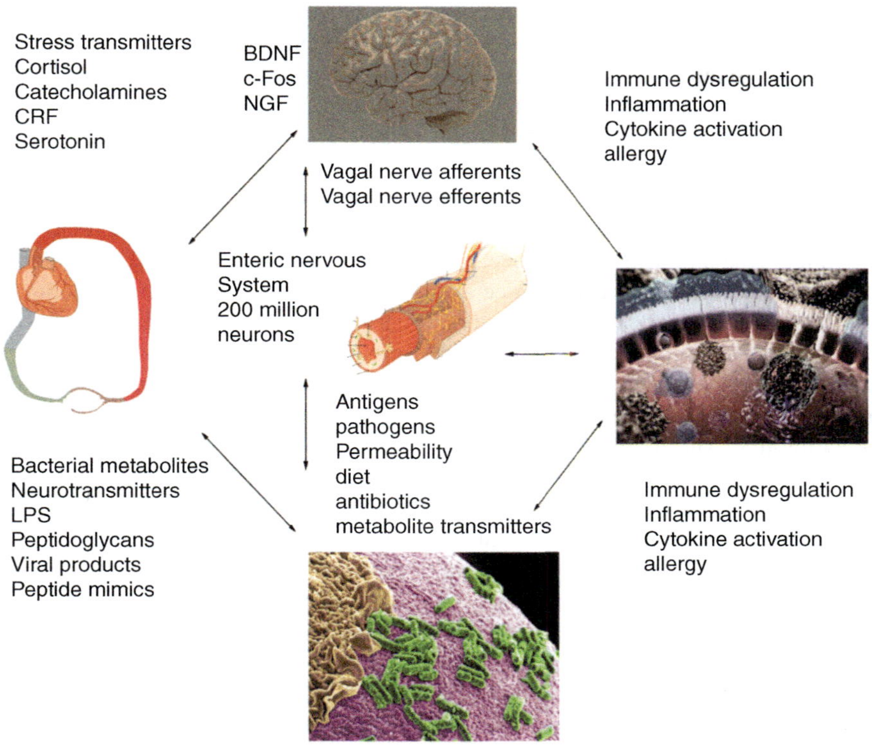

FIGURE 42.1 **Components of the gut–brain axis.**

The microbiota produces humoral transmitters carried by the circulatory system to the other control centers. Examples of the various intermediary substances and conveyance mechanism are shown.

EXAMPLES OF MICROBIOME ACTIONS ON THE SUPERORGANISM

The microbiota produces changes at every level of the superorganism. They affect brain function through direct release of humoral messengers, such as neurotransmitters, neurotransmitter precursors, regulatory hormones, and a variety of neuroactive metabolites, such as short chain fatty acids. They also produce effects on the local gut, the enteric nervous system, the autonomic nervous system, and the immune system, all of which can then secondarily change brain function. There is a question of how much the synthesis of bioactive and precursor compounds by gut bacteria have effects on the brain, but we do know that certain *Lactobacillus* and *Bifidobacterium* species produce gamma-aminobutyric acid (GABA) one of the major transmitters in the CNS. *Escherichia*, *Bacillus*, and *Saccharomyces* spp. produce noradrenaline, an important neurotransmitter in the central and peripheral nervous system. Varied species of *Candida*, *Streptococcus*, *Escherichia*, and *Enterococcus* produce serotonin (5-HT) a transmitter involved extensively in depression and anxiety, as well as impulsive behaviors. *Bacillus* produces dopamine, a critical transmitter in the motor and reward systems of the brain. *Lactobacillus* produces acetylcholine, the major transmitter involved in the parasympathetic system, as well as cognition in the CNS.

Most importantly, the bacteria of the gut directly compete to change and regulate the local environment. A well-characterized example of this is the response of the gut to stool transplants in patients with recurrent *Clostridium difficile* (C diff) infection. The actions of the transplanted stool microbiota will usually suppress the overexpression of C diff and the diarrhea will stabilize [5].

Examples of local actions of the gut microbiome include actions on nerve transmission. Rats fed with *Lactobacillus reuteri* were found to have increased enteric nerve excitability (in sensory AH neurons) and decreased calcium-dependent potassium channel activation. This will change the sensory signals from the gut to the enteric nervous system. The microbiome has been implicated in pain regulation and sensitization early in life and plays a regulatory role in pain sensitivity, particularly visceral sensitivity.

There are several exciting examples of microbiota effects transduced by the vagus nerve, such as a study in which mice fed *Lactobacillus rhamnosus* develop altered GABA receptor expression in certain brain regions that correlates with reduced anxiety- and depression-like symptoms in response to stress. This model of probiotic benefit is abolished by vagotomy [6].

Other effects produced by microbiota on brain and behavior are not transduced by the vagus nerve but may be transduced by humoral transmitters elaborated by the bacteria or by actions on the immune system. An example of a profound behavioral effect unrelated to vagal transmission is the study by Bercik et al. who looked at model behaviors in two strains of mice. One strain, BALB/c mice, are more timid, while the second strain, NIH Swiss mice are less timid in models of exploratory and other behaviors. In the study, mice with intact microbiota were compared with germ-free mice (with sterile GI tracts) in these tasks, and then the GI tracts of the intact mice were sterilized with nonabsorbable antibiotics. The more naturally timid BALB/c mice became less timid when colonized with the microbiota taken from the guts of the NIH Swiss mice, and the NIH Swiss mice became more timid when colonized by the microbiota from the BALB/c mice. The effects were not abolished by vagotomy and did not appear related to inflammation, but did correlate with altered brain-derived neurotrophic factor (BDNF) synthesis, suggesting a complex humoral pathway that is both specific and produces a change in a complex behavioral pattern [7]. It is still unclear how the change in BDNF occurs but this is an area of extremely active research.

The microbiome also regulates the set point for the immune system. Normal immune function is profoundly affected by altered gut microbiome. Immune defects in germ-free mice include decreased Peyer's patches, lamina propria, and isolated lymphoid follicles. There are also decreased intestinal CD8+ T cells and CD4+ T helper 17 cells and reduced B-cell production of secretory IgA [8]. IgA maintains a critical element in endothelial barrier function. *Bacteroides fragilis* introduced into germ-free mice restored immune maturation at gut-associated lymphoid tissues [9]. The microbiome provides diverse signals for tuning host immune status toward either effector or regulator direction, and is thus critical to peripheral immune education and homeostasis. Chronic inflammation, chronic immune activation, and sensitization of afferent sensory pathways can all play a role in brain function.

EXAMPLES OF HUMAN BRAIN EFFECTS ON THE MICROBIOME

One key feature of the organism is that it can survive in the absence of gut bacteria, albeit in a clearly less functional way and in a way which is far less environmentally adapted. The superorganism, deprived of gut bacteria, requires a variety of vitamins and nutritional supplements, and has coevolved to such an extent that there are gross distortions of gut function, immune function, and behavioral and brain dysfunction.

There are numerous ways in which the brain, the gut, and the immune system shape and effect the gut microbiome. The most obvious is food. The choices made in feeding profoundly shape the microbiome. Dietary intake has been shown to effect the bacterial make up of the microbiome. In a study by Wu et al., high animal protein and fat consumption was associated with higher levels of *Bacteroides* enterotype, while carbohydrates were associated with higher levels of *Prevotella* enterotype. A diet change resulted in a rapid shift in bacterial composition, but the base enterotypes were conserved [10]. A study comparing animal product–rich diets with plant-based diets showed that the animal-based diet increased the abundance of bile-tolerant microorganisms (*Alistipes*, *Bilophila*, and *Bacteroides*) and decreased the levels of Firmicutes that metabolize dietary plant polysaccharides (*Roseburia*, *Eubacterium rectale*, and *Ruminococcus bromii*) [11]. These findings (and others) show that brain choices strongly influence the microbiome. These choices appear to us to be a result of "free choice," but there is also evidence that the gut microbiome, the immune system, and the enteric nervous system may all influence decisions about feeding that seem to be brain based.

The microbiome is also sensitive to the release of a variety of neurohormones and transmitters released by the nervous system. These may originate in brain or the enteric nervous system. The microbiome is also influenced strongly by the immune system, a subject worthy of an entire review. Separation of infant rats from their mother, a rat model of early life stress, increases systemic corticosterone level and immune responses, changes that are associated with altered microbiota [12]. Host catecholamines have profound effects on the microbiota, such as a six order increase in bacterial host pathogens in response to norepinephrine and dopamine [13]. Stress-related hormones have been shown to increase conjugative transfer of antibiotic-resistant genes between enteric bacteria thereby contributing to the increased prevalence of antibiotic-resistant foodborne bacterial pathogens in the food supply [14,15]. Additionally, the ability of monoamines, such as norepinephrine and dopamine to alter gene expression has now been shown for a number of pathogenic microorganisms including *Mycoplasma hyopneumoniae* [16], *Salmonella enterica* serovar *Typhimurium* [17], and *Vibrio parahaemolyticus* [18].

ROLE OF THE GUT–BRAIN AXIS IN SPECIFIC DISEASE STATES

This chapter has focused on the complexity of a four-arm model of behavioral regulation of the "superorganism." This concept is supported by role for the gut, the brain, the immune system, and the microbiome in a number of psychiatric disease states. The complexity described earlier of what came first is difficult to tease out. Additionally, many of the studies I will present have methodological flaws and present some conflicting findings. Despite these caveats, the idea of foods as medications (phagopharmacology), prebiotics (foods that alter the microbiome), probiotics (eating live bacterial cultures), antibiotics, and fecal transplants directed at psychiatric disorders, provide new targets and therapeutic options that are quite exciting.

MOOD DISORDERS

Both anxiety and depression have been implicated as conditions related to GI dysfunction and dysbiosis. The concept of major depression has undergone considerable revision. Evidence from genetic studies and clinical studies suggest that anxiety and depression overlap in numerous ways and they are increasingly studied under the rubric of "mood disorders." There are several lines of evidence linking the gut microbiota to mood disorders. They include altered microbiota in subjects with major depression, increased inflammation associated with major depression, and metabolic products of the microbiota associated with major depression. No single study gives a compelling uniform causal relationship, but accumulating evidence is quite exciting.

Altered gut microbiota have been observed in association with depression including a study showing increased levels of Enterobacteriaceae and *Alistipes* but reduced levels of *Faecalibacterium* [19], as well as a study showing increase presence of Bacteroidales family and a decreased presence of the Lachnospiraceae family, as well as selected *Oscillibacter* and *Alistipes* that showed a significant association with depression. The latter study noted that *Oscillibacter* produces valeric acid as a metabolite [20], which has been associated with an anxiety model of depression in mice. In an unrelated study, a similar metabolite, isovaleric acid, was increased in the stool of depressed patients [21]. There is increasing evidence that antibiotics alter the gut microbiome and that exposure to antibiotics is associated with numerous health-related problems. A very recent study looking at a very large British database and using a nested case controlled model showed that exposure to even a single course of antibiotics increases the risk for major depression as well as anxiety [22]. Specific bacteria can alter behavior associated with mood disorders, such as resilience, reward sensitivity, and anxiety. The pathogen *Campylobacter jejuni* elevated anxiety-like behavior, apparently through activation of neuronal c-Fos protein, in both the autonomic and central nervous system [23,24], an effect also shown for *Citrobacter rodentium* [25].

Specific bacteria also seem to be able to attenuate mood disorders or reverse model mood disorders in animals. Bercik et al. showed that induced gut inflammation was associated with decreased BDNF and corresponding anxiety-like behavior. Drugs that decreased gut inflammation decreased the anxiety like behavior, while introduction of the

bacteria *Bifidobacterium longum* decreased the anxiety and increased the BDNF [26]. In a related study they found that the reversal of anxiety-like behavior by *B. longum* required an intact vagus nerve [27]. (In a fascinating reverse experiment, a brain lesion mouse model of depression resulted in altered GI motility and a change in the microbiome of the mouse [28].) In a mouse model in which stress was used to produce anxiety and depression, feeding with *L. rhamnosus* (JB-1) reduced stress-induced corticosterone and anxiety- and depression-related behavior, and also produced specific changes in GABA synthesis in specific brain areas. This finding was also dependent on an intact vagus nerve [29]. A less specific but more clinical study showed that a probiotic formulation (*Lactobacillus helveticus* R0052 and *B. longum* R0175) administered to rats decreased anxiety-like behavior in rats and improved mood and decreased anxiety based on checklist responses in humans. The human part of the study was double blinded and placebo controlled, making the study finding significantly more powerful than other open studies, but the study used normal volunteers and did not use patients with defined mood disorders [30].

Major depression (and possibly bipolar disorder) have been increasingly linked to inflammation [31]. Conditions associated with CNS inflammation, such as HIV, hepatitis C, multiple sclerosis, transverse myelitis, and a variety of rheumatologic conditions are associated with dramatic increases in major depression. Numerous basic models of depression can be provoked by inflammation, and treatment with antidepressants can improve the symptoms. The "leaky gut" syndrome, in which there is a decreased integrity in the gut endothelium and increased bacterial translocation across the barrier is associated with chronic immune activation, increased general inflammation, and increased activation of the HPA–stress axis. Animal models of irritable bowel syndrome have also been associated with depression [32]. Stress and the HPA axis also play a role in mood disorders. There is increasing evidence from germ-free mice and other experimental models that the HPA–stress axis illness (or symptom) set-point may be determined by the gut microbiome [33]. Gut-related chronic immune activation may be a primary mechanism for some types of depression. A link between hippocampal serotonin system and early life microbiome manipulations supports this idea. The finding was also sex sensitive, and as depression may be a disorder with sex specificity, this is again an exciting finding [34]. Exposure to stress alters the microbiota and induces increased inflammation. In a mouse model of early stress, adult mice showed decreased abundance of *Bacteroides*, increased abundance of *Clostridium*, and changes of other bacteria genera, which were concurrent with enhanced circulatory proinflammatory cytokines [35]. Effects of the "leaky gut" on inflammation and stress, as well as depressive behavior was reversed by *L. helveticus* and *rhamnosus* in a probiotic formulation. The gut "leakiness" also improved [36]. A similar effect was shown for *Lactobacillus farciminis* (decreased stress-induced gut leakiness and decreased HPA axis stress response) [37].

These actions of gut inflammation are not unidirectional, but are bidirectional and may even form amplification loops. There are numerous effects of stress and sympathetic nervous system transmitters on the gut microbiome, some of which were mentioned in the previous section, but there is little doubt that as much as we are sensing the messages from our microbiome, they are sensing the messages from their animal hosts. The identification of bacterial "receptors" for host transmitters is beginning to delineate the mechanisms of the bidirectional communications in our complex system [38].

AUTISM

Autism was described by Leo Kanner in 1943, but the definition has been broadened considerably since that time. It appears to occur on a spectrum of severity and may represent several overlapping disorders with similar symptom clusters. The high incidence of comorbid GI pathology has been seen in numerous descriptive studies and resulted in investigations into GI pathology on these patients. There are clearly abnormalities in the microbiome, the endothelial wall [39], and the enteric nervous system, as well as the brain and behavioral changes that characterize the condition. Altered bacterial flora include elevated levels of *Clostridia* species [40] as well as other species [15]. Finegold et al. reported an altered balance between Firmicutes and Bacteroidetes with an increased presence of Bacteroidetes in the severely autistic group and a predominant presence of Firmicutes in healthy controls [41]. There are several studies showing altered microbial flora in autism, but these studies have methodological limitations and should be viewed with a critical eye [42]. These changes may be part of the pathology behind autism or a consequence of it. There is a lot of evidence for chronic inflammation, with local infiltration within the gut, as well as systemic immune activation, elevated cytokines, and elevated levels of LPS [43]. Other CNS disorders, such as multiple sclerosis, appear to be the result of immune dysregulation, and are often accompanied by major depression and other mood disorders. Some late onset forms of autism have a partial response to antibiotic treatment with vancomycin [44] suggesting a role for bacterial dysbiosis but it is unclear if there is benefit in other cases, further supporting the idea of heterogeneity in the condition. A recent study suggested that a probiotic introduction of the bacteria *B. fragilis* may improve autism-like behavior in an animal model of autism [45]. An intriguing line of evidence implicates propionic acid, a short chain fatty acid fermentation product of clostridia (and other species), as playing a role in autism as suggested by actions in several models of the condition [46]. The primary pathology

in autism remains elusive, with genetic, environmental, and gut–brain axis pathology all playing roles in the disorder, but of unclear contribution.

SCHIZOPHRENIA

Another disorder with considerable heterogeneity, schizophrenia is a life long neurodegenerative disorder that profoundly affects executive function. Schizophrenia is associated with households containing cats, with the latitude of birth and the time of the year of birth and with numerous other correlates that seem beyond explanation. There are several lines of evidence that implicate the microbiome in schizophrenia, but the most compelling is the role of immune dysregulation in schizophrenia. Elevated serum levels of proinflammatory cytokines are found in schizophrenia patients compared to controls [47,48]. The inflammatory markers and severity of clinical symptoms appear to correlate in schizophrenia as well [49]. These findings are correlative and by no means definitive. Schizophrenia is a particularly human disease, and animal models are flawed by nonspecificity, although the exact phenotype defining schizophrenia is also difficult to define. Models of chronic immune activation in mice have several schizophrenia-like behaviors, including stereotypical behavior, distorted social behaviors, and disturbed relationships within developing groups. Chronically immune activated animals (mice, rats, and primates) have been used as models for schizophrenia [50]. In one model, the metabolomics profile in chronic immune-activated mice were treated with a *Bacteroides* species which reversed the abnormal metabolomics as well as some of the behavioral correlates of schizophrenia [51]. Germ-free mice are socially abnormal in several tests and if they are recolonized early in life these can be reversed. More interestingly, this effect is amplified in male mice, which in part correlates with the earlier onset of schizophrenia in males [52]. Most patients with schizophrenia have cognitive deficits, and germ-free mice have model cognitive defects that are thought to be a model of schizophrenia-related cognitive deficits [53]. There are other clues that suggest that the microbiome is involved in schizophrenia, such as alterations in the microbiome [54], a report of alterations in the bacteriophage library [55], and a report of altered metabolites associated with excess clostridia [56], but all of these need much more extensive investigation to be compelling.

THE BEHAVIORAL EFFECTS OF TOXOPLASMOSIS

While the unraveling of the complex nature of our interactions with the microbiome have only begun, many studies of extreme behavioral responses to host–microbe colonization have been described. Many of these involve the actions of "parasitic" fungi and protozoans that have been observed and described. This discussion would not be complete without a discussion on the effects of the obligate intracellular protozoan organism, *Toxoplasma gondii*. While long been considered a benign infection that is only clinically apparent in those with immunosuppression and the babies of woman infected during pregnancy (where congenital infection causes a variety of birth defects), new discoveries make us reconsider the role of microbiota in human behavior.

In general, the immune system was thought to make this common infection (approximately one-third of the worlds population) benign in the absence of immunosuppression. The organism has a preference for the CNS, and in immunoincompetent patients can produce fatal CNS lesions. In recent years there is increasing evidence of more subtle effects. These include evidence of a link to the risk for schizophrenia and mood disorders in the offspring of women who have had toxoplasmosis. There is also an extremely high prevalence of toxoplasmosis in mothers of children with Down syndrome. In patients with acute psychosis, there is an increase prevalence of IgM for Toxoplasma. Numerous studies have shown an increased prevalence of toxoplasmosis among schizophrenic patients. Toxoplasma-infected schizophrenic patients differ from Toxoplasma-free schizophrenic patients by brain anatomy and by a higher intensity of the positive symptoms of the disease.

The core description of behavioral reprogramming of the animal host shows that mice and rats lose their normal aversion to the scent of predatory feline urine once they are infected. Presumably, this makes them more likely to be eaten and pass the parasite on to the feline predator, helping to complete the parasite life cycle. This is a remarkable behavioral reprogramming, as it is very specific. The animals do not lose their fear of dog urine, they can still learn fear, and they respond to learned fear in a normal way. There is increasing evidence that latent infection in humans has subtle but remarkable effects in humans. As compared to uninfected people, latent toxoplasmosis is associated with increased traffic accidents and work accidents, prolonged reaction times, increased suicide and self-inflicted violence, increased intelligence and sociability in women, decreased intelligence and sociability in men, increased avoidance of cat urine odor in women, decreased avoidance of cat urine odor in men, and an increase in "neuroticism" (a trait associated with increased emotional lability) in populations with increased prevalence of latent infection. Women shown pictures of men with latent toxoplasmosis and control men rated infected men as more masculine and dominant, probably related to an increased concentration of testosterone in latently infected men, who are also taller than uninfected men (for review see Flegr [57]).

The subtleties of these effects of a single-celled organism are remarkable and must make us rethink the control of human behavior as far more complex than only nature and nurture from our own genes and environment, but we must now consider nature and nurture of the entire set of microbiota that accompany each superorganism and the interacting nature of the genome, metabolomics, competition, and even viral infection (both eukaryotic virus and bacterial phage) that determine behavior and disorder for each of us.

REFERENCES

[1] Schwartz MW, Seeley RJ, Tschöp MH, et al. Cooperation between brain and islet in glucose homeostasis and diabetes. Nature 2013;503(7474):59–66.

[2] Azevedo FA, Carvalho LR, Grinberg LT, et al. Equal numbers of neuronal and nonneuronalcells make the human brain an isometrically scaled-up primate brain. J Comp Neurol 2009;513(5):532–41.

[3] De Filippo C, Cavalieri D, Di Paola M, et al. Impact of diet in shaping gut microbiota revealed by a comparative study in children from Europe and rural Africa. Proc Natl Acad Sci USA 2010;107(33):14691–6.

[4] Hehemann JH, Correc G, Barbeyron T, Helbert W, Czjzek M, Michel G. Transfer of carbohydrate-active enzymes from marine bacteria to Japanese gut microbiota. Nature 2010;464(7290):908–12.

[5] Dutta SK, Girotra M, Garg S, et al. Efficacy of combined jejunal and colonic fecal microbiota transplantation for recurrent *Clostridium difficile* infection. Clin Gastroenterol Hepatol 2014;12(9):1572–6.

[6] Bravo JA, Forsythe P, Chew MV, et al. Ingestion of *Lactobacillus* strain regulates emotional behavior and central GABA receptor expression in a mouse via the vagus nerve. Proc Natl Acad Sci USA 2011;108(38):16050–5.

[7] Bercik P, Denou E, Collins J, et al. The intestinal microbiota affect central levels of brain-derived neurotropic factor and behavior in mice. Gastroenterology 2011;141(2):599–609.

[8] Round JL, Mazmanian SK. The gut microbiota shapes intestinal immune responses during health and disease. Nat Rev Immunol 2009;9:313–23.

[9] Round JL, Mazmanian SK. Inducible Foxp3+ regulatory T-cell development by a commensal bacterium of the intestinal microbiota. Proc Natl Acad Sci USA 2010;107:12204–9.

[10] Wu GD, Chen J, Hoffmann C, et al. Linking long-term dietary patterns with gut microbial enterotypes. Science 2011;334(6052):105–8.

[11] Maurice RN, Carmody DB, Gootenberg JE, et al. Diet rapidly and reproducibly alters the human gut microbiome. Nature 2014;505(7484):559–63.

[12] O'Mahony SM, Marchesi JR, Scully P, et al. Early life stress alters behavior, immunity, and microbiota in rats: implications for irritable bowel syndrome and psychiatric illnesses. Biol Psychiatry 2009;65:263–7.

[13] Lyte M, Ernst S. Catecholamine induced growth of gram negative bacteria. Life Sci 1992;50(3):203–12.

[14] Peterson G, Kumar A, Gart E, Narayanan S. Catecholamines increase conjugative gene transfer between enteric bacteria. Microb Pathog 2011;51(1–2):1–8.

[15] Williams BL, Hornig M, Parekh T, Lipkin WI. Application of novel PCR-based methods for detection, quantitation, and phylogenetic characterization of *Sutterella* species in intestinal biopsy samples from children with autism and gastrointestinal disturbances. MBio 2012;3(1):e00261-11.

[16] Oneal MJ, Schafer ER, Madsen ML, Minion FC. Global transcriptional analysis of *Mycoplasma hyopneumoniae* following exposure to norepinephrine. Microbiology 2008;154(Pt. 9):2581–8.

[17] Bearson BL, Bearson SM, Uthe JJ, et al. Iron regulated genes of *Salmonella enterica* serovar *Typhimurium* in response to norepinephrine and the requirement of fepDGC for norepinephrine-enhanced growth. Microbes Infect 2008;10(7):807–16.

[18] Nakano M, Takahashi A, Sakai Y, Nakaya Y. Modulation of pathogenicity with norepinephrine related to the type III secretion system of *Vibrio parahaemolyticus*. J Infect Dis 2007;195(9):1353–60.

[19] Jiang H, Ling Z, Zhang Y, Mao H, Ma Z, Yin Y, Wang W, Tang W, Tan Z, Shi J, Li L, Ruan B. Altered fecal microbiota composition in patients with major depressive disorder. Brain Behav Immun 2015;48:186–94.

[20] Naseribafrouei A, Hestad K, Avershina E, et al. Correlation between the human fecal microbiota and depression. Neurogastroenterol Motil 2014;26(8):1155–62.

[21] Szczesniak O, A Hestad K, Hanssen JF, Rudi K. 2015. Isovaleric acid in stool correlates with human depression. Nutr Neurosci.

[22] Lurie I, Yang YX, Haynes K, Mamtani R, Boursi B. Antibiotic exposure and the risk for depression, anxiety, or psychosis: a nested case–control study. J Clin Psychiatry 2015;76(11):1522–8.

[23] Goehler LE, Park SM, Opitz N, Lyte M, Gaykema RP. *Campylobacter jejuni* infection increases anxiety-like behavior in the holeboard: possible anatomical substrates for viscerosensory modulation of exploratory behavior. Brain Behav Immun 2008;22:354–66.

[24] Gaykema RP, Goehler LE, Lyte M. Brain response to cecal infection with *Campylobacter jejuni*: analysis with Fos immunohistochemistry. Brain Behav Immun 2004;18:238–45.

[25] Bercik P, Verdu EF, Foster JA, et al. Chronic gastrointestinal inflammation induces anxiety-like behavior and alters central nervous system biochemistry in mice. Gastroenterology 2010;139. 2102–2112.e1.

[26] Bercik P, Denou E, Collins J, et al. The intestinal microbiota affect central levels of brain-derived neurotropic factor and behavior in mice. Gastroenterology 2011;141:599–609. e591–e593.

[27] Bercik P, Park AJ, Sinclair D. The anxiolytic effect of *Bifidobacterium longum* NCC3001 involves vagal pathways for gut–brain communication. Neurogastroenterol Motil 2011;23:1132–9.

[28] Park AJ, Collins J, Blennerhassett PA, et al. Altered colonic function and microbiota profile in a mouse model of chronic depression. Neurogastroenterol Motil 2013;25(9):733. e575.

[29] Bravo JA, Forsythe P, Chew MV, et al. Ingestion of *Lactobacillus* strain regulates emotional behavior and central GABA receptor expression in a mouse via the vagus nerve. Proc Natl Acad Sci USA 2011;108:16050–5.

[30] Messaoudi M, Lalonde R, Violle N, et al. Assessment of psychotropic-like properties of a probiotic formulation (*Lactobacillus helveticus* R0052 and *Bifidobacterium longum* R0175) in rats and human subjects. Br J Nutr 2011;105:755–64.

[31] Miller AH, Maletic V, Raison CL. Inflammation and its discontents: the role of cytokines in the pathophysiology of major depression. Biol Psychiatry 2009;65(9):732–41.

[32] Liu L, Li Q, Sapolsky R, et al. Transient gastric irritation in the neonatal rats leads to changes in hypothalamic CRF expression, depression- and anxiety-like behavior as adults. PLoS One 2011;6(5): e19498.

[33] Dinan TG, Cryan JF. Regulation of the stress response by the gut microbiota: implications for psychoneuroendocrinology. Psychoneuroendocrinology 2012;37:1369–78.

[34] Clarke G, Grenham S, Scully P, et al. The microbiome–gut–brain axis during early life regulates the hippocampal serotonergic system in a sex-dependent manner. Mol Psychiatry 2013;18:666–73.

[35] Bailey MT, Dowd SE, Galley JD, et al. Exposure to a social stressor alters the structure of the intestinal microbiota: implications for stressor-induced immunomodulation. Brain Behav Immun 2011;25:397–407.

[36] Zareie M, Johnson-Henry K, Jury J, et al. Probiotics prevent bacterial translocation and improve intestinal barrier function in rats following chronic psychological stress. Gut 2006;55:1553–60.

[37] Ait-Belgnaoui A, Durand H, Cartier C, et al. Prevention of gut leakiness by a probiotic treatment leads to attenuated HPA response to an acute psychological stress in rats. Psychoneuroendocrinology 2012;37:1885–95.

[38] Clarke MB, Hughes DT, Zhu C, Boedeker EC, Sperandio V. The QseC sensor kinase: a bacterial adrenergic receptor. Proc Natl Acad Sci USA 2006;103:10420–5.

[39] de Magistris L, Familiari V, Pascotto A, et al. Alterations of the intestinal barrier in patients with autism spectrum disorders and in their first-degree relatives. J Pediatr Gastroenterol Nutr 2010;51(4): 418–24.

[40] Song Y, Liu C, Finegold SM. Real-time PCR quantitation of clostridia in feces of autistic children. Appl Environ Microbiol 2004;70(11):6459–65.

[41] Finegold SM, Molitoris D, Song Y, et al. Gastrointestinal microflora studies in late-onset autism. Clin Infect Dis 2002;35:S6–S16.

[42] Gondalia SV, Palombo EA, Knowles SR, Cox SB, Meyer D, Austin DW. Molecular characterisation of gastrointestinal microbiota of children with autism (with and without gastrointestinal dysfunction) and their neurotypical siblings. Autism Res 2012;5:419–27.

[43] Emanuele E, Orsi P, Boso D, et al. Low-grade endotoxemia in patients with severe autism. Neurosci Lett 2012;471(3):162–5.

[44] Sandler RH, Finegold SM, Bolte ER, et al. Short-term benefit from oral vancomycin treatment of regressive-onset autism. J Child Neurol 2000;15(7):429–35.

[45] Hsiao EY, McBride SW, Hsien S, et al. Microbiota modulate behavioral and physiological abnormalities associated with neurodevelopmental disorders. Cell 2013;155(7):1451–63.

[46] MacFabe DF. Enteric short-chain fatty acids: microbial messengers of metabolism, mitochondria, and mind: implications in autism spectrum disorders. Microb Ecol Health Dis 2015;26:28177.

[47] Song X, Fan X, Song X, et al. Elevated levels of adiponectin and other cytokines in drug naïve, first episode schizophrenia patients with normal weight. Schizophr Res 2013;150(1):269–73.

[48] Pedrini M, Massuda R, Fries GR, et al. Similarities in serum oxidative stress markers and inflammatory cytokines in patients with overt schizophrenia at early and late stages of chronicity. J Psychiatr Res 2012;46(6):819–22.

[49] Fan X, Liu EY, Freudenreich O, et al. Higher white blood cell counts are associated with an increased risk for metabolic syndrome and more severe psychopathology in non-diabetic patients with schizophrenia. Schizophr Res 2010;118(1–3):211–7.

[50] Bauman MD, Iosif AM, Smith SE, Bregere C, Amaral DG, Patterson PH. Activation of the maternal immune system during pregnancy alters behavioral development of rhesus monkey offspring. Biol Psychiatry 2014;75:332–41.

[51] Hsiao EY, McBride SW, Hsien S, et al. Microbiota modulate behavioral and physiological abnormalities associated with neurodevelopmental disorders. Cell 2013;155:1451–63.

[52] Desbonnet L, Clarke G, Shanahan F, Dinan TG, Cryan JF. Microbiota is essential for social development in the mouse. Mol Psychiatry 2014;19:146–8.

[53] Gareau MG, Wine E, Rodrigues DM, et al. Bacterial infection causes stress-induced memory dysfunction in mice. Gut 2011;60:307–17.

[54] Castro-Nallar E, Bendall ML, Pérez-Losada M, et al. Composition, taxonomy and functional diversity of the oropharynx microbiome in individuals with schizophrenia and controls. PeerJ 2015;3:e1140.

[55] Yolken RH, Severance EG, Sabunciyan S, et al. Metagenomic sequencing indicates that the oropharyngeal phageome of individuals with schizophrenia differs from that of controls. Schizophr Bull 2015;41(5):1153–61.

[56] Shaw W. Increased urinary excretion of a 3-(3-hydroxyphenyl)-3-hydroxypropionic acid (HPHPA), an abnormal phenylalanine metabolite of Clostridia spp. in the gastrointestinal tract, in urine samples from patients with autism and schizophrenia. Nutr Neurosci 2010;13:135–43.

[57] Flegr J. Influence of latent toxoplasma infection on human personality, physiology and morphology: pros and cons of the toxoplasma-human model in studying the manipulation hypothesis. J Exp Biol 2013;216(Pt. 1):127–33.

Chapter 43

Management of Disease and Disorders by Prebiotics and Probiotic Therapy: Probiotics in Bacterial Vaginosis

B. Vitali*, A. Abruzzo* and P. Mastromarino**

*Department of Pharmacy and Biotechnology, University of Bologna, Bologna, Italy; **Department of Public Health and Infectious Diseases, Sapienza University, Rome, Italy

INTRODUCTION

The human body harbors an enormous number of microorganisms that inhabit surfaces and cavities exposed or connected to the external environment [1]. As one of these human-microbe habitats, the female genital tract is colonized by bacterial communities that are known to confer antimicrobial protection to the vagina and play a crucial role in health [2,3]. This ecosystem is dynamic with changes in structure and composition being influenced by stage of life cycle, hormone levels, immune responses, nutritional status and disease states. The vaginal microbiota can also be altered by external factors, such as environmental exposures, microbial interspecies competition or commensalism, and hygiene behaviors [4].

In healthy women, the vaginal ecosystem is dominated by *Lactobacillus* species. Five distinct vaginal bacterial biotypes, characterized by the dominance of *Lactobacillus crispatus*, *Lactobacillus gasseri*, *Lactobacillus iners*, *Lactobacillus jensenii*, or an increased proportion of other strictly anaerobic bacteria, were described [5]. The vaginal vault is colonized within 24 h of a female child's birth and remains colonized until death. Lactobacilli become the predominant inhabitants of the vagina at the time of puberty, because of the effect of estrogens on the glycogen content of vaginal epithelial cells. Menopause is marked by a dramatic reduction in estrogen production, resulting in a drop of glycogen content in the vaginal epithelium. This leads to depletion of lactobacilli with a subsequent rise in vaginal pH, since glucose is not converted to lactic acid. High pH values promote growth of opportunistic pathogens, particularly colonization by enteric bacteria. Therefore, the vaginal microbiota suffers significant structural changes at various stages in a woman's life that are closely associated to the level of estrogens in the body [4].

Lactobacilli form a critical line of defense against potential pathogens through different mechanisms, such as the production of various antibacterial compounds (lactic acid, hydrogen peroxide, and bacteriocins), coaggregation, competitive exclusion, and immunomodulation [6,7]. Alterations in the types and relative proportions of the microbial species in the vagina can be associated with the development of infectious conditions, such as bacterial vaginosis (BV), aerobic vaginitis, candidiasis, and sexually transmitted infections [8,9].

BACTERIAL VAGINOSIS

BV, largely defined as a dysbiosis of the vaginal microbiome, is the most common and enigmatic vaginal condition in reproductive-age women with unknown etiology and poorly understood pathogenesis [10,11].

BV represents an imbalance in the ecology of the normal vaginal microbiota, characterized by a decrease in the prevalence and concentration of *Lactobacillus* species, and an increase in the prevalence and concentration of several pathogenic bacteria, mainly anaerobes [12]. Conventional culture-dependent analysis of the vaginal microbiota of women with BV identified a typical spectrum of anaerobes, including *Gardnerella vaginalis*, *Atopobium vaginae*, *Mobiluncus mulieris*, *Prevotella bivia*, *Fusobacterium nucleatum*, *Ureaplasma urealyticum*, and *Mycoplasma hominis* [13]. In recent years, culture-independent techniques based on the analysis of rRNA gene sequences have been developed, providing powerful tools to reveal the phylogenetic diversity of the microorganisms found within the vaginal ecosystem and to understand community dynamics [8,14–16]. These molecular studies indicate that the vaginal bacterial communities are dramatically different between women with and without BV. BV is associated with increased taxonomic richness and diversity. The microbiota composition is highly variable among subjects at a fine taxonomic scale (species or genus level), but, at the phylum level, Actinobacteria and Bacteroidetes are strongly associated with BV, while higher proportions of Firmicutes are found in healthy subjects. With the advances in molecular techniques, the spectrum of anaerobes detected in BV-affected women was greatly expanded with the addition of *Eggerthella, Megasphaera, Leptotrichia, Dialister,* and *Sneathia* [17]. Also, the Vaginal Human Microbiome Project has detected several newly described bacteria in the Clostridiales order, which are currently designated BVAB1, BVAB2, and BVAB3 [18]. All these organisms are of relatively low virulence. Therefore, BV is not caused by the mere presence of the potential pathogens but rather by their unrestrained increase in number, reaching cell counts that are 100- to 1000-fold above the normal bacterial levels in healthy vagina [19].

Recently, attention has been directed not only to the microbiome associated with BV, but also to the characterization of the vaginal proteome and metabolome. It has been demonstrated that BV is marked by profound changes in the proteome of vaginal fluids, mainly regarding the abundance of proteins involved in the innate immune response [20]. Studies investigating the metabolites produced by lactobacilli-dominated microbiota and polymicrobial BV have shown a striking loss of lactic acid and a shift toward mixed short-chain fatty acids production during BV state [21–23].

Clinical presentation of BV is typical. Women have an unpleasant, "fishy-smelling" discharge that is more noticeable after unprotected intercourse. The discharge is off-white, thin, and homogeneous. Erythema and inflammation are usually absent, and most patients are asymptomatic [24]. The overgrowth of vaginal anaerobes determines an increased production of amines (putrescine, cadaverine, and trimethylamine) that become volatile at alkaline pH, that is, after sexual intercourse and during menstrual cycle, and contribute to the typical malodour of the vaginal discharge [25].

The diagnosis of BV is based on clinical criteria (Amsel method) or Gram staining (Nugent method). Amsel method [26] implies the presence of at least three of the following criteria: (1) thin, homogeneous, milky vaginal discharge; (2) vaginal pH higher than 4.5; (3) "fishy" odor of vaginal fluid after addition of 10% KOH (whiff test); and (4) presence of "clue cells" on microscopic evaluation, that are vaginal epithelial cells heavily coated with bacteria of saline wet preparations. Amsel's criteria, although widely accepted as the best available means

to diagnose BV in the clinical setting, may fail to identify women with asymptomatic BV. The second method, the Gram staining score of vaginal smears according to Nugent [27], involves the microscopic quantitation of bacterial morphotypes yielding a score between 0 and 10 (normal microflora: score $\leq$ 3; intermediate microflora: score 4–6; BV: score $\geq$ 7). Nugent score system is generally preferred in the scientific community [28].

A large body of literature confirms that BV may lead to potentially severe complications and sequelae. Several studies have demonstrated an increased risk of abortion in the first trimester of pregnancy, premature ruptures of the membranes, chorioamnionitis, and preterm birth in women with BV [29–31]. A causal relationship has been established also between BV, pelvic inflammatory disease [32,33] and cervicitis [34]. Moreover, alterations in the vaginal microenvironment have been associated with urinary tract infections [35]. Increasing data also indicates that BV facilitates the acquisition of sexually transmitted diseases, such as *Neisseria gonorrhoeae, Chlamydia trachomatis,* HIV, and Herpes simplex virus type-2 infection (HSV-2) [36–42]. In vitro experiments using immortalized vaginal epithelial cells cocultured with common vaginal bacterial species offered some insight into the host–microbe interactions and mucosal immune response to BV. These studies suggest that BV-associated bacteria can induce an innate immune response from the genital epithelium characterized by upregulation of cytokines associated with increased risk of HIV-1 transmission, while commensal lactobacilli exert an antiinflammatory influence [43].

Current therapy for BV involves oral or intravaginal administration of metronidazole or clindamycin [44], but long-term follow-up suggests high recurrence rates [45]. The reasons for recurrence may include the failure to eradicate the offending organisms, due to the formation of a prolific bacterial biofilm adhering to the vaginal epithelium. The main components of this polymicrobial biofilm are *G. vaginalis* and *A. vaginae* [46]. Using mathematical models, it has been demonstrated that the human microbiome is dynamic within a single anatomic niche and can undergo complete reorganization in less than a day during antibiotic treatment. Within 24 h of metronidazole treatment initiation, most BV-associated bacteria undergo rapid exponential depletion. Levels of *Lactobacillus* species, particularly *L. iners*, often surge to fill the transient microbial vacuum. When treatment is stopped anaerobic species quickly re-emerge, suggesting a possible role for intermittent prophylactic treatment [47].

The high recurrence rates, which result in repeated exposure to antibiotics and the emergence of drug-resistant strains, suggest a need for alternative therapeutic tools. Novel antibiotic treatments are progressively being studied, like rifaximin [48], in order to have more options for switching therapy, combining therapies and long-term prophylactic use to prevent recurrences. In this panorama, probiotics are emerging as good candidates to improve the efficacy of BV therapy as well as to prevent recurrences.

RATIONALE FOR USING PROBIOTICS IN BACTERIAL VAGINOSIS

Evidence exists of the beneficial functions of the human microbiota, and prompted the selection of bacterial strains, recognized as probiotics, with health-promoting activities for the treatment of conditions in which the microbiota composition or functionality is perturbed. Probiotics are defined as "live microorganisms which when administered in adequate amounts confer health benefit on the host" [49], and therefore can represent an alternative approach for the prophylaxis and therapy of vaginal infections. *Lactobacillus* are the commonest bacteria used as probiotics. The rationale for the use of probiotics in BV is based on the regulatory role played by lactobacilli in the vaginal ecosystem and the need for restoration of the microbial homeostasis after insult [7].

Probiotics used in the treatment and prevention of BV may be administered both by oral and vaginal formulations. Orally consumed probiotics are believed to ascend to the vaginal tract after they are excreted from the rectum; vaginal administration allows for direct replacement of the probiotics for unhealthy vaginal microbiota and occupation of specific adhesion sites at the epithelial surface of the vagina, which consequently results in maintenance of a low pH and production of antimicrobial substances like acids, hydrogen peroxide, and biosurfactants [50].

The next paragraph reviews the most relevant current literature and investigations on probiotics' potential to prevent and treat BV.

CLINICAL TRIALS ON PROBIOTICS USE IN BACTERIAL VAGINOSIS

Various studies have been carried out to assess the efficacy of a single strain or a combination of different probiotic strains in the treatment of BV. The probiotic products contained a range of *Lactobacillus* species (*Lactobacillus acidophilus, Lactobacillus rhamnosus, Lactobacillus reuteri, Lactobacillus brevis, Lactobacillus salivarius, Lactobacillus. plantarum, L. gasseri*) and were administered orally or intravaginally. Two approaches for the treatment of BV have been used: treatment with probiotics alone or administration of probiotics following a conventional antibiotic therapy (adjuvant therapy).

Treatment of Bacterial Vaginosis Using Only Probiotics

The relevant randomized controlled trials using the first type of approach on women affected by BV are reported in Table 43.1.

The first clinical trials were conducted in the nineties using various strains of *L. acidophilus*. A product containing H_2O_2-producing *L. acidophilus* turned out to be ineffective for treatment of BV assessed according to Amsel criteria [51]. Nevertheless, it is difficult to evaluate the real efficacy of the product tested in this study since 50% of the patients in the active group and 86% of the placebo group did not complete the trial. In comparison, a high BV cure rate (88%) was observed in a placebo-controlled study using a pharmaceutical product (containing a H_2O_2-producing *L. acidophilus* strain plus estriol) that included both pregnant and nonpregnant women [52]. However, the results reported in this trial may have been biased by the enrolment criteria

TABLE 43.1 Clinical Trials on Probiotics Use for Treatment of Bacterial Vaginosis (BV)

Authors	No. of patients	Type of Study/ Duration	Intervention	BV Cure Rate
Hallén et al. [51]	57	R, DB, PC; 20–40 days	Twice daily vaginal suppository containing *L. acidophilus* 10^{8-9} CFU or placebo for 6 days	21% compared to 0% control ($P = $ NS)
Parent et al. [52]	32	R, PC; 4 weeks	1–2 daily vaginal tablet containing *L. acidophilus* $\geq 10^7$ CFU and 0.03 mg estriol for 6 days	88% compared to 22% control ($P < 0.05$)
Anukam et al. [53]	40	R, OB, AC; 30 days	Twice daily vaginal capsule containing *L. rhamnosus* GR-1 (10^9 CFU) and *L. reuteri* RC-14 (10^9 CFU), or 0.75% metronidazole gel for 5 days	65% compared to 33% metronidazole ($P = 0.056$)
Mastromarino et al. [54]	34	R, DB, PC; 3 weeks	Daily vaginal tablet containing $\geq 10^9$ CFU of *L. brevis* CD2, *L. salivarius* FV2, and *L. plantarum* FV9 for 7 days	50% compared to 6% control ($P = 0.017$)

AC, Active controlled; CFU, colony forming units; DB, double blind; OB, observer blind; PC, placebo controlled; R, randomized.

in which only two of the four Amsel criteria were required for a positive definition of BV status.

Both studies described previously do not use well-characterized probiotic *Lactobacillus* strains. Two more recent clinical trials employing different species of lactobacilli have been performed using well-characterized and well-selected strains specific for treatment of genitourinary infections [53,54]. Both studies used a combination of different species of lactobacilli with different biological properties on fertile nonpregnant women. *L. rhamnosus* GR-1 and *Lactobacillus fermentum* RC-14 strains used in the first study [53] adhere to uroepithelial cells, inhibit pathogen binding [55,56] and can be recovered from the vagina after oral administration [57]. In addition, *L. fermentum* RC-14 produces biosurfactants [58] and significant amounts of hydrogen peroxide [56]. A single-blind comparison of intravaginal probiotics (*L. rhamnosus* GR-1 and *L. fermentum* RC-14) and metronidazole gel for the treatment of BV was carried out on a group of Nigerian women [53]. Cure of BV was based on a Nugent score $\leq$ 3 at 30 days. A BV cure rate of 65% was achieved after probiotic treatment compared to 33% of the metronidazole therapy ($P = 0.056$).

The second study also used a product containing well-characterized and rationally selected *Lactobacillus* strains (*L. brevis* CD2, *L. salivarius* FV2, and *L. plantarum* FV9) [54]. *L. salivarius* FV2 and *L. plantarum* FV9 produce anti-infective agents, including hydrogen peroxide, and are able to coaggregate efficiently with vaginal pathogens [59]. *L. plantarum* and *L. brevis* strains are able to adhere at high levels to human epithelial cells, displacing vaginal pathogens [59,60]. The strains are able to temporarily colonize the human vagina [61], reduce vaginal proinflammatory cytokines IL-1β and IL-6 [62], and showed inhibitory activity toward HSV-2 and *C. trachomatis* replication in cell cultures [63–65]. In the double-blind, placebo-controlled trial [54] both the Amsel criteria and Nugent scores to assess BV cure were utilized as recommended by FDA (US Department of Health and Human Services, Food and Drug Administration, Center for Drug Evaluation and Research. Guidance for industry: BV—developing antimicrobial drugs for treatment 1998). The intravaginal probiotic-treated group showed a BV cure rate of 50% compared to 6% in the placebo-treated group with the combined test methods, whereas 67% versus 12% cure rate was obtained when considering only the Amsel criteria.

Treatment of Bacterial Vaginosis Using Probiotics as Adjuvant Therapy

Eight randomized controlled trials used lactobacilli following conventional antibiotic treatment (Table 43.2) to evaluate BV cure rate after 1 month [66–68] or BV recurrence after 2–6 months [69–73].

A trial in Nigeria evaluated augmentation of antimicrobial metronidazole therapy for BV by a 30-day oral probiotic treatment (*L. rhamnosus* GR-1 and *L. fermentum* RC-14) compared to placebo-treated control [66]. At the end of treatment a significantly greater number of women in the probiotic group compared to the placebo group were BV-free (Nugent score $\leq$ 3). The same *Lactobacillus* strains were used in a subsequent trial after treatment with a single dose of tinidazole [68]. Probiotic oral capsules were administered for 28 days starting on the first day of antibiotic treatment. At the end of the treatment the probiotic group had a significantly higher cure rate of BV than the placebo group.

In the study of Petricevic and Witt [67] a 7-day *Lactobacillus* treatment was performed after clindamycin therapy. Intravaginal *L. casei rhamnosus* (Lcr35) was used in the intervention group, whereas women in the control group did not receive Lcr35. The BV cure rate, evaluated by Nugent method 4 weeks after the last administration of medication, showed a significantly higher cure rate in the intervention group.

The efficacy of *Lactobacillus* supplementation after clindamycin or metronidazole treatment on the recurrence rate of BV has been evaluated in five trials [69–73]. Administration of tampons impregnated with *L. gasseri, L. casei* subsp. *rhamnosus* and *L. fermentum,* or placebo tampons during the menstrual period following clindamycin treatment was exploited [70]. Cure rates assessed by Amsel criteria after the second menstrual period did not show a significant difference between the two groups. Possible explanations for the lack of effects could be the low amount of lactobacilli in tampons at the end of the study (10^6 CFU, colony forming units) or the unfavorable period of administration, that is, during the menstrual flow.

A 10-day repeated treatment with *L. gasseri* Lba EB01-DSM 14869 and *L. rhamnosus* Lbp PB01-DSM 14870 during three menstrual cycles was compared with a placebo treatment on BV-affected women enrolled according to Amsel criteria [71]. The cure was evaluated by the Hay/Ison score [74]. Probiotic use did not improve the efficacy of BV therapy after the first month of treatment, but it significantly reduced the recurrence rate of BV at 6 months from initiation of treatment.

Even two studies carried out by the same group [72,73] indicated that vaginal application of a *L. rhamnosus* strain significantly prevented BV recurrence. Thirty days after the end of a 2 months once weekly treatment with the probiotic a significantly higher number of women in the treated group were BV-free in comparison to control [72]. Consistently, a higher percentage of treated patients showed no BV recurrences after another 3 months compared to controls, although the difference was not statistically significant ($P = 0.07$). A similar 6 months probiotic treatment indicated that during the first 6 months

TABLE 43.2 Clinical Trials on Probiotics Use Combined With Antibiotic Therapy for Treatment of Bacterial Vaginosis (BV)

Authors	No. of Patients	Type of Study/ Duration	Intervention	BV Cure Rate
Eriksson et al. [70]	187	R, DB, PC; 2 menstrual periods	Vaginal 100 mg clindamycin ovules for 3 days, then tampons containing 10^8 CFU of *L. gasseri, Lactobacillus casei rhamnosus, L. fermentum* or placebo tampons during the next menstrual period	56% compared to 62% control (P = NS)
Anukam et al. [66]	125	R, DB, PC; 30 days	Twice daily oral metronidazole 500 mg for 7 days and twice daily oral capsules containing *L. rhamnosus* GR-1 (10^9 CFU), and *L. reuteri* RC-14 (10^9 CFU) or placebo for 30 days starting on day 1 of metronidazole treatment	88% compared to 40% control (P < 0.001)
Petricevic and Witt [67]	190	R, OB, PC; 4 weeks	Twice daily oral clindamycin 300 mg for 7 days, then vaginal capsules containing 10^9 CFU of *L. casei rhamnosus* (Lcr35) for 7 days	83% compared to 35% control (P < 0.001)
Marcone et al. [72]	84	R; 3–6 months	Twice daily oral metronidazole 500 mg for 7 days followed by vaginal tablet containing > 10^4 CFU freeze-dried *L. rhamnosus* once a week for 2 months starting 1 week after the last antibiotic administration	88% compared to 71% control (P = 0.05) at 3 months 83% compared to 67% control (P = 0.07) at 6 months
Larsson et al. [71]	100	R, DB, PC; 6 menstrual periods	Daily 2% vaginal clindamycin cream directly followed by vaginal capsules containing *L. gasseri* Lba EB01-DSM 14869 (10^8–10^9 CFU) and *L. rhamnosus* Lbp PB01-DSM 14870 (10^8–10^9 CFU) for 10 days, probiotic treatment repeated for 10 days after each menstruation during three menstrual cycles	65% compared to 46% control (P = 0.042)
Martinez et al. [68]	64	R, DB, PC; 28 days	Single dose of tinidazole (2 g) followed by two oral capsules containing *L. rhamnosus* GR-1 (10^9 CFU) and *L. reuteri* RC-14 (10^9 CFU) or placebo for 28 days starting on the first day 1 of tinidazole treatment	87.5% compared to 50% control (P < 0.05)
Marcone et al. [73]	49	R; 12 months	Twice-daily oral metronidazole 500 mg for 7 days followed by once-weekly vaginal tablet containing > 10^4 CFU freeze-dried *L. rhamnosus* for 6 months.	91% compared to 69% control (P < 0.05)
Bradshaw et al. [69]	268	R, DB, PC; 6 months	Twice daily oral metronidazole 400 mg for 7 days followed by vaginal pessary containing *L. acidophilus* KS400 ≥ 10^7 CFU and 0.03 mg estriol for 12 days	72% compared to 73% control (P = NS)

CFU, Colony forming units; DB, double blind; OB, observer blind; PC, placebo controlled; R, randomized.

of follow-up, a constant percentage (96%) of patients in probiotic treated group had a balanced vaginal ecosystem. Follow-up over 12 months showed no statistically significant difference among vaginal ecosystems in patients in the treated group, while in the control group there was a significant increase in the number of women with abnormal flora over time [73].

Pessaries containing *L. acidophilus* KS400 were used in a recent trial to evaluate the efficacy of probiotics on the recurrence rate of BV following oral metronidazole treatment [69]. A 12 days course of probiotic pessary did not achieve higher cure rates for BV compared with placebo pessary over 6 months of follow-up as assessed by Nugent score.

A recent prospective, randomized, placebo-controlled, double-blinded study evaluated the efficacy of vaginal probiotic capsules for BV prophylaxis in healthy women with a history of recurrent BV [75]. One hundred twenty healthy Chinese women with a history of recurrent BV (≥2 BV episodes in the previous year), were assigned randomly to daily vaginal prophylaxis with 1 capsule that contained 8×10^9 CFU of *L. rhamnosus* (6.8×10^9 CFU), *L. acidophilus* (0.4×10^9 CFU), and *Streptococcus thermophilus* (0.8×10^9 CFU) or 1 placebo capsule for 7 days on, 7 days off, and 7 days on. Probiotic prophylaxis resulted in lower recurrence rates for BV (15.8% vs. 45.0%; P < 0.001) through 2 months as assessed according to Amsel criteria. Between the 2- and 11-month follow-up periods, women who received probiotics reported a lower incidence of BV (10.6% vs. 27.7%; $P = 0.04$). However, a limitation of this study was that 11-month outcomes were collected by telephone follow-up interview.

FORMULATIONS OF PROBIOTIC BACTERIA FOR THE TREATMENT OF BACTERIAL VAGINOSIS

A suitable vaginal probiotic delivery system should (1) ensure the inclusion of a high number of microorganisms, (2) guarantee their viability during the production and the storage period, (3) maintain their ability to produce active and beneficial substances, (4) modulate their release into the vaginal environment [76]. Thus, the selection of specific technological processes, dosage forms, and excipients is a critical point in order to design a functional and successful probiotic delivery system. In particular, encapsulation and freeze-drying are widely applied methods for preserving and protecting probiotic cells against stressing physical-chemical factors [77–80]. Probiotics have been mainly formulated in different traditional vaginal forms, such as douches, gels, ovules, tablets, or capsules [81–83]. Generally, douches or aqueous solutions are not suitable formulations for bacteria viability, due to the high water activity, which hinders the bacteria survival. Semisolid formulations, ovules, tablets, and capsules, able to dissolve in situ or gelify and characterized by a lower moisture content, are rapidly removed from the washing action of vaginal fluids, thus requiring multiple daily doses and consequently leading to poor patient compliance [84]. For this reason, several mucoadhesive polymers can be used in order to prolong the residence time of the formulation into vaginal cavity [85]. Furthermore, prebiotics can be also used thanks to their ability to promote probiotic viability and growth [77,86].

Although different probiotic delivery systems were widely investigated, only a few studies reported the use of probiotic-based formulation for BV treatment. Mastromarino et al. [59] formulated effervescent vaginal tablets containing different lactobacilli strains. The authors demonstrated that tablet production did not influence cell viability; moreover, tablets showed a significant bactericidal activity toward *G. vaginalis*.

CONCLUSIONS

The well-known physiological role played by a normal and stable *Lactobacillus*-dominated microbiota in preventing vaginal infections have raised interest for the potential of bacteriotherapy with probiotic lactobacilli in the setting of urogenital infections, including BV. In the past years several studies have evaluated the effect of probiotics in the treatment of vaginal infections. The studies using lactobacilli to treat BV showed the potential of probiotic bacteria

to cure this dysbiosis. When probiotics were used following antibiotic treatment, BV cure rate was increased and recurrence rates were reduced.

The preferred route of delivery for probiotic lactobacilli is intravaginal. However, some authors delivered lactobacilli orally to repopulate the vagina, based on the observation that *Lactobacillus* strains can pass from the gut into the urogenital system and can be recovered from the vagina. It is noteworthy that the capability of the lactobacilli to colonize the vagina after oral ingestion is strictly dependent on their viability and capability to survive gastric acid and bile salts. Obviously, the timing of vaginal colonization after oral administration is longer compared to direct administration at the vaginal level. Also, the load of lactobacilli that can be delivered orally to the vagina is clearly lower than direct vaginal administration. However, the oral administration of probiotics has the advantage of producing systemic effects, such as promoting intestinal health and modulates the immune response, which may contribute to improving the overall clinical picture of the woman.

Anyway, probiotic administration requires the design of a suitable delivery system able to ensure microorganism viability and to maintain their beneficial properties during its preparation and storage period. In particular, the selection of specific technological processes, dosage forms and excipients represented a key requirement in order to obtain a final successful probiotic formulation.

A systematic review conducted in 2009 [87] only based on four studies [52,53,66,70] did not provide conclusive evidence that probiotics are superior to or enhance the effectiveness of antibiotics in the treatment of BV. However, a more recent metaanalysis based on 12 clinical trials described here [51,52,54,66–73,75] concluded that probiotics supplementation can significantly improve the cure rate of BV [88]. Moreover, the pooled results of these trials also showed that probiotics were associated with a significant improvement in the cure rate of BV either in orally intervention or in vaginal application [88].

In conclusion, lactobacilli use in BV is supported by scientific results. However, further clinical trials, including larger samples of BV-affected women, in which lactobacilli are compared either with placebo or conventional antibiotics, need to be conducted before drawing definitive conclusions about the actual efficacy of probiotics as a strategy to treat women with BV.

REFERENCES

[1] Turnbaugh PJ, Ley RE, Hamady M, Fraser-Liggett CM, Knight R, Gordon JI. The human microbiome project. Nature 2007;449: 804–10.

[2] Mirmonsef P, Gilbert D, Zariffard MR, et al. The effects of commensal bacteria on innate immune responses in the female genital tract. Am J Reprod Immunol 2011;65:190–5.

[3] Petrova MI, Lievens E, Malik S, Imholz N, Lebeer S. *Lactobacillus* species as biomarkers and agents that can promote various aspects of vaginal health. Front Physiol 2015;6:81.

[4] Borges S, Silva J, Teixeira P. The role of lactobacilli and probiotics in maintaining vaginal health. Arch Gynecol Obstet 2014;289:479–89.

[5] Ravel J, Gajer P, Abdo Z, et al. Vaginal microbiome of reproductive-age women. Proc Natl Acad Sci USA 2011;108(Suppl. 1):4680–7.

[6] Kaewsrichan J, Peeyananjarassri K, Kongprasertkit J. Selection and identification of anaerobic lactobacilli producing inhibitory compounds against vaginal pathogens. FEMS Immunol Med Microbiol 2006;48:75–83.

[7] Reid G, Younes JA, Van der Mei HC, Gloor GB, Knight R, Busscher HJ. Microbiota restoration: natural and supplemented recovery of human microbial communities. Nat Rev Microbiol 2011;9:27–38.

[8] Fredricks DN, Fiedler TL, Marrazzo JM. Molecular identification of bacteria associated with bacterial vaginosis. N Engl J Med 2005;353:1899–911.

[9] Donders GG. Definition and classification of abnormal vaginal flora. Best Pract Res Clin Obstet Gynaecol 2007;21:355–73.

[10] Sobel JD. Bacterial vaginosis. Annu Rev Med 2000;51:349–56.

[11] Aldunate M, Srbinovski D, Hearps AC, et al. Antimicrobial and immune modulatory effects of lactic acid and short chain fatty acids produced by vaginal microbiota associated with eubiosis and bacterial vaginosis. Front Physiol 2015;6:164.

[12] Machado A, Cerca N. Influence of biofilm formation by *Gardnerella vaginalis* and other anaerobes on bacterial vaginosis. J Infect Dis 2015;212:1856–61.

[13] Livengood CH. Bacterial vaginosis: an overview for 2009. Rev Obstet Gynecol 2009;2:28–37.

[14] Vitali B, Pugliese C, Biagi E, et al. Dynamics of vaginal bacterial communities in women developing bacterial vaginosis, candidiasis, or no infection, analyzed by PCR-denaturing gradient gel electrophoresis and real-time PCR. Appl Environ Microbiol 2007;73:5731–41.

[15] Shipitsyna E, Roos A, Datcu R, et al. Composition of the vaginal microbiota in women of reproductive age—sensitive and specific molecular diagnosis of bacterial vaginosis is possible? PLoS One 2013;8:e60670.

[16] Cruciani F, Biagi E, Severgnini M, et al. Development of a microarray-based tool to characterize vaginal bacterial fluctuations and application to a novel antibiotic treatment for bacterial vaginosis. Antimicrob Agents Chemother 2015;59:2825–34.

[17] Romero R, Hassan SS, Gajer P, et al. The composition and stability of the vaginal microbiota of normal pregnant women is different from that of non-pregnant women. Microbiome 2014;2:4.

[18] Huang B, Fettweis JM, Brooks JP, Jefferson KK, Buck GA. The changing landscape of the vaginal microbiome. Clin Lab Med 2014;34:747–61.

[19] Forsum U, Holst E, Larsson PG, Vasquez A, Jakobsson T, Mattsby-Baltzer I. Bacterial vaginosis—a microbiological and immunological enigma. APMIS 2005;113:81–90.

[20] Cruciani F, Wasinger V, Turroni SB, et al. Proteome profiles of vaginal fluids from women affected by bacterial vaginosis and healthy controls: outcomes of rifaximin treatment. J Antimicrob Chemother 2013;68:2648–59.

[21] Yeoman CJ, Thomas SM, Miller ME, et al. A multi-omic systems-based approach reveals metabolic markers of bacterial vaginosis and insight into the disease. PLoS One 2013;8:e56111.

[22] McMillan A, Rulisa S, Sumarah M, et al. A multi-platform metabolomics approach identifies highly specific biomarkers of bacterial diversity in the vagina of pregnant and non-pregnant women. Sci Rep 2015;5:14174.

[23] Vitali B, Cruciani F, Picone G, Parolin C, Donders G, Laghi L. Vaginal microbiome and metabolome highlight specific signatures of bacterial vaginosis. Eur J Clin Microbiol Infect Dis 2015;34:2367–76.

[24] Sobel JD. Vaginitis. N Engl J Med 1997;337:1896–903.

[25] Nelson TM, Borgogna JL, Brotman RM, Ravel J, Walk ST, Yeoman CJ. Vaginal biogenic amines: biomarkers of bacterial vaginosis or precursors to vaginal dysbiosis? Front Physiol 2015;6:253.

[26] Amsel R, Totten PA, Spiegel CA, Chen KCS, Eschenbach D, Holmes KK. Non specific vaginitis: diagnostic criteria and microbial and epidemiologic associations. Am J Med 1983;74:14–22.

[27] Nugent RP, Krohn MA, Hillier SL. Reliability of diagnosing bacterial vaginosis is improved by a standardized method of Gram stain interpretation. J Clin Microbiol 1991;29:297–301.

[28] Schwiertz A, Taras D, Rusch K, Rusch V. Throwing the dice for the diagnosis of vaginal complaints? Ann Clin Microbiol Antimicrob 2006;5:4.

[29] Guaschino S, De Seta F, Piccoli M, Maso G, Alberico S. Aetiology of preterm labour: bacterial vaginosis. BJOG 2006;113(Suppl. 3):46–51.

[30] Donati L, Di Vico A, Nucci M, et al. Vaginal microbial flora and outcome of pregnancy. Arch Gynecol Obstet 2010;281:589–600.

[31] Redelinghuys MJ, Ehlers MM, Dreyer AW, Kock MM. Normal flora and bacterial vaginosis in pregnancy: an overview. Crit Rev Microbiol 2016;42:352–63.

[32] Ness RB, Hillier SL, Ki KE, et al. Bacterial vaginosis and risk of pelvic inflammatory disease. Obstet Gynecol 2004;104:761–9.

[33] Taylor BD, Darville T, Haggerty CL. Does bacterial vaginosis cause pelvic inflammatory disease? Sex Transm Dis 2013;40:117–22.

[34] Gorgos LM, Sycuro LK, Srinivasan S, et al. Relationship of specific bacteria in the cervical and vaginal microbiotas with cervicitis. Sex Transm Dis 2015;42:475–81.

[35] Harmanli OH, Cheng GY, Nyirjesy P, Chatwani A, Gaughan JP. Urinary tract infections in women with bacterial vaginosis. Obstet Gynecol 2000;95:710–2.

[36] Cherpes TL, Meyn LA, Krohn MA, Lurie JG, Hillier SL. Association between acquisition of herpes simplex virus type 2 in women and bacterial vaginosis. Clin Infect Dis 2003;37:319–25.

[37] Wiesenfeld H, Hillier S, Krohn MA, Landers D, Sweet R. Bacterial vaginosis is a strong predictor of *Neisseria gonorrhoeae* and *Chlamydia trachomatis* infection. Clin Infect Dis 2003;36:663–8.

[38] Brotman RM. Vaginal microbiome and sexually transmitted infections: an epidemiologic perspective. J Clin Invest 2011;121:4610–7.

[39] Nardis C, Mosca L, Mastromarino P. Vaginal microbiota and viral sexually transmitted diseases. Ann Ig 2013;25:443–56.

[40] Petrova MI, van den Broek M, Balzarini J, Vanderleyden J, Lebeer S. Vaginal microbiota and its role in HIV transmission and infection. FEMS Microbiol Rev 2013;37:762–92.

[41] Fastring DR, Amedee A, Gatski M, et al. Co-occurrence of *Trichomonas vaginalis* and bacterial vaginosis and vaginal shedding of HIV-1 RNA. Sex Transm Dis 2014;41:173–9.

[42] Esber A, Vicetti Miguel RD, Cherpes TL, et al. Risk of bacterial vaginosis among women with herpes simplex virus type 2 infection: a systematic review and meta-analysis. J Infect Dis 2015;212:8–17.

[43] Mitchell C, Marrazzo J. Bacterial vaginosis and the cervicovaginal immune response. Am J Reprod Immunol 2014;71:555–63.

[44] Donders GG, Zodzika J, Rezeberga D. Treatment of bacterial vaginosis: what we have and what we miss. Expert Opin Pharmacother 2014;15:645–57.

[45] Bradshaw CS, Morton AN, Hocking J, et al. High recurrence rates of bacterial vaginosis over the course of 12 months after oral metronidazole therapy and factors associated with recurrence. J Infect Dis 2006;193:1478–86.

[46] Muzny CA, Schwebke JR. Biofilms: an underappreciated mechanism of treatment failure and recurrence in vaginal infections. Clin Infect Dis 2015;61:601–6.

[47] Mayer BT, Srinivasan S, Fiedler TL, Marrazzo JM, Fredricks DN, Schiffer JT. Rapid and profound shifts in the vaginal microbiota following antibiotic treatment for bacterial vaginosis. J Infect Dis 2015;212:793–802.

[48] Cruciani F, Brigidi P, Calanni F, et al. Efficacy of rifaximin vaginal tablets in treatment of bacterial vaginosis: a molecular characterization of the vaginal microbiota. Antimicrob Agents Chemother 2012;56:4062–70.

[49] Joint FAO/WHO Working Group Report on Drafting Guidelines for the Evaluation of Probiotics in Food. London, Ontario, Canada, April 30 and May 1 (2002), http://www.who.int/foodsafety/fs_management/en/probiotic_guidelines.pdf; 2002.

[50] Homayouni A, Bastani P, Ziyadi S, et al. Effects of probiotics on the recurrence of bacterial vaginosis: a review. J Low Genit Tract Dis 2014;18:79–86.

[51] Hallén A, Jarstrand C, Påhlson C. Treatment of bacterial vaginosis with lactobacilli. Sex Transm Dis 1992;19:146–8.

[52] Parent D, Bossens M, Bayot D, et al. Therapy of bacterial vaginosis using exogenously-applied Lactobacilli acidophili and a low dose of estriol: a placebo-controlled multicentric clinical trial. Arzneimittelforschung 1996;46:68–73.

[53] Anukam KC, Osazuwa E, Osemene GI, Ehigiagbe F, Bruce AW, Reid G. Clinical study comparing probiotic Lactobacillus GR-1 and RC-14 with metronidazole vaginal gel to treat symptomatic bacterial vaginosis. Microbes Infect 2006;8:2772–6.

[54] Mastromarino P, Macchia S, Meggiorini L, et al. Effectiveness of Lactobacillus-containing vaginal tablets in the treatment of symptomatic bacterial vaginosis. Clin Microbiol Infect 2009;15:67–74.

[55] Reid G, Cook RL, Bruce AW. Examination of strains of lactobacilli for properties that may influence bacterial interference in the urinary tract. J Urol 1987;138:330–5.

[56] Reid G, Bruce AW. Selection of Lactobacillus strains for urogenital probiotic applications. J Infect Dis 2001;183(Suppl. 1):S77–80.

[57] Reid G, Bruce AW, Fraser N, Heinemann C, Owen J, Henning B. Oral probiotics can resolve urogenital infections. FEMS Immunol Med Microbiol 2001;30:49–52.

[58] Velraeds MC, van der Belt B, van der Mei HC, Reid G, Busscher HJ. Interference in initial adhesion of uropathogenic bacteria and yeasts silicone rubber by a Lactobacillus acidophilus biosurfactant. J Med Microbiol 1998;49:790–4.

[59] Mastromarino P, Brigidi P, Macchia S, et al. Characterization and selection of vaginal Lactobacillus strains for the preparation of vaginal tablets. J Appl Microbiol 2002;93:884–93.

[60] Maggi L, Mastromarino P, Macchia S, et al. Technological and biological evaluation of tablets containing different strains of lactobacilli for vaginal administration. Eur J Pharm Biopharm 2000;50:389–95.

[61] Massi M, Vitali B, Federici F, Matteuzzi D, Brigidi P. Identification method based on PCR combined with automated ribotyping for tracking probiotic Lactobacillus strains colonizing the human gut and vagina. J Appl Microbiol 2004;96:777–86.

[62] Hemalatha R, Mastromarino P, Ramalaxmi BA, Balakrishna NV, Sesikeran B. Effectiveness of vaginal tablets containing lactobacilli versus pH tablets on vaginal health and inflammatory cytokines: a randomized, double-blind study. Eur J Clin Microbiol Infect Dis 2012;31:3097–105.

[63] Conti C, Malacrino C, Mastromarino P. Inhibition of herpes simplex virus type 2 by vaginal lactobacilli. J. Physiol. Pharmacol 2009;6:19–26.

[64] Mastromarino P, Cacciotti F, Masci A, Mosca L. Antiviral activity of Lactobacillus brevis towards herpes simplex virus type 2: role of cell wall associated components. Anaerobe 2011;17:334–6.

[65] Mastromarino P, Di Pietro M, Schiavoni G, Nardis C, Gentile M, Sessa R. Effects of vaginal lactobacilli in Chlamydia trachomatis infection. Int J Med Microbiol 2014;304:654–61.

[66] Anukam K, Osazuwa E, Ahonkhai I, et al. Augmentation of antimicrobial metronidazole therapy of bacterial vaginosis with oral probiotic Lactobacillus rhamnosus GR-1 and Lactobacillus reuteri RC-14: randomized, double-blind, placebo-controlled trial. Microbes Infect 2006;8:1450–4.

[67] Petricevic L, Witt A. The role of Lactobacillus casei rhamnosus Lcr35 in restoring the normal vaginal flora after antibiotic treatment of bacterial vaginosis. BJOG 2008;115:1369–74.

[68] Martinez RC, Franceschini SA, Patta MC, et al. Improved cure of bacterial vaginosis with single dose of tinidazole (2 g), Lactobacillus rhamnosus GR-1, and Lactobacillus reuteri RC-14: a randomized, double-blind, placebo-controlled trial. Can J Microbiol 2009;55:133–8.

[69] Bradshaw CS, Pirotta M, De Guingand D, et al. Efficacy of oral metronidazole with vaginal clindamycin or vaginal probiotic for bacterial vaginosis: randomised placebo-controlled double-blind trial. PLoS One 2012;7(4):e34540.

[70] Eriksson K, Carlsson B, Forsum U, Larsson PG. A double-blind treatment study of bacterial vaginosis with normal vaginal lactobacilli after an open treatment with vaginal clindamycin ovules. Acta Derm Venereol 2005;85:42–6.

[71] Larsson PG, Stray-Pedersen B, Ryttig KR, Larsen S. Human lactobacilli as supplementation of clindamycin to patients with bacterial vaginosis reduce the recurrence rate; a 6-month, double-blind, randomized, placebo-controlled study. BMC Women's Health 2008;8:3.

[72] Marcone V, Calzolari E, Bertini M. Effectiveness of vaginal administration of Lactobacillus rhamnosus following conventional metronidazole therapy: how to lower the rate of bacterial vaginosis recurrences. New Microbiol 2008;31:429–33.

[73] Marcone V, Rocca G, Lichtner M, Calzolari E. Long-term vaginal administration of Lactobacillus rhamnosus as a complementary approach to management of bacterial vaginosis. Int J Gynaecol Obstet 2010;110:223–6.

[74] Ison CA, Hay PE. Validation of a simplified grading of Gram stained vaginal smears for use in genitourinary medicine clinics. Sex Transm Infect 2002;78:413–5.

[75] Ya W, Reifer C, Miller LE. Efficacy of vaginal probiotic capsules for recurrent bacterial vaginosis: a double-blind, randomized, placebo-controlled study. Am J Obstet Gynecol 2010;203. 120.e1–126.e1.

[76] Domig KJ, Kiss H, Petricevic L, Viernstein H, Unger F, Kneifel W. Strategies for the evaluation and selection of potential vaginal probiotics from human sources: an exemplary study. Benef Microbes 2014;5:263–72.

[77] Pliszczak D, Bourgeois S, Bordes C, et al. Improvement of an encapsulation process for the preparation of pro- and prebiotics-loaded bioadhesive microparticles by using experimental design. Eur J Pharm Sci 2011;44:83–92.

[78] Verdenelli MC, Coman MM, Cecchini C, Silvi S, Orpianesi C, Cresci A. Evaluation of antipathogenic activity and adherence properties of human *Lactobacillus* strains for vaginal formulations. J Appl Microbiol 2014;116:1297–307.

[79] Borges S, Costa P, Silva J, Teixeira P. Effects of processing and storage on *Pediococcus pentosaceus* SB83 in vaginal formulations: lyophilized powder and tablets. Biomed Res Int 2013;2013:680767.

[80] Juárez Tomás MS, De Gregorio PR, Leccese Terraf MC, Nader-Macías ME. Encapsulation and subsequent freeze-drying of *Lactobacillus reuteri* CRL 1324 for its potential inclusion in vaginal probiotic formulations. Eur J Pharm Sci 2015;79:87–95.

[81] Palmeira-de-Oliveira R, Palmeira-de-Oliveira A, Martinez-de-Oliveira J. New strategies for local treatment of vaginal infections. Adv Drug Deliv Rev 2015;92:105–22.

[82] Zárate G, Tomás MS, Nader-Macias ME. Effect of some pharmaceutical excipients on the survival of probiotic vaginal lactobacilli. Can J Microbiol 2005;51:483–9.

[83] Fazeli MR, Toliyat T, Samadi N, Hajjaran S, Jamalifar H. Viability of *Lactobacillus acidophilus* in various vaginal tablet formulations. Daru 2006;14(4):172–7.

[84] Abruzzo A, Bigucci F, Cerchiara T, et al. Chitosan/alginate complexes for vaginal delivery of chlorhexidinedigluconate. Carb Polym 2013;91:651–8.

[85] Valenta C. The use of mucoadhesive polymers in vaginal delivery. Adv Drug Deliv Rev 2005;57:1692–712.

[86] Vitali B, Abruzzo A, Parolin C, et al. Association of *Lactobacillus crispatus* with fructo-oligosaccharides and ascorbic acid in hydroxypropyl methylcellulose vaginal insert. Carbohydr Polym 2016;136:1161–9.

[87] Senok AC, Verstraelen H, Temmerman M, Botta GA. Probiotics for the treatment of bacterial vaginosis. Cochrane Database Syst Rev 2009;4:CD006289.

[88] Huang H, Song L, Zhao W. Effects of probiotics for the treatment of bacterial vaginosis in adult women: a meta-analysis of randomized clinical trials. Arch Gynecol Obstet 2014;289:1225–34.

Index

9780128040249